Quantum Information Processing and Quantum Error Correction

This page intentionally left blank

Quantum Information Processing and Quantum Error Correction

An Engineering Approach

Ivan Djordjevic

AMSTERDAM • BOSTON • HEIDELBERG • LONDON
NEW YORK • OXFORD • PARIS • SAN DIEGO
SAN FRANCISCO • SINGAPORE • SYDNEY • TOKYO

ELSEVIER

Academic Press is an imprint of Elsevier

Academic Press is an imprint of Elsevier
The Boulevard, Langford Lane, Kidlington, Oxford OX5 1GB, UK
225 Wyman Street, Waltham, MA 02451, USA

First edition 2012

British Library Cataloguing in Publication Data
A catalogue record for this book is available from the British Library

Library of Congress Cataloging-in-Publication Data
A catalog record for this book is available from the Library of Congress

ISBN: 978-0-12-385491-9

For information on all Academic Press publications
visit our web site at books.elsevier.com

To Milena

This page intentionally left blank

Contents

This page intentionally left blank

Preface

Quantum information relates to the use of quantum mechanics concepts to perform information processing and transmission of information. Quantum information processing (QIP) is an exciting research area with numerous applications, including quantum key distribution (QKD), quantum teleportation, quantum computing, quantum lithography, and quantum memories. This area is currently experiencing rapid development, which can be judged from the number of published books on QIP and by the number of conferences devoted solely to QIP concepts. Given the novelty of these underlying QIP concepts, it is expected that this topic will be of interest to a broad range of scientists, not just those involved in QIP research. Moreover, based on Moore's law, which claims that the number of transistors that can be etched on a single chip doubles every 18 months, leading to doubling of the memory and doubling of the computational speed, it is expected that the feature size might fall below 10 nm around 2020 with the current trend. At this point, as the individual properties of atoms and electrons start to dominate, Moore's law will cease to be valid. Therefore, the ever-increasing demands of miniaturization of electronics will eventually lead us to the point when quantum effects become important. Given this fact, it seems that a much broader range of scientist will need to turn to the study of QIP much sooner than is apparent at present. Note that, on the other hand, as multicore architecture is becoming a prevailing high-performance chip design approach, improvement of computational speed can be achieved even without reducing the feature size through parallelization. Therefore, due to multicore processor architectures, the need for QIP may be prolonged for a period of time. Another turning down point might be the fact that, despite intensive development of quantum algorithms, the number of available quantum algorithms is still small compared to that of classical algorithms. Further, the current quantum gates can operate only on several tens of quantum bits (also known as qubits), which is too low for any meaningful quantum computation operation. Until recently, it was widely believed that quantum computation would never become a reality. However, recent advances in various quantum gate implementations, as well as the proof of the accuracy threshold theorem, give rise to optimism that quantum computers might become a reality soon.

Because of the interdisciplinary nature of the fields of quantum information processing, quantum computing, and quantum error correction, this book aims to provide the right balance between quantum mechanics, quantum error correction, quantum computing, and quantum communication. The main *objectives* of the book can be listed as follows:

1. To describe the trends in quantum information processing, quantum error correction, and quantum computing.
2. To provide a self-contained introduction to quantum information processing and quantum error correction.

3. It targets a very wide range of readers – electrical engineers, optical engineers, applied mathematicians, computer scientists, and physicists.
4. It does not require prior knowledge of quantum mechanics. The basic concepts of quantum mechanics are given in Chapter 2.
5. It does not require any prerequisite background, except for an understanding of the basic concepts of vector algebra at undergraduate level. An appendix is provided containing a basic description of abstract algebra at a level sufficient to follow the book easily.
6. For readers not familiar with the concepts of channel coding, Chapter 6 provides the basic concepts and definitions of coding theory. Only the concepts of coding theory of importance in quantum error correction are covered in this chapter.
7. Readers not interested in quantum error correction can approach only the chapters relating to quantum information processing and quantum computing.
8. Readers interested only in quantum error correction do not need to read the chapters on quantum computing, except the chapter on quantum circuits. An effort has been made to create independent chapters, while ensuring proper flow between chapters.
9. The book introduces concepts gradually, starting from intuitive and basic concepts, through topics of medium difficulty to highly mathematically involved topics.
10. At the end of each chapter, a set of problems is provided so that the reader can gain a deeper understanding of the underlying concepts.
11. Several different courses can be offered by using this book. Examples include: (i) a course on quantum error correction; (ii) a course on quantum information processing and quantum computing; and (iii) a course integrating the concepts of quantum information processing and quantum error correction.
12. The reader completing the book and problems provided at the end of each chapter will be able to perform independent research in quantum error correction and/or quantum computing.

This book covers various topics of quantum information processing and quantum error correction. It is a self-contained introduction to quantum information, quantum computation, and quantum error correction. Readers of this book will be ready for further study in this area, and will be prepared to perform independent research. Having completed the book, readers should be able to design information processing circuits, stabilizer codes, Calderbank–Shor–Steane (CSS) codes, subsystem codes, topological codes and entanglement-assisted quantum error correction codes, and propose corresponding physical implementations. They should also be proficient in quantum fault-tolerant design as well. The book starts with basic principles of quantum mechanics, including state vectors, operators, density operators, measurements, and dynamics of a quantum system. It continues with fundamental principles of quantum computation, quantum gates, quantum algorithms, quantum teleportation, and fault-tolerant quantum computing. There is significant coverage of quantum error correction codes (QECCs), in particular stabilizer codes, CSS codes, quantum low-density parity-check (LDPC) codes, subsystem codes (also known as operator QECCs), topological codes, and entanglement-assisted QECCs. The next topic considered is fault-tolerant QECC and fault-tolerant quantum computing. The book continues with quantum information theory. The quantum information theory chapter is a basic chapter for topics to be included in the next edition, such as quantum communications and multidimensional quantum key distribution. It is also recommended that readers with a strong mathematical background read this chapter before Chapter 8, devoted to stabilizer codes (and beyond). The next part of the book is spent investigating physical realizations of quantum computers, encoders, and decoders, including nuclear magnetic resonance (NMR), ion traps, photonic quantum realization, cavity quantum electrodynamics, and quantum dots.

Some *unique features* of the book can be summarized as follows:

1. This book integrates quantum information processing, quantum computing, and quantum error correction.
2. It does not require prior knowledge of quantum mechanics.
3. It does not require any prerequisite material except for basic concepts of vector algebra at undergraduate level.
4. It offers in-depth exposition on the design and realization of quantum information processing and quantum error correction circuits.
5. The successful student should be prepared for further study in this area, and should be qualified to perform independent research.
6. Readers completing this book should be able to design information processing circuits, stabilizer codes, CSS codes, topological codes, subsystem codes, entanglement-assisted quantum error correction codes, and should be proficient in fault-tolerant design.
7. Successful students should be able to propose physical implementations of quantum information processing and quantum error correction circuits.
8. Extra material to support the book will be provided on an accompanying website.

The author would like to thank his colleague Dr Shikhar Uttam, who contributed Chapters 4 and 12, and would also like to acknowledge the National Science Foundation (NSF) for its support.

Finally, special thanks are extended to Charlotte Kent, Tim Pitts, and Pauline Wilkinson of Elsevier for their tremendous effort in organizing the logistics of the book, including editing and promotion, which was indispensible in making this book happen.

This page intentionally left blank

About the Author

Dr Djordjevic is Assistant Professor at the Department of Electrical and Computer Engineering of College of Engineering, University of Arizona, with a joint appointment in the College of Optical Sciences. Prior to this appointment in August 2006, he worked at the University of Arizona, Tucson, USA (as a Research Assistant Professor); the University of the West of England, Bristol, UK; the University of Bristol, Bristol, UK; Tyco Telecommunications, Eatontown, USA; and the National Technical University of Athens, Athens, Greece. His current research interests include optical networks, error control coding, constrained coding, coded modulation, turbo equalization, OFDM applications, and quantum error correction. He presently directs the Optical Communications Systems Laboratory (OCSL) within the ECE Department at the University of Arizona.

Dr Djordjevic is an author, together with Dr William Shieh, of the book *OFDM for Optical Communications*, Elsevier, October 2009. He is also an author, together with Professors Ryan and Vasic, of the book *Coding for Optical Channels*, Springer, March 2010. Dr Djordjevic is the author of over 130 journal publications and over 150 conference papers. He serves as an Associate Editor for the *Frequenz* and the *International Journal of Optics*. Dr Djordjevic is an IEEE Senior Member and an OSA Member.

This page intentionally left blank

CHAPTER

Introduction

1

CHAPTER OUTLINE

This chapter introduces quantum information processing (QIP) and quantum error correction coding (QECC) [1−11, 13−16] concepts. *Quantum information* is related to the use of quantum mechanics concepts to perform information processing and transmission of information. QIP is an exciting research area with numerous applications, including quantum key distribution (QKD), quantum teleportation, quantum computing, quantum lithography, and quantum memories. This area is currently experiencing intensive development and, given the novelty of underlying concepts, it could become of interest to a broad range of scientists, not only those involved in research. Moore's law claims that the number of chips that can be etched on a single chip doubles every 18 months, leading to doubling of memory and doubling of computational speed. On this basis, extrapolation of this law until 2020 indicates that feature size might fall below 10 nm, and at this point, as the individual properties of atoms and electrons start to dominate, Moore's law will cease to be valid. Therefore, the ever-increasing demands in miniaturization of electronics will eventually lead to a point when quantum effects become important. Given this fact, it seems that a much broader range of scientists will have to study QIP much sooner than it appears right now. Note that as the multicore architecture is becoming a prevailing high-performance chip design approach, improving computational speed can be achieved even without reducing the feature size through parallelization. Therefore, due to multicore processor architectures, the need for QIP could be prolonged for a certain amount of time. Another turning down point could be the fact that, despite intensive development of quantum algorithms, the number of available quantum algorithms is still small compared to that of classical algorithms. Until recently, it was widely believed that quantum computation would never become a reality. However, recent advances in various quantum gate implementations, as

Quantum Information Processing and Quantum Error Correction. DOI: 10.1016/B978-0-12-385491-9.00001-0

well as the proof of the accuracy threshold theorem, have given rise to optimism that quantum computers will soon become reality.

The *fundamental features* of QIP are different from those of classical computing and can be broken down into three categories: (1) linear superposition; (2) entanglement; (3) quantum parallelism. Below we provide some basic details of these features:

1. **Linear superposition.** Contrary to the classical bit, a quantum bit or *qubit* can take not only two discrete values 0 and 1, but also *all* possible *linear combinations* of them. This is a consequence of a fundamental property of quantum states: It is possible to construct a *linear superposition* of quantum state $|0\rangle$ and quantum state $|1\rangle$.
2. **Entanglement.** At a quantum level it appears that two quantum objects can form a single entity, even when they are well separated from each other. Any attempt to consider this entity as a combination of two independent quantum objects given by the tensor product of quantum states fails, unless the possibility of signal propagation at superluminal speed is allowed. These quantum objects that cannot be decomposed into a tensor product of individual independent quantum objects are called *entangled* quantum objects. Given the fact that arbitrary quantum states cannot be copied, which is the consequence of the no-cloning theorem, communication at superluminal speed is not possible, and as a consequence the entangled quantum states cannot be written as the tensor product of independent quantum states. Moreover, it can be shown that the amount of information contained in an entangled state of N qubits grows exponentially instead of linearly, which is the case for classical bits.
3. **Quantum parallelism.** This makes it possible to perform a large number of operations in parallel, which represents a key difference from classical computing. Namely, in classical computing it is possible to know the internal status of the computer. On the other hand, because of the no-cloning theorem, it is not possible to know the current state of a quantum computer. This property has led to the development of the Shor factorization algorithm, which can be used to crack the Rivest–Shamir–Adleman (RSA) encryption protocol. Some other important quantum algorithms include: the Grover search algorithm, which is used to perform a search for an entry in an unstructured database; the quantum Fourier transform, which is the basis for a number of different algorithms; and Simon's algorithm. These algorithms are the subject of Chapter 5. A quantum computer is able to encode all input strings of length N simultaneously into a single computational step. In other words, the quantum computer is able simultaneously to pursue 2^N classical paths, indicating that a quantum computer is significantly more powerful than a classical one.

Although the QIP has opened up some fascinating perspectives, as indicated above, there are certain limitations that need to be overcome before QIP becomes a commercial reality. The first is related to the number of existing quantum algorithms, whose number is significantly lower than that of classical algorithms. The second problem is related to physical implementation issues. There are many potential technologies, such as nuclear magnetic resonance (NMR), ion traps, cavity quantum electrodynamics, photonics, quantum dots, and superconducting technologies, to mention just a few. Nevertheless, it is not clear which technology will prevail. Regarding quantum teleportation, most probably the photonic implementation will prevail. On the other hand, for quantum computing applications there are many potential technologies that compete with each other. Moreover, presently the number of qubits that can be manipulated is of the order of tens, well below that needed for meaningful quantum computation, which is of the order of thousands. Another problem, which can be considered as the major difficulty, is related to *decoherence*. Decoherence is related to the interaction of qubits with environments that blur the fragile superposition states. It also introduces errors, indicating that the quantum register should be sufficiently isolated from the environment so that only few random errors occur occasionally, which can be corrected by QECC techniques. One of the most powerful applications of quantum error correction is the protection of quantum information as it dynamically undergoes quantum computation. Imperfect quantum gates affect quantum computation by introducing errors in computed data. Moreover, the imperfect control gates introduce errors in processed sequences since wrong operations can be applied. The QECC scheme now needs to deal not only with errors introduced by quantum channels, but also with errors introduced by imperfect quantum gates during the encoding/decoding process. Because of this, the reliability of data processed by quantum computers is not *a priori* guaranteed by QECC. The reason is threefold: (i) the gates used for encoders and decoders are composed of imperfect gates, including controlled imperfect gates; (ii) the syndrome extraction applies unitary operators to entangle ancillary qubits with code block; and (iii) the error recovery action requires the use of controlled operation to correct for these errors. Nevertheless, it can be shown that arbitrary good quantum error protection can be achieved even with imperfect gates, provided that the error probability per gate is below a certain *threshold*; this claim is known as the accuracy threshold theorem and will be discussed in Chapter 11, together with various fault-tolerant concepts.

This introductory chapter is organized as follows. In Section 1.1, photon polarization is described as it represents the simplest and most natural connection to QIP. In the same section, some basic concepts of quantum mechanics are introduced, such as the concept of state. The Dirac notation is also introduced, and will be used throughout the book. In Section 1.2, the concept of the qubit is formally introduced and its geometric interpretation given. In Section 1.3, another interesting example of a representation of qubits, the spin-1/2 system, is provided. Section 1.4 covers basic quantum gates and QIP fundamentals. The basic concepts of quantum teleportation are introduced in Section 1.5. Section 1.6 considers the basic QECC concepts. Finally, Section 1.7 is devoted to the quantum key distribution (QKD), also known as quantum cryptography.

1.1 PHOTON POLARIZATION

The electric/magnetic field of plane linearly polarized waves is described as follows [12]:

$$\boldsymbol{A}(\boldsymbol{r},t) = \boldsymbol{p}A_0 \exp[j(\omega t - \boldsymbol{k}\cdot\boldsymbol{r})], \quad \boldsymbol{A} \in \{\boldsymbol{E}, \boldsymbol{H}\}, \tag{1.1}$$

where $\boldsymbol{E}$ ($\boldsymbol{H}$) denotes electric (magnetic) field, $\boldsymbol{p}$ denotes the polarization orientation, $\boldsymbol{r} = x\boldsymbol{e}_x + y\boldsymbol{e}_y + z\boldsymbol{e}_z$ is the position vector, and $\boldsymbol{k} = k_x\boldsymbol{e}_x + k_y\boldsymbol{e}_y + k_z\boldsymbol{e}_z$ denotes the wave propagation vector whose magnitude is $k = 2\pi/\lambda$ (λ is the operating wavelength). For the x-polarization waves ($\boldsymbol{p} = \boldsymbol{e}_x$, $\boldsymbol{k} = k\boldsymbol{e}_z$), Eq. (1.1) becomes

$$\boldsymbol{E}_x(z,t) = \boldsymbol{e}_x E_{0x} \cos(\omega t - kz), \tag{1.2}$$

while for y-polarization ($\boldsymbol{p} = \boldsymbol{e}_y$, $\boldsymbol{k} = k\boldsymbol{e}_z$) it becomes

$$\boldsymbol{E}_y(z,t) = \boldsymbol{e}_y E_{0y} \cos(\omega t - kz + \delta), \tag{1.3}$$

where δ is the relative phase difference between the two orthogonal waves. The resultant wave can be obtained by combining (1.2) and (1.3) as follows:

$$\boldsymbol{E}(z,t) = \boldsymbol{E}_x(z,t) + \boldsymbol{E}_y(z,t) = \boldsymbol{e}_x E_{0x}\cos(\omega t - kz) + \boldsymbol{e}_y E_{0y}\cos(\omega t - kz + \delta). \tag{1.4}$$

The *linearly polarized* wave is obtained by setting the phase difference to an integer multiple of 2π, namely $\delta = m\cdot 2\pi$:

$$\boldsymbol{E}(z,t) = (\boldsymbol{e}_x E\cos\theta + \boldsymbol{e}_y E\sin\theta)\cos(\omega t - kz); \quad E = \sqrt{E_{0x}^2 + E_{0y}^2}, \quad \theta = \tan^{-1}\frac{E_{0y}}{E_{0x}}. \tag{1.5}$$

By ignoring the time-dependent term, we can represent the linear polarization as shown in Figure 1.1a. On the other hand, if $\delta \neq m\cdot 2\pi$, the *elliptical polarization* is obtained. From Eqs (1.2) and (1.3), by eliminating the time-dependent term we obtain the following equation of the ellipse:

$$\left(\frac{E_x}{E_{0x}}\right)^2 + \left(\frac{E_y}{E_{0y}}\right)^2 - 2\frac{E_x}{E_{0x}}\frac{E_y}{E_{0y}}\cos\delta = \sin^2\delta, \quad \tan 2\phi = \frac{2E_{0x}E_{0y}\cos\delta}{E_{0x}^2 - E_{0y}^2}, \tag{1.6}$$

which is shown in Figure 1.1b. By setting $\delta = \pm\pi/2, \pm 3\pi/2, \ldots$, the equation of the ellipse becomes

$$\left(\frac{E_x}{E_{0x}}\right)^2 + \left(\frac{E_y}{E_{0y}}\right)^2 = 1. \tag{1.7}$$

By setting further $E_{0x} = E_{0y} = E_0$; $\delta = \pm\pi/2, \pm 3\pi/2, \ldots$, the equation of the ellipse becomes the circle:

$$E_x^2 + E_y^2 = 1, \tag{1.8}$$

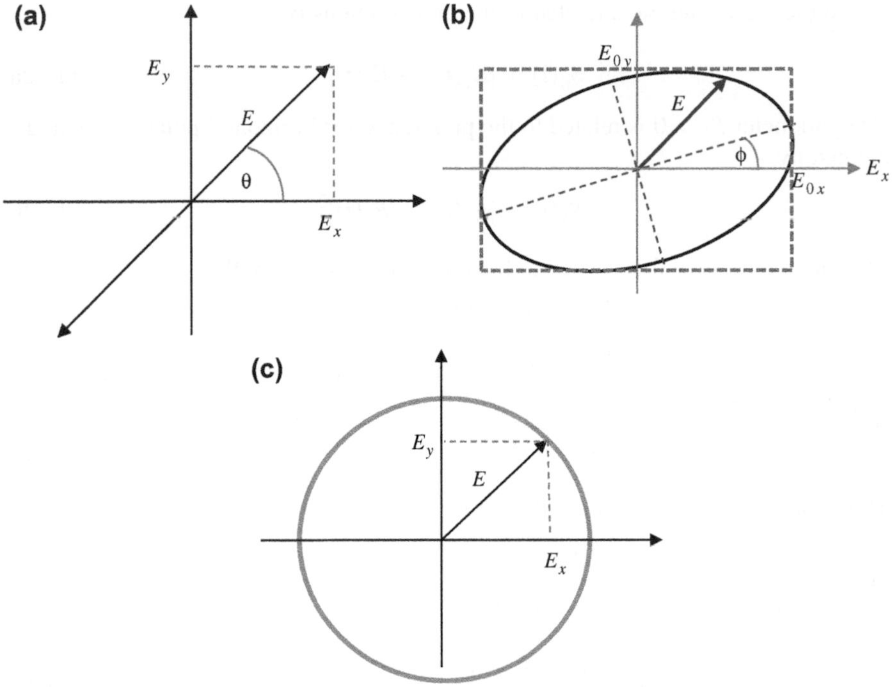

FIGURE 1.1

Various forms of polarization: (a) linear polarization; (b) elliptic polarization; (c) circular polarization.

and the corresponding polarization is known as *circular polarization* (see Figure 1.1c). A *right circularly polarized* wave is obtained for $\delta = \pi/2 + 2m\pi$:

$$\boldsymbol{E} = E_0[\boldsymbol{e}_x \cos(\omega t - kz) - \boldsymbol{e}_y \sin(\omega t - kz)], \tag{1.9}$$

Otherwise, for $\delta = -\pi/2 + 2m\pi$, the polarization is known as left circularly polarized.

Very often, the *Jones vector representation* of a polarization wave is used:

$$\boldsymbol{E}(t) = \begin{bmatrix} E_x(t) \\ E_y(t) \end{bmatrix} = E \begin{bmatrix} \sqrt{1-\kappa} \\ \sqrt{k} e^{j\delta} \end{bmatrix} e^{j(\omega t - kz)}, \tag{1.10}$$

where k is the power-splitting ratio between states of polarizations (SOPs), with the complex phasor term being typically omitted in practice.

Another interesting representation is the *Stokes vector representation*:

$$\boldsymbol{S}(t) = \begin{bmatrix} S_0(t) \\ S_1(t) \\ S_2(t) \\ S_3(t) \end{bmatrix}, \tag{1.11}$$

where the parameter S_0 is related to the optical intensity by

$$S_0(t) = |E_x(t)|^2 + |E_y(t)|^2. \tag{1.12a}$$

The parameter $S_1 > 0$ is related to the preference for horizontal polarization and is defined by

$$S_1(t) = |E_x(t)|^2 - |E_y(t)|^2. \tag{1.12b}$$

The parameter $S_2 > 0$ is related to the preference for $\pi/4$ SOP:

$$S_2(t) = E_x(t)E_y^*(t) + E_x^*(t)E_y(t). \tag{1.12c}$$

Finally, the parameter $S_3 > 0$ is related to the preference for right-circular polarization and is defined by

$$S_3(t) = j[E_x(t)E_y^*(t) - E_x^*(t)E_y(t)]. \tag{1.12d}$$

The parameter S_0 is related to other Stokes parameters by

$$S_0^2(t) = S_1^2(t) + S_2^2(t) + S_3^2(t). \tag{1.13}$$

The degree of polarization is defined by

$$p = \frac{\left[S_1^2 + S_2^2 + S_3^2\right]^{1/2}}{S_0}, \qquad 0 \leq p \leq 1 \tag{1.14}$$

For $p = 1$ the polarization does not change with time. The Stokes vector can be represented in terms of Jones vector parameters as

$$s_1 = 1 - 2\kappa, \quad s_2 = 2\sqrt{\kappa(1-\kappa)}\cos\delta, \quad s_3 = 2\sqrt{\kappa(1-\kappa)}\sin\delta. \tag{1.15}$$

After the *normalization* with respect to S_0, the normalized Stokes parameters are given by

$$s_i = \frac{S_i}{S_0}; \quad i = 0, 1, 2, 3. \tag{1.16}$$

If the normalized Stokes parameters are used, the polarization state can be represented as a point on a *Poincaré sphere*, as shown in Figure 1.2. The points located at the opposite sides of the line crossing the center represent the orthogonal polarizations.

The polarization ellipse is very often represented in terms of the ellipticity and the azimuth, which are illustrated in Figure 1.3. The ellipticity is defined by the ratio of half-axis lengths. The corresponding angle is called the ellipticity angle and is denoted by ε. Small ellipticity means that the polarization ellipse is highly elongated, while for zero elipticity the polarization is linear. For $\varepsilon = \pm\pi/4$, the polarization is circular. For $\varepsilon > 0$ the polarization is right-elliptical. On the other hand, the azimuth angle η defines the orientation of the main axis of the ellipse with respect to E_x.

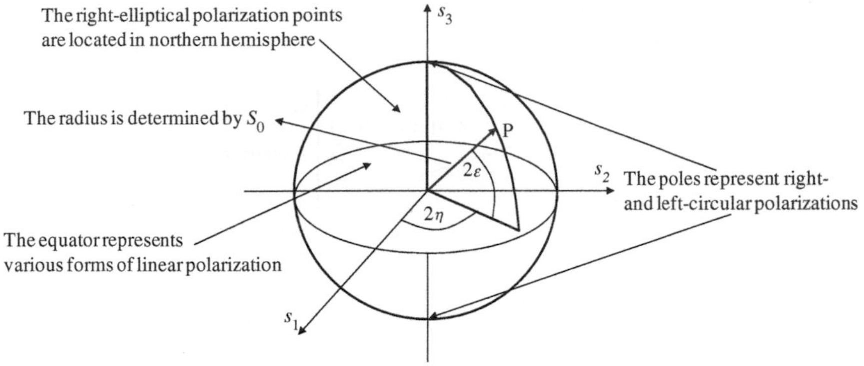
The right-elliptical polarization points
are located in northern hemisphere
s_3
The radius is determined by S_0
P
2ε
s_2 The poles represent right-
and left-circular polarizations
2η
The equator represents
various forms of linear polarization
s_1

FIGURE 1.2

Representation of polarization state as a point on a Poincaré sphere.

The polarization ellipse parameters can be related to the Jones vector parameters by

$$\sin 2\varepsilon = 2\sqrt{\kappa(1-\kappa)}\,\sin\delta, \quad \tan 2\eta = \frac{2\sqrt{\kappa(1-\kappa)}\sin\delta}{1-2\kappa}. \tag{1.17}$$

Finally, the parameters of the polarization ellipse can be related to the Stokes vector parameters by

$$s_1 = \cos 2\eta \cos 2\varepsilon, \quad s_2 = \sin 2\eta \cos 2\varepsilon, \quad s_3 = \sin 2\varepsilon, \tag{1.18}$$

and the corresponding geometrical interpretation is provided in Figures 1.2 and 1.3.

Let us now observe the *polarizer–analyzer ensemble*, shown in Figure 1.4. When an electromagnetic wave passes through the polarizer, it can be represented as a vector in the xOy plane transverse to the propagation direction, as given by Eq. (1.5), where the angle θ depends on the filter orientation. By introducing the unit vector $\hat{p} = (\cos\theta, \sin\theta)$, Eq. (1.5) can be rewritten as

$$\boldsymbol{E} = E_0\hat{p}\cos(\omega t - kz). \tag{1.19}$$

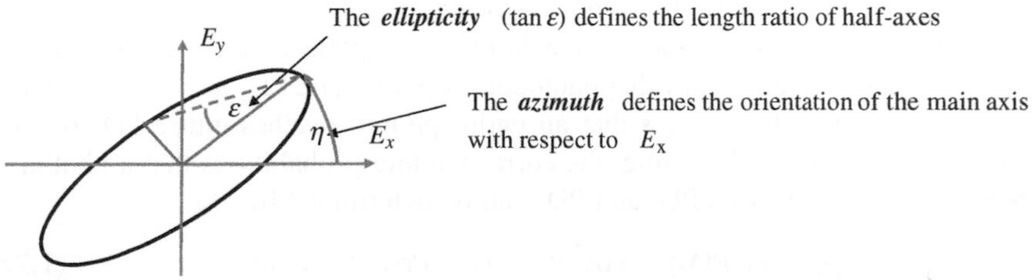
The *ellipticity* (tan ε) defines the length ratio of half-axes
E_y
ε
η
E_x
The *azimuth* defines the orientation of the main axis
with respect to E_x

FIGURE 1.3

The ellipticity and azimuth of the polarization ellipse.

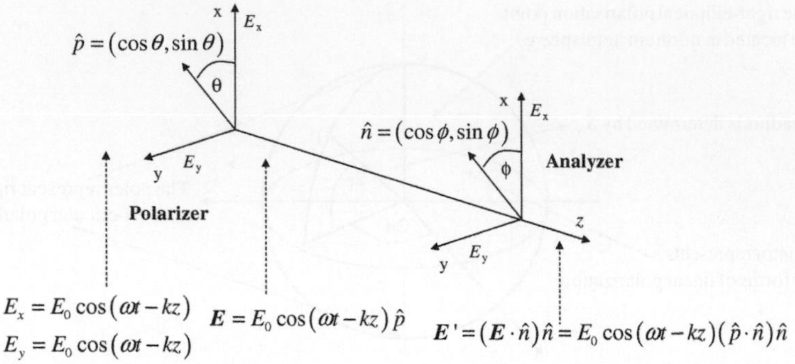

FIGURE 1.4

The polarizer–analyzer ensemble for study of photon polarization.

If $\theta = 0$ rad, the light is polarized along the x-axis, while for $\theta = \pi/2$ rad it is polarized along the y-axis. *Natural light* is *unpolarized* as it represents *incoherent superposition* of 50% of the light polarized along the x-axis and 50% of light polarized along the y-axis. After the analyzer, whose axis makes an angle ϕ with respect to the x-axis, which can be represented by unit vector $\hat{n} = (\cos\phi, \sin\phi)$, the output electric field is given by

$$\begin{aligned} \boldsymbol{E}' = (\boldsymbol{E}\cdot\hat{n})\hat{n} &= E_0\cos(\omega t - kz)(\hat{p}\cdot\hat{n})\hat{n} \\ &= E_0\cos(\omega t - kz)[(\cos\theta, \sin\theta)\cdot(\cos\phi, \sin\phi)]\hat{n} \\ &= E_0\cos(\omega t - kz)[\cos\theta\cos\phi + \sin\theta\sin\phi]\hat{n} \\ &= E_0\cos(\omega t - kz)\cos(\theta - \phi)\hat{n}. \end{aligned} \tag{1.20}$$

The intensity of the analyzer output field is given by

$$I' = |\boldsymbol{E}'|^2 = I\cos^2(\theta - \phi), \tag{1.21}$$

which is commonly referred to as the *Malus law*.

Decomposition of polarization by a birefringent plate is now considered (see Figure 1.5). Experiments show that photodetectors PD_x and PD_y are never triggered simultaneously, which indicates that an entire photon reaches either PD_x or PD_y (a photon never splits). Therefore, the corresponding probabilities that a photon is detected by photodetectors PD_x and PD_y can be determined by

$$p_x = \Pr(PD_x) = \cos^2\theta, \quad p_y = \Pr(PD_y) = \sin^2\theta. \tag{1.22}$$

If the total number of photons is N, the number of detected photons in x-polarization will be $N_x \cong N\cos^2\theta$ and the number of detected photons in y-polarization

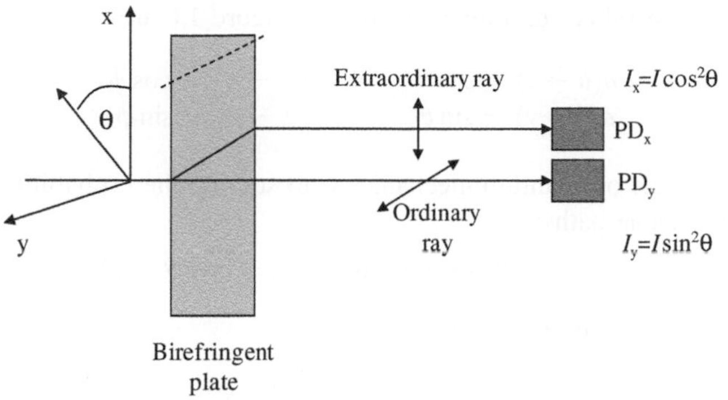

FIGURE 1.5

Polarization decomposition by a birefringent plate. PD, photodetector.

will be $N_y \cong N\sin^2\theta$. In the limit, as $N \to \infty$ we would expect the Malus law to be obtained.

Let us now study the polarization decomposition and recombination by means of birefringent plates, as illustrated in Figure 1.6. Classical physics prediction of the total probability of a photon passing the polarizer—analyzer ensemble is given by

$$p_{\text{tot}} = \cos^2\theta \cos^2\phi + \sin^2\theta \sin^2\phi \neq \cos^2(\theta - \phi), \tag{1.23}$$

which is inconsistent with the Malus law, given by Eq. (1.21). In order to reconstruct the results from wave optics, it is necessary to introduce into quantum mechanics the concept of *probability amplitude*, that α is detected as β, which is denoted as $a(\alpha \to \beta)$, and it is a complex number. The probability is obtained as the squared magnitude of probability amplitude:

$$p(\alpha \to \beta) = |a(\alpha \to \beta)|^2. \tag{1.24}$$

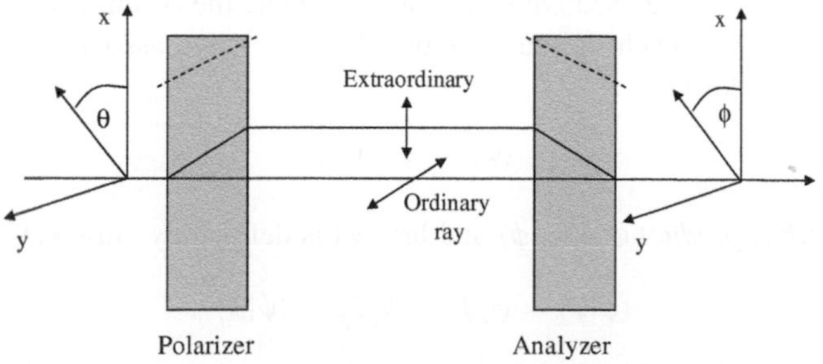

FIGURE 1.6

Polarization decomposition and recombination by a birefringent plate.

The relevant probability amplitudes relating to Figure 1.6 are

$$\begin{aligned} a(\theta \to x) &= \cos\theta \qquad & a(x \to \phi) &= \cos\phi \\ a(\theta \to y) &= \sin\theta \qquad & a(x \to \phi) &= \sin\phi. \end{aligned} \tag{1.25}$$

The basic principle of quantum mechanics is to sum up the probability amplitudes for indistinguishable paths:

$$a_{\text{tot}} = \cos\theta\cos\phi + \sin\theta\sin\phi = \cos(\theta - \phi). \tag{1.26}$$

The corresponding total probability is

$$p_{\text{tot}} = |a_{\text{tot}}|^2 = \cos^2(\theta - \phi), \tag{1.27}$$

and this result is consistent with the Malus law!

Based on the previous discussion, the *state vector* of the photon polarization is given by

$$|\psi\rangle = \begin{pmatrix} \psi_x \\ \psi_y \end{pmatrix}, \tag{1.28}$$

where ψ_x is related to x-polarization and ψ_y to y-polarization, with the normalization condition as follows:

$$|\psi_x|^2 + |\psi_y|^2 = 1. \tag{1.29}$$

In this representation, the x- and y-polarization photons can be represented by

$$|x\rangle = \begin{pmatrix} 1 \\ 0 \end{pmatrix}, \quad |y\rangle = \begin{pmatrix} 0 \\ 1 \end{pmatrix}, \tag{1.30}$$

and the right and left circular polarization photons are represented by

$$|R\rangle = \frac{1}{\sqrt{2}}\begin{pmatrix} 1 \\ j \end{pmatrix}, \quad |L\rangle = \frac{1}{\sqrt{2}}\begin{pmatrix} 1 \\ -j \end{pmatrix}. \tag{1.31}$$

In Eqs (1.28)–(1.31) we used Dirac notation to denote the column vectors (kets). In Dirac notation, with each column vector ("ket") $|\psi\rangle$, we associate a row vector ("bra") $\langle\psi|$ as follows:

$$\langle\psi| = (\psi_x^* \;\; \psi_y^*). \tag{1.32}$$

The *scalar (dot) product* of ket $|\phi\rangle$ and bra $\langle\psi|$ is defined by "bracket" as follows:

$$\langle\phi|\psi\rangle = \phi_x^*\psi_x + \phi_y^*\psi_y = \langle\psi|\phi\rangle^*. \tag{1.33}$$

The normalization condition can be expressed in terms of scalar product by

$$\langle\psi|\psi\rangle = 1. \tag{1.34}$$

Based on (1.30) and (1.31), it is evident that

$$\langle x|x\rangle = \langle y|y\rangle = 1, \qquad \langle R|R\rangle = \langle L|L\rangle = 1.$$

Because the vectors $|x\rangle$ and $|y\rangle$ are *orthogonal*, their dot product is zero:

$$\langle x|y\rangle = 0 \tag{1.35}$$

and they form the *basis*. Any state vector $|\psi\rangle$ can be written as a *linear superposition* of basis kets as follows:

$$|\psi\rangle = \begin{pmatrix} \psi_x \\ \psi_y \end{pmatrix} = \psi_x|x\rangle + \psi_y|y\rangle. \tag{1.36}$$

We can use now (1.34) and (1.36) to derive an important relation in quantum mechanics, known as the *completeness relation*. The projections of state vector $|\psi\rangle$ along basis vectors $|x\rangle$ and $|y\rangle$ are given by

$$\langle x|\psi\rangle = \psi_x \underbrace{\langle x|x\rangle}_{1} + \psi_y \underbrace{\langle x|y\rangle}_{0} = \psi_x, \quad \langle y|\psi\rangle = \psi_x\langle y|x\rangle + \psi_y\langle y|y\rangle = \psi_y. \tag{1.37}$$

By substituting (1.37) into (1.36) we obtain:

$$|\psi\rangle = |x\rangle\langle x|\psi\rangle + |y\rangle\langle y|\psi\rangle = \underbrace{(|x\rangle\langle x| + |y\rangle\langle y|)}_{I}|\psi\rangle, \tag{1.38}$$

and from the right side of Eq. (1.38) we derive the completeness relation:

$$|x\rangle\langle x| + |y\rangle\langle y| = I. \tag{1.39}$$

The probability that the photon in state $|\psi\rangle$ will pass the x-polaroid is given by

$$p_x = \frac{|\psi_x|^2}{|\psi_x|^2 + |\psi_y|^2} = |\psi_x|^2 = |\langle x|\psi\rangle|^2. \tag{1.40}$$

The probability amplitude of the photon in state $|\psi\rangle$ to pass the x-polaroid is

$$a_x = \langle x|\psi\rangle. \tag{1.41}$$

Let $|\phi\rangle$ and $|\psi\rangle$ be two physical states. The probability amplitude of finding ϕ in ψ, denoted as $a(\phi \rightarrow \psi)$, is given by

$$a(\phi \rightarrow \psi) = \langle \phi|\psi\rangle, \tag{1.42}$$

and the probability of ϕ passing the ψ test is given by

$$p(\phi \rightarrow \psi) = |\langle \phi|\psi\rangle|^2. \tag{1.43}$$

1.2 THE CONCEPT OF THE QUBIT

Based on the previous section, it can be concluded that the quantum bit, also known as the *qubit*, lies in a two-dimensional Hilbert space H, isomorphic to C^2 space, where C is the complex number space, and can be represented as

$$|\psi\rangle = \alpha|0\rangle + \beta|1\rangle = \begin{pmatrix} \alpha \\ \beta \end{pmatrix}; \quad \alpha, \beta \in C; \quad |\alpha|^2 + |\beta|^2 = 1, \tag{1.44}$$

where the $|0\rangle$ and $|1\rangle$ states are computational basis (CB) states, and $|\psi\rangle$ is a superposition state. If we perform the measurement of a qubit, we will get $|0\rangle$ with probability $|\alpha|^2$ and $|1\rangle$ with probability of $|\beta|^2$. Measurement changes the state of a qubit from a superposition of $|0\rangle$ and $|1\rangle$ to the specific state consistent with the measurement result. If we parametrize the probability amplitudes α and β as follows:

$$\alpha = \cos\left(\frac{\theta}{2}\right), \quad \beta = e^{j\phi} \sin\left(\frac{\theta}{2}\right), \tag{1.45}$$

where θ is a polar angle and ϕ is an azimuthal angle, we can geometrically represent the qubit by a Bloch sphere (or a Poincaré sphere for the photon), as illustrated in Figure 1.7. (Note that the Bloch sphere in Figure 1.7 is a little different from the Poincaré sphere in Figure 1.2.) Bloch vector coordinates are given by ($\cos\phi \sin\theta$, $\sin\phi \sin\theta$, $\cos\theta$). This Bloch vector representation is related to the CB by

$$|\psi(\theta,\phi)\rangle = \cos(\theta/2)|0\rangle + e^{j\phi}\sin(\theta/2)|1\rangle \doteq \begin{pmatrix} \cos(\theta/2) \\ e^{j\phi}\sin(\theta/2) \end{pmatrix}, \tag{1.46}$$

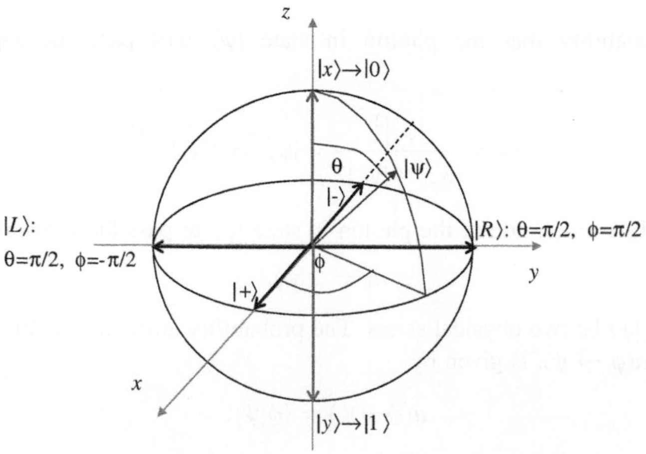

FIGURE 1.7

Block (Poincaré) sphere representation of a single qubit.

where $0 \leq \theta \leq \pi$ and $0 \leq \phi < 2\pi$. The north and south poles correspond to computational $|0\rangle$ ($|x\rangle$-polarization) and $|1\rangle$ ($|y\rangle$-polarization) basis kets respectively. Other important bases are the *diagonal basis* $\{|+\rangle, |-\rangle\}$, very often denoted as $\{|\nearrow\rangle, |\searrow\rangle\}$, related to the CB by

$$|+\rangle = |\nearrow\rangle = \frac{1}{\sqrt{2}}(|0\rangle + |1\rangle), \qquad |-\rangle = |\searrow\rangle = \frac{1}{\sqrt{2}}(|0\rangle - |1\rangle) \tag{1.47}$$

and the *circular basis* $\{|R\rangle, |L\rangle\}$, related to the CB as follows:

$$|R\rangle = \frac{1}{\sqrt{2}}(|0\rangle + j|1\rangle), \qquad |L\rangle = \frac{1}{\sqrt{2}}(|0\rangle - j|1\rangle). \tag{1.48}$$

1.3 SPIN-1/2 SYSTEMS

In addition to the photon, an important realization of the qubit is a spin-1/2 system. Nuclear magnetic resonance (NMR) is based on the fact that the proton possesses a magnetic moment μ that can take only two values along the direction of the magnetic field. That is, the projection along the $\hat{n}$ axis, obtained by $\boldsymbol{\mu}\cdot\hat{n}$, takes only two values, and this property characterizes a spin-1/2 particle. Experimentally this was confirmed by the *Stern–Gerlach experiment*, shown in Figure 1.8. A beam of protons is sent into a nonhomogenic magnetic field of a direction $\hat{n}$ orthogonal to the beam direction. It can be observed that the beam splits into two subbeams, one deflected in the direction $+\hat{n}$ and the other in the opposite direction $-\hat{n}$. Therefore, the proton never splits. The basis in spin-1/2 systems can be written as $\{|+\rangle, |-\rangle\}$, where the corresponding basis kets represent the spin-up and spin-down states. The superposition state can be represented in terms of these bases as follows:

$$|\psi\rangle = \begin{pmatrix} \psi_+ \\ \psi_- \end{pmatrix} = \psi_+|+\rangle + \psi_-|-\rangle, \quad |\psi_-|^2 + |\psi_+|^2 = 1, \tag{1.49}$$

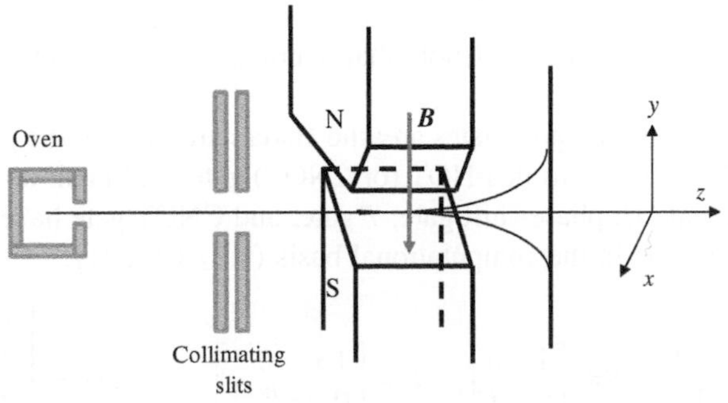

FIGURE 1.8

The Stern–Gerlach experiment.

where $|\psi_+|^2$ ($|\psi_-|^2$) denotes the probability of finding the system in the spin-up (spin-down) state. By using the trigonometric identity $\sin^2(\theta/2)+\cos^2(\theta/2)=1$, the expansion coefficients ψ_+ and ψ_- can be expressed as follows:

$$\psi_+ = \mathrm{e}^{-j\phi/2}\cos(\theta/2) \qquad \psi_- = \mathrm{e}^{j\phi/2}\sin(\theta/2), \tag{1.50}$$

so that

$$|\psi\rangle = \mathrm{e}^{-j\phi/2}\cos(\theta/2)|+\rangle + \mathrm{e}^{j\phi/2}\sin(\theta/2)|-\rangle. \tag{1.51}$$

Therefore, the single superposition state of a spin-1/2 system can also be visualized as the point (θ,ϕ) on a unit sphere (Bloch sphere).

1.4 QUANTUM GATES AND QUANTUM INFORMATION PROCESSING

In quantum mechanics, the primitive undefined concepts are *physical system*, *observable*, and *state*. The concept of state has been introduced in the previous sections. An observable, such as momentum and spin, can be represented by an *operator*, such as A, in the vector space in question. An operator, or gate, acts on a ket from the left, $(A)\cdot|\alpha\rangle = A|\alpha\rangle$, and results in another ket. A linear operator (gate) B can be expressed in terms of eigenkets $\{|a^{(n)}\rangle\}$ of a Hermitian operator A. (An operator A is said to be *Hermitian* if $A^\dagger = A,\ A^\dagger = (A^{\mathrm{T}})^*$.) The *operator* X is associated with a *square matrix* (albeit infinite in extent), whose elements are

$$X_{mn} = \langle a^{(m)}|X|a^{(n)}\rangle, \tag{1.52}$$

and can explicitly be written as

$$X \doteq \begin{pmatrix} \langle a^{(1)}|X|a^{(1)}\rangle & \langle a^{(1)}|X|a^{(2)}\rangle & \cdots \\ \langle a^{(2)}|X|a^{(1)}\rangle & \langle a^{(2)}|X|a^{(2)}\rangle & \cdots \\ \vdots & \vdots & \ddots \end{pmatrix}, \tag{1.53}$$

where we use the notation $\doteq$ to denote that operator X is represented by the matrix above.

Very important single-qubit gates are: the Hadamard gate H, the phase shift gate S, the $\pi/8$ (or T) gate, controlled-NOT (or CNOT) gate, and Pauli operators X, Y, Z. The Hadamard gate H, phase-shift gate, T gate, and CNOT gate have the following matrix representation in the computational basis (CB) $\{|0\rangle,|1\rangle\}$:

$$H \doteq \frac{1}{\sqrt{2}}\begin{bmatrix} 1 & 1 \\ 1 & -1 \end{bmatrix},\quad S \doteq \begin{bmatrix} 1 & 0 \\ 0 & j \end{bmatrix},\quad T \doteq \begin{bmatrix} 1 & 0 \\ 0 & \mathrm{e}^{j\pi/4} \end{bmatrix},\quad \mathrm{CNOT} \doteq \begin{bmatrix} 1 & 0 & 0 & 0 \\ 0 & 1 & 0 & 0 \\ 0 & 0 & 0 & 1 \\ 0 & 0 & 1 & 0 \end{bmatrix}. \tag{1.54}$$

The Pauli operators, on the other hand, have the following matrix representation in the CB:

$$X \doteq \begin{bmatrix} 0 & 1 \\ 1 & 0 \end{bmatrix}, \quad Y \doteq \begin{bmatrix} 0 & -j \\ j & 0 \end{bmatrix}, \quad Z \doteq \begin{bmatrix} 1 & 0 \\ 0 & -1 \end{bmatrix}. \tag{1.55}$$

The action of Pauli gates on an arbitrary qubit $|\psi\rangle = a|0\rangle + b|1\rangle$ is given as follows:

$$X(a|0\rangle + b|1\rangle) = a|1\rangle + b|0\rangle, \quad Y(a|0\rangle + b|1\rangle) = j(a|1\rangle - b|0\rangle), \quad Z(a|0\rangle + b|1\rangle) = a|0\rangle - b|1\rangle. \tag{1.56}$$

So the action of an X gate is to introduce the bit flip, the action of a Z gate is to introduce the phase flip, and the action of a Y gate is to simultaneously introduce the bit and phase flips.

Several important single-, double-, and three-qubit gates are shown in Figure 1.9. The action of a single-qubit gate is to apply the operator U on qubit $|\psi\rangle$, which results in another qubit. A controlled-U gate conditionally applies the operator U on target qubit $|\psi\rangle$, when the control qubit $|c\rangle$ is in the $|1\rangle$ state. One particularly important controlled U-gate is the controlled-NOT (CNOT) gate. This gate flips the content of target qubit $|t\rangle$ when the control qubit $|c\rangle$ is in the $|1\rangle$ state. The purpose of the SWAP gate is to interchange the positions of two qubits, and can be implemented by using three CNOT gates, as shown in Figure 1.9d. Finally, the Toffoli gate represents the generalization of the CNOT gate, where two control qubits are used. The minimum set of gates that can be used to perform an arbitrary quantum computation algorithm

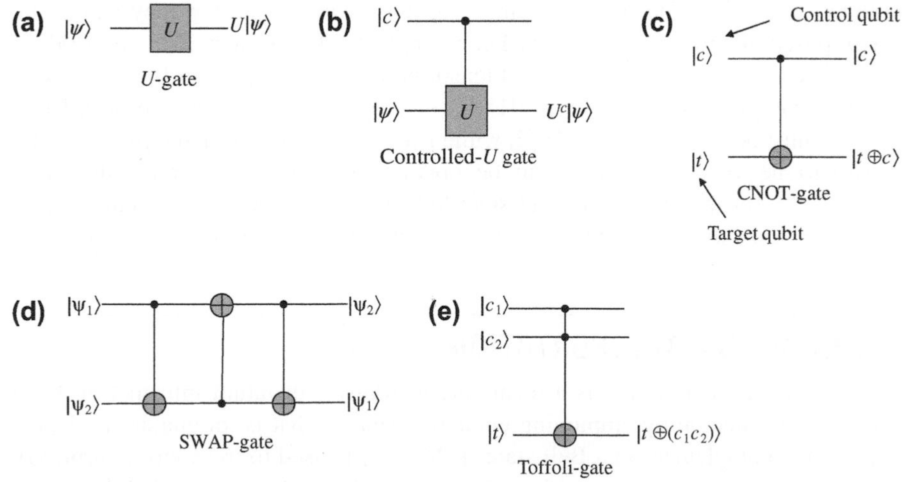

FIGURE 1.9

Important quantum gates and their actions: (a) single-qubit gate; (b) controlled-U gate; (c) CNOT gate; (d) SWAP gate; (e) Toffoli gate.

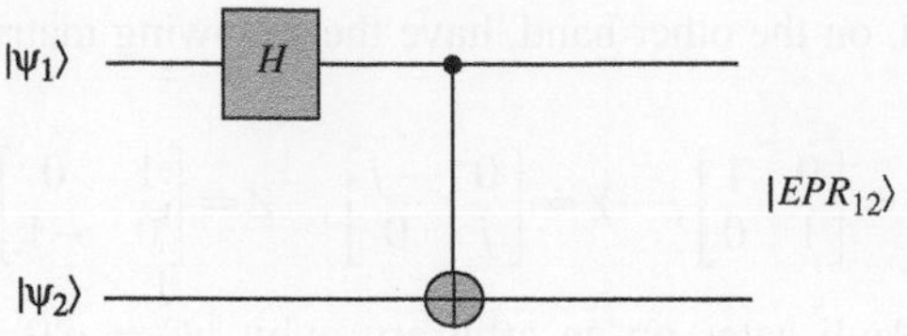

FIGURE 1.10

Bell states (EPR pairs) preparation circuit.

is known as the *universal set of gates*. The most popular sets of universal quantum gates are $\{H, S, \text{CNOT, Toffoli}\}$ gates, $\{H, S, \pi/8\ (T), \text{CNOT}\}$ gates, the Barenco gate [13], and the Deutsch gate [14]. By using these universal quantum gates, more complicated operations can be performed. As an illustration, in Figure 1.10 the Bell states (Einstein–Podolsky–Rosen (EPR) pairs) preparation circuit is shown; this is very important in quantum teleportation and QKD applications.

So far, single-, double-, and triple-qubit quantum gates have been considered. An arbitrary quantum state of K qubits has the form $\Sigma_s \alpha_s |s\rangle$, where s runs over all binary strings of length K. Therefore, there are 2^K complex coefficients, all independent except for the normalization constraint:

$$\sum_{s=00...00}^{11...11} |\alpha_s|^2 = 1. \tag{1.57}$$

For example, the state $\alpha_{00}|00\rangle + \alpha_{01}|01\rangle + \alpha_{10}|10\rangle + \alpha_{11}|11\rangle$ (with $|\alpha_{00}|^2 + |\alpha_{01}|^2 + |\alpha_{10}|^2 + |\alpha_{11}|^2 = 1$) is the general two-qubit state (we use $|00\rangle$ to denote the tensor product $|0\rangle \otimes |0\rangle$). The multiple qubits can be *entangled* so that they cannot be decomposed into two separate states. For example, the Bell state or EPR pair $(|00\rangle + |11\rangle)/\sqrt{2}$ cannot be written in terms of tensor product $|\psi_1\rangle|\psi_2\rangle = (\alpha_1|0\rangle + \beta_1|1\rangle) \otimes (\alpha_2|0\rangle + \beta_2|1\rangle) = \alpha_1\alpha_2|00\rangle + \alpha_1\beta_2|01\rangle + \beta_1\alpha_2|10\rangle + \beta_1\beta_2|11\rangle$, because in order to do so it must be $\alpha_1\alpha_2 = \beta_1\beta_2 = 1/\sqrt{2}$, while $\alpha_1\beta_2 = \beta_1\alpha_2 = 0$, which *a priori* has no reason to be valid. This state can be obtained by using the circuit shown in Figure 1.10, for the two-qubit input state $|00\rangle$. For more details on quantum gates and algorithms, the interested reader is referred to Chapters 3 and 5 respectively.

1.5 QUANTUM TELEPORTATION

Quantum teleportation [17] is a technique to transfer quantum information from source to destination by employing entangled states. That is, in quantum teleportation, the entanglement in a Bell state (EPR pair) is used to transport an arbitrary quantum state $|\psi\rangle$ between two distant observers A and B (often called Alice and Bob), as illustrated in Figure 1.11. The quantum teleportation system employs three qubits: qubit 1 is an arbitrary state to be teleported, while qubits 2 and 3 are in a Bell state $|B_{00}\rangle = (|00\rangle + |11\rangle)/\sqrt{2}$. Let the state to be teleported be denoted by

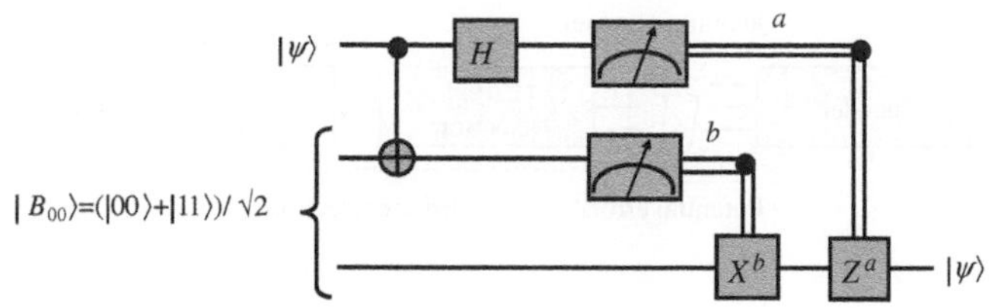

FIGURE 1.11

The quantum teleportation principle.

$|\psi\rangle = a|0\rangle + b|1\rangle$. The input to the circuit shown in Figure 1.11 is therefore $|\psi\rangle|B_{00}\rangle$, and can be rewritten as

$$\begin{aligned}|\psi\rangle|B_{00}\rangle &= (a|0\rangle + b|1\rangle)(|00\rangle + |11\rangle)/\sqrt{2} \\ &= (a|000\rangle + a|011\rangle + b|100\rangle + b|111\rangle)/\sqrt{2}. \end{aligned} \tag{1.58}$$

The CNOT gate is then applied with the first qubit serving as control and the second qubit as target, which transforms (1.58) into

$$\begin{aligned}\mathrm{CNOT}^{(12)}|\psi\rangle|B_{00}\rangle &= \mathrm{CNOT}^{(12)}(a|000\rangle + a|011\rangle + b|100\rangle + b|111\rangle)/\sqrt{2} \\ &= (a|000\rangle + a|011\rangle + b|110\rangle + b|101\rangle)/\sqrt{2}. \end{aligned} \tag{1.59}$$

In the next stage, the Hadamard gate is applied to the first qubit, which maps $|0\rangle$ to $(|0\rangle + |1\rangle)/\sqrt{2}$ and $|1\rangle$ to $(|0\rangle - |1\rangle)/\sqrt{2}$, so that the overall transformation of (1.59) is as follows:

$$\begin{aligned}H^{(1)}\mathrm{CNOT}^{(12)}|\psi\rangle|B_{00}\rangle = \frac{1}{2}(&a|000\rangle + a|100\rangle + a|011\rangle + a|111\rangle \\ &+ b|010\rangle - b|110\rangle + b|001\rangle - b|101\rangle). \end{aligned} \tag{1.60}$$

The measurements are performed on qubits 1 and 2, and based on the results of measurements, denoted respectively as a and b, the controlled-X (CNOT) and controlled-Z gates are applied conditionally to lead to the following content on qubit 3:

$$\frac{1}{2}(2a|0\rangle + 2b|1\rangle) = a|0\rangle + b|1\rangle = |\psi\rangle, \tag{1.61}$$

indicating that the arbitrary state $|\psi\rangle$ is teleported to the remote destination and can be found at the qubit 3 position.

1.6 QUANTUM ERROR CORRECTION CONCEPTS

The QIP relies on delicate superposition states, which are sensitive to interactions with the environment, resulting in decoherence. Moreover, the quantum gates are

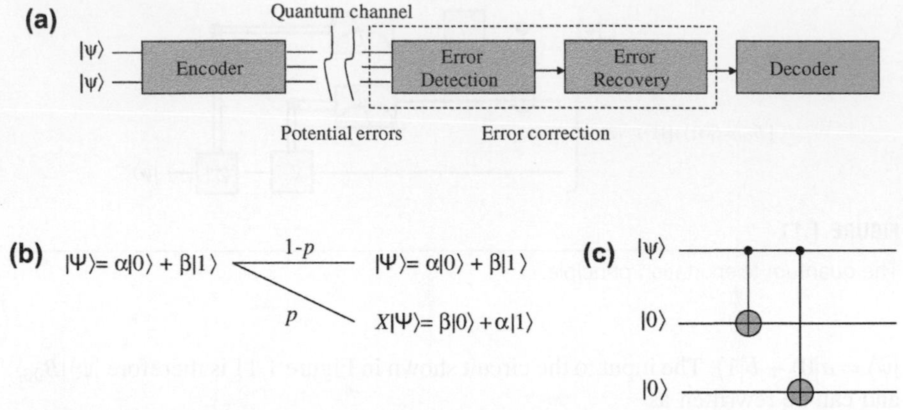

FIGURE 1.12

(a) Quantum error-correction principle. (b) The bit-flipping quantum channel model. (c) Three-qubit flip-error correction code encoder.

imperfect and the use of quantum error correction coding (QECC) is necessary to enable fault-tolerant computing and to deal with quantum errors [18–23]. QECC is also essential in quantum communication and quantum teleportation applications. The elements of quantum error correction codes are shown in Figure1.12a. The (*N*,*K*) QECC code performs encoding of the quantum state of *K* qubits, specified by 2^K complex coefficients α_s, into a quantum state of *N* qubits, in such a way that errors can be detected and corrected, and all 2^K complex coefficients can be perfectly restored, up to the global phase shift. This means that, from quantum mechanics (see Chapter 2), we know that two states $|\psi\rangle$ and $e^{j\theta}|\psi\rangle$ are equal up to a *global phase shift* as the results of measurement on both states are the same. A quantum error correction consists of four major steps: encoding, error detection, error recovery, and decoding. The sender (Alice) encodes quantum information in state $|\psi\rangle$ with the help of local ancillary qubits $|0\rangle$, and then sends the encoded qubits over a noisy quantum channel (say, a free-space optical channel or optical fiber). The receiver (Bob) performs multi-qubit measurement on all qubits to diagnose the channel error and performs a recovery unitary operation *R* to reverse the action of the channel. Quantum error correction is essentially more complicated than classical error correction. The difficulties in quantum error correction can be summarized as follows: (i) the no-cloning theorem indicates that it is impossible to make a copy of an arbitrary quantum state; (ii) quantum errors are continuous and a qubit can be in any superposition of the two bases states; and (iii) the measurements destroy the quantum information. The quantum error correction principles will be more evident after a simple example given below.

Assume we want to send a single qubit $|\psi\rangle = \alpha|0\rangle + \beta|1\rangle$ through the quantum channel in which during transmission the transmitted qubit can be flipped to

$X|\psi\rangle = \beta|0\rangle + \alpha|1\rangle$ with probability p. Such a quantum channel is called a *bit-flip channel* and it can be described as shown in Figure 1.12b. The three-qubit flip code sends the same qubit three times, and therefore represents the *repetition code* equivalent. The corresponding codewords in this code are $|\bar{0}\rangle = |000\rangle$ and $|\bar{1}\rangle = |111\rangle$. The three-qubit flip-code encoder is shown in Figure 1.12c. One input qubit and two ancillaries are used at the input encoder, which can be represented by $|\psi_{123}\rangle = \alpha|000\rangle + \beta|100\rangle$. The first ancillary qubit (the second qubit at the encoder input) is controlled by the information qubit (the first qubit at encoder input) so that its output can be represented by $\text{CNOT}_{12}(\alpha|000\rangle + \beta|100\rangle) = \alpha|000\rangle + \beta|110\rangle$ (if the control qubit is $|1\rangle$ the target qubit gets flipped, otherwise it stays unchanged). The output of the first CNOT gate is used as input to the second CNOT gate in which the second ancillary qubit (the third qubit) is controlled by the information qubit (the first qubit), so that the corresponding encoder output is obtained as $\text{CNOT}_{13}(\alpha|000\rangle + \beta|110\rangle) = \alpha|000\rangle + \beta|111\rangle$, which indicates that basis codewords are indeed $|\bar{0}\rangle$ and $|\bar{1}\rangle$. With this code, we are able to correct a single qubit-flip error, which occurs with probability $(1-p)^3 + 3p(1-p)^2 = 1 - 3p^2 + 2p^3$. Therefore, the probability of an error remaining uncorrected or wrongly corrected with this code is $3p^2 - 2p^3$. It is clear from Figure 1.12c that the three-qubit flip-code encoder is a *systematic encoder* in which the information qubit is unchanged, and the ancillary qubits are used to impose the encoding operation and create the parity qubits (the output qubits 2 and 3).

Let us assume that a qubit flip occurred on the first qubit, leading to received quantum word $|\psi_r\rangle = \alpha|100\rangle + \beta|011\rangle$. In order to identify the error it is necessary to perform the measurements on the observables Z_1Z_2 and Z_2Z_3, where the subscript denotes the index of qubit on which a given Pauli gate is applied. The result of the measurement is the eigenvalue ± 1, and corresponding eigenvectors are two valid codewords, namely $|000\rangle$ and $|111\rangle$. The observables can be represented as follows:

$$\begin{aligned} Z_1Z_2 &= (|00\rangle\langle 11| + |11\rangle\langle 11|)\otimes I - (|01\rangle\langle 01| + |10\rangle\langle 10|)\otimes I \\ Z_2Z_3 &= I\otimes(|00\rangle\langle 11| + |11\rangle\langle 11|) - I\otimes(|01\rangle\langle 01| + |10\rangle\langle 10|). \end{aligned} \tag{1.62}$$

It can be shown that $\langle\psi_r|Z_1Z_2|\psi_r\rangle = -1$, $\langle\psi_r|Z_2Z_3|\psi_r\rangle = +1$, indicating that an error occurred on either the first or second qubit, but not on the second or third qubit. The intersection reveals that the first qubit was in error. By using this approach we can create the three-qubit look-up table (LUT), given in Table 1.1.

Table 1.1 The three-qubit flip-code LUT

Z_1Z_2	Z_2Z_3	Error
+1	+1	I
+1	−1	X_3
−1	+1	X_1
−1	−1	X_2

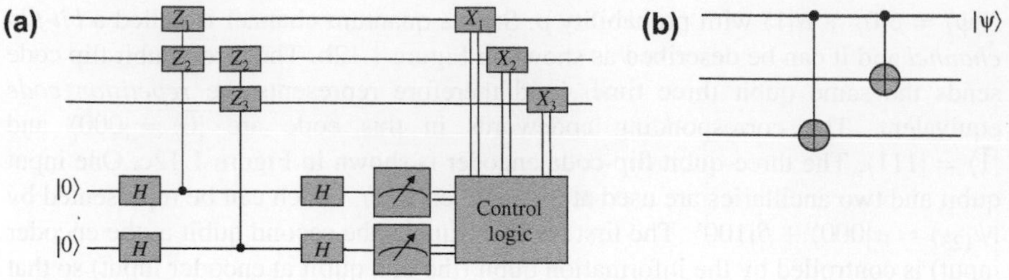

FIGURE 1.13

(a) Three-qubit flip-code error detection and error correction circuit. (b) Decoder circuit configuration.

Three-qubit flip-code error detection and error correction circuits are shown in Figure 1.13. The results of measurements on ancillaries (see Figure 1.13a) will determine the error syndrome $[\pm 1 \pm 1]$, and based on the LUT given in Table 1.1, we identify the error event and apply the corresponding X_i gate on the ith qubit being in error, and the error is corrected since $X^2 = I$. The control logic operation is described in Table 1.1. For example, if both outputs of the measurements circuits are -1, the operator X_2 is activated. The last step is to perform decoding as shown in Figure 1.13b by simply reversing the order of elements in the corresponding encoder.

1.7 QUANTUM KEY DISTRIBUTION (QKD)

The QKD exploits the principle of quantum mechanics in order to enable to demonstrably secure distribution of a private key between two remote destinations. Private key cryptography is much older than public key cryptosystems, commonly used today. In a private key cryptosystem Alice (sender) must have an *encoding key*, while Bob (receiver) must have a matching *decoding key* to decrypt the encoded message. The simplest private key cryptosystem is the *Vernam cipher* (*one time pad*), which operates as follows [1]: (i) Alice and Bob share n-bit key strings; (ii) Alice encodes her n-bit message by adding the message and the key together; and (iii) Bob decodes the information by subtracting the key from the received message. There are several *drawbacks* in this scheme: (i) secure distribution of the key as well as the length of the key must be at least as long as the message length; (ii) the key bits cannot be reused; and (iii) the keys must be delivered in advance, securely stored until use, and be destroyed after their use.

On the other hand, if the classical information, such as key, is transmitted over the quantum channel, because of the no-cloning theorem, which states that a device cannot be constructed to produce an exact copy of an arbitrary quantum state, an eavesdropper cannot get an exact copy of the key in a QKD system. That is, in an attempt to distinguish between two non-orthogonal states, information gain is only possible at the expense of introducing disturbance to the signal. This observation can be proved as follows. Let $|\psi_1\rangle$ and $|\psi_2\rangle$ be two non-orthogonal states Eve is interested to learn about and $|\alpha\rangle$ be the standard state prepared by Eve. Eve tries to interact with these states without disturbing them, and as a result of this interaction the following transformation is performed [1]:

$$|\psi_1\rangle \otimes |\alpha\rangle \rightarrow |\psi_1\rangle \otimes |\alpha\rangle, \quad |\psi_2\rangle \otimes |\alpha\rangle \rightarrow |\psi_2\rangle \otimes |\beta\rangle. \tag{1.63}$$

Eve hopes that the states $|\alpha\rangle$ and $|\beta\rangle$ are different, in an attempt to learn something about the states. However, any unitary transformation must preserve the dot product, so from (1.63) we obtain:

$$\langle\alpha|\beta\rangle \, \langle\psi_1|\psi_2\rangle = \langle\alpha|\alpha\rangle\langle\psi_1|\psi_2\rangle, \tag{1.64}$$

which indicates that $\langle\alpha|\alpha\rangle = \langle\alpha|\beta\rangle = 1$, and consequently $|\alpha\rangle = |\beta\rangle$. Therefore, in attempting to distinguish between $|\psi_1\rangle$ and $|\psi_2\rangle$, Eve will disturb them. The key idea of QKD is therefore to transmit non-orthogonal qubit states between Alice and Bob, and by checking for disturbances in the transmitted state, they can establish an upper bound on the noise/eavesdropping level in their communication channel. Once the QKD protocol is completed, transmitter A and receiver B perform a series of classical steps, which are illustrated in Figure 1.14. As the transmission distance increases and for higher key distribution speeds, the error correction becomes increasingly important. By performing information reconciliation by low-density parity-check (LDPC) codes, the higher input bit error rates (BERs) can be tolerated compared to other coding schemes. The *privacy amplification* is further performed to eliminate any information obtained by an eavesdropper. Privacy amplification must be performed between Alice and Bob to distill from the generated key a smaller set of bits whose correlation with Eve's string is below the desired threshold. One way to accomplish privacy amplification is through the use of universal hash functions. The simplest family of hash functions is based on multiplication with a random element of the Galois field GF(2^m) ($m > n$), where n denotes the number of bits remaining after information reconciliation is completed. The threshold for maximum tolerable error rate is

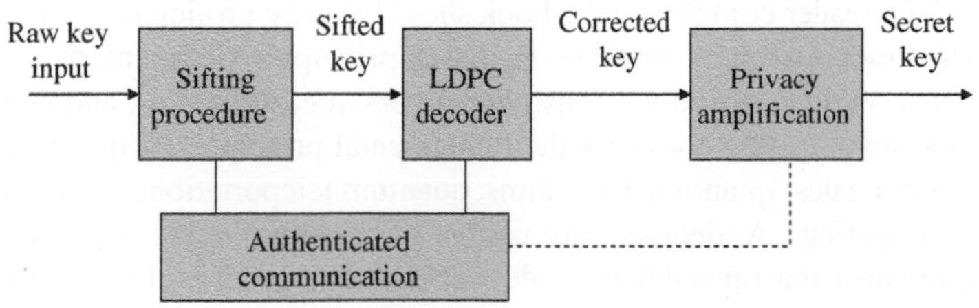

FIGURE 1.14

Classical postprocessing steps.

dictated by the efficiency of the implemented information reconciliation and privacy amplification protocols. The BB84 protocol [24, 25] is described next as an illustration of the QKD procedure.

The BB84 protocol consists of the following steps [1]:

1. Alice chooses $(4+\delta)n$ random data bits $\boldsymbol{a}$. Alice also chooses at random a $(4+\delta)n$-long bit string $\boldsymbol{b}$ $(\delta > 0)$ and encodes each data bit (in $\boldsymbol{a}$) as $\{|0\rangle, |1\rangle\}$ if the corresponding bit in $\boldsymbol{b}$ is 0 or $\{|+\rangle, |-\rangle\}$ if the corresponding bit in $\boldsymbol{b}$ is 1. Alice sends the resulting quantum state to Bob.
2. Bob receives the $(4+\delta)n$ qubits, and measures each qubit in either the $\{|0\rangle, |1\rangle\}$ or $\{|+\rangle, |-\rangle\}$ basis at random.
3. Alice announces the vector $\boldsymbol{b}$ over a public channel.
4. Alice and Bob discard any bits where Bob measured in a different basis from Alice's prepared basis. With high probability there are at least $2n$ bits left (otherwise the protocol should be aborted), and they keep $2n$ bits.
5. Alice then selects a subset of n bits, to be used against Eve's interference, and provides Bob the information about which ones get selected.
6. Alice and Bob compare the values of n check bits. If there is disagreement in the number of locations exceeding the error correction capability of the error correction scheme, they abort the protocol.
7. Otherwise, Alice and Bob perform information reconciliation and privacy amplification on the remaining n bits to obtain m $(m < n)$ shared key bits.

1.8 ORGANIZATION OF THE BOOK

This section is related to the organization of the book. The book covers various topics of quantum information processing and quantum error correction. It is a self-contained introduction to quantum information, quantum computation, and quantum error correction. Readers of the book will be ready for further study in this area, and will be prepared to perform independent research. On completing the book the reader should be able design information processing circuits, stabilizer codes, Calderbank–Shor–Steane (CSS) codes, subsystem codes, topological codes, and entanglement-assisted quantum error correction codes, and propose corresponding physical implementation. The reader completing the book should also be proficient in fault-tolerant design. The book starts by introducing basic principles of quantum mechanics, including state vectors, operators, density operators, measurements, and dynamics of a quantum system. It then considers the fundamental principles of quantum computation, quantum gates, quantum algorithms, quantum teleportation, and fault-tolerant quantum computing. A detailed discussion of quantum error correction codes (QECCs), in particular on stabilizer codes, CSS codes, quantum low-density parity-check (LDPC) codes, subsystem codes (also known as operator QECCs), topological codes and entanglement-assisted QECCs, is then undertaken. The book goes on to consider fault-tolerant QECC and fault-tolerant quantum computing, then continues with quantum information theory. The next part of the book is spent investigating physical realizations of quantum computers, encoders, and decoders, including NMR, ion traps, photonic quantum realization, cavity quantum electrodynamics, and quantum dots. The chapters of the book are organized as follows.

Chapter 2 covers the quantum mechanics fundamentals. The basic topics of quantum mechanics are covered, including state vectors, operators, density operators, and dynamics of a quantum system. Quantum measurements (Section 2.4), the uncertainty principle (Section 2.5), and change of basis (Section 2.7) are then described. In addition, the harmonic oscillator (Section 2.9), orbital angular momentum (Section 2.10), spin-1/2 systems (Section 2.11), and hydrogen-like atoms (Section 2.12) are discussed. For a deeper understanding of this material, a set of problems to be used for self-study is included in Section 2.14. Readers not interested in physical implementation, but only in either the quantum computing or quantum error correction aspects, can skip Sections 2.9–2.12.

Chapter 3 covers quantum circuits and quantum information processing fundamentals. In this chapter, basic single-qubit (Section 3.1), two-qubit (Section 3.2) and many-qubit gates (Section 3.3) are described. The fundaments of quantum information processing (Section 3.3) are also discussed. The various sets of universal quantum gates are discussed in Section 3.5. Quantum measurements and the principles of deferred implicit measurements are discussed in Section 3.4. The Bloch sphere representation of a single qubit is provided in Section 3.1. The Gottesman–Knill theorem is given in Section 3.5. The Bell state preparation circuit and quantum relay are described in Sections 3.2 and 3.6 respectively. The basic concepts of quantum teleportation are described in Section 3.6. In addition to the topics above, the concepts of quantum parallelism and entanglement are introduced (see Section 3.2). We also discuss the no-cloning theorem, in Section 3.4, and consider its consequences. For a deeper understanding of the underlying concepts, in Section 3.8 we provide a series of problems for self-study.

Chapter 4 is contributed by Dr Shikhar Uttam from University of Pittsburgh. This chapter is concerned with basic concepts of quantum information processing. The chapter starts by considering the superposition principle and the concept of quantum parallelism. It continues with a formulation and proof of the no-cloning theorem. Another important theorem is then formulated and proved, which claims that it is impossible to unambiguously distinguish non-orthogonal quantum states. The next section is devoted to entanglement, in which an important theorem and corresponding corollaries are defined. It considers how entanglement can be used in detecting and correcting the quantum errors. Finally, the operator-sum representation of a quantum operation is discussed and the concept applied to describe quantum errors and decoherence. Several illustrative examples are given; namely, bit-flip, phase-flip and depolarizing channels.

Chapter 5 describes various quantum algorithms: (i) Deutsch and Deutsch–Jozsa algorithms, presented in Section 5.2; (ii) the Grover search algorithm, presented in Section 5.3; (iii) quantum Fourier transform, presented in Section 5.4; (iv) period-finding and Shor's factorization algorithms, presented in Section 5.5; and

(v) Simon's algorithm, presented in Section 5.6. Section 5.1 reviews the quantum parallelism concept. Section 5.5 describes how to crack the RSA encryption protocol. Section 5.7 covers classical/quantum computing complexities and Turing machines. In Section 5.9, a set of problems is given that will help the reader to gain a deeper understanding of these concepts. In addition, several problems have been devoted to finding phase algorithm, quantum discrete logarithm algorithm, Kitaev's algorithm, and quantum simulation.

Chapter 6 details the basic concepts of coding theory. Only topics from classical error correction important in quantum error correction are covered. This chapter is organized as follows. In Section 6.1 channel coding preliminaries are introduced, namely basic definitions, channel models, the concept of channel capacity, and the channel coding theorem. Section 6.2 is devoted to the basics of linear block codes (LBCs), such as definition of generator and parity-check matrices, syndrome decoding, distance properties of LBCs, and some important coding bounds. In Section 6.3 cyclic codes are introduced. The BCH codes are described in Section 6.4. The RS codes, concatenated and product codes are described in Section 6.5. After a short summary section, a set of problems is provided to enable readers to gain a deeper understanding of classical error correction.

Chapter 7 considers quantum error correction concepts, ranging from an intuitive description to a rigorous mathematical framework. After introduction of the Pauli operators, basic quantum codes are described, such as three-qubit flip code, three-qubit phase flip code, Shor's nine-qubit code, stabilizer codes, and CSS codes. Quantum error correction is then formally introduced, including quantum error correction mapping, quantum error representation, stabilizer group definition, quantum-check matrix representation, and the quantum syndrome equation. The necessary and sufficient conditions for quantum error correction are included, distance properties and error correction capability are discussed, and the CSS codes are revisited. Section 7.4 describes important quantum coding bounds, including the quantum Hamming bound, Gilbert–Varshamov bound, and Singleton bound. This section also discusses quantum weight enumerators and quantum MacWilliams identities. Quantum superoperators and various quantum channels, including the quantum depolarizing, amplitude damping, and generalized amplitude damping channels, are then considered. For a deeper understanding of this quantum error correction material, in Section 7.7 a set of problems is included.

Chapter 8 covers stabilizer codes and related issues. The stabilizer codes are introduced in Section 8.1. Their basic properties are discussed in Section 8.2. The encoded operations are introduced in the same section. In Section 8.3 finite geometry interpretation of stabilizer codes is introduced. This representation is used in Section 8.4 to introduce the so-called standard form of stabilizer code. The standard form is a basic representation for the efficient encoder and decoder implementations (Section 8.5). Section 8.6 discusses nonbinary stabilizer codes, which generalizes the previous sections. The subsystem codes are introduced in Section 8.7. In the same section efficient encoding and decoding of subsystem codes are discussed as well. The entanglement-assisted codes are introduced in Section 8.8. An important class of quantum codes, namely the topological codes, is discussed in Section 8.9. Finally, for better understanding of material presented in the chapter, a set of problems is provided in Section 8.11.

Chapter 9 concerns entanglement-assisted (EA) quantum error correction codes, which use pre-existing entanglement between transmitter and receiver to improve the reliability of transmission. A key advantage of EA quantum codes compared to CSS codes is that EA quantum codes do not do not impose the dual-containing constraint. Therefore, arbitrary classical codes can be used to design EA quantum codes. The number of required entangled qubits, also known as ebits, is determined by $e = \mathrm{rank}(\boldsymbol{H}\boldsymbol{H}^{\mathrm{T}})$, where $\boldsymbol{H}$ is the parity-check matrix of the corresponding classical code used to design EA code. A general description of EA quantum error correction is provided in Section 9.1. In Section 9.2 the EA canonical code and its error correction capability are considered. In Section 9.3 the concept of generalization of EA canonical code to arbitrary EA code is covered. It also describes how to design EA codes from classical codes, in particular classical quaternary codes. In Section 9.4 the encoding for EA quantum codes is discussed. In Section 9.5 the concept of operator quantum error correction, also known as subsystem codes, is introduced. EA operator quantum error correction is discussed in Section 9.6. Finally, in Section 9.8 a set of problems that will help the reader better understand the material in the chapter is included.

Chapter 10 covers quantum LDPC codes. Section 10.1 describes classical LDPC codes, their design, and decoding algorithms. Section 10.2 considers the dual-containing quantum LDPC codes. Various design algorithms are introduced. Section 10.3 concerns EA quantum LDPC codes. Various classes of finite geometry codes are described. Section 10.4 describes the probabilistic sum-product algorithm based on the quantum-check matrix. Section 10.6 includes a set of problems that will help the reader better understand the material in the chapter.

Chapter 11 relates to fault-tolerant quantum computation concepts. After the introduction of fault-tolerance basics and traversal operations (Section 11.1), the basics of fault-tolerant quantum computation concepts and procedures are given in Section 11.2, using Steane's code as an illustrative example. Section 11.3 provides a rigorous description of fault-tolerant QECC. Section 11.4 discusses fault-tolerant computing, in particular the fault-tolerant implementation of the Toffoli gate. Section 11.5 formulates and proves the accuracy threshold theorem, then Section 11.7 provides a set of problems for the interested reader to gain deeper understanding of the fault-tolerant theory.

Chapter 12 is contributed by Dr Shikhar Uttam from University of Pittsburgh. This chapter covers quantum information theory. The chapter starts with classical and von Neumann entropy definitions and properties, followed by quantum representation of classical information. Further is given an introduction to accessible information as the maximum mutual information over all possible measurement schemes. An important bound, the Holevo bound, is introduced and proven. Next typical sequence, typical state, typical subspace, and projector on the typical subspace are discussed. Then the typical subspace theorem is proven, followed by formulation and proof of Schumacher's source coding theorem, the quantum equivalent of Shannon's source coding theorem. The concept of quantum compression and decompression, along with the concept of reliable compression, is introduced. The final section of this chapter gives a formulation and proof of the Holevo–Schumacher–Westmoreland (HSW) theorem, the quantum equivalent of Shannon's channel capacity theorem. Several common quantum channel models are considered and their channel capacity determined. At the end of the chapter, a set of problems is provided for self-study.

Chapter 13 covers several promising physical implementations of quantum information processing. The introductory section discusses di Vincenzo criteria and gives an overview of physical implementation concepts. Section 13.2 relates to nuclear magnetic resonance implementation. In Section 13.3 the use of ion traps in quantum computing is described. Section 13.4 gives various photonic implementations, including bulky optics implementation (subsection 13.4.1) and integrated optics implementation (subsection 13.4.2). Section 13.5 considers quantum relay implementation. The implementation of quantum encoders and decoders is discussed in Section 13.6. Further, the implementation of quantum computing based on optical cavity electrodynamics is discussed in Section 13.7. The use of quantum dots in quantum computing is discussed in Section 13.8. In Section 13.10 several interesting problems are provided for self-study.

References

[1] M.A. Neilsen, I.L. Chuang, Quantum Computation and Quantum Information, Cambridge University Press, Cambridge, 2000.

[2] M. Le Bellac, An Introduction to Quantum Information and Quantum Computation, Cambridge University Press, Cambridge, 2006.

[3] F. Gaitan, Quantum Error Correction and Fault Tolerant Quantum Computing, CRC Press, 2008.

[4] G. Jaeger, Quantum Information: An Overview, Springer, 2007.

[5] D. Petz, Quantum Information Theory and Quantum Statistics, Theoretical and Mathematical Physics, Springer, Berlin, 2008.

[6] P. Lambropoulos, D. Petrosyan, Fundamentals of Quantum Optics and Quantum Information, Springer, Berlin, 2007.

[7] G. Johnson, A Shortcut Through Time: The Path to the Quantum Computer, Knopf, New York, 2003.

[8] J. Preskill, Quantum Computing (1999). Available at: http://www.theory.caltech.edu/~preskill/.

[9] J. Stolze, D. Suter, Quantum Computing, Wiley, New York, 2004.

[10] R. Landauer, Information is physical, Phys. Today 44 (5) (May 1991) 23–29.

[11] R. Landauer, The physical nature of information, Phys. Lett. A 217 (1991) 188–193.

[12] G. Keiser, Optical Fiber Communications, McGraw-Hill, 2000.

[13] A. Barenco, A universal two-bit quantum computation, Proc. R. Soc. Lond. A 449 (1937) 679–683, 1995.

[14] D. Deutsch, Quantum computational networks, Proc. R. Soc. Lond. A 425 (1868) 73−90, 1989.

[15] D.P. DiVincenzo, Two-bit gates are universal for quantum computation, Phys. Rev. A 51 (2) (1995) 1015−1022.

[16] A. Barenco, C.H. Bennett, R. Cleve, D.P. DiVincenzo, N. Margolus, P. Shor, T. Sleator, J.A. Smolin, H. Weinfurter, Elementary gates for quantum computation, Phys. Rev. A 52 (5) (1995) 3457−3467.

[17] C.H. Bennett, G. Brassard, C. Crépeau, R. Jozsa, A. Peres, W.K. Wootters, Teleporting an unknown quantum state via dual classical and Einstein−Podolsky−Rosen channels, Phys. Rev. Lett. 70 (13) (1993) 1895−1899.

[18] D.J.C. MacKay, G. Mitchison, P.L. McFadden, Sparse-graph codes for quantum error correction, IEEE Trans. Inform. Theory 50 (2004) 2315−2330.

[19] I.B. Djordjevic, Photonic implementation of quantum relay and encoders/decoders for sparse-graph quantum codes based on optical hybrid, IEEE Photon. Technol. Lett. 22 (19) (2010) 1449−1451.

[20] I.B. Djordjevic, Photonic entanglement-assisted quantum low-density parity-check encoders and decoders, Opt. Lett. 35 (9) (2010) 1464−1466.

[21] I.B. Djordjevic, Quantum LDPC codes from balanced incomplete block designs, IEEE Commun. Lett. 12 (2008) 389−391.

[22] I.B. Djordjevic, Cavity quantum electrodynamics (CQED) based quantum LDPC encoders and decoders, IEEE Photon. J. 3 (4) (2011) 727−738.

[23] I.B. Djordjevic, Photonic quantum dual-containing LDPC encoders and decoders, IEEE Photon. Technol. Lett. 21 (13) (1 July 2009) 842−844.

[24] C.H. Bennet, G. Brassard, Quantum cryptography: public key distribution and coin tossing, in: Proc. IEEE International Conference on Computers, Systems, and Signal Processing, Bangalore, India, 1984, 175−179.

[25] C.H. Bennett, Quantum cryptography: uncertainty in the service of privacy, Science 257 (August 1992) 752−753.

This page intentionally left blank

CHAPTER

Quantum Mechanics Fundamentals

2

CHAPTER OUTLINE

This chapter covers the quantum mechanics fundamentals and is based on Refs [1–25]. The following topics from quantum mechanics are considered: state vectors, operators, density operators, measurements, and dynamics of a quantum system. In addition, harmonic oscillators, orbital angular momentum, spin-1/2 systems, and hydrogen-like systems are described. These systems are of great importance in various quantum information processing systems and in quantum communications.

Quantum Information Processing and Quantum Error Correction. DOI: 10.1016/B978-0-12-385491-9.00002-2

The chapter is organized as follows. In Section 2.1 the basic concepts in quantum mechanics are introduced. Section 2.2 provides a description of how eigenkets can be used as basis kets. The matrix representation of operators, kets, and bras is described in Section 2.3. Section 2.4 covers quantum measurements, commutators, and Pauli operators, while Section 2.5 discusses the uncertainty principle. Density operators are described in Section 2.6. Section 2.7 deals with change of basis and diagonalization. Section 2.8 considers the dynamics of quantum systems and the Schrödinger equation. The harmonic oscillator theory is used in many quantum computing applications, and is considered in Section 2.9. The orbital angular momentum is described in Section 2.10, while spin-1/2 systems are covered in Section 2.11. In Section 2.12 hydrogen-like atoms are described. Finally, Section 2.13 concludes the chapter. Some problems for self-study are provided at the end of the chapter.

2.1 INTRODUCTION

In quantum mechanics, the primitive undefined concepts are *physical system*, *observable*, and *state* [1–4]. A physical system is any sufficiently isolated quantum object, say an electron, a photon, or a molecule. An observable will be associated with a measurable property of a physical system, say energy or the z-component of the spin. The state of a physical system is a trickier concept in quantum mechanics compared to classical mechanics. The problem arises when considering composite physical systems. In particular, states exist, known as *entangled states*, for a bipartite physical system in which neither of the subsystems is in a definite state. Even in cases where physical systems can be described as being in a state, two classes of states are possible: pure and mixed. The condition of a quantum mechanical system is completely specified by its *state vector* $|\psi\rangle$ in a Hilbert space H (a vector space on which a positive-definite scalar product is defined) over the field of complex numbers. Any state vector $|\alpha\rangle$, also known as a *ket*, can be expressed in terms of basis vectors $|\phi_n\rangle$ by

$$|\alpha\rangle = \sum_{n=1}^{\infty} a_n|\phi_n\rangle. \tag{2.1}$$

An *observable*, such as momentum and spin, can be represented by an *operator*, such as A, in the vector space in question. Generally, an operator acts on a ket from the left: $(A) \cdot |\alpha\rangle = A|\alpha\rangle$, which results in another ket. An operator A is said to *Hermitian* if

$$A^{\dagger} = A, \qquad A^{\dagger} = (A^{\mathrm{T}})^{*}. \tag{2.2}$$

Suppose that the Hermitian operator A has a discrete set of eigenvalues $a^{(1)}$, ..., $a^{(n)}$, ... The associated eigenvectors (eigenkets) $|a^{(1)}\rangle$, ..., $|a^{(n)}\rangle$, ... can be obtained from

$$A\left|a^{(n)}\right\rangle = a^{(n)}\left|a^{(n)}\right\rangle. \tag{2.3}$$

The Hermitian conjugate of a ket $|\alpha\rangle$ is denoted by $\langle\alpha|$ and called the "bra". The space dual to ket space is known as *bra* space. There exists a one-to-one correspondence, dual correspondence (D.C.), between a ket space and a bra space:

$$\begin{aligned}
&|\alpha\rangle \overset{\text{D.C.}}{\leftrightarrow} \langle\alpha| \\
&|a^{(1)}\rangle, |a^{(2)}\rangle, \ldots \overset{\text{D.C.}}{\leftrightarrow} \langle a^{(1)}|, \langle a^{(2)}|, \ldots \\
&|\alpha\rangle + |\beta\rangle \overset{\text{D.C.}}{\leftrightarrow} \langle\alpha| + \langle\beta| \\
&c_\alpha|\alpha\rangle + c_b|\beta\rangle \overset{\text{D.C.}}{\leftrightarrow} c_\alpha^*\langle\alpha| + c_\beta^*\langle\beta|.
\end{aligned} \tag{2.4}$$

The *scalar (inner) product* of two state vectors $|\phi\rangle$ and $|\psi\rangle$ is defined by

$$\langle\beta|\alpha\rangle = \sum_{n=1}^{\infty} a_n b_n^*. \tag{2.5}$$

Let $|\alpha\rangle$, $|\beta\rangle$, and $|\gamma\rangle$ be the state kets. The following *properties of inner product* are valid:

$$\begin{aligned}
&1.\ \langle\beta|\alpha\rangle = \langle\alpha|\beta\rangle^* \\
&2.\ \langle\beta|(a|\alpha\rangle + b|\gamma\rangle) = a\langle\beta|\alpha\rangle + b\langle\beta|\gamma\rangle \\
&3.\ (\langle a\alpha| + \langle b\beta|)|\gamma\rangle = a^*\langle\alpha|\gamma\rangle + b^*\langle\beta|\gamma\rangle \\
&4.\ \langle\alpha|\alpha\rangle \geq 0 \\
&5.\ \sqrt{\langle\alpha+\beta|\alpha+\beta\rangle} \leq \sqrt{\langle\alpha|\alpha\rangle} + \sqrt{\langle\beta|\beta\rangle} \quad \text{(Triangle inequality)} \\
&6.\ |\langle\beta|\alpha\rangle|^2 \leq \langle\alpha|\alpha\rangle\langle\beta|\beta\rangle \qquad \text{(Cauchy – Schwartz inequality).}
\end{aligned} \tag{2.6}$$

Property 5 is known as the *triangle inequality* and proerty 6 as the *Cauchy–Schwartz inequality*, which is important in proving the Heisenberg uncertainty relationship, and as such will be proved here. The starting point in the proof is to observe the linear combination of two kets $|\alpha\rangle$ and $|\beta\rangle$ – that is, $|\alpha\rangle + \lambda|\beta\rangle$ – and apply property 4 to obtain:

$$((\langle\alpha| + \lambda^*\langle\beta|)(|\alpha\rangle + \lambda|\beta\rangle) \geq 0 \Leftrightarrow \langle\alpha|\alpha\rangle + \lambda\langle\alpha|\beta\rangle + \lambda^*\langle\beta|\alpha\rangle + |\lambda|^2\langle\beta|\beta\rangle \geq 0. \tag{2.7}$$

The inequality (2.7) is satisfied for any complex λ, including $\lambda = -\langle\beta|\alpha\rangle/\langle\beta|\beta\rangle$:

$$\langle\alpha|\alpha\rangle - \frac{\langle\beta|\alpha\rangle}{\langle\beta|\beta\rangle}\langle\alpha|\beta\rangle - \frac{\langle\alpha|\beta\rangle}{\langle\beta|\beta\rangle}\langle\beta|\alpha\rangle + \frac{|\langle\beta|\alpha\rangle|^2}{|\langle\beta|\beta\rangle|^2}\langle\alpha|\alpha\rangle \geq 0. \tag{2.8}$$

By multiplying (2.8) by $\langle\beta|\beta\rangle$ we obtain:

$$|\langle\beta|\alpha\rangle|^2 \leq \langle\alpha|\alpha\rangle\langle\beta|\beta\rangle, \tag{2.9}$$

therefore proving the Cauchy–Schwartz inequality.

The norm of ket $|\alpha\rangle$ is defined by $\sqrt{\langle\alpha|\alpha\rangle}$, and the normalized ket by $|\tilde{\alpha}\rangle = |\alpha\rangle/\sqrt{\langle\alpha|\alpha\rangle}$. Clearly, $\langle\tilde{\alpha}|\tilde{\alpha}\rangle = 1$. Two kets are said to be *orthogonal* if their dot product is zero – that is, $\langle\alpha|\beta\rangle = 0$. Two operators X and Y are equal, $X = Y$, if their action on a ket is the same: $X|\alpha\rangle = Y|\alpha\rangle$. The identity operator I is defined by $I|\psi\rangle = |\psi\rangle$. Operators can be *added*; addition operations are commutative and associative. Operators can be *multiplied*; multiplication operations are in general *noncommutative*, but associative. The associative property in quantum mechanics, known as the *associative axiom*, is postulated to hold quite generally as long as we are dealing with "legal" multiplications among kets, bras, and operators as follows:

$$
\begin{aligned}
&(|\beta\rangle\langle\alpha|)\cdot|\gamma\rangle = |\beta\rangle\cdot(\langle\alpha|\gamma\rangle) \qquad && (\langle\beta|)\cdot X|\alpha\rangle = (\langle\beta|X)\cdot|\alpha\rangle = \{\langle\alpha|X^\dagger|\beta\rangle\}^* \\
&X = |\beta\rangle\langle\alpha| \Rightarrow X^\dagger = |\alpha\rangle\langle\beta| && \langle\beta|X|\alpha\rangle = \langle\alpha|X^\dagger|\beta\rangle^* \\
& && X^\dagger = X:\ \ \langle\beta|X|\alpha\rangle = \langle\alpha|X|\beta\rangle^*.
\end{aligned} \tag{2.10}
$$

2.2 EIGENKETS AS BASE KETS

The eigenkets $\{|\xi^{(n)}\rangle\}$ of operator Ξ form the basis so that arbitrary ket $|\psi\rangle$ can be expressed in terms of eigenkets by

$$|\psi\rangle = \sum_{n=1}^{\infty} c_n \left|\xi^{(n)}\right\rangle. \tag{2.11}$$

By multiplying (2.11) by $\langle\xi^{(n)}|$ from the left we obtain:

$$\left\langle\xi^{(n)}|\psi\right\rangle = \sum_{j=1}^{\infty} c_j\left\langle\xi^{(n)}\middle|\xi^{(j)}\right\rangle = c_n\left\langle\xi^{(n)}\middle|\xi^{(n)}\right\rangle + \sum_{j=1,\, j\neq n}^{\infty} c_j\left\langle\xi^{(n)}\middle|\xi^{(j)}\right\rangle. \tag{2.12}$$

Since the eigenkets $\{|\xi^{(n)}\rangle\}$ form the basis, the principle of orthonormality is satisfied, $\langle\xi^{(n)}|\xi^{(j)}\rangle = \delta_{nj},\ \ \delta_{nj} = \begin{cases} 1, & n = j \\ 0, & n \neq j \end{cases}$, so that Eq. (2.12) becomes

$$c_n = \left\langle\xi^{(n)}\middle|\psi\right\rangle. \tag{2.13}$$

By substituting (2.13) into (2.11) we obtain:

$$|\psi\rangle = \sum_{n=1}^{\infty}\left\langle\xi^{(n)}\middle|\psi\right\rangle\left|\xi^{(n)}\right\rangle = \sum_{n=1}^{\infty}\left|\xi^{(n)}\right\rangle\left\langle\xi^{(n)}\middle|\psi\right\rangle. \tag{2.14}$$

Because $|\psi\rangle = I|\psi\rangle$, from (2.14) it is clear that

$$\sum_{n=1}^{\infty}\left|\xi^{(n)}\right\rangle\left\langle\xi^{(n)}\right| = I, \tag{2.15}$$

and the relation above is known as the *completeness relation*. The operators under summation in (2.15) are known as *projection* operators, P_n:

$$P_n = \left|\xi^{(n)}\right\rangle\left\langle\xi^{(n)}\right| \tag{2.16}$$

which satisfy the relationship $\sum_{n=1}^{\infty} P_n = I$. It is easy to show that the ket (2.11) with c_n determined from (2.13) is of unit length:

$$\langle\psi|\psi\rangle = \sum_{n=1}^{\infty}\left\langle\psi\middle|\xi^{(n)}\right\rangle\left\langle\xi^{(n)}\middle|\psi\right\rangle = \sum_{n=1}^{\infty}\left|\left\langle\psi|\xi^{(n)}\right\rangle\right|^2 = 1. \tag{2.17}$$

The following is an important theorem that will be used throughout the chapter.

Theorem. The eigenvalues of a Hermitian operator A are real and the eigenkets are orthogonal:

$$\left\langle a^{(m)}\middle|a^{(n)}\right\rangle = \delta_{nm}.$$

This theorem is quite straightforward to prove starting from the eigenvalue equation

$$A\left|a^{(i)}\right\rangle = a^{(i)}\left|a^{(i)}\right\rangle \tag{2.18a}$$

and its dual conjugate

$$\left\langle a^{(j)}\right|A^{\dagger} = a^{(j)^*}\left\langle a^{(j)}\right|. \tag{2.18b}$$

By multiplying (2.18a) by $\langle a^{(j)}|$ from the left and (2.18b) by $|a^{(i)}\rangle$ from the right and subtracting them, we obtain:

$$\left(a^{(i)} - a^{(j)^*}\right)\left\langle a^{(j)}|a^{(i)}\right\rangle = 0. \tag{2.19}$$

For $i = j$ it is clear from (2.19) that $a^{(i)} = a^{(j)*}$, proving that eigenvalues are real. For $i \neq j$ it is clear from (2.19) that $\langle a^{(j)}|a^{(i)}\rangle = 0$, proving that the eigenkets are orthogonal. We can now represent the operator A in terms of its eigenkets as follows:

$$A = \sum_i a^{(i)}\left|a^{(i)}\right\rangle\left\langle a^{(i)}\right|, \tag{2.20}$$

which is known as spectral decomposition.

2.3 MATRIX REPRESENTATIONS

A linear operator X can be expressed in terms of eigenkets $\{|a^{(n)}\rangle\}$ by applying the completeness relation twice, as follows:

$$X = IXI = \sum_{m=1}^{\infty}\left|a^{(m)}\right\rangle\left\langle a^{(m)}\right|X\sum_{n=1}^{\infty}\left|a^{(n)}\right\rangle\left\langle a^{(n)}\right|$$

$$= \sum_{n=1}^{\infty}\sum_{m=1}^{\infty}\left|a^{(m)}\right\rangle\left\langle a^{(m)}\middle|X\middle|a^{(n)}\right\rangle\left\langle a^{(n)}\right|. \tag{2.21}$$

Therefore, the *operator* X is associated with a *square matrix* (albeit infinite in extent), whose elements are

$$X_{mn} = \left\langle a^{(m)} \middle| X \middle| a^{(n)} \right\rangle \tag{2.22a}$$

and can be written explicitly as

$$X \doteq \begin{pmatrix} \langle a^{(1)}|X|a^{(1)}\rangle & \langle a^{(1)}|X|a^{(2)}\rangle & \cdots \\ \langle a^{(2)}|X|a^{(1)}\rangle & \langle a^{(2)}|X|a^{(2)}\rangle & \cdots \\ \vdots & \vdots & \ddots \end{pmatrix}, \tag{2.22b}$$

where we use the notation $\doteq$ to denote that operator X is represented by the matrix above.

Multiplication operator $Z = XY$ can be represented in matrix form by employing (2.22) and the completeness relation to obtain:

$$Z_{mn} = \left\langle a^{(m)} \middle| Z \middle| a^{(n)} \right\rangle = \left\langle a^{(m)} \middle| XY \middle| a^{(n)} \right\rangle = \sum_k \left\langle a^{(m)} \middle| X \middle| a^{(k)} \right\rangle \left\langle a^{(k)} \middle| Y \middle| a^{(n)} \right\rangle. \tag{2.23}$$

The *kets* are represented as *column vectors* and the corresponding representation can be obtained by multiplying the ket $|\gamma\rangle = X|\alpha\rangle$ from the left by $|a^{(i)}\rangle$ and by applying the completeness relation as follows:

$$\left\langle a^{(i)} \middle| \gamma \right\rangle = \left\langle a^{(i)} \middle| X \middle| \alpha \right\rangle = \sum_j \underbrace{\left\langle a^{(i)} \middle| X \middle| a^{(j)} \right\rangle}_{\text{matrix}} \underbrace{\left\langle a^{(j)} \middle| \alpha \right\rangle}_{\text{column vector}}. \tag{2.24}$$

Therefore, the ket $|\alpha\rangle$ is a column vector based on (2.24):

$$|\alpha\rangle = \begin{pmatrix} \langle a^{(1)}|\alpha\rangle \\ \langle a^{(2)}|\alpha\rangle \\ \vdots \end{pmatrix}. \tag{2.25}$$

The *bras* are represented as *row vectors* and and the corresponding representation can be obtained by multiplying the bra $\langle\gamma| = \langle\alpha|X$ from the right by $|a^{(i)}\rangle$ and by applying the completeness relation as follows:

$$\left\langle \gamma \middle| a^{(i)} \right\rangle = \left\langle \alpha \middle| X \middle| a^{(i)} \right\rangle = \sum_j \underbrace{\left\langle \alpha \middle| a^{(j)} \right\rangle}_{\langle a^{(j)}|\alpha\rangle^*,\ \text{row vector}} \underbrace{\left\langle a^{(j)} \middle| X \middle| a^{(i)} \right\rangle}_{\text{matrix}}. \tag{2.26}$$

From (2.26) it is clear that the bra $\langle\alpha|$ can be represented as a row vector by

$$\langle\alpha| \doteq \left(\left\langle a^{(1)} \middle| \alpha \right\rangle^* \quad \left\langle a^{(2)} \middle| \alpha \right\rangle^* \quad \cdots \right). \tag{2.27}$$

The *inner product* of the ket $|\beta\rangle$ and bra $\langle\alpha|$ can be represented by applying the completeness representation as a conventional vector dot-product as follows:

$$\langle\alpha|\beta\rangle = \sum_i \left\langle\alpha\middle|a^{(i)}\right\rangle\left\langle a^{(i)}\middle|\beta\right\rangle = \sum_i \left\langle a^{(i)}|\alpha\right\rangle^* \left\langle a^{(i)}\middle|\beta\right\rangle$$
$$= \left(\left\langle a^{(1)}\middle|\alpha\right\rangle^* \quad \left\langle a^{(2)}\middle|\alpha\right\rangle^* \quad \ldots\right)\begin{pmatrix}\langle a^{(1)}|\beta\rangle \\ \langle a^{(2)}|\beta\rangle \\ \vdots\end{pmatrix}. \tag{2.28}$$

The *outer product* $|\beta\rangle\langle\alpha|$ is represented in the matrix form by using twice the completeness relation, as follows:

$$|\beta\rangle\langle\alpha| = \sum_i\sum_j |a^{(i)}\rangle\langle a^{(i)}|\beta\rangle\langle\alpha|a^{(j)}\rangle\langle a^{(j)}| = \sum_i\sum_j |a^{(i)}\rangle\langle a^{(j)}|\langle a^{(i)}|\beta\rangle\langle a^{(j)}|\alpha\rangle^*$$
$$= \begin{pmatrix}\langle a^{(1)}|\beta\rangle\langle a^{(1)}|\alpha\rangle^* & \langle a^{(1)}|\beta\rangle\langle a^{(2)}|\alpha\rangle^* & \cdots \\ \langle a^{(2)}|\beta\rangle\langle a^{(1)}|\alpha\rangle^* & \langle a^{(2)}|\beta\rangle\langle a^{(2)}|\alpha\rangle^* & \cdots \\ \vdots & \vdots & \ddots\end{pmatrix}. \tag{2.29}$$

Finally, the *operator* can be represented in terms of *outer products* by

$$X = \sum_{n=1}^{\infty}\sum_{m=1}^{\infty}\left|a^{(m)}\right\rangle\left\langle a^{(m)}\middle|X\middle|a^{(n)}\right\rangle\left\langle a^{(n)}\right| = \sum_{n=1}^{\infty}\sum_{m=1}^{\infty} X_{mn}\left|a^{(m)}\right\rangle\left\langle a^{(n)}\right|,$$
$$X_{mn} = \left\langle a^{(m)}\middle|X\middle|a^{(n)}\right\rangle. \tag{2.30}$$

In order to illustrate the various representations above, we study the photons, spin-1/2 systems, and Hadamard operator below.

2.3.1 Photons

The x- and y-polarizations can be represented by

$$|E_x\rangle = \begin{pmatrix}1\\0\end{pmatrix} \qquad |E_y\rangle = \begin{pmatrix}0\\1\end{pmatrix}.$$

On the other hand, the right- and left-circular polarizations can be represented by

$$|E_R\rangle = \frac{1}{\sqrt{2}}\begin{pmatrix}1\\j\end{pmatrix} \qquad |E_L\rangle = \frac{1}{\sqrt{2}}\begin{pmatrix}1\\-j\end{pmatrix}.$$

The 45° polarization ket can be represented as follows:

$$|E_{45^\circ}\rangle = \cos\left(\frac{\pi}{4}\right)|E_x\rangle + \sin\left(\frac{\pi}{4}\right)|E_y\rangle = \frac{1}{\sqrt{2}}(|E_x\rangle + |E_y\rangle) = \frac{1}{\sqrt{2}}\begin{pmatrix}1\\1\end{pmatrix}.$$

The bras corresponding to the left and right polarizations can be written as

$$\langle E_R| = \frac{1}{\sqrt{2}}(1 \quad -j) \qquad \langle E_L| = \frac{1}{\sqrt{2}}(1 \quad j).$$

It can be easily verified that the left and right states are orthogonal and that the right polarization state is of unit length:

$$\langle E_R|E_L\rangle = \frac{1}{2}(1 \quad -j)\begin{pmatrix} 1 \\ -j \end{pmatrix} = 0 \qquad \langle E_R|E_R\rangle = \frac{1}{2}(1 \quad -j)\begin{pmatrix} 1 \\ j \end{pmatrix} = 1.$$

The completeness relation is clearly satisfied because

$$|E_x\rangle\langle E_x| + |E_y\rangle\langle E_y| = \begin{pmatrix} 1 \\ 0 \end{pmatrix}(1 \quad 0) + \begin{pmatrix} 0 \\ 1 \end{pmatrix}(0 \quad 1) = \begin{pmatrix} 1 & 0 \\ 0 & 1 \end{pmatrix} = I_2.$$

An arbitrary polarization state can be represented as

$$|E\rangle = |E_R\rangle\langle E_R|E\rangle + |E_L\rangle\langle E_L|E\rangle.$$

For example, for $E = E_x$ we obtain:

$$|E_x\rangle = |E_R\rangle\langle E_R|E_x\rangle + |E_L\rangle\langle E_L|E_x\rangle.$$

For the photon spin operator S matrix representation, we have to solve the following eigenvalue equation:

$$S|\psi\rangle = \lambda|\psi\rangle.$$

The photon spin operator satisfies $S^2 = I$, so that we can write

$$|\psi\rangle = S^2|\psi\rangle = S(S|\psi\rangle) = S(\lambda|\psi\rangle) = \lambda S|\psi\rangle = \lambda^2|\psi\rangle.$$

It is clear from the previous equation that $\lambda^2 = 1$, so that the corresponding eigenvalues are $\lambda = \pm 1$. By substituting the eigenvalues into the eigenvalue equation we find that the corresponding eigenkets are the left and right polarization states:

$$S|E_R\rangle = |E_R\rangle \qquad S|E_L\rangle = -|E_L\rangle.$$

The photon spin represented in the $\{|E_x\rangle, |E_y\rangle\}$ basis can be obtained by

$$S \doteq \begin{pmatrix} S_{xx} & S_{xy} \\ S_{yx} & S_{yy} \end{pmatrix} = \begin{pmatrix} \langle E_x|S|E_x\rangle & \langle E_x|S|E_y\rangle \\ \langle E_y|S|E_x\rangle & \langle E_y|S|E_y\rangle \end{pmatrix} = \begin{pmatrix} 0 & -j \\ j & 0 \end{pmatrix}.$$

2.3.2 Spin-1/2 Systems

The S_z-basis in spin-1/2 systems can be written as $\{|S_z;+\rangle, |S_z;-\rangle\}$, where the corresponding basis kets represent the spin-up and spin-down states. The eigenvalues are $\{\hbar/2, -\hbar/2\}$, and the corresponding eigenket–eigenvalue relation is

$$S_z|S_z;\pm\rangle = \pm\frac{\hbar}{2}|S_z;\pm\rangle,$$

where S_z is the spin operator that can be represented in the basis above as follows:

$$S_z = \sum_{i=+,-}\sum_{j=+,-} \underbrace{|i\rangle\langle j|S_z|i\rangle\langle j|}_{i\frac{\hbar}{2}|i\rangle} = \sum_{i=+,-} i\frac{\hbar}{2}|i\rangle\langle i| = \frac{\hbar}{2}(|+\rangle\langle +|-|-\rangle\langle -|).$$

The matrix representation of spin-1/2 systems is obtained as

$$|S_z;+\rangle = \begin{pmatrix} \langle S_z;+|S_z;+\rangle \\ \langle S_z;-|S_z;+\rangle \end{pmatrix} \doteq \begin{pmatrix} 1 \\ 0 \end{pmatrix} \qquad |S_z;-\rangle \doteq \begin{pmatrix} 0 \\ 1 \end{pmatrix}$$

$$S_z \doteq \begin{pmatrix} \langle S_z;+|S_z|S_z;+\rangle & \langle S_z;+|S_z|S_z;-\rangle \\ \langle S_z;-|S_z|S_z;+\rangle & \langle S_z;-|S_z|S_z;-\rangle \end{pmatrix} = \frac{\hbar}{2}\begin{pmatrix} 1 & 0 \\ 0 & -1 \end{pmatrix}.$$

2.3.3 Hadamard Gate

The matrix representation of the Hadamard operator (gate) is given by

$$H = \frac{1}{\sqrt{2}}\begin{bmatrix} 1 & 1 \\ 1 & -1 \end{bmatrix}.$$

It can easily be shown that the Hadamard gate is Hermitian and unitary as follows:

$$H^{\dagger} = \frac{1}{\sqrt{2}}\begin{bmatrix} 1 & 1 \\ 1 & -1 \end{bmatrix} = H$$

$$H^{\dagger}H = \frac{1}{\sqrt{2}}\begin{bmatrix} 1 & 1 \\ 1 & -1 \end{bmatrix}\frac{1}{\sqrt{2}}\begin{bmatrix} 1 & 1 \\ 1 & -1 \end{bmatrix} = \begin{bmatrix} 1 & 0 \\ 0 & 1 \end{bmatrix} = I.$$

The eigenvalues for the Hadamard gate can be obtained from $\det(H - \lambda I) = 0$ to be $\lambda_{1,2} = \pm 1$. By substituting the eigenvalues into the eigenvalue equation, namely $H|\Psi_{1,2}\rangle = \pm|\Psi_{1,2}\rangle$, the corresponding eigenkets are obtained as follows:

$$|\Psi_1\rangle = \begin{bmatrix} \dfrac{1}{\sqrt{4-2\sqrt{2}}} \\ \dfrac{1}{\sqrt{2\sqrt{2}}} \end{bmatrix} \qquad |\Psi_2\rangle = \begin{bmatrix} \dfrac{1}{\sqrt{4+2\sqrt{2}}} \\ -\dfrac{1}{\sqrt{2\sqrt{2}}} \end{bmatrix}.$$

2.4 QUANTUM MEASUREMENTS, COMMUTATORS, AND PAULI OPERATORS

Each measurable physical quantity – observable (such as position, momentum, or angular momentum) – is associated with a Hermitian operator that has a complete set of eigenkets. According to P. A. Dirac, "A measurement always causes the system to jump into an eigenstate of the dynamical variable that is being measured" [3]. Dirac's statement can be formulated as the following *postulate*: An exact measurement of an observable with operator A always yields as a result one of the eigenvalues $a^{(n)}$ of A. Thus, the measurement changes the state, with the measurement system "thrown into" one of its eigenstates, which can be represented as $|\alpha\rangle \xrightarrow{A\text{ measurement}} |a^{(j)}\rangle$. If, before measurement, the system was in state $|\alpha\rangle$, the probability that the result of a measurement will be the eigenvalue $a^{(i)}$ is given by

$$\Pr\left(a^{(i)}\right) = \left|\left\langle a^{(i)} \middle| \alpha \right\rangle\right|^2. \tag{2.31}$$

Since at least one of the eigenvalues must occur as the result of the measurements, this probability satisfies

$$\sum_i \Pr\left(a^{(i)}\right) = \sum_i \left|\left\langle a^{(i)} \middle| \alpha \right\rangle\right|^2 = 1. \tag{2.32}$$

The expected value of the outcome of the measurement of A is given by

$$\langle A \rangle = \sum_i a^{(i)} \Pr\left(a^{(i)}\right) = \sum_i a^{(i)} \left|\left\langle a^{(i)} \middle| \alpha \right\rangle\right|^2 = \sum_i a^{(i)} \left\langle \alpha \middle| a^{(i)} \right\rangle \left\langle a^{(i)} \middle| \alpha \right\rangle. \tag{2.33}$$

By applying the eigenvalue equation $a^{(i)}\left|a^{(i)}\right\rangle = A\left|a^{(i)}\right\rangle$, Eq. (2.33) becomes

$$\langle A \rangle = \sum_i \left\langle \alpha \middle| A \middle| a^{(i)} \right\rangle \left\langle a^{(i)} \middle| \alpha \right\rangle. \tag{2.34}$$

By using further the completeness relation $\sum_i |a^{(i)}\rangle\langle a^{(i)}| = I$, we obtain the expected value of the measurement of A to be simply

$$\langle A \rangle = \langle \alpha | A | \alpha \rangle. \tag{2.35}$$

In various situations, like initial state preparations for quantum information processing applications, we need to select one particular outcome of the measurement. This procedure is known as *selective measurement* (or filtration) and it can be conducted as shown in Figure 2.1.

The result of the selective measurement can be interpreted as applying the *projection operator* $P_{a'}$ to $|\alpha\rangle$ to obtain:

$$P_{a'}|\alpha\rangle = |a'\rangle\langle a'|\alpha\rangle. \tag{2.36}$$

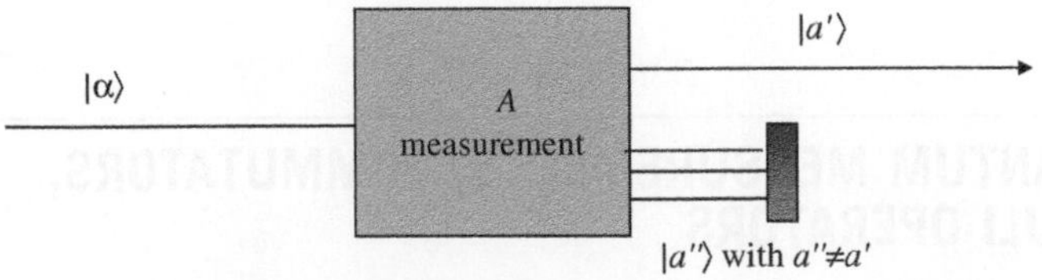

FIGURE 2.1

The concept of selective measurement (filtration).

The probability that the outcome of the measurement of observable Ξ with eigenvalues $\xi^{(n)}$ lies between (a,b) is given by

$$\Pr(\xi \in R(a,b)) = \sum_{\xi^{(n)} \in R(a,b)} \left| \left\langle \xi^{(n)} \middle| \alpha \right\rangle \right|^2 = \sum_{\xi^{(n)} \in R(a,b)} \left\langle \alpha \middle| \xi^{(n)} \right\rangle \left\langle \xi^{(n)} \middle| \alpha \right\rangle$$
$$= \langle \alpha | P_{ab} | \alpha \rangle = \langle P_{ab} \rangle, \tag{2.37}$$

where P_{ab} denotes the following projection operator:

$$P_{ab} = \sum_{\xi^{(n)} \in R(a,b)} \left| \xi^{(n)} \right\rangle \left\langle \xi^{(n)} \right|. \tag{2.38}$$

It is straightforward to show that the projection operator P_{ab} satisfies

$$P_{ab}^2 = P_{ab} \Leftrightarrow P_{ab}(P_{ab} - I) = 0. \tag{2.39}$$

Therefore, the eigenvalues of projection operator P_{ab} are either 0 (corresponding to the "false proposition") or 1 (corresponding to the "true proposition"); this is important in *quantum detection theory* [5].

In terms of projection operators, the state of the system after the measurement is given by

$$|\alpha\rangle \xrightarrow{A \text{ measurement}} \frac{1}{\sqrt{\langle \alpha | P_j | \alpha \rangle}} P_j |\alpha\rangle, \quad P_j = |a^{(j)}\rangle \langle a^{(j)}|. \tag{2.40}$$

In the case where operator A has *degenerate* eigenvalues corresponding to the eigenkets $\left\{ \left| a_i^{(j)} \right\rangle \right\}_{j=1}^{d_i}$, where each eigenket corresponds to the same eigenvalue a_i:

$$A \left| a_i^{(j)} \right\rangle = a_i \left| a_i^{(j)} \right\rangle; \quad j = 1, \ldots, d_i; \tag{2.41}$$

we say that eigenvalue a_i is *degenerate* of order d_i. The corresponding probability of obtaining the measurement result a_i can be found by

$$\Pr(a_i) = \sum_{j=1}^{d_i} \left| \left\langle a_i^{(j)} \middle| \alpha \right\rangle \right|^2. \tag{2.42}$$

2.4.1 Commutators

Let A and B be two operators, which in general do not commute, i.e. $AB \neq BA$. The quantity $[A,B] = AB - BA$ is called the *commutator* of A and B, while the quantity $\{A,B\} = AB + BA$ is called the *anticommutator*. Two observables A and B are said to be *compatible* when their corresponding operators commute: $[A,B] = 0$. Two observables A and B are said to be *incompatible* when $[A,B] \neq 0$. If in the set of operators $\{A,B,C,\ldots\}$ all operators commute in pairs, namely $[A,B] = [A,C] = [B,C] = \ldots = 0$, we say the set is a *complete set of commuting observables* (CSCO).

Let A, B, and C be the operators, then the following *properties* of the commutator are valid:

$$\begin{aligned}
&1.\ [A,B] = -[B,A] \\
&2.\ [A+B,C] = [A,C] + [B,C] \\
&3.\ [A,BC] = B[A,C] + [A,B]C \\
&4.\ [X,P] = j\hbar,\ \ [X,X] = [P,P] = 0,
\end{aligned} \tag{2.43}$$

where X and P are the position and momentum operators. Properties 1–3 are quite straightforward to prove by using the definition of commutator, and as such will not be considered at this point. Here we provide the proof of property 4. It is convenient to establish a *connection* between *wave quantum mechanics* and *matrix quantum mechanics*. In wave mechanics, information about the state of a particle is described by the corresponding *wave function* $\psi(x,t) = \langle x|\psi\rangle$. The wave function gives information about the location of the particle – namely, the magnitude squared of the wave functions $|\psi(x,t)|^2$ is related to the *probability density*. The probability of finding the particle within the interval x and $x + \mathrm{d}x$ is given by

$$\mathrm{d}P(x,t) = |\psi(x,t)|^2 \mathrm{d}x. \tag{2.44}$$

In wave quantum mechanics, the position X and the momentum P operators are defined by

$$X\psi(x,t) = x\psi(x,t) \qquad P\psi(x,t) = -j\hbar\frac{\partial}{\partial x}\psi(x,t). \tag{2.45}$$

We apply the commutator to the test wave function $\psi(x,t)$ and obtain:

$$\begin{aligned}
[X,P]\psi(x,t) &= (XP - PX)\psi(x,t) \\
&= XP\psi(x,t) - PX\psi(x,t) \\
&= -j\hbar x\frac{\partial}{\partial x}\psi(x,t) + j\hbar\frac{\partial}{\partial x}(X\psi(x,t)) \\
&= -j\hbar x\frac{\partial}{\partial x}\psi(x,t) + j\hbar\frac{\partial}{\partial x}(x\psi(x,t)) \\
&= -j\hbar x\frac{\partial}{\partial x}\psi(x,t) + j\hbar\left(\psi(x,t) + x\frac{\partial}{\partial x}\psi(x,t)\right) = j\hbar\psi(x,t),
\end{aligned} \tag{2.46}$$

therefore proving that $[X,P] = j\hbar$.

The Pauli operators occur very often in quantum information processing and quantum error correction, and as such are very suitable to be used to illustrate the basic ideas above.

2.4.2 Pauli Operators

The basic unit of information in a quantum computer is known as a *quantum bit* or *qubit*. The corresponding state space is two-dimensional with the following basis: $\{|0\rangle = [1\ 0]^{\mathrm{T}}, |1\rangle = [0\ 1]^{\mathrm{T}}\}$. The arbitrary state ket can be written in terms of basis kets as follows:

$$|\psi\rangle = \alpha|0\rangle + \beta|1\rangle, \tag{2.47}$$

with α and β satisfying the following normalization condition: $|\alpha|^2 + |\beta|^2 = 1$. The probability that a qubit upon measurement is in a state $|0\rangle$ is determined by $|\langle 0|\psi\rangle|^2 = |\alpha|^2$.

The *Pauli operators* X, Y, Z (very often denoted as σ_x, σ_y, and σ_z or σ_1, σ_2, and σ_3) correspond to the measurement of the spin along the x-, y-, and z-axes respectively. Their actions on basis states are given by

$$\begin{aligned} &X|0\rangle = |1\rangle, \quad X|1\rangle = |0\rangle \\ &Y|0\rangle = -j|1\rangle, \quad Y|1\rangle = j|0\rangle \\ &Z|0\rangle = |0\rangle, \quad Z|1\rangle = -|1\rangle. \end{aligned} \tag{2.48}$$

It is clear that basis states are eigenkets of Z. We have shown earlier that an operator Ξ can be represented in matrix form in $\{|a^{(k)}\rangle\}$ basis with matrix elements given by $\Xi_{ij} = \langle a^{(i)}|\Xi|a^{(j)}\rangle$. Based on the action of the Pauli X operator in (2.48), it can be represented in matrix form as

$$X \doteq \begin{pmatrix} \langle 0|X|0\rangle & \langle 0|X|1\rangle \\ \langle 1|X|0\rangle & \langle 1|X|1\rangle \end{pmatrix} = \begin{pmatrix} \langle 0|1\rangle & \langle 0|0\rangle \\ \langle 1|1\rangle & \langle 1|0\rangle \end{pmatrix} = \begin{pmatrix} 0 & 1 \\ 1 & 0 \end{pmatrix}. \tag{2.49}$$

In similar fashion, the Pauli X and Y operators can be represented as

$$Y \doteq \begin{pmatrix} 0 & -j \\ j & 0 \end{pmatrix}, \quad Z \doteq \begin{pmatrix} 1 & 0 \\ 0 & -1 \end{pmatrix}. \tag{2.50}$$

Since any operator Ξ can be written in terms of outer product as $\Xi = \sum_{n=1}^{\infty}\sum_{m=1}^{\infty} \Xi_{mn}|a^{(m)}\rangle\langle a^{(n)}|$, the Pauli X operator can be written as

$$X \doteq \begin{pmatrix} 0 & X_{01} \\ X_{10} & 0 \end{pmatrix} = \begin{pmatrix} 0 & 1 \\ 1 & 0 \end{pmatrix} = X_{01}|0\rangle\langle 1| + X_{10}|1\rangle\langle 0| = |0\rangle\langle 1| + |1\rangle\langle 0|. \tag{2.51}$$

We have shown that if Ξ has eigenvalues $\xi^{(k)}$ and eigenkets $|\xi^{(k)}\rangle$ determined from $\Xi|\xi^{(k)}\rangle = \xi^{(k)}|\xi^{(k)}\rangle$, the spectral decomposition Ξ is given by $\Xi = \sum_k \xi^{(k)}|\xi^{(k)}\rangle\langle\xi^{(k)}|$, so that the spectral decomposition of the Pauli Z operator will be

$$Z = |0\rangle\langle 0| - |1\rangle\langle 1|. \tag{2.52}$$

The projection operators corresponding to measurements of 1 and −1 can be defined as

$$P_0 = |0\rangle\langle 0|, \qquad P_1 = |1\rangle\langle 1|. \tag{2.53}$$

The projector P_0 (P_1) performs projection of an arbitrary state to the $|0\rangle$ ($|1\rangle$) state:

$$\begin{aligned} &|\psi\rangle = \alpha|0\rangle + \beta|1\rangle \\ &P_0|\psi\rangle = (|0\rangle\langle 0|)|\psi\rangle = \alpha|0\rangle, \qquad P_1|\psi\rangle = (|1\rangle\langle 1|)|\psi\rangle = \beta|1\rangle. \end{aligned} \tag{2.54}$$

The sum of projection operators clearly results in the identity operator as shown below:

$$\begin{aligned} P_0 + P_1 &= |0\rangle\langle 0| + |1\rangle\langle 1| = \begin{pmatrix} 1 \\ 0 \end{pmatrix}\begin{pmatrix} 1 & 0 \end{pmatrix} + \begin{pmatrix} 0 \\ 1 \end{pmatrix}\begin{pmatrix} 0 & 1 \end{pmatrix} \\ &= \begin{pmatrix} 1 & 0 \\ 0 & 0 \end{pmatrix} + \begin{pmatrix} 0 & 0 \\ 0 & 1 \end{pmatrix} = \begin{pmatrix} 1 & 0 \\ 0 & 1 \end{pmatrix} = I. \end{aligned} \tag{2.55}$$

We have shown earlier that we can express the state ket $|\psi\rangle$ in terms of operator Ξ as $|\psi\rangle = \sum_k \alpha_k |\xi^{(k)}\rangle$. We have also shown that the probability of obtaining the measurement result $\xi^{(k)}$ can be expressed in terms of projection operator $P_k = |\xi^{(k)}\rangle\langle\xi^{(k)}|$ by

$$\begin{aligned} |\alpha_k|^2 &= \left|\left\langle \xi^{(k)} \middle| \psi \right\rangle\right|^2 = \left\langle \xi^{(k)} \middle| \psi \right\rangle \left\langle \xi^{(k)} \middle| \psi \right\rangle^* = \left\langle \xi^{(k)} \middle| \psi \right\rangle \left\langle \psi \middle| \xi^{(k)} \right\rangle \\ &= \left\langle \psi \middle| \xi^{(k)} \right\rangle \left\langle \xi^{(k)} \middle| \psi \right\rangle = \langle \psi | P_k | \psi \rangle. \end{aligned}$$

We have also shown that the final state of the system of the measurement will be $P_k|\psi\rangle / \sqrt{\langle\psi|P_k|\psi\rangle}$. For a two-dimensional system, if the measurement result +1 is obtained the result after the measurement will be

$$|\psi'\rangle = \frac{1}{\sqrt{\langle\psi|P_0|\psi\rangle}} P_0|\psi\rangle \overset{|\psi\rangle = \alpha|0\rangle + \beta|1\rangle}{=} \frac{\alpha}{|\alpha|}|0\rangle, \tag{2.56}$$

and if measurement result −1 is obtained the corresponding state will be

$$|\psi'\rangle = \frac{1}{\sqrt{\langle\psi|P_1|\psi\rangle}} P_1|\psi\rangle = \frac{\beta}{|\beta|}|1\rangle. \tag{2.57}$$

2.5 UNCERTAINTY PRINCIPLE

If two observables, say A and B, are to be measured simultaneously and exactly on the same system, the system after the measurement must be left in the state $|a^{(n)};b^{(n)}\rangle$, which is an eigenstate of both observables:

$$A\left|a^{(n)};b^{(n)}\right\rangle = a^{(n)}\left|a^{(n)};b^{(n)}\right\rangle$$

$$B\left|a^{(n)};b^{(n)}\right\rangle = b^{(n)}\left|a^{(n)};b^{(n)}\right\rangle. \tag{2.58}$$

This will be true only if $AB = BA$ or equivalently the commutator $[A,B] = AB - BA = 0$ – that is, when two operators *commute* as shown below:

$$\begin{aligned} AB\left|a^{(n)};b^{(n)}\right\rangle &= A\left(B\left|a^{(n)};b^{(n)}\right\rangle\right) = Ab^{(n)}\left|a^{(n)};b^{(n)}\right\rangle = b^{(n)}\cdot A\left|a^{(n)};b^{(n)}\right\rangle \\ &= a^{(n)}b^{(n)}\left|a^{(n)};b^{(n)}\right\rangle \\ BA\left|a^{(n)};b^{(n)}\right\rangle &= a^{(n)}b^{(n)}\left|a^{(n)};b^{(n)}\right\rangle \Rightarrow AB = BA. \end{aligned} \tag{2.59}$$

When two operators do not commute, they cannot be simultaneously measured with complete precision. Given an observable A, we define the operator $\Delta A = A - \langle A\rangle$, and the corresponding expectation value of $(\Delta A)^2$ that is known as the *dispersion* of A:

$$\left\langle(\Delta A)^2\right\rangle = \left\langle A^2 - 2A\langle A\rangle + \langle A\rangle^2\right\rangle = \left\langle A^2\right\rangle - \left\langle A\right\rangle^2. \tag{2.60}$$

Then for any state the following inequality is valid:

$$\left\langle(\Delta A)^2\right\rangle\left\langle(\Delta B)^2\right\rangle \geq \frac{1}{4}|\langle[\Delta A, \Delta B]\rangle|^2, \tag{2.61}$$

which is known as the *Heisenberg* uncertainty principle.

Example. The commutation relation for coordinate X and momentum P observables is $[X, P] = j\hbar$, as shown above. By substituting this commutation relation into (2.61) we obtain:

$$\langle X^2\rangle\langle P^2\rangle \geq \frac{\hbar^2}{4}.$$

Say we observe a large ensemble of N independent systems, all of them being in the state $|\psi\rangle$. On some systems X is measured, and on some systems P is measured. The uncertainty principle asserts that for any state the product of dispersions (variances) cannot be less than $\hbar^2/4$.

For the derivation of the uncertainty principle, the following two lemmas are important.

Lemma 1 (L1). The expectation operator of a Hermitian operator B is purely real.

This lemma can easily be proved by using the definition of Hermitian operator $B^\dagger = B$ and the definition of the expectation of operator B, $\langle B\rangle = \langle\alpha|B|\alpha\rangle$. By using Eq. (2.10) we obtain $\langle\alpha|B|\alpha\rangle^* = \langle\alpha|B^\dagger|\alpha\rangle^* = \langle\alpha|B|\alpha\rangle$, which proves that the expectation of a Hermitian operator is indeed real.

Lemma 2 (L2). The expectation operator of an anti-Hermitian operator, defined by $C^\dagger = -C$, is purely imaginary.

This lemma can also be proved by using the definition of anti-Hermitian operator $C^\dagger = -C$ and the definition of the expectation of operator C, $\langle C\rangle = \langle\alpha|C|\alpha\rangle$. By again using Eq. (2.10) we obtain $\langle\alpha|C|\alpha\rangle^* = \langle\alpha| - C^\dagger|\alpha\rangle^* = -\langle\alpha|C^\dagger|\alpha\rangle^* = -\langle\alpha|C|\alpha\rangle$, which proves that the expectation of an anti-Hermitian operator is indeed imaginary.

By assuming that A and B are Hermitian operators, the uncertainty principle can now be derived starting from the *Cauchy–Schwartz inequality* (see Eq. (2.6)) as follows. Let $|\alpha\rangle = \Delta A|\psi\rangle$ and $|\beta\rangle = \Delta B|\psi\rangle$. From the Cauchy–Schwartz inequality we know that

$$\langle\alpha|\alpha\rangle\langle\beta|\beta\rangle = \langle(\Delta A)^2\rangle\langle(\Delta B)^2\rangle \geq |\langle\Delta A\Delta B\rangle|^2. \tag{2.62}$$

It is easy to show that $\langle\Delta A\Delta B\rangle = (\langle[\Delta A,\Delta B]\rangle + \langle\{\Delta A,\Delta B\}\rangle)/2$. Since $[A,B]^\dagger = (AB-BA)^\dagger = BA - AB = -[A,B]$ from (L2) is clear that the expectation of $[A,B]$ is purely imaginary. In similar fashion we show that $\{A,B\}^\dagger = (AB+BA)^\dagger = BA + AB = \{A,B\}$, indicating that the expectation of $\{A,B\}$ is purely real (based on (L1)). Therefore, the first term in expectation of $\Delta A\Delta B$ is purely imaginary and the second term is purely real, as indicated below:

$$\langle\Delta A\Delta B\rangle = \underbrace{\frac{1}{2}\langle[\Delta A,\Delta B]\rangle}_{\text{purely imaginery}} + \underbrace{\frac{1}{2}\langle\{\Delta A,\Delta B\}\rangle}_{\text{purely real}}. \tag{2.63}$$

From complex numbers theory we know that the magnitude squared of a complex number $z = x + jy$ (j is the imaginary unit) is given by $|z|^2 = x^2 + y^2$, so that the magnitude squared of $|\langle\Delta A\Delta B\rangle|^2$ is given by

$$|\langle\Delta A\Delta B\rangle|^2 = \frac{1}{4}|\langle[\Delta A,\Delta B]\rangle|^2 + \frac{1}{4}|\langle\{\Delta A,\Delta B\}\rangle|^2. \tag{2.64}$$

If we omit the anticommutator term in (2.64) we obtain the uncertainty relation given by Eq. (2.61).

2.6 DENSITY OPERATORS

Let a large number of quantum systems of the same kind be prepared, each in a set of orthonormal states $|\phi_n\rangle$, and let the fraction of the system being in state $|\phi_n\rangle$ be denoted by probability p_n ($n = 1,2,\ldots$):

$$\langle\phi_m|\phi_n\rangle = \delta_{mn}, \qquad \sum_n p_n = 1. \tag{2.65}$$

Therefore, this ensemble of quantum states represents a classical *statistical mixture* of kets. The probability of obtaining ξ_n from the measurement of Ξ will be

$$\Pr(\xi_k) = \sum_{n=1}^{\infty} p_n|\langle\xi_k|\phi_n\rangle|^2 = \sum_{n=1}^{\infty} p_n\langle\xi_k|\phi_n\rangle\langle\phi_n|\xi_k\rangle = \langle\xi_k|\rho|\xi_k\rangle, \tag{2.66}$$

where the operator ρ is known as a *density operator* and is defined as

$$\rho = \sum_{n=1}^{\infty} p_n|\phi_n\rangle\langle\phi_n|. \tag{2.67}$$

The expected value of the density operator is given by

$$\langle\rho\rangle=\sum_{k=1}^{\infty}\xi_k\Pr(\xi_k)=\sum_{k=1}^{\infty}\xi_k\langle\xi_k|\rho|\xi_k\rangle=\sum_{k=1}^{\infty}\langle\xi_k|\rho\Xi|\xi_k\rangle=\mathrm{Tr}(\rho\Xi). \tag{2.68}$$

The *properties* of the density operator can be summarized as follows:

1. The density operator is Hermitian ($\rho^+=\rho$), with the set of orthonormal eigenkets $|\phi_n\rangle$ corresponding to the non-negative eigenvalues p_n and $\mathrm{Tr}(\rho)=1$.
2. Any Hermitian operator with non-negative eigenvalues and trace 1 may be considered as a density operator.
3. The density operator is positive definite: $\langle\psi|\rho|\psi\rangle\geq 0$ for all $|\psi\rangle$.
4. The density operator has the property $\mathrm{Tr}(\rho^2)\leq 1$, with equality iff one of the prior probabilities is 1, and all the rest 0: $\rho=|\phi_n\rangle\langle\phi_n|$, and the density operator is then a projection operator.
5. The eigenvalues of a density operator satisfy $0\leq\lambda_i\leq 1$.

The proof of these properties is quite straightforward, and is left for the reader. When ρ is the projection operator, we say that it represents the system in a *pure state*; otherwise, with $\mathrm{Tr}(\rho^2)<1$, it represents a *mixed state*. A mixed state in which all eigenkets occur with the same probability is known as a *completely mixed state* and can be represented by

$$\rho=\sum_{k=1}^{\infty}\frac{1}{n}|\phi_n\rangle\langle\phi_n|=\frac{1}{n}I\Rightarrow\mathrm{Tr}(\rho^2)=\frac{1}{n}\Rightarrow\frac{1}{n}\leq\mathrm{Tr}(\rho^2)\leq 1. \tag{2.69}$$

If the density matrix has off-diagonal elements different from zero, we say that it exhibits *quantum interference*, which means that the state terms can interfere with each other. Let us observe the following pure state:

$$\begin{aligned}|\psi\rangle=\sum_{i=1}^{n}\alpha_i|\xi_i\rangle\Rightarrow\rho=|\psi\rangle\langle\psi|&=\sum_{i=1}^{n}|\alpha_i|^2|\psi_i\rangle\langle\psi_i|+\sum_{i=1}^{n}\sum_{j=1,j\neq i}^{n}\alpha_i\alpha_j^*|\xi_i\rangle\langle\xi_j|\\&=\sum_{i=1}^{n}\langle\xi_i|\rho|\xi_i\rangle|\xi_i\rangle\langle\xi_i|+\sum_{i=1}^{n}\sum_{j=1,j\neq i}^{n}\langle\xi_i|\rho|\xi_j\rangle|\xi_i\rangle\langle\xi_j|.\end{aligned} \tag{2.70}$$

The first term in (2.70) relates to the probability of the system being in state $|\psi_i\rangle$, and the second term relates to the quantum interference. It appears that the off-diagonal elements of a mixed state will be zero, while those of a pure state will be nonzero. Notice that the existence of off-diagonal elements is base dependent; therefore, to check for purity it is a good idea to compute $\mathrm{Tr}(\rho^2)$ instead.

In quantum information theory, the density matrix can be used to determine the amount of information conveyed by the quantum state, i.e. to compute the von Neumann entropy:

$$S=\mathrm{Tr}(\rho\log\rho)=-\sum_i\lambda_i\log_2\lambda_i, \tag{2.71}$$

where λ_i are the eigenvalues of the density matrix. The corresponding Shannon entropy can be calculated by

$$H = -\sum_i p_i \log_2 p_i. \tag{2.72}$$

Let us further consider the observable Ξ with the set of eigenvalues $\{\xi_k;d\}$, where d is the *degeneracy* index. For simplicity of derivation without losing generality, let us assume that $\rho = |\psi\rangle\langle\psi|$ (the pure state). The probability that the measurement of Ξ will lead to eigenvalue ξ_k is given by

$$\begin{aligned}
\Pr(\xi_k) &= \sum_d |\langle \xi_k;d|\psi\rangle|^2 = \sum_d \langle \xi_k;d|\psi\rangle\langle\psi|\xi_k;d\rangle \\
&= \sum_d \langle \xi_k;d| \left[\sum_n |\phi_n\rangle\langle\phi_n|\right] |\psi\rangle\langle\psi|\xi_k;d\rangle = \sum_n \langle\phi_n|\psi\rangle \sum_d \langle\psi|\xi_k;d\rangle\langle\xi_k;d|\phi_n\rangle \\
&= \sum_n \langle\phi_n|\psi\rangle\langle\psi| \left[\sum_d |\xi_k;d\rangle\langle\xi_k;d|\right] |\phi_n\rangle = \sum_n \langle\phi_n|\rho P_k|\phi_n\rangle, \\
\rho &= |\psi\rangle\langle\psi|, P_k = \sum_d |\xi_k;d\rangle\langle\xi_k;d|.
\end{aligned} \tag{2.73}$$

From the last line of the previous expression it is clear that

$$\Pr(\xi_k) = \mathrm{Tr}(\rho P_k). \tag{2.74}$$

For a *two-level system*, the density operator can be written as

$$\rho = \frac{1}{2}(I + \vec{r}\cdot\vec{\sigma}), \quad \vec{\sigma} = [X \;\; Y \;\; Z], \tag{2.75}$$

where X, Y, Z are the Pauli operators and $\vec{r} = [r_x \;\; r_y \;\; r_z]$ is the Bloch vector, whose components can be determined by

$$r_x = \mathrm{Tr}(\rho X), \quad r_y = \mathrm{Tr}(\rho Y), \quad r_z = \mathrm{Tr}(\rho Z). \tag{2.76}$$

The magnitude of the Bloch vector is $\|\vec{r}\| \leq 1$, with equality corresponding to the pure state.

Suppose now that S is a *bipartite composite system* with component subsystems A and B. For example, the subsystem A can represent the quantum register Q and subsystem B the environment E. The composite system can be represented by $AB = A \otimes B$, where $\otimes$ stands for the tensor product. If the dimensionality of Hilbert space H_A is m and the dimensionality of Hilbert space H_B is n, then the dimensionality of Hilbert space H_{AB} will be mn. Let $|\alpha\rangle \in A$ and $|\beta\rangle \in B$, then $|\alpha\rangle|\beta\rangle = |\alpha\rangle \otimes |\beta\rangle \in AB$. If the operator A acts on kets from H_A and the operator B on kets from H_B, then the action of AB on $|\alpha\rangle|\beta\rangle$ can be described as follows:

$$(AB)|\alpha\rangle|\beta\rangle = (A|\alpha\rangle)(B|\beta\rangle). \tag{2.77}$$

The norm of state $|\psi\rangle = |\alpha\rangle|\beta\rangle \in AB$ is determined by

$$\langle\psi|\psi\rangle = \langle\alpha|\alpha\rangle\langle\beta|\beta\rangle. \tag{2.78}$$

Let $\{|\alpha_i\rangle\}$ $(\{|\beta_i\rangle\})$ be a basis for the Hilbert space H_A (H_B) and let E be an ensemble of physical systems S described by the density operator ρ. The *reduced density operator* ρ_A for subsystem A is defined to be the partial trace of ρ over B:

$$\rho_A = \mathrm{Tr}_B(\rho) = \sum_j \langle\beta_j|\rho|\beta_j\rangle. \tag{2.79}$$

Similarly, the *reduced density operator* ρ_B for subsystem B is defined to be the partial trace of ρ over A:

$$\rho_B = \mathrm{Tr}_A(\rho) = \sum_i \langle\alpha_j|\rho|\alpha_j\rangle. \tag{2.80}$$

2.7 CHANGE OF BASIS

Suppose that we have two incompatible observables A and B. The ket space can be viewed as being spanned by $|a^{(i)}\rangle$ or $|b^{(i)}\rangle$. The two different sets of base kets span the same ket. We are interested in finding out how these two descriptions can be related. The following theorem will provide the answer.

Theorem. Given two sets of base kets, $\{|a^{(i)}\rangle\}$ and $\{|b^{(i)}\rangle\}$, both satisfying orthonormality and completeness, there exists a unitary operator U (the unitary operator is defined by $U^\dagger U = I$) such that

$$|b^{(i)}\rangle = U|a^{(i)}\rangle. \tag{2.81}$$

This theorem can be proved by construction. Let the unitary operator be chosen as follows:

$$U = \sum_k |b^{(k)}\rangle\langle a^{(k)}|. \tag{2.82}$$

Since $U|a^{(l)}\rangle = |b^{(l)}\rangle$, we can show that such a chosen operator U is unitary as follows:

$$U^\dagger U = \sum_k \sum_l \left|a^{(l)}\right\rangle\left\langle b^{(l)}\middle|b^{(k)}\right\rangle\left\langle a^{(k)}\right| \overset{\langle b^{(l)}|b^{(k)}\rangle=\delta_{lk}}{=} \sum_k \left|a^{(k)}\right\rangle\left\langle a^{(k)}\right| = I. \tag{2.83}$$

The matrix representation of the U operator in the initial basis can be obtained, based on (2.81), as follows:

$$\left\langle a^{(k)}\middle|U\middle|a^{(l)}\right\rangle = \left\langle a^{(k)}\middle|b^{(l)}\right\rangle. \tag{2.84}$$

The ket $|\alpha\rangle$ in the old basis can be represented by:

$$|\alpha\rangle = \sum_i |a^{(i)}\rangle\langle a^{(i)}|\alpha\rangle. \tag{2.85}$$

By multiplying (2.85) by $\langle b^{(k)}|$ from the left we obtain:

$$\langle b^{(k)}|\alpha\rangle = \sum_i \langle b^{(k)}|a^{(i)}\rangle\langle a^{(i)}|\alpha\rangle \overset{\langle b^{(k)}|=\langle a^{(k)}|U^\dagger}{=} \sum_i \langle a^{(k)}|U^\dagger|a^{(i)}\rangle\langle a^{(i)}|\alpha\rangle. \tag{2.86}$$

Therefore, the column vector for $|\alpha\rangle$ in the new basis can be obtained by applying U^+ to the column vector in the old basis:

$$(\text{New}) = (U^\dagger)(\text{Old}).$$

The relationship between the old matrix elements and the new matrix elements can be established by

$$\begin{aligned}\langle b^{(k)}|X|b^{(l)}\rangle &= \sum_m \sum_n \langle b^{(k)}|a^{(m)}\rangle\langle a^{(m)}|X|a^{(n)}\rangle\langle a^{(n)}|b^{(l)}\rangle \\ &\underset{\langle b^{(k)}|=\langle a^{(k)}|U^\dagger}{\overset{|b^{(k)}\rangle=U|a^{(k)}\rangle}{=}} \sum_m \sum_n \langle a^{(k)}|U^\dagger|a^{(m)}\rangle\langle a^{(m)}|X|a^{(n)}\rangle\langle a^{(n)}|U|a^{(l)}\rangle.\end{aligned} \tag{2.87}$$

Therefore, the matrix representation X' in the new basis can be obtained from the matrix representation in the old basis X as follows:

$$X' = U^\dagger X U, \tag{2.88}$$

where the transformation $U^\dagger X U$ is known as the *similarity transformation.*

The trace of an operator X, Tr(X), is independent of the basis, as shown below:

$$\begin{aligned}\text{Tr}(X) &= \sum_i \langle a^{(i)}|X|a^{(i)}\rangle = \sum_i \sum_l \sum_k \langle a^{(i)}|b^{(l)}\rangle\langle b^{(l)}|X|b^{(k)}\rangle\langle b^{(k)}|a^{(i)}\rangle \\ &= \sum_i \sum_l \sum_k \langle b^{(k)}|a^{(i)}\rangle\langle a^{(i)}|b^{(l)}\rangle\langle b^{(l)}|X|b^{(k)}\rangle = \sum_k \langle b^{(k)}|X|b^{(k)}\rangle.\end{aligned} \tag{2.89}$$

The properties of the trace can be summarized as follows:

$$\begin{aligned}&\text{Tr}(XY) = \text{Tr}(YX) \qquad \text{Tr}(XYZ) = \text{Tr}(ZXY) = \text{Tr}(YZX) \quad \text{(the trace is cyclic)} \\ &\text{Tr}(U^\dagger XU) = \text{Tr}(X) \\ &\text{Tr}(|a^{(i)}\rangle\langle a^{(j)}|) = \delta_{ij} \qquad \text{Tr}(|b^{(i)}\rangle\langle a^{(j)}|) = \langle a^{(j)}|b^{(i)}\rangle.\end{aligned} \tag{2.90}$$

2.7.1 Photon Polarization (Revisited)

Figure 2.2 illustrates the transformation of bases using photon polarization. In Figure 2.2a the coordinate system is rotated for θ counterclockwise, while in Figure 2.2b the ket $|\psi\rangle$ is rotated for θ in the clockwise direction.

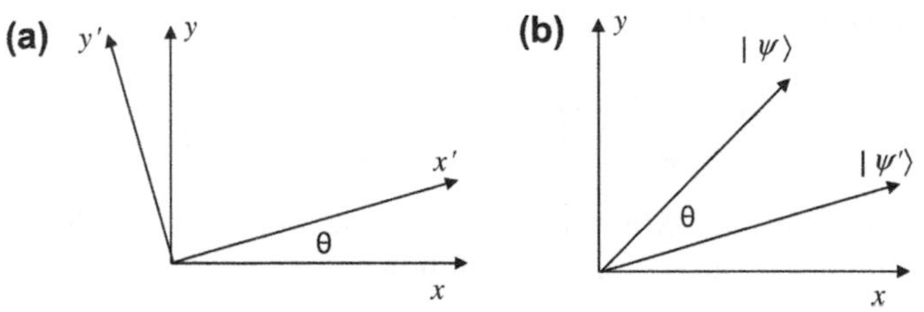

FIGURE 2.2

The transformation bases using photon polarization: (a) the rotation of a coordinate system; (b) the rotation of the ket.

The ket $|\psi\rangle$ can be represented in the original basis by

$$|\psi\rangle = |x\rangle\langle x|\psi\rangle + |y\rangle\langle y|\psi\rangle, \tag{2.91}$$

where $\langle x|\psi\rangle$ ($\langle y|\psi\rangle$) denotes the projection along x-polarization (y-polarization). By multiplying (2.91) by $\langle x'|$ and $\langle y'|$ from the left side we obtain:

$$\langle x'|\psi\rangle = \langle x'|x\rangle\langle x|\psi\rangle + \langle x'|y\rangle\langle y|\psi\rangle \qquad \langle y'|\psi\rangle = \langle y'|x\rangle\langle x|\psi\rangle + \langle y'|y\rangle\langle y|\psi\rangle, \tag{2.92}$$

or by expressing the equations (2.92) in matrix form:

$$\begin{pmatrix} \langle x'|\psi\rangle \\ \langle y'|\psi\rangle \end{pmatrix} = \begin{pmatrix} \langle x'|x\rangle & \langle x'|y\rangle \\ \langle y'|x\rangle & \langle y'|y\rangle \end{pmatrix} \begin{pmatrix} \langle x|\psi\rangle \\ \langle y|\psi\rangle \end{pmatrix}, \tag{2.93}$$

which is consistent with (2.86). From Figure 2.2a is clear that the old basis $\{|x\rangle, |y\rangle\}$ is related to the new basis $\{|x'\rangle, |y'\rangle\}$ by

$$|x\rangle = \cos\theta|x'\rangle - \sin\theta|y'\rangle \qquad |y\rangle = \sin\theta|x'\rangle + \cos\theta|y'\rangle. \tag{2.94}$$

The corresponding projections are

$$\langle x'|x\rangle = \cos\theta \qquad \langle x'|y\rangle = \sin\theta \qquad \langle y'|x\rangle = -\sin\theta \qquad \langle y'|y\rangle = \cos\theta. \tag{2.95}$$

From Figure 2.2 it is clear that rotation of the basis counterclockwise for θ is equivalent to the rotation of the ket $|\psi\rangle$ for the same angle but clockwise, so that the ket in the new basis can be expressed as follows:

$$|\psi'\rangle = R(\theta)|\psi\rangle, \tag{2.96}$$

where $R(\theta)$ is the rotation operator, which can be represented in matrix form, based on (2.93) and (2.95), by

$$R(\theta) = \begin{pmatrix} \cos\theta & \sin\theta \\ -\sin\theta & \cos\theta \end{pmatrix}. \tag{2.97}$$

In other words, Eq. (2.96) can be written as

$$\begin{pmatrix} \langle x'|\psi\rangle \\ \langle y'|\psi\rangle \end{pmatrix} = \begin{pmatrix} \cos\theta & \sin\theta \\ -\sin\theta & \cos\theta \end{pmatrix} \begin{pmatrix} \langle x|\psi\rangle \\ \langle y|\psi\rangle \end{pmatrix}. \tag{2.98}$$

The rotation operator $R(\theta)$ can be represented in terms of the photon spin operator S, introduced in Section 2.3, as follows:

$$R(\theta) = \cos\theta I + j\sin\theta \begin{pmatrix} 0 & -j \\ j & 0 \end{pmatrix} = R(\theta) = \cos\theta I + jS\sin\theta, \quad S = \begin{pmatrix} 0 & -j \\ j & 0 \end{pmatrix}. \tag{2.99}$$

We have shown earlier (see Section 2.3) that the left-circular $|E_L\rangle$ and the right-circular $|E_R\rangle$ polarizations are eigenvectors of the spin operator, namely $S|E_L\rangle = -|E_L\rangle$ and $S|E_R\rangle = |E_R\rangle$. The action of the rotation operator on eigenkets of S is then

$$\begin{aligned} R(\theta)|E_R\rangle &= (\cos\theta I + j\sin\theta S)|E_R\rangle = (\cos\theta I + j\sin\theta)|E_R\rangle = e^{j\theta}|E_R\rangle \\ R(\theta)|E_L\rangle &= (\cos\theta I + j\sin\theta S)|E_L\rangle = (\cos\theta I - j\sin\theta)|E_L\rangle = e^{-j\theta}|E_L\rangle. \end{aligned} \tag{2.100}$$

2.7.2 Diagonalization and Unitary Equivalent Observables

We are often concerned in quantum mechanics with the problem of determining the eigenvalues and eigenkets of an operator B, whose matrix elements in the old basis $|a^{(j)}\rangle$ are known. Let the eigenvalue equation of B be given as

$$B|b^{(i)}\rangle = b^{(i)}|b^{(i)}\rangle. \tag{2.101}$$

By multiplying (2.101) by $\langle a^{(j)}|$ from the left we obtain:

$$\langle a^{(j)}|B|b^{(i)}\rangle = b^{(i)}\langle a^{(j)}|b^{(i)}\rangle. \tag{2.102}$$

By applying further the completeness relation we obtain:

$$\sum_k \langle a^{(j)}|B|a^{(k)}\rangle\langle a^{(k)}|b^{(i)}\rangle = b^{(i)}\langle a^{(j)}|b^{(i)}\rangle, \tag{2.103}$$

which can be expressed in matrix form as

$$\begin{pmatrix} \langle a^{(1)}|B|a^{(1)}\rangle & \langle a^{(1)}|B|a^{(2)}\rangle & \cdots \\ \langle a^{(2)}|B|a^{(1)}\rangle & \langle a^{(2)}|B|a^{(2)}\rangle & \cdots \\ \vdots & \vdots & \ddots \end{pmatrix} \begin{pmatrix} \langle a^{(1)}|b^{(i)}\rangle \\ \langle a^{(2)}|b^{(i)}\rangle \\ \vdots \end{pmatrix} = b^{(i)} \begin{pmatrix} \langle a^{(1)}|b^{(i)}\rangle \\ \langle a^{(2)}|b^{(i)}\rangle \\ \vdots \end{pmatrix}. \tag{2.104}$$

We can perform the *diagonalization* using the following matrix, in which eigenkets are used as columns:

$$U = \begin{pmatrix} \langle a^{(1)}|b^{(1)}\rangle & \langle a^{(1)}|b^{(2)}\rangle & \cdots \\ \langle a^{(2)}|b^{(1)}\rangle & \langle a^{(2)}|b^{(2)}\rangle & \cdots \\ \vdots & \vdots & \ddots \end{pmatrix}. \tag{2.105}$$

The nontrivial solutions are possible if the characteristic cquation

$$\det(B - \lambda I) = 0 \tag{2.106}$$

is solvable. Namely, from matrix theory we know that if an $n \times n$ matrix A has n linearly independent eigenkets, and eigenkets are used as columns of U, then $U^{-1}AU$ is diagonal. If A is Hermitian,

$$(U^{-1}AU)^{\dagger} = U^{\dagger}A^{\dagger}(U^{-1})^{\dagger} = U^{\dagger}A(U^{-1})^{\dagger},$$

then the matrix U is a unitary matrix, since $U^{\dagger} = U^{-1}$.

Example. Consider the 2×2 rotation matrix $R = \begin{bmatrix} \cos\theta & \sin\theta \\ -\sin\theta & \cos\theta \end{bmatrix}$. We can determine the unitary transformation that diagonalizes R as follows. We first find that the roots of the characteristic polynomial $\det(R - \lambda I) = 0$ are $\lambda_{1,2} = \cos\theta \pm j\sin\theta$. The eigenvectors can be found from the eigenvalue relation $R|\psi_{1,2}\rangle = \lambda_{1,2}|\psi_{1,2}\rangle$ as follows:

$$|\psi_1\rangle = \frac{1}{\sqrt{2}}\begin{bmatrix} 1 \\ j \end{bmatrix} \qquad |\psi_2\rangle = \frac{1}{\sqrt{2}}\begin{bmatrix} 1 \\ -j \end{bmatrix}.$$

The unitary matrix that diagonalizes R is then given by

$$U = [\psi_1 \quad \psi_2] = \frac{1}{\sqrt{2}}\begin{bmatrix} 1 & 1 \\ j & -j \end{bmatrix}.$$

We finally demonstrate that the unitary matrix indeed diagonalizes R by

$$U^{\dagger}RU = \frac{1}{\sqrt{2}}\begin{bmatrix} 1 & -j \\ 1 & j \end{bmatrix}\begin{bmatrix} \cos\theta & \sin\theta \\ -\sin\theta & \cos\theta \end{bmatrix}\frac{1}{\sqrt{2}}\begin{bmatrix} 1 & 1 \\ j & -j \end{bmatrix} = \begin{bmatrix} e^{j\theta} & 0 \\ 0 & e^{-j\theta} \end{bmatrix}.$$

Consider two sets of orthonormal basis $\{|a'\rangle\}$ and $\{|b'\rangle\}$ connected by the U operator. Knowing U we can construct a unitary transform of A, UAU^{-1}; then A and UAU^{-1} are said to be *unitary equivalent observables*. For the unitary equivalent observables the following theorem is important.

Theorem. The unitary equivalent observables, A and UAU^{-1}, have identical spectra.

This theorem can be proved starting from the eigenvalue relation $A|a^{(l)}\rangle = a^{(l)}|a^{(l)}\rangle$ and base transformation $|b^{(i)}\rangle = U|a^{(i)}\rangle$. By applying the U

operator from the left on the eigenvalue relation and by inserting an identity operator $I = UU^{\dagger}$, we obtain:

$$\underbrace{UAU^{-1}\, U|a^{(l)}\rangle}_{b^{(l)}} = a^{(l)} U|a^{(l)}\rangle \Leftrightarrow (UAU^{-1})|b^{(l)}\rangle = a^{(l)}|b^{(l)}\rangle, \tag{2.107}$$

which confirms that the unitary equivalent objects have the same spectrum $\{a^{(l)}\}$.

2.8 TIME EVOLUTION – SCHRÖDINGER EQUATION

The time-evolution operator $U(t,t_0)$ operator transforms the initial ket at time t_0, $|\alpha,t_0\rangle$, into the final ket at time t:

$$|\alpha, t_0;t\rangle = U(t, t_0)|\alpha, t_0\rangle. \tag{2.108}$$

The time-evolution operators must satisfy the following two properties:

1. Unitary property: $U^{+}(t,t_0)\, U(t,t_0) = I$.
2. Composition property: $U(t_2,t_0) = U(t_2,t_1)U(t_1,t_0)$, $t_2 > t_1 > t_0$.

Following Eq. (2.108), the action of the infinitesimal time-evolution operator $U(t_0 + dt,t_0)$ can be described by

$$|\alpha, t_0;t_0 + dt\rangle = U(t_0 + dt, t_0)|\alpha, t_0\rangle. \tag{2.109}$$

The following operator satisfies all the propositions above, when $dt \to 0$:

$$U(t_0 + dt, t_0) = 1 - j\Omega\, dt, \qquad \Omega^{\dagger} = \Omega, \tag{2.110}$$

where the operator Ω is related to the Hamiltonian H by $H = \hbar\Omega$, and the Hamiltonian eigenvalues correspond to the energy $E = \hbar\omega$. For the infinitesimal time-evolution operator $U(t_0 + dt,t_0)$ we can derive the time-evolution equation as follows. The starting point in the derivation is the composition property:

$$U(t + dt, t_0) = U(t + dt, t)U(t, t_0) = \left(1 - \frac{j}{\hbar}H\, dt\right)U(t, t_0). \tag{2.111}$$

Equation (2.111) can be rewritten in the following form:

$$\lim_{dt\to 0} \frac{U(t + dt, t_0) - U(t, t_0)}{dt} = -\frac{j}{\hbar}HU(t, t_0), \tag{2.112}$$

which by taking the partial derivative definition into account becomes

$$j\hbar\frac{\partial}{\partial t}U(t, t_0) = HU(t, t_0). \tag{2.113}$$

This equation is known as the Schrödinger equation for the time-evolution operator.

The Schrödinger equation for a state ket can be obtained by applying the time-evolution operator on the initial ket:

$$j\hbar\frac{\partial}{\partial t}U(t,t_0)|\alpha,t_0\rangle = HU(t,t_0)|\alpha,t_0\rangle, \tag{2.114}$$

which, based on (2.108), can be rewritten as

$$j\hbar\frac{\partial}{\partial t}|\alpha,t_0;t\rangle = H|\alpha,t_0;t\rangle. \tag{2.115}$$

For *conservative systems*, for which the Hamiltonian is time invariant, we can easily solve Eq. (2.113) to obtain:

$$U(t,t_0) = \mathrm{e}^{-\frac{j}{\hbar}H(t-t_0)}. \tag{2.116}$$

The time evolution of kets in conservative systems can therefore be described by applying (2.116) in (2.108), which yields

$$|\alpha(t)\rangle = \mathrm{e}^{-\frac{j}{\hbar}H(t-t_0)}|\alpha(t_0)\rangle. \tag{2.117}$$

Therefore, the operators do not explicitly depend on time – this concept is known as the *Schrödinger picture.*

In the *Heisenberg picture*, the state vector is independent of time, but operators depend on time:

$$A(t) = \mathrm{e}^{\frac{j}{\hbar}H(t-t_0)}A\mathrm{e}^{-\frac{j}{\hbar}H(t-t_0)}. \tag{2.118}$$

The time-evolution equation in the Heisenberg picture is given by

$$j\hbar\frac{\mathrm{d}A(t)}{\mathrm{d}t} = [A(t),H] + j\hbar\frac{\partial A(t)}{\partial t}. \tag{2.119}$$

The density operator ρ, representing the statistical mixture of states, is independent of time in the Heisenberg picture. The expectation value of a measurement of an observable $\Xi(t)$ at time t is given by

$$\begin{aligned} E_t[\Xi] &= \mathrm{Tr}[\rho\Xi(t)] \\ E_t[\Xi] &= \mathrm{Tr}[\rho(t)\Xi], \quad \rho(t) = \mathrm{e}^{-\frac{j}{\hbar}H(t-t_0)}\rho\mathrm{e}^{-\frac{j}{\hbar}H(t-t_0)}. \end{aligned} \tag{2.120}$$

For a *time-variant Hamiltonian* $H(t)$, the integration leads to a time-ordered exponential:

$$\begin{aligned} U(t,t_0) &= T\left\{\exp\left[-\frac{j}{\hbar}\int_{t_0}^{t} H(t')\mathrm{d}t'\right]\right\} \\ &= \lim_{N\to\infty}\prod_{n=0}^{N-1}\exp\left[-\frac{j}{\hbar}H(t_0+n\Delta_N)\Delta_N\right], \quad \Delta_N = (t-t_0)/N. \end{aligned} \tag{2.121}$$

Example. The Hamiltonian for a two-state system is given by

$$H = \begin{bmatrix} \omega_1 & \omega_2 \\ \omega_2 & \omega_1 \end{bmatrix}.$$

The basis for this system is given by $\{|0\rangle = [1\ 0]^{\mathrm{T}}, |1\rangle = [0\ 1]^{\mathrm{T}}\}$.

(a) Determine the eigenvalues and eigenkets of H, and express the eigenkets in terms of basis.

(b) Determine the time evolution of the system described by the Schrödinger equation:

$$j\hbar \frac{\partial}{\partial t}|\psi\rangle = H|\psi\rangle, \quad |\psi(0)\rangle = |0\rangle.$$

To determine the eigenkets of H we start from the characteristic equation $\det(H - \lambda I) = 0$ and find that the eigenvalues are $\lambda_{1,2} = \omega_1 \pm \omega_2$. The corresponding eigenvectors are

$$|\lambda_1\rangle = \frac{1}{\sqrt{2}}\begin{bmatrix} 1 \\ 1 \end{bmatrix} = \frac{1}{\sqrt{2}}(|0\rangle + |1\rangle) \qquad |\lambda_1\rangle = \frac{1}{\sqrt{2}}\begin{bmatrix} 1 \\ -1 \end{bmatrix} = \frac{1}{\sqrt{2}}(|0\rangle - |1\rangle).$$

We now have to determine the time evolution of arbitrary ket $|\psi(t)\rangle = [\alpha(t)\ \beta(t)]^{\mathrm{T}}$. The starting point is the Schrödinger equation:

$$j\hbar \frac{\partial}{\partial t}|\psi\rangle = j\hbar \begin{bmatrix} \dot{\alpha}(t) \\ \dot{\beta}(t) \end{bmatrix}, \quad H|\psi\rangle = \begin{bmatrix} \omega_1 & \omega_2 \\ \omega_2 & \omega_1 \end{bmatrix}\begin{bmatrix} \alpha(t) \\ \beta(t) \end{bmatrix}$$

$$= \begin{bmatrix} \omega_1\alpha(t) + \omega_2\beta(t) \\ \omega_2\alpha(t) + \omega_1\beta(t) \end{bmatrix} \Rightarrow j\hbar \begin{bmatrix} \dot{\alpha}(t) \\ \dot{\beta}(t) \end{bmatrix} = \begin{bmatrix} \omega_1\alpha(t) + \omega_2\beta(t) \\ \omega_2\alpha(t) + \omega_1\beta(t) \end{bmatrix}.$$

By substituting $\alpha(t) + \beta(t) = \gamma(t)$ and $\alpha(t) - \beta(t) = \delta(t)$, we obtain the ordinary set of differential equations:

$$j\hbar \frac{\mathrm{d}\gamma(t)}{\mathrm{d}t} = (\omega_1 + \omega_2)\,\gamma(t) \quad j\hbar \frac{\mathrm{d}\delta(t)}{\mathrm{d}t} = (\omega_1 - \omega_2)\delta(t),$$

whose solution is $\gamma(t) = C\exp(\frac{\omega_1 + \omega_2}{j\hbar}t)$ and $\delta(t) = D\exp(\frac{\omega_1 - \omega_2}{j\hbar}t)$. From the initial state $|\psi(0)\rangle = |0\rangle = [1 \quad 0]^{\mathrm{T}}$, we obtain the unknown constants $C = D = 1$, so that state time evolution is given by:

$$|\psi(t)\rangle = \exp\left(-\frac{j}{\hbar}\omega_1 t\right)\begin{bmatrix} \cos\left(\frac{\omega_2 t}{\hbar}\right) \\ -j\sin\left(\frac{\omega_2 t}{\hbar}\right) \end{bmatrix}.$$

Before concluding this section, it is interesting to see how the base kets evolve over time. This is particularly simple to determine if the base kets are chosen to be eigenkets of A that commutes with Hamiltonian H, i.e. $[A,H] = 0$. Then both A and H

have simultaneous eigenkets, called *energy eigenkets*, with eigenvalues denoted by $E_a{}'$, that satisfy the eigenvalue equation:

$$H|a'\rangle = E_{a'}|a'\rangle \tag{2.122}$$

The time-evolution operator can now be expanded in terms of projection operators as follows:

$$e^{-\frac{i}{\hbar}Ht} = \sum_{a'}\sum_{a''}|a''\rangle\langle a''|e^{-\frac{i}{\hbar}Ht}|a'\rangle\langle a'| = \sum_{a'}|a'\rangle e^{-\frac{i}{\hbar}E_{a'}t}|a'\rangle. \tag{2.123}$$

By applying this time-evolution operator on the initial state,

$$|\alpha, t_0 = 0\rangle = \sum_{a'}|a'\rangle\langle a'|\alpha, t_0 = 0\rangle = \sum_{a'}c_{a'}|a'\rangle, \tag{2.124}$$

we obtain:

$$|\alpha, t_0 = 0;t\rangle = e^{-\frac{i}{\hbar}Ht}|\alpha, t_0 = 0\rangle = \sum_{a'}|a'\rangle\langle a'|\alpha\rangle e^{-\frac{i}{\hbar}E_{a'}t}. \tag{2.125}$$

The evolution of expansion coefficient is therefore as follows:

$$c_{a'}(t=0)\rightarrow c_{a'}(t) = c_{a'}(t=0)e^{-\frac{i}{\hbar}E_{a'}t}. \tag{2.126}$$

If the initial state was a base state, then from (2.125) it is clear that

$$|\alpha, t_0 = 0\rangle = |a'\rangle\rightarrow|\alpha, t_0 = 0;t\rangle = |a'\rangle e^{-\frac{i}{\hbar}E_{a'}t}. \tag{2.127}$$

Therefore, with the system initially being in a simultaneous eigenstate of A and H, it remains so all the time!

By completing this section, we conclude the introduction to quantum mechanics. In the next several sections we study different quantum systems that are important in various quantum information processing systems and in quantum communications. These examples can also be used to gain a deeper understanding of the underlying concepts of quantum mechanics presented above.

2.9 HARMONIC OSCILLATOR

The Hamiltonian of a particle in a one-dimensional parabolic potential well $V(x) = m\omega^2x^2/2$ is given by

$$H = \frac{p^2}{2m} + \frac{m\omega^2x^2}{2}, \tag{2.128}$$

where p is the momentum operator and x is the position operator. The *annihilation* a and *creation* $a^\dagger$ operators, often used in quantum mechanics, are related to the momentum and position operators as follows:

$$a = \sqrt{\frac{m\omega}{2\hbar}}\left(x + \frac{jp}{m\omega}\right) \qquad a^\dagger = \sqrt{\frac{m\omega}{2\hbar}}\left(x - \frac{jp}{m\omega}\right). \tag{2.129}$$

It can be shown that these two operators satisfy the following commutation relation:

$$[a, a^\dagger] = \frac{1}{2\hbar}(-j[x, p] + j[p, x]) \overset{[x_i, p_j] = j\hbar\delta_{ij}I}{=} I. \tag{2.130}$$

Another important operator is the *number* operator defined by

$$N = a^\dagger a \tag{2.131}$$

From (2.129) it is clear that

$$a^\dagger a = \left(\frac{m\omega}{2\hbar}\right)\left(x^2 + \frac{p^2}{m^2\omega^2}\right) + \frac{j}{2\hbar}[x, p] = \frac{H}{\hbar\omega} - \frac{1}{2}, \tag{2.132}$$

so that we can relate the Hamiltonian and the number operators as follows:

$$H = \hbar\omega\left(N + \frac{1}{2}\right). \tag{2.133}$$

By denoting the energy eigenket of N by its eigenvalue n, i.e. $N|n\rangle = n|n\rangle$, the energy eigenvalues E_n can be determined from

$$H|n\rangle = \hbar\omega\left(N + \frac{1}{2}\right)|n\rangle = \hbar\omega\left(n + \frac{1}{2}\right)|n\rangle = E_n|n\rangle, \tag{2.134}$$

by

$$E_n = \left(n + \frac{1}{2}\right)\hbar\omega \tag{2.135}$$

To understand the physical significance of a, a^+, and N, we first prove the following commutation relation:

$$[N, a] = [aa^\dagger, a] = a^\dagger[a, a] + [a^\dagger, a]a = -a. \tag{2.136}$$

In similar fashion we can prove the commutation relation $[N, a^\dagger] = a^\dagger$. We now observe the action of operators $Na^\dagger$ and Na on $|n\rangle$ and by using the commutation relations just derived we obtain:

$$\begin{aligned} Na^\dagger|n\rangle &= ([N, a^\dagger] + a^\dagger N)|n\rangle = (n+1)a^\dagger|n\rangle \\ Na|n\rangle &= ([N, a] + aN)|n\rangle = (n-1)a|n\rangle. \end{aligned} \tag{2.137}$$

Therefore, the creation (annihilation) operator increases (decreases) the quantum energy for one unit. By using (2.135) and the commutation relations just derived, it is easy to show that the commutation relation $[H, a^\dagger] = \hbar\omega a^\dagger$ is valid, which can be used to determine the action of the operators $Ha^\dagger$ and Ha on ket $|\psi\rangle$ as follows:

$$\begin{aligned} Ha^\dagger|\psi\rangle &= ([H, a^\dagger] + a^\dagger H)|\psi\rangle = (E + \hbar\omega)a^\dagger|\psi\rangle \\ Ha|\psi\rangle &= (E - \hbar\omega)a|\psi\rangle. \end{aligned} \tag{2.138}$$

This is the reason why the creation and annihilation operators are also known as *rising* and *lowering* operators respectively.

What remains is to determine the simultaneous eigenkets of N and H. Equation (2.137) shows that $a|n\rangle$ and $|n-1\rangle$ differ only in a multiplicative constant as follows:

$$a|n\rangle = c|n-1\rangle. \tag{2.139}$$

From $N|n\rangle = n|n\rangle$, it is clear that $\langle n|N|n\rangle = n$, while from (2.139) it is clear that

$$\langle n|a^\dagger a|n\rangle = |c|^2 \geq 0, \tag{2.140}$$

and by combining these two equations we find that $c = \sqrt{n}$ and that n is never negative, which indicates that the ground state of the harmonic oscillator has energy $E_0 = \hbar\omega/2$. Therefore, the action of annihilation and creation operators on $|n\rangle$ can be described as follows:

$$a|n\rangle = \sqrt{n}|n-1\rangle \qquad a^\dagger|n\rangle = \sqrt{n+1}|n+1\rangle. \tag{2.141}$$

If we continue applying the annihilation operator we obtain the following sequence:

$$\begin{aligned} a^2|n\rangle &= \sqrt{n(n-1)}|n-2\rangle \\ a^3|n\rangle &= \sqrt{n(n-1)(n-2)}|n-3\rangle \\ &\vdots \\ a^k|n\rangle &= \sqrt{n(n-1)\ldots(n-k+1)}|n-k\rangle. \end{aligned} \tag{2.142}$$

In similar fashion, if we keep applying the creation operator on the ground state we obtain:

$$\begin{aligned} a^\dagger|0\rangle &= |1\rangle \\ (a^\dagger)^2|0\rangle &= \sqrt{1\cdot 2}|2\rangle \\ (a^\dagger)^3|0\rangle &= \sqrt{1\cdot 2\cdot 3}|3\rangle \\ &\vdots \\ (a^\dagger)^n|0\rangle &= \sqrt{n!}|n\rangle. \end{aligned} \tag{2.143}$$

In other words, we can express the state $|n\rangle$ in terms of ground state as

$$|n\rangle = \frac{(a^\dagger)^n}{\sqrt{n!}}|0\rangle. \tag{2.144}$$

The matrix elements of annihilation and creation operators can be obtained by using (2.141) and employing the orthogonality principle for basis $\{|n\rangle\}$ as follows:

$$\langle m|a|n\rangle = \sqrt{n}\delta_{m,\,n-1} \quad \langle m|a^\dagger|n\rangle = \sqrt{n+1}\delta_{m,\,n+1}. \tag{2.145}$$

Time evolution of eigenkets, based on the previous section, can be described by

$$|\psi(t)\rangle = \mathrm{e}^{-\frac{j}{\hbar}H}|\psi(0)\rangle. \tag{2.146}$$

We can express the initial state $|\psi(0)\rangle$ in basis $\{|n\rangle\}$ as follows:

$$|\psi(0)\rangle = \sum_n c_n(0)|n\rangle, \tag{2.147}$$

where $c_n(0)$ is the expansion coefficient. Based on (2.126) the evolution of expansion coefficient can be described as

$$c_n(t) = c_n(0)\mathrm{e}^{-\frac{j}{\hbar}(E_n-E_0)t} \overset{E_n=(n+1/2)\hbar\omega}{=} c_n(0)\mathrm{e}^{-jn\omega t}, \tag{2.148}$$

so that Eq. (2.146) becomes

$$|\psi(t)\rangle = \sum_n c_n(0)\mathrm{e}^{-jn\omega t}|n\rangle. \tag{2.149}$$

By using (2.129) we can express the position x and momentum p operators in terms of annihilation and creation operators as follows:

$$x = \sqrt{\frac{\hbar}{2m\omega}}(a^\dagger + a) \qquad p = j\sqrt{\frac{m\hbar\omega}{2}}(a^\dagger - a). \tag{2.150}$$

Based on (2.145) we can determine the matrix elements of x and p operators:

$$\begin{aligned}\langle k|x|l\rangle &= \sqrt{\frac{\hbar}{2m\omega}}(\sqrt{l+1}\delta_{k,\,l+1} + \sqrt{l}\delta_{k,\,l-1})\\ \langle k|p|l\rangle &= j\sqrt{\frac{m\hbar\omega}{2}}(\sqrt{l+1}\delta_{k,\,l+1} - \sqrt{l}\delta_{k,\,l-1}).\end{aligned} \tag{2.151}$$

Clearly, neither x nor p are diagonal in the number operator representation, which is consistent with the fact that neither annihilation nor creation operators commute with N. Therefore, the matrix representations of x and p, based on (2.151), are as follows:

$$x \doteq \sqrt{\frac{\hbar}{2m\omega}}\begin{bmatrix} 0 & 1 & 0 & 0 & \cdots \\ 1 & 0 & \sqrt{2} & 0 & \cdots \\ 0 & \sqrt{2} & 0 & \sqrt{3} & \cdots \\ 0 & 0 & \sqrt{3} & 0 & \ddots \\ \vdots & \vdots & & \ddots & \ddots \end{bmatrix} \quad p \doteq j\sqrt{\frac{m\hbar\omega}{2}}\begin{bmatrix} 0 & -1 & 0 & 0 & \cdots \\ 1 & 0 & -\sqrt{2} & 0 & \cdots \\ 0 & \sqrt{2} & 0 & -\sqrt{3} & \cdots \\ 0 & 0 & \sqrt{3} & 0 & \ddots \\ \vdots & \vdots & & \ddots & \ddots \end{bmatrix}.$$

Before concluding this section, we determine the uncertainty product for x and p. From (2.145) it is clear that $\langle x\rangle = \langle p\rangle = 0$ for the ground state. On the other hand, from (2.150) we obtain $x^2 = ((a^\dagger)^2 + a^\dagger a + aa^\dagger + a^2)\hbar/(2m\omega)$, so that $\langle x^2\rangle = \hbar/(2m\omega)$. In similar fashion, we obtain the expectation of p^2 for the ground state to be $\langle p^2\rangle = \hbar m\omega/2$, so that the uncertainty product is

$$\left\langle(\Delta x)^2\right\rangle\left\langle(\Delta p)^2\right\rangle = \left(\frac{\hbar}{2}\right)^2. \tag{2.152}$$

Following a similar procedure we can determine the uncertainty product of the nth excited state as follows:

$$\left\langle(\Delta x)^2\right\rangle\left\langle(\Delta p)^2\right\rangle = \left(n+\frac{1}{2}\right)^2\left(\frac{\hbar}{2}\right)^2. \tag{2.153}$$

2.10 ANGULAR MOMENTUM

The angular momentum in classical physics is defined as

$$\begin{aligned} \boldsymbol{L} = \boldsymbol{r}\times\boldsymbol{p} &= \begin{bmatrix} \hat{x} & \hat{y} & \hat{z} \\ x & y & z \\ p_x & p_y & p_z \end{bmatrix} = (yp_z - zp_y)\hat{x} + (-xp_z + zp_x)\hat{y} + (xp_y - yp_x)\hat{z} \\ &= L_x\hat{x} + L_y\hat{y} + L_z\hat{z};\ L_x = yp_z - zp_y,\quad L_y = zp_x - xp_z,\quad L_z = xp_y - yp_x. \end{aligned} \tag{2.154}$$

The position operator x can be defined as follows (see Eq. (2.45)):

$$x|x'\rangle = x'|x'\rangle, \tag{2.155}$$

where x' is the eigenvalue. We assume that eigenkets form the complete set, and we can represent an arbitrary continuous ket $|\alpha\rangle$ as follows:

$$|\alpha\rangle = \int_{-\infty}^{\infty} \langle x'|\alpha\rangle|x'\rangle \mathrm{d}x'. \tag{2.156}$$

The corresponding three-dimensional generalization leads to

$$|\alpha\rangle = \int_{-\infty}^{\infty} \langle \boldsymbol{r}'|\alpha\rangle|\boldsymbol{r}'\rangle \mathrm{d}x'\,\mathrm{d}y'\,dz',\ \ \boldsymbol{r}' = [x', y', z'], \tag{2.157}$$

where $|\boldsymbol{r}'\rangle$ is the simultaneous eigenket of observables x, y, and z, i.e.

$$x|\boldsymbol{r}'\rangle = x'|\boldsymbol{r}'\rangle, \qquad y|\boldsymbol{r}'\rangle = y'|\boldsymbol{r}'\rangle, \qquad z|\boldsymbol{r}'\rangle = z'|\boldsymbol{r}'\rangle. \tag{2.158}$$

In (2.158), we implicitly assumed that all three coordinates can be measured simultaneously with arbitrary small accuracy, indicating that

$$[x_m, x_n] = 0;\ x_m, x_n = x, y, z. \tag{2.159}$$

We now introduce the *infinitesimal translation operator* $T(\mathrm{d}\boldsymbol{r}')$ as follows:

$$T(\mathrm{d}\boldsymbol{r}')|\boldsymbol{r}'\rangle = |\boldsymbol{r}' + \mathrm{d}\boldsymbol{r}'\rangle, \tag{2.160}$$

which changes the position from $\boldsymbol{r}'$ to $\boldsymbol{r}' + \mathrm{d}\boldsymbol{r}'$. This operator must satisfy the following *properties*:

1. *Unitarity property*: $T(\mathrm{d}\boldsymbol{r}')T^{\dagger}(\mathrm{d}\boldsymbol{r}') = I$.
2. *Composition property*: $T(\mathrm{d}\boldsymbol{r}')T(\mathrm{d}\boldsymbol{r}'') = T(\mathrm{d}\boldsymbol{r}' + \mathrm{d}\boldsymbol{r}'')$.
3. *Opposite direction translation property*: $T^{-1}(\mathrm{d}\boldsymbol{r}') = T(-\mathrm{d}\boldsymbol{r}')$.
4. $\lim_{\mathrm{d}r' \to 0} T(\mathrm{d}\boldsymbol{r}') = I$.

The following translation operator:

$$T(\mathrm{d}\boldsymbol{r}') = 1 - \frac{j}{\hbar}\boldsymbol{p}\cdot\mathrm{d}\boldsymbol{r}', \quad \boldsymbol{p} = [p_x, p_y, p_z] \tag{2.161}$$

satisfies all the properties from the above. In (2.161), p_j denotes the momentum operator along the jth axis ($j = x,y,z$) introduced by (2.45). Therefore, we can establish the relationship between classical and quantum momentum by the following mapping:

$$p_i \rightarrow -j\hbar\frac{\partial}{\partial i}; \ i = x, y, z, \tag{2.162}$$

which can be used in (2.154) to determine the quantum angular momentum operator. For example, the quantum angular momentum along the z-axis is given by

$$L_z = xp_y - yp_x = x\left(-j\hbar\frac{\partial}{\partial y}\right) - y\left(-j\hbar\frac{\partial}{\partial x}\right) = -j\hbar\left(x\frac{\partial}{\partial y} - y\frac{\partial}{\partial x}\right). \tag{2.163}$$

2.10.1 Commutation Relations

Given that angular momentum operators can be expressed in terms of position x and linear momentum operators p_i ($i = x,y,z$), and that x and p_i obey the following commutation relation:

$$[x_m, p_n] = j\hbar\delta_{mn}; \ x_m, x_n = x, y, z, \tag{2.164}$$

it can be shown that components of angular momentum satisfy the following commutation relations:

$$[L_x, L_y] = j\hbar L_z \quad [L_z, L_x] = j\hbar L_y \quad [L_y, L_z] = j\hbar L_x\,. \tag{2.165}$$

The first commutation relation can be proved as follows. Since $L_y = zp_x - xp_z$, we obtain:

$$[L_x, L_y] = [L_x, zp_x - xp_z] = [L_x, zp_x] - [L_x, xp_z]. \tag{2.166}$$

By applying property 3 of (2.43), $[A, BC] = [A, B]C + B[A, C]$, $[L_x, zp_x]$ can be written as

$$[L_x, zp_x] = [L_x, z]p_x + z[L_x, p_x]. \tag{2.167}$$

By using $L_x = yp_z - zp_y$, we obtain:

$$\begin{aligned}[L_x, z] &= [yp_z - zp_y, z] = [yp_z, z] - [zp_y, z] = -[z, yp_z] + [z, zp_y] \\ &= -[z, y]p_z - y[z, p_z] + [z, z]p_y + z[z, p_y] = -j\hbar y.\end{aligned} \tag{2.168}$$

Similarly, we show that

$$[L_x, p_x] = [yp_z - zp_y, p_x] = 0 \tag{2.169}$$

By substituting (2.169) and (2.168) into (2.167) we obtain:

$$[L_x, zp_x] = -j\hbar y. \tag{2.170}$$

Applying again property 3 of (2.43) on the second term in (2.166), we obtain:

$$[L_x, xp_z] = [L_x, x]p_z + x[L_x, p_z]. \tag{2.171}$$

Since $[L_x, x] = [yp_z - zp_y, x] = 0$, it follows that

$$[L_x, p_z] = [yp_z - zp_y, p_z] = -j\hbar p_y, \tag{2.172}$$

which upon substitution in (2.171) yields

$$[L_x, xp_z] = -j\hbar xp_y. \tag{2.173}$$

Now by substituting (2.173) and (2.170) into (2.166), we obtain:

$$\begin{aligned}[L_x, L_y] &= [L_x, z]p_x + z[L_x, p_x] - ([L_x, x]p_z + x[L_x, p_z]) = -j\hbar yp_x + j\hbar xp_y \\ &= j\hbar(xp_y - yp_x) = j\hbar L_z,\end{aligned} \tag{2.174}$$

which proves the fist commutation relation in (2.165). The other commutation relations can be proved in similar fashion. Because the components of angular momentum do not commute, we can specify only one component at the time. It is straightforward to show that every component of angular momentum commutes with $L^2 = L_x^2 + L_y^2 + L_z^2$. Therefore, in order to specify a state of angular momentum, it is sufficient to use L^2 and L_z. In quantum mechanics, in addition to *orbital angular momentum* (OAM) $\boldsymbol{L}$, the *spin angular momentum* (SAM) $\boldsymbol{S}$ is also present. We can therefore define the total angular momentum (TAM) as $\boldsymbol{J} = \boldsymbol{L} + \boldsymbol{S}$. The components of TAM satisfy the same set of commutation relations as OAM:

$$[J_x, J_y] = i\hbar J_z \quad [J_z, J_x] = i\hbar J_y \quad [J_y, J_z] = i\hbar J_x \tag{2.175}$$

(the imaginary unit is denoted by i to avoid any confusion with the corresponding quantum number j to be introduced soon).

The states of OAM are specified by two quantum numbers: (i) the *orbital quantum number* l and (ii) the *azimuthal (magnetic) quantum number* m. Similarly, the states of TAM are specified by two quantum numbers j and m. The eigenvalues of J^2 are labeled as j, while the eigenvalues of J_z are labeled as m. The corresponding eigenvalue equations are given by

$$J^2|j,m\rangle = \hbar^2 j(j+1)|j,m\rangle \qquad J_z|j,m\rangle = m\hbar|j,m\rangle$$
$$m = -j, -j+1, \cdots, -1, 0, 1, \cdots j-1, j. \tag{2.176}$$

In the case of OAM both quantum numbers l and m are integers. For example, for $l = 3$, the possible values of m are $\{-3,-2,-1,0,1,2,3\}$, and the corresponding "angular momentum" is $\hbar\sqrt{l(l+1)} = 2\hbar\sqrt{3}$. The spin, on the other hand, can have half-odd-integral values. Since $j = l + s$, the same is true for TAM. For example, the electron has spin $s = 1/2$ and for $l = 1$ we obtain $j = 3/2$, so that possible values for j are $\{-3/2,-1/2,1/2,3/2\}$.

To move up and down in states of TAM, we introduce the *ladder* operators J_+ and $J-$, also known as *raising* and *lowering* operators respectively. The ladder operators are defined as follows:

$$J_+ = J_x + iJ_y \qquad J_- = J_x - iJ_y. \tag{2.177}$$

The ladder operators satisfy the following commuting relations:

$$[J_z, J_+] = \hbar J_+ \quad [J_z, J_-] = -\hbar J_- \quad [J_+, J_-] = 2\hbar J_z. \tag{2.178}$$

It is often very useful to express the J^2, J_x, and J_y operators in terms of ladder operators as follows:

$$J^2 = J_-J_+ + J_z^2 + \hbar J_z = J_z^2 + (J_+J_- + J_-J_+)/2$$
$$J_x = (J_+ + J_-)/2 \quad J_y = (J_+ - J_-)/2i. \tag{2.179}$$

The ladder operators act on TAM states as follows:

$$J_+|j,m\rangle = \hbar\sqrt{j(j+1) - m(m+1)}|j,m+1\rangle$$
$$J_-|j,m\rangle = \hbar\sqrt{j(j+1) - m(m-1)}|j,m-1\rangle. \tag{2.180}$$

Because $m = -j,\ldots,0,\ldots,+j$, meaning that the maximum possible value for m is j and minimum possible value for m is $-j$, we require

$$J_+|j,m=j\rangle = 0 \qquad J_-|j,m=-j\rangle = 0. \tag{2.181}$$

The TAM states $|j,m\rangle$ satisfy the following orthonormality relation:

$$\langle j_1, m_1 | j_2, m_2\rangle = \delta_{j_1,j_2}\delta_{m_1,m_2} \tag{2.182}$$

and the completeness relation as follows:

$$\sum_{j=0}^{\infty}\sum_{m=-j}^{j} |j,m\rangle\langle j,m| = I. \tag{2.183}$$

2.10.2 Matrix Representation of Angular Momentum

We have shown in Section 2.3 that an operator A is associated with a square matrix, and this representation is dependent on eigenkets of choice. For TAM, it is natural to use a standard basis so that the A_{mn} element satisfies

$$A_{mn} = \langle j,m|A|j,n\rangle. \tag{2.184}$$

For every constant j we can associate a $(2j+1)\times(2j+1)$ matrix with J^2 and J_z as follows:

$$\begin{aligned}(J^2)_{mn} &= \langle j,m|J^2|j,n\rangle = \langle j,m|\hbar^2 j(j+1)|j,n\rangle = \hbar^2 j(j+1)\delta_{mn}\\ (J_z)_{mn} &= \langle j,m|J_z|j,n\rangle = \langle j,m|n\hbar|j,n\rangle = n\hbar\delta_{mn}.\end{aligned} \tag{2.185}$$

For example, for $l=1$ the matrix representations for operators L^2 and L_z can be written as follows:

$$L^2 = \begin{bmatrix} \langle 1,1|L^2|1,1\rangle & \langle 1,1|L^2|1,0\rangle & \langle 1,1|L^2|1,-1\rangle \\ \langle 1,0|L^2|1,1\rangle & \langle 1,0|L^2|1,0\rangle & \langle 1,0|L^2|1,-1\rangle \\ \langle 1,-1|L^2|1,1\rangle & \langle 1,-1|L^2|1,0\rangle & \langle 1,-1|L^2|1,-1\rangle \end{bmatrix} = 2\hbar^2\begin{bmatrix}1&0&0\\0&1&0\\0&0&1\end{bmatrix}$$

$$L_z = \begin{bmatrix} \langle 1,1|L_z|1,1\rangle & \langle 1,1|L_z|1,0\rangle & \langle 1,1|L_z|1,-1\rangle \\ \langle 1,0|L_z|1,1\rangle & \langle 1,0|L_z|1,0\rangle & \langle 1,0|L_z|1,-1\rangle \\ \langle 1,-1|L_z|1,1\rangle & \langle 1,-1|L_z|1,0\rangle & \langle 1,-1|L_z|1,-1\rangle \end{bmatrix} = \hbar\begin{bmatrix}1&0&0\\0&0&0\\0&0&-1\end{bmatrix}. \tag{2.186}$$

Example. Let us consider the quantum system with angular momentum $l=1$. Determine the matrix representation in the basis of simultaneous eigenkets of L_z and L^2.

The eigenkets of the system with $l=1$ are

$$|1\rangle = \begin{pmatrix}1\\0\\0\end{pmatrix} \equiv |l=1,m=1\rangle,\quad |0\rangle = \begin{pmatrix}0\\1\\0\end{pmatrix} \equiv |l=1,m=0\rangle,$$

$$|-1\rangle = \begin{pmatrix}0\\0\\1\end{pmatrix} \equiv |l=1,m=-1\rangle.$$

Based on (2.176), (2.177), and (2.180), we derive the following set of relations to be used in the determination of the matrix representation of L_x:

$$\begin{aligned}&L^2|l,m\rangle = \hbar^2 l(l+1)|l,m\rangle \qquad\qquad L_z|l,m\rangle = m\hbar|l,m\rangle\\ &L_+|l,m\rangle = \hbar\sqrt{l(l+1)-m(m+1)}|l,m+1\rangle\\ &L_-|j,m\rangle = \hbar\sqrt{l(l+1)-m(m-1)}|l,m-1\rangle;\quad L_x = \frac{1}{2}(L_+ + L_-).\end{aligned}$$

In order to determine the matrix representation of L_x, let us observe the action of L_x on base kets:

$$L_x|1\rangle = \frac{1}{2}(L_+ + L_-)\,|1\rangle = \frac{1}{2}L_-|1\rangle = \frac{1}{2}\hbar|0\rangle,\; L_x\,|-1\rangle = \frac{1}{2}(L_+ + L_-)|-1\rangle$$
$$= \frac{1}{2}L_+|-1\rangle = \frac{1}{2}\hbar\sqrt{2}|0\rangle;$$
$$L_x|0\rangle = \frac{1}{2}(L_+ + L_-)\,|0\rangle = \frac{1}{2}L_+|0\rangle + \frac{1}{2}L_-|0\rangle = \frac{1}{2}\hbar\sqrt{2}|1\rangle + \hbar\sqrt{2}|-1\rangle$$
$$= \frac{\hbar}{\sqrt{2}}(|1\rangle + |-1\rangle).$$

The matrix representation of L_x, based on previous equations, is as follows:

$$L_x \doteq \begin{bmatrix} \langle 1|L_x|1\rangle & \langle 1|L_x|0\rangle & \langle 1|L_x|-1\rangle \\ \langle 0|L_x|1\rangle & \langle 0|L_x|0\rangle & \langle 0|L_x|-1\rangle \\ \langle -1|L_x|1\rangle & \langle -1|L_x|0\rangle & \langle -1|L_x|-1\rangle \end{bmatrix} = \frac{\hbar}{\sqrt{2}}\begin{bmatrix} 0 & 1 & 0 \\ 1 & 0 & 1 \\ 0 & 1 & 0 \end{bmatrix}.$$

Example. Let the system with angular momentum $l = 1$ be initially in the state $[1\ 2\ 3]^{\mathrm{T}}$. Determine the probability that a measurement of L_x will be equal to zero.

First of all, the initial state is not normalized; upon normalization the initial state is $[1\ 2\ 3]^{\mathrm{T}}/\sqrt{14}$. Based on the matrix representation of L_x from the previous example, we determine the eigenvalues from

$$\frac{\hbar}{\sqrt{2}}\begin{bmatrix} -\lambda & 1 & 0 \\ 1 & -\lambda & 1 \\ 0 & 1 & -\lambda \end{bmatrix} = 0$$

to be $\lambda_{1,2,3} = \hbar, -\hbar, 0$. The corresponding eigenkets can be found from $L_x|\lambda_{1,2,3}\rangle = \lambda_{1,2,3}|\lambda_{1,2,3}\rangle$ as follows:

$$|1\rangle_x = \frac{1}{2}\begin{bmatrix} 1 \\ \sqrt{2} \\ 1 \end{bmatrix}, \quad |-1\rangle_x = \frac{1}{2}\begin{bmatrix} 1 \\ -\sqrt{2} \\ 1 \end{bmatrix}, \quad |0\rangle_x = \frac{1}{2}\begin{bmatrix} 1 \\ 0 \\ -1 \end{bmatrix}.$$

The initial state in L_x basis is now

$$|\psi\rangle_x = {}_x\langle 1|\psi\rangle|1\rangle_x + {}_x\langle 0|\psi\rangle|0\rangle_x + {}_x\langle -1|\psi\rangle|-1\rangle_x, \quad |\psi\rangle = \frac{1}{\sqrt{14}}\begin{bmatrix} 1 \\ 2 \\ 3 \end{bmatrix}.$$

Since

$${}_x\langle 1|\psi\rangle = (2+\sqrt{2})/\sqrt{14}, \quad {}_x\langle 0|\psi\rangle = -1/\sqrt{7}, \quad {}_x\langle -1|\psi\rangle = (2-\sqrt{2})/\sqrt{14},$$

we obtain that

$$\Pr(|0\rangle_x) = |{}_x\langle 0|\psi\rangle|^2 = 1/7.$$

2.10.3 Coordinate Representation of OAM

We are concerned here with coordinate representation of OAM. Because of spherical symmetry, it is convenient to work with spherical coordinates, which are related to Cartesian coordinates as follows:

$$x = r\sin\theta\cos\phi \qquad y = r\sin\theta\sin\phi \qquad z = r\cos\theta. \tag{2.187}$$

With these substitutions, the components of OAM can be represented in spherical coordinates as

$$L_x = j\hbar\left(\sin\phi\frac{\partial}{\partial\theta} + \frac{\cos\phi}{\tan\theta}\frac{\partial}{\partial\phi}\right) \qquad L_y = j\hbar\left(-\cos\phi\frac{\partial}{\partial\theta} + \frac{\sin\phi}{\tan\theta}\frac{\partial}{\partial\phi}\right)$$
$$L_z = -j\hbar\frac{\partial}{\partial\phi}. \tag{2.188}$$

The corresponding L^2 and leader operators can be expressed in spherical coordinates as follows:

$$L^2 = -\hbar^2\left(\frac{\partial^2}{\partial\theta^2} + \frac{1}{\tan\theta}\frac{\partial}{\partial\theta} + \frac{1}{\sin^2\theta}\frac{\partial^2}{\partial\phi^2}\right)$$
$$L_+ = \hbar e^{j\phi}\left(\frac{\partial}{\partial\theta} + j\frac{1}{\tan\theta}\frac{\partial}{\partial\phi}\right) \qquad L_- = -\hbar e^{-j\phi}\left(\frac{\partial}{\partial\theta} - j\frac{1}{\tan\theta}\frac{\partial}{\partial\phi}\right). \tag{2.189}$$

Because the various operators above are only functions of angular variables, the corresponding eigenfunctions of OAM will also be functions of θ and ϕ only:

$$\langle\theta,\phi|l,m\rangle = Y_l^m(\theta,\phi), \tag{2.190}$$

where $Y_l^m(\theta,\phi)$ are spherical harmonics, defined by

$$Y_l^m(\theta,\phi) = \begin{cases} (-1)^m\sqrt{\dfrac{(2l+1)(l-m)!}{4\pi(l+m)!}}P_l^m(\cos\theta)e^{jm\phi}, & m>0 \\ (-1)^{|m|}\sqrt{\dfrac{(2l+1)(l-|m|)!}{4\pi(l+m)!}}P_l^{|m|}(\cos\theta)e^{jm\phi}, & m<0 \end{cases}. \tag{2.191}$$

The associated Legendre polynomials, denoted as $P_l^m(x)$, are defined as:

$$P_l^m(x) = (1-x^2)^{m/2}\frac{d^m}{dx^m}P_l(x), \tag{2.192}$$

where $P_l(x)$ are the Legendre polynomials, defined by

$$P_l(x) = \frac{1}{2^l l!}\frac{d^l}{dx^l}\left[\left(x^2-1\right)^l\right]. \tag{2.193}$$

The Legendre polynomials can be determined recursively as follows:

$$(l+1)P_{l+1}(x) = (2l+1)xP_l(x) - lP_{l-1}(x); \quad P_0(x) = 1, \quad P_1(x) = x. \tag{2.194}$$

Example. A symmetrical top with moments of inertia I_x, I_y, and I_z in the body of axes frame is described by the following Hamiltonian:

$$H = \frac{1}{2I_x}\left(L_x^2 + L_y^2\right) + \frac{1}{2I_z}L_z^2.$$

Determine the eigenvalues and eigenkets of the Hamiltonian. Determine the expected measurement of $L_x + L_y + L_z$ for any state. Finally, if initially the top was in state $|l = 3, m = 0\rangle$, determine the probability that the result of a measurement of L_x at time $t = 4\pi I_x/\hbar$ is $\hbar$.

We first express the Hamiltonian in terms of L^2 and L_z:

$$\begin{aligned} H &= \frac{1}{2I_x}\left(L_x^2 + L_y^2\right) + \frac{1}{2I_z}L_z^2 = \frac{1}{2I_x}\left(L_x^2 + L_y^2 + L_z^2 - L_z^2\right) + \frac{1}{2I_z}L_z^2 \\ &= \frac{1}{2I_x}L^2 + \left(\frac{1}{2I_z} - \frac{1}{2I_x}\right)L_z^2. \end{aligned}$$

We already know that when A is an operator with eigenvalues λ_i, then the eigenvalues for the function of A, $f(A)$, are $f(\lambda_i)$. Therefore, based on (2.176) we obtain the following energy eigenvalues:

$$E_{lm} = \frac{1}{2I_x}\hbar^2 l(l+1) + \left(\frac{1}{2I_z} - \frac{1}{2I_x}\right)m^2\hbar^2.$$

Furthermore, since L^2 and L_z have simultaneous eigenkets, the eigenkets of H will be $|Y_l^m(\theta,\phi)\rangle$ with eigenenergies being E_{lm}. The measurement of $L_x + L_y + L_z$ yields

$$\begin{aligned} \langle Y_l^m(\theta,\phi)|L_x + L_y + L_z|Y_l^m(\theta,\phi)\rangle &= \left\langle Y_l^m(\theta,\phi)\left|\frac{L_+ + L_-}{2} + \frac{L_+ - L_-}{2j} + L_z\right|Y_l^m(\theta,\phi)\right\rangle \\ &= \langle Y_l^m(\theta,\phi)|L_z|Y_l^m(\theta,\phi)\rangle = m\hbar. \end{aligned}$$

Since the state at $t = 0$, $|Y_0^3(\theta,\phi)\rangle$, is an eigenket of H, it will remain so all the time. The measurement of L_z for $m = 0$ is zero, so that the probability of getting $\hbar$ is zero.

2.10.4 Angular Momentum and Rotations

The angular momentum operator L is often called the generator of rotations. The state ket $|\psi\rangle$ in a coordinate system O can be rotated to a new coordinate system O' by means of rotation operator U_R as follows:

$$|\psi'\rangle = U_R|\psi\rangle. \tag{2.195}$$

The infinitesimal rotation operator can be defined as

$$U_{\mathrm{R}}(\mathrm{d}\theta,\hat{n}) = I - \frac{j}{\hbar}\,\mathrm{d}\theta\,\boldsymbol{L}\cdot\hat{n}. \tag{2.196}$$

This operator can be used to describe the rotation of the state ket $|\psi\rangle$ in coordinate system O' around $\hat{n}$ for an angle θ (relative to O):

$$\begin{aligned} U_{\mathrm{R}}(\theta,\hat{n}) &= \lim_{N\to\infty} U_{\mathrm{R}}(\mathrm{d}\theta=\theta/N,\hat{n}) \\ &= \lim_{N\to\infty}\left(I-\frac{j}{\hbar}\frac{\theta}{N}\boldsymbol{L}\cdot\hat{n}\right)^{N} \overset{\lim_{N\to\infty}(1+a/N)^N=\mathrm{e}^a}{=} \exp\left(-\frac{j}{\hbar}\theta\boldsymbol{L}\cdot\hat{n}\right). \end{aligned} \tag{2.197}$$

For example, the rotation around the z-axis by an angle φ can be described by

$$U_{\mathrm{R}}(\varphi,z) = \exp\left(-\frac{j}{\hbar}\varphi L_z\right). \tag{2.198}$$

The rotation operator must satisfy the following property

$$U_{\mathrm{R}}(0,\hat{n}) = U_{\mathrm{R}}(2\pi,\hat{n}) = I. \tag{2.199}$$

An observable A in coordinate system O is transformed into A' in coordinate system O' by

$$A' = U_{\mathrm{R}} A U_{\mathrm{R}}^{\dagger}. \tag{2.200}$$

Example. Determine the rotation operator around $\hat{n}=\hat{y}$ for $l=1$. By using this operator determine the representation of L_x in standard L_z basis.

The rotation operator (2.197) for $\hat{n}=\hat{y}$ is as follows:

$$U_{\mathrm{R}}(\theta,\hat{y}) = \exp\left(-\frac{j\theta}{\hbar}L_y\right) = \sum_{n=0}^{\infty}\frac{(-j\theta)^n}{n!}\left(\frac{L_y}{\hbar}\right)^n.$$

The matrix representation of L_y in the L_z basis is given by

$$\begin{aligned} L_y &\doteq \begin{bmatrix} \langle 1|L_y|1\rangle & \langle 1|L_y|0\rangle & \langle 1|L_y|-1\rangle \\ \langle 0|L_y|1\rangle & \langle 0|L_y|0\rangle & \langle 0|L_y|-1\rangle \\ \langle -1|L_y|1\rangle & \langle -1|L_y|0\rangle & \langle -1|L_y|-1\rangle \end{bmatrix} = \frac{\hbar}{\sqrt{2}}\begin{bmatrix} 0 & -j & 0 \\ j & 0 & -j \\ 0 & j & 0 \end{bmatrix} \\ &= \frac{j\hbar}{\sqrt{2}}\begin{bmatrix} 0 & -1 & 0 \\ 1 & 0 & -1 \\ 0 & 1 & 0 \end{bmatrix}. \end{aligned}$$

The first three powers of $L_y/\hbar$ are therefore

$$\frac{L_y}{\hbar} = \frac{j}{\sqrt{2}}\begin{bmatrix} 0 & -1 & 0 \\ 1 & 0 & -1 \\ 0 & 1 & 0 \end{bmatrix}, \quad \left(\frac{L_y}{\hbar}\right)^2 = -\frac{1}{2}\begin{bmatrix} -1 & 0 & 1 \\ 0 & -2 & 0 \\ 1 & 0 & -1 \end{bmatrix},$$

$$\left(\frac{L_y}{\hbar}\right)^3 = \frac{j}{\sqrt{2}}\begin{bmatrix} 0 & -1 & 0 \\ 1 & 0 & -1 \\ 0 & 1 & 0 \end{bmatrix}.$$

It is clear that the third power $L_y/\hbar$ is the same as the first power. The rotation operator can therefore be written as

$$U_{\rm R}(\theta,\hat{y}) = \sum_{n=0}^{\infty}\frac{(-j\theta)^n}{n!}\left(\frac{L_y}{\hbar}\right)^n = I + \sum_{n=0}^{\infty}\frac{(-j\theta)^{2n+1}}{(2n+1)!}\left(\frac{L_y}{\hbar}\right) + \sum_{n=0}^{\infty}\frac{(-j\theta)^{2n}}{(2n)!}\left(\frac{L_y}{\hbar}\right)^2$$
$$= I - j\sin\theta\frac{L_y}{\hbar} - (1-\cos\theta)\left(\frac{L_y}{\hbar}\right)^2$$

The corresponding representation in matrix form is as follows:

$$U_{\rm R}(\theta,\hat{y}) = \begin{bmatrix} 1 & 0 & 0 \\ 0 & 1 & 0 \\ 0 & 0 & 1 \end{bmatrix} + \frac{\sin\theta}{\sqrt{2}}\begin{bmatrix} 0 & -1 & 0 \\ 1 & 0 & -1 \\ 0 & 1 & 0 \end{bmatrix} + \frac{1-\cos\theta}{2}\begin{bmatrix} -1 & 0 & 1 \\ 0 & -2 & 0 \\ 1 & 0 & -1 \end{bmatrix}$$
$$= \begin{bmatrix} \frac{1+\cos\theta}{2} & -\frac{\sin\theta}{\sqrt{2}} & \frac{1-\cos\theta}{2} \\ \frac{\sin\theta}{\sqrt{2}} & \cos\theta & -\frac{\sin\theta}{\sqrt{2}} \\ \frac{1-\cos\theta}{2} & \frac{\sin\theta}{\sqrt{2}} & \frac{1+\cos\theta}{2} \end{bmatrix}.$$

To obtain the eigenvectors of L_x by using the eigenkets of L_z we have to rotate the eigenkets of L_z by $\theta = \pi/2$ – that is, to apply the rotation operator

$$U_{\rm R}(\pi/2,\hat{y}) = \begin{bmatrix} \frac{1}{2} & -\frac{1}{\sqrt{2}} & \frac{1}{2} \\ \frac{1}{\sqrt{2}} & 0 & -\frac{1}{\sqrt{2}} \\ \frac{1}{2} & \frac{1}{\sqrt{2}} & \frac{1}{2} \end{bmatrix}$$

on base kets to obtain:

$$|1\rangle_x = U_{\rm R}(\pi/2,\hat{y})|1\rangle = \begin{bmatrix} 1/2 \\ 1/\sqrt{2} \\ 1/2 \end{bmatrix}, \quad |0\rangle_x = U_{\rm R}(\pi/2,\hat{y})|0\rangle = \begin{bmatrix} -1/\sqrt{2} \\ 0 \\ 1/\sqrt{2} \end{bmatrix},$$

$$|-1\rangle_x = U_{\rm R}(\pi/2,\hat{y})|-1\rangle = \begin{bmatrix} 1/2 \\ -1/\sqrt{2} \\ 1/2 \end{bmatrix}.$$

2.11 SPIN-1/2 SYSTEMS

The spin is an intrinsic property of particles, which was deduced from the Stern–Gerlach experiment. Similarly to OAM, to completely characterize the state of the spin, it is sufficient to observe operators S^2 and S_z, whose simultaneous eigenkets can be denoted by $|s,m\rangle$. The corresponding eigenvalue equations are given as follows:

$$S^2|s,m\rangle = \hbar^2 s(s+1)|s,m\rangle \qquad S_z|s,m\rangle = m\hbar|s,m\rangle$$
$$m = -s, \; -s+1, \ldots, \; -1, 0, 1, \ldots, s-1, s. \tag{2.201}$$

Unlike the OAM, spin is not a function of spatial coordinates. To move up and down in states of SAM, we introduce the *ladder* operators S_+ and $S-$, also known as *raising* and *lowering* operators:

$$S_+ = S_x + iS_y \qquad S_- = S_x - iS_y. \tag{2.202}$$

The ladder operators act on SAM states as follows:

$$S_\pm|s,m\rangle = \hbar\sqrt{s(s+1) - m(m\pm 1)}|s,m\pm 1\rangle \qquad S_+|s,s\rangle = 0 \quad S_-|s,-s\rangle = 0. \tag{2.203}$$

The components of the spin operator satisfy commutation relations similar to those of OAM:

$$[S_x, S_y] = j\hbar S_z \quad [S_z, S_x] = j\hbar S_y \quad [S_y, S_z] = j\hbar S_x\,. \tag{2.204}$$

Unlike OAM, the spin for a given particle is fixed. For example, the force-carrying boson has spin $s = 1$, the graviton has spin $s = 2$, while electrons and quarks have spin $s = 1/2$. The eigenvalue equation for spin-1/2 particles is

$$S^2|1/2,m\rangle = \frac{3}{4}\hbar^2|1/2,m\rangle \qquad S_z|1/2,m\rangle = m\hbar|1/2,m\rangle \qquad m = -1/2, \; 1/2. \tag{2.205}$$

Therefore, the measurement of the spin of any spin-1/2 system can have only two possible results: spin-up and spin-down. The basis states for spin-1/2 systems can be labeled as

$$|+\rangle = |1/2, +1/2\rangle = \begin{bmatrix} 1 \\ 0 \end{bmatrix} \qquad |-\rangle = |1/2, -1/2\rangle = \begin{bmatrix} 0 \\ 1 \end{bmatrix}. \tag{2.206}$$

These states are orthonormal, since

$$\langle +|+\rangle = \langle -|-\rangle = 1 \qquad \langle +|-\rangle = \langle -|+\rangle = 0. \tag{2.207}$$

The superposition state $|\psi\rangle$ can be written in terms of base states as

$$|\psi\rangle = \alpha|+\rangle + \beta|-\rangle = \begin{bmatrix} \alpha \\ \beta \end{bmatrix}, \qquad |\alpha|^2 + |\beta|^2 = 1, \tag{2.208}$$

where $|\alpha|^2$ ($|\beta|^2$) denotes the probability of finding the system in the spin-up (spin-down) state. By using the trigonometric identity $\sin^2(\theta/2) + \cos^2(\theta/2) = 1$, the expansion coefficients α and β can be expressed as follows:

$$\alpha = e^{-j\phi/2}\cos(\theta/2) \qquad \beta = e^{j\phi/2}\sin(\theta/2), \tag{2.209}$$

so that

$$|\psi\rangle = e^{-j\phi/2}\cos(\theta/2)|+\rangle + e^{j\phi/2}\sin(\theta/2)|-\rangle. \tag{2.210}$$

Therefore, the single superposition state, also known as the quantum bit (qubit), can be visualized as the point (θ,ϕ) on a unit sphere (Bloch or Poincaré sphere).

The arbitrary operator A can be represented in this basis by

$$A \doteq \begin{bmatrix} \langle +|A|+\rangle & \langle +|A|-\rangle \\ \langle -|A|+\rangle & \langle -|A|-\rangle \end{bmatrix}. \tag{2.211}$$

For example, the matrix representation of the S_z operator is given by

$$S_z \doteq \begin{bmatrix} \langle +|S_z|+\rangle & \langle +|S_z|-\rangle \\ \langle -|S_z|+\rangle & \langle -|S_z|-\rangle \end{bmatrix} = \begin{bmatrix} \langle +|\frac{\hbar}{2}|+\rangle & \langle +|-\frac{\hbar}{2}|-\rangle \\ \langle -|\frac{\hbar}{2}|+\rangle & \langle -|-\frac{\hbar}{2}|-\rangle \end{bmatrix}$$
$$= \frac{\hbar}{2}\begin{bmatrix} 1 & 0 \\ 0 & -1 \end{bmatrix} \tag{2.212}$$

The action of ladder operators for spin-1/2 systems can be described as follows:

$$S_+|-\rangle = \hbar|+\rangle \qquad S_-|+\rangle = \hbar|-\rangle, \tag{2.213}$$

which can be used for matrix representation in the $\{|+\rangle,|-\rangle\}$ basis:

$$S_+ \doteq \begin{bmatrix} \langle +|S_+|+\rangle & \langle +|S_+|-\rangle \\ \langle -|S_+|+\rangle & \langle -|S_+|-\rangle \end{bmatrix} = \begin{bmatrix} 0 & \langle +|\hbar|+\rangle \\ 0 & \langle -|\hbar|+\rangle \end{bmatrix} = \hbar\begin{bmatrix} 0 & 1 \\ 0 & 0 \end{bmatrix}$$
$$S_- \doteq \begin{bmatrix} \langle +|S_-|+\rangle & \langle +|S_-|-\rangle \\ \langle -|S_-|+\rangle & \langle -|S_-|-\rangle \end{bmatrix} = \begin{bmatrix} \langle +|\hbar|-\rangle & 0 \\ \langle -|\hbar|-\rangle & 0 \end{bmatrix} = \hbar\begin{bmatrix} 0 & 0 \\ 1 & 0 \end{bmatrix}. \tag{2.214}$$

The S_x and S_y operators can be represented in the same basis by

$$S_x = (S_+ + S_-)/2 = \frac{\hbar}{2}\begin{bmatrix} 0 & 1 \\ 1 & 0 \end{bmatrix} \qquad S_y = (S_+ - S_-)/2j = \frac{\hbar}{2}\begin{bmatrix} 0 & -j \\ j & 0 \end{bmatrix}. \tag{2.215}$$

The action of the S_x operator on basis kets is given by

$$S_x|+\rangle = \frac{S_+ + S_-}{2}|+\rangle = \frac{1}{2}S_+|+\rangle + \frac{1}{2}S_-|+\rangle = \frac{\hbar}{2}|+\rangle$$
$$S_x|-\rangle = \frac{S_+ + S_-}{2}|-\rangle = \frac{1}{2}S_+|-\rangle + \frac{1}{2}S_-|-\rangle = \frac{\hbar}{2}|+\rangle. \tag{2.216}$$

The unitary matrix that can be used to transform the S_z basis to the S_x basis is given by

$$U = \begin{bmatrix} \langle +_x|+\rangle & \langle +_x|-\rangle \\ \langle -_x|+\rangle & \langle -_x|-\rangle \end{bmatrix}, \tag{2.217}$$

where

$$|+_x\rangle = \frac{1}{\sqrt{2}}(|+\rangle + |-\rangle) \qquad |-_x\rangle = \frac{1}{\sqrt{2}}(|+\rangle - |-\rangle). \tag{2.218}$$

Since $\langle \pm_x | \pm \rangle = \pm 1/\sqrt{2}$ and $\langle \pm_x | \mp \rangle = 1/\sqrt{2}$, we obtain:

$$U = \frac{1}{\sqrt{2}} \begin{bmatrix} 1 & 1 \\ 1 & -1 \end{bmatrix}. \tag{2.219}$$

The application of U to an arbitrary state gives

$$U|\psi\rangle = U \begin{bmatrix} \alpha \\ \beta \end{bmatrix} = \frac{1}{\sqrt{2}} \begin{bmatrix} \alpha + \beta \\ \alpha - \beta \end{bmatrix}. \tag{2.220}$$

The operator U, also known as the Hadamard gate in quantum computing, is clearly Hermitian since $UU^\dagger = I$, and can be used to diagonalize S_x as follows:

$$US_xU^\dagger = \frac{1}{\sqrt{2}} \begin{bmatrix} 1 & 1 \\ 1 & -1 \end{bmatrix} \frac{\hbar}{2} \begin{bmatrix} 0 & 1 \\ 1 & 0 \end{bmatrix} \frac{1}{\sqrt{2}} \begin{bmatrix} 1 & 1 \\ 1 & -1 \end{bmatrix} = \frac{\hbar}{2} \begin{bmatrix} 1 & 0 \\ 0 & -1 \end{bmatrix}.$$

From the matrix representation of the spin operators S_x, S_y, S_z and the matrix representation of an operator A, we can represent the spin operators in outer product representation as follows:

$$S_x = \frac{\hbar}{2} \begin{bmatrix} 0 & 1 \\ 1 & 0 \end{bmatrix} = \frac{\hbar}{2}(|+\rangle\langle -| + |-\rangle\langle +|), \quad S_y = \frac{\hbar}{2} \begin{bmatrix} 0 & -j \\ j & 0 \end{bmatrix} = \frac{j\hbar}{2}(-|+\rangle\langle -| + |-\rangle\langle +|),$$
$$S_z = \frac{\hbar}{2} \begin{bmatrix} 1 & 0 \\ 0 & -1 \end{bmatrix} = \frac{\hbar}{2}(|+\rangle\langle +| - |-\rangle\langle -|). \tag{2.221}$$

The projection operators $P_+ = |+\rangle\langle +|$ and $P_- = |-\rangle\langle -|$ can be represented in matrix form as follows:

$$P_+ \doteq \begin{bmatrix} \langle +|(|+\rangle\langle +|)|+\rangle & \langle +|(|+\rangle\langle +|)|-\rangle \\ \langle -|(|+\rangle\langle +|)|+\rangle & \langle -|(|+\rangle\langle +|)|-\rangle \end{bmatrix} = \begin{bmatrix} 1 & 0 \\ 0 & 0 \end{bmatrix}$$
$$P_- \doteq \begin{bmatrix} \langle +|(|-\rangle\langle -|)|+\rangle & \langle +|(|-\rangle\langle -|)|-\rangle \\ \langle -|(|-\rangle\langle -|)|+\rangle & \langle -|(|-\rangle\langle -|)|-\rangle \end{bmatrix} = \begin{bmatrix} 0 & 0 \\ 0 & 1 \end{bmatrix}. \tag{2.222}$$

Clearly, the projection operators satisfy the completeness relation:

$$P_+ + P_- = I.$$

2.11.1 Pauli Operators (Revisited)

By using the standard basis of the S_z operator, we have shown that spin operators can be represented as follows:

$$S_x = \frac{\hbar}{2} \begin{bmatrix} 0 & 1 \\ 1 & 0 \end{bmatrix} = \frac{\hbar}{2} X, \quad S_y = \frac{\hbar}{2} \begin{bmatrix} 0 & -j \\ j & 0 \end{bmatrix} = \frac{\hbar}{2} Y, \quad S_z = \frac{\hbar}{2} \begin{bmatrix} 1 & 0 \\ 0 & -1 \end{bmatrix} = \frac{\hbar}{2} Z;$$
$$X = \begin{bmatrix} 0 & 1 \\ 1 & 0 \end{bmatrix}, \quad Y = \begin{bmatrix} 0 & -j \\ j & 0 \end{bmatrix}, \quad Z = \begin{bmatrix} 1 & 0 \\ 0 & -1 \end{bmatrix}, \tag{2.223}$$

where X, Y, and Z denote the Pauli matrices, which are often denoted in the literature by σ_x, σ_y, and σ_z respectively. It is straightforward to show that the Pauli matrices satisfy the following commutation and anticommutation relations:

$$[\sigma_k, \sigma_l] = 2j\varepsilon_{klm}\sigma_m, \quad \varepsilon_{klm} = \begin{cases} 1, & \text{for cyclic permutations of } k, l, \text{ and } m \\ -1, & \text{for anticyclic permutations of } k, l, \text{ and } m \\ 0, & \text{otherwise} \end{cases}$$
$$\{\sigma_k, \sigma_l\} = 2\delta_{kl}. \tag{2.224}$$

By simple inspection of the Pauli matrices we conclude that $\mathrm{Tr}(\sigma_k) = 0$. Further, we show that $\mathrm{Tr}(\sigma_k\sigma_l) = 2\delta_{kl}$. This property can be proved by summing the commutation and anticommutation relations to obtain:

$$[\sigma_k, \sigma_l] + \{\sigma_k, \sigma_l\} = \sigma_k\sigma_l + \sigma_l\sigma_k + \sigma_k\sigma_l - \sigma_l\sigma_k = 2\sigma_k\sigma_l = 2j\varepsilon_{klm}\sigma_m$$
$$+2\delta_{kl} \Rightarrow \sigma_k\sigma_l = j\varepsilon_{klm}\sigma_m + \delta_{kl}. \tag{2.225}$$

By applying the Tr operator on the last equality in (2.225), we obtain:

$$\mathrm{Tr}(\sigma_k\sigma_l) = \mathrm{Tr}(j\varepsilon_{klm}\sigma_m + \delta_{kl}) = j\varepsilon_{klm}\mathrm{Tr}(\sigma_m) + \mathrm{Tr}(\delta_{kl}I) = \begin{cases} 2, & k = l \\ 0, & k \neq l \end{cases} = 2\delta_{kl}. \tag{2.226}$$

Because the eigenvalues of spin operators S_k ($k = x,y,z$) are $\pm\hbar/2$ and $S_k = \hbar\sigma_k/2$, it is clear that eigenvalues of Pauli matrices are ±1, while the eigenkets are the same as those of S_k, namely $|+\rangle$ and $|-\rangle$:

$$Z|+\rangle = |+\rangle \qquad Z|-\rangle = -|-\rangle \tag{2.227}$$

From the definition of the Pauli matrices it is clear that $X^2 = Y^2 = Z^2 = I$. By defining the ladder operator as $\sigma_\pm = (\sigma_x \pm \sigma_y)/2$, it is obvious that the action of ladder operators on basis states is as follows:

$$\sigma_+|+\rangle = 0 \qquad \sigma_+|-\rangle = |+\rangle \qquad \sigma_-|+\rangle = |-\rangle \quad \sigma_-|-\rangle = 0. \tag{2.228}$$

The matrix representation of ladder operators can easily be obtained from definition formulas and the matrix representation of Pauli operators:

$$\sigma_+ \doteq \begin{bmatrix} 0 & 1 \\ 0 & 0 \end{bmatrix} \qquad \sigma_- \doteq \begin{bmatrix} 0 & 0 \\ 1 & 0 \end{bmatrix}. \tag{2.229}$$

An arbitrary operator A from a two-dimensional Hilbert space can be represented in terms of Pauli operators as follows:

$$A = \frac{1}{2}(a_0 I + \boldsymbol{a}\cdot\boldsymbol{\sigma}), \quad a_0 = \mathrm{Tr}(A), \quad \boldsymbol{a} = \mathrm{Tr}(A\boldsymbol{\sigma}), \quad \boldsymbol{\sigma} = (X, Y, Z). \tag{2.230}$$

This property can be proved starting from the matrix representation of operator A:

$$A=\begin{bmatrix} a & b \\ c & d \end{bmatrix}.$$

By calculating the dot product of $\boldsymbol{a}$ and σ we obtain:

$$\begin{aligned}
\boldsymbol{a}\cdot\boldsymbol{\sigma} &= \mathrm{Tr}(AX)X+\mathrm{Tr}(AY)Y+\mathrm{Tr}(AZ)Z = \mathrm{Tr}\left(\begin{bmatrix} a & b \\ c & d \end{bmatrix}\begin{bmatrix} 0 & 1 \\ 1 & 0 \end{bmatrix}\right)\sigma_X \\
&+\mathrm{Tr}\left(\begin{bmatrix} a & b \\ c & d \end{bmatrix}\begin{bmatrix} 0 & -j \\ j & 0 \end{bmatrix}\right)\sigma_Y+\mathrm{Tr}\left(\begin{bmatrix} a & b \\ c & d \end{bmatrix}\begin{bmatrix} 1 & 0 \\ 0 & -1 \end{bmatrix}\right)\sigma_Z \\
&= \mathrm{Tr}\left(\begin{bmatrix} b & a \\ d & c \end{bmatrix}\right)\sigma_X+\mathrm{Tr}\left(\begin{bmatrix} jb & -ja \\ jd & -jc \end{bmatrix}\right)\sigma_Y+\mathrm{Tr}\left(\begin{bmatrix} a & -b \\ c & -d \end{bmatrix}\right)\sigma_Z \\
&= (b+c)\sigma_X+(jb-jc)\sigma_Y+(a-d)\sigma_Z.
\end{aligned} \tag{2.231}$$

By using (2.231) it follows that

$$\begin{aligned}
\frac{1}{2}(a_0 I+\boldsymbol{a}\cdot\boldsymbol{\sigma}) &= \frac{1}{2}\left(\begin{bmatrix} a+d & 0 \\ 0 & a+d \end{bmatrix}+\begin{bmatrix} a-d & b+c+b-c \\ b+c-b+c & -a+d \end{bmatrix}\right) \\
&= \begin{bmatrix} a & b \\ c & d \end{bmatrix}=A,
\end{aligned} \tag{2.232}$$

which proves Eq. (2.230).

The projection operators can be represented in this form as follows:

$$P_+=\frac{1}{2}(I+Z) \qquad P_-=\frac{1}{2}(I-Z). \tag{2.233}$$

Another interesting result that will be used later in the quantum circuits chapter is

$$\mathrm{e}^{j\theta X}=I\cos\theta+jX\sin\theta. \tag{2.234}$$

This identity can be proved starting from the following expansions:

$$\begin{aligned}
e^x &= 1+x+\frac{x^2}{2!}+\frac{x^3}{3!}+\ldots \qquad \sin x = x-\frac{x^3}{3!}+\frac{x^5}{5!}+\ldots \\
\cos x &= 1-\frac{x^2}{2!}+\frac{x^4}{4!}-\ldots.
\end{aligned} \tag{2.235}$$

By applying (2.235), we obtain:

$$\begin{aligned}
\mathrm{e}^{j\theta X} &= I+j\theta X+\frac{(j\theta X)^2}{2!}+\frac{(j\theta X)^3}{3!}+\frac{(j\theta X)^4}{4!}+\frac{(j\theta X)^5}{5!}+\ldots \\
&= I\left(1-\frac{\theta^2}{2!}+\frac{\theta^4}{4!}-\ldots\right)+jX\left(\theta-\frac{\theta^3}{3!}+\frac{\theta^5}{5!}+\ldots\right) \\
&= I\cos\theta+jX\sin\theta,
\end{aligned} \tag{2.236}$$

where we have used the identity $X^{2n}=(X^2)^n=I$.

2.11.2 Density Operator for Spin-1/2 Systems

Let us observe the pure state given by the density operator $\rho_1 = |+_y\rangle\langle +_y|$ and a completely mixed state given by the density operator $\rho_2 = 0.5|+\rangle\langle +|+0.5|-\rangle\langle -|$. The corresponding matrix representations are given by

$$\begin{aligned}\rho_1 &= |+_y\rangle\langle +_y| = \frac{1}{\sqrt{2}}(|+\rangle + j|-\rangle)\,\frac{1}{\sqrt{2}}(\langle +| - j\langle -|) \\ &= \frac{1}{2}(|+\rangle\langle +| - j|+\rangle\langle -| + j|-\rangle\langle +| + |-\rangle\langle -|) = \begin{bmatrix} 1/2 & -j/2 \\ j/2 & 1/2 \end{bmatrix} \\ \rho_2 &= \frac{1}{2}|+\rangle\langle +| + \frac{1}{2}|-\rangle\langle -| = \begin{bmatrix} 1/2 & 0 \\ 0 & 1/2 \end{bmatrix}.\end{aligned} \tag{2.237}$$

It can easily be verified that both density operators are Hermitian and have unit trace. The corresponding traces for ρ^2 can be found by

$$\begin{aligned}\mathrm{Tr}(\rho_1^2) &= \mathrm{Tr}\left(\begin{bmatrix} 1/2 & -j/2 \\ j/2 & 1/2 \end{bmatrix}\begin{bmatrix} 1/2 & -j/2 \\ j/2 & 1/2 \end{bmatrix}\right) = 1 \\ \mathrm{Tr}(\rho_2^2) &= \mathrm{Tr}\left(\begin{bmatrix} 1/2 & 0 \\ 0 & 1/2 \end{bmatrix}\begin{bmatrix} 1/2 & 0 \\ 0 & 1/2 \end{bmatrix}\right) = \frac{1}{2} < 1.\end{aligned} \tag{2.238}$$

If we measure S_z the probability of finding the result $\hbar/2$ is

$$\begin{aligned}\mathrm{Tr}(\rho_1|+\rangle\langle +|) &= \langle +|\rho_1|+\rangle = [1 \quad 0]\begin{bmatrix} 1/2 & -j/2 \\ j/2 & 1/2 \end{bmatrix}\begin{bmatrix} 1 \\ 0 \end{bmatrix} = \frac{1}{2} \\ \mathrm{Tr}(\rho_2|+\rangle\langle +|) &= \langle +|\rho_2|+\rangle = [1 \quad 0]\begin{bmatrix} 1/2 & 0 \\ 0 & 1/2 \end{bmatrix}\begin{bmatrix} 1 \\ 0 \end{bmatrix} = \frac{1}{2}.\end{aligned} \tag{2.239}$$

If, on the other hand, we measure S_y, the probability of finding the result $\hbar/2$ is

$$\begin{aligned}\mathrm{Tr}(\rho_1|+_y\rangle\langle +_y|) &= \langle +_y|\rho_1|+_y\rangle = \langle +_y|(|+_y\rangle\langle +_y|)|+_y\rangle = 1 \\ \mathrm{Tr}(\rho_2|+_y\rangle\langle +_y|) &= \langle +_y|\rho_2|+_y\rangle = \frac{1}{\sqrt{2}}[1 \quad -j]\begin{bmatrix} 1/2 & 0 \\ 0 & 1/2 \end{bmatrix}\frac{1}{\sqrt{2}}\begin{bmatrix} 1 \\ j \end{bmatrix} = \frac{1}{2}.\end{aligned} \tag{2.240}$$

2.11.3 Time Evolution of Spin-1/2 Systems

The Hamiltonian of a particle of spin $\boldsymbol{S}$ in a magnetic field $\boldsymbol{B}$ is given by

$$H = -\gamma \boldsymbol{S}\cdot\boldsymbol{B}, \quad \gamma = 2e/mc. \tag{2.241}$$

If a constant magnetic field is applied in the z-direction, the corresponding Hamiltonian is given by

$$H = -\gamma S_z B_z. \tag{2.242}$$

Because B_z and gyromagnetic ratio γ are constant, the eigenkets of Hamiltonian H are the same as the eigenkets of S_z:

$$H|+\rangle = -\gamma B_z S_z|+\rangle = -\gamma B_z \frac{\hbar}{2}|+\rangle = E_+|+\rangle,\quad E_+ = -\gamma B_z \frac{\hbar}{2}$$
$$H|-\rangle = -\gamma B_z S_z|-\rangle = \gamma B_z \frac{\hbar}{2}|-\rangle = E_-|+\rangle,\quad E_- = \gamma B_z \frac{\hbar}{2}. \tag{2.243}$$

The time evolution of spin kets is governed by the Schrödinger equation:

$$j\hbar \frac{\mathrm{d}}{\mathrm{d}t}|\psi\rangle = H|\psi\rangle,\quad |\psi\rangle = \begin{bmatrix}\alpha\\ \beta\end{bmatrix}, \tag{2.244}$$

and by solving it we obtain:

$$|\psi(t)\rangle = \begin{bmatrix}\alpha \mathrm{e}^{-\frac{j}{\hbar}E_+ t}\\ \beta \mathrm{e}^{-\frac{j}{\hbar}E_- t}\end{bmatrix}. \tag{2.245}$$

For example, if the initial state was $|+_y\rangle$, it would interesting to determine the probability of finding $\pm\hbar/2$ if we measure S_x or S_z at time t. For the initial state $|\psi\rangle = |+_y\rangle = [1/\sqrt{2}\quad 1/\sqrt{2}]^{\mathrm{T}}$, and by applying (2.245) we obtain:

$$|\psi(t)\rangle = \frac{1}{\sqrt{2}}\begin{bmatrix}\mathrm{e}^{-\frac{j}{\hbar}E_+ t}\\ \mathrm{e}^{-\frac{j}{\hbar}E_- t}\end{bmatrix} = \frac{1}{\sqrt{2}}\begin{bmatrix}\mathrm{e}^{j\gamma B_z t/2}\\ \mathrm{e}^{-j\gamma B_z t/2}\end{bmatrix}, \tag{2.246}$$

so that the probability of finding $\pm\hbar/2$ is

$$\mathrm{Pr}_y(\hbar/2) = |\langle +_y|\psi(t)\rangle|^2 = \cos^2\left(\frac{\gamma B_z t}{2}\right),$$
$$\mathrm{Pr}_y(-\hbar/2) = |\langle -_y|\psi(t)\rangle|^2 = \sin^2\left(\frac{\gamma B_z t}{2}\right). \tag{2.247}$$

In similar fashion,

$$\langle +|\psi(t)\rangle = \frac{1}{\sqrt{2}}\mathrm{e}^{j\gamma B_z/t}, \tag{2.248}$$

so that

$$\mathrm{Pr}_z(\hbar/2) = |\langle +|\psi(t)\rangle|^2 = \frac{1}{2}, \mathrm{Pr}_z(-\hbar/2) = |\langle -|\psi(t)\rangle|^2 = \frac{1}{2}. \tag{2.249}$$

For the state $|+_y\rangle$ we can determine $\langle S_x\rangle$ and $\langle S_z\rangle$ by

$$\langle S_x\rangle = \sum_i p_i\lambda_i = \cos^2\left(\frac{\gamma B_z t}{2}\right)\cdot\frac{\hbar}{2} + \sin^2\left(\frac{\gamma B_z t}{2}\right)\cdot\left(-\frac{\hbar}{2}\right) = \frac{\hbar}{2}\cos(\gamma B_z t)$$
$$\langle S_z\rangle = \sum_i p_i\lambda_i = \frac{1}{2}\cdot\frac{\hbar}{2} + \frac{1}{2}\cdot\left(-\frac{\hbar}{2}\right) = 0. \tag{2.250}$$

In the rest of this section, we provide several examples that will be of interest in various physical implementations of quantum computers.

Example. A particle is under the influence of a periodic magnetic field $\boldsymbol{B} = B_0 \sin(\omega t)\hat{z}$. Determine the Hamiltonian and the state of the particle at time t. If the initial state was $|+_y\rangle$, determine the probability that the result of measurement of S_y at t is $\hbar/2$.

To determine the Hamiltonian we use (2.241) to obtain:

$$H = -\gamma \boldsymbol{S}\cdot\boldsymbol{B} = -\gamma B_0 \sin(\omega t) S_z. \tag{2.251}$$

We now solve the Schrödinger equation:

$$j\hbar \frac{\mathrm{d}}{\mathrm{d}t}|\psi\rangle = H|\psi\rangle,$$

for arbitrary state $|\psi\rangle = [\alpha \quad \beta]^{\mathrm{T}}$ as follows:

$$j\hbar \begin{bmatrix} \dot{\alpha}(t) \\ \dot{\beta}(t) \end{bmatrix} = -\gamma B_0 \sin(\omega t)\frac{\hbar}{2}\begin{bmatrix} \alpha(t) \\ -\beta(t) \end{bmatrix}. \tag{2.252}$$

The first differential equation

$$\frac{\mathrm{d}\alpha}{\mathrm{d}t} = j\frac{\gamma B_0}{2}\sin(\omega t)$$

can simply be solved by the method of separation variables:

$$\frac{\mathrm{d}\alpha}{\alpha} = j\frac{\gamma B_0}{2}\sin(\omega t)\mathrm{d}t$$

as follows:

$$\alpha(t) = \alpha(0)\exp\left(-j\frac{\gamma B_0}{2\omega}\cos(\omega t)\right) \tag{2.253}$$

In similar fashion, by solving the second differential equation we obtain:

$$\beta(t) = \beta(0)\exp\left(j\frac{\gamma B_0}{2\omega}\cos(\omega t)\right) \tag{2.254}$$

For initial state $|\psi(0)\rangle = |+_y\rangle = [1 \quad j]^{\mathrm{T}}$, meaning that $\alpha(0) = 1/\sqrt{2}$, $\beta(0) = j/\sqrt{2}$, we obtain from (2.253) and (2.254):

$$\begin{aligned}
|\psi(t)\rangle &= \frac{1}{\sqrt{2}}\begin{bmatrix} \exp\left(-j\frac{\gamma B_0}{2\omega}\cos(\omega t)\right) \\ -j\exp\left(j\frac{\gamma B_0}{2\omega}\cos(\omega t)\right) \end{bmatrix} \\
&= \frac{1}{\sqrt{2}}\begin{bmatrix} \cos\left(\frac{\gamma B_0}{2\omega}\cos(\omega t)\right) - j\sin\left(\frac{\gamma B_0}{2\omega}\cos(\omega t)\right) \\ \sin\left(\frac{\gamma B_0}{2\omega}\cos(\omega t)\right) - j\cos\left(j\frac{\gamma B_0}{2\omega}\cos(\omega t)\right) \end{bmatrix} \\
&= \cos\left(\frac{\gamma B_0}{2\omega}\cos(\omega t)\right)\frac{1}{\sqrt{2}}\begin{bmatrix} 1 \\ j \end{bmatrix} - j\sin\left(\frac{\gamma B_0}{2\omega}\cos(\omega t)\right)\frac{1}{\sqrt{2}}\begin{bmatrix} 1 \\ -j \end{bmatrix} \\
&= \cos\left(\frac{\gamma B_0}{2\omega}\cos(\omega t)\right)|+_y\rangle - j\sin\left(\frac{\gamma B_0}{2\omega}\cos(\omega t)\right)|-_y\rangle
\end{aligned}$$

and the corresponding probability is

$$|\langle +_y|\psi(t)\rangle|^2 = \cos^2\left(\frac{\gamma B_0}{2\omega}\cos(\omega t)\right).$$

Example. A particle is under the influence of a periodic magnetic field $\boldsymbol{B} = B_0\cos(\omega t)\hat{x} + B_0\sin(\omega t)\hat{y}$. Determine the Hamiltonian of the system. Determine the transformation that will allow us to write a time-invariant Hamiltonian. If the initial state was $|+\rangle$, determine the probability that at time t the system is in state $|-\rangle$. Finally, determine the first time $t > 0$ for which the system is with certainty in state $|-\rangle$.

To determine the Hamiltonian we again use (2.241) to obtain:

$$\begin{aligned} H = -\gamma \boldsymbol{S}\cdot\boldsymbol{B} &= -\gamma B_0\cos(\omega t)S_x - \gamma B_0\sin(\omega t)S_y \\ &= \omega_0(\cos(\omega t)S_x + \sin(\omega t)S_y), \quad \omega_0 = -\gamma B_0. \end{aligned} \tag{2.255}$$

Based on the matrix representation of S_x and S_y, the Hamiltonian can be represented by

$$\begin{aligned} H &= \frac{\hbar\omega_0}{2}\left(\begin{bmatrix} 0 & \cos(\omega t) \\ \cos(\omega t) & 0 \end{bmatrix} + \begin{bmatrix} 0 & -j\sin(\omega t) \\ j\sin(\omega t) & 0 \end{bmatrix}\right) \\ &= \frac{\hbar\omega_0}{2}\begin{bmatrix} 0 & \exp(-j\omega t) \\ \exp(j\omega t) & 0 \end{bmatrix}. \end{aligned} \tag{2.256}$$

We now need to solve the Schrödinger equation:

$$j\hbar\frac{\mathrm{d}}{\mathrm{d}t}|\psi\rangle = H|\psi\rangle,$$

which for arbitrary state $|\psi\rangle = [\alpha \quad \beta]^{\mathrm{T}}$ can be written as follows:

$$j\hbar\begin{bmatrix} \dot{\alpha}(t) \\ \dot{\beta}(t) \end{bmatrix} = \frac{\hbar\omega_0}{2}\begin{bmatrix} 0 & \exp(-j\omega t) \\ \exp(j\omega t) & 0 \end{bmatrix}\begin{bmatrix} \alpha(t) \\ -\beta(t) \end{bmatrix} \tag{2.257}$$

and the corresponding differential equations resulting from (2.257) are

$$\frac{\mathrm{d}\alpha}{\mathrm{d}t} = \frac{-j\omega_0}{2}\exp(-j\omega t)\beta \qquad \frac{\mathrm{d}\beta}{\mathrm{d}t} = \frac{-j\omega_0}{2}\exp(j\omega t)\alpha, \tag{2.258}$$

which are mutually dependent. By substituting $\alpha = \exp(-j\omega t/2)\alpha'$, $\beta = \exp(j\omega t/2)\beta'$, from (2.258) we obtain:

$$\begin{aligned} \dot{\alpha} &= -\frac{j\omega}{2}\exp(-j\omega t/2)\alpha' + \exp(-j\omega t/2)\dot{\alpha}', \\ \dot{\beta} &= \frac{j\omega}{2}\exp(j\omega t/2)\beta' + \exp(j\omega t/2)\dot{\beta}', \end{aligned} \tag{2.259}$$

which is equivalent to

$$\frac{j\omega_0}{2}\exp(-j\omega t/2)\beta = -\frac{j\omega}{2}\alpha' + \dot{\alpha}', \qquad \frac{j\omega_0}{2}\exp(j\omega t/2)\alpha = \frac{j\omega}{2}\beta' + \dot{\beta}' \tag{2.260}$$

or, expressed in more compact form:

$$j\dot{\alpha}' = -\frac{\omega}{2}\alpha' + \frac{\omega_0}{2}\beta' \qquad \frac{\omega_0}{2}\alpha' + \frac{\omega}{2}\beta' = j\dot{\beta}'. \tag{2.261}$$

The Hamiltonian corresponding to (2.261) is

$$H' = \begin{bmatrix} -\frac{\omega}{2} & \frac{\omega_0}{2} \\ \frac{\omega_0}{2} & \frac{\omega}{2} \end{bmatrix} \tag{2.262}$$

In order to solve the Schrödinger equation in transform coordinates, we differentiate the corresponding component equations to obtain:

$$\ddot{\alpha}' = \frac{\omega}{2}\left(-\frac{\omega}{2}\alpha' + \frac{\omega_0}{2}\beta'\right) - \frac{\omega_0}{2}\left(\frac{\omega_0}{2}\alpha' + \frac{\omega}{2}\beta'\right),$$

$$\ddot{\beta}' = -\frac{\omega_0}{2}\left(-\frac{\omega}{2}\alpha' + \frac{\omega_0}{2}\beta'\right) - \frac{\omega}{2}\left(\frac{\omega_0}{2}\alpha' + \frac{\omega}{2}\beta'\right),$$

which is equivalent to

$$\ddot{\alpha}' = -\frac{\omega^2 + \omega_0^2}{4}\alpha' \qquad \ddot{\beta}' = -\frac{\omega^2 + \omega_0^2}{4}\beta'. \tag{2.263}$$

The corresponding solutions for (2.263) are

$$\alpha'(t) = A_1\cos(\Gamma t) + A_2\sin(\Gamma t), \quad \Gamma = \sqrt{\frac{\omega^2 + \omega_0^2}{4}}. \tag{2.264}$$

Since $\alpha'(0) = \alpha(t)\exp(j\omega t/2)|_{t=0} = 1$, we obtain $A_1 = 1$, $A_2 = 0$, so that

$$\alpha'(t) = \cos(\Gamma t) \Rightarrow \alpha(t) = \exp(-j\omega t/2)\alpha'(t) = \exp(-j\omega t/2)\cos(\Gamma t). \tag{2.265}$$

Because $|\alpha(t)|^2 + |\beta(t)|^2 = 1$, it follows that $\beta(t) = \exp(j\omega t/2)\sin(\Gamma t)$ and the state evolution is described by

$$|\psi(t)\rangle = \begin{bmatrix} \exp(-j\omega t/2)\cos(\Gamma t) \\ \exp(j\omega t/2)\sin(\Gamma t) \end{bmatrix}, \tag{2.266}$$

and the corresponding probability is

$$|\langle -|\psi\rangle|^2 = \sin^2(\Gamma t). \tag{2.267}$$

Since $|\langle -|\psi\rangle|^2 = 1$, it follows that $\Gamma t = \pi/2$ and the first time instance at which the system is in state $|-\rangle$ is

$$t = \frac{4\pi}{\sqrt{\omega^2 + \omega_0^2}}.$$

2.12 HYDROGEN-LIKE ATOMS AND BEYOND

A hydrogen atom is a bound system, consisting of a proton and a neutron, with potential given by

$$V(r) = -\frac{1}{4\pi\varepsilon_0}\frac{e^2}{r}, \tag{2.268}$$

where e is electron charge. Therefore, the potential is only a function of the radial coordinate, and because of spherical symmetry it is convenient to use a spherical coordinate system in which the Laplacian is defined by

$$\nabla^2 = \frac{1}{r^2}\frac{\partial}{\partial r}\left(r^2\frac{\partial}{\partial r}\right) + \frac{1}{r^2 \sin\theta}\frac{\partial}{\partial\theta}\left(\sin\theta\frac{\partial}{\partial\theta}\right) + \frac{1}{r^2\sin^2\theta}\frac{\partial^2}{\partial\phi^2}. \tag{2.269}$$

The angular momentum operator L^2 in spherical coordinates is given by

$$L^2 = -\hbar^2\left[\frac{1}{\sin\theta}\frac{\partial}{\partial\theta}\left(\sin\theta\frac{\partial}{\partial\theta}\right) + \frac{1}{\sin^2\theta}\frac{\partial^2}{\partial\phi^2}\right]. \tag{2.270}$$

The Hamiltonian can be written as

$$H = -\frac{\hbar^2}{2m}\frac{1}{r^2}\frac{\partial}{\partial r}\left(r^2\frac{\partial}{\partial r}\right) + \frac{1}{2mr^2}L^2 + V(r). \tag{2.271}$$

Because the operators L^2 and L_z have common eigenkets, the Hamiltonian leads to the following three equations:

$$\begin{aligned} &H\Psi(r,\theta,\phi) = E\Psi(r,\theta,\phi) \\ &L^2\Psi(r,\theta,\phi) = \hbar^2 l(l+1)\Psi(r,\theta,\phi) \\ &L_z\Psi(r,\theta,\phi) = m\hbar\Psi(r,\theta,\phi). \end{aligned} \tag{2.272}$$

We can associate three quantum numbers with this problem: (i) the *principal quantum number*, n, corresponding to the energy (originating from H); (ii) the *azimuthal quantum number*, l, representing the angular momentum (originating from L^2); and (iii) the *magnetic quantum number*, m, originating from L_z. In order to solve (2.272), we can use the method of separation of variables: $\Psi(r,\theta,\phi) = R(r)\Theta(\theta)\Phi(\phi)$. Since the $\Theta(\theta)\Phi(\phi)$ term has already been found in the section on OAM to be related to the spherical harmonics, $\Theta(\theta)\Phi(\phi) = Y_l^m(\theta,\phi)$, we are left with a radial equation to solve:

$$-\frac{\hbar^2}{2mr}\frac{d^2}{dr^2}[rR_{nl}(r)] + \left[\frac{l(l+1)\hbar^2}{2mr^2} + V(r)\right]R_{nl}(r) = ER_{nl}(r). \tag{2.273}$$

By substituting the Coulomb potential into the radial equation we obtain:

$$-\frac{\hbar^2}{2mr}\frac{d^2}{dr^2}[rR_{nl}(r)] + \left[\frac{l(l+1)\hbar^2}{2mr^2} - \frac{1}{4\pi\varepsilon_0}\frac{e^2}{r}\right]R_{nl}(r) = ER_{nl}(r). \tag{2.274}$$

The solution of the radial equation can be written as

$$R_{nl}(r) = \sqrt{\left(\frac{2}{na_{\mathrm{H}}}\right)\frac{(n-l-1)!}{2n[(n-1)!]^2}}\,\mathrm{e}^{-r/2a_{\mathrm{H}}}\left(\frac{r}{a_{\mathrm{H}}}\right)^l L_{n+1}^{2l+1}\left(\frac{r}{a_{\mathrm{H}}}\right), \tag{2.275}$$

where a_{H} is the first Bohr radius (the lowest energy orbit radius; $a_{\mathrm{H}} = 0.0529$ nm) and $L_{n+1}^{2l+1}(r/a_{\mathrm{H}})$ are the corresponding associated Laguerre polynomials, defined by

$$L_n^{\alpha}(x) = \frac{x^{-\alpha}\mathrm{e}^x}{n!}\frac{\mathrm{d}^n}{\mathrm{d}x^n}(\mathrm{e}^{-x}x^{n+\alpha}). \tag{2.276}$$

The radial portion of the wave function is typically normalized as follows:

$$\int_0^{\infty} r^2|R(r)|^2\mathrm{d}r = 1, \tag{2.277}$$

while the expectation value is defined by

$$\langle r^k\rangle = \int_0^{\infty} r^{2+k}|R(r)|^2\mathrm{d}r. \tag{2.278}$$

The following expectation values of the hydrogen atom can be determined in analytic form:

$$\langle r\rangle = \frac{a_{\mathrm{H}}}{2}[3n^2 - l(l+1)] \qquad \langle r^2\rangle = \frac{a_{\mathrm{H}}^2 n^2}{2}[5n^2 + 1 - 3l(l+1)] \qquad \langle r^{-1}\rangle = \frac{1}{a_0 n^2}. \tag{2.279}$$

The complete wave function is the product of the radial wave function and spherical harmonics:

$$\Psi(r,\theta,\phi) = \sqrt{\left(\frac{2}{na_{\mathrm{H}}}\right)\frac{(n-l-1)!}{2n[(n-1)!]^2}}\,\mathrm{e}^{-r/2a_H}\left(\frac{r}{a_{\mathrm{H}}}\right)^l L_{n+1}^{2l+1}\left(\frac{r}{a_{\mathrm{H}}}\right)Y_l^m(\theta,\phi). \tag{2.280}$$

By substituting (2.191) into (2.280), we obtain:

$$\Psi(\rho,\theta,\phi) = C_{n,\,l,\,m}\mathrm{e}^{-\rho/2}(\rho)^l L_{n+1}^{2l+1}(\rho)P_l^{|m|}(\cos\theta)\mathrm{e}^{jm\phi}, \tag{2.281}$$

where $\rho = r/a_{\mathrm{H}}$, and $C_{n,m,l}$ is the normalization constant obtained as a product of normalization constants in (2.191) and (2.280). The energy levels in the hydrogen atom are only functions of n and are given by

$$E_n = \frac{E_1}{n^2}, \quad E_1 = -\frac{m^2e^4}{32\pi^2\varepsilon_0^2\hbar^2} = -13.6 \text{ eV}. \tag{2.282}$$

Because the values that l can take are $\{0,1,\ldots,n-1\}$, while the values that m can take are $\{-l,-l+1,\ldots,l-1,l\}$, and since the radial component is not a function of

m, the number of states with the same energy (the total *degeneracy* of the energy level E_n) is

$$2\sum_{l=0}^{n-1}(2l+1) = 2n^2. \tag{2.283}$$

The states in which $l = 0, 1, 2, 3, 4$ are traditionally called s, p, d, f, and g respectively. The simpler eigenfunctions of the hydrogen atom, ignoring the normalization constant, are given in Table 2.1, which is obtained based on (2.281).

Example. The electron in a hydrogen atom is in the state $\psi_{nlm} = R_{32}(\frac{1}{\sqrt{6}}Y_2^1 + \frac{1}{\sqrt{2}}Y_2^0 + \frac{1}{\sqrt{3}}Y_2^{-1})$. Determine the energy of the electron. If L^2 is measured, determine the possible results of measurement. If, on the other hand, L_z is measured, determine the results of measurements and corresponding probabilities. Finally, determine the expectation of L_z.

The energy level is only a function of n, so that for $n = 3$ we obtain:

$$E_3 = \frac{E_1}{3^2} = -13.6/9 \text{ eV} = -1.51 \text{ eV}.$$

The measurements of L^2 will give the result $\hbar^2 l(l+1)|_{l=2} = 6\hbar^2$. The measurements results for L_z will only be a function of the angular portion of the wave function: $\frac{1}{\sqrt{6}}Y_2^1 + \frac{1}{\sqrt{2}}Y_2^0 + \frac{1}{\sqrt{3}}Y_2^{-1}$. Because L_z satisfies the eigenvalue equation $L_z Y_l^m = m\hbar Y_l^m$, it follows that possible results of measurements will be $\hbar$, 0, and $-\hbar$, which occur with probabilities 1/6, 1/2, and 1/3 respectively. Therefore, the expectation value of L_z is found to be $\langle L_z\rangle = \frac{1}{6}\hbar + \frac{1}{2}0 + \frac{1}{3}(-\hbar) = -\frac{\hbar}{6}$.

Table 2.1 Hydrogen Atom Eigenfunctions

State	n	l	m	Eigenfunction
1s	1	0	0	$e^{-\rho/2}$
2s	2	0	0	$e^{-\rho/2}(1-\rho)$
2p	2	1	−1 0 1	$e^{-\rho/2}\rho\begin{cases}\sin\theta\, e^{-j\phi}\\ \cos\theta\\ \sin\theta\, e^{j\phi}\end{cases}$
3s	3	0	0	$e^{-\rho/2}(\rho^2-4\rho+2)$
3p	3	1	−1 0 1	$e^{-\rho/2}(\rho^2-2\rho)\begin{cases}\sin\theta\, e^{-j\phi}\\ \cos\theta\\ \sin\theta\, e^{j\phi}\end{cases}$
3d	3	2	−2 −1 0 1 2	$e^{-\rho/2}\rho^2\begin{cases}\sin^2\theta\, e^{-j2\phi}\\ \sin\theta\cos\theta\, e^{-j\phi}\\ 1-3\cos^2\theta\\ \sin\theta\cos\theta\, e^{j\phi}\\ \sin^2\theta\, e^{j2\phi}\end{cases}$

The results above are applicable to many two-particle systems, with attraction energy being inversely proportional to the distance between them, providing that parameters are properly chosen. For instance, if the charge of the nucleus is Z, then in the calculations above we need to substitute e^2 by Ze^2. Examples include deuterium and tritium ions that contain only one electron, positronium, and muonic atoms.

The total angular momentum of an electron $\boldsymbol{j}$ in an atom can be found as the vector sum of orbital angular momentum $\boldsymbol{l}$ and spin $\boldsymbol{s}$ as

$$\boldsymbol{j} = \boldsymbol{l} + \boldsymbol{s}. \tag{2.284}$$

For a given value of azimuthal quantum number l, there exist two values of total angular momentum quantum number of an electron: $j = l + 1/2$ and $j = l - 1/2$. Namely, as the electron undergoes orbital motion around the nucleus, it experiences a magnetic field – this interaction is known as spin-orbit interaction. This interaction results in two states $j = l + s$ and $j = l - s$ with slightly different energies, as shown in Figure 2.3.

For atoms containing more than one electron, the total angular momentum $\boldsymbol{J}$ is given by the sum of individual orbital momenta $\boldsymbol{L} = \boldsymbol{l}_1 + \boldsymbol{l}_2 + \ldots$ and spins $\boldsymbol{S} = \boldsymbol{s}_1 + \boldsymbol{s}_2 + \ldots$, so that we can write

$$\boldsymbol{J} = \boldsymbol{L} + \boldsymbol{S}; \quad \boldsymbol{L} = \boldsymbol{l}_1 + \boldsymbol{l}_2 + \ldots, \quad \boldsymbol{S} = \boldsymbol{s}_1 + \boldsymbol{s}_2 + \ldots. \tag{2.285}$$

This type of coupling is known as *LS* coupling. Another type of coupling, namely *JJ* coupling, occurs when individual ***j***s add together to produce the resulting $\boldsymbol{J}$. *LS* coupling typically occurs in lighter elements, while *JJ* coupling typically occurs in heavy elements. In *LS* coupling, the magnitudes of $\boldsymbol{L}$, $\boldsymbol{S}$, and $\boldsymbol{J}$ are given by

$$|\boldsymbol{L}| = \hbar\sqrt{L(L+1)}, \quad |\boldsymbol{S}| = \hbar\sqrt{S(S+1)}, \quad |\boldsymbol{J}| = \hbar\sqrt{J(J+1)}, \tag{2.286}$$

where L, S, and J are quantum numbers satisfying the following properties: (i) L is always a non-negative integer; (ii) the spin quantum number S is either integral or half-integral depending whether the number of electrons is even or odd; and (iii) the

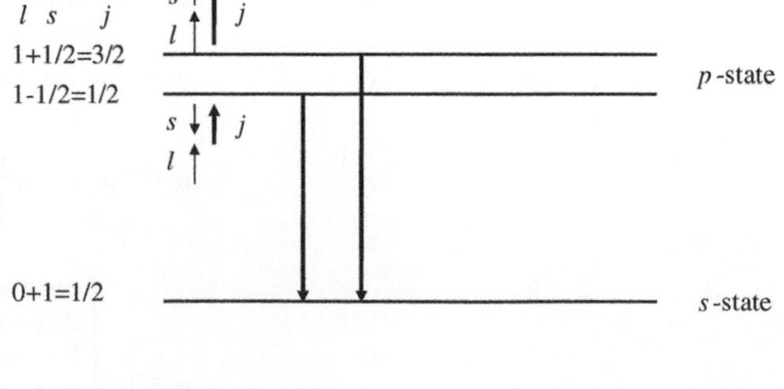

FIGURE 2.3

Spin-orbit interaction.

total angular momentum quantum number J is either integral or half-integral depending whether the number of electrons is even or odd respectively. The *spectroscopic notation* of a state characterized by the quantum numbers L, S, and J is as follows:

$$^{2S+1}L_J, \tag{2.287}$$

where the quantity $2S+1$ is known as the *multiplicity*, and determines the numbers of different Js for a given value of L. If $L << S$, different values of J are $L+S, L+S-1, \ldots, L-S$, meaning that there are $2S+1$ possible values of J. If, on the other hand, $L < S$, then the possible values of J are $L+S, L+S-1, \ldots, |L-S|$, meaning only $2L+1$ different values for J exist. The states in which $L = 0, 1, 2, 3, 4, 5, 6, 7, 8, \ldots$ are traditionally called S, P, D, F, G, H, I, K, M, ... respectively. The states with multiplicity $2S+1 = 1, 2, 3, 4, 5$, and 6 are typically called singlet, doublet, triplet, quartet, quintet, and sextet states respectively.

Example. Let us observe the hydrogen-like atom whose nucleus charge is Ze and which contains only one electron. The wave function of the electron in this atom is given by $\psi(r) = Ce^{-r/a}$, $a = a_H/Z$, $a_H = 0.05$ nm. If the nucleolus number is $A = 173$ and $Z = 70$, determine the probability that the electron is in the nucleolus. By assuming that the radius of the nucleolus is $R = 1.2 \times A^{1/3}$ fm, determine the probability that the electron is in the region $x, y, z > 0$.

We first determine the normalization constant from

$$\iiint \psi^* \psi d^3 r = C^2 \underbrace{\int_0^\infty r^2 e^{-2r/a} dr}_{(a/2)^3 \Gamma(3)} \int_0^{2\pi} d\phi \int_0^\pi \sin\theta d\theta = 1$$

as $C = (\pi a^3)^{-1/2}$. The probability that the electron is found in the nucleolus of radius R is

$$P = C^2 \int_0^R r^2 e^{-2r/a} dr \int_0^{2\pi} d\phi \int_0^\pi \sin\theta d\theta = 4\pi C^2 \int_0^R r^2 e^{-2r/a} dr \approx \frac{4}{a^3} \int_0^R r^2 dr$$

$$= \frac{4}{3}\left(\frac{R}{a}\right)^3 = \frac{4}{3}\left(\frac{Zr_0}{a_H}\right)^3 A = 1.1 \times 10^{-6}.$$

Since the wave function is independent of θ and ϕ, the probability of the electron being in the region $x, y, z > 0$ is 1/8.

2.13 SUMMARY

This chapter has considered the quantum mechanics fundamentals. The basic topics of quantum mechanics have been covered, including state vectors, operators, density operators, and dynamics of a quantum system. Quantum measurements (Section 2.4), the uncertainty principle (Section 2.5), and change of basis (Section 2.7) have also been described. In addition, the harmonic oscillator (Section 2.9), orbital angular momentum (Section 2.10), spin-1/2 systems (Section 2.11), and hydrogen-like atoms (Section 2.12) have been considered. For a deeper understanding of this material, a set of problems to be used for self-study is given in Section 2.14. Students not interested in physical implementation, but only in either the quantum computing or quantum error correction aspects, can skip Sections 2.9–2.12.

2.14 PROBLEMS

1. Show that states $|R\rangle$ and $|L\rangle$, representing right- and left-handed polarized photons, can be represented in terms of $|x\rangle$ and $|y\rangle$ state vectors (representing x- and y-polarized photons) as follows:

$$|R\rangle = \frac{1}{\sqrt{2}}(|x\rangle + j|y\rangle) \qquad |L\rangle = \frac{1}{\sqrt{2}}(|x\rangle - j|y\rangle).$$

2. Let the states $|\theta\rangle$ and $|\theta_\perp\rangle$, representing photons linearly polarized along directions making an angle θ and $\theta_\perp$ with Ox and Oy respectively, be defined as

$$|\theta\rangle = \cos\theta|x\rangle + \sin\theta|y\rangle \qquad |\theta_\perp\rangle = -\sin\theta|x\rangle + \cos\theta|y\rangle.$$

How are $|R'\rangle = (|\theta\rangle + j|\theta_\perp\rangle)/\sqrt{2}$ and $|L'\rangle = (|\theta\rangle - j|\theta_\perp\rangle)/\sqrt{2}$ related to the $|R\rangle$ and $|L\rangle$ states?

3. Let the Hermitian operator Σ be defined as

$$\Sigma = P_{\mathrm{R}} - P_{\mathrm{L}},$$

where P_{R} and P_{L} are projection operators with respect to right- and left-handed polarized photons respectively. What is the action of Σ on $|R\rangle$ and $|L\rangle$? Determine the action of $\exp(-j\theta\Sigma)$ on these kets.

4. Using the matrix representation of Σ in the basis $\{|x\rangle, |y\rangle\}$, show that $\Sigma^2 = \mathrm{I}$ and express $\exp(-j\theta\Sigma)$ in terms of Σ. What is the action of $\exp(-j\theta\Sigma)$ on $|x\rangle$ and $|y\rangle$?

5. **(a)** Write down the basis corresponding to 45° and 135° polarizations.
   (b) Write down the basis that is neither plane nor circularly polarized.

6. **(a)** Show that the matrix $|\phi\rangle\langle\phi|$ is Hermitian.
   (b) Show that the photon spin operator S is Hermitian.

7. Show that the transformation matrix from one base to another is unitary.

8. **(a)** Determine the transformation matrix from the x, y basis to the R, L basis.
   (b) Determine the transformation matrix from the R, L basis to the basis devised in Problem 5b.
   (c) Determine the transformation matrix from the x, y basis to the basis devised in Problem 5b and show that it can be written as a product of the matrix determined in Problems 8a and 8b.

(d) Show that the product of transformation matrix from basis 1 to basis 2 with the transformation matrix from basis 2 to basis 3 is the same as the transformation matrix from basis 1 to basis 3.

9. The probability that a photon in state $|\psi\rangle$ passes through an x-polaroid is the average value of a physical observable called "x-polarizedness". Write down the operator P_x corresponding to this observable. What are its eigenstates? Write down its expression in terms of its eigenvalues and eigenstates. Verify that the probability that a photon in state $|\psi\rangle$ passes through the x-polaroid is $\langle\psi|P_x|\psi\rangle$.

10. Consider a system of two photons, both traveling along the z-axis. The photons are entangled, and their state vector is

$$|\psi\rangle = \frac{1}{\sqrt{2}}(|xx\rangle + |yy\rangle), \tag{2.P10}$$

where $|xx\rangle$ denotes the state in which both photons are polarized along the x-axis, $|xy\rangle$ is the state in which the first photon is polarized along x and the second along y, etc.

(a) Show that $|xx\rangle$, $|xy\rangle$, $|yx\rangle$, $|yy\rangle$ form an orthonormal basis.

(b) Let the axes x', y' be rotated from x, y by an angle θ. Show that the state $|\psi\rangle$ is rotationally invariant by showing that it has the same form as Eq. (2.P10) in the $|x'x'\rangle$, $|x'y'\rangle$, $|y'x'\rangle$, $|y'y'\rangle$ basis.

(c) Write $|\psi\rangle$ in the circularly polarized basis: $|RR\rangle$, $|RL\rangle$, $|LR\rangle$, $|LL\rangle$. How does this transform under the rotation of part (b)?

(d) The two photons are absorbed by a piece of matter. What are all the possible values of the total angular momentum, L_z, transferred to the matter?

11. This problem relates to the properties of a density operator, defined as

$$\rho = \sum_i p_i |i\rangle\langle i|, \qquad \sum_i p_i = 1.$$

Prove that the most general density operator must satisfy the following properties:

(a) It must be Hermitian: $\rho^\dagger = \rho$.

(b) It must have unit trace: $\mathrm{Tr}\rho = 1$.

(c) It must be positive: $\langle\phi|\rho|\phi\rangle \geq 0 \ \ \forall \ \ |\phi\rangle$.

Show that the expectation value of operator M is given by $\langle M\rangle = \mathrm{Tr}(\rho M)$. Show that if $\rho^2 = \rho$ then all p_i are zero except one, that is equal to unity. Prove that condition $\rho^2 = \rho$ is the necessary and sufficient condition for a state to be pure. Also show that $\mathrm{Tr}\rho^2 = 1$ is a necessary and sufficient condition for the density operator to describe a pure state.

12. We would like to determine the most general form of density operator ρ for a qubit.

(a) Show that the most general Hermitian matrix of unit trace has the form:

$$\rho = \begin{pmatrix} a & c \\ c^* & 1-a \end{pmatrix},$$

where a is a real number and c is a complex number. Show that positivity of eigenvalues of ρ introduces the following constraint on matrix elements:

$$0 \le a(1-a) - |c|^2 \le 1/4.$$

Show that necessary and sufficient condition for the quantum state described by ρ to be represented by a vector in Hilbert space H is $a(1-a) = |c|^2$. Determine a and c for the matrix ρ describing the normalized state vector $|\psi\rangle = \lambda|0\rangle + \mu|1\rangle$, $|\lambda|^2 + |\mu|^2 = 1$, and show that in this case $a(1-a) = |c|^2$.

(b) Show that ρ can be written as a function of Bloch vector $\vec{b}$ as follows:

$$\rho = \frac{1}{2}(I + \vec{b}\cdot\vec{\sigma}) = \frac{1}{2}\begin{pmatrix} 1+b_z & b_x - jb_y \\ b_x + jb_y & 1-b_z \end{pmatrix}, \quad |\vec{b}| \le 1.$$

Show that a quantum state represented by a vector in Hilbert space H corresponds to the case $|\vec{b}|^2 = 1$. To interpret the vector $\vec{b}$ physically, we calculate the expectation of $\vec{\sigma}$: $\langle\sigma_i\rangle = \text{Tr}(\rho\sigma_i)$. Show that the expectation value of $\vec{\sigma}$ is $\vec{b}$.

13. Let the composite system S be composed of two subsystems A and B, and let A and B themselves be composite systems with corresponding Hilbert spaces H_A and H_B respectively. The orthonormal bases for H_A and H_B are given as $\{|e_i: i = 1,\ldots,N\rangle\}$ and $\{|f_j: j = 1,\ldots,M\rangle\}$ respectively.

(a) Let $|\psi\rangle = |\phi\rangle \otimes |\chi\rangle$, where $|\phi\rangle$ and $|\chi\rangle$ are normalized states in H_A and H_B respectively. Show that $E(|\psi\rangle) = 0$.

(b) Let $|ij\rangle$ be the computational basis for $H_E \otimes H_F$ and let

$$|\psi\rangle = \frac{1}{\sqrt{N}}\sum_i |ii\rangle.$$

Show that $E(|\psi\rangle) = \log(N)$.

14. Show that the operator

$$\frac{1}{2}(I + \vec{\sigma_A}\cdot\vec{\sigma_B})$$

permutes the qubits A and B ($\vec{\sigma_A}\cdot\vec{\sigma_B}$ stands for both scalar and tensor product). This operator is also known as the SWAP operator. Its matrix representation in the basis $\{|00\rangle, |01\rangle, |10\rangle, |11\rangle\}$ is given by

$$U_{\text{SWAP}} = \begin{pmatrix} 1 & 0 & 0 & 0 \\ 0 & 0 & 1 & 0 \\ 0 & 1 & 0 & 0 \\ 0 & 0 & 0 & 1 \end{pmatrix}.$$

15. Show that for a two-level system the density operator can be written as

$$\rho = \frac{1}{2}(I + \vec{r} \cdot \vec{\sigma}), \quad \vec{\sigma} = [X \ \ Y \ \ Z],$$

where X, Y, Z are the Pauli operators, and $\vec{r} = [r_x \ \ r_y \ \ r_z]$ is the Bloch vector, whose components can be determined by

$$r_x = \mathrm{Tr}(\rho X), \quad r_y = \mathrm{Tr}(\rho Y), \quad r_z = \mathrm{Tr}(\rho Z).$$

16. Using the Pauli matrices prove:
(a) $(\boldsymbol{\sigma} \cdot \boldsymbol{A})(\boldsymbol{\sigma} \cdot \boldsymbol{B}) = (\boldsymbol{A} \cdot \boldsymbol{B})I + j\boldsymbol{\sigma} \cdot (\boldsymbol{A} \times \boldsymbol{B}); \boldsymbol{\sigma} = (X, Y, Z)$, $\boldsymbol{A} = (A_x, A_y, A_z), \boldsymbol{B} = (B_x, B_y, B_z)$.
(b) $\exp(-\frac{j\theta}{2}\boldsymbol{n} \cdot \boldsymbol{\sigma}) = \cos(\theta/2)I - j\boldsymbol{n} \cdot \boldsymbol{\sigma}\sin(\theta/2), \quad \boldsymbol{n} = (n_x, n_y, n_z)$.
(c) $(\boldsymbol{n} \cdot \boldsymbol{\sigma})^2 = I$.

17. Consider the rotation operator $U_R(\theta, \hat{u}) = \exp\left(\frac{j\theta}{2}\hat{u} \cdot \boldsymbol{\sigma}\right)$, $\boldsymbol{\sigma} = (X, Y, Z)$. By rotating the eigenkets of S_z determine the eigenkets of S_x and S_y in the standard basis.

18. Determine the normalized momentum distribution for a hydrogen atom electron in states $1s$, $2s$, and $2p$.

19. The parity operator is obtained by replacing $\boldsymbol{r}$ with $-\boldsymbol{r}$. How does this operator affect the wave function of an electron in a hydrogen atom?

20. Let the hydrogen atom be in a state specified by quantum numbers n and l. Determine the dispersion for the distance of the electron from the nucleus, defined as $\sqrt{\langle r^2 \rangle - \langle r \rangle^2}$.

21. For a two-dimensional hydrogen-like atom the Schrödinger equation in atomic units is given by $(-\nabla^2 - 2Z/r)\psi = E\psi$. Using the cylindrical coordinate system and the method of separation of variables, determine the differential equations for $R(r)$ and $\Phi(\phi)$.

22. Consider the Hamiltonian of a three-dimensional isotropic harmonic oscillator:

$$H = \frac{1}{2m}\left(p_x^2 + p_y^2 + p_z^2\right) + \frac{m\omega^2}{2}\left(x^2 + y^2 + z^2\right).$$

Represent the Hamiltonian in spherical coordinates and determine the eigenfunctions and energy eigenvalues.

23. In a quantum system with an angular momentum $l = 1$, the eigenkets of L_z are given by $|1\rangle, |0\rangle, |-1\rangle$ and the action of L_z on base kets is given by $L_z|1\rangle = \hbar, L_z|0\rangle = 0, L_z|-1\rangle = -\hbar|-1\rangle$. The Hamiltonian is given by $H = \omega_0(L_x^2 - L_y^2)/\hbar$, where ω_0 is a constant. Determine the matrix representation of H, the corresponding eigenvalues and eigenkets in the bases above.

24. Compute $\langle lm|L_x^2|lm\rangle$ and $\langle lm|L_xL_y|lm\rangle$ in the standard angular momentum basis.

25. Consider a particle in a central potential. Given that $|lm\rangle$ is an eigenket of L_z and L^2 determine the sum $\Delta L_x^2 + \Delta L_y^2$. For which values of l and m does this sum vanish?

26. The classical equation of motion for a particle with a mass m and charge q in the presence of electric field $\boldsymbol{E}$ and magnetic field $\boldsymbol{B}$ is given by

$$m\boldsymbol{a} = q\boldsymbol{E} + \frac{q}{c}\boldsymbol{v} \times \boldsymbol{B}, \quad \boldsymbol{v} = \frac{d\boldsymbol{r}}{dt} = \dot{\boldsymbol{r}}, \quad \boldsymbol{a} = \frac{d\boldsymbol{v}}{dt} = \ddot{\boldsymbol{r}}.$$

$\boldsymbol{E}$ and $\boldsymbol{B}$ must satisfy Maxwell's equation so that it is possible to find the vector potential $\boldsymbol{A}(\boldsymbol{r},t)$ and scalar potential $\phi(\boldsymbol{r},t)$ such that

$$\boldsymbol{E} = -\nabla\phi - \frac{1}{c}\frac{\partial \boldsymbol{A}}{\partial t} \qquad \boldsymbol{B} = \nabla \times \boldsymbol{A}.$$

By using the Hamilton equations $\dot{\boldsymbol{r}} = \partial H/\partial \boldsymbol{p}$ and $\boldsymbol{p} = -\partial H/\partial \boldsymbol{r}$, show that the following Hamiltonian leads to the equation of motion:

$$H = \frac{1}{2m}\left(\boldsymbol{p} - \frac{q}{c}\boldsymbol{A}\right) \cdot \left(\boldsymbol{p} - \frac{q}{c}\boldsymbol{A}\right) + q\phi.$$

References

[1] J.J. Sakurai, Modern Quantum Mechanics, Addison-Wesley, 1994.

[2] G. Baym, Lectures on Quantum Mechanics, Westview Press, 1990.

[3] P.A.M. Dirac, Quantum Mechanics, fourth ed., Oxford University Press, London, 1958.

[4] F. Gaitan, Quantum Error Correction and Fault Tolerant Quantum Computing, CRC Press, 2008.

[5] C.W. Helstrom, Quantum Detection and Estimation Theory, Academic, New York, 1976.

[6] Y. Peleg, R. Pnini, E. Zaarur, Schaum's Outline of Theory and Problems of Quantum Mechanics, McGraw-Hill, New York, 1998.

[7] D. McMahon, Quantum Mechanics Demystified, McGraw-Hill, New York, 2006.

[8] D. McMahon, Quantum Field Theory Demystified, McGraw-Hill, New York, 2008.

[9] C.W. Helstrom, J.W.S. Liu, J.P. Gordon, Quantum Mechanical Communication Theory, Proc. IEEE 58 (October 1970) 1578–1598.

[10] M.A. Neilsen, I.L. Chuang, Quantum Computation and Quantum Information, Cambridge University Press, Cambridge, 2000.

[11] S. Fleming, PHYS 570: Quantum Mechanics (Lecture Notes), University of Arizona, 2008.

[12] G.R. Fowles, Introduction to Modern Optics, Dover Publications, New York, 1989.

[13] M.O. Scully, M.S. Zubairy, Quantum Optics, Cambridge University Press, Cambridge, 1997.

[14] M. Le Bellac, A Short Introduction to Quantum Information and Quantum Computation, Cambridge University Press, Cambridge, 2006.

[15] C.C. Pinter, A Book of Abstract Algebra, Dover Publications, New York, 2010.

[16] D. Griffiths, Introduction to Quantum Mechanics, Prentice-Hall, Englewood Cliffs, NJ, 1995.

[17] N. Zettilli, Quantum Mechanics, Concepts and Applications, John Wiley, New York, 2001.
[18] J. von Neumann, Mathematical Foundations of Quantum Mechanics, Princeton University Press, Princeton, NJ, 1955.
[19] J.M. Jauch, Foundations of Quantum Mechanics, Addison-Wesley, Reading, MA, 1968.
[20] A. Peres, Quantum Theory: Concepts and Methods, Kluwer, Boston, 1995.
[21] A. Goswami, Quantum Mechanics, McGraw-Hill, New York, 1996.
[22] M. Jammer, Philosophy of Quantum Mechanics, John Wiley, New York, 1974.
[23] R. McWeeny, Quantum Mechanics, Principles and Formalism, Dover Publications, Mineola, NY, 2003.
[24] H. Weyl, Theory of Groups and Quantum Mechanics, Dover Publications, New York, 1950.
[25] R.L. Liboff, Introductory Quantum Mechanics, Addison-Wesley, Reading, MA, 1997.

This page intentionally left blank

CHAPTER

Quantum Circuits and Quantum Information Processing Fundamentals

3

CHAPTER OUTLINE

In this chapter, basic single-qubit, two-qubit, and many-qubit gates [1−13] are described. The fundamentals of quantum information processing are provided as well. Various sets of universal quantum gates are also described. Further, the Bloch sphere representation of a single qubit is provided, the Gottesman−Knill theorem is formulated, and the Bell state preparation circuit and the quantum relay are described. The basics concepts of quantum teleportation and quantum computation are introduced. Also, quantum measurement circuits are provided, and the principles of deferred measurement and implicit measurement are formulated. In addition to the topics mentioned above, the concepts of quantum parallelism and entanglement are introduced as well. Finally, the no-cloning theorem is provided and its consequences are discussed.

3.1 SINGLE-QUBIT OPERATIONS

A single qubit is a state ket (vector),

$$|\psi\rangle = a|0\rangle + b|1\rangle, \tag{3.1}$$

parameterized by two complex numbers a and b satisfying the normalization condition $|a|^2 + |b|^2 = 1$. The parameters a and b are known as (quantum) *probability amplitudes* as their magnitudes squared, $|a|^2$ and $|b|^2$, represent probabilities of finding the state $|\psi\rangle$ in basis states $|0\rangle$ and $|1\rangle$ respectively. The computational basis (CB) is represented in matrix form as follows:

$$|0\rangle \doteq \begin{pmatrix} 1 \\ 0 \end{pmatrix}, \quad |1\rangle \doteq \begin{pmatrix} 0 \\ 1 \end{pmatrix}. \tag{3.2}$$

Quantum Information Processing and Quantum Error Correction. DOI: 10.1016/B978-0-12-385491-9.00003-4

Another important basis is the *diagonal basis* $\{|+\rangle, |-\rangle\}$, very often denoted as $\{|\nearrow\rangle, |\searrow\rangle\}$, which is related to the CB by

$$|+\rangle = |\nearrow\rangle = \frac{1}{\sqrt{2}}(|0\rangle + |1\rangle), \quad |-\rangle = |\searrow\rangle = \frac{1}{\sqrt{2}}(|0\rangle - |1\rangle). \tag{3.3}$$

The circular basis $\{|R\rangle, |L\rangle\}$, is related to CB as follows:

$$|R\rangle = \frac{1}{\sqrt{2}}(|0\rangle + j|1\rangle), \quad |L\rangle = \frac{1}{\sqrt{2}}(|0\rangle - j|1\rangle). \tag{3.4}$$

The most important single-qubit quantum gates are Pauli operators X, Y, Z; the Hadamard gate H, the phase shift gate S (sometimes denoted by P), and the $\pi/8$ (or T) gate. The Pauli operators can be represented in matrix form in the CB by

$$X = \sigma_x = \sigma_1 \doteq \begin{bmatrix} 0 & 1 \\ 1 & 0 \end{bmatrix}, \quad Y = \sigma_y = \sigma_2 \doteq \begin{bmatrix} 0 & -j \\ j & 0 \end{bmatrix},$$
$$Z = \sigma_z = \sigma_3 \doteq \begin{bmatrix} 1 & 0 \\ 0 & -1 \end{bmatrix}, \quad I = \sigma_0 \doteq \begin{bmatrix} 1 & 0 \\ 0 & 1 \end{bmatrix}. \tag{3.5}$$

Based on their representation in (3.5), the following properties of Pauli operators (gates) can easily be proved:

$$\begin{aligned} &X(a|0\rangle + b|1\rangle) = a|1\rangle + b|0\rangle, \quad Y(a|0\rangle + b|1\rangle) = j(a|1\rangle - b|0\rangle), \\ &Z(a|0\rangle + b|1\rangle) = a|0\rangle - b|1\rangle \\ &X^2 = I, \quad Y^2 = I, \quad Z^2 = I \\ &XY = jZ \quad YX = -jZ. \end{aligned} \tag{3.6}$$

The Hadamard gate H, phase shift gate S, and T gate in the CB are represented by

$$H \doteq \frac{1}{\sqrt{2}}\begin{bmatrix} 1 & 1 \\ 1 & -1 \end{bmatrix}, \quad S \doteq \begin{bmatrix} 1 & 0 \\ 0 & j \end{bmatrix}, \quad T \doteq \begin{bmatrix} 1 & 0 \\ 0 & e^{j\pi/4} \end{bmatrix} = e^{j\pi/8}\begin{bmatrix} e^{-j\pi/8} & 0 \\ 0 & e^{j\pi/8} \end{bmatrix} \tag{3.7}$$

respectively. The action of the Hadamard gate on a single qubit $|\psi\rangle$ is to perform the following transformation:

$$\begin{aligned} H(a|0\rangle + b|1\rangle) &= a\frac{1}{\sqrt{2}}(|0\rangle + |1\rangle) + b\frac{1}{\sqrt{2}}(|0\rangle - |1\rangle) \\ &= \frac{1}{\sqrt{2}}[(a+b)|0\rangle + (a-b)|1\rangle]. \end{aligned} \tag{3.8}$$

The Hadamard gate therefore interchanges the CB and diagonal basis: $|0\rangle \leftrightarrow |+\rangle$ and $|1\rangle \leftrightarrow |-\rangle$.

A single qubit in the state $a|0\rangle + b|1\rangle$ can be visualized as the point (θ,ϕ) on a unit sphere (*Bloch or Poincaré sphere*) where $a = \cos(\theta/2)$ and $b = e^{j\phi}\sin(\theta/2)$; the corresponding *Bloch vector* coordinates are: $(\cos\phi\sin\theta, \sin\phi\sin\theta, \cos\theta)$. A Bloch sphere representation of a single qubit (or the Poincaré sphere for the photon) is shown in Figure 3.1. This spinor representation is related to the CB by

$$|\psi(\theta,\phi)\rangle = \cos(\theta/2)|0\rangle + e^{j\phi}\sin(\theta/2)|1\rangle \doteq \begin{pmatrix} \cos(\theta/2) \\ e^{j\phi}\sin(\theta/2) \end{pmatrix}, \tag{3.9}$$

where $0 \le \theta \le \pi$ and $0 \le \phi < 2\pi$. The north and south poles correspond to computational $|0\rangle$ ($|x\rangle$-polarization) and $|1\rangle$ ($|y\rangle$-polarization) basis kets respectively. The pure qubit states lie on the Bloch sphere, while the mixed qubit states lie in the interior of the Block sphere. The maximally mixed state ($I/2$) lies in the center of the Bloch sphere. The orthogonal states are antipodal. From Figure 3.1 we see that the CB, diagonal basis, and circular bases are 90° apart from each other, and we often say that these three bases are mutually *conjugate bases*. These bases are used as three pairs of signal states for the six-state quantum key distribution (QKD) protocol. Another important basis used in QKD and for eavesdropping is the *Breidbart basis* given by $\{\cos(\pi/8)|0\rangle + \sin(\pi/8)|1\rangle, -\sin(\pi/8)|0\rangle + \cos(\pi/8)|1\rangle\}$.

The common single-qubit gates discussed above and their matrix representation are summarized in Figure 3.2.

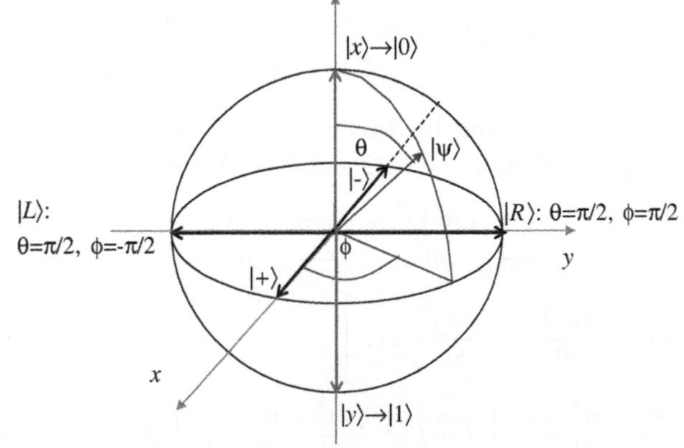

FIGURE 3.1

Block (Poincaré) sphere representation of a single qubit.

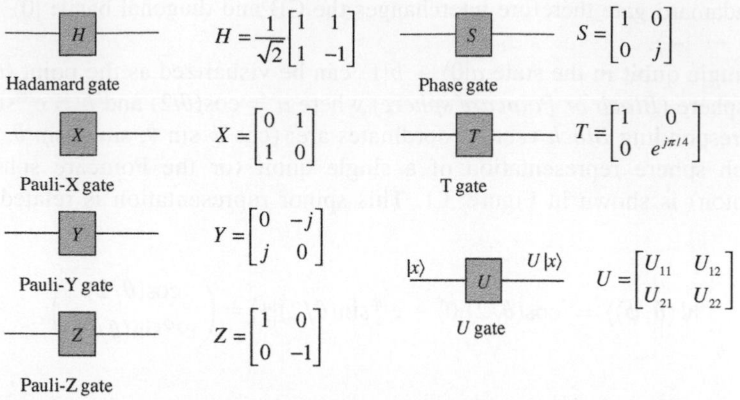

FIGURE 3.2

Common single-qubit gates and their matrix representations.

The *rotation operators* for θ about the *x*-, *y*-, and *z*-axes, denoted as $R_x(\theta)$, $R_y(\theta)$, and $R_z(\theta)$ respectively, can be represented in terms of Pauli operators as follows:

$$R_x(\theta) = e^{-j\theta X/2} = \cos\frac{\theta}{2}I - j\sin\frac{\theta}{2}X = \begin{bmatrix} \cos\frac{\theta}{2} & -j\sin\frac{\theta}{2} \\ -j\sin\frac{\theta}{2} & \cos\frac{\theta}{2} \end{bmatrix},$$

$$R_y(\theta) = e^{-j\theta Y/2} = \cos\frac{\theta}{2}I - j\sin\frac{\theta}{2}Y = \begin{bmatrix} \cos\frac{\theta}{2} & -\sin\frac{\theta}{2} \\ \sin\frac{\theta}{2} & \cos\frac{\theta}{2} \end{bmatrix},$$

$$R_z(\theta) = e^{-j\theta Z/2} = \cos\frac{\theta}{2}I - j\sin\frac{\theta}{2}Z = \begin{bmatrix} e^{-j\frac{\theta}{2}} & 0 \\ 0 & e^{j\frac{\theta}{2}} \end{bmatrix}. \tag{3.10}$$

Equations (3.10) can be proved by Taylor series expansion. Namely, the Taylor expansion of $\exp(j\alpha A)$, where $\alpha = -\theta/2$ and $A \in \{X,Y,Z\}$, is given by

$$\begin{aligned} e^{j\alpha A} &= \sum_{n=0}^{\infty}\frac{(j\alpha A)^n}{n!} = \sum_{n-\text{even}}\frac{(j\alpha A)^n}{n!} + \sum_{n-\text{odd}}\frac{(j\alpha A)^n}{n!} = 1 - \frac{(\alpha A)^2}{2!} + \frac{(\alpha A)^4}{4!} - \dots \\ &+ j\left[(\alpha A) - \frac{(\alpha A)^3}{3!} + \frac{(\alpha A)^5}{5!} - \dots\right] \\ &\overset{A^2=I}{=} \underbrace{\left[1 - \frac{\alpha^2}{2!} + \frac{\alpha^4}{4!} - \dots\right]}_{\cos(\alpha)} I + j\underbrace{\left[\alpha - \frac{\alpha^3}{3!} + \frac{\alpha^5}{5!} - \dots\right]}_{\sin(\alpha)} A = \cos(\alpha)I + j\sin(\alpha)A. \end{aligned} \tag{3.11}$$

By substituting $\alpha = -\theta/2$ and $A = X,Y,Z$ we obtain equations (3.10).

The *rotation by* θ around unit vector $\hat{\boldsymbol{n}} = (n_x, n_y, n_z)$ can be described by

$$R_{\hat{n}}(\theta) = \exp(-j\theta\hat{\boldsymbol{n}}\cdot\boldsymbol{\sigma}/2) = \cos\left(\frac{\theta}{2}\right)I - j\sin\left(\frac{\theta}{2}\right)(n_xX + n_yY + n_zZ)$$

$$\hat{\boldsymbol{n}} = (n_x, n_y, n_z), \quad \boldsymbol{\sigma} = (X, Y, Z). \tag{3.12}$$

Notice that Eq. (3.12) is just a special case of (3.11), obtained by setting $\alpha = -\theta/2$, $A = \hat{\boldsymbol{n}}\cdot\boldsymbol{\sigma}$. Using this substitution, since (3.11) is valid for $A^2 = I$, clearly, it must be proved that the following property holds as well:

$$(\hat{\boldsymbol{n}}\cdot\boldsymbol{\sigma})^2 = I, \tag{3.13}$$

which is left as a homework problem. By performing the matrix multiplication $R_z(\beta)R_y(\gamma)R_z(\delta)$ we obtain:

$$R_z(\beta)R_y(\gamma)R_z(\delta) = \begin{bmatrix} e^{-j\frac{\beta}{2}} & 0 \\ 0 & e^{j\frac{\beta}{2}} \end{bmatrix} \begin{bmatrix} \cos\frac{\gamma}{2} & -\sin\frac{\gamma}{2} \\ \sin\frac{\gamma}{2} & \cos\frac{\gamma}{2} \end{bmatrix} \begin{bmatrix} e^{-j\frac{\delta}{2}} & 0 \\ 0 & e^{j\frac{\delta}{2}} \end{bmatrix}$$

$$= \begin{bmatrix} e^{j(-\beta/2-\delta/2)}\cos\frac{\gamma}{2} & -e^{j(-\beta/2+\delta/2)}\sin\frac{\gamma}{2} \\ e^{j(+\beta/2-\delta/2)}\sin\frac{\gamma}{2} & e^{j(+\beta/2+\delta/2)}\cos\frac{\gamma}{2} \end{bmatrix}, \tag{3.14}$$

which is clearly a unitary matrix. The unitarity will not be changed if we insert a global phase factor $\exp(j\alpha)$. Therefore, an arbitrary single-qubit operation can be decomposed as

$$U = \begin{bmatrix} e^{j(\alpha-\beta/2-\delta/2)}\cos\frac{\gamma}{2} & -e^{j(\alpha-\beta/2+\delta/2)}\sin\frac{\gamma}{2} \\ e^{j(\alpha+\beta/2-\delta/2)}\sin\frac{\gamma}{2} & e^{j(\alpha+\beta/2+\delta/2)}\cos\frac{\gamma}{2} \end{bmatrix} = e^{j\alpha}R_z(\beta)R_y(\gamma)R_z(\delta), \tag{3.15}$$

and this decomposition is sometimes called the single-qubit *Z–Y decomposition theorem* [1]. By simple matrix multiplication we can prove the following important *circuit identities*:

$$HXH = Z, \quad HYH = -Y, \quad HZH = X, \quad HTH = R_x(\pi/4), \tag{3.16}$$

where by $R_x(\theta)$ we denote the rotation operator for θ about the x-axis.

The *quantum state purity* is specified by the density operator ρ (in some books this is called the state operator [3]), which was defined in Chapter 2, as follows:

$$\mathscr{P}(\rho) = \mathrm{Tr}\rho^2, \quad 1/D \le \mathscr{P}(\rho) \le 1, \tag{3.17}$$

where D is the dimensionality of the corresponding Hilbert space H. If $\mathscr{P}(\rho) = 1$ we say that the state is *pure*; on the other hand, if $\mathscr{P}(\rho) < 1$ we say that the quantum state is *mixed*. Alternatively, instead of purity the term *mixedness* may be used, defined as $\mathscr{M}(\rho) = 1 - \mathscr{P}(\rho)$, which represents the complement of purity. What is interesting is the fact that the purity does not change under *conjugation mapping*: $\rho \rightarrow U\rho U^\dagger$, where U is a unitary operator. The is also true for the time-evolution operator $U = \exp[-(j/\hbar)H(t - t_0)]$, where H is the Hamiltonian operator. We have shown in the previous chapter that for pure states $\rho^2 = \rho$, indicating that the density operators serve as projection operators for pure states. The projector operator is defined as

$$P|\psi_i\rangle \equiv |\psi_i\rangle\langle\psi_i|, \tag{3.18}$$

where $|\psi_i\rangle$ is a pure state. Let a finite state of projectors $\{P|\psi_i\rangle\}$ be given. Any state ρ' can be written in terms of projectors as follows:

$$\rho' = \sum_i p_i P|\psi_i\rangle;\ 0 < p_i < 1,\ \sum_i p_i = 1. \tag{3.19}$$

A quantum state is said to be in a (partially) *coherent superposition* of states $|a_i\rangle$, where $|a_i\rangle$ are eigenkets of Hermitian operator A, if the corresponding density matrix is not diagonal. In discussing Figure 3.1, we said that mixed qubit states lie inside of the Bloch sphere and can be represented as weighted convex combinations of pure states. We can use *Stokes parameters* to represent the density operator:

$$\rho = \frac{1}{2}\sum_{i=0}^{3} S_i\sigma_i = \frac{1}{2}(S_0 I + S_1\sigma_1 + S_2\sigma_2 + S_3\sigma_3) = \frac{1}{2}\begin{pmatrix} S_0 + S_3 & S_1 - jS_2 \\ S_1 - jS_2 & S_0 - S_3 \end{pmatrix}, \tag{3.20}$$

where S_i $(i = 0,1,2,3)$ are Stokes parameters that can be obtained as

$$S_i = \mathrm{Tr}(\rho\sigma_i);\ i = 0, 1, 2, 3, \tag{3.21}$$

as we have shown in the previous chapter. The Euclidean distance with respect to the origin is given by

$$r = \sqrt{S_1^2 + S_2^2 + S_3^2}, \tag{3.22}$$

and the corresponding three-component vector $\mathbf{S} = (S_1, S_2, S_3)$ is known as the Stokes (Bloch) vector. In optical applications, the vector $\mathbf{S} = (S_1, S_2, S_3)$ describes the polarization state of a photon. The ratio r/S_0 is known as the *degree of polarization*. When the state is normalized $S_0 = 1$ represents the total quantum probability. Going back to Figure 3.1, we conclude that Stokes coordinates are given by $S_0 = 1$, $S_1 = \sin\theta\cos\phi$, $S_2 = \sin\theta\sin\phi$, and $S_3 = \cos\theta$.

3.2 TWO-QUBIT OPERATIONS

The mathematical foundation for two-qubit states relies on the tensor product concept. The first qubit occupies the Hilbert space H_A with orthonormal basis $\{|0_A\rangle, |1_A\rangle\}$ and the second qubit occupies the Hilbert space H_B with orthonormal basis

$\{|0_B\rangle,|1_B\rangle\}$. The bipartite state representing the fact that subsystem A is in the state $|m_A\rangle$ ($m=0,1$), while at the same time the subsystem B is in the state $|n_A\rangle$ ($n=0,1$) can be denoted as $|m_A m_B\rangle = |m_A\rangle \otimes |m_B\rangle$, where $\otimes$ stands for tensor product, which is often omitted for brevity. The qubits from subsystems A and B can be represented as

$$\begin{aligned} |\psi_A\rangle &= \alpha_A|0_A\rangle + \beta_A|1_A\rangle, \quad |\alpha_A|^2 + |\beta_A|^2 = 1; \\ |\psi_B\rangle &= \alpha_B|0_B\rangle + \beta_B|1_B\rangle, \quad |\alpha_B|^2 + |\beta_B|^2 = 1 \end{aligned} \tag{3.23}$$

respectively. The bipartite state will then be given by

$$\begin{aligned} |\psi_A\psi_B\rangle &= |\psi_A\rangle \otimes |\psi_B\rangle = (\alpha_A|0_A\rangle + \beta_A|1_A\rangle) \otimes (\alpha_B|0_B\rangle + \beta_B|1_B\rangle) \\ &= \alpha_A\alpha_B|0_A0_B\rangle + \alpha_A\beta_B|0_A1_B\rangle + \beta_A\alpha_B|1_A0_B\rangle + \beta_A\beta_B|1_A1_B\rangle. \end{aligned} \tag{3.24}$$

In Eq. (3.24) we defined the space $H_A \otimes H_B$ as the tensor product of subspaces H_A and H_B. Unfortunately, the state (3.24) is not the most general state of $H_A \otimes H_B$. Interestingly enough, states of the form (3.24) do not even create the subspace of $H_A \otimes H_B$. The most general state in space $H_A \otimes H_B$ has the form:

$$|\psi\rangle = \alpha_{00}|0_A0_B\rangle + \alpha_{01}|0_A1_B\rangle + \alpha_{10}|1_A0_B\rangle + \alpha_{11}|1_A1_B\rangle. \tag{3.25}$$

The state given by (3.25) has the form given by (3.24) only when the following condition, which is necessary and sufficient, is satisfied:

$$\alpha_{00}\alpha_{11} = \alpha_{01}\alpha_{10}, \tag{3.26}$$

which *a priori* has no reason to be satisfied. The two-qubit state $|\psi\rangle$ that cannot be written in the form $|\psi_A\rangle \otimes |\psi_B\rangle$ is called an *entangled state*. Important examples of two-qubit states are *Bell states*, also known as Einstein–Podolsky–Rosen (EPR) states (pairs):

$$\begin{aligned} |B_{00}\rangle &= \frac{1}{\sqrt{2}}(|00\rangle + |11\rangle), \quad |B_{01}\rangle = \frac{1}{\sqrt{2}}(|01\rangle + |10\rangle), \\ |B_{10}\rangle &= \frac{1}{\sqrt{2}}(|00\rangle - |11\rangle), \quad |B_{11}\rangle = \frac{1}{\sqrt{2}}(|01\rangle - |10\rangle). \end{aligned} \tag{3.27}$$

The Bell states are *maximally entangled states* as if we disregard the information on one qubit and perform the measurement on the second qubit, the result of the measurement will be purely random. Notice that in Eq. (3.27) we omitted the subscripts, which is commonly done in quantum information processing. The fundamental property of an entangled state is that the subsystem A (B) cannot be in a definite quantum state $|\psi_A\rangle$ ($|\psi_B\rangle$). Let us observe the Bell state $|B_{01}\rangle$, and let M be the physical property of subsystem A, which in space H_AH_B is denoted by $M \otimes I_B$. The expectation value of M is given by

$$\begin{aligned} \langle M\rangle &= \langle B_{01}|M|B_{01}\rangle = \frac{1}{2}(\langle 01| + \langle 10|)M(|01\rangle + |10\rangle) \\ &= \frac{1}{2}(\langle 01| + \langle 10|)\,[(M|0\rangle)|1\rangle + (M|1\rangle)|0\rangle] = \frac{1}{2}[\langle 0_A|M|0_A\rangle + \langle 1_A|M|1_A\rangle], \end{aligned} \tag{3.28}$$

where we used $\langle 0_B|0_B\rangle = \langle 1_B|1_B\rangle = 1$ and $\langle 0_B|1_B\rangle = \langle 1_B|0_B\rangle = 0$. Now we have to prove that there is no state $|\psi_A\rangle$ such that

$$\langle \psi_A|M|\psi_A\rangle = \langle B_{01}|M|B_{01}\rangle. \tag{3.29}$$

Let us now evaluate the expected value of M with respect to state $|\psi_A\rangle$:

$$\begin{aligned}\langle \psi_A|M|\psi_A\rangle &= (\alpha_A^*\langle 0_A| + \beta_A^*\langle 1_A|)M(\alpha_A|0_A\rangle + \beta_A|1_A\rangle) \\ &= |\alpha_A|^2\langle 0_A|M|0_A\rangle + \alpha_A^*\beta_A\langle 0_A|M|1_A\rangle + \alpha_A\beta_A^*\langle 1_A|M|0_A\rangle \\ &\quad + |\beta_A|^2\langle 1_A|M|1_A\rangle.\end{aligned} \tag{3.30}$$

For (3.30) to be in the same form as (3.28), it is required that $|\alpha_A|^2 = |\beta_A|^2 = 1/2$, but the terms $\alpha_A^*\beta_A$ and $\alpha_A\beta_A^*$ will not vanish. So, it appears that Eq. (3.28) is an incoherent mixture of 50% of state $|0_A\rangle$ and 50% of state $|1_A\rangle$, not a linear superposition. In conclusion, in general, it is not possible to represent the subsystem of an entangled state by a state ket. An example of an incoherent mixture is unpolarized (natural) light, which is a mixture of 50% of light along the x-axis and 50% of light along the y-axis.

An n-qubit register has 2^n mutually orthogonal states, which in the CB can be represented by $|x_1x_2\ldots x_n\rangle$ $(x_i = 0,1)$. Any state of the quantum register can be specified by 2^n complex amplitudes c_x $(x \equiv x_1x_2\ldots x_n)$ by

$$|\Psi\rangle = \sum_x c_x|x\rangle, \quad \sum_x |c_x|^2 = 1. \tag{3.31}$$

Even for moderate n (close to 1000), the number of complex amplitudes c_x specifying the state of the quantum register is enormous. A classical computer will require enormous resources to store and manipulate such large amount of complex coefficients. The n-qubit $(n > 2)$ analogs of Bell states will be now briefly reviewed. One popular family of entangled multiqubit states is Greenberger–Horne–Zeilinger (GHZ) states:

$$|\text{GHZ}\rangle = \frac{1}{\sqrt{2}}(|00\ldots 0\rangle \pm |11\ldots 1\rangle). \tag{3.32}$$

Another popular family of multiqubit entangled states is known as W-states:

$$|\text{W}\rangle = \frac{1}{\sqrt{N}}(|00\ldots 01\rangle + |00\ldots 10\rangle + \ldots + |01\ldots 00\rangle + |10\ldots 00\rangle). \tag{3.33}$$

The W-state of n qubits represents a superposition of single-weighted CB states, each occurring with a probability amplitude of $N^{-1/2}$.

In the rest of this section, we describe several important two-qubit gates. We start our description with the controlled-NOT (CNOT) gate. The CNOT gate is a quantum gate, with circuit representation and operating principle shown in Figure 3.3, and has two input qubits, known as *control qubit* $|c\rangle$ and *target qubit* $|t\rangle$.

The action of the CNOT gate can be described by $|c,t\rangle \to |c,t \oplus c\rangle$. Therefore, if the control qubit is set to 1, then the target qubit is flipped. The matrix representation of the CNOT gate is given by

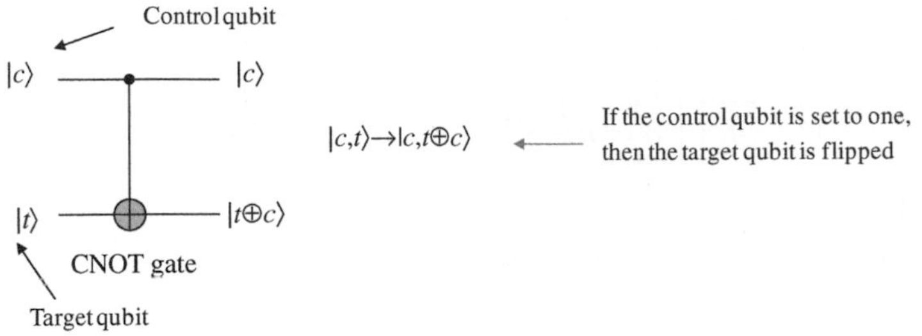

FIGURE 3.3

The circuit representation and operating principle of a CNOT gate.

$$\text{CNOT} = \begin{bmatrix} 1 & 0 & 0 & 0 \\ 0 & 1 & 0 & 0 \\ 0 & 0 & 0 & 1 \\ 0 & 0 & 1 & 0 \end{bmatrix} = \begin{bmatrix} I & 0 \\ 0 & X \end{bmatrix}. \tag{3.34}$$

Another important two-qubit gate is the quantum swapping circuit (SWAP gate), whose circuit representation and operating principle is provided in Figure 3.4. The action of the SWAP gate is to interchange the states of two input qubits:

$$U_{\text{SWAP}}^{(12)}|\psi_1\rangle|\psi_2\rangle = |\psi_2\rangle|\psi_1\rangle. \tag{3.35}$$

In certain technologies, the controlled-Z gate, $C(Z)$, is easier to implement than the CNOT gate. Its operating principle and circuit representation is shown in Figure 3.5. The corresponding matrix representation is given by

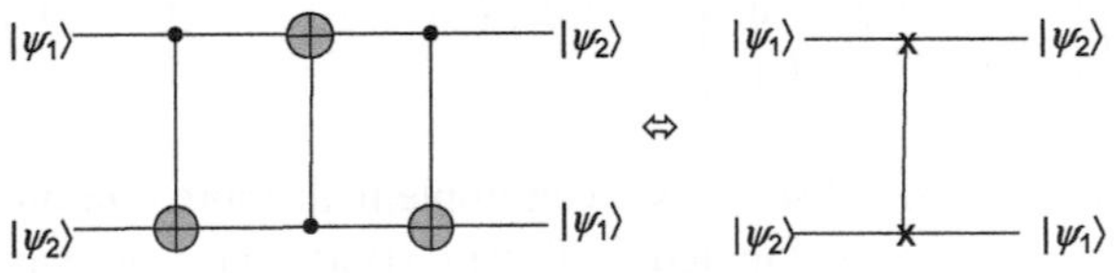

FIGURE 3.4

The operating principle and circuit representation of a SWAP gate.

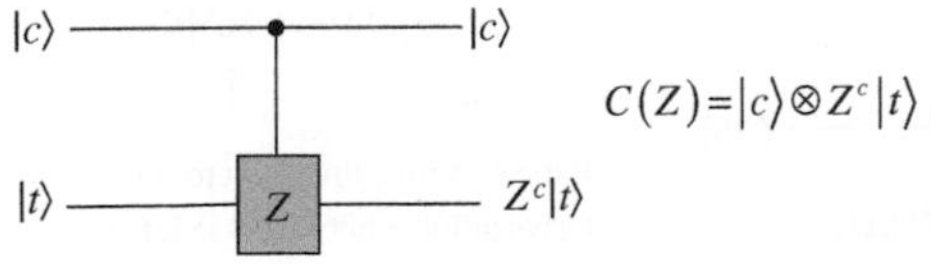

FIGURE 3.5

The operating principle and circuit representation of a controlled-Z gate, $C(Z)$.

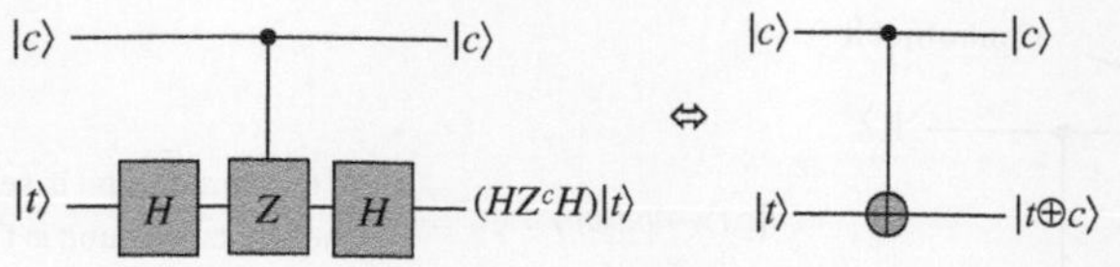

FIGURE 3.6

Implementation of a CNOT gate with the help of a controlled-Z gate.

$$C(Z) = \begin{bmatrix} 1 & 0 & 0 & 0 \\ 0 & 1 & 0 & 0 \\ 0 & 0 & 1 & 0 \\ 0 & 0 & 0 & -1 \end{bmatrix} = \begin{bmatrix} I & 0 \\ 0 & Z \end{bmatrix}. \tag{3.36}$$

We can establish the connection between the CNOT gate and the controlled-Z gate as shown in Figure 3.6. This equivalence can easily be proved as follows:

$$\begin{aligned} \text{CNOT}|c,t\rangle &= |c\rangle \otimes X^c|t\rangle = |c\rangle \otimes (HZH)^c|t\rangle = |c\rangle \otimes HZ^cH|t\rangle \\ &= I\otimes H\cdot C(Z)\cdot I\otimes H|c,t\rangle. \end{aligned} \tag{3.37}$$

In general, an arbitrary controlled-U gate can be implemented as shown in Figure 3.7. The action of the U gate can be described by $|c\rangle|t\rangle \to |c\rangle U^c|t\rangle$. Therefore, if the control qubit is set to 1, then U is applied to the target qubit, otherwise it is left unchanged. The corresponding matrix representation of the controlled-U gate is given by

$$C(U) = \begin{bmatrix} 1 & 0 & 0 & 0 \\ 0 & 1 & 0 & 0 \\ 0 & 0 & U_{00} & U_{01} \\ 0 & 0 & U_{10} & U_{11} \end{bmatrix} = \begin{bmatrix} I & 0 \\ 0 & U \end{bmatrix}, \quad U = \begin{bmatrix} U_{00} & U_{01} \\ U_{10} & U_{11} \end{bmatrix}. \tag{3.38}$$

By using the gates described so far, we can implement more complicated quantum circuits. For example, the entangled EPR pair (Bell state) preparation circuit is

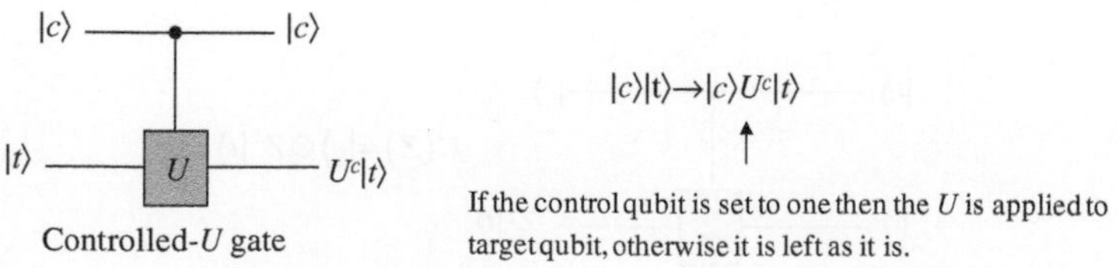

FIGURE 3.7

Controlled-U gate implementation and operating principle.

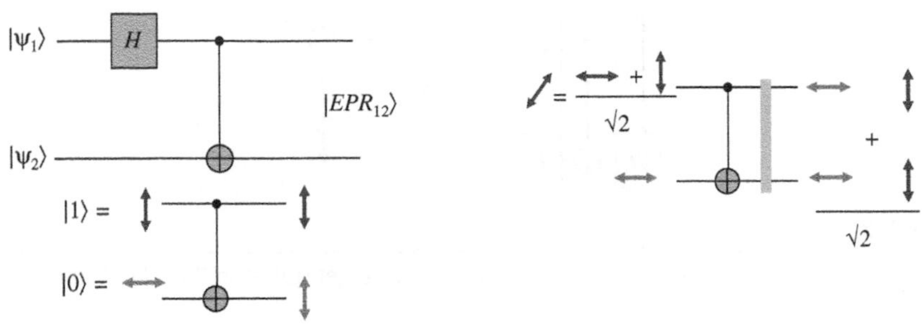

FIGURE 3.8

Entangled EPR pair (Bell state) preparation circuit.

shown in Figure 3.8. In the same figure we provide the corresponding implementation in the optical domain. The operating principle of this circuit can be described as follows:

$$\begin{aligned}\mathrm{CNOT}(H\otimes I)|\psi_1\rangle|\psi_2\rangle &= \begin{bmatrix}1&0&0&0\\0&1&0&0\\0&0&0&1\\0&0&1&0\end{bmatrix}\otimes\left(\frac{1}{\sqrt{2}}\begin{bmatrix}1&1\\1&-1\end{bmatrix}\otimes\begin{bmatrix}1&0\\0&1\end{bmatrix}\right)\\ &\quad\times\left(\begin{bmatrix}a_1\\b_1\end{bmatrix}\otimes\begin{bmatrix}a_2\\b_2\end{bmatrix}\right)\\ &=\frac{1}{\sqrt{2}}\begin{bmatrix}1&0&0&0\\0&1&0&0\\0&0&0&1\\0&0&1&0\end{bmatrix}\begin{bmatrix}1&0&1&0\\0&1&0&1\\1&0&-1&0\\0&1&0&-1\end{bmatrix}\begin{bmatrix}a_1a_2\\a_1b_2\\b_1a_2\\b_1b_2\end{bmatrix}\\ &=\frac{1}{\sqrt{2}}\begin{bmatrix}a_1a_2+b_1b_2\\a_1b_2+b_1b_2\\b_1a_2-b_1b_2\\a_1a_2-b_1a_2\end{bmatrix}.\end{aligned}\tag{3.39}$$

For example, by setting $a_1 = a_2 = 1$ and $b_1 = b_2 = 0$, we obtain:

$$\mathrm{EPR}_{12}|0\rangle|0\rangle = \mathrm{CNOT}(H\otimes I)\begin{bmatrix}1\\0\\0\\0\end{bmatrix} = \frac{|00\rangle+|11\rangle}{\sqrt{2}} = |B_{00}\rangle.$$

Therefore, by setting $|\psi_1\rangle = |a\rangle$ and $|\psi_1\rangle = |b\rangle$, where $a, b \in \{0,1\}$, the output of the circuit shown in Figure 3.8 will be the Bell state $|B_{ab}\rangle$. Regarding the optical implementation, the vertical photon corresponds to state $|1\rangle$, while the horizontal

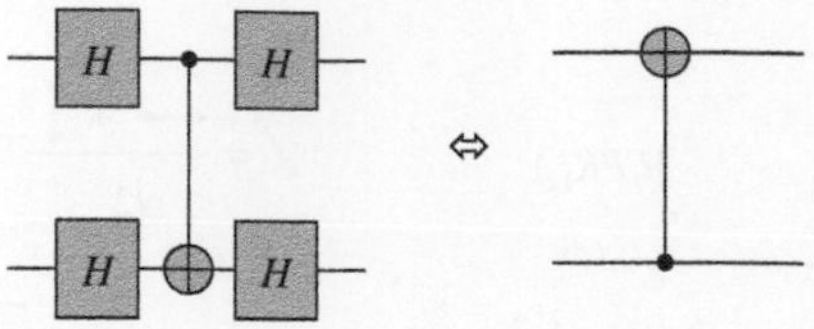

FIGURE 3.9

The quantum circuit to swap the roles of control and target qubits in the CNOT gate.

photon corresponds to state $|0\rangle$. If the control input to the CNOT gate is the vertical photon and the target input is the horizontal photon, both outputs of the CNOT gate will be vertical photons. If both inputs of the EPR circuit are horizontal photons, then the control photon will be transformed by the Hadamard gate to the $|45°\rangle$ photon, and after the CNOT gate at the output we will obtain the Bell $|B_{00}\rangle$ state.

By using one CNOT gate and four Hadamard gates we can change the roles of control and target qubits in the CNOT gate, which is illustrated in Figure 3.9. We learned in the previous section that the Hadamard gate interchanges the CB and diagonal basis, $|0\rangle \to |+\rangle$ and $|1\rangle \to |-\rangle$. After this basis transformation, the action of the CNOT gate is to perform the following mappings:

$$\begin{array}{ll} |+\rangle|+\rangle \to |+\rangle|+\rangle & |-\rangle|+\rangle \to |-\rangle|+\rangle \\ |+\rangle|-\rangle \to |-\rangle|-\rangle & |-\rangle|-\rangle \to |+\rangle|-\rangle. \end{array} \tag{3.40}$$

Therefore, the second qubit now behaves as the control and the first as the target qubit. The second pair of Hadmard gates maps the diagonal basis back to the computational basis, and the roles of input control and target qubits are interchanged.

As another illustrative example, in Figure 3.10 we provide the control phase-shift gate and its representation. If the control qubit is 1, we introduce the phase shift $\exp(j\phi)$; otherwise we leave the target qubit unchanged. On the other hand, the circuit on the right-hand side of Figure 3.10 performs the following mapping:

$$\begin{array}{ll} |00\rangle \to |00\rangle & |01\rangle \to |01\rangle \\ |10\rangle \to e^{j\phi}|10\rangle & |11\rangle \to e^{j\phi}|11\rangle, \end{array} \tag{3.41}$$

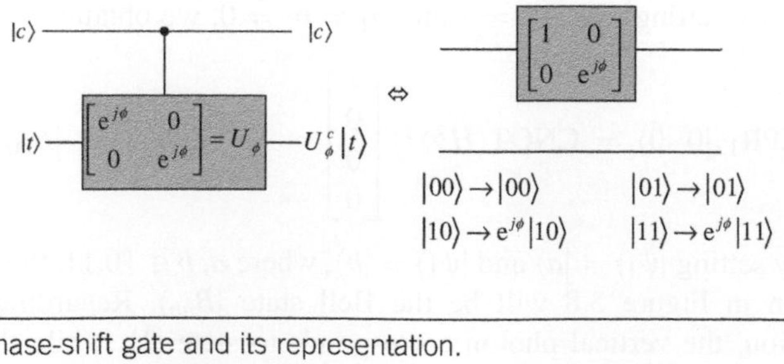

FIGURE 3.10

The control phase-shift gate and its representation.

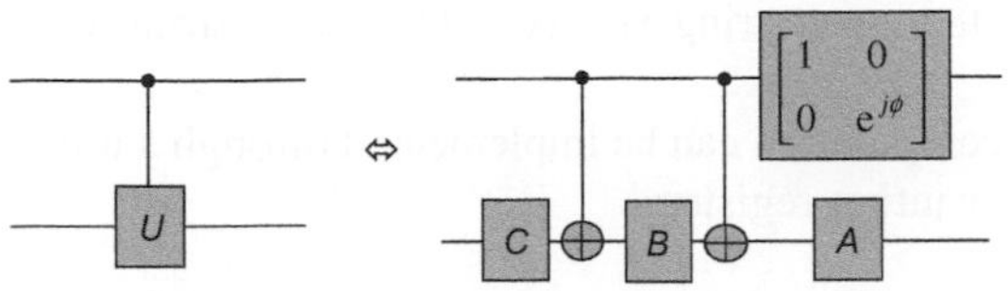

FIGURE 3.11

The quantum circuit to implement the controlled-U gate by means of single-qubit gates A, B, C satisfying the property $ABC = I$ and the following equality $U = e^{j\phi}AXBXC$.

which is the same as that of the circuit on the left-hand side of Figure 3.10.

Before concluding this section, we provide the *corollary of the Z–Y single-qubit decomposition theorem*, which is useful in the physical implementation of various controlled-U circuits. The proof of this corollary is left to the reader (see Problem 1). Let U be a unitary gate on a single qubit. Then there exist unitary operators A, B, C on a single-qubit such that $ABC = I$ and $U = e^{j\phi}AXBXC$, where ϕ is the overall phase factor. This equivalence is illustrated in Figure 3.11.

3.3 GENERALIZATION TO N-QUBIT GATES AND QUANTUM COMPUTATION FUNDAMENTALS

A generic N-qubit gate is shown in Figure 3.12. This quantum gate performs the following computation:

$$\begin{pmatrix} O_1(x) \\ \vdots \\ O_N(x) \end{pmatrix} = U \begin{pmatrix} x_1 \\ \vdots \\ x_N \end{pmatrix}; \quad (x)_{10} = (x_1 \ldots x_N)_2. \tag{3.42}$$

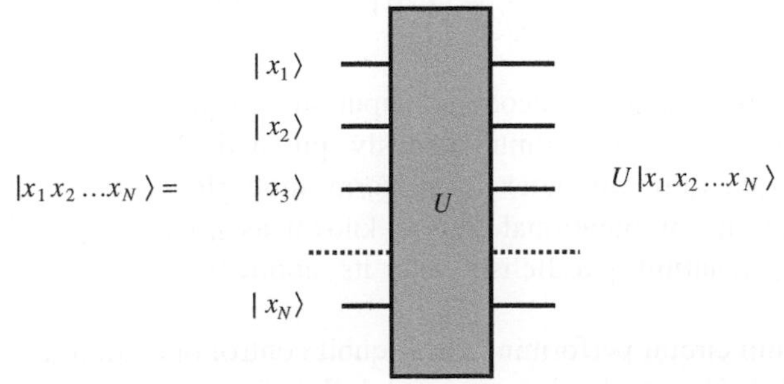

FIGURE 3.12

Representation of a generic N-qubit gate.

This circuit maps the input string $x_1 \ldots x_N$ to the output string $O_1(x) \ldots O_N(x)$, where $x \equiv x_1 \ldots x_N$.

The quantum computation can be implemented through a unitary transformation U on a $2N$-qubit quantum register:

$$U|x_1 \ldots x_N\rangle \otimes |0 \ldots 0\rangle = |x_1 \ldots x_N\rangle \otimes |O_1(x) \ldots O_N(x)\rangle, \tag{3.43}$$

where we added an additional N ancillary qubits in state $|0\rangle$ in order to preserve the unitarity of operator U at the output of the quantum circuit. That is, when computation is performed according to (3.42), with N qubits, when $x \neq y$ if

$$O_1(x) \ldots O_N(x) = O_1(y) \ldots O_N(y), \tag{3.44}$$

then

$$\langle O_1(x) \ldots O_N(x)|O_1(y) \ldots O_N(y)\rangle = 1, \tag{3.45}$$

indicating that unitarity is not preserved upon quantum computation. On the other hand, when computation is performed based on (3.43), in situations when $O_1(x) \ldots O_N(x) = O_1(y) \ldots O_N(y)$ $(x \neq y)$, we obtain:

$$\begin{aligned} &\langle x_1 \ldots x_N O_1(x) \ldots O_N(x)|y_1 \ldots y_N O_1(y) \ldots O_N(y)\rangle \\ &= \langle x_1 \ldots x_N|y_1 \ldots y_N\rangle \langle O_1(x) \ldots O_N(x)|O_1(y) \ldots O_N(y)\rangle = 0 \cdot 1 = 0, \end{aligned} \tag{3.46}$$

and the unitarity of the output is preserved. The linear superposition allows us to create the following $2N$-qubit state:

$$|\psi_{\text{in}}\rangle = \left[\frac{1}{\sqrt{2^N}} \sum_x |x_1 \ldots x_N\rangle\right] \otimes |0 \ldots 0\rangle, \tag{3.47}$$

and upon the application of quantum operation U, the output can be represented by

$$|\psi_{\text{out}}\rangle = U(C)|\psi_{\text{in}}\rangle = \frac{1}{\sqrt{2^N}} \sum_x |x_1 \ldots x_N\rangle \otimes |O_1(x) \ldots O_N(x)\rangle. \tag{3.48}$$

Therefore, we were able to encode all input strings generated by U into $|\psi_{\text{out}}\rangle$; in other words, we have simultaneously pursued 2^N classical paths. This ability of a quantum computer to perform multiple function evaluations in a single quantum computational step is known as *quantum parallelism*. More details about quantum parallelism and its applications will be provided in Chapter 5.

The quantum circuit performing the N-qubit control operation, $C^N(U)$, is shown in Figure 3.13. If N control qubits are set to 1, then the U gate is applied to quantum state $|y\rangle$. The action of the U gate on the multiqubit state $|y\rangle$ has already been discussed above.

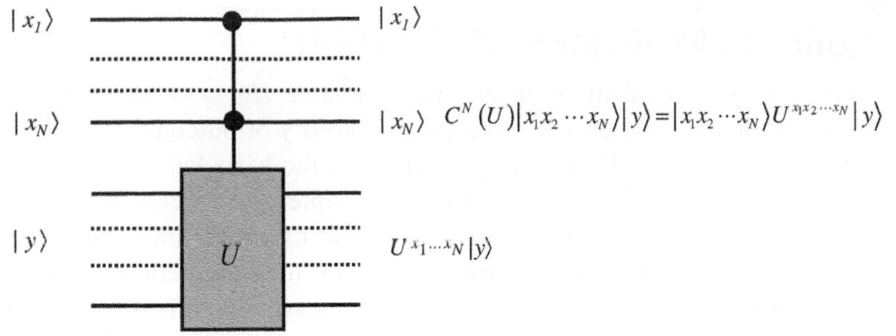

FIGURE 3.13

Quantum circuit to perform N-qubit control operation, $C^N(U)$.

For example, the special case for $N = 2$, the $C^2(U)$ gate, is shown in Figure 3.14, together with its equivalent representation.

When $U = X$, then the $C^2(X)$ gate is called the *Toffoli gate*, and its representation is shown in Figure 3.15. The Toffoli gate can be implemented based on Figure 3.14 for $V = (1 - j)(1 + jX/2)$. Its implementation is shown on the right in Figure 3.15.

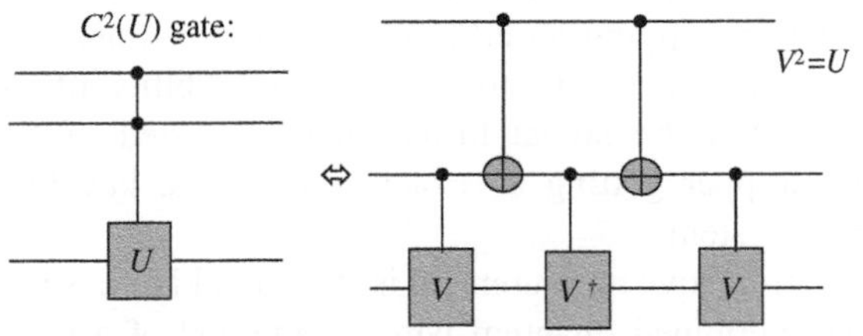

FIGURE 3.14

Quantum circuit to perform two-qubit controlled U operation, $C^2(U)$, and its equivalent representation.

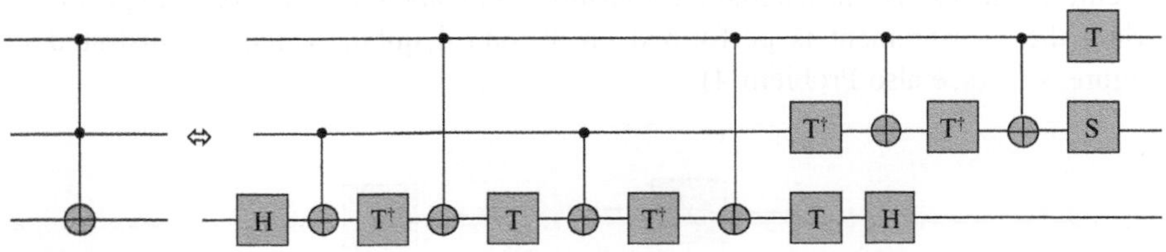

FIGURE 3.15

Toffoli gate representation and its implementation.

3.4 QUBIT MEASUREMENT (REVISITED)

When measuring a single qubit in an arbitrary state $|\psi\rangle = a\ |0\rangle + b\ |1\rangle$, the probability of the outcome 0 is $|a|^2$ and the probability of outcome 1 is $|b|^2$. Such a measurement is done by Hermitian projection on the basis kets $|0\rangle$ and $|1\rangle$. The axioms of quantum mechanics, described in Chapter 2, indicate that after the measurement an arbitrary qubit is collapsed to the measured basis state, so that a qubit will be destroyed in this measurement. The measurement is therefore an *irreversible* operation. For example, if during the measurement it is found that the result of the measurement is 0, then the post-measurement state will be $|\psi_{\text{out}}\rangle = P_0|\psi\rangle/\sqrt{\langle\psi|P_0|\psi\rangle}$. On the other hand, if the result of the measurement is 1, then the post-measurement state will be $|\psi_{\text{out}}\rangle = P_1|\psi\rangle/\sqrt{\langle\psi|P_1|\psi\rangle}$. We denote by P_i $(i = 0{,}1)$ the projection operators, namely $P_i = |i\rangle\langle i|$ $(i = 0{,}1)$. The density operator upon measurement will be $\rho_{\text{out}} = P_0\rho P_0 + P_1\rho P_1$, where ρ is the density operator of the input state. This problem cannot be solved by copying because of the *no-cloning theorem*. The no-cloning theorem claims that a quantum device cannot be constructed to generate an exact copy of an arbitrary quantum state $|\psi\rangle$, namely to perform the following mapping $|\psi\rangle \otimes |0\rangle \to |\psi\rangle \otimes |\psi\rangle$. This theorem can easily be proved by contradiction (see Problem 3). Notice that this theorem does not rule out the possibility of building a device that can copy a particular set of orthonormal quantum states. Important consequences of the no-cloning theorem can be summarized as follows [2]: (i) it does not allow in quantum computing copying an arbitrary number of times as a way of backing data in case of errors; (ii) it rules out the possibility of using Bell states (EPR pairs) to transmit the signal faster than the speed of light; and (iii) it prevents an eavesdropper getting an exact copy of the key in a quantum key distribution (QKD) system.

The symbol for projective measurement is shown in Figure 3.16. Without loss of generality, any undetermined quantum wires at the end of a quantum circuit are assumed to be measured. This claim is sometimes called the *principle of implicit measurement*. It is also clear that classically conditioned quantum operations can be replaced by quantum conditioned ones. The claim is known as the *principle of differed measurement*, and can also be formulated as: Measurements can always be moved from an intermediate stage of a quantum circuit to the end of the quantum circuit. It can also be shown that the *measurement commutes with control operation*, when the measurement is performed on a control qubit, which is illustrated in Figure 3.17 (see also Problem 4).

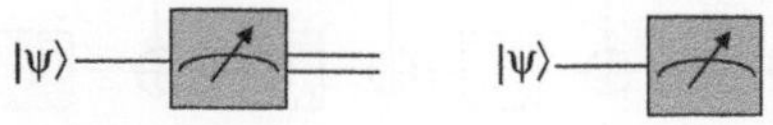

FIGURE 3.16

The symbol for projective measurement. The double line denotes the classical bit.

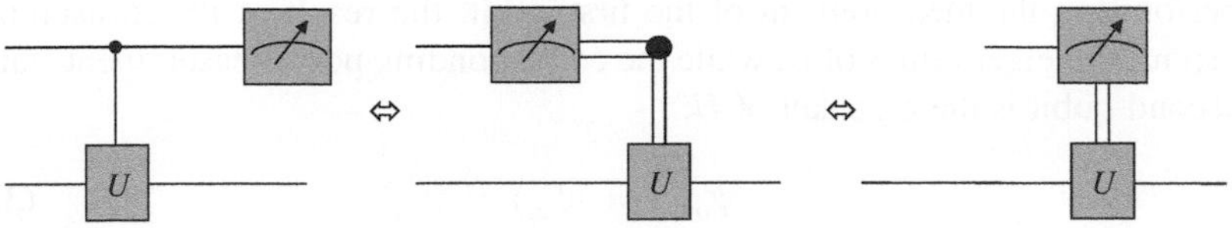

FIGURE 3.17

The measurement and control commute, when the measurement is performed on a control qubit.

Suppose now that we are concerned with the measurement of an observable associated with an operator U, which is a Hermitian and a unitary with eigenvalues ± 1, acting on the quantum state $|\psi_{\text{in}}\rangle$. The quantum circuit shown in Figure 3.18 can be used for measuring its eigenvalues without completely destroying the measured qubit. (After the measurement, the post-measurement qubit will be in the corresponding eigenket.)

As an illustration, let us observe the measurement of an observable of operator U with eigenvalues ± 1 and corresponding eigenkets $|\psi_{\pm}\rangle$. The sequence of states at different points of quantum circuits from Figure 3.18 is given as follows:

$$
\begin{aligned}
|\psi_1\rangle &= |0\rangle|\psi_{\text{in}}\rangle = a|0\rangle|0\rangle + b|0\rangle|1\rangle \\
|\psi_2\rangle &= \frac{1}{\sqrt{2}}(|0\rangle + |1\rangle)|\psi_{\text{in}}\rangle = \frac{1}{\sqrt{2}}(a|0\rangle|0\rangle + a|1\rangle|0\rangle + b|0\rangle|1\rangle + b|1\rangle|1\rangle) \\
|\psi_3\rangle &= \frac{1}{\sqrt{2}}(a|0\rangle|0\rangle + a|1\rangle|0\rangle + b|0\rangle|1\rangle - b|1\rangle|1\rangle) \\
|\psi_4\rangle &= \frac{1}{2}(a(|0\rangle + |1\rangle)|0\rangle + a(|0\rangle - |1\rangle)|0\rangle + b(|0\rangle + |1\rangle)|1\rangle - b(|0\rangle - |1\rangle)|1\rangle) \\
&= \frac{1}{2}(|0\rangle(a|0\rangle + a|0\rangle + b|1\rangle - b|1\rangle) + |1\rangle(a|0\rangle - a|0\rangle + b|1\rangle + b|1\rangle)) \\
&= a|0\rangle|0\rangle + b|1\rangle|1\rangle. \qquad (3.49)
\end{aligned}
$$

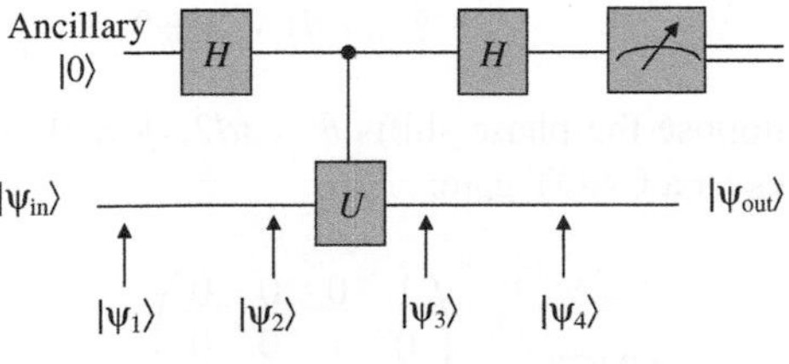

FIGURE 3.18

Measurement of an observable of the operator U.

By performing the measurement of the first qubit, the result of the measurement corresponds to eigenvalues of U, while the corresponding post-measurement state of the second qubit is the eigenket of U:

$$|\psi_{\text{out}}\rangle = |\psi_{\pm}\rangle. \tag{3.50}$$

Notice that we have not explicitly said anything about the operator U, in the sense of whether it is applied to single qubits or many qubits. Therefore, this measurement circuit can be applied to observables U acting on any number of qubits with two eigenvalues ± 1.

3.5 UNIVERSAL QUANTUM GATES

In order to perform an arbitrary quantum computation, a minimum number of gates known as *universal quantum gates* [12–15] is needed. A set of quantum gates is said to be universal if it contains a minimum finite number of gates so that arbitrary unitary operation can be performed with arbitrary small error probability by using only gates from that set. The most popular sets of universal quantum gates are: {Hadamard (H), phase (S), CNOT, Toffoli (U_{T})} gates, {H, S, $\pi/8$ (T), CNOT} gates, Barenco gate [12], and Deutsch gate [13]. The *Gottesman–Knill theorem* (see Problem 5) has shown that a quantum computation that solely uses gates from the Clifford group $N_U(G_N) = \{H, S, \text{CNOT}\}$ and the measurements of observables that belong to the Pauli group G_N (N-qubit tensor product of Pauli operators) can be simulated efficiently on a classical computer. Typically, for quantum error correction, the use of CNOT, S, and H gates is sufficient. Unfortunately, for arbitrary quantum computation the Clifford group gates are not sufficient. However, if the set N_U is extended using either the Toffoli or $\pi/8$ gate, we obtain a universal set of quantum gates. Barenco [12] has shown that the following two-qubit gate (with matrix elements relative to the computational basis $\{|ij\rangle: i,j = 0,1\}$) is universal:

$$A(\phi, \alpha, \theta) = \begin{pmatrix} 1 & 0 & 0 & 0 \\ 0 & 1 & 0 & 0 \\ 0 & 0 & e^{j\alpha}\cos(\theta) & -je^{j(\alpha+\phi)}\sin(\theta) \\ 0 & 0 & -je^{j(\alpha-\phi)}\sin(\theta) & e^{j\alpha}\cos(\theta) \end{pmatrix}. \tag{3.51}$$

For example, if we impose the phase shifts $\theta = \pi/2$, $\phi = 0$ rad, and $\alpha = \pi/2$, the Barenco gate operates as a CNOT gate:

$$\text{CNOT} = \begin{pmatrix} 1 & 0 & 0 & 0 \\ 0 & 1 & 0 & 0 \\ 0 & 0 & 0 & 1 \\ 0 & 0 & 1 & 0 \end{pmatrix}.$$

The controlled-Y gate, $C(Y)$, can be obtained from the Barenco gate by setting $\theta = \pi/2$, $\phi = 3\pi/2$, and $\alpha = \pi/2$:

$$C(Y) = \begin{pmatrix} 1 & 0 & 0 & 0 \\ 0 & 1 & 0 & 0 \\ 0 & 0 & 0 & -j \\ 0 & 0 & j & 0 \end{pmatrix}.$$

Various single-qubit gates can be obtained from the Barenco gate by omitting the control qubit and by appropriate setting of the parameters α, θ, ϕ.

The three-qubit Deutsch gate [13], denoted by $D(\theta)$, represents the generalization of the Barenco gate, and its matrix representation is given by

$$D(\theta) = \begin{pmatrix} 1 & 0 & 0 & 0 & 0 & 0 & 0 & 0 \\ 0 & 1 & 0 & 0 & 0 & 0 & 0 & 0 \\ 0 & 0 & 1 & 0 & 0 & 0 & 0 & 0 \\ 0 & 0 & 0 & 1 & 0 & 0 & 0 & 0 \\ 0 & 0 & 0 & 0 & 1 & 0 & 0 & 0 \\ 0 & 0 & 0 & 0 & 0 & 1 & 0 & 0 \\ 0 & 0 & 0 & 0 & 0 & 0 & j\cos(\theta) & \sin(\theta) \\ 0 & 0 & 0 & 0 & 0 & 0 & \sin(\theta) & j\cos(\theta) \end{pmatrix}. \quad (3.52)$$

By using the Z–Y decomposition theorem:

$$U = \begin{pmatrix} e^{j(\alpha-\beta/2-\delta/2)}\cos\left(\frac{\gamma}{2}\right) & -e^{j(\alpha-\beta/2+\delta/2)}\sin\left(\frac{\gamma}{2}\right) \\ e^{j(\alpha+\beta/2-\delta/2)}\sin\left(\frac{\gamma}{2}\right) & \cos\left(\frac{\gamma}{2}\right)e^{j(\alpha+\beta/2+\delta/2)} \end{pmatrix}, \quad (3.53)$$

where α, β, γ, and δ are phase shifts, we can implement an arbitrary single-qubit gate. By setting $\gamma = \delta = 0$ rad, $\alpha = \pi/4$, and $\beta = \pi/2$ rad, the U gate above operates as a phase gate:

$$S = \begin{pmatrix} 1 & 0 \\ 0 & j \end{pmatrix}.$$

By setting $\gamma = \delta = 0$ rad, $\alpha = \pi/8$, and $\beta = \pi/4$ rad, the U gate operates as a $\pi/8$ gate:

$$T = \begin{pmatrix} 1 & 0 \\ 0 & e^{j\pi/4} \end{pmatrix}.$$

By setting $\gamma = +\pi/2$, $\alpha = \pi/2$ rad, and $\beta = 0$, $\delta = \pi$, the U gate given by Eq. (3.53) operates as a Hadamard gate:

$$H = \frac{1}{\sqrt{2}}\begin{pmatrix} 1 & 1 \\ 1 & -1 \end{pmatrix}.$$

The Pauli gates X, Y, and Z can be obtained from the U gate given by Eq. (3.53) by appropriately setting the phase shifts α, β, γ, and δ. The X gate is obtained by setting $\gamma = \pi$, $\beta = \delta = 0$ rad, and $\alpha = \pi/2$:

$$X = \begin{pmatrix} 0 & 1 \\ 1 & 0 \end{pmatrix}.$$

The Z gate is obtained by setting $\gamma = \delta = 0$ rad, $\alpha = \pi/2$, and $\beta = \pi$:

$$Z = \begin{pmatrix} 1 & 0 \\ 0 & -1 \end{pmatrix}.$$

The Y gate is obtained by setting $\gamma = \pi$, $\delta = 0$ rad, $\alpha = \pi/4$, and $\beta = -\pi/2$:

$$Y = \begin{pmatrix} 0 & -j \\ j & 0 \end{pmatrix}.$$

3.6 QUANTUM TELEPORTATION

Quantum teleportation [16] is a technique to transfer quantum information from source to destination by employing entangled states. Namely, in quantum teleportation, the entanglement in the Bell state (EPR pair) is used to transport an arbitrary quantum state $|\psi\rangle$ between two distant observers A and B (often called Alice and Bob). The quantum teleportation system employs three qubits: qubit 1 is an arbitrary state to be teleported, while qubits 2 and 3 are in Bell state $|B_{00}\rangle = (|00\rangle + |11\rangle)/\sqrt{2}$. The operating principle is shown in Figure 3.19, for an input state $|x\rangle$, where $x = 0,1$. The evaluation of the three-qubit state in five characteristic points of a quantum circuit is shown as well. The input three-qubit state at point 1 can be written as

$$|\psi_1\rangle = \frac{|x00\rangle + |x11\rangle}{\sqrt{2}}. \tag{3.54}$$

The first qubit is then used as control qubit and the second qubit is the target qubit for the CNOT gate, so that the state at point 2 becomes

$$|\psi_2\rangle = \frac{|xx0\rangle + |x(1-x)1\rangle}{\sqrt{2}}. \tag{3.55}$$

The Hadamard gate performs the following mapping on the first qubit: $H|x\rangle = (|0\rangle + (-1)^x|1\rangle)/\sqrt{2}$, so that the state in point 3 can be represented by

$$|\psi_3\rangle = \frac{|0x0\rangle + (-1)^x|1x0\rangle + |0(1-x)1\rangle + (-1)^x|1(1-x)1\rangle}{2}. \tag{3.56}$$

The second qubit is then used as control qubit and the third qubit as target in the control-X gate (or CNOT gate) so that quantum state at point 4 can be written as

$$|\psi_4\rangle = \frac{|0xx\rangle + (-1)^x|1xx\rangle + |0(1-x)x\rangle + (-1)^x|1(1-x)x\rangle}{2}. \tag{3.57}$$

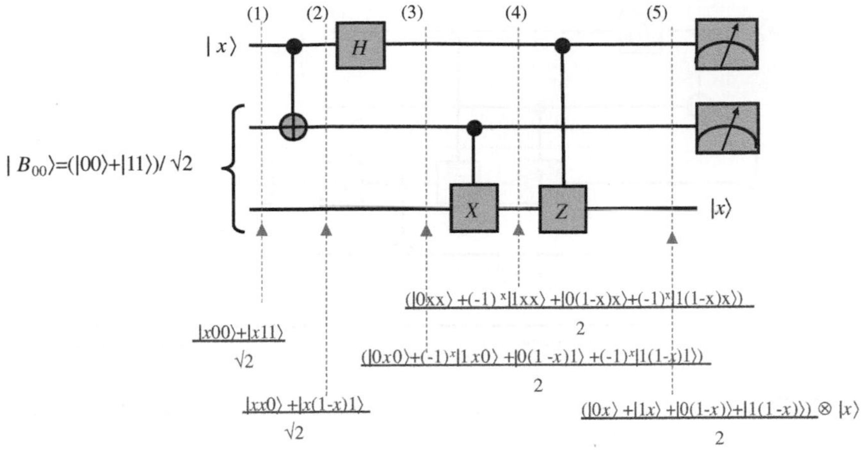

FIGURE 3.19

The quantum teleportation principle.

Further, the first qubit is used as control and the third qubit as target for the controlled-Z gate, so that the state at point 5 is given by

$$|\psi_5\rangle = \frac{|0xx\rangle + |1xx\rangle + |0(1-x)x\rangle + |1(1-x)x\rangle}{2}$$
$$= \frac{|0x\rangle + |1x\rangle + |0(1-x)\rangle + |1(1-x)\rangle}{2} \otimes |x\rangle. \quad (3.58)$$

Finally, we perform measurements on the first two qubits that are destroyed, and the third qubit is delivered to the destination. By comparing the destination and source qubits we can conclude that the correct quantum state is teleported. Notice that in this analysis we assume there is no error introduced by the quantum channel. Another interesting point is that the teleported state was not a superposition state. Following the same procedure we can show that an arbitrary quantum state $|\psi\rangle$ can be teleported by using the scheme from Figure 3.19 (see Problem 6).

Another alternative quantum teleportation scheme is shown in Figure 3.20. In this version, we perform the measurement in the middle of the circuit and based on

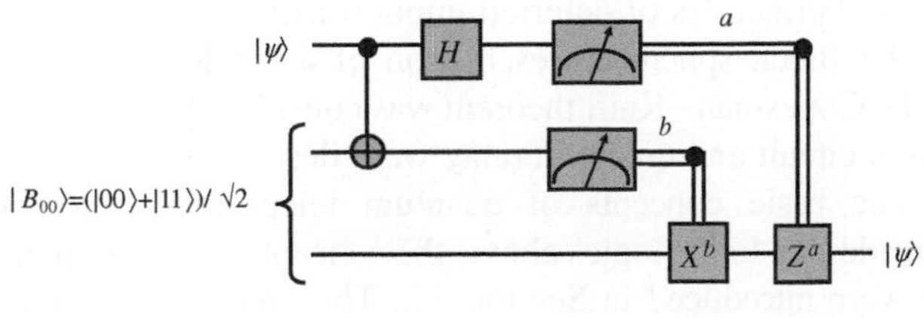

FIGURE 3.20

The quantum teleportation scheme performing measurement in the middle of the circuit.

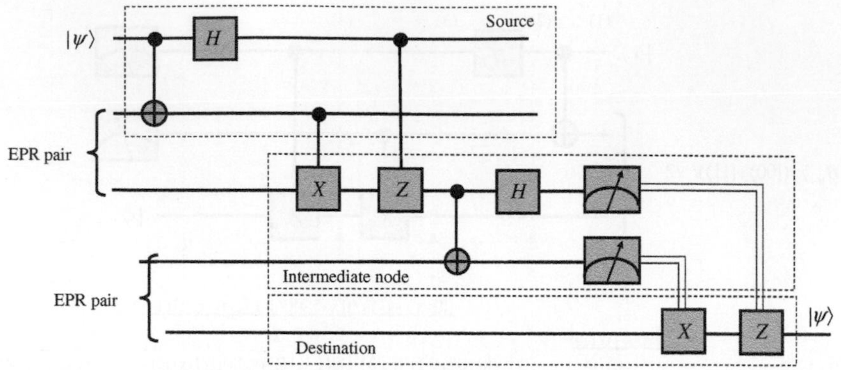

FIGURE 3.21

The operating principle of quantum relay.

results of the measurement we conditionally execute X and Z operators. We can interpret this circuit as the application of the principle of deferred measurement, but now in the opposite direction.

In order to extend the total teleportation distance, *quantum repeaters* [17], entanglement purification [18], and *quantum relay* [19–21] can be used. The key obstacle for quantum repeater and relay for implementation is the no-cloning theorem. However, we can use the quantum teleportation principle described above to implement the quantum relay. One such quantum relay, based on the quantum teleportation circuit from Figure 3.19, is shown in Figure 3.21. Its operating principle is the subject of Problem 8.

3.7 SUMMARY

This chapter has covered quantum gates and quantum information processing fundamentals. In this chapter, basic single-qubit (Section 3.1), two-qubit (Section 3.2), and many-qubit gates (Section 3.3) have been described. The fundamentals of quantum information processing (Section 3.3) have been discussed as well. The various sets of universal quantum gates were considered in Section 3.5. The quantum measurements and principles of deferred implicit measurements were discussed in Section 3.4. The Bloch sphere representation of a single qubit was provided in Section 3.1. The Gottesman–Knill theorem was considered in Section 3.5. The Bell state preparation circuit and quantum relay were described in Sections 3.2 and 3.6 respectively. The basic concepts of quantum teleportation were described in Section 3.6. In addition to the topics above, the concepts of quantum parallelism and entanglement were introduced in Section 3.2. The no-cloning theorem was introduced in Section 3.4, and its consequences discussed. For a deeper understanding of underlying concepts, in Section 3.8 a series of problems for self-study are given.

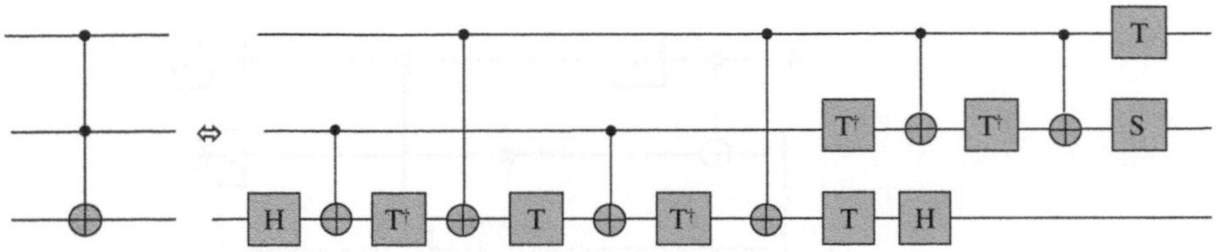

FIGURE 3.P2

Toffoli gate.

3.8 PROBLEMS

1. Prove the corollary of the Z–Y single-qubit decomposition theorem, which can be formulated as follows. Let U be a unitary gate on a single qubit. Then there exist unitary operators A, B, C on a single qubit such that $ABC = I$ and $U = e^{j\phi}AXBXC$, where ϕ is the overall phase factor.
2. Verify that the quantum circuit shown in Figure 3.P2 implements the Toffoli gate.
3. This problem is related to the *no-cloning theorem*, which states that a quantum device cannot be constructed with the action of creating an exact copy of an arbitrary quantum state. Prove this claim. *Hint*: use the contradiction approach.
4. A consequence of the principle of deferred measurements is that measurements commute with quantum gates when the measurement is performed on a control qubit, which is shown in Figure 3.P4. Prove this claim.
5. This problem concerns the *Gottesman–Knill theorem* stating that a quantum computation that solely uses gates from the Clifford group $N_U(G_N) = \{H, S, \text{CNOT}\}$ and the measurements of observables that belong to the Pauli group G_N (N-qubit tensor product of Pauli operators) can be simulated efficiently on a classical computer. Prove the theorem.
6. The quantum teleportation circuit is shown in Figure 3.P6. By using a similar procedure to that described in Section 3.6, show that this circuit can be used to teleport an arbitrary quantum state to a remote destination.
7. Prove that the quantum circuit shown in Figure 3.P7 can be used as a quantum teleportation circuit. Are the circuits from Figures 3.P6 and 3.P7 equivalent?

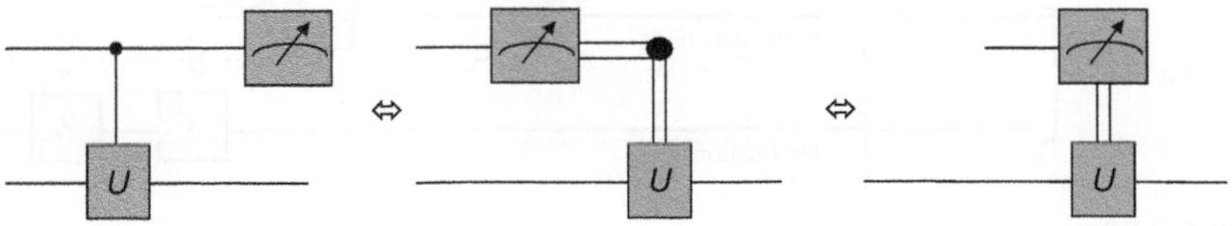

FIGURE 3.P4

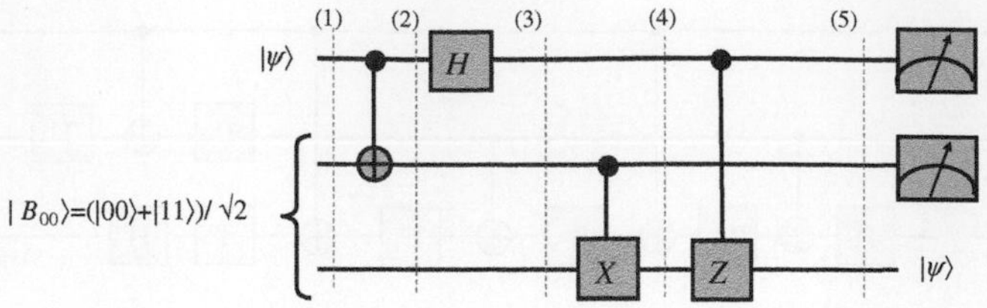

FIGURE 3.P6

The quantum teleportation circuit.

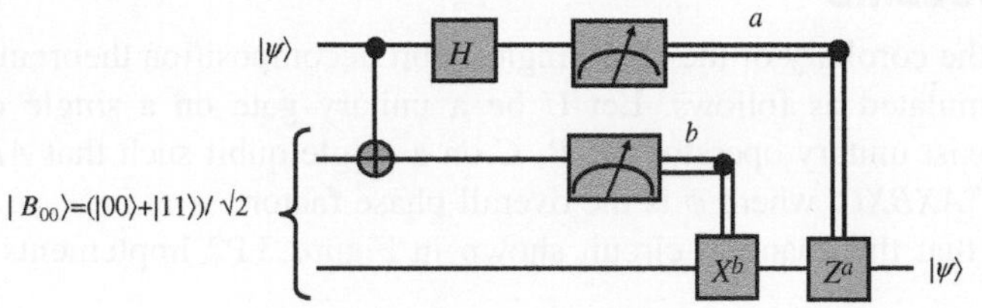

FIGURE 3.P7

The quantum teleportation circuit (version 2).

8. By using a similar approach to that in the previous two problems, describe the operating principle of *quantum relay* shown in Figure 3.P8. Prove that an arbitrary quantum state can indeed be teleported to a remote destination using this circuit. Can this circuit be implemented without any measurement? If yes, how?

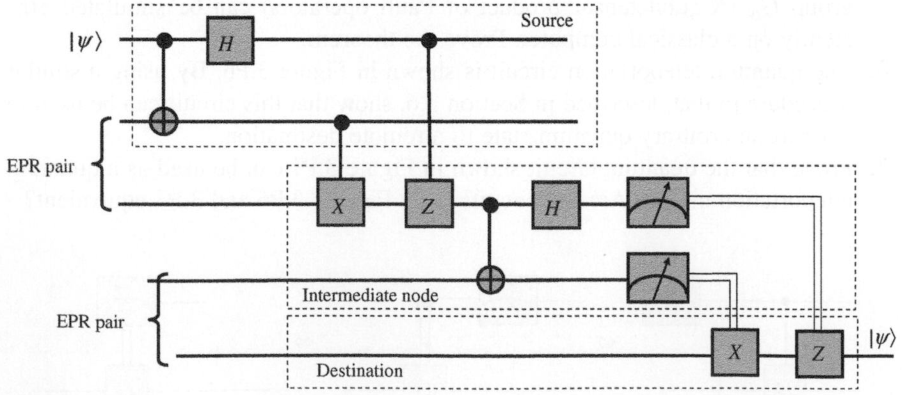

FIGURE 3.P8

Quantum relay.

9. Show that the following operator:

$$\frac{1}{2}(I + \overrightarrow{\sigma_A} \cdot \overrightarrow{\sigma_B}),$$

where $\overrightarrow{\sigma_A} \cdot \overrightarrow{\sigma_B}$ stands for simultaneous scalar and tensor product, swaps the qubits A and B. This operator is known as the SWAP operator. Determine its matrix representation in the basis $\{|00\rangle,|01\rangle,|10\rangle,|11\rangle\}$. Also determine $U_{\text{SWAP}}^{1/2}$. Show further how the controlled-X gate can be constructed from the SWAP gate.

10. Consider a two-dimensional Hilbert space and define the state as follows:

$$|\psi\rangle = \cos(\theta/2)|0\rangle + \sin(\theta/2)|1\rangle,$$

and let the density matrix be given by

$$\rho = p|0\rangle\langle 0| + (1-p)|1\rangle\langle 1|, \ \ 0 < p < 1.$$

Determine the Shannon H_{Sh} and von Neumann H_{vN} entropies and show that $H_{\text{Sh}} \geq H_{\text{vN}}$. The Shannon entropy is defined as $H_{\text{Sh}} = -\Sigma_i p_i \log p_i$, while the von Neumann entropy is defined as $H_{\text{vN}} = -\text{Tr}\ \rho \log \rho$.

11. The mutual information is defined as $I(X,Y) = H_{\text{Sh}}(X) - H_{\text{Sh}}(X|Y)$, where X is the channel input and Y is the channel output. Let us evaluate the amount of information that Eve can get from Alice, denoted as $I(A,E)$, for the following two situations:

- Let i represent the bit sent in computational basis (CB) or in the diagonal basis. Therefore, the bit i can take different values with probability $p(i) =$ 1/4. Let Eve use the CB to measure result r, which takes two different values. Create a table of conditional probabilities $p(r|i)$ and determine $p(i|r)$. What is Eve's mutual information in this case?
- Let Eve now use the Breidbart basis instead. What is Eve's mutual information in this case? Which one is higher?

12. Show that if the perfect quantum copying machine exists, then two distant parties (say Alice and Bob), sharing the pair of entangled qubits in state $|B_{11}\rangle$, could communicate at superluminal velocity.

13. This problem is related to the *superdense coding* [22]. Two distant parties (say Alice and Bob) share a pair of entagled qubits in state $|B_{00}\rangle$. Alice intends to send to Bob two classical bits i,j $(i,j = 0,1)$, while using a single qubit. In order to do so, she applies the operator A_{ij} acting on her half of $|B_{00}\rangle$ as $A_{ij} = X^i Z^j$, where i and j are exponents. She then sends this transformed qubit to Bob, who actually receives $A_{ij}|B_{00}\rangle$.

(a) Provide explicit expressions for $A_{ij}|B_{00}\rangle$ $(i,j = 0,1)$ in terms of states $|ij\rangle$ $(i,j = 0,1)$.

(b) Let us assume that Bob has the circuit shown in Figure 3.P13 available. Show that CNOT gates transform $A_{ij}|B_{00}\rangle$ into a tensor product, and that the measurement on the second qubit reveals the transmitted bit i. Finally,

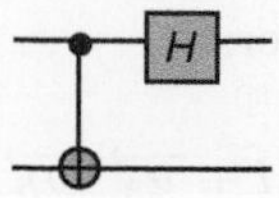

FIGURE 3.P13

show that the second measurement on the first qubit reveals the value of bit j and therefore that the procedure indeed allows two classical bits to be transmitted over one entangled state.

References

[1] M.A. Neilsen, I.L. Chuang, Quantum Computation and Quantum Information, Cambridge University Press, Cambridge, 2000.

[2] F. Gaitan, Quantum Error Correction and Fault Tolerant Quantum Computing, CRC Press, 2008.

[3] M. Le Bellac, An Introduction to Quantum Information and Quantum Computation, Cambridge University Press, 2006.

[4] G. Jaeger, Quantum Information: An Overview, Springer, 2007.

[5] D. Petz, Quantum Information Theory and Quantum Statistics, Theoretical and Mathematical Physics, Springer, Berlin, 2008.

[6] P. Lambropoulos, D. Petrosyan, Fundamentals of Quantum Optics and Quantum Information, Springer-Verlag, Berlin, 2007.

[7] G. Johnson, A Shortcut Through Time: The Path to the Quantum Computer, Knopf, New York, 2003.

[8] J. Preskill, Quantum Computing, 1999. Available at: http://www.theory.caltech.edu/~preskill/.

[9] J. Stolze, D. Suter, Quantum Computing, Wiley, New York, 2004.

[10] R. Landauer, Information is physical, Phys. Today 44 (5) (May 1991) 23–29.

[11] R. Landauer, The physical nature of information, Phys. Lett. A 217 (1991) 188–193.

[12] A. Barenco, A universal two-bit quantum computation, Proc. R. Soc. Lond. A 449 (1937) 679–683, 1995.

[13] D. Deutsch, Quantum computational networks, Proc. R. Soc. Lond. A 425 (1868) 73–90, 1989.

[14] D.P. DiVincenzo, Two-bit gates are universal for quantum computation, Phys. Rev. A 51 (2) (1995) 1015–1022.

[15] A. Barenco, C.H. Bennett, R. Cleve, D.P. DiVincenzo, N. Margolus, P. Shor, T. Sleator, J.A. Smolin, H. Weinfurter, Elementary gates for quantum computation, Phys. Rev. A 52 (5) (1995) 3457–3467.

[16] C.H. Bennett, G. Brassard, C. Crépeau, R. Jozsa, A. Peres, W.K. Wootters, Teleporting an unknown quantum state via dual classical and Einstein–Podolsky–Rosen channels, Phys. Rev. Lett. 70 (13) (1993) 1895–1899.

[17] H.-J. Briegel, W. Dür, J.I. Cirac, P. Zoller, Quantum repeaters: the role of imperfect local operations in quantum communication, Phys. Rev. Lett. 81 (26) (1998) 5932–5935.

[18] J.-W. Pan, C. Simon, Č. Brunker, A. Zeilinger, Entanglement purification for quantum communication, Nature 410 (26 April 2001) 1067–1070.

[19] I.B. Djordjevic, Photonic implementation of quantum relay and encoders/decoders for sparse-graph quantum codes based on optical hybrid, IEEE Photon. Technol. Lett. 22 (19) (1 October 2010) 1449–1451.

[20] I.B. Djordjevic, On the photonic implementation of universal quantum gates, Bell states preparation circuit and quantum LDPC encoders and decoders based on directional couplers and HNLF, Opt. Express 18 (8) (1 April 2010) 8115–8122.

[21] I.B. Djordjevic, On the photonic implementation of universal quantum gates, Bell states preparation circuit, quantum relay and quantum LDPC encoders and decoders, IEEE Photon. J. 2 (1) (February 2010) 81–91.

[22] A. Harrow, Superdense coding of quantum states, Phys. Rev. Lett. 92 (18) (7 May 2004) 187901-1–187901-4.

This page intentionally left blank

CHAPTER

Introduction to Quantum Information Processing

4

Shikhar Uttam
University of Pittsburgh (shf28@pitt.edu)

CHAPTER OUTLINE

In this chapter the fundamental ideas that lay the foundations of quantum information processing are introduced. These ideas do not have a counterpart in the classical realm and in this context show the richness and weirdness of quantum mechanics. They underlie all algorithms and protocols developed, and those being developed, within quantum information processing such as quantum computing, quantum communications, and the development of a quantum computer, a computational machine that processes information by exploiting quantum effects.

The elemental roots of quantum information processing lie with information theory [1–5], computer science [6–8], and quantum mechanics [9–18]. Shannon's development of information theory [1,2] allowed us to formally define "information", while Alan Turing's seminal work [6–8], in laying the foundation of computer science, provided us with a computational mechanism to process information. Together they have led to breakthroughs in efficient and secure data communication over communication channels such as memory storage devices and wire (or wireless) communication links. These breakthroughs, however, are based on the classical unit of

Quantum Information Processing and Quantum Error Correction. DOI: 10.1016/B978-0-12-385491-9.00004-6

information, a binary digit or a bit, that can be either 0 or 1. In this classical setup, to increase the computational speed of a classical computing machine we need to proportionally increase the processing power. For example, given classical computers that can perform a certain number of computations per second, the number of computations can be approximately doubled by adding a similar computer. This picture dramatically changes with the introduction of the computational consequences of quantum mechanics [19]. Now, instead of working with bits, we work with quantum bits or qubits that represent quantum states (such as photon polarization or electron spin) in a two-dimensional Hilbert space. Based on the notion of a qubit, the quantum ideas of superposition and parallelism are introduced in Section 4.1. Quantum superposition deals with coherent superposition of qubits or, more generally, of quantum states in higher dimensional Hilbert space. Quantum parallelism is a consequence of superposition and we discuss the resulting increase in computational speed exponentially with size instead of linearly as is the case in classical computing. Another consequence of superposition is the no-cloning theorem that plays an important role in quantum key distribution (QKD) protocols. The no-cloning theorem and accompanying proof are presented in Section 4.2. A result closely related to the no-cloning theorem is that quantum mechanics puts strict restrictions on reliably distinguishing non-orthogonal states. This result has far-reaching consequences in quantum cryptography. This result is presented in Section 4.3.

The next major quantum mechanical idea introduced in this chapter is quantum entanglement (Section 4.4). Quantum entanglement has played a key role in the debate about the validity of quantum mechanics. This is discussed in the context of the well-known EPR pair. The section, however, begins with the mathematical formulation of entanglement and then goes on to consider Schmidt decomposition as a mathematical tool for identifying entanglement. The physical context of entanglement is also presented. Entanglement has wide-ranging consequences for quantum measurements, teleportation, superdense coding, and quantum error correction codes. These topics form the core of this book, and understanding entanglement will help the reader to grasp these ideas more thoroughly as they are discussed in later chapters.

After introducing these main ideas, in Section 4.5 the mechanism through which interaction between quantum states and the environment can be efficiently formulated for describing quantum information processing tasks is introduced. This mechanism is known as the operator-sum representation. In Section 4.6 this representation is derived and then used to describe decoherence and quantum errors. The very important Pauli matrices are also briefly introduced. They will be further discussed in detail in later chapters in relation to Bloch vectors and unitary operations.

4.1 SUPERPOSITION PRINCIPLE AND QUANTUM PARALLELISM

The superposition principle is a direct consequence of the linearity of the Schrödinger wave equation. The Schrödinger wave equation is a linear, second-order, partial differential equation, and therefore admits as a solution the superposition of all linearly independent solutions. (For a quantum particle with an associated quantum state, the Schrödinger wave equation describes the evolution of its quantum state given the initial state.) Quantum superposition is a highly non-intuitive and paradoxical result with no corresponding analog in the classical realm. Here we eschew a rigorous derivation of the Schrödinger wave equation. Instead, we rely on Schrödinger's cat to provide an intuitive illustration of the idea of quantum superposition.

Suppose there is a cat inside a steel chamber along with a certain amount of radioactive material that may or may not decay within a fixed time being considered. If the radioactive material decays then a poisonous gas is released inside the steel chamber, killing the cat. If, however, there is no radioactive decay, the cat lives. Schrödinger reasoned that assuming that we cannot look inside the chamber we can associate a certain likelihood with the event the cat is alive and also with the event the cat is dead. We therefore have a highly non-intuitive situation where Schrödinger's cat is both alive and dead at the same time! In other words, the quantum state of Schrödinger's cat is a superposition of the two states of being alive and dead. It is only when we open the chamber – that is, make a measurement – that we can introduce the classical notion that the cat is either alive or dead.

This highly paradoxical notion of quantum superposition made explicit by Schrödinger's thought experiment is practically manifested in Young's double-slit experiment [20]. Let us consider a stream of photons in the form of a light beam impinging on an opaque screen with two parallel vertical slits. The photons passing through these slits are recorded on a photographic plate. If we close any one of the slits, the result on the photographic plate is an intensity band corresponding to the slit that was open. (Note that we are not considering classical diffraction effects at the edges of the slit.) Applying the classical notion of a particle we deduce that if both slits are open we would get two intensity bands corresponding to the two slits. However, on performing the experiment with both slits open we get an interference pattern, completely different from the classical prediction. This interference can be explained by the quantum superposition principle: Each photon that impinges on the screen does not have only one specific path to the photographic plate through a single slit. Instead, the photon simultaneously passes through both the slits and follows every conceivable path to the photographic plate, resulting in a wave-like interference with itself. This situation is analogous to Schrödinger's cat being both dead and alive. It is important to note that here, as in there, the superposition exists until the time we do not measure it. The moment we attempt a measurement by trying to ascertain which slit the electron passes through the superposition collapses. What we end up measuring is either of the following two situations:

- A photon passes through slit 1 with probability $|\psi_1|^2$
- A photon passes through slit 2 with probability $|\psi_2|^2$.

Here ψ is the *probability amplitude* function, called the Schrödinger wave function, that completely describes the quantum behavior of the photon. In quantum mechanics the wave function can be defined as a finite or infinite vector, or a function of a set of continuous real variables depending on the quantum mechanical system under consideration. In this chapter we will be primarily concerned with a two-state quantum system, which can be described as a vector in a complex two-dimensional Hilbert space. We refer to this wave-function representation as a qubit. We formalize the definition as follows. Let $|\mu\rangle$ and $|\nu\rangle$ form the basis for a complex two-dimensional Hilbert space. Then the qubit $|\psi\rangle$ is given by

$$|\psi\rangle = \alpha|\mu\rangle + \beta|\nu\rangle, \tag{4.1}$$

where α and β are complex probability amplitudes. In quantum information processing $|\mu\rangle$ and $|\nu\rangle$ are referred to as the computational basis. Without any loss in generality we assume the computational basis $|\mu\rangle$and $|\nu\rangle$ to be the orthogonal canonical states $|0\rangle = \begin{pmatrix}1\\0\end{pmatrix}$ and $|1\rangle = \begin{pmatrix}0\\1\end{pmatrix}$ respectively, resulting in

$$|\psi\rangle = \alpha|0\rangle + \beta|1\rangle = \begin{pmatrix}\alpha\\\beta\end{pmatrix}. \tag{4.2}$$

Equation (4.2) makes explicit the connection between wave function and probability amplitude, and it illustrates qubit $|\psi\rangle$ being represented by a quantum superposition of $|0\rangle$ and $|1\rangle$. However, as with Schrödinger's cat, if we attempted to measure this quantum state we would get a classical result, 0 or 1 with probabilities proportional to $|\alpha|^2$ and $|\beta|^2$ respectively. In other words, any attempt to measure the quantum bit results in the measurement of a classical bit; whether it is 0 or 1, however, is a decision we cannot make, but instead is made for us by the quantum world. It is important to note that because 0 and 1 are the only two measurement outcomes the relation

$$|\alpha|^2+|\beta|^2 = 1 \tag{4.3}$$

is satisfied.

It is superposition that makes quantum parallelism possible. To see this, let us juxtapose n qubits lying in n distinct two-dimensional Hilbert spaces H_0, H_1, ..., H_{n-1} that are isomorphic to each other. In practice this means the qubits have been prepared separately, without any interaction. This juxtaposition is mathematically described by the tensor product

$$|\psi_0\rangle \otimes |\psi_1\rangle \otimes \ldots \otimes |\psi_{n-1}\rangle \in H_0 \otimes H_1 \otimes \ldots \otimes H_{n-1}. \tag{4.4}$$

Any arbitrary basis can be selected as the computational basis for H_i, $i = 0, 1, \ldots, n-1$. However, for ease of exposition, we assume the computational basis to be $|0_i\rangle$ and $|1_i\rangle$. Consequently, we can represent the ith qubit as

$$|\psi_i\rangle = \alpha_i|0_i\rangle + \beta_i|1_i\rangle. \tag{4.5}$$

Introducing a further assumption, $\alpha_i = \frac{1}{\sqrt{2}}$ and $\beta_i = \frac{1}{\sqrt{2}}$, we now have

$$
\begin{aligned}
&|\psi_0\rangle \otimes |\psi_1\rangle \otimes \ldots \otimes |\psi_{n-1}\rangle \\
&\quad = \frac{1}{\sqrt{2}}(|0_0\rangle + |1_0\rangle) \otimes \frac{1}{\sqrt{2}}(|0_1\rangle + |1_1\rangle) \otimes \ldots \otimes \frac{1}{\sqrt{2}}(|0_{n-1}\rangle + |1_{n-1}\rangle) \qquad (4.6)
\end{aligned}
$$

$$
\begin{aligned}
&= \frac{1}{\sqrt{2^n}}(|0_0\rangle \otimes |0_1\rangle \otimes \ldots \otimes |0_{n-1}\rangle + |0_0\rangle \otimes |0_1\rangle \otimes \ldots \otimes |1_{n-1}\rangle \\
&\quad + |0_0\rangle \otimes |0_1\rangle \otimes \ldots \otimes |1_{n-2}\rangle \otimes |0_{n-1}\rangle \ldots + |1_0\rangle \otimes |1_1\rangle \otimes \ldots \otimes |1_{n-1}\rangle). \qquad (4.7)
\end{aligned}
$$

This composite quantum system is called the n-qubit register and, as can be seen from (4.7), is a superposition of 2^n quantum states that exist simultaneously! This is an example of quantum parallelism. In the classical realm a linear increase in size corresponds roughly to a linear increase in processing power. In the quantum world, due to the power of quantum parallelism, a linear increase in size corresponds to an exponential increase in processing power. The downside, however, is the accessibility to this parallelism. Remember that superposition collapses the moment we attempt to measure it. This raises the question if quantum parallelism has practical uses. The first insight into this question was provided by the seminal works of Deutsch and Jozsa [21]. In 1985 Deutsch proposed the first quantum algorithm to determine whether or not a Boolean function is constant by exploiting quantum parallelism [22]. In 1992 Deutsch and Jozsa generalized the Deutsch algorithm to n qubits [23]. Both these algorithm only showed the feasibility of quantum computing without any practical applications. Shor's groundbreaking work [24] changed that when he showed how quantum parallelism could be exploited to achieve a polynomial–time quantum algorithm for prime factorization that achieved the promise of exponential speed-up over classical factorization algorithms. Furthermore, Grover proposed a quantum algorithm [25] that achieved a quadratic speed-up over classical algorithms to perform unstructured database searches. These algorithms are presented in detail in later chapters.

4.2 NO-CLONING THEOREM

Just like parallelism, quantum superposition is also the key idea behind our inability to clone arbitrary quantum states. To see this, let us think of a quantum copier that takes as input an arbitrary quantum state and outputs two copies of that state, resulting in a clone of the original state. For example, if the input state is $|\psi\rangle$, then the output of the copier is $|\psi\rangle \otimes |\psi\rangle$. For an arbitrary quantum state, such a copier raises a fundamental contradiction. Consider two arbitrary states $|\psi\rangle$ and $|\chi\rangle$ that are input to the copier. When they are inputted individually we expect to get $|\psi\rangle \otimes |\psi\rangle$ and $|\chi\rangle \otimes |\chi\rangle$. Now consider a superposition of these two states given by

$$|\varphi\rangle = \alpha|\psi\rangle + \beta|\chi\rangle \qquad (4.8)$$

Based on the above description that the quantum copier clones the original state, we expect the output

$$|\varphi\rangle \otimes |\varphi\rangle = (\alpha|\psi\rangle + \beta|\chi\rangle) \otimes (\alpha|\psi\rangle + \beta|\chi\rangle), \tag{4.9}$$

that can be expanded to

$$|\varphi\rangle \otimes |\varphi\rangle = \alpha^2|\psi\rangle \otimes |\psi\rangle + \beta^2|\chi\rangle \otimes |\chi\rangle + \alpha\beta|\psi\rangle \otimes |\chi\rangle + \beta\alpha|\chi\rangle \otimes |\psi\rangle. \tag{4.10}$$

On the other hand, linearity of quantum mechanics, as evidenced by the Schrödinger wave equation, tells us that the quantum copier can be represented by a unitary operator that performs the cloning. If such a unitary operator were to act on the superposition state $|\varphi\rangle$, the output would be a superposition of $|\psi\rangle \otimes |\psi\rangle$ and $|\chi\rangle \otimes |\chi\rangle$:

$$|\varphi'\rangle = \alpha|\psi\rangle \otimes |\psi\rangle + \beta|\chi\rangle \otimes |\chi\rangle. \tag{4.11}$$

As is clearly evident, the difference between (4.10) and (4.11) leads to the contradiction mentioned above. As a consequence, there is no unitary operator that can clone $|\varphi\rangle$ as described by (4.10). We therefore have the no-cloning theorem: No quantum copier can clone an arbitrary quantum state.

This result raises a related question: Do there exist some specific states for which cloning is possible? The answer to this question is surprisingly yes. Remember, a key result of quantum mechanics is that unitary operators preserve probabilities. This implies that inner products $\langle\varphi|\varphi\rangle$ and $\langle\varphi'|\varphi'\rangle$ should be identical. Using (4.8) and (4.11), the inner products $\langle\varphi|\varphi\rangle$ and $\langle\varphi'|\varphi'\rangle$ are respectively given by

$$\langle\varphi|\varphi\rangle = \alpha^2|\langle\psi|\psi\rangle| + \beta^2|\langle\chi|\chi\rangle| + \alpha^*\beta|\langle\psi|\chi\rangle| + \alpha\beta^*|\langle\chi|\psi\rangle| \tag{4.12}$$

and

$$\langle\varphi'|\varphi'\rangle = \alpha^2|\langle\psi|\psi\rangle|^2 + \beta^2|\langle\chi|\chi\rangle|^2 + \alpha^*\beta|\langle\psi|\chi\rangle|^2 + \alpha\beta^*|\langle\chi|\psi\rangle|^2. \tag{4.13}$$

We know that $|\langle\psi|\psi\rangle| = |\langle\psi|\psi\rangle|^2 = 1$ and $|\langle\chi|\chi\rangle| = |\langle\chi|\chi\rangle|^2 = 1$. Therefore, the discrepancy lies in the cross terms. Specifically, to avoid the contradiction that resulted in the no-cloning theorem we require that $|\langle\chi|\psi\rangle|^2 = |\langle\chi|\psi\rangle|$. This condition is satisfied when the states are orthogonal. Thus, cloning is possible only for mutually orthogonal states. It is, however, important to remember a subtle point here. Even if we have a mutually orthogonal set of states we need a quantum copier (or unitary operator) specifically for those states. If the unitary operator is specific to a different set of mutually orthogonal states, cloning will fail.

It would seem that the no-cloning theorem would prevent us from exploiting the richness of quantum mechanics. It turns out that this is not the case. A key example is quantum key distribution (QKD), that with very high probability guarantees secure communication. QKD protocols will be discussed in detail in later chapters.

4.3 DISTINGUISHING QUANTUM STATES

Not only can non-orthogonal quantum states not be cloned, they also cannot be reliably distinguished. There is no measurement we can perform that can reliably distinguish non-orthogonal states. This fundamental result plays an important role in quantum cryptography. Its proof is based on contradiction.

Let us suppose Alice prepares two quantum states $|\psi_1\rangle$ and $|\psi_2\rangle$ that are not orthogonal to each other and gives them to Bob to see if he can distinguish between them. Bob asserts that he can design a measurement that can distinguish the two states. He has a set of measurement operators $\{\boldsymbol{P}_i\}$ and designs a decision rule d that, based on the measurement outcome, makes the decision if the quantum state is 1 or 2. Toward that end he designs

$$\boldsymbol{O}_1 = \sum_{\{i|d(i)=1\}} \boldsymbol{P}_i^\dagger \boldsymbol{P}_i, \tag{4.14}$$

and

$$\boldsymbol{O}_2 = \sum_{\{i|d(i)=2\}} \boldsymbol{P}_i^\dagger \boldsymbol{P}_i, \tag{4.15}$$

such that

$$\langle\psi_1|\boldsymbol{O}_1|\psi_1\rangle = \langle\psi_1|\sqrt{\boldsymbol{O}_1}\sqrt{\boldsymbol{O}_1}|\psi_1\rangle = 1, \tag{4.16}$$

and

$$\langle\psi_2|\boldsymbol{O}_2|\psi_2\rangle = \langle\psi_2|\sqrt{\boldsymbol{O}_2}\sqrt{\boldsymbol{O}_2}|\psi_2\rangle = 1. \tag{4.17}$$

Equation (4.16) states that for state $|\psi_1\rangle$ the outcome using the measurement operator $\sqrt{\boldsymbol{O}_1}$ occurs with complete certainty. Equation (4.17) states the same for $|\psi_2\rangle$. As a consequence of (4.16) Bob can state that

$$\langle\psi_2|\sqrt{\boldsymbol{O}_1}\sqrt{\boldsymbol{O}_1}|\psi_2\rangle = 0. \tag{4.18}$$

The validity of (4.18) requires that $\sqrt{\boldsymbol{O}_1}|\psi_2\rangle = 0$. Bob then decomposes $|\psi_1\rangle$ along $|\psi_2\rangle$ and a quantum state $|\xi\rangle$ orthogonal to $|\psi_2\rangle$ to get

$$|\psi_1\rangle = c_1|\psi_2\rangle + c_2|\xi\rangle, \tag{4.19}$$

such that

$$|c_1|^2 + |c_2|^2 = 1. \tag{4.20}$$

Bob notes that since $|\psi_1\rangle$ and $|\psi_2\rangle$ are non-orthogonal $|c_1|^2 \neq 0$, and consequently $|c_2|^2 < 1$. He further notes that

$$\sqrt{O_1}|\psi_1\rangle = \sqrt{O_1}(c_1|\psi_2\rangle + c_2|\xi\rangle) \tag{4.21}$$

$$= c_1\sqrt{O_1}|\psi_2\rangle + c_2\sqrt{O_1}|\xi\rangle \tag{4.22}$$

$$= 0 + c_2\sqrt{O_1}|\xi\rangle \tag{4.23}$$

$$\Rightarrow \sqrt{O_1}|\psi_1\rangle = c_2\sqrt{O_1}|\xi\rangle \tag{4.24}$$

Equation (4.23) follows from (4.18). Bob then substitutes (4.24) in $\langle\psi_1|\sqrt{O_1}\sqrt{O_1}|\psi_1\rangle$ to get

$$\langle\psi_1|\sqrt{O_1}\sqrt{O_1}|\psi_1\rangle = c_2c_2^*\langle\xi|\sqrt{O_1}\sqrt{O_1}|\xi\rangle, \tag{4.25}$$

$$\Rightarrow \langle\psi_1|\sqrt{O_1}\sqrt{O_1}|\psi_1\rangle = |c_2|^2\langle\xi|\sqrt{O_1}\sqrt{O_1}|\xi\rangle. \tag{4.26}$$

Bob knows that the completeness relation tells him

$$\sum_k O_k = \sum_k \sum_{\{i|d(i)=k\}} P_i^\dagger P_i = 1, \tag{4.27}$$

resulting in

$$1 = \langle\xi|\xi\rangle = \left\langle\xi\middle|\sum_k O_k\middle|\xi\right\rangle \geq \langle\xi|O_1|\xi\rangle = \left\langle\xi\middle|\sqrt{O_1}\sqrt{O_1}\middle|\xi\right\rangle. \tag{4.28}$$

Substituting (4.28) in (4.26), Bob gets

$$\left\langle\psi_1\middle|\sqrt{O_1}\sqrt{O_1}\middle|\psi_1\right\rangle = |c_2|^2\left\langle\xi\middle|\sqrt{O_1}\sqrt{O_1}\middle|\xi\right\rangle \leq |c_2|^2 < 1, \tag{4.29}$$

leading to a contradiction with (4.16). Thus, Bob's initial assertion is wrong, resulting in his inability to distinguish non-orthogonal quantum states. Note that if the $|\psi_1\rangle$ and $|\psi_2\rangle$ are orthogonal we have $|c_1|^2 = 0$, and $|c_2|^2 = 1$ and Bob can indeed design measurements that are able to distinguish the orthogonal states.

4.4 QUANTUM ENTANGLEMENT

Quantum entanglement is another fascinating and key idea from the quantum realm that has no counterpart in the classical world. It is a subject of intense current research and is seen by many physicists as the holy grail of quantum mechanics. Quantum entanglement is not completely understood and efforts are underway to remedy this. In this section we explain quantum entanglement through the well-understood pure state entanglement of a composite system.

Let $|\psi_0\rangle$ and $|\psi_1\rangle$ be two qubits lying in the two-dimensional Hilbert spaces H_0 and H_1 respectively, and let the state of the joint quantum system lying in $H_0 \otimes H_1$ be denoted by $|\psi\rangle$. The qubit $|\psi\rangle$ is said to be entangled if it cannot be written in the product state form

$$|\psi\rangle = |\psi_0\rangle \otimes |\psi_1\rangle. \tag{4.30}$$

As an example, the state

$$|\psi\rangle = \frac{1}{\sqrt{2}}(|0_0\rangle \otimes |0_1\rangle + |1_0\rangle \otimes |1_1\rangle) \tag{4.31}$$

is an entangled state because it cannot be expressed in a product state form.

The definition of entanglement can be extended from a two-qubit case to an n-qubit case. Let $|\psi_0\rangle, \ldots, |\psi_{n-1}\rangle$ be n qubits lying in the Hilbert spaces $H_0, \ldots, H_{n-1}$ respectively, and let the state of the joint quantum system lying in $H_0 \otimes \ldots \otimes H_{n-1}$ be denoted by $|\psi\rangle$. The qubit $|\psi\rangle$ is then said to be entangled if it cannot be written in the product state form:

$$|\psi\rangle = |\psi_0\rangle \otimes |\psi_1\rangle \otimes \ldots \otimes |\psi_{n-1}\rangle. \tag{4.32}$$

4.4.1 Schmidt Decomposition

For a bipartite system, it can be elegantly determined whether or not the qubit $|\psi\rangle$ is in a product state or an entangled one. The Schmidt decomposition states that a pure state $|\psi\rangle$ of the composite system AB can be represented as

$$|\psi\rangle = \sum_i c_i |i_A\rangle \otimes |i_B\rangle, \tag{4.33}$$

where $|i_A\rangle$ and $|i_B\rangle$ are orthonormal bases of the subsystems A and B respectively, and $c_i \in \mathbb{R}_+$ are Schmidt coefficients that satisfy $\sum_i c_i^2 = 1$.

Let us prove this theorem for the general case where the respective Hilbert spaces of the quantum systems A and B are not two-dimensional as we have considered till now, but are instead m-dimensional. For each of these m-dimensional Hilbert spaces we can associate an orthonormal basis. Let us label the two bases as $\{|a_0\rangle, |a_1\rangle, \ldots, |a_{m-1}\rangle\}$ and $\{|b_0\rangle, |b_1\rangle, \ldots, |b_{m-1}\rangle\}$. The state of the composite system can now be synthesized from these bases as

$$|\psi\rangle = \sum_k \sum_l \gamma_{kl} |a_k\rangle \otimes |b_l\rangle. \tag{4.34}$$

The coefficients γ_{kl} can be expressed as a matrix $\Gamma = [\gamma_{kl}]$. To factorize this matrix we employ the well-known factorization method of singular valued decomposition that expresses Γ as

$$\Gamma = U \Lambda V^{\dagger}, \tag{4.35}$$

where U is a matrix of eigenvectors of $\Gamma\Gamma^\dagger$, V is a matrix of eigenvectors of $\Gamma^\dagger\Gamma$, and Λ is a diagonal matrix of singular values. Based on this decomposition the elements of Γ can be expressed as

$$\gamma_{kl} = \sum_i u_{ki}\lambda_{ii}v_{li}^*. \tag{4.36}$$

Substituting (4.36) into (4.34) we get

$$|\psi\rangle = \sum_k \sum_l \sum_i u_{ki}\lambda_{ii}v_{li}^*|a_k\rangle \otimes |b_l\rangle. \tag{4.37}$$

We now consider two marginal summations and define them as

$$|i_A\rangle = \sum_k u_{ki}|a_k\rangle \tag{4.38}$$

and

$$|i_B\rangle = \sum_l v_{li}^*|b_l\rangle. \tag{4.39}$$

Substituting these marginal sums in (4.37) we get

$$|\psi\rangle = \sum_i c_i|i_A\rangle \otimes |i_B\rangle, \tag{4.40}$$

where we have $c_i = \lambda_{ii}$. Since we know $|\psi\rangle$ is a pure state, it is evident that $\sum_i c_i^2 = \sum_i \lambda_{ii}^2 = 1$. Let us look further at the orthonormality of $|i_A\rangle$ by considering any two of its bases $|q_A\rangle$ and $|r_A\rangle$, $q, r \in \{0, 1, \ldots, m-1\}$:

$$|q_A\rangle = \sum_k u_{kq}|a_k\rangle, \tag{4.41}$$

$$|r_A\rangle = \sum_k u_{kr}|a_k\rangle. \tag{4.42}$$

The inner product of these two states is

$$\langle q_A|r_A\rangle = \sum_s \sum_t \langle a_s|u_{sq}^*u_{tr}|a_t\rangle. \tag{4.43}$$

From orthonormality of $|a_s\rangle$ and $|a_t\rangle$, $t, s \in \{0, 1, \ldots, m-1\}$, Eq. (4.43) reduces to

$$\langle q_A|r_A\rangle = \sum_k \langle a_k|u_{kq}^*u_{kr}|a_k\rangle. \tag{4.44}$$

$$= \sum_k u_{kq}^*u_{kr}\langle a_k|a_k\rangle \tag{4.45}$$

$$= \sum_k u_{kq}^*u_{kr} \qquad \text{(due to orthonormality } \langle a_k|a_k\rangle = 1\text{)}. \tag{4.46}$$

Finally, due to the orthonormality of eigenvectors $\sum_k u^*_{kq} u_{kr} = 1$, if and only if $q = r$. Therefore, $|i_A\rangle, i = \{0, 1, \ldots, m-1\}$ is a set of orthonormal states. We can get similar results for $|i_B\rangle$, thus proving the Schmidt decomposition theorem.

The number c_i is known as the Schmidt basis and the number of c_is is called the Schmidt rank. A corollary of the Schmidt decomposition theorem is that a pure state in a composite system is a product state if and only if the Schmidt rank is 1, and is an entangled state if and only if the Schmidt rank is greater than 1. This can be deduced from (4.40). If the Schmidt rank is 1 then we have a product state. On the other hand, if it is greater than 1 we get entanglement.

Example 4.1. Let us consider an example of two qubits in a bipartite system. The state we consider is

$$|\psi\rangle = \frac{1}{2}(|0_A\rangle \otimes |0_B\rangle + |0_A\rangle \otimes |1_B\rangle + |1_A\rangle \otimes |0_B\rangle + |1_A\rangle \otimes |1_B\rangle). \tag{4.47}$$

For ease of exposition we drop the subscripts and write the bipartite state as

$$|\psi\rangle = \frac{1}{2}(|0\rangle \otimes |0\rangle + |0\rangle \otimes |1\rangle + |1\rangle \otimes |0\rangle + |1\rangle \otimes |1\rangle), \tag{4.48}$$

where, as before, $|0\rangle = \begin{pmatrix} 1 \\ 0 \end{pmatrix}$ and $|1\rangle = \begin{pmatrix} 0 \\ 1 \end{pmatrix}$. We can now expand $|\psi\rangle$ as

$$|\psi\rangle = \frac{1}{2}\begin{pmatrix} 1 \\ 1 \\ 1 \\ 1 \end{pmatrix}. \tag{4.49}$$

If we consider the orthonormal basis

$$|0_A\rangle = |0_B\rangle = \frac{1}{\sqrt{2}}\begin{pmatrix} 1 \\ 1 \end{pmatrix} \tag{4.50}$$

and

$$|1_A\rangle = |1_B\rangle = \frac{1}{\sqrt{2}}\begin{pmatrix} 1 \\ -1 \end{pmatrix}, \tag{4.51}$$

we can then represent $|\psi\rangle$ as the product state

$$|\psi\rangle = |0_A\rangle \otimes |0_B\rangle. \tag{4.52}$$

As can be seen, the Schmidt rank is 1 and the value of the Schmidt basis is 1.

Example 4.2. Let us now turn to our second example that we gave as an example of the entangled state:

$$|\psi\rangle = \frac{1}{\sqrt{2}}(|0_A\rangle \otimes |0_B\rangle + |1_A\rangle \otimes |1_B\rangle), \tag{4.53}$$

$$= \frac{1}{\sqrt{2}}\begin{pmatrix} 1 \\ 0 \\ 0 \\ 1 \end{pmatrix}. \tag{4.54}$$

Following the proof of the Schmidt decomposition theorem, we first define $\{|a_0\rangle, |a_1\rangle\}$ and $\{|b_0\rangle, |b_1\rangle\}$. Let them be

$$|a_0\rangle = |b_0\rangle = \begin{pmatrix} 1 \\ 0 \end{pmatrix}, \tag{4.55}$$

$$|a_1\rangle = |b_1\rangle = \begin{pmatrix} 0 \\ 1 \end{pmatrix}. \tag{4.56}$$

As a result the Λ matrix

$$\Lambda = \begin{pmatrix} 1/\sqrt{2} & 0 \\ 0 & 1/\sqrt{2} \end{pmatrix} \tag{4.57}$$

is already diagonalized, telling us that the Schmidt decomposition of $|\psi\rangle$ is $|\psi\rangle$ itself! Also, the Schmidt rank is 2, proving that the quantum state of the bipartite system is entangled. Both Schmidt bases are equal to $\frac{1}{\sqrt{2}}$.

The entangled state discussed above is famously called an EPR pair of qubits, after Einstein, Podolsky and Rosen, who raised doubts about quantum mechanics itself [26]. Their fundamental objection was about the entanglement of qubits and their apparent nonlocal interaction, which intrigued the quantum physics community for quite some time. The objection raised by them is referred to as the hidden variable theory. To put the debate to rest, Bell developed a set of inequalities that, if violated, would validate quantum mechanics [27,28]. Bell's theoretical predictions were experimentally validated [29–31], leading to the establishment of quantum mechanics as the best physical description of nature around us.

Until now we have given the mathematical definition of entanglement and how to identify it. Let us now look at what it physically means. Going back to (4.53), let us suppose we want to measure qubit A. Our measurement outcome would then either be 0 or 1, each with probability 0.5, and consequently this measurement would send qubit A to either state $|0_A\rangle$ or $|1_A\rangle$. Herein lies the weirdness of entanglement. Due to this measurement on qubit A the state of our composite system is either $|0_A\rangle \otimes |0_B\rangle$or $|1_A\rangle \otimes |1_B\rangle$. Thus, we have changed the state of qubit B by performing measurement of qubit A because the two qubits are correlated, coupled, or entangled: A change in state of one causes change of state in the other. In fact, if we have prior knowledge of the EPR pair, a single measurement of one entangled qubit gives us complete information about the state of the other entangled qubit! This idea is exploited, for example, in quantum superdense coding to send two classical bits of information per qubit over a quantum channel. Superdense coding will be discussed in a later chapter, where this idea will be further extended to explain entanglement-assisted quantum error correction codes.

We end this section by stating the complete set of EPR pairs:

$$|\psi_1\rangle = \frac{1}{\sqrt{2}}(|0_A\rangle \otimes |0_B\rangle + |1_A\rangle \otimes |1_B\rangle) \tag{4.58}$$

$$|\psi_2\rangle = \frac{1}{\sqrt{2}}(|0_A\rangle \otimes |1_B\rangle + |1_A\rangle \otimes |0_B\rangle) \tag{4.59}$$

$$|\psi_3\rangle = \frac{1}{\sqrt{2}}(|0_A\rangle \otimes |0_B\rangle - |1_A\rangle \otimes |1_B\rangle) \tag{4.60}$$

$$|\psi_4\rangle = \frac{1}{\sqrt{2}}(|0_A\rangle \otimes |1_B\rangle - |1_A\rangle \otimes |0_B\rangle). \tag{4.61}$$

We leave it as an exercise for the reader to show that these states form an orthonormal basis for the Hilbert space of any bipartite quantum system. In the next chapter it is shown how these EPR pairs can be created using only two quantum gates: the Hadamard and CNOT gates.

4.5 OPERATOR-SUM REPRESENTATION

In this section we introduce the operator-sum representation that provides an elegant way to describe the dynamic evolution of an open quantum system. Closed quantum systems dynamics are described by a unitary transformation. For an open system, though, the initial and the final states are not generally related through a unitary transformation. To develop the operator-sum relation, the trick is to realize that the open system interacts with its surrounding environment. We can therefore describe a closed system as a composite system of the original quantum system and the environment. This closed composite system can then be described by a unitary evolution. To get the final state of the original quantum system we are interested in, we perform a partial trace over the environment.

Let us begin by considering an open quantum system Q that is in the state $\boldsymbol{\rho}_Q$ at some initial time t_i. Here $\boldsymbol{\rho}_Q$ is the density matrix describing the quantum state of Q. The density matrix description is a more efficient representation of a quantum system Q in either pure or mixed state with minimum mathematical clutter. Let us also assume that at time t_i the state of the environment is described by the density matrix $\boldsymbol{\rho}_E = |e_0\rangle\langle e_0|$. The initial state of the composite system is then described by the product state

$$\boldsymbol{\rho}_{QE} = \boldsymbol{\rho}_Q \otimes \boldsymbol{\rho}_E. \tag{4.62}$$

Let us denote by $\boldsymbol{U}$ the unitary transform on the Hilbert space $H_Q \otimes H_E$ in which the closed composite system lies. Here, H_Q and H_E are the Hilbert spaces where the quantum system Q and the environment E respectively lie. In general, any operator acting on the Hilbert space $H_Q \otimes H_E$ of the composite system can be represented as a linear combination of tensor products of linear operators. Therefore, for our case, the unitary transform U can be mathematically described by

$$U = \sum_j k_j T_{Q_j} \otimes T_{E_j}, \tag{4.63}$$

where T_{Q_i} and T_{E_i} are some linear transforms defined on the respective Hilbert spaces H_Q and H_E. The action of U on ρ_{QE} results in

$$\rho'_{QE} = U \rho_{QE} U^\dagger. \tag{4.64}$$

We now perform a partial trace over the environment E to get the desired reduced state:

$$\varepsilon(\rho_Q) = \mathrm{tr}_E(U \rho_{QE} U^\dagger). \tag{4.65}$$

Here $\varepsilon(\rho_Q)$ is the final state of the quantum system Q after quantum operation ε as described by (4.65) is performed on ρ_Q. To get the operator-sum representation let us substitute (4.62) in (4.65). We get

$$\varepsilon(\rho_Q) = \mathrm{tr}_E(U \rho_Q \otimes \rho_E U^\dagger) \tag{4.66}$$

$$= \sum_i \langle e_i | U \rho_Q \otimes \rho_E U^\dagger | e_i \rangle. \tag{4.67}$$

Further substituting (4.63) in (4.67) we get

$$\varepsilon(\rho_Q) = \sum_i \langle e_i | \sum_j k_j T_{Q_j} \otimes T_{E_j} (\rho_Q \otimes \rho_E) \sum_k k_k^* (T_{Q_k} \otimes T_{E_k})^\dagger | e_i \rangle \tag{4.68}$$

$$= \sum_i \sum_j \sum_k k_j k_k^* \langle e_i | T_{Q_j} \rho_Q T_{Q_k}^\dagger \otimes T_{E_j} \rho_E T_{E_k}^\dagger | e_i \rangle \tag{4.69}$$

$$= \sum_i \sum_j \sum_k k_j k_k^* \langle e_i | T_{Q_j} \rho_Q T_{Q_k}^\dagger \otimes T_{E_j} | e_0 \rangle \langle e_0 | T_{E_k}^\dagger | e_i \rangle \tag{4.70}$$

$$= \sum_i \sum_j \sum_k k_j k_k^* T_{Q_j} \rho_Q T_{Q_k}^\dagger \langle e_i | T_{E_j} | e_0 \rangle \langle e_0 | T_{E_k}^\dagger | e_i \rangle \tag{4.71}$$

$$= \sum_i \sum_j \sum_k (k_j T_{Q_j} \langle e_i | T_{E_j} | e_0 \rangle) \rho_Q (k_k^* T_{Q_k}^\dagger \langle e_0 | T_{E_k}^\dagger | e_i \rangle) \tag{4.72}$$

$$= \sum_i \left(\sum_j k_j T_{Q_j} \langle e_i | T_{E_j} | e_0 \rangle \right) \rho_Q \left(\sum_k k_k^* T_{Q_k}^\dagger \langle e_0 | T_{E_k}^\dagger | e_i \rangle \right) \tag{4.73}$$

$$= \sum_i \left(\left\langle e_i \middle| \sum_j k_j T_{Q_j} \otimes T_{E_j} \middle| e_0 \right\rangle \right) \rho_Q \left(\left\langle e_0 \middle| \left(\sum_k k_k T_{Q_k} \otimes T_{E_k} \right)^\dagger \middle| e_i \right\rangle \right) \quad (4.74)$$

$$= \sum_i (\langle e_i|U|e_0\rangle) \rho_Q (\langle e_0|U^\dagger|e_i\rangle) \quad (4.75)$$

$$= \sum_i (\langle e_i|U|e_0\rangle) \rho_Q (\langle e_i|U|e_0\rangle^\dagger). \quad (4.76)$$

Denoting

$$E_i = \langle e_i|U|e_0\rangle, \quad (4.77)$$

we have

$$\varepsilon(\rho_Q) = \sum_i E_i \rho_Q E_i^\dagger. \quad (4.78)$$

Equation (4.78) is the operator-sum representation. The individual operators E_i are variously referred to as Kraus operators, operation elements, or decomposition operators. Kraus operators provide an elegant way to summarize the effect of the environment on the quantum system Q without explicitly considering the physical interactions between the two. This fact can be made explicit by rewriting (4.78) as

$$\varepsilon(\rho_Q) = \sum_i \mathrm{tr}\left(E_i \rho_Q E_i^\dagger\right) \left(\frac{E_i \rho_Q E_i^\dagger}{\mathrm{tr}\left(E_i \rho_Q E_i^\dagger\right)} \right). \quad (4.79)$$

Using the idea of implicit measurements, we can think of the second term in the summation as the state of the quantum system Q as a result of the action of E_i, with normalized probability. The first term is the probability that the quantum system would be acted upon by E_i. Thus, we have defined the effect of the quantum system Q interacting with the environment: With probability $\mathrm{tr}(E_i \rho_Q E_i^\dagger)$, the quantum system goes to the state $\left(\frac{E_i \rho_Q E_i^\dagger}{\mathrm{tr}(E_i \rho_Q E_i^\dagger)} \right)$. This interaction with the environment is very useful in characterizing noisy quantum channels and closely follows the effect of noise on classical channels. In the classical realm we can describe the channel through conditional probabilities: Given that the input is x, the probability that the output is y is given by the channel characteristic $p(y|x)$. This is the same idea behind quantum noise as expressed by (4.79). We will use this idea behind the operator-sum representation to look at quantum errors and decoherence in the following section.

Under the trace-preserving condition that states that

$$\mathrm{tr}(\varepsilon(\rho_Q)) = \mathrm{tr}(\rho_Q) = 1, \quad (4.80)$$

the Kraus operators satisfy the completeness relation:

$$\sum_i \boldsymbol{E}_i^\dagger \boldsymbol{E}_i = \boldsymbol{I}, \tag{4.81}$$

where $\boldsymbol{I}$ is the identity operator. Equation (4.81) is a simple consequence of (4.78) and (4.80):

$$\mathrm{tr}\left(\sum_i \boldsymbol{E}_i \boldsymbol{\rho}_Q \boldsymbol{E}_i^\dagger\right) = 1 \tag{4.82}$$

$$\Rightarrow \mathrm{tr}\left(\sum_i \boldsymbol{E}_i^\dagger \boldsymbol{E}_i \boldsymbol{\rho}_Q\right) = 1 \tag{4.83}$$

$$\Rightarrow \mathrm{tr}\left(\left(\sum_i \boldsymbol{E}_i^\dagger \boldsymbol{E}_i\right) \boldsymbol{\rho}_Q\right) = 1. \tag{4.84}$$

Since we know $\mathrm{tr}(\boldsymbol{\rho}_Q) = 1$, Eq. (4.84) gives us the desired completeness relation $\sum_i \boldsymbol{E}_i^\dagger \boldsymbol{E}_i = 1$.

Let us now look at the nontrace-preserving condition: $\sum_i \boldsymbol{E}_i^\dagger \boldsymbol{E}_i < 1$. This condition arises when we perform an explicit measurement on the composite system QE after its unitary evolution. As an example let us consider a projective measurement operator $\boldsymbol{P}_a$ on the evolved composite system. We can incorporate this projective within the quantum operation. Consequently, in a manner similar to (4.65), the quantum operation is described by

$$\varepsilon_a(\boldsymbol{\rho}_Q) = \mathrm{tr}_E(\boldsymbol{P}_a \boldsymbol{U} \boldsymbol{\rho}_{QE} \boldsymbol{U}^\dagger \boldsymbol{P}_a). \tag{4.85}$$

We leave it as an exercise for the reader to show that (4.85) can be expressed in an operator-sum representation. The trace of $\varepsilon_a(\boldsymbol{\rho}_Q)$ is the probability that the measurement outcome a occurs. Since it is the probability of one particular outcome, we get the nontrace-preserving condition $\mathrm{tr}(\varepsilon_a(\boldsymbol{\rho}_Q)) < 1$.

As a final point, we note that the operator-sum representation describing a particular quantum operation is not unique. Specifically, we state without proof that two sets of Kraus operators that are related to each other through unitary transformations represent the same quantum operation.

4.6 DECOHERENCE AND QUANTUM ERRORS

Quantum computation works by manipulating the quantum interference effect. We discussed at the beginning of this chapter that quantum interference, a manifestation of coherent superposition of quantum states, is the cornerstone behind all quantum information tasks such as quantum computation and quantum communication.

A major source of problems is our inability to prevent our quantum system of interest from interacting with the surrounding environment. This interaction results in an entanglement between the quantum system and the environment, leading to decoherence.

To understand this system–environment entanglement and decoherence, let us consider an example where the quantum system Q of interest is in the quantum state

$$|\psi_Q\rangle = \alpha|0\rangle + \beta|1\rangle, \tag{4.86}$$

where, as before, $|0\rangle$ and $|1\rangle$ are the computational basis. We assume the initial state of the environment E to be $|\psi_E\rangle$. Thus, the initial state of the composite system QE is

$$|\psi_{QE}\rangle = |\psi_Q\rangle \otimes |\psi_E\rangle = \alpha|0\rangle \otimes |\psi_E\rangle + \beta|1\rangle \otimes |\psi_E\rangle. \tag{4.87}$$

Let this composite state undergo a unitary evolution $\boldsymbol{U}$ such that the resulting state is

$$\boldsymbol{U}|\psi_{QE}\rangle = \alpha|0\rangle \otimes |\psi_{E_1}\rangle + \beta|1\rangle \otimes |\psi_{E_2}\rangle. \tag{4.88}$$

The corresponding density operator for this state is

$$\begin{aligned} \boldsymbol{\rho}'_{QE} = {} & |\alpha|^2(|0\rangle \otimes |\psi_{E_1}\rangle)(\langle 0| \otimes \langle \psi_{E_1}|) + |\beta|^2(|1\rangle \otimes |\psi_{E_2}\rangle)(\langle 1| \otimes \langle \psi_{E_2}|) \\ & + \alpha^*\beta(|1\rangle \otimes |\psi_{E_2}\rangle)(\langle 0| \otimes \langle \psi_{E_1}|) + \alpha\beta^*(|0\rangle \otimes |\psi_{E_1}\rangle)(\langle 1| \otimes \langle \psi_{E_2}|). \end{aligned} \tag{4.89}$$

Taking the partial trace over the environment E results in

$$\begin{aligned} \boldsymbol{\rho}'_Q = {} & \mathrm{tr}_E\Big(|\alpha|^2(|0\rangle \otimes |\psi_{E_1}\rangle\Big)(\langle 0| \otimes \langle \psi_{E_1}|) + |\beta|^2(|1\rangle \otimes |\psi_{E_2}\rangle)(\langle 1| \otimes \langle \psi_{E_2}|) \\ & + \alpha^*\beta(|1\rangle \otimes |\psi_{E_2}\rangle)(\langle 0| \otimes \langle \psi_{E_1}|) + \alpha\beta^*(|0\rangle \otimes |\psi_{E_1}\rangle)(\langle 1| \otimes \langle \psi_{E_2}|)\Big) \end{aligned} \tag{4.90}$$

$$\begin{aligned} \Rightarrow \boldsymbol{\rho}'_Q = {} & |\alpha|^2\langle \psi_{E_1}|\psi_{E_1}\rangle|0\rangle \otimes \langle 0| + |\beta|^2\langle \psi_{E_2}|\psi_{E_2}\rangle|1\rangle \otimes \langle 1| + \alpha^*\beta\langle \psi_{E_2}|\psi_{E_1}\rangle|1\rangle \otimes \langle 0| \\ & + \alpha\beta^*\langle \psi_{E_1}|\psi_{E_2}\rangle|0\rangle \otimes \langle 1|. \end{aligned} \tag{4.91}$$

In matrix form $\boldsymbol{\rho}'_Q$ is given by

$$\boldsymbol{\rho}'_Q = \begin{pmatrix} |\alpha|^2\langle \psi_{E_1}|\psi_{E_1}\rangle & \alpha\beta^*\langle \psi_{E_1}|\psi_{E_2}\rangle \\ \alpha^*\beta\langle \psi_{E_2}|\psi_{E_1}\rangle & |\beta|^2\langle \psi_{E_2}|\psi_{E_2}\rangle \end{pmatrix}. \tag{4.92}$$

We can simplify (4.92) by noting that $\langle \psi_{E_1}|\psi_{E_1}\rangle = 1$ and $\langle \psi_{E_2}|\psi_{E_2}\rangle = 1$. We therefore have

$$\boldsymbol{\rho}'_Q = \begin{pmatrix} |\alpha|^2 & \alpha\beta^*\langle \psi_{E_1}|\psi_{E_2}\rangle \\ \alpha^*\beta\langle \psi_{E_2}|\psi_{E_1}\rangle & |\beta|^2 \end{pmatrix}. \tag{4.93}$$

Note that in the off-diagonal terms $\alpha^*\beta$ and $\alpha\beta^*$ indicate the quantum interference effect of coherent superposition. However, due to the interaction between the quantum system Q and the environment E we have the additional components $\langle\psi_{E_1}|\psi_{E_2}\rangle$ and $\langle\psi_{E_2}|\psi_{E_1}\rangle$. These terms indicate an entanglement of (or coupling between) the quantum system Q with the environment E, and in the following sense a loss in coherence and loss of information to the environment. The two environment-related terms are in reality time dependent and can be usefully described by an exponential decay function,

$$\langle\psi_{E_1}|\psi_{E_2}\rangle = \mathrm{e}^{-\gamma(t)}, \tag{4.94}$$

where the exponent $\gamma(t)$ is a positive real number. The larger the decay, the more off-diagonal terms $\alpha^*\beta$ and $\alpha\beta^*$ are suppressed. As a consequence, $\gamma(t)$ parameterizes the system—environment entanglement, and is an example of what is known as amplitude damping. Furthermore, it also characterizes the loss in coherent (relative phase) information. This is known as the dephasing effect of decoherence. A more general example of dephasing is depolarization, which will be considered later in this section.

In the above example, we have considered the coupling between a single-qubit quantum system and the environment, and discussed the resulting loss of interference or coherent superposition. In general, for multiple qubit systems decoherence also results in loss of coupling between the qubits, which is highly desirable in any quantum computer. In fact, with increasing complexity and size of the computer, the decoherence effect becomes worse. Additionally, the quantum system lies in some complex Hilbert space where there are infinite variations of errors that can cause decoherence. How is it ever possible to correct for each of these quantum errors? The first step was the realization that the effect of any of the infinite variations of quantum errors on a quantum system of interest can be completely characterized by four quantum operations. We describe these operations by considering their effect on our favorite qubit

$$|\psi\rangle = \alpha|0\rangle + \beta|1\rangle. \tag{4.95}$$

1. Identity operation — This operation keeps the qubit as it is.
2. Bit-flip operation — This operation flips the computational bases, resulting in $|\psi'\rangle = \alpha|1\rangle + \beta|0\rangle$.
3. Phase-flip operation — This operation flips the phase of the second computational basis, resulting in $|\psi'\rangle = \alpha|0\rangle - \beta|1\rangle$.
4. Bit-and-phase-flip operation — This operation flips the computational bases and changes the phase of the second basis, resulting in $|\psi'\rangle = \alpha|1\rangle - \beta|0\rangle$.

These four operations are performed by the following matrices:

1. Identity operation (**I**): $\begin{pmatrix} 1 & 0 \\ 0 & 1 \end{pmatrix}$
2. Bit-flip operation (**X**): $\begin{pmatrix} 0 & 1 \\ 1 & 0 \end{pmatrix}$

3. Phase-flip operation (**Z**): $\begin{pmatrix} 1 & 0 \\ 0 & -1 \end{pmatrix}$

4. Bit-and-phase-flip operation (**Y**): $\begin{pmatrix} 0 & -i \\ i & 0 \end{pmatrix}$.

These matrices **X**, **Y**, and **Z** are called the Pauli matrices. Pauli matrices will be discussed in greater detail in a later chapter, as they play a key role in quantum computing and quantum communication. Here, we prove the equivalence between the above four quantum operations completely characterizing quantum errors and the Pauli matrices along with the identity operator forming an orthogonal basis for a complex Hilbert space $H_{2\times 2}$ of 2×2 matrices. Toward that end we use the Hilbert–Schmidt inner product, defined as

$$\langle \boldsymbol{A}, \boldsymbol{B} \rangle_{\mathrm{HS}} = \frac{1}{2}\mathrm{tr}(\boldsymbol{A}^{\dagger}\boldsymbol{B}), \tag{4.96}$$

where $\boldsymbol{A}, \boldsymbol{B} \in H_{2\times 2}$, to first show that all four matrices are orthogonal:

$$\begin{aligned} \frac{1}{2}\mathrm{tr}(\mathbf{I}^{\dagger}\mathbf{X}) &= \frac{1}{2}\mathrm{tr}\left(\begin{pmatrix} 1 & 0 \\ 0 & 1 \end{pmatrix}\begin{pmatrix} 0 & 1 \\ 1 & 0 \end{pmatrix}\right) = \frac{1}{2}\mathrm{tr}\left(\begin{pmatrix} 0 & 1 \\ 1 & 0 \end{pmatrix}\right) \\ &= 0 = \frac{1}{2}\mathrm{tr}(\mathbf{I}^{\dagger}\mathbf{Z}) = \frac{1}{2}\mathrm{tr}(\mathbf{I}^{\dagger}\mathbf{Y}) \end{aligned} \tag{4.97}$$

$$\begin{aligned} \frac{1}{2}\mathrm{tr}(\mathbf{X}^{\dagger}\mathbf{Y}) &= \frac{1}{2}\mathrm{tr}\left(\begin{pmatrix} 0 & 1 \\ 1 & 0 \end{pmatrix}\begin{pmatrix} 0 & -i \\ i & 0 \end{pmatrix}\right) = \frac{1}{2}\mathrm{tr}\left(\begin{pmatrix} i & 0 \\ 0 & -i \end{pmatrix}\right) \\ &= 0 = \frac{1}{2}\mathrm{tr}(\mathbf{Y}^{\dagger}\mathbf{X}) \end{aligned} \tag{4.98}$$

$$\begin{aligned} \frac{1}{2}\mathrm{tr}(\mathbf{X}^{\dagger}\mathbf{Z}) &= \frac{1}{2}\mathrm{tr}\left(\begin{pmatrix} 0 & 1 \\ 1 & 0 \end{pmatrix}\begin{pmatrix} 1 & 0 \\ 0 & -1 \end{pmatrix}\right) = \frac{1}{2}\mathrm{tr}\left(\begin{pmatrix} 0 & -1 \\ 1 & 0 \end{pmatrix}\right) \\ &= 0 = \frac{1}{2}\mathrm{tr}(\mathbf{Z}^{\dagger}\mathbf{X}) \end{aligned} \tag{4.99}$$

$$\begin{aligned} \frac{1}{2}\mathrm{tr}(\mathbf{Y}^{\dagger}\mathbf{Z}) &= \frac{1}{2}\mathrm{tr}\left(\begin{pmatrix} 0 & -i \\ i & 0 \end{pmatrix}\begin{pmatrix} 1 & 0 \\ 0 & -1 \end{pmatrix}\right) = \frac{1}{2}\mathrm{tr}\left(\begin{pmatrix} 0 & i \\ i & 0 \end{pmatrix}\right) \\ &= 0 = \frac{1}{2}\mathrm{tr}(\mathbf{Z}^{\dagger}\mathbf{Y}) \end{aligned} \tag{4.100}$$

Based on the Hilbert–Schmidt inner product, we can go a step further and show that **I**, **X**, **Y**, **Z** are not only orthogonal but also orthonormal. This is simple to do and is left as an exercise for the reader.

Next, we need to prove that **I, X, Y, Z** spans $H_{2\times 2}$, i.e. we can express any matrix $P \in H_{2\times 2}$ as

$$P = c_1 I + c_2 X + c_3 Y + c_4 Z, \tag{4.101}$$

such that $c_1, c_2, c_3, c_4 \in \mathbb{C}$. To show this, let us first define P as

$$P = \begin{pmatrix} r & s \\ t & u \end{pmatrix}, \tag{4.102}$$

where $r, s, t, u \in \mathbb{C}$. Due to the orthonormality of **I, X, Y, Z**, the coefficients can be easily calculated using the Hilbert–Schmidt inner product:

$$c_1 = \frac{1}{2}\mathrm{tr}(\mathbf{I}^\dagger \mathbf{P}) = \frac{1}{2}\mathrm{tr}\left(\begin{pmatrix} 1 & 0 \\ 0 & 1 \end{pmatrix}\begin{pmatrix} r & s \\ t & u \end{pmatrix}\right) = \frac{1}{2}(r+u) \in \mathbb{C} \tag{4.103}$$

$$c_2 = \frac{1}{2}\mathrm{tr}(\mathbf{X}^\dagger \mathbf{P}) = \frac{1}{2}\mathrm{tr}\left(\begin{pmatrix} 0 & 1 \\ 1 & 0 \end{pmatrix}\begin{pmatrix} r & s \\ t & u \end{pmatrix}\right) = \frac{1}{2}(t+s) \in \mathbb{C} \tag{4.104}$$

$$c_3 = \frac{1}{2}\mathrm{tr}(\mathbf{Y}^\dagger \mathbf{P}) = \frac{1}{2}\mathrm{tr}\left(\begin{pmatrix} 0 & -i \\ i & 0 \end{pmatrix}\begin{pmatrix} r & s \\ t & u \end{pmatrix}\right) = \frac{i}{2}(s-t) \in \mathbb{C} \tag{4.105}$$

$$c_4 = \frac{1}{2}\mathrm{tr}(\mathbf{Z}^\dagger \mathbf{P}) = \frac{1}{2}\mathrm{tr}\left(\begin{pmatrix} 1 & 0 \\ 0 & -1 \end{pmatrix}\begin{pmatrix} r & s \\ t & u \end{pmatrix}\right) = \frac{1}{2}(r-u) \in \mathbb{C}. \tag{4.106}$$

Substituting (4.103)–(4.106) in (4.101) verifies our claim, thereby completing the proof.

4.6.1 Quantum Errors and Noisy Quantum Channels

Based on the four quantum operations, we now briefly consider a few examples of noisy quantum channels. In a manner similar to classical channels, quantum channels can be described as acting on the input qubit according to a probability rule: With certain probability they leave the state alone while with some other probabilities (always summing to 1) they perform a quantum operation on the qubit. Here we give a few examples of quantum noisy channels using the operator-sum representation. We assume that the initial state of our quantum system is ρ_Q.

Bit-Flip Channel

Let us begin by looking at the operation of this noisy quantum channel:

$$\varepsilon(\rho_Q) = pX\rho_Q X^\dagger + (1-p)\rho_Q. \tag{4.107}$$

The above equation describes the quantum operation of the bit-flip channel on the input qubit. It says that with probability p it will flip the computational basis states of a qubit, while with probability $1-p$ it will leave the qubit alone. This, however, is not the strict operator-sum representation. To get there we define

$$\boldsymbol{E}_1 = \sqrt{p}\boldsymbol{X} \tag{4.108}$$

$$\boldsymbol{E}_2 = \sqrt{1-p}\boldsymbol{I}. \tag{4.109}$$

We can now define the operator-sum representation of the bit-flip channel as

$$\varepsilon(\boldsymbol{\rho}_Q) = \boldsymbol{E}_1\boldsymbol{\rho}_Q\boldsymbol{E}_1{}^{\dagger} + \boldsymbol{E}_2\boldsymbol{\rho}_Q\boldsymbol{E}_2{}^{\dagger}. \tag{4.110}$$

Noting that $\boldsymbol{X}^{\dagger} = \boldsymbol{X}$, we can reduce (4.109) to

$$\varepsilon(\boldsymbol{\rho}_Q) = \boldsymbol{E}_1\boldsymbol{\rho}_Q\boldsymbol{E}_1 + \boldsymbol{E}_2\boldsymbol{\rho}_Q\boldsymbol{E}_2. \tag{4.111}$$

Example 4.3. As an example, for an input qubit in the computational basis state $|1\rangle$ the output quantum state would then be

$$|\psi_Q\rangle = p|0\rangle + (1-p)|1\rangle. \tag{4.112}$$

Phase-Flip Channel

The operation of this noisy quantum channel can be described by

$$\varepsilon(\boldsymbol{\rho}_Q) = p\boldsymbol{Z}\boldsymbol{\rho}_Q\boldsymbol{Z}^{\dagger} + (1-p)\boldsymbol{\rho}_Q. \tag{4.113}$$

Following the steps for the bit-flip channel the corresponding operator-sum representation is

$$\varepsilon(\boldsymbol{\rho}_Q) = \boldsymbol{E}_1\boldsymbol{\rho}_Q\boldsymbol{E}_1 + \boldsymbol{E}_2\boldsymbol{\rho}_Q\boldsymbol{E}_2, \tag{4.114}$$

where

$$\boldsymbol{E}_1 = \sqrt{p}\boldsymbol{Z} \tag{4.115}$$

$$\boldsymbol{E}_2 = \sqrt{1-p}\boldsymbol{I}, \tag{4.116}$$

while noting that $\boldsymbol{Z}^{\dagger} = \boldsymbol{Z}$.

Example 4.4. Consider, as an example, the input qubit to be

$$|\psi_Q\rangle = \begin{pmatrix}\alpha \\ \beta\end{pmatrix} = \begin{pmatrix}\frac{1}{\sqrt{2}} \\ \frac{1}{\sqrt{2}}\end{pmatrix} = \frac{1}{\sqrt{2}}\begin{pmatrix}1 \\ 1\end{pmatrix}. \tag{4.117}$$

The output of the phase-flip channel is

$$|\psi_Q\rangle = \frac{p}{\sqrt{2}}\begin{pmatrix}1\\-1\end{pmatrix} + \frac{1-p}{\sqrt{2}}\begin{pmatrix}1\\1\end{pmatrix}. \tag{4.118}$$

Thus, with probability p the channel introduces a relative phase-shift of π between the probability amplitudes α and β, while there is a $1-p$ probability that the channel leaves the qubit alone. We encourage the reader to relate the *ket* picture employed in the two examples to the reduced density matrix picture in the operator-sum representation.

Depolarization Channel

This noisy channel is described as

$$\varepsilon(\boldsymbol{\rho}_Q) = p\frac{\boldsymbol{I}}{2} + (1-p)\boldsymbol{\rho}_Q. \tag{4.119}$$

This operation tells us that with probability p we get complete depolarization (dephasing) due to entanglement with the environment. The result is a completely mixed state with all relative phase information lost to the environment. As before, there is $1-p$ probability that the channel leaves the qubit alone.

The operator-sum representation of this method is not evident from (4.119). This is because using Pauli operators we can get different realizations. An example is

$$\varepsilon(\boldsymbol{\rho}_Q) = (1-p)\boldsymbol{\rho}_Q + \frac{p}{3}(\boldsymbol{X}\boldsymbol{\rho}_Q\boldsymbol{X}^\dagger + \boldsymbol{Y}\boldsymbol{\rho}_Q\boldsymbol{Y}^\dagger + \boldsymbol{Z}\boldsymbol{\rho}_Q\boldsymbol{Z}^\dagger) \tag{4.120}$$

$$= \boldsymbol{E}_1\boldsymbol{\rho}_Q\boldsymbol{E}_1 + \boldsymbol{E}_2\boldsymbol{\rho}_Q\boldsymbol{E}_2 + \boldsymbol{E}_3\boldsymbol{\rho}_Q\boldsymbol{E}_3 + \boldsymbol{E}_4\boldsymbol{\rho}_Q\boldsymbol{E}_4), \tag{4.121}$$

where

$$\boldsymbol{E}_1 = \sqrt{1-p\boldsymbol{I}} \tag{4.122}$$

$$\boldsymbol{E}_2 = \sqrt{\frac{p}{3}}\boldsymbol{X} \tag{4.123}$$

$$\boldsymbol{E}_3 = \sqrt{\frac{p}{3}}\boldsymbol{Y}, \tag{4.124}$$

and

$$\boldsymbol{E}_4 = \sqrt{\frac{p}{3}}\boldsymbol{Z}. \tag{4.125}$$

This operator-sum representation leaves the qubit alone with probability $1-p$, and with equal probability of $p/3$, the qubit is acted upon by **X**, **Y**, **Z**. The latter actions result in a depolarization effect. This representation is close to the strict definition in (4.119), but it is not identical to it because for any $\boldsymbol{\rho}_Q$, $p\frac{\boldsymbol{I}}{2}$ is not equal to

$\frac{p}{3} = (X\rho_Q X^\dagger + Y\rho_Q Y^\dagger + Z\rho_Q Z^\dagger)$. In the problems that follow we ask the reader to verify the exact form of the operator-sum representation for the depolarizing channel.

4.7 CONCLUSION

In this chapter the key ideas and concepts that lie at the heart of information processing tasks and quantum mechanics in general have been introduced. We have presented the quantum idea of superposition of states and how it leads to quantum parallelism, the no-cloning theorem and our inability to reliably distinguish non-orthogonal quantum states. The mathematical formalism and the physical idea behind quantum entanglement have also been presented. The reader will see that all these concepts resurface in all later chapters as they underpin the algorithms and protocols discussed therein. The operator-sum representation is also crucial for understanding communication over quantum channels. We have discussed some of these quantum channels here. They will arise again in later chapters when quantum error correction codes are discussed.

This introductory chapter is by no means a thorough examination of the concepts that lie herein. In fact, we have barely scratched the surface. But the ideas outlined here are crucial to understanding the rest of this book. If the reader is interested in delving into the details of each individual topic, we would encourage the reader to consult the references. References [1–5] are seminal works on information theory both from Shannon's probabilistic/statistical point of view [1–3] and Kolmogorov's complexity point of view [3,5]. Landauer's important work [4] provides an important connection between classic physics, information theory, and reversible and irreversible computation. References [6–8] detail Alan Turing's groundbreaking work that founded the field of computer science. References [9–17] list a series of papers that established the theoretical foundations of quantum mechanics. Reference [18] is J. J. Sakurai's eloquent exposition of the field of quantum mechanics. The jump from quantum mechanics to quantum computation began when Feynman proposed the idea of a quantum computer in his keynote address in 1981 at the California Institute of Technology [19]. References [21–25] are papers that discuss, for the first time, the practical feasibility of quantum computing.

Finally, Refs [26–31] present the fundamental debate regarding nonlocal interactions and qubit entanglement and how it was settled, thereby establishing quantum mechanics as a legitimate physical theory. Quantum entanglement is a complex field and the subject of current research. Reference [32] discusses the various aspects of quantum entanglement.

4.8 PROBLEMS

1. Which of the following are legitimate quantum states?
   (a) $|\psi\rangle = 0.5|0\rangle + 0.5|1\rangle$
   (b) $|\psi\rangle = \cos\theta\,|0\rangle + \sin\theta\,|1\rangle$

(c) $|\psi\rangle = \frac{\sqrt{3}}{2}|0\rangle + \frac{i}{2}|1\rangle$

(d) $|\psi\rangle = \cos^2\theta\,|0\rangle + \sin^2\theta\,|1\rangle$.

2. We presented the ket vector to describe the quantum states resulting in the noisy channel Examples 4.3 and 4.4. Use the operator-sum notation to verify them.
3. The ket vectors and the corresponding density matrices in the previous problem are expressed in the computational basis $|0\rangle$ and $|1\rangle$. Express each of them in the basis

$$|+\rangle = \frac{1}{\sqrt{2}}(|0\rangle + |1\rangle) \quad \text{and} \quad |-\rangle = \frac{1}{\sqrt{2}}(|0\rangle + |1\rangle).$$

4.

(a) Show that for some quantum state ρ_Q

$$I = \frac{I\rho_Q I + X\rho_Q X + Y\rho_Q Y + Z\rho_Q Z}{2}.$$

(b) Based on (a), derive the operator-sum relation for

$$\varepsilon(\rho_Q) = p\frac{I}{2} + (1-p)\rho_Q.$$

5. Derive the operator-sum representation of (4.85).
6. Consider a qubit initially in the quantum state

$$|\psi\rangle = \alpha|0\rangle + \beta|1\rangle,$$

with $|\alpha|^2 + |\beta|^2 = 1$, that interacts with a generalized damping channel with the Kraus operators given by

$$E_1 = \sqrt{p}\begin{pmatrix} 1 & 0 \\ 0 & \sqrt{1-\gamma} \end{pmatrix}, \; E_2 = \sqrt{p}\begin{pmatrix} 0 & \sqrt{\gamma} \\ 0 & 0 \end{pmatrix},$$

$$E_3 = \sqrt{1-p}\begin{pmatrix} \sqrt{1-\gamma} & 0 \\ 0 & 1 \end{pmatrix}, \quad E_4 = \sqrt{1-p}\begin{pmatrix} 0 & 0 \\ \sqrt{\gamma} & 0 \end{pmatrix}.$$

(a) Show that the quantum operation associated with this channel is trace preserving.

(b) Determine $\varepsilon(\rho_Q)$.

(c) Determine the probabilities $P(i)$ that processes $\mathbf{E}_i$ $(i = 1,2,3,4)$ occur.

7.

(a) Show that the Pauli matrix **X** performs bit-flip operation.

(b) Show that the Pauli matrix **Z** performs phase-flip operation.

(c) Show that the Pauli matrix **Y** performs bit-phase-flip operation.

8. If the no-cloning theorem could be violated, would it be possible to reliably distinguish two non-orthogonal quantum states?

9. Find the Schmidt decomposition of the following quantum states:
   (a) All the EPR pairs.
   (b) $|\psi\rangle = \dfrac{|0\rangle \otimes |1\rangle + |1\rangle \otimes |0\rangle + |1\rangle \otimes |1\rangle}{\sqrt{3}}$.
10. Suppose we have a quantum state $|\psi\rangle$ in a bipartite quantum system AB (i.e. $|\psi\rangle \in H_A \otimes H_B$). Show that the partial traces $\mathrm{tr}_A(|\psi\rangle\langle\psi|)$ and $\mathrm{tr}_B(|\psi\rangle\langle\psi|)$ have the same eigenvalues. (Assume that the two Hilbert spaces have the same dimension.)
11. Show that the EPR pair states form an orthonormal basis for the Hilbert space of any bipartite quantum system.
12. Compute:
    (a) The trace of the Pauli matrices.
    (b) The determinant of the Pauli matrices.
    (c) The square of the Pauli matrices.

References

[1] C.E. Shannon, A mathematical theory of communication, Bell Syst. Tech. J. 27 (1948) 379–423, 623–656.

[2] C.E. Shannon, W. Weaver, The Mathematical Theory of Communication, University of Illinois Press, Urbana, IL, 1949.

[3] T. Cover, J. Thomas, Elements of Information Theory, Wiley-Interscience, 1991.

[4] R. Landauer, Irreversibility and heat generation in the computing process, IBM J. Res. Dev. 5 (3) (1961).

[5] A.N. Kolmogorov, Three approaches to the quantitative definition of information, Prob. Inform. Transm. 1 (1) (1965) 1–7.

[6] A. Turing, On computable numbers, with an application to the Entscheidungs problem, Proc. Lond. Math. Soc. 2 (42) (1936) 230–265.

[7] A. Turing, Computability and lambda-definability, J. Symbolic Logic 2 (1937) 153–163.

[8] A. Turing, Systems of logic based on ordinals, Proc. Lond. Math. Soc. 3 (45) (1939) 161–228.

[9] M. Planck, On the law of distribution of energy in the normal spectrum, Ann. Phys. 4 (1901) 553.

[10] A. Einstein, On a heuristic viewpoint concerning the production and transformation of light, Ann. Phys. (1905).

[11] L. de Broglie, Waves and quanta, Compt. Rend. 177 (1923) 507.

[12] W. Pauli, On the connection between the completion of electron groups in an atom with the complex structure of spectra, Z. Phys. 31 (1925) 765.

[13] W. Heisenberg, Quantum-theoretical re-interpretation of kinematic and mechanical relations, Z. Phys. 33 (1925) 879.

[14] M. Born, P. Jordan, On quantum mechanics, Z. Phys. 34 (1925) 858.

[15] E. Schrödinger, Quantization as a problem of proper values, Part I, Ann. Phys. 79 (1926) 361.

[16] E. Schrödinger, On the relation between the quantum mechanics of Heisenberg, Born, and Jordan, and that of Schrödinger, Ann. Phys. 79 (1926) 734.

[17] W. Heisenberg, The actual content of quantum theoretical kinematics and mechanics, Z. Phys. 43 (1927) 172.

[18] J.J. Sakurai, Modern Quantum Mechanics, Addison-Wesley, 1993.

[19] R.P. Feynman, Simulating physics with computers, Int. J. Theor. Phys. 21 (1982) 467–488.

[20] M. Born, E. Wolf, Principles of Optics, Cambridge University Press, 1999.

[21] D. Deutsch, The Fabric of Reality, Penguin Press, New York, 1997.

[22] D. Deutsch, Quantum theory, the Church–Turing principle and the universal quantum computer, Proc. R. Soc. Lond. A 400 (1985) 97–117.

[23] D. Deutsch, R. Jozsa, Rapid solution of problems by quantum computation, Proc. Math. Phys. Sci. 439 (1907) 553–558, 1992.

[24] P. Shor, Algorithms for Quantum Computation: Discrete Logarithms and Factoring, IEEE Comput. Soc. Press, November 1994, pp. 124–134.

[25] L.K. Grover, A fast quantum mechanical algorithm for database search, Proc. 28th Annual ACM Symp. on the Theory of Computing, May 1996, p. 212.

[26] A. Einstein, B. Podolsky, N. Rosen, Can quantum-mechanical description of physical reality be considered complete? Phys. Rev. 47 (15 May 1935) 777–780.

[27] J.S. Bell, On the Einstein–Podolsky–Rosen paradox, Physics 1 (1964) 195–200.

[28] J.S. Bell, Speakable and Unspeakable in Quantum Mechanics, Cambridge University Press, 1987.

[29] J.F. Clauser, A. Shimony, Bell's theorem Experimental Tests and implications, Rep. Prog. Phys. 41 (1978) 1881.

[30] A. Aspect, P. Grangier, G. Roger, Experimental Tasks of Realistic Local Theories via Bell's Theorem, Phys. Rev. Lett. 47 (1981) 460.

[31] P.G. Kwiat, et al., New High-Intensity Source of Polarization-Entangled Photon Pairs, Phys. Rev. Lett. 75 (1995) 4337.

[32] R. Horodecki, P. Horodecki, M. Horodecki, K. Horodecki, Quantum entanglement, Rev. Mod. Phys. 81 (2009) 865.

CHAPTER

Quantum Algorithms

5

CHAPTER OUTLINE

This chapter considers basic quantum algorithms [1–16]. The chapter starts by revisiting the quantum parallelism concept and describing its power in calculating the global property of a certain function by performing only one evaluation of that function, namely the Deutsch and Deutsch–Jozsa algorithms. Further, the Grover search algorithm to perform a search for an entry in an unstructured database is described. Next, the quantum Fourier transform is described, which is a basic algorithm used in many other quantum algorithms. An algorithm to evaluate the period of a function is also provided. It is then described how to solve the Rivest–Shamir–Adleman (RSA) encryption protocol. Then, Shor's factorization algorithm is given. Finally, Simon's algorithm is described. The quantum computational complexity and Turing machine representation are discussed as well. After summarizing the chapter, a set of problems is given to help getting a deeper understanding of the material presented in this chapter.

Quantum Information Processing and Quantum Error Correction. DOI: 10.1016/B978-0-12-385491-9.00005-8

5.1 QUANTUM PARALLELISM (REVISITED)

The quantum computation C implemented on a quantum register maps the input string $i_1 \ldots i_N$ to the output string $O_1(i) \ldots O_N(i)$:

$$\begin{pmatrix} O_1(i) \\ \vdots \\ O_N(i) \end{pmatrix} = U(C) \begin{pmatrix} i_1 \\ \vdots \\ i_N \end{pmatrix}; \quad (i)_{10} = (i_1 \ldots i_N)_2. \tag{5.1}$$

The computation basis (CB) states are denoted by

$$|i_1 \ldots i_N\rangle = |i_1\rangle \otimes \ldots \otimes |i_N\rangle; \quad i_1 \ldots i_N \in \{0, 1\}. \tag{5.2}$$

The linear superposition allows us to form the following $2N$-qubit state:

$$|\psi_{\text{in}}\rangle = \left[\frac{1}{\sqrt{2^N}} \sum_i |i_1 \ldots i_N\rangle\right] \otimes |0 \ldots 0\rangle, \tag{5.3}$$

and upon the application of quantum operation $U(C)$, the output can be represented by

$$|\psi_{\text{out}}\rangle = U(C)|\psi_{\text{in}}\rangle = \frac{1}{\sqrt{2^N}} \sum_i |i_1 \ldots i_N\rangle \otimes |O_1(i) \ldots O_N(i)\rangle. \tag{5.4}$$

The quantum computer has been able to encode all input strings generated by C into $|\psi_{\text{out}}\rangle$; in other words, it has simultaneously pursued 2^N classical paths. This ability of a quantum computer to encode multiple computational results into a quantum state in a single quantum computational step is known as *quantum parallelism.*

The quantum circuit to simultaneously evaluate $f(0)$ and $f(1)$ of the mapping $f(x)$: $\{0,1\} \rightarrow \{0,1\}$ is shown in Figure 5.1. Its output state contains information about both $f(0)$ and $f(1)$.

We can show that the quantum operation U_f is unitary as follows:

$$|x, y \oplus f(x)\rangle \xrightarrow{U_f} |x, [y \oplus f(x)] \oplus f(x)\rangle = |x, y\rangle \Rightarrow U_f^2 = I. \tag{5.5}$$

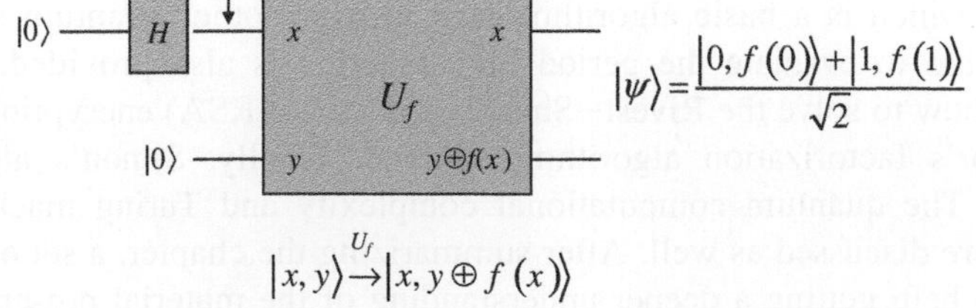

FIGURE 5.1

Quantum circuit to simultaneously compute $f(0)$ and $f(1)$.

(a) $|0\rangle$ ——[H]——

$|0\rangle$ ——[H]——

$|\psi\rangle = \dfrac{|0\rangle+|1\rangle}{\sqrt{2}}\dfrac{|0\rangle+|1\rangle}{\sqrt{2}} = \dfrac{|00\rangle+|01\rangle+|10\rangle+|11\rangle}{2}$

(b) $|0\rangle$ ——[H]——
⋮
$|0\rangle$ ——[H]——

$|\psi\rangle = \dfrac{1}{\sqrt{2^n}}\sum_x |x\rangle$

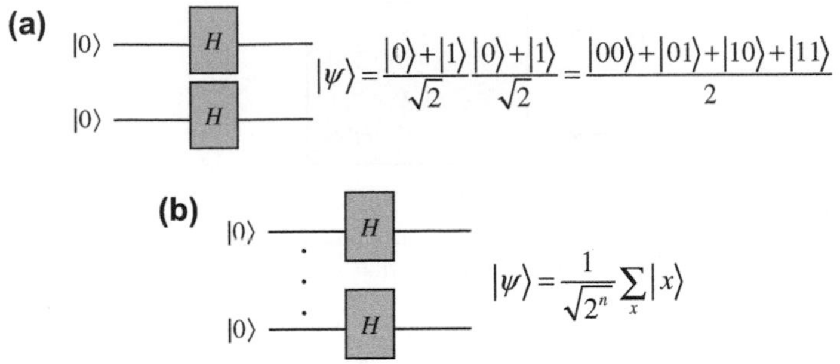

FIGURE 5.2

The Walsh–Hadamard transform: (a) on two qubits; (b) on n qubits.

The action of operation U_f in operator notation can be represented by

$$U_f|x,0\rangle = |x,f(x)\rangle \qquad U_f|x,y\rangle = |x,y\oplus f(x)\rangle. \tag{5.6}$$

The output of the circuit shown in Figure 5.1 can now be obtained as

$$|\psi\rangle = U_f|H0\otimes 0\rangle = U_f\frac{1}{\sqrt{2}}(|00\rangle + |10\rangle) = \frac{1}{\sqrt{2}}(|0,f(0)\rangle + |1,f(1)\rangle), \tag{5.7}$$

proving that the circuit in Figure 5.1 can indeed simultaneously evaluate $f(0)$ and $f(1)$.

We would like to generalize the computation of $f(x)$ on $n + m$ qubits. Before we provide the quantum circuit to perform this computation, we need the quantum circuit to create the superposition state from Eq. (5.3). The Walsh–Hadamard transform on two ancillary qubits in state $|00\rangle$ can be implemented by applying the Hadamard gates on ancillary qubits as shown in Figure 5.2a.

The Walsh–Hadamard transform on n qubits can be implemented as shown in Figure 5.2b. Now, by using the Walsh–Hadamard transform circuit, the quantum circuit to evaluate $f(x)$: $\{0,1\}^n \to \{0,1\}$ can be implemented as shown in Figure 5.3.

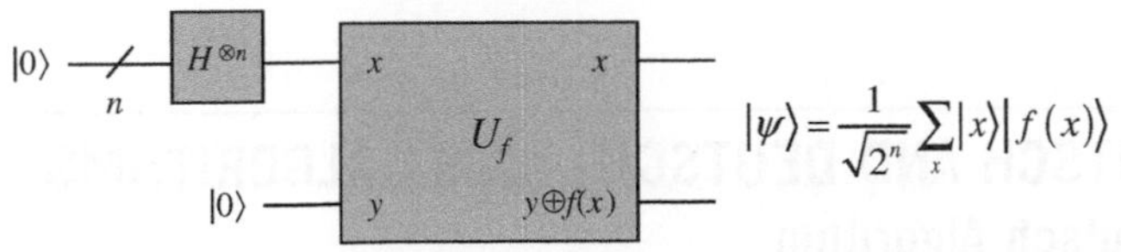

FIGURE 5.3

The quantum circuit to evaluate $f(x)$: $\{0,1\}^n \to \{0,1\}$.

$$|x\rangle \; \frac{\quad}{n} \quad \boxed{\begin{array}{c} x \qquad\qquad x \\ U_f \\ z \qquad\quad z \oplus f(x) \end{array}} \quad |\psi\rangle$$

$$|z\rangle \; \frac{\quad}{m}$$

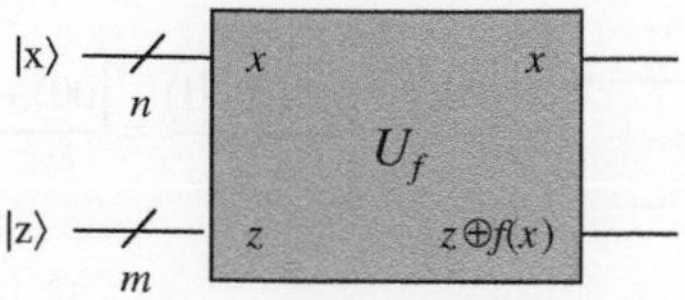

FIGURE 5.4

The generalization of quantum operation U_f on $n + m$ qubits.

$$H^{\otimes n}|0^{\otimes n}\rangle = \frac{1}{2^{n/2}} \sum_{x=0}^{2^n-1} |x\rangle, \quad |x\rangle = |x_{n-1} \cdots x_0\rangle, \; x_i = 0, 1$$

$$|\psi\rangle = U_f \left| (H^{\otimes n} 0^{\otimes n}) \otimes 0^{\otimes m} \right\rangle = \frac{1}{2^{n/2}} \sum_{x=0}^{2^n-1} |x, f(x)\rangle$$

$$U_f |x, z\rangle = |x, z \oplus f(x)\rangle, \quad |z\rangle = |z_{n-1} \cdots z_0\rangle, \; z_i = 0, 1$$

$$U_f |x, 0^{\otimes m}\rangle = |x, f(x)\rangle$$

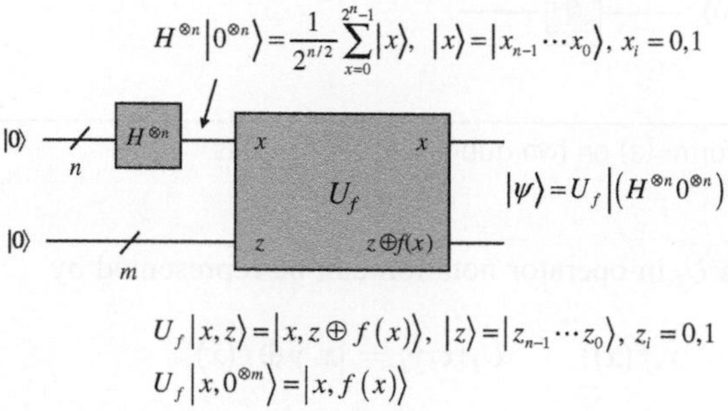

FIGURE 5.5

The quantum circuit to evaluate $f(x)$ on $n + m$ qubits.

The quantum circuit performing U_f on $n + m$ qubits is shown in Figure 5.4. Its operation can be described as follows:

$$U_f |x, z\rangle = |x, z \oplus f(x)\rangle, \quad |z\rangle = |z_{n-1} \dots z_0\rangle, \quad z_i = 0, 1. \tag{5.8}$$

By setting $z = 0 \dots 0$, the action of this gate is

$$U_f |x, 0^{\otimes m}\rangle = |x, f(x)\rangle, \tag{5.9}$$

which is the same as the action of the gate shown in Figure 5.3. By using this circuit, we can generalize the evaluation $f(x)$ to $n + m$ qubits as illustrated in Figure 5.5. The output state of the quantum circuit shown in Figure 5.5 can be determined as

$$|\psi\rangle = U_f \left| (H^{\otimes n} 0^{\otimes n}) \otimes 0^{\otimes m} \right\rangle = \frac{1}{2^{n/2}} U_f \left(\sum_{x=0}^{2^n-1} |x\rangle \otimes |0^{\otimes m}\rangle \right) = \frac{1}{2^{n/2}} \sum_{x=0}^{2^n-1} |x, f(x)\rangle. \tag{5.10}$$

5.2 DEUTSCH AND DEUTSCH–JOZSA ALGORITHMS

5.2.1 Deutsch Algorithm

Although $|\psi\rangle = (|0, f(0)\rangle + |1, f(1)\rangle)/\sqrt{2}$ contains information about both $f(0)$ and $f(1)$, there is no advantage to classic computation if a table of $f(x)$ is desired.

On the other hand, quantum computation allows us to evaluate some *global property*, say $f(0) \oplus f(1)$, by performing only one evaluation of $f(x)$. Although this example is trivial and not very practical, it allows us to describe the basic concepts of quantum computing. The quantum circuit implementing the Deutsch algorithm is shown in Figure 5.6. The state $|\psi\rangle$ after the Hadamard gate stage is given by

$$|\psi\rangle = H|0\rangle \otimes H|1\rangle = \frac{1}{2}(|0\rangle + |1\rangle)\,(|0\rangle - |1\rangle)$$

$$= \frac{1}{2}\left(\sum_{x=0}^{1}|x\rangle\right)(|0\rangle - |1\rangle). \tag{5.11}$$

The application of the operator U_f to $|\psi\rangle$ performs the following mapping:

- Option 1: $f(x) = 0 \Rightarrow (|0\rangle - |1\rangle) \rightarrow (|0\rangle - |1\rangle)$
- Option 2: $f(x) = 1 \Rightarrow (|0\rangle - |1\rangle) \rightarrow (|1\rangle - |0\rangle) = -(|0\rangle - |1\rangle)$.

By combining these two options, the same mapping can be represented by $(|0\rangle - |1\rangle) \rightarrow (-1)^{f(x)}(|0\rangle - |1\rangle)$, so that the action of U_f on $|\psi\rangle$ is as follows:

$$U_f|\psi\rangle = H|0\rangle \otimes H|1\rangle = \frac{1}{2}\left(\sum_{x=0}^{1}(-1)^{f(x)}|x\rangle\right)(|0\rangle - |1\rangle). \tag{5.12}$$

The operator that performs the mapping $|x\rangle \xrightarrow{U_f} (-1)^{f(x)}|x\rangle$ is known as an *oracle* operator. The upper and lower registers on the right-hand side of the circuit shown in Figure 5.6 are unentangled after the oracle operator. This operator is important in the Grover search algorithm, which will be described later. The function operator output state (see Figure 5.6) is given by $|\varphi\rangle = \left[(-1)^{f(0)}|0\rangle + (-1)^{f(1)}|1\rangle\right]/\sqrt{2}$, and by applying the Hadamard gate we obtain:

$$H|\varphi\rangle = \frac{1}{2}\left[(-1)^{f(0)} + (-1)^{f(1)}\right]|0\rangle + \frac{1}{2}\left[(-1)^{f(0)} - (-1)^{f(1)}\right]|1\rangle. \tag{5.13}$$

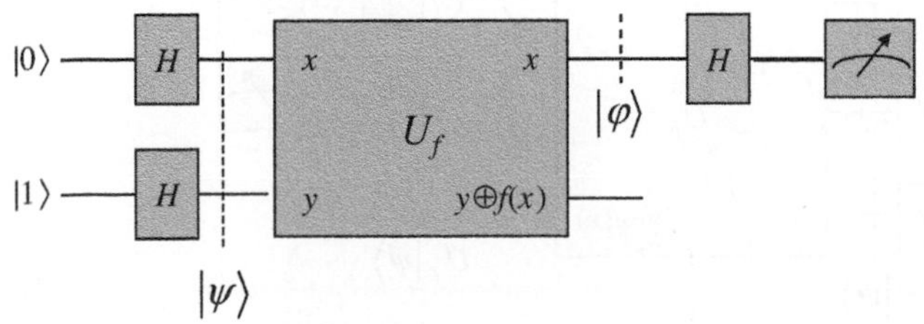

FIGURE 5.6

The quantum circuit implementing the Deutsch algorithm.

By performing the measurement on upper qubit, we can obtain two possible states:

- Option 1: Qubit $|0\rangle$ is the final state upon measurement, meaning that $f(0) = f(1)$; in other words, the function $f(x)$ is constant.
- Option 2: Qubit $|1\rangle$ is obtained, meaning that $f(0) \neq f(1)$, indicating that the function $f(x)$ is balanced.

Therefore, in this example, we employed quantum parallelism to bypass the explicit calculation of $f(x)$.

5.2.2 Deutsch–Jozsa Algorithm

The Deutsch–Jozsa algorithm is a generalization of Deutsch's algorithm. Namely, it verifies if the mapping $f(x)$: $\{0,1\}^n \rightarrow \{0,1\}$ is constant or balanced for all values of $x \in \{0, 2^n - 1\}$. The quantum circuit implementing the Deutsch–Jozsa algorithm is shown in Figure 5.7. We will describe the Deutsch–Jozsa algorithm for $n = 2$ (see Figure 5.8), while the description for arbitrary n is left to the reader. Based on Figure 5.8 we conclude that the action of U_f on $|\psi\rangle$ is given by

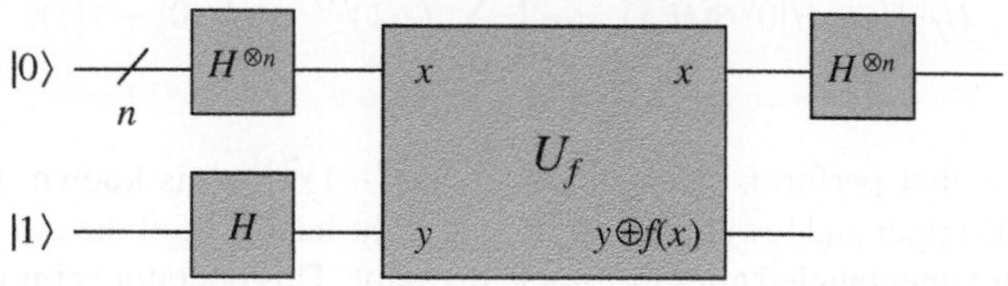

FIGURE 5.7

Deutsch–Jozsa algorithm implementation.

$$H^{\otimes 2}|00\rangle = \frac{1}{\sqrt{2}}\left(|0\rangle+|1\rangle\right)\frac{1}{\sqrt{2}}\left(|0\rangle+|1\rangle\right) = \frac{1}{2}\left(|00\rangle+|01\rangle+|10\rangle+|11\rangle\right) = \frac{1}{\sqrt{2^2}}\sum_{x=0}^{3}|x\rangle$$

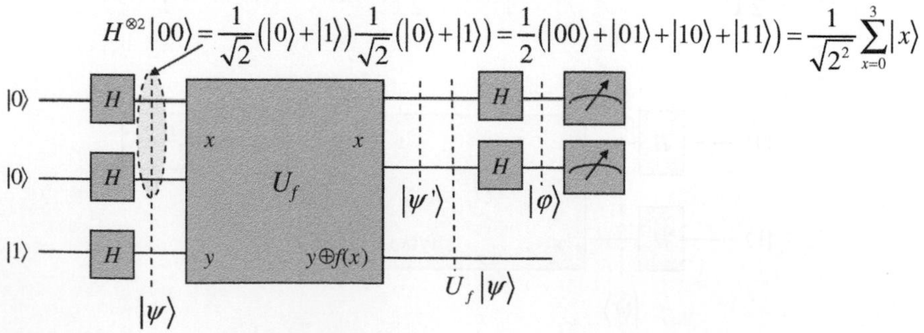

FIGURE 5.8

Deutsch–Jozsa algorithm implementation for $n = 2$.

$$U_f|\psi\rangle = U_f\left[\frac{1}{2}\sum_{x=0}^{3}|x\rangle\right]\otimes\frac{1}{\sqrt{2}}(|0\rangle-|1\rangle)$$
$$=\left[\frac{1}{2}\sum_{x=0}^{3}(-1)^{f(x)}|x\rangle\right]\otimes\frac{1}{\sqrt{2}}(|0\rangle-|1\rangle). \quad (5.14)$$

Clearly the upper and lower registers on the right-hand side of gate U_f are unentangled.

By applying the Hadamard gates on upper qubits $|\psi'\rangle$ and performing measurement we obtain the following options:

- Option 1: $f(x) = \text{const}$, in which case the upper qubits $|\psi'\rangle$ can be represented as

$$|\psi'\rangle = \pm\frac{1}{2}(|00\rangle+|01\rangle+|10\rangle+|11\rangle) = \pm\frac{1}{\sqrt{2}}(|0\rangle+|1\rangle)\frac{1}{\sqrt{2}}(|0\rangle+|1\rangle). \quad (5.15)$$

By applying the corresponding Hadamard gates, the upper qubits $|\psi'\rangle$ are transformed to

$$|\varphi\rangle = H^{\otimes 2}|\psi'\rangle = \pm|00\rangle. \quad (5.16)$$

- Option 2: $f(x) = x \bmod 2$, in which case the upper qubits (based on (5.14)) are in fact

$$|\psi'\rangle = \frac{1}{2}(|00\rangle-|01\rangle+|10\rangle-|11\rangle) = \frac{1}{2}[|0\rangle(|0\rangle-|1\rangle)+(|0\rangle-|1\rangle)|1\rangle]. \quad (5.17)$$

Now, by applying the Hadamard gates, the upper qubits $|\psi'\rangle$ are transformed to:

$$H^{\otimes 2}|\psi'\rangle = H^{\otimes 2}\frac{1}{2}[(|0\rangle+|1\rangle)\otimes(|0\rangle-|1\rangle)] = |01\rangle. \quad (5.18)$$

The result in upper qubits is unambiguous only if $|\psi'\rangle$ is a non-entangled state, indicating that it must be

$$(-1)^{f(0)+f(1)} = (-1)^{f(1)+f(2)}. \quad (5.19)$$

5.3 GROVER SEARCH ALGORITHM

The Grover search algorithm performs a *search* for an entry in an *unstructured database*. In order to efficiently describe the Grover search algorithm, we use in interpretation similar to that of Le Bellac [1]. If N is the number of entries, a classical algorithm will take on average $N/2$ attempts, while the Grover search algorithm solves this problem in $\sim\sqrt{N}$ operations. The search is conducted under the following assumptions: (1) the database address contains n qubits and (2) the *search function* is defined by

$$f(x) = \begin{cases} 0, & x \neq y \\ 1, & x = y \end{cases} = \delta_{xy}, \quad x \in \{0, 1, \ldots, 2^n - 1\}, \tag{5.20}$$

meaning that the search function result is 1 if the correct item $|y\rangle$ is found. In the previous section, we defined the *oracle* operator as follows:

$$O|x\rangle = (-1)^{f(x)}|x\rangle. \tag{5.21}$$

Based on the oracle operator, we define the *Grover* operator as

$$G = H^{\otimes n} X H^{\otimes n} O, \tag{5.22}$$

where the action of operator X (not related to the Pauli X operator) on state $|x\rangle$ is given by

$$X|x\rangle = -(-1)^{\delta_{x0}}|x\rangle, \tag{5.23}$$

while the corresponding matrix representation of X is

$$X \doteq \begin{bmatrix} 1 & 0 & 0 & 0 & \ldots & 0 \\ 0 & -1 & 0 & 0 & \ldots & 0 \\ 0 & 0 & -1 & 0 & \ldots & 0 \\ \ldots & & \vdots & & & \\ 0 & 0 & 0 & 0 & \ldots & -1 \end{bmatrix}. \tag{5.24}$$

From Eq. (5.24) it is clear that the X operator can be represented as $X = 2|0\rangle\langle 0| - I$, and upon substitution into (5.22), the Grover search operator becomes

$$G = H^{\otimes n}(2|0\rangle\langle 0| - I)H^{\otimes n}O. \tag{5.25}$$

To simplify further the representation of the Grover operator, we represent it in terms of superposition state

$$|\Psi\rangle = H^{\otimes n}|0^{\otimes n}\rangle = \frac{1}{2^{n/2}} \sum_{x=0}^{2^n-1} |x\rangle, \tag{5.26}$$

as follows:

$$\begin{aligned} G &= H^{\otimes n}(2|0\rangle\langle 0| - I)H^{\otimes n}O \\ &= \Bigg(2\underbrace{H^{\otimes n}|0\rangle}_{|\Psi\rangle}\underbrace{\langle 0|H^{\otimes n}}_{\langle\Psi|} - \underbrace{H^{\otimes n}H^{\otimes n}}_{I}\Bigg)O \overset{H^2=I}{=} (2|\Psi\rangle\langle\Psi| - I)O. \end{aligned} \tag{5.27}$$

The Grover operator implementation circuit for $n = 3$ is shown in Figure 5.9. The geometric interpretation of the Grover search algorithm, in terms of reflection operator O, is essentially the rotation in a 2D plane as shown in Figure 5.10. The key idea of the Grover search algorithm is to determine the desired entry in an iterative

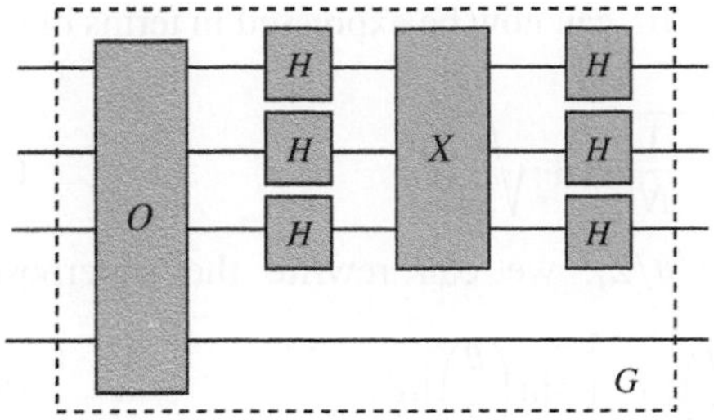

$$X = \begin{bmatrix} 1 & 0 & 0 & 0 & \cdots & 0 \\ 0 & -1 & 0 & 0 & \cdots & 0 \\ 0 & 0 & -1 & 0 & \cdots & 0 \\ \cdots & & \vdots & & & \\ 0 & 0 & 0 & 0 & \cdots & -1 \end{bmatrix}$$

$X = 2|0\rangle\langle 0| - I$ $\qquad$ $O|x\rangle = (-1)^{\delta_{xy}}|x\rangle$

FIGURE 5.9

The quantum circuit to perform the Grover search.

fashion, by rotating a current state $|\alpha\rangle$ in small angles θ until we reach the $|y\rangle$ state. The rotation of superposition state $|\Psi\rangle$ for an angle θ in counterclockwise fashion can be implemented by two reflections (or equivalently by applying the oracle operator twice), the first reflection of $|\Psi\rangle$ with respect to the state $|\alpha\rangle$ to get $O|\Psi\rangle$, and the second reflection of $O|\Psi\rangle$ with respect to the superposition state $|\Psi\rangle$ to get the resulting state $G|\Psi\rangle$, which is just the counterclockwise rotation of $|\Psi\rangle$ for θ.

The state $|\alpha\rangle$ is essentially the superposition of all other database entries $|x\rangle$ different from $|y\rangle$; in other words, we can write

$$|\alpha\rangle = \frac{1}{\sqrt{N-1}} \sum_{x \neq y} |x\rangle. \tag{5.28}$$

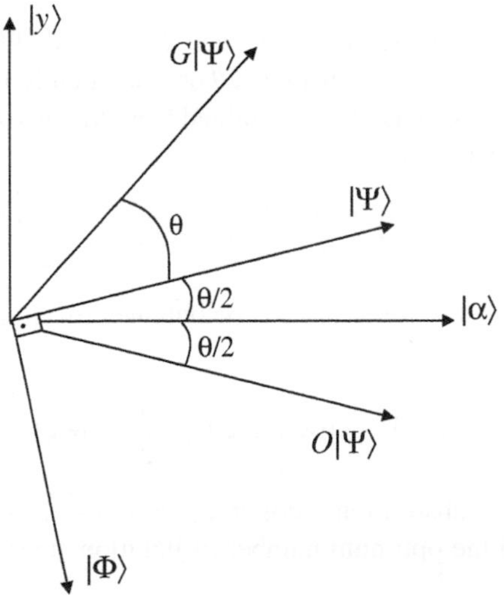

FIGURE 5.10

The geometric interpretation of the Grover search algorithm.

The superposition state, given by (5.26), can now be expressed in terms of (5.28) as follows:

$$|\Psi\rangle = \sqrt{1-\frac{1}{N}}|\alpha\rangle + \sqrt{\frac{1}{N}}|y\rangle. \tag{5.29}$$

By substituting $\sqrt{1-1/N} = \cos(\theta/2)$, we can rewrite the superposition state $|\Psi\rangle$ as

$$|\Psi\rangle = \cos\left(\frac{\theta}{2}\right)|\alpha\rangle + \sin\left(\frac{\theta}{2}\right)|\alpha\rangle. \tag{5.30}$$

Let us now apply the oracle operator on $a|\alpha\rangle + b|y\rangle$ to obtain:

$$O(a|\alpha\rangle + b|y\rangle) = a(-1)^{\delta_{\alpha y}}|\alpha\rangle + b(-1)^{\delta_{yy}}|y\rangle = a|\alpha\rangle - b|y\rangle, \tag{5.31}$$

therefore proving that the action of operator O is a reflection with respect to the direction of $|\alpha\rangle$. The Grover operator action on state $a|\Psi\rangle + b|\Phi\rangle$ is given by

$$(2|\Psi\rangle\langle\Psi| - I)(a|\Psi\rangle + b|\Phi\rangle) = 2a|\Psi\rangle \underbrace{\langle\Psi|\Psi\rangle}_{=1} - a|\Psi\rangle + 2b|\Psi\rangle \underbrace{\langle\Psi|\Phi\rangle}_{=0} - b|\Phi\rangle$$
$$= a|\Psi\rangle - b|\Phi\rangle, \tag{5.32}$$

which is just the reflection with respect to the direction of $|\Psi\rangle$. Therefore, the product of two reflections is a rotation, and from Figure 5.10 we conclude that the angle between $G|\Psi\rangle$ and $|\alpha\rangle$ is $\theta + \theta/2 = 3\theta/2$, so that we can write

$$G|\Psi\rangle = \cos\left(\frac{3\theta}{2}\right)|\alpha\rangle + \sin\left(\frac{3\theta}{2}\right)|y\rangle. \tag{5.33}$$

In a similar fashion, since the angle between $G|\Psi\rangle$ and $|\Psi\rangle$ is θ, $G^2|\Psi\rangle$ can be obtained from $|\Psi\rangle$ by rotation by an angle 2θ or equivalently from $|\alpha\rangle$ by rotation by $(2\theta + \theta/2)$. After k iterations, $G^k|\Psi\rangle$ is obtained from $|\alpha\rangle$ by rotation by $(k\theta + \theta/2) = (2k+1)\theta/2$, and we can write

$$G^k|\Psi\rangle = \cos\left(\frac{(2k+1)\theta}{2}\right)|\alpha\rangle + \sin\left(\frac{(2k+1)\theta}{2}\right)|y\rangle. \tag{5.34}$$

The optimum number of iterations can be obtained by aligning $G^k|\Psi\rangle$ with $|y\rangle$ or equivalently by setting

$$\cos\left(\frac{(2k+1)\theta}{2}\right) = 0 \Leftrightarrow \cos k\theta \cos\frac{\theta}{2} - \sin k\theta \sin\frac{\theta}{2} = 0. \tag{5.35}$$

By using the substitution $\cos(\theta/2) = \sqrt{1-1/N}$, we obtain that $\cos k\theta = 1/\sqrt{N}$, and the optimum number of iteration steps is then

$$k_o = \left[\frac{1}{\theta}\cos^{-1}\sqrt{\frac{1}{N}}\right] + 1. \tag{5.36}$$

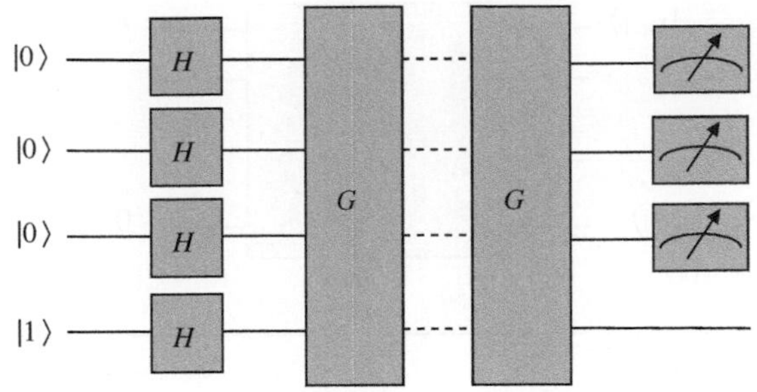

FIGURE 5.11

The quantum circuit to perform the Grover search for $n = 3$.

If $N >> 1$ we can use the Taylor expansion and keep the first term to obtain $\theta \simeq 2/\sqrt{N}$ and by substituting it into (5.36) we obtain the following approximation for optimum number of interations:

$$k_o \simeq \frac{\sqrt{N}}{2} \cos^{-1}\sqrt{\frac{1}{N}} \simeq \frac{\pi\sqrt{N}}{4}. \tag{5.37}$$

Therefore, in order to successfully complete the Grover search, we need to apply the oracle operator $\sim\sqrt{N}$ times. The quantum circuit to perform the Grover search for $n = 3$ is shown in Figure 5.11.

After the optimum number of steps k_o, the angle between $G^{ko}|\Psi\rangle$ and $|y\rangle$ is less than $\theta/2$, so that the probability of error is less than $O(1/N)$. It can be shown that the Grover algorithm is optimal. By taking all gates into account, the total number of operations in the Grover search algorithm is proportional to $\sqrt{(N)}\log N$.

5.4 QUANTUM FOURIER TRANSFORM

In this section we are concerned with the quantum Fourier transform. The unitary Fourier transform U_{FT} can be represented by the following matrix elements:

$$\langle y|U_{\mathrm{FT}}|x\rangle = (U_{\mathrm{FT}})_{yx} = \frac{1}{2^{n/2}} e^{j2\pi xy/2^n};$$

$$|x\rangle = |x_{n-1}...x_0\rangle, \quad x_i = 0, 1. \tag{5.38}$$

We can show that the U_{FT} operator is unitary as follows:

$$\sum_{y=0}^{2^n-1} (U_{\mathrm{FT}}^{\dagger})_{x'y}(U_{\mathrm{FT}})_{yx} = \sum_{y=0}^{2^n-1} (U_{\mathrm{FT}}^{*})_{x'y}(U_{\mathrm{FT}})_{yx} = \frac{1}{2^{n/2}} \sum_{y=0}^{2^n-1} e^{j2\pi(x-x')y/2^n} = \delta_{x'x}. \tag{5.39}$$

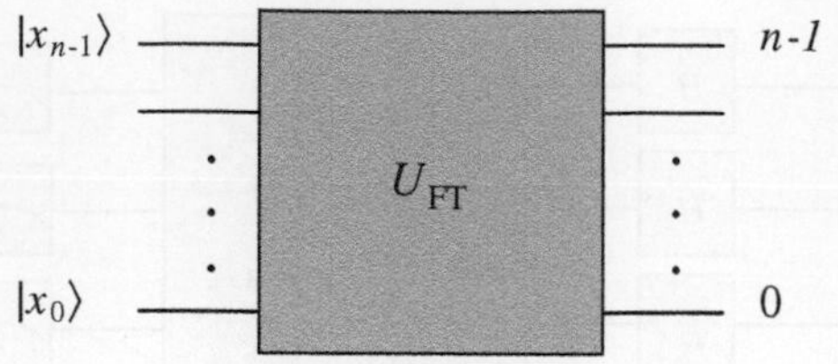

FIGURE 5.12

The notation of a quantum FT circuit.

The quantum FT notation is provided in Figure 5.12.

Let $|\Psi\rangle$ be a linear superposition of vectors $|x\rangle$:

$$|\Psi\rangle = \sum_{x=0}^{2^n-1} f(x)|x\rangle, \tag{5.40}$$

where $f(x) = \langle x|\Psi\rangle$ is a projection of $|\Psi\rangle$ along $|x\rangle$, while the probability amplitudes satisfy the following normalization condition: $\sum_{x=0}^{2^n-1}|f(x)|^2 = 1$. By applying the U_{FT} operator on the superposition state we obtain:

$$|\Phi\rangle = U_{FT}|\Psi\rangle = U_{FT}\sum_{x=0}^{2^n-1}\langle x|\Psi\rangle|x\rangle. \tag{5.41}$$

The probability amplitude of finding the state $|y\rangle$ of the computational basis at the output of the FT block is given by

$$\langle y|\Phi\rangle = \sum_{x=0}^{2^n-1}\langle y|U_{FT}|x\rangle\langle x|\Psi\rangle = \frac{1}{2^{n/2}}\sum_{x=0}^{2^n-1} e^{j2\pi xy/2^n} f(x), \tag{5.42}$$

which is just the Fourier transform of $f(x)$:

$$\tilde{f}(y) = \frac{1}{2^{n/2}}\sum_{x=0}^{2^n-1} f(x)e^{j2\pi xy/2^n}. \tag{5.43}$$

The action of U_{FT} on $|x\rangle$ in operator notation is given by

$$U_{FT}|x\rangle = \sum_{y=0}^{2^n-1}|y\rangle\langle y|U_{FT}|x\rangle = \frac{1}{2^{n/2}}\sum_{y=0}^{2^n-1} e^{j2\pi xy/2^n}|y\rangle, \tag{5.44}$$

where

$$x = x_0 + 2x_1 + 2^2x_2 + \ldots + 2^{n-1}x_{n-1}$$

$$y = y_0 + 2y_1 + 2^2y_2 + \ldots + 2^{n-1}y_{n-1}.$$

For $n = 3$, $N = 2^n = 8$, we can represent xy as

$$\frac{xy}{8} = y_0\left(\frac{x_2}{2}+\frac{x_1}{4}+\frac{x_0}{8}\right)+y_1\left(\frac{x_1}{2}+\frac{x_0}{4}\right)+y_2\frac{x_0}{2}. \tag{5.45}$$

We also know that

$$0\cdot x_m x_{m-1}\ldots x_1 x_0 = \frac{x_m}{2}+\frac{x_{m-1}}{2^2}+\ldots+\frac{x_0}{2^m}, \tag{5.46}$$

and $\exp(j2\pi m) = 1$, when m is an integer. By representing $|y\rangle = |y_{n-1}\ldots y_0\rangle$, from (5.44) we obtain:

$$\begin{aligned} U_{FT}|x\rangle &= \frac{1}{2^{n/2}}\sum_{y_0,\ldots,y_{n-1}} e^{j2\pi y_{n-1}.x_0}\ldots e^{j2\pi y_0.x_{n-1}\ldots x_0}|y_{n-1}\ldots y_0\rangle \\ &= \frac{1}{2^{n/2}}\left(\sum_{y_{n-1}} e^{j2\pi y_{n-1}.x_0}|y_{n-1}\rangle\right)\ldots\left(\sum_{y_1} e^{j2\pi y_1.x_{n-2}\ldots x_0}|y_1\rangle\right) \\ &\quad\times\left(\sum_{y_0} e^{j2\pi y_0.x_{n-1}\ldots x_0}|y_0\rangle\right). \end{aligned} \tag{5.47}$$

By representing (5.47) in the developed form we obtain:

$$\begin{aligned} U_{FT}|x\rangle &= \frac{1}{2^{n/2}}(|0_{n-1}\rangle + e^{j2\pi.x_0}|1_{n-1}\rangle)\ldots(|0_1\rangle + e^{j2\pi.x_{n-2}\ldots x_0}|1_1\rangle) \\ &\quad\times(|0_0\rangle + e^{j2\pi.x_{n-1}\ldots x_0}|1_0\rangle). \end{aligned} \tag{5.48}$$

By following a similar procedure, for $n = 2$, $N = 2^n = 4$, the application of U_{FT} on $|x\rangle$ yields

$$\begin{aligned} U_{FT}|x\rangle = U_{FT}|x_1x_0\rangle &= \frac{1}{2}(|0_1\rangle + e^{j2\pi.x_0}|1_1\rangle)\,(|0_0\rangle + e^{j2\pi.x_1x_0}|1_0\rangle) \\ &= \frac{1}{2}(|00\rangle + e^{j2\pi.x_1x_0}|01\rangle + e^{j2\pi.x_0}|10\rangle + e^{j2\pi(.x_1x_0+.x_0)}|11\rangle). \end{aligned} \tag{5.49}$$

The quantum circuit to perform FT for $n = 3$ is shown in Figure 5.13.

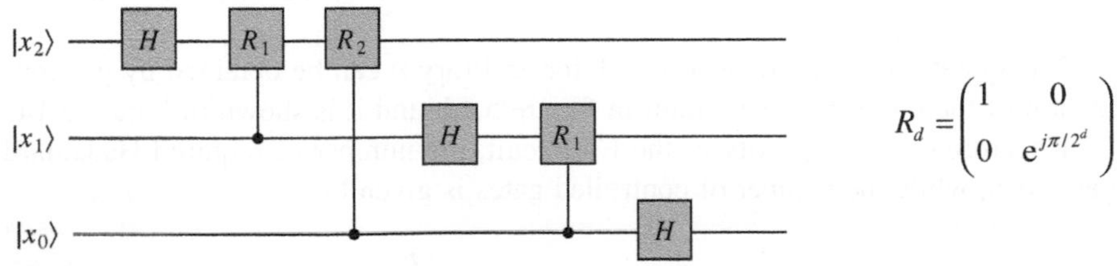

FIGURE 5.13

The quantum circuit to perform the FT for $n = 3$.

Since $H|0\rangle_2 = (|0\rangle_2 + |1\rangle_2)/\sqrt{2}$, $H|1\rangle_2 = (|0\rangle_2 - |1\rangle_2)/\sqrt{2}$, the action of the Hadamard gate on qubit $|x_2\rangle$ (see Figure 5.13) is given by

$$H|x\rangle_2 = \frac{1}{\sqrt{2}}(|0\rangle_2 + e^{j2\pi . x_2}|1\rangle_2). \tag{5.50}$$

In Figure 5.13 we denote by R_d the discrete-phase shift gate whose action in matrix representation is given by

$$R_d = \begin{bmatrix} 1 & 0 \\ 0 & e^{j\pi/2^d} \end{bmatrix}. \tag{5.51}$$

Let $C_{ij}(R_d)$ denote the control-R_d gate, where i is the control qubit and j the target qubit. If $x_1 = 0$, the action of $C_{12}(R_d)$ is given by

$$C_{12}(R_1)H|x_2\rangle = \frac{1}{\sqrt{2}}(|0\rangle_2 + e^{j2\pi . x_2}|1\rangle_2) \tag{5.52}$$

On the other hand, if $x_1 = 1$ then the action of $C_{12}(R_d)$ is given by

$$C_{12}(R_1)H|x_2\rangle = \frac{1}{\sqrt{2}}(|0\rangle_2 + e^{j\pi/2}e^{j2\pi . x_2}|1\rangle_2). \tag{5.53}$$

Equations (5.52) and (5.53) can be jointly combined into

$$C_{12}(R_1)H|x_2\rangle = \frac{1}{\sqrt{2}}(|0\rangle_2 + e^{j2\pi . x_2 x_1}|1\rangle_2). \tag{5.54}$$

To determine the final state in the top branch of Figure 5.13, we further apply $C_{02}(R_2)$ on $C_{12}(R_1)|x_2\rangle$ and obtain:

$$C_{02}(R_2)C_{12}(R_1)H|x_2\rangle = \frac{1}{\sqrt{2}}(|0\rangle_2 + e^{j2\pi . x_2 x_1 x_0}|1\rangle_2). \tag{5.55}$$

The final state of the FT circuit in Figure 5.13 is then given by

$$|\Psi'\rangle = \frac{1}{\sqrt{8}}(|0\rangle_0 + e^{j2\pi \cdot x_0}|1\rangle_0)(|0\rangle_1 + e^{j2\pi . x_1 x_0}|1\rangle_1)(|0\rangle_2 + e^{j2\pi \cdot x_2 x_1 x_0}|1\rangle_2). \tag{5.56}$$

We can see that the qubits are arranged in the opposite order so that additional swap gates are needed. Another option would be to redefine the CB states as $|x\rangle = |x_0 x_1 \ldots x_{n-1}\rangle$.

The quantum circuit to perform FT for arbitrary n can be obtained by generalization of the corresponding circuit in Figure 5.13, and it is shown in Figure 5.14.

Regarding the complexity of the FT circuit, the number of required Hadamard gates is n, while the number of controlled gates is given by

$$n + (n-1) + \ldots + 1 \simeq \frac{n^2}{2}. \tag{5.57}$$

Therefore, the overall complexity of the FT computation circuit is $O(n^2)$.

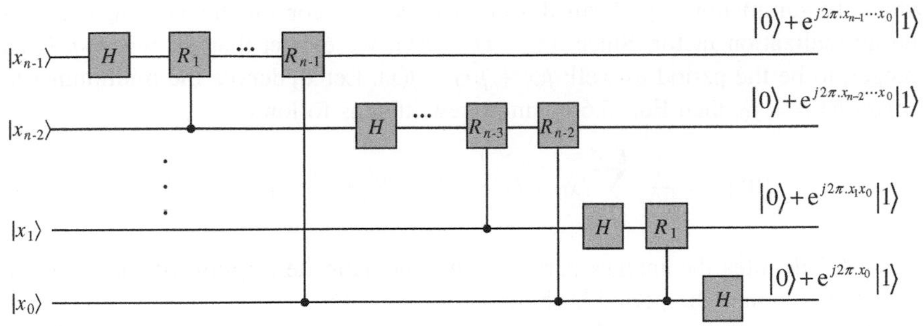

FIGURE 5.14

The quantum circuit to perform FT for arbitrary *n*.

5.5 THE PERIOD OF A FUNCTION AND SHOR FACTORING ALGORITHM

5.5.1 Period of a Function

The Shor factorization algorithm [12–15] is based on the possibility of determining in polynomial time the period of a given function $f(x)$. The function $f(x)$ in Shor's algorithm is defined as $f(x) = b^x \bmod N$. This function has a period r: $f(x) = f(x + r)$, $x \in \{0,1,\ldots,2^n - 1\}$. For the algorithm to be successful it is important that $2^n > N^2$. A classical algorithm requires $O(N)$ elementary operations. The corresponding quantum algorithm requires $O(n^3)$ elementary operations. The function $b^x \bmod N$ behaves as random noise over a period in the classical case. To describe the algorithm for determination of the period of function, we use an interpretation similar to that of Le Bellac [1]. The initial state of $n + m$ qubits N is given by

$$|\Phi\rangle = \frac{1}{2^{n/2}}\left(\sum_{x=0}^{2^n-1}|x\rangle\right)\otimes|0\ldots0\rangle. \tag{5.58}$$

By applying the function operator U_f on initial state $|\Phi\rangle$, the final state is

$$|\Phi_f\rangle = U_f|\Phi\rangle = \frac{1}{2^{n/2}}\sum_{x=0}^{2^n-1}|x,f(x)\rangle. \tag{5.59}$$

If we now perform the measurement of the output register and obtain the result f_0, the corresponding final state of the first register will be "the state vector collapse":

$$|\Psi_0\rangle = \frac{1}{\mathcal{N}}\sum_{x:\ f(x)=f_0}|x\rangle \tag{5.60}$$

where the summation is performed over all values of x for which $f(x) = f_0$, and $\mathcal{N}$ is the normalization factor. Since $f(x + r) = f(x)$, we expect that pr, where p is an integer, to be the period as well: $f(x + pr) = f(x)$. Let x_0 denote the minimum x to satisfy $f(x_0) = f_0$, then Eq. (5.60) can be rewritten as follows:

$$|\Psi_0\rangle = \frac{1}{\sqrt{K}} \sum_{k=0}^{K-1} |x_0 + kr\rangle; \quad K = [2^n/r] \text{ or } [2^n/r] + 1, \tag{5.61}$$

where $[z]$ denotes the integer part of z. A schematic description of the quantum circuit to calculate the period is shown in Figure 5.15.

The state of the input register is an incoherent superposition of vectors $|\Psi_i\rangle$:

$$|\Psi_i\rangle = \frac{1}{\sqrt{K_i}} \sum_{k=0}^{K_i-1} |x_i + kr\rangle, \tag{5.62}$$

where the minimum x to satisfy $f(x_i) = f_i$ is denoted by x_i. The input register density operator can be written, based on (5.59), as

$$\rho_{\text{tot}} = |\Psi_f\rangle\langle\Psi_f| = \frac{1}{2^n} \sum_{x,z} |x, f(x)\rangle\langle z, f(z)|, \tag{5.63}$$

and the density operator of the input register can be obtained by applying the trace operator over output register states:

$$\rho_{\text{in}} = \text{Tr}_{\text{out}}\rho_{\text{tot}} = \frac{1}{2^n} \sum_{x,z} |x\rangle\langle z| \langle f(x)|f(z)\rangle. \tag{5.64}$$

As an illustration, let the function $f(x)$ take the value f_0 exactly N_0 times and the value f_1 exactly N_1 times, so that $N_0 + N_1 = 2^n$, then the following is valid:

$$\langle f(x)|f(z)\rangle = \begin{cases} 0, & f(x) \neq f(z) \\ 1, & f(x) = f(z) \end{cases}. \tag{5.65}$$

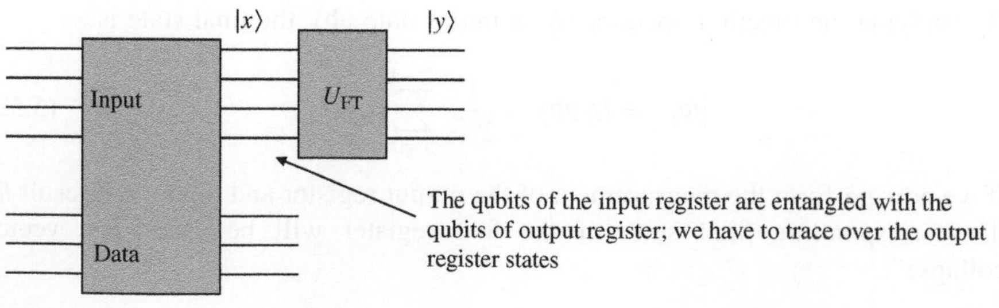
|x⟩
|y⟩
Input
U_FT
Data
The qubits of the input register are entangled with the qubits of output register; we have to trace over the output register states

FIGURE 5.15

The quantum circuit to determine the period of $f(x)$.

Upon substitution of (5.65) into (5.64), we obtain:

$$\rho_{\text{in}} = \frac{1}{2^n}\left(\sum_{x,z:\ f(x)=f(z)=f_0} |x\rangle\langle z| + \sum_{x,z:\ f(x)=f(z)=f_1} |x\rangle\langle z|\right). \tag{5.66}$$

Therefore, the density operator ρ_{in} represents the incoherent superposition of states $|\Psi_0\rangle$ and $|\Psi_1\rangle$, each occurring with probability $p_i = N_0/2^n$ $(i = 0,1)$, defined as

$$|\Psi_0\rangle = \frac{1}{\sqrt{N_0}}\sum_{x:\, f(x)=f_0} |x\rangle \qquad |\Psi_1\rangle = \frac{1}{\sqrt{N_1}}\sum_{x:\, f(x)=f_1} |x\rangle. \tag{5.67}$$

Based on (5.61), we can rewrite (5.66) as follows:

$$\rho_{\text{in}} = \frac{1}{2^n}\sum_{i=0}^{r-1}\sum_{k_i,k_j=0}^{K_i-1} |x_i + k_i r\rangle\langle x_i + k_j r|. \tag{5.68}$$

In order to determine the probability amplitude of finding the CB state $|y\rangle$, we can employ the FT of state:

$$|\Psi\rangle = \sum_{x=0}^{2^n-1} f(x)|x\rangle,\ f(x) = \langle x|\Psi\rangle,\ \sum_{x=0}^{2^n-1}|f(x)|^2 = 1, \tag{5.69}$$

which becomes $|\Phi_0\rangle$ for $f(x) = 1/\sqrt{K},\ x = x_0 + kr$. Since $|\Phi_0\rangle = U_{\text{FT}}|\Psi_0\rangle$, the probability amplitude of finding the CB state $|y\rangle$ is given by

$$a(\Phi_0 \rightarrow y) = \langle y|\Phi_0\rangle = \frac{1}{2^{n/2}}\frac{1}{\sqrt{K}}\sum_{k=0}^{K-1} e^{j2\pi y(x_0+kr)/2^n}, \tag{5.70}$$

and the corresponding probability is

$$p(y) = |a(\Phi_0 \rightarrow y)|^2 = \frac{1}{2^n K}\left|\sum_{k=0}^{K-1} e^{j2\pi y(x_0+kr)/2^n}\right|^2. \tag{5.71}$$

By using the geometric series formula we obtain:

$$\sum_{k=0}^{K-1} e^{j2\pi ykr/2^n} = \frac{1 - e^{j2\pi yKr/2^n}}{1 - e^{j2\pi yr/2^n}} = e^{j\pi(K-1)r/2^n}\frac{\sin(\pi yKr/2^n)}{\sin(\pi yr/2^n)}, \tag{5.72}$$

By using (5.72), the probability $p(y)$ for $2^n/r = K$ can be written as

$$p(y) = \frac{1}{2^n K}\frac{\sin^2(\pi y)}{\sin^2(\pi y/K)} = \begin{cases} 1/r, & y = iK \\ 0, & \text{otherwise} \end{cases}. \tag{5.73}$$

By substituting $y_i = i2^n/r + \delta_i$, we can rewrite Eq. (5.73) as

$$p(y_i) = \frac{1}{2^n K}\frac{\sin^2(\pi\delta_i Kr/2^n)}{\sin^2(\pi\delta_i r/2^n)}. \tag{5.74}$$

It can be shown that δ_i satisfies the following inequality:

$$|\delta_i| = \left| y_i - i\frac{2^n}{r} \right| < 1/2, \tag{5.75}$$

so that the probability $p(y)$ can be lower bounded by

$$p(y_i) \geq \frac{4}{\pi^2}\frac{K}{2^n} \simeq \frac{4}{\pi^2}\frac{1}{r}. \tag{5.76}$$

It can be shown that the function $p(y)$ has sharp maxima when the value of y is close to $i2^n/r$. The value i/r can be determined by developing $y/2^n$ in continued fractions, repeatedly applying the split and invert method. For example, 5/13 can be represented by using the continued fractions expansion as

$$\cfrac{1}{2 + \cfrac{1}{1 + \cfrac{1}{1 + \cfrac{1}{2}}}}.$$

The fraction i/r is then obtained as an irreducible fraction i_0/r_0. If i and r do not have a common factor, then we can immediately obtain $r = r_0$. The probability that two large numbers do not have a common factor is larger than 60%. Therefore, with probability 0.4×0.6 (0.4 comes from the probability of finding y_i close to $i2^n/r$), the protocol above will directly give the period $r = r_0$. If $f(x) \neq f(x + r_0)$, we try $f(x) = f(x + kr_0)$ with small multiples of r_0 (such as $2r_0$, $3r_0$, ...). If all these trials are unsuccessful, it means $|\delta_i| > 1/2$ and the protocol needs to be repeated. This procedure requires $O(n^3)$ elementary operations, $O(n^2)$ for FT, and $O(n)$ for the calculation of b^x.

5.5.2 Solving the RSA Encryption Protocol

Once we determine the period r, we can solve the *Rivest–Shamir–Adleman (RSA) encryption protocol*. Before we describe how to do so, we briefly overview the RSA encryption protocol, which is illustrated in Figure 5.16. Bob chooses two primes p and q, determines $N = pq$, and a number c that does not have a common divisor with the product $(p - 1)(q - 1)$. He further calculates d, the inverse for mod $(p - 1)(q - 1)$ multiplication, by

$$cd \equiv 1 \bmod (p - 1)(q - 1) \tag{5.77}$$

By a nonsecure path, using a classical communication channel, he sends Alice the values of N and c (but not p and q of course). Alice then sends Bob an encrypted message, which must be represented as a number $a < N$. Alice calculates

$$b \equiv a^c \bmod N \tag{5.78}$$

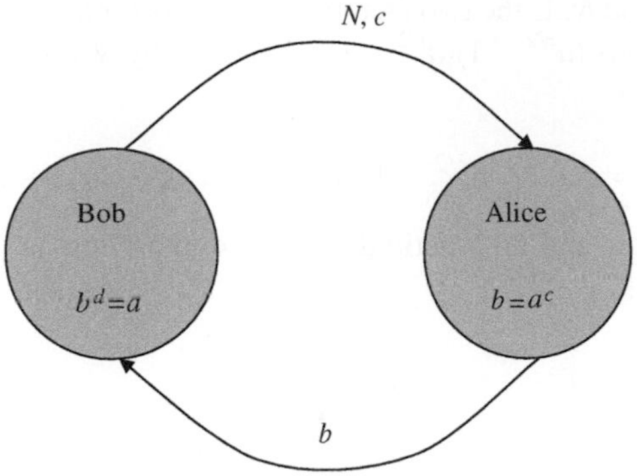

FIGURE 5.16

The RSA encryption protocol.

and sends the result b over the nonsecure channel. Bob, upon receiving b, calculates

$$b^d \bmod N = a, \tag{5.79}$$

and recovers the encrypted number.

Example. Let $p=3$ and $q=7$, meaning that $N=21$ and $(p-1)(q-1)=12$. We further choose $c=5$, which does not have a common divisor with 12. Next, d is determined as the inverse per mod 12 as 5, from $cd \bmod 12 = 25 \bmod 12 = 1$. Let us assume that Alice wants to send $a=4$. She first computes $a^c \bmod 12 = 4^5 \bmod 21 = (1024 = 21 \times 28 + 16) = 16 = b$, and sends it to Bob. Bob computes $b^5 \bmod 21 = 16^5 \bmod 21 = (49932 \times 21 + 4) \bmod 21 = 4 \bmod 21$ and therefore recovers the encrypted number a.

After this short overview of RSA protocol, we now describe how to solve it. Eve obtains b, N, c from the public channel, and calculates d' from

$$cd' \equiv 1 \bmod r. \tag{5.80}$$

She further calculates $b^{d'} \bmod N$ as follows:

$$\begin{aligned} b^{d'} \bmod N &= a^{cd'} \bmod N = a^{1+mr} \bmod N = a(a^r)^m \bmod N \\ &= a \bmod N, \quad a^r \equiv 1 \bmod N. \end{aligned} \tag{5.81}$$

5.5.3 Shor Factoring Algorithm

In order to *factorize* N we must write $a^r - 1$ as

$$a^r - 1 = (a^{r/2} - 1)\,(a^{r/2} + 1), \tag{5.82}$$

while $a^{r/2} \neq \pm 1 \bmod N$. If the two factors are integers (and $a^{r/2} \neq \pm 1 \bmod N$), then the product of integers $(a^{r/2}-1)(a^{r/2}+1)$ is divisible by $N=pq$. The values of p and q are obtained by

$$p = \gcd(N,\ a^{r/2}-1) \text{ and } q = \gcd(N,\ a^{r/2}+1). \tag{5.83}$$

From the fundamental arithmetic theorem we know that any positive integer N can be factorized as

$$N = \prod_{i=1}^{k} p_i^{l_i}, \tag{5.84}$$

where p_i are the primes and l_i are their corresponding powers. To factorize N we can generate an integer b between 1 and $N-1$ randomly, and apply the classical Euclidean algorithm to determine if $\gcd(b,N)=1$, which requires $O[(\log_2 N)^3]$ (classical steps). If $\gcd(b,N)>1$, then b is a nontrivial factor of N. If, on the other hand, $\gcd(b,N)=1$, we choose the order of b to be r, i.e. $b^r \bmod N = 1$, and N divides b^r-1. Further, if r is even we can factorize b^r-1 as $(b^{r/2}-1)(b^{r/2}+1)$, and we can find the factors of N using (5.83). Based on the previous discussion, we can summarize the *Shor factorization algorithm* as follows:

1. Randomly generate an integer b between 0 and $N-1$, and determine $\gcd(b,N)$. If $\gcd(b,N)>1$, then b is a nontrivial factor of N; otherwise go to the next step.
2. Prepare two quantum registers initialized to the $|0\rangle$ state as follows:

$$|\Psi_{\text{in}}\rangle = |0\rangle^{\otimes m}|0\rangle^{\otimes n},\quad n=\lceil \log_2 N\rceil. \tag{5.85}$$

3. By applying the m-qubit Hadamard transform on the first register, create the superposition state:

$$|\Psi_1\rangle = \frac{1}{2^{m/2}}\sum_{x=0}^{2^m-1}|x\rangle|0\rangle^{\otimes n}. \tag{5.86}$$

4. Apply the unitary transformation performing the following mapping:

$$|\Psi_2\rangle = C\sum_{x=0}^{2^m-1}|x\rangle|b^x \bmod N\rangle, \tag{5.87}$$

where C is a normalization constant.
5. Perform measurement on the second register in the CB to obtain f_0, while leaving the first register in uniform superposition of all states for which $f(x)=f_0$:

$$|\Psi_3\rangle = C'\sum_{k=0}^{K-1}|x_0+kr\rangle|f_0\rangle,\quad K=\lceil 2^m/r\rceil. \tag{5.88}$$

6. Apply the inverse quantum FT on the first register to obtain:

$$|\Psi_4\rangle = C'' \sum_{l=0}^{2^m-1} \sum_{k=0}^{K-1} e^{j2\pi(x_0+kr)l/2^m} |l\rangle |f_0\rangle. \tag{5.89}$$

7. Perform the measurement on the first register to obtain l/r, as described in the discussion for inequality (5.75).
8. Apply the continuous fractions expansion algorithm to determine an irreducible fraction l_0/r_0, followed by the procedure described in the paragraph below Eq. (5.76) that determines the period r.
9. If r is even and $b^{r/2} \neq 1 \bmod N$, then compute $\gcd(b^{r/2} - 1, N) = p$ and $\gcd(b^{r/2} + 1, N) = q$ to check if either p or q is a nontrivial factor.

Repeat the procedure above until full prime factorization in the form of Eq. (5.84) is obtained.

5.6 SIMON'S ALGORITHM

The Shor factorization algorithm requires the determination of the period of a function. An alternative to this algorithm is Simon's algorithm [16], which represents a particular instance of the more general problem known as the *Abelian hidden subgroup* problem. Let us observe the function f from $(F_2)^N$ to itself, represented by the following unitary map:

$$|x\rangle|0\rangle \xrightarrow{U_f} |x\rangle|f(x)\rangle \quad \forall x. \tag{5.90}$$

Let H' be a subgroup of $(F_2)^N$ for which the function f is a unique constant for every right coset of H'. Simon's algorithm performs $O(N)$ evaluations of the oracle f in combination with classical computation to determine the *generators* of subgroup H'. Simon's algorithm can be formulated as follows:

0. *Initialization.* Prepare two N-qubit quantum registers initialized to the $|0\rangle$ state:

$$|\Psi_{in}\rangle = |0\rangle^{\otimes N} |0\rangle^{\otimes N}. \tag{5.91}$$

1. *Superposition state.* By applying the N-qubit Hadamard transform on the first register, create the superposition state:

$$|\Psi_1\rangle = \frac{1}{2^{N/2}} \sum_{x \in F_2^N} |x\rangle |0\rangle^{\otimes N}. \tag{5.92}$$

2. *Unitary operation U_f application.* Upon applying the U_f operator on the second register we obtain:

$$|\Psi_2\rangle = \frac{1}{2^{N/2}} \sum_{x \in F_2^N} |x\rangle |f(x)\rangle. \tag{5.93}$$

3. *Measurement step.* Perform the measurement on the second register to obtain a value y, which yields the coset of the hidden subgroup in the first register:

$$|\Psi_3\rangle = \frac{1}{|H'|^{1/2}} \sum_{x:\, f(x)=y} |x\rangle|y\rangle. \tag{5.94}$$

4. *Hadamard transform followed by measurement.* Apply the N-qubit Hadamard transform to the first register, followed by measurement on the first register in the computational basis.
5. Repeat steps 1–4 approximately N times. The result of this algorithm will be the orthogonal complement of H' with respect to the scalar product in $(F_2)^N$.
6. Solve the system of linear equations by *Gauss elimination* to obtain the kernel of a square matrix, which contains the generators of H'.

5.7 CLASSICAL/QUANTUM COMPUTING COMPLEXITIES AND TURING MACHINES

Quantum algorithms have raised some questions on the accuracy of some statements from classical algorithm theory where algorithm complexity is concerned. Namely, some problems that appear to be "intractable" (the complexity is exponential in the number of bits n) in classical computing can be solved polynomially in n by means of quantum computers. For example, if prime factorization is an intractable problem, which was suggested by practice but not proven, then Shor's algorithm discussed above contradicts this observation, as it can represent a composite number in terms of primes in a number of steps that is polynomial in n. The key advantage of quantum algorithms is that they can explore all branches of a nondeterministic algorithm in parallel, through the concept of quantum parallelism.

Classical algorithms can be classified as *effective* when the number of steps is a *polynomial* function of size n. The computational complexity of these algorithms is typically denoted as **P**. The class of problems that are solvable *nondeterministically* in *polynomial* time are called **NP**. The subclass of these problems, which are the most difficult ones, are **NP-complete** problems. For example, the Traveling Salesman Problem belongs to this subclass. If one of these problems can be solved efficiently, then all of them can be solved. The class of problems that can be solved with the amount of *memory polynomial in input size* is called **PSPACE**. Further, the class of problems that can be solved with high probability by the use of a random generator is known as **BPP**, originating from "bounded error probability polynomial in time." Finally, the class of problems that can be solved in polynomial time, if sums of exponentially many contributors are themselves computable in polynomial time, is denoted as $\mathbf{P}^{\#\mathbf{P}}$. The different classes of problems discussed above can be related as follows:

$$\mathrm{P} \subset \mathrm{BPP},\ \mathrm{P} \subset \mathrm{NP} \subset \mathrm{P}^{\#\mathrm{P}} \subset \mathrm{PSPACE}. \tag{5.95}$$

Turing defined a class of machines, known as *Turing machines*, which can be used to study the complexity of a computational algorithm. In particular, there exist so-called *universal machines*, which can be used to simulate any other Turing machine. Interestingly enough, the Turing machine can be used to simulate all operations performed on a modern computer. This led to the formulation of the *Church–Turing* thesis as follows: The class of functions that can be computed by a Turing machine corresponds exactly to the class of functions that one would naturally consider to be computable by an algorithm. This thesis essentially establishes the equivalence between rigorous mathematical description (Turing machine) and an intuitive concept. There exist problems that are not calculable and there is no known algorithm to solve them. For example, the halting problem of a Turing machine belongs to this class. The Church–Turing thesis is also applicable to quantum algorithms. In the previous paragraph we said that an algorithm is effective if it can be solved in a polynomial number of steps. We also learned that Turing machines can be used to describe these effective algorithms. These two observations can be used to formulate a *strong* version of the *Church–Turing thesis*: Any computational model can be simulated on a probabilistic Turing machine with a polynomial number of computational steps. There exist several types of Turing machines depending on the type of computation. The *deterministic*, *probabilistic*, and *multitape Turing machines* are described briefly next.

The deterministic Turing machine is described by an alphabet A, a set of control states Q, and by a transition function δ:

$$\delta: \; Q \times A \to Q \times A \times D, \;\; D = \{-1, 0, 1\}. \tag{5.96}$$

The elements of the alphabet are called *letters*, and by concatenating the letters we obtain *words*. The set D is related to the *read-write* head, with the elements -1, $+1$, and 0 denoting the movement of the head to the left, right, and standing respectively. The deterministic Turing machine can be *defined* as $(Q, A, \delta, q_0, q_a, q_r)$, where the *state* of the machine is specified by $q \in Q$. In particular, $q_0, q_a, q_r \in Q$ denote the initial state, the accepting state, and rejecting state respectively. The *configuration* of the Turing machine is given by $c = (q, x, y)$, where $x, y \in A'$, with A' being the set of all words obtained by concatenating the letters from A. The Turing machine has a *tape* (memory), specified by xy with x being the *scanning* (*reading*). A *computation* is a sequence of configurations beginning with an *initial configuration* c_0, until we reach the *halting configuration*. The computation halts after t computation steps when either one of the configurations does not have a successor or if its state is q_a or q_r.

The *probabilistic Turing machine* is more general, as the transition function assigns *probabilities* to possible operations:

$$\delta: \; Q \times A \times Q \times A \times D \to [0, 1]. \tag{5.97}$$

In other words, a probabilistic Turing machine is a nondeterministic Turing machine that randomly selects possible transitions according to some probability distribution. As a consequence, the machine-state transitions can be described by a *stochastic matrix*. A given configuration is a successor configuration with probability δ. A terminal configuration can be computed from an initial configuration with a probability given by the product of probabilities of intermediate configurations leading to it by a particular computation, defined by a sequence of states. The deterministic Turing machine is just a particular instance of a probabilistic Turing machine.

An *m-type deterministic Turing machine* is characterized by *m* tapes, an alphabet *A*, a finite state of control states *Q*, and the following transition function:

$$\delta: \; Q \times A^m \rightarrow Q \times (A \times D)^{\times m}, \tag{5.98}$$

and defined by (Q,A,δ,q_0,q_a,q_r). The *configuration* of an *m*-type machine is given by $(q,x_1,y_1,\ldots,x_m,y_m)$, where q is the current state of the machine, $(x_i,y_i) \in A' \times A'$, and x_iy_i denotes content of the *i*th type. The *m*-type machines are suitable for problems involving parallelism. If the computational time needed for a one-type machine is *t*, then the computation time needed for an *m*-type machine is $O(t^{1/2})$, while in terms of computational complexity they are comparable.

The quantum Turing machine is characterized by the following transition function:

$$\delta: \; Q \times A \times Q \times A \times D \rightarrow \mathscr{C}, \tag{5.99}$$

which moves a given configuration to a range of successor configurations, each occurring with a quantum *probability amplitude*, which corresponds to a unitary transformation of a quantum state (we denote the set of complex numbers by $\mathscr{C}$). Bennett has shown that *m*-type Turing machines can be simulated by *reversible Turing machines*, with a certain reduction in efficiency [17]. Further, Toffoli has shown that arbitrary finite mapping can be computed reversibly by padding strings with zeros, permuting them, and projecting some of the bit strings to other bit strings [18]. The elementary reversible gates can be used to implement permutations of bit strings. Finally, Benioff has shown that unitary quantum state evolution (that is reversible) is at least as powerful as a Turing machine [19,20]. The probabilistic classical process can be represented by a tree, which grows exponentially with possible outcomes. The key difference in quantum computing is that we assign the quantum probability amplitudes to the branches in a tree, which can interfere with each other.

5.8 SUMMARY

This chapter has presented various quantum algorithms: (i) Deutsch and Deutsch–Jozsa algorithms (Section 5.2); (ii) Grover search algorithm (Section 5.3); (iii) quantum Fourier transform (Section 5.4); (iv) period-finding and Shor's factorization algorithms (Section 5.5); and (v) Simon's algorithm (Section 5.6). Section 5.1 contains a review of the quantum parallelism concept. In Section 5.5 a procedure to solve the RSA encryption protocol was also described. In Section 5.7 classical/quantum computing complexities and Turing machines are considered.

In Section 5.9 a set of problems to help the reader gain a deeper understanding of these topics is provided. In addition, problems have been included as regards finding the phase algorithm, the quantum discrete logarithm algorithm, Kitaev's algorithm, and quantum simulation.

5.9 PROBLEMS

1. By using mathematical induction prove that the circuit in Figure 5.P1 can be used to implement the Deutsch–Jozsa algorithm – that is, to verify whether the mapping $\{0,1\}^n \rightarrow \{0,1\}$ is constant or balanced.
2. By using mathematical induction, prove that the circuit in Figure 5.P2 can be used to calculate the FT for arbitrary *n*.
3. Assume that a unitary operator U has an eigenket $|u\rangle$ with eigenvalue $\exp(j2\pi\varphi_u)$. The goal of *phase estimation* is to estimate the phase φ_u. Describe how the quantum Fourier transform can be used for phase estimation.
4. This problem relates to the *shift-invariance property* of quantum Fourier transform. Let G be a group and H be a subgroup of G. If a function f on G is constant on cosets of H, then the FT of f is invariant over cosets of H. Prove the claim.

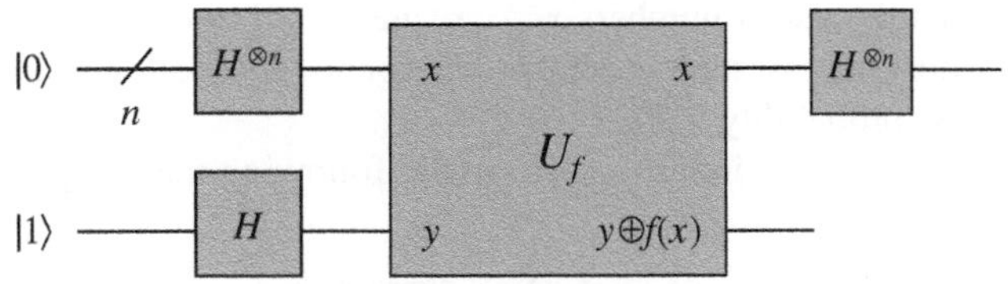

FIGURE 5.P1

Deutsch-Jozsa algorithm implentation circuit

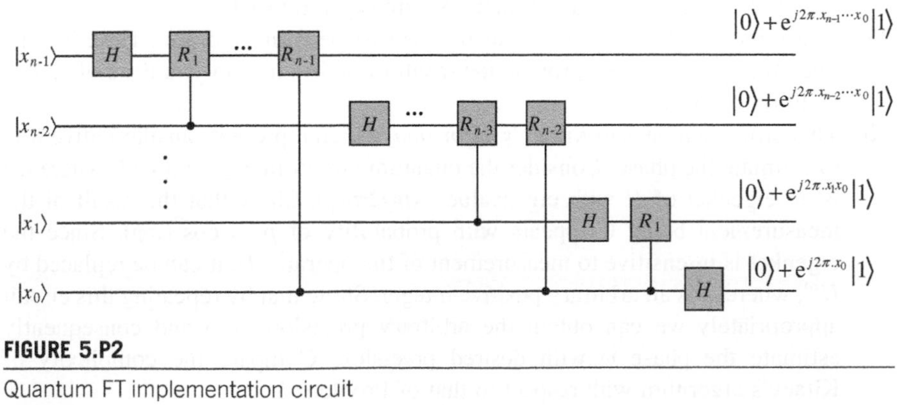

FIGURE 5.P2

Quantum FT implementation circuit

5. The quantum circuit to perform the Grover search for $n = 3$ was shown in Figure 5.11. Provide the quantum circuit that can be used to perform a Grover search for arbitrary n. Explain the operating principle and prove that this circuit can indeed be used for any n.
6. Provide the quantum circuit to implement the Shor factorization algorithm for $n = 15$. Describe the operating principle of this circuit.
7. This problem is devoted to *quantum discrete logarithms*. Let us consider the following function $f(x_1, x_2) = a^{bx_1+x_2} \bmod N$, where all variables are integers. Let r be the smallest positive integer for which $a^r \bmod N = 1$. This integer can be determined by using the order-finding algorithm. Clearly, this function is periodic as $f(x_1 + i, x_2 - ib) = f(x_1, x_2)$, where i is an integer. The discrete logarithm problem can be formulated as follows: given a and $c = a^b$, determine b. This problem is important in solving the RSA encryption protocol, as discussed in Section 5.5. Your task is to provide a quantum algorithm that can solve this problem by using one query of a quantum block U that performs the following unitary mapping: $U|x_1\rangle|x_2\rangle|y\rangle \to |x_1\rangle|x_2\rangle|y \oplus f(x_1, x_2)\rangle$.
8. In Problem 6 you were asked to provide the quantum circuit to implement the Shor factorization algorithm for $n = 15$. By using Simon's algorithm, describe how to perform the same task. Provide the corresponding quantum circuit. Analyze the complexity of both algorithms.
9. Suppose that the list of numbers $x_1, \ldots, x_n$ are stored in quantum memory. How many memory accesses are needed to determine the smallest number in the list with success probability $\geq 1/2$?
10. Provide the quantum circuit to perform the following mapping:

$$|m\rangle \to \frac{1}{\sqrt{p}} \sum_{n=0}^{p-1} e^{j2\pi mn/p} |n\rangle,$$

where p is a prime.
11. Design a quantum circuit to perform the mapping $|x\rangle \to |x + c \bmod 2^n\rangle$, where $x \in [0, 2^n - 1]$ and c is a constant, by using quantum FT.
12. The circuit in Figure 5.P12 can be used to perform addition. Describe the operating principle and provide the result of addition. Can you generalize this addition problem?
13. This problem relates to *Kitaev's algorithm*, which represents an alternative way to estimate the phase. Consider the quantum circuit in Figure 5.P13, where $|u\rangle$ is an eigenket of U with eigenvalue $\exp(j2\pi\varphi_u)$. Show that the result of the measurement being 0 appears with probability of $p = \cos^2(\pi\varphi)$. Since the eigenket is insensitive to measurement of the operator U, it can be replaced by U^m, where m is an arbitrary positive integer. Show that by repeating this circuit appropriately we can obtain the arbitrary precision of p and consequently estimate the phase φ with desired precision. Compare the complexity of Kitaev's algorithm with respect to that of Problem 3.

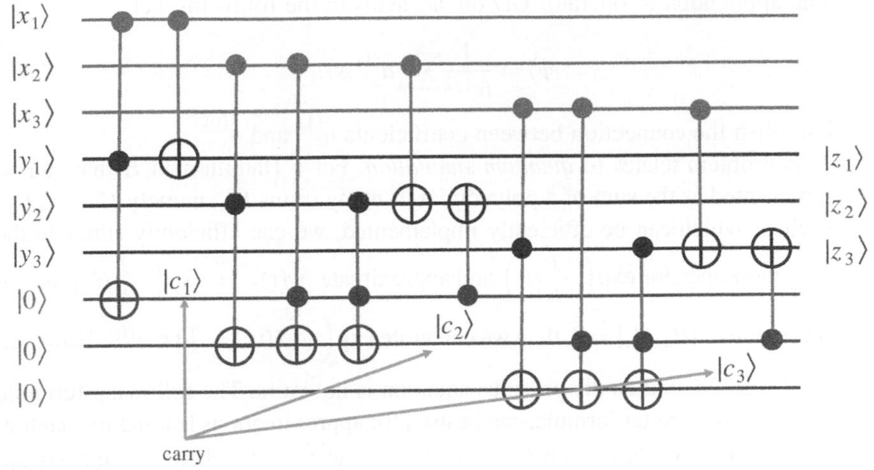

FIGURE 5.P12

Circuit.

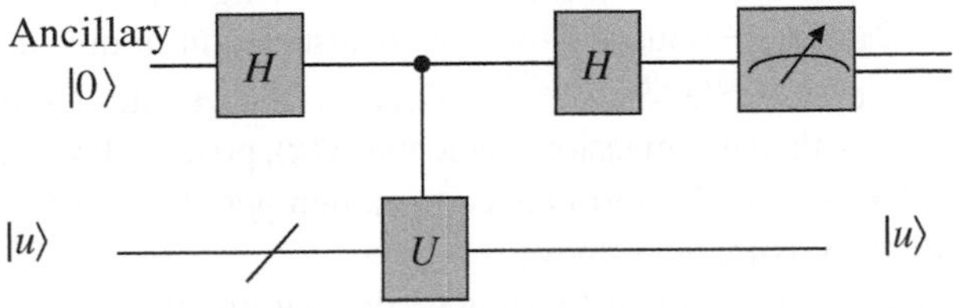

FIGURE 5.P13

Kitaev's algorithm implentation

14. The state ket after application of Hadamard gates of the circuit in Figure 5.P14 can be written in the following form:

$$|\psi\rangle = \frac{1}{N^{1/2}} \sum_x a_x^{(0)} |x\rangle, \; a_x^{(0)} = 1.$$

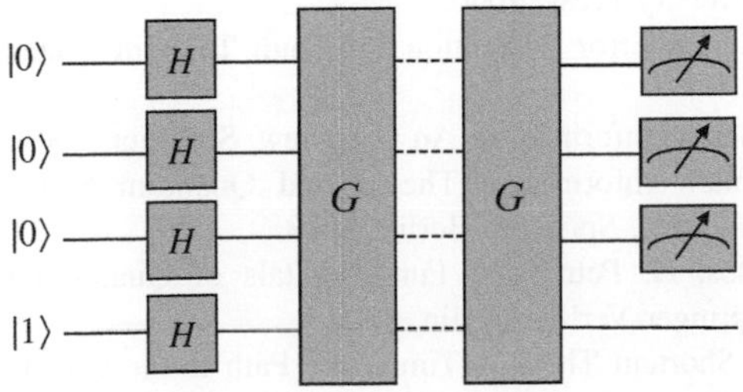

FIGURE 5.P14

Quantum circuit to perform Grover search (n = 3)

The application of operator GO on $|\psi\rangle$ leads to the following ket:

$$|\psi\rangle = \frac{1}{N^{1/2}}\sum\nolimits_x a_x^{(1)}|x\rangle.$$

Establish the connection between coefficients $a_x^{(1)}$ and $a_x^{(0)}$.

15. This problem relates to *quantum simulation.* For a Hamiltonian H that can be represented as the sum of a polynomial of many terms H_m, namely $H = \Sigma_m H_m$, each of which can be efficiently implemented, we can efficiently simulate the evolution operator $\exp\left(-\frac{j}{\hbar}Ht\right)$ and approximate $|\psi(t)\rangle = \exp\left(-\frac{j}{\hbar}Ht\right)|\psi(0)\rangle$. If for all m,n $[H_m,H_n] = 0$, then we can write $\exp\left(-\frac{j}{\hbar}Ht\right) = \prod_k e^{-\frac{j}{\hbar}H_k t}$. However, if $[H_m,H_n] \neq 0$, then the previous equation is not valid. The following formula, known as the Trotter formula, can be used for approximations leading to quantum simulation algorithms, $\lim_{n\to\infty}(e^{jU_1t/n}e^{jU_2t/n}) = e^{j(U_1+U_2)t}$, where U_1 and U_2 are Hermitian operators. Prove the Trotter formula. Prove also the following useful approximations: $e^{j(U_1+U_2)\Delta t} = e^{jU_1\Delta t}e^{jU_2\Delta t} + O(\Delta t^2)$, $e^{j(U_1+U_2)\Delta t} = e^{jU_1\Delta t/2}e^{jU_2\Delta t}e^{jU_1\Delta t/2} + O(\Delta t^3)$. Finally, the following approximation, known as the Baker–Campbell–Hausdorf formula, is also useful in quantum simulation: $e^{(U_1+U_2)\Delta t} = e^{U_1\Delta t}e^{U_2\Delta t}e^{-[U_1,U_2]\Delta t^2/2} + O(\Delta t^3)$. Prove it. Consider now a single particle with one-dimensional potential $V(x)$, governed by the Hamiltonian $H = p^2/(2m) + V(x)$. Perform the computation $|\psi(t)\rangle = \exp(-jHt/\hbar)|\psi(0)\rangle$ by using the approximations above.

16. Construct the quantum circuit to simulate the Hamiltonian $H = Z_1Z_2\ldots Z_n$, performing the unitary transform $|\psi(t)\rangle = \exp(-jHt/\hbar)|\psi(0)\rangle$ for arbitrary Δt.

References

[1] M. Le Bellac, An Introduction to Quantum Information and Quantum Computation, Cambridge University Press, 2006.

[2] M.A. Neilsen, I.L. Chuang, Quantum Computation and Quantum Information, Cambridge University Press, 2000.

[3] F. Gaitan, Quantum Error Correction and Fault Tolerant Quantum Computing, CRC Press, 2008.

[4] G. Jaeger, Quantum Information: An Overview, Springer, 2007.

[5] D. Petz, Quantum Information Theory and Quantum Statistics, Theoretical and Mathematical Physics, Springer, Berlin, 2008.

[6] P. Lambropoulos, D. Petrosyan, Fundamentals of Quantum Optics and Quantum Information, Springer-Verlag, Berlin, 2007.

[7] G. Johnson, A Shortcut Through Time: The Path to the Quantum Computer, Knopf, New York, 2003.

[8] J. Preskill, Quantum Computing (1999). Available at: http://www.theory.caltech.edu/~preskill/.

[9] J. Stolze, D. Suter, Quantum Computing, Wiley, New York, 2004.
[10] R. Landauer, Information is physical, Phys. Today 44 (5) (May 1991) 23–29.
[11] R. Landauer, The physical nature of information, Phys. Lett. A 217 (1991) 188–193.
[12] P.W. Shor, Algorithms for quantum computation: discrete logarithm and factoring, Proc. IEEE 35th Annual Symposium on Foundations of Computer Science (1994) p. 124.
[13] P.W. Shor, Polynomial-time algorithms for prime number factorization and discrete logarithms on a quantum computer, SIAM J. Comput. 26 (1997) 1484.
[14] A. Ekert, R. Josza, Shor's factoring algorithm, Rev. Mod. Phys. 68 (1996) 733–753.
[15] T. Beth, M. Rötteler, Quantum algorithms: Applicable algebra and quantum physics, in: G. Alber, T. Beth, M. Horodečki, P. Horodečki, R. Horodečki, M. Rötteler, H. Weinfurter, R. Werner, A. Zeilinger (Eds.), Quantum Information, Springer-Verlag, Berlin, 2001, pp. 96–150.
[16] D.R. Simon, On the power of quantum computation, in: S. Goldwasser (Ed.), Proc. IEEE 35th Annual Symposium on the Foundations of Computer Science, Los Alamitos, CA, 1994, p. 116.
[17] C.H. Bennett, Time/space trade-offs for reversible computation, SIAM J. Comput. 18 (1989) 766–776.
[18] T. Toffoli, Reversible computing, in: G. Goos, J. Hartmanis (Eds.), Automata, Languages and Programming, Lecture Notes in Computer Science 85, Springer-Verlag, Berlin, 1980, p. 632.
[19] P. Benioff, The computer as a physical system: A microscopic quantum mechanical Hamiltonian model of computers as represented by Turing machines, J. Stat. Phys. 22 (5) (1980) 563–591.
[20] P. Benioff, Models of quantum Turing machines, Fortschr. Phys. 46 (1998) 423.

CHAPTER

Classical Error Correcting Codes

6

CHAPTER OUTLINE

In this chapter, which is based on Refs [1–39], the basic concepts of coding theory are provided. Only topics from classical error correction important in quantum error correction are covered. This chapter is organized as follows. In Section 6.1, the channel coding preliminaries are introduced, in particular basic definitions, channel models, the concept of channel capacity, and statement of the channel coding theorem. Section 6.2 covers the basics of linear block codes (LBCs), such as definitions of generator and parity-check matrices, syndrome decoding, distance properties of LBCs, and some important coding bounds. In Section 6.3 cyclic codes are introduced. The Bose–Chaudhuri–Hocquenghem (BCH) codes are described in Section 6.4. The Reed–Solomon (RS), concatenated, and product codes are described in Section 6.5. After a short summary section, a set of problems is provided for readers to gain a deeper understanding of classical error correction.

6.1 CHANNEL CODING PRELIMINARIES

Two key system parameters are transmitted power and channel bandwidth, which together with additive noise sources determine the signal-to-noise ratio (SNR) and consequently the bit error rate (BER). In practice, we very often encounter

Quantum Information Processing and Quantum Error Correction. DOI: 10.1016/B978-0-12-385491-9.00006-X

a situation where the target BER cannot be achieved with a given modulation format. For fixed SNR, the only practical option to change the data quality transmission from unacceptable to acceptable is through the use of *channel coding*. Another practical motivation of introducing channel coding is to reduce the required SNR for a given target BER. The energy that can be saved by coding is commonly described as coding gain. *Coding gain* refers to the savings attainable in the energy per information bit to noise spectral density ratio (E_b/N_0) required to achieve a given bit error probability when coding is used compared to that with no coding. A typical digital communication system employing channel coding is shown in Figure 6.1. The discrete source generates information in the form of a sequence of symbols. The channel encoder accepts the message symbols and adds redundant symbols according to a corresponding prescribed rule. Channel coding is the act of transforming a length-k sequence into a length-n codeword. The set of rules specifying this transformation are called the channel code, which can be represented as the following mapping:

$$C: M \rightarrow X,$$

where C is the channel code, M is the set of information sequences of length k, and X is the set of codewords of length n. The decoder exploits these redundant symbols to determine which message symbol was actually transmitted. The encoder and decoder consider the whole digital transmission system as a discrete channel. Different classes of channel codes can be classified into three broad categories: (i) *error detection*, in which we are concerned only with detecting the errors occurring during transmission (examples include automatic requests for transmission, ARQ); (ii) *forward error correction* (FEC), where we are interested in correcting errors that occur during transmission; and (iii) hybrid channel codes that combine the previous two approaches. In this chapter we are concerned only with FEC.

The key idea behind the forward error correcting codes is to add extra redundant symbols to the message to be transmitted, and to use those redundant symbols in the decoding procedure to correct the errors introduced by the channel. The redundancy can be introduced in the time, frequency, or space domain. For example, redundancy in the time domain is introduced if the same message is transmitted at least twice; this is known as the *repetition code*. Space redundancy is used as a means to achieve high spectrally efficient transmission, in which the modulation is combined with error control.

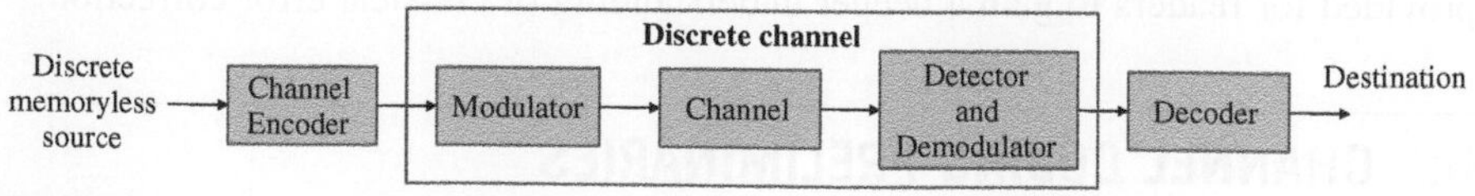

FIGURE 6.1

Block diagram of a point-to-point digital communication system.

The codes commonly considered in digital communications and storage applications belong either to the class of *block codes* or to the class of *convolutional codes*. In an *(n,k) block code* the channel encoder accepts information in successive k-symbol blocks and adds $n-k$ redundant symbols that are algebraically related to the k message symbols, thereby producing an overall encoded block of n symbols ($n>k$), known as a codeword. If the block code is systematic, the information symbols remain unchanged during the encoding operation, and the encoding operation may be considered as adding the $n-k$ generalized parity checks to k information symbols. Since the information symbols are statistically independent (a consequence of source coding or scrambling), the next codeword is independent of the content of the current codeword. The code rate of an (n,k) block code is defined as $R=k/n$ and overhead by $OH=(1/R-1)\bullet 100\%$. In convolutional code, however, the encoding operation may be considered as the discrete-time convolution of the input sequence with the impulse response of the encoder. Therefore, the $n-k$ generalized parity checks are functions not only of k information symbols, but also of m previous k-tuples, with $m+1$ being the encoder impulse response length. The statistical dependence is introduced to a window of length $n(m+1)$, a parameter known as the constraint length of convolutional codes.

As mentioned above, the channel code considers the whole transmission system as a discrete channel in which the sizes of input and output alphabets are finite. Two examples of such channels are shown in Figure 6.2. In Figure 6.2a an example of a discrete memoryless channel (DMC) is shown, which is characterized by channel (transition) probabilities. Let $X=\{x_1, x_1,\ldots,x_M\}$ denote the channel input alphabet and $Y=\{y_1, y_2,\ldots,y_N\}$ denote the channel output alphabet. This channel is completely characterized by the following set of transition probabilities:

$$p(y_j|x_i)=P(Y=y_j|X=x_i),\quad 0\le p(y_j|x_i)\le 1,\\ i\in\{1,\ldots,M\},\quad j\in\{1,\ldots,N\}, \tag{6.1}$$

where M and N denote the sizes of input and output alphabets respectively. The transition probability $p(y_j|x_i)$ represents the conditional probability that channel output $Y=y_j$ given the channel input $X=x_i$. The channel introduces errors, and if $j\neq i$ the corresponding $p(y_j|x_i)$ represents the conditional probability of error, while for $j=i$ it represents the conditional probability of correct reception. For $M=N$, the *average symbol error probability* is defined as the probability that output random variable Y_j is different from input random variable X_i, with averaging being performed for all $j\neq i$:

$$P_e=\sum_{i=1}^{M}p(x_i)\sum_{j=1,j\neq i}^{N}p(y_j|x_i), \tag{6.2}$$

where the inputs are selected from the distribution $\{p(x_i)=P(X=x_i);\ i=1,\ldots,M\}$, with $p(x_i)$ being known as the *a priori* probability of

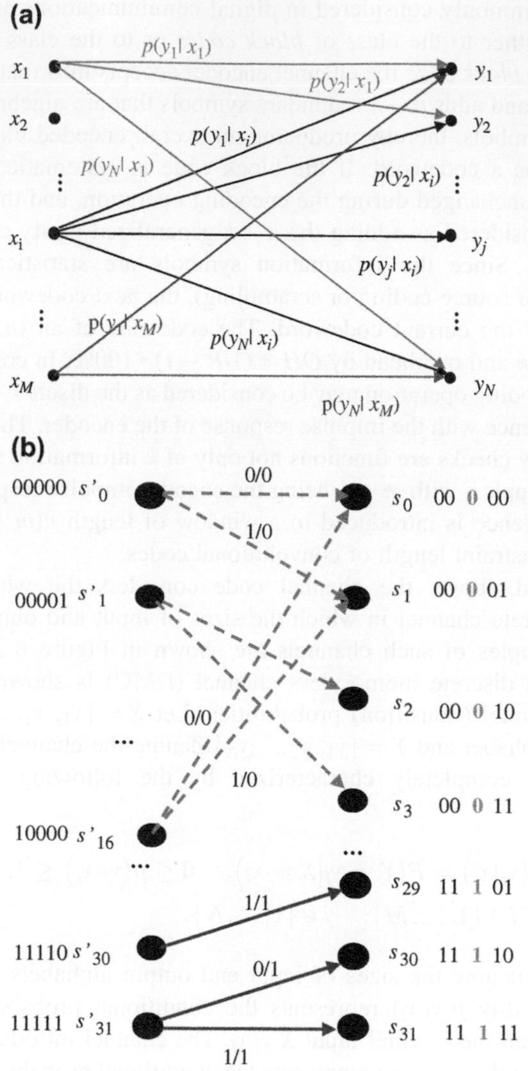

FIGURE 6.2

Two examples of discrete channels: (a) discrete memoryless channel (DMC); (b) discrete channel with memory described as dynamic trellis [14].

input symbol x_i. The corresponding probabilities of output symbols can be calculated by

$$p(y_j) = \sum_{i=1}^{M} P(Y = y_j | X = x_i) P(X = x_i) = \sum_{i=1}^{M} p(y_j | x_i) p(x_i); \quad j = 1, \ldots, N. \tag{6.3}$$

The decision rule that minimizes average symbol error probability (6.2), denoted as $D(y_j) = x^*$, is known as the maximum *a posteriori* (MAP) rule, and can be formulated as follows:

$$D(y_j) = x^* : P(x^* | y_j) \geq P(x_i | y_j), \quad i = 1, \ldots, M. \tag{6.4}$$

Therefore, the symbol error probability P_e will be minimal when to every output symbol y_j the input symbol x^* is assigned that has the largest *a posteriori* probability $P(x^*|y_j)$. By using the Bayes rule, Eq. (6.4) can be rewritten as

$$D(y_j) = x^* : \frac{P(y_j | x^*) P(x^*)}{P(y_j)} \geq \frac{P(y_j | x_i) P(x_i)}{P(y_j)}, \quad i = 1, \ldots, M. \tag{6.5}$$

If all input symbols are equally likely, $P(x_i) = 1/M$ $(i = 1,\ldots,M)$, the corresponding decision rule is known as the maximum-likelihood (ML) decision rule:

$$D(y_j) = x^* : P(y_j | x^*) \geq P(y_j | x_i), \quad i = 1, \ldots, M. \tag{6.6}$$

Figure 6.2b shows a discrete channel model with memory [14]. We assume that the optical channel has memory equal to $2m + 1$, with $2m$ being the number of bits that influence the observed bit from both sides. This dynamical trellis is uniquely defined by the set of the previous state, the next state, in addition to the channel output. The state (the bit-pattern configuration) in the trellis is defined as $\mathbf{s}_j = (x_{j-m}, x_{j-m+1},\ldots,x_j,x_{j+1},\ldots,x_{j+m}) = \mathbf{x}[j-m,j+m]$, where $x_k \in \boldsymbol{X} = \{0,1\}$. An example trellis of memory $2m + 1 = 5$ is shown in Figure 6.2b. The trellis has $2^5 = 32$ states $(\mathbf{s}_0,\mathbf{s}_1,\ldots,\mathbf{s}_{31})$, each of which corresponds to a different five-bit pattern. For a complete description of the trellis, the transition probability density functions (PDFs) $p(y_j|x_j) = p(y_j|\mathbf{s})$, $\mathbf{s} \in \boldsymbol{S}$ can be determined from *collected histograms*, where y_j represents the sample that corresponds to the transmitted bit x_j, and $\boldsymbol{S}$ is the set of states in the trellis.

One important figure of merit in the channel is the *channel capacity*, which is obtained by maximization of mutual information $I(\boldsymbol{X};\boldsymbol{Y})$ over all possible input distributions:

$$C = \max_{\{p(x_i)\}} I(\boldsymbol{X}; \boldsymbol{Y}), \quad I(\boldsymbol{X}; \boldsymbol{Y}) = H(\boldsymbol{X}) - H(\boldsymbol{X}|\boldsymbol{Y}), \tag{6.7a}$$

where $H(U) = E(\log_2 P(U))$ denotes the entropy of a random variable U and E(•) denotes the mathematical expectation operator. The mutual information can be determined by

$$I(X;Y) = H(X) - H(X|Y) = \sum_{i=1}^{M} p(x_i)\log_2\left[\frac{1}{p(x_i)}\right] - \sum_{j=1}^{N} p(y_j)\sum_{i=1}^{M} p(x_i|y_j)\log_2\left[\frac{1}{p(x_i|y_j)}\right]. \tag{6.7b}$$

$H(X)$ represents the uncertainty about the channel input before observing the channel output, also known as *entropy*; $H(X|Y)$ denotes the conditional entropy or the amount of uncertainty remaining about the channel input after the channel output has been received. Therefore, the mutual information represents the amount of information (per symbol) that is conveyed by the channel, which represents the uncertainty about the channel input that is resolved by observing the channel output. The mutual information can be interpreted by means of the Venn diagram shown in Figure 6.3a. The left circle represents the entropy of channel input, the right circle represents the entropy of channel output, and the mutual information is obtained in the intersection of these two circles. Another interpretation, due to Ingels [28], is shown in Figure 6.3b. The mutual information, i.e. the information conveyed by the channel, is obtained as the output information minus information lost in the channel.

Since for an M-ary input M-ary output symmetric channel, such as M-ary pulse position (PPM), $p(y_j|x_i) = P_s/(M-1)$ and $p(y_j|x_j) = 1 - P_s$, where P_s is symbol error probability, the channel capacity (in bits/symbol) can be found by

$$C = \log_2 M + (1 - P_s)\log_2(1 - P_s) + P_s \log_2\left(\frac{P_s}{M-1}\right). \tag{6.8}$$

The channel capacity represents an important bound on data rates achievable by any modulation and coding schemes. It can also be used to compare different coded modulation schemes in terms of their distance to the channel capacity curve. Figure 6.4 shows the channel capacity for a Poisson M-ary PPM channel against the average number of signal photons per slot, expressed on a dB scale, for the average number of background photons of $K_b = 1$.

Now we have built enough knowledge to formulate the *channel coding theorem* [24,27]:

Let a discrete memoryless source with an alphabet S have entropy $H(S)$ and emit the symbols every T_s seconds. Let a discrete memoryless channel have capacity C and be used once in T_c seconds. Then, if

$$H(S)/T_s \leq C/T_c \tag{6.9}$$

there exists a coding scheme for which the source output can be transmitted over the channel and reconstructed with an arbitrary small probability of error. The parameter

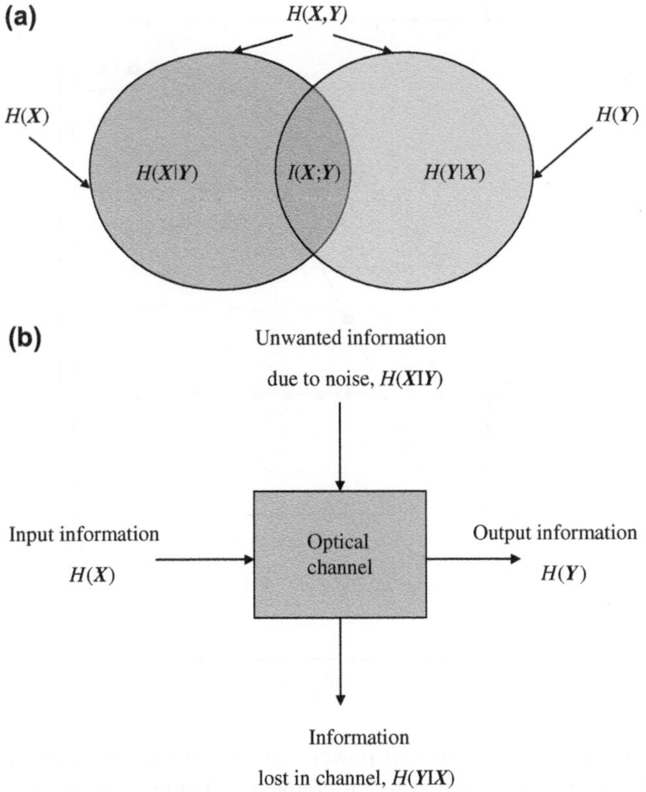

FIGURE 6.3

Interpretation of the mutual information: (a) using Venn diagrams; (b) using the approach of Ingels.

$H(S)/T_s$ is related to the average information rate, while the parameter C/T_c is related to the channel capacity per unit time. For a binary symmetric channel ($M = N = 2$) the inequality (6.9) simply becomes

$$R \leq C, \tag{6.10}$$

where R is the code rate introduced above.

Another very important theorem is Shannon's third theorem, also known as the *information capacity theorem*, which can be formulated as follows [24,27].

The information capacity of a continuous channel of bandwidth B Hz, perturbed by AWGN of PSD $N_0/2$ and limited in bandwidth B, is given by

$$C = B \log_2\left(1 + \frac{P}{N_0 B}\right) \quad \text{(bits/s)}, \tag{6.11}$$

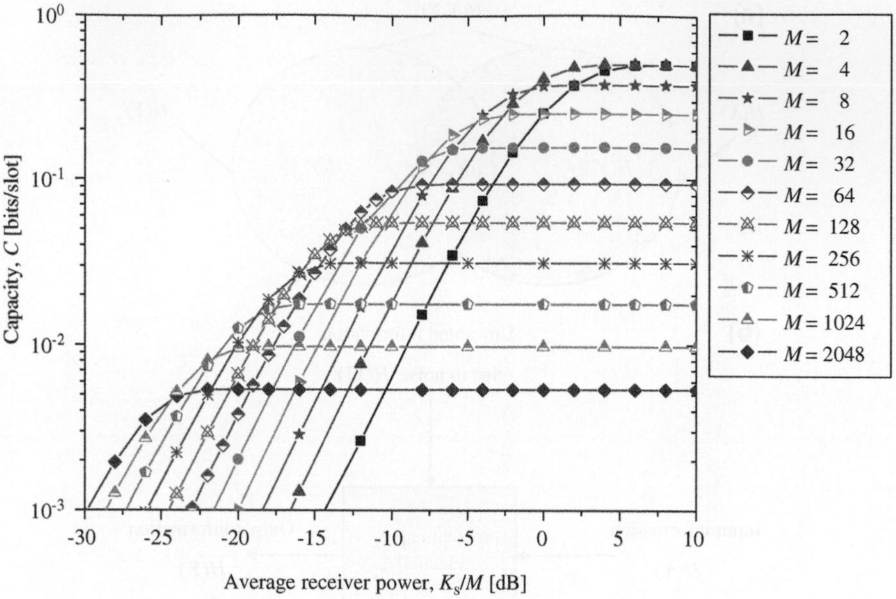

FIGURE 6.4

Channel capacity for *M*-ary PPM on a Poisson channel for $K_b = 1$.

where P is the average transmitted power. This theorem represents a remarkable result of information theory, because it connects all important system parameters (transmitted power, channel bandwidth, and noise power spectral density) in only one formula. What is also interesting is that LDPC codes can approach Shannon's limit within 0.0045 dB [13]. By using Eq. (6.11) and Fano's inequality [27],

$$\begin{aligned} H(\mathbf{X}|\mathbf{Y}) &\le H(P_e) + P_e \log_2(M-1), \\ H(P_e) &= -P_e \log_2 P_e - (1-P_e)\log_2(1-P_e) \end{aligned} \tag{6.12}$$

for an optical channel, when amplified spontaneous emission (ASE) noise-dominated scenario is observed and binary phase-shift keying (BPSK) at 40 Gb/s, in Figure 6.5a we report the minimum BERs against optical SNR for different code rates. On the other hand, Figure 6.5b shows the minimum channel capacity BERs for different code rates for 64-PPM over a Poisson channel in the presence of background radiation with an average number of background photons of $K_b = 0.2$, also obtained by using Fano's inequality. For example, it has been shown in Ref. [39] that for 64-PPM over a Poisson channel with serial concatenated turbo code (8184,4092) we are about 1 dB away from channel capacity (for $K_b = 0.2$ at BER of 10^{-5}), while with LDPC (8192,4096) code, of fixed column weight 6 and irregular row weight, we are 1.4 dB away from channel capacity.

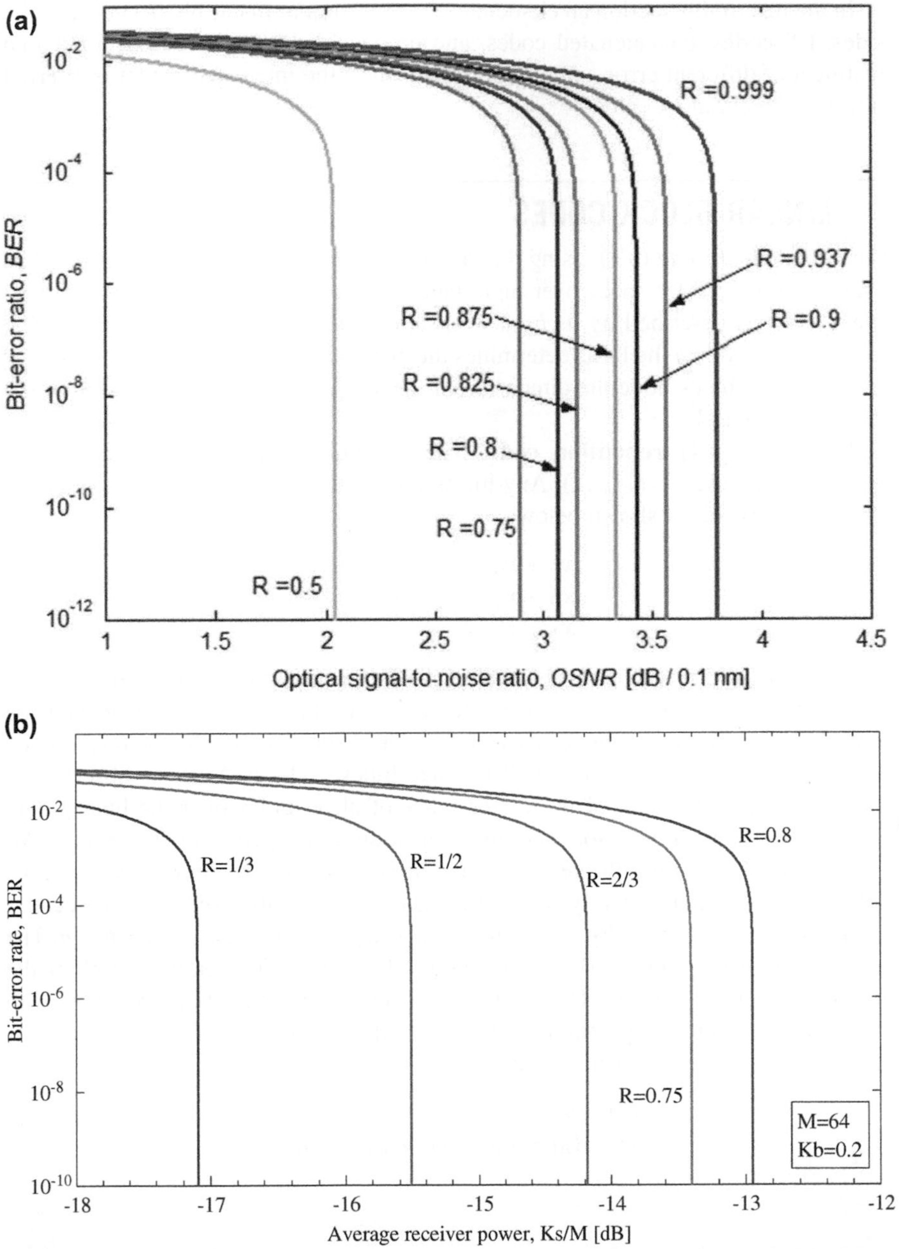

FIGURE 6.5

(a) Minimum BER against optical SNR for different code rate values (for BPSK at 40 Gb/s). (b) Channel capacity BER curves against average receiver power for 64-PPM over a Poisson channel with background radiation of $K_b = 0.2$ and different code rates.

In the rest of this section an elementary introduction to linear block codes, cyclic codes, RS codes, concatenated codes, and product codes is given. For a detailed treatment of different error-control coding schemes, the interested reader is referred to Refs [15–20,26,36–38].

6.2 LINEAR BLOCK CODES

The *linear block code* (n,k), using the language of vector spaces, can be defined as a subspace of a vector space over finite field GF(q), with q being the prime power. Every space is described by its *basis* – a set of linearly independent vectors. The number of vectors in the basis determines the dimension of the space. Therefore, for an (n,k) linear block code the dimension of the space is n and the dimension of the code subspace is k.

Example: $(n,1)$ repetition code. The repetition code has two codewords $\boldsymbol{x}_0 = (00\ldots0)$ and $\boldsymbol{x}_1 = (11\ldots1)$. Any linear combination of these two codewords is another codeword, as shown below:

$$\boldsymbol{x}_0 + \boldsymbol{x}_0 = \boldsymbol{x}_0$$
$$\boldsymbol{x}_0 + \boldsymbol{x}_1 = \boldsymbol{x}_1 + \boldsymbol{x}_0 = \boldsymbol{x}_1$$
$$\boldsymbol{x}_1 + \boldsymbol{x}_1 = \boldsymbol{x}_0.$$

The set of codewords from a linear block code forms a group under the addition operation, because an all-zero codeword serves as the identity element, and the codeword itself serves as the inverse element. This is the reason why the linear block codes are also called group codes. The linear block code (n,k) can be observed as a k-dimensional subspace of the vector space of all n-tuples over the binary field GF(2) = {0,1}, with addition and multiplication rules given in Table 6.1. All n-tuples over GF(2) form the vector space. The sum of two n-tuples $\boldsymbol{a} = (a_1\ a_2\ \ldots\ a_n)$ and $\boldsymbol{b} = (b_1\ b_2\ \ldots\ b_n)$ is clearly an n-tuple and the commutative rule is valid because $\boldsymbol{c} = \boldsymbol{a} + \boldsymbol{b} = (a_1 + b_1\ a_2 + b_2\ \ldots\ a_n + b_n) = (b_1 + a_1\ b_2 + a_2\ \ldots\ b_n + a_n) = \boldsymbol{b} + \boldsymbol{a}$. The all-zero vector $\boldsymbol{0} = (0\ 0\ \ldots\ 0)$ is the identity element, while n-tuple $\boldsymbol{a}$ itself is the inverse element $\boldsymbol{a} + \boldsymbol{a} = 0$. Therefore, the n-tuples form the Abelian group with respect to the addition operation. The scalar multiplication is defined as $\alpha\boldsymbol{a} = (\alpha a_1\ \alpha a_2\ \ldots\ \alpha a_n)$, $\alpha \in$ GF(2). The distributive laws

$$\alpha(\boldsymbol{a} + \boldsymbol{b}) = \alpha\boldsymbol{a} + \alpha\boldsymbol{b}$$
$$(\alpha + \beta)\boldsymbol{a} = \alpha\boldsymbol{a} + \beta\boldsymbol{a},\ \forall\ \alpha, \beta \in \mathrm{GF}(2)$$

Table 6.1 Addition (+) and Multiplication (•) Rules

+	0	1	•	0	1
0	0	1	0	0	0
1	1	0	1	0	1

are also valid. The associate law $(\alpha \cdot \beta)\boldsymbol{a} = \alpha \cdot (\beta \boldsymbol{a})$ is clearly satisfied. Therefore, the set of all n-tuples is a vector space over GF(2). The set of all codewords from an (n,k) linear block code forms an Abelian group under the addition operation. It can be shown, in a fashion similar to that above, that all codewords of an (n,k) linear block code form a vector space of dimensionality k. There exists k basis vectors (codewords) such that every codeword is a linear combination of these codewords.

Example: $(n,1)$ repetition code, $C = \{(0\ 0 \ldots 0), (1\ 1 \ldots 1)\}$. Two codewords in C can be represented as a linear combination of an all-ones basis vector: $(1\ 1 \ldots 1) = 1 \cdot (1\ 1 \ldots 1)$, $(0\ 0 \ldots 0) = 1 \cdot (1\ 1 \ldots 1) + 1 \cdot (1\ 1 \ldots 1)$.

6.2.1 Generator Matrix for Linear Block Code

Any codeword $\boldsymbol{x}$ from the (n,k) linear block code can be represented as a linear combination of k basis vectors $\boldsymbol{g}_i$ $(i = 0,1,\ldots, k-1)$ as given below:

$$\boldsymbol{x} = m_0\boldsymbol{g}_0 + m_1\boldsymbol{g}_1 + \ldots + m_{k-1}\boldsymbol{g}_{k-1} = \boldsymbol{m}\begin{bmatrix} \boldsymbol{g}_0 \\ \boldsymbol{g}_1 \\ \ldots \\ \boldsymbol{g}_{k-1} \end{bmatrix} = \boldsymbol{mG}; \quad \boldsymbol{G} = \begin{bmatrix} \boldsymbol{g}_0 \\ \boldsymbol{g}_1 \\ \ldots \\ \boldsymbol{g}_{k-1} \end{bmatrix},$$
$$\boldsymbol{m} = (m_0 \quad m_1 \quad \ldots \quad m_{k-1}), \tag{6.13}$$

where $\boldsymbol{m}$ is the message vector and $\boldsymbol{G}$ is the generator matrix (of dimensions $k \times n$), in which every row represents a basis vector from the coding subspace. Therefore, in order to encode, the message vector $\boldsymbol{m}(m_0,m_1,\ldots,m_{k-1})$ has to be multiplied by a generator matrix $\boldsymbol{G}$ to get $\boldsymbol{x} = \boldsymbol{mG}$, where $\boldsymbol{x}(x_0,x_1,\ldots,x_{n-1})$ is a codeword.

Example. Generator matrices for repetition $(n,1)$ code $\boldsymbol{G}_{\text{rep}}$ and $(n,n-1)$ single-parity-check code $\boldsymbol{G}_{\text{par}}$ are given respectively as

$$\boldsymbol{G}_{\text{rep}} = [11\ldots1] \quad \boldsymbol{G}_{\text{par}} = \begin{bmatrix} 100\ldots01 \\ 010\ldots01 \\ \ldots \\ 000\ldots11 \end{bmatrix}.$$

By elementary operations on rows in the generator matrix, the code may be transformed into *systematic* form:

$$\boldsymbol{G}_{\text{s}} = [\boldsymbol{I}_k | \boldsymbol{P}], \tag{6.14}$$

where $\boldsymbol{I}_k$ is the identity matrix of dimensions $k \times k$, and $\boldsymbol{P}$ is a matrix of dimensions $k \times (n-k)$ with columns denoting the positions of parity checks:

$$\boldsymbol{P} = \begin{bmatrix} p_{00} & p_{01} & \ldots & p_{0,n-k-1} \\ p_{10} & p_{11} & \ldots & p_{1,n-k-1} \\ \ldots & & \ldots & \ldots \\ p_{k-1,0} & p_{k-1,1} & \ldots & p_{k-1,n-k-1} \end{bmatrix}.$$

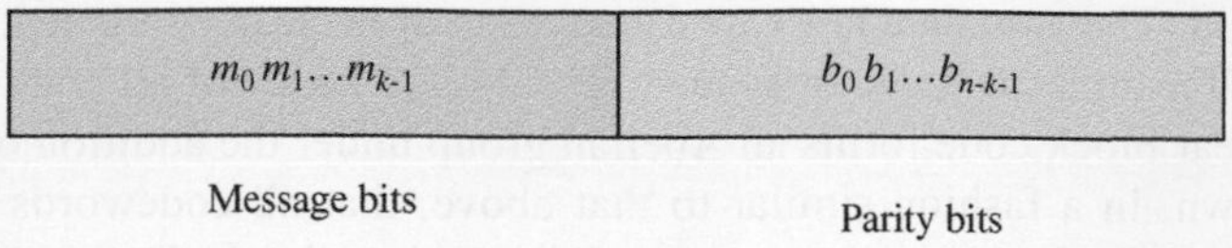

FIGURE 6.6

Structure of systematic codeword.

The codeword of a systematic code is obtained by

$$x = [m|b] = m[I_k|P] = mG, \quad G = [I_k|P], \tag{6.15}$$

and the structure of a systematic codeword is shown in Figure 6.6.

Therefore, during encoding the message vector is unchanged and the elements of the vector of parity checks b are obtained by

$$b_i = p_{0i}m_0 + p_{1i}m_1 + \ldots + p_{k-1,i}m_{k-1}, \tag{6.16}$$

where

$$p_{ij} = \begin{cases} 1, & \text{if} \quad b_i \quad \text{depends on} \quad m_j \\ 0, & \text{otherwise} \end{cases}.$$

During transmission the channel introduces errors so that the received vector r can be written as $r = x + e$, where e is the error vector (pattern) with components determined by

$$e_i = \begin{cases} 1 & \text{if an error occurred in the } i\text{th location} \\ 0 & \text{otherwise.} \end{cases}$$

To determine whether the received vector r is a codeword vector, we introduce the concept of a *parity-check matrix.*

6.2.2 Parity-Check Matrix for Linear Block Code

Another useful matrix associated with the linear block codes is the parity-check matrix. We expand the matrix equation $x = mG$ in scalar form as follows:

$$\begin{aligned}
x_0 &= m_0 \\
x_1 &= m_1 \\
&\ldots \\
x_{k-1} &= m_{k-1} \\
x_k &= m_0 p_{00} + m_1 p_{10} + \ldots + m_{k-1} p_{k-1,0} \\
x_{k+1} &= m_0 p_{01} + m_1 p_{11} + \ldots + m_{k-1} p_{k-1,1} \\
&\ldots \\
x_{n-1} &= m_0 p_{0,n-k-1} + m_1 p_{1,n-k-1} + \ldots + m_{k-1} p_{k-1,n-k-1}.
\end{aligned} \tag{6.17}$$

By using the first k equalities, the last $n-k$ equations can be rewritten as follows:

$$\begin{aligned} & x_0 p_{00} + x_1 p_{10} + \ldots + x_{k-1} p_{k-1,0} + x_k = 0 \\ & x_0 p_{01} + x_1 p_{11} + \ldots + x_{k-1} p_{k-1,0} + x_{k+1} = 0 \\ & \ldots \\ & x_0 p_{0,n-k+1} + x_1 p_{1,n-k-1} + \ldots + x_{k-1} p_{k-1,n-k+1} + x_{n-1} = 0. \end{aligned} \tag{6.18}$$

The matrix representation of (6.18) is

$$\begin{bmatrix} x_0 & x_1 & \ldots & x_{n-1} \end{bmatrix} \begin{bmatrix} p_{00} & p_{10} & \cdots & p_{k-1,0} & 1 & 0 & \ldots & 0 \\ p_{01} & p_{11} & \cdots & p_{k-1,1} & 0 & 1 & \ldots & 0 \\ \cdots & & \cdots & & \cdots & & & \cdots \\ p_{0,n-k-1} & p_{1,n-k-1} & \cdots & p_{k-1,n-k-1} & 0 & 0 & \ldots & 1 \end{bmatrix}^{\mathrm{T}}$$
$$= \boldsymbol{x}\begin{bmatrix} \boldsymbol{P}^{\mathrm{T}} & \boldsymbol{I}_{n-k} \end{bmatrix} = \boldsymbol{x}\boldsymbol{H}^{\mathrm{T}} = \boldsymbol{0}, \quad \boldsymbol{H} = \begin{bmatrix} \boldsymbol{P}^{\mathrm{T}} & \boldsymbol{I}_{n-k} \end{bmatrix}_{(n-k)\times n}. \tag{6.19}$$

The matrix $\boldsymbol{H}$ in (6.19) is known as the parity-check matrix. We can easily verify that:

$$\boldsymbol{G}\boldsymbol{H}^{\mathrm{T}} = [\boldsymbol{I}_k \quad \boldsymbol{P}] \begin{bmatrix} \boldsymbol{P} \\ \boldsymbol{I}_{n-k} \end{bmatrix} = \boldsymbol{P} + \boldsymbol{P} = \boldsymbol{0}, \tag{6.20}$$

meaning that the parity-check matrix of an (n,k) linear block code $\boldsymbol{H}$ is a matrix of rank $n-k$ and dimensions $(n-k) \times n$ whose null space is a k-dimensional vector with basis forming the generator matrix $\boldsymbol{G}$.

Example. Parity-check matrices for $(n,1)$ repetition code $\boldsymbol{H}_{\text{rep}}$ and $(n,n-1)$ single-parity check code $\boldsymbol{H}_{\text{par}}$ are given respectively as

$$\boldsymbol{H}_{\text{rep}} = \begin{bmatrix} 100\ldots01 \\ 010\ldots01 \\ \ldots \\ 000\ldots11 \end{bmatrix} \qquad \boldsymbol{H}_{\text{par}} = [11\ldots1].$$

Example. For Hamming (7,4) code, the generator $\boldsymbol{G}$ and parity-check $\boldsymbol{H}$ matrices are given respectively as

$$\boldsymbol{G} = \begin{bmatrix} 1000|110 \\ 0100|011 \\ 0010|111 \\ 0001|101 \end{bmatrix} \qquad \boldsymbol{H} = \begin{bmatrix} 1011|100 \\ 1110|010 \\ 0111|001 \end{bmatrix}.$$

Every (n,k) linear block code with generator matrix $\boldsymbol{G}$ and parity-check matrix $\boldsymbol{H}$ has a dual code with generator matrix $\boldsymbol{H}$ and parity-check matrix $\boldsymbol{G}$. For example, $(n,1)$ repetition and $(n,n-1)$ single-parity check codes are dual.

6.2.3 Distance Properties of Linear Block Codes

To determine the *error correction capability* of the code we have to introduce the concept of Hamming distance and Hamming weight. *Hamming distance* between two codewords $\boldsymbol{x}_1$ and $\boldsymbol{x}_2$, $d(\boldsymbol{x}_1,\boldsymbol{x}_2)$, is defined as the number of locations in which their respective elements differ. *Hamming weight*, $w(\boldsymbol{x})$, of a codeword vector $\boldsymbol{x}$ is defined as the number of nonzero elements in the vectors. The *minimum distance*, $d_{\min}$, of a linear block code is defined as the smallest Hamming distance between any pair of code vectors in the code. Since the zero vector is a codeword, the minimum distance of a linear block code can be determined simply as the smallest Hamming weight of the nonzero code vectors in the code. Let the parity-check matrix be written as $\boldsymbol{H}=[\boldsymbol{h}_1\ \boldsymbol{h}_2 \ldots \boldsymbol{h}_n]$, where $\boldsymbol{h}_i$ is the ith column in $\boldsymbol{H}$. Since every codeword $\boldsymbol{x}$ must satisfy the syndrome equation, $\boldsymbol{x}\boldsymbol{H}^{\mathrm{T}}=\boldsymbol{0}$ (see Eq. (6.19)), the minimum distance of a linear block code is determined by the minimum number of columns of the matrix $\boldsymbol{H}$ whose sum is equal to the zero vector. For example, (7,4) Hamming code in the example above has a minimum distance $d_{\min}=3$ since the addition of the first, fifth, and sixth columns leads to a zero vector. The codewords can be represented as points in n-dimensional space, as shown in Figure 6.7. The decoding process can be visualized by creating spheres of radius t around codeword points. The received word vector $\boldsymbol{r}$ in Figure 6.7a will be decoded as a codeword $\boldsymbol{x}_i$ because its Hamming distance $d(\boldsymbol{x}_i,\boldsymbol{r}) \leq t$ is closest to the codeword $\boldsymbol{x}_i$. On the other hand, in the example shown in Figure 6.7b the Hamming distance $d(\boldsymbol{x}_i,\boldsymbol{x}_j) \leq 2t$ and the received vector $\boldsymbol{r}$ that falls in the intersection area of the two spheres cannot be uniquely decoded.

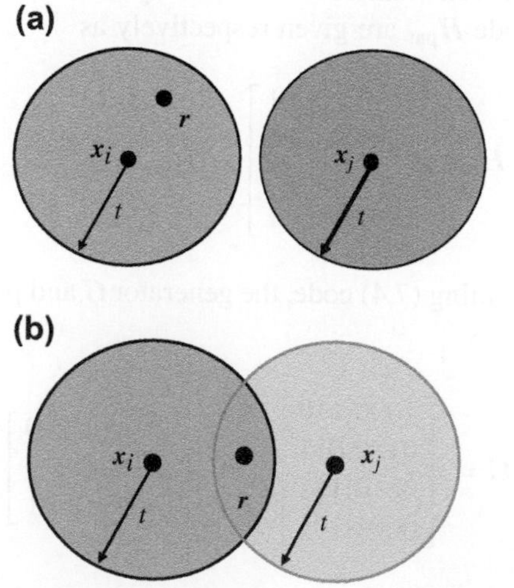

FIGURE 6.7

Illustration of Hamming distance: (a) $d(\boldsymbol{x}_i,\boldsymbol{x}_j) \geq 2t+1$; (b) $d(\boldsymbol{x}_i,\boldsymbol{x}_j) < 2t+1$.

Therefore, an (n,k) linear block code of minimum distance $d_{\min}$ can correct up to t errors if, and only if, $t \le \lfloor 1/2(d_{\min}-1) \rfloor$ (where $\lfloor\ \rfloor$ denotes the largest integer less than or equal to the enclosed quantity) or equivalently $d_{\min} \ge 2t+1$. If we are only interested in detecting e_{d} errors then $d_{\min} \ge e_{\mathrm{d}}+1$. Finally, if we are interested in detecting e_{d} errors and correcting e_{c} errors then $d_{\min} \ge e_{\mathrm{d}}+e_{\mathrm{c}}+1$. The Hamming (7,4) code is therefore a single error correcting and double error detecting code. More generally, families of (n,k) linear block codes with the following parameters:

- Block length: $n = 2^m - 1$
- Number of message bits: $k = 2^m - m - 1$
- Number of parity bits: $n - k = m$
- $d_{\min} = 3$

where $m \ge 3$, are known as *Hamming* codes. Hamming codes belong to the class of perfect codes, codes that satisfy the Hamming inequality below with equality sign [24,27]:

$$2^{n-k} \ge \sum_{i=0}^{t} \binom{n}{i}. \tag{6.21}$$

This bound gives how many errors t can be corrected with an (n,k) linear block code by using syndrome decoding procedure (described in Section 6.2.5).

6.2.4 Coding Gain

A very important characteristics of an (n,k) linear block code is so-called coding gain, which was introduced at the beginning of this chapter as being the savings attainable in the energy per information bit to noise spectral density ratio (E_b/N_0) required to achieve a given bit error probability when coding is used compared to that with no coding. Let E_{c} denote the transmitted bit energy and E_{b} denote the information bit energy. Since the total information word energy kE_{b} must be the same as the total codeword energy nE_{c}, we obtain the following relationship between E_{c} and E_{b}:

$$E_c = (k/n)E_b = RE_b. \tag{6.22}$$

The probability of error for BPSK on an AWGN channel, when a coherent hard decision (bit-by-bit) demodulator is used, can be obtained as follows:

$$p = \frac{1}{2}\operatorname{erfc}\left(\sqrt{\frac{E_{\mathrm{c}}}{N_0}}\right) = \frac{1}{2}\operatorname{erfc}\left(\sqrt{\frac{RE_{\mathrm{b}}}{N_0}}\right), \tag{6.23}$$

where the $\operatorname{erfc}(x)$ function is defined by

$$\operatorname{erfc}(x) = \frac{2}{\sqrt{\pi}} \int_x^{+\infty} e^{-z^2}\,dz.$$

For high SNRs the word error probability (remained upon decoding) of a t-error correcting code is dominated by a $t+1$ error event:

$$P_w(e) \approx \binom{n}{t+1} p^{t+1}(1-p)^{n-t-1} \approx \binom{n}{t+1} p^{t+1}. \tag{6.24}$$

The bit error probability P_b is related to the word error probability by

$$P_b \approx \frac{2t+1}{n} P_w(e) \approx c(n,t)p^{t+1}, \tag{6.25}$$

because $2t+1$ and more errors per codeword cannot be corrected and they can be located anywhere on n codeword locations, and $c(n,t)$ is a parameter dependent on the error correcting capability t and codeword length n. By using the upper bound on erfc(x) we obtain:

$$P_b \approx \frac{c(n,t)}{2}\left[\exp\left(\frac{-RE_b}{N_0}\right)\right]^{t+1}. \tag{6.26}$$

The corresponding approximation for the uncoded case is

$$P_{b,\text{uncoded}} \approx \frac{1}{2}\exp\left(-\frac{E_b}{N_0}\right). \tag{6.27}$$

By equating (6.26) and (6.27) and ignoring the parameter $c(n,t)$, we obtain the following expression for hard decision decoding coding gain:

$$\frac{(E_b/N_0)_{\text{uncoded}}}{(E_b/N_0)_{\text{coded}}} \approx R(t+1). \tag{6.28}$$

The corresponding soft decision asymptotic coding gain of convolutional codes is [15,17–19,26]

$$\frac{(E_b/N_0)_{\text{uncoded}}}{(E_b/N_0)_{\text{coded}}} \approx Rd_{\text{min}}, \tag{6.29}$$

and it is about 3 dB better than hard decision decoding (because $d_{\text{min}} \geq 2t+1$).

In optical communications it is very common to use the Q-factor as the figure of merit instead of the SNR, which is related to the BER on an AWGN as follows:

$$\text{BER} = \frac{1}{2}\text{erfc}\left(\frac{Q}{\sqrt{2}}\right) \tag{6.30}$$

Let BER_{in} denote the BER at the input of an FEC decoder and BER_{out} denote the BER at the output of the FEC decoder, and let BER_{ref} denote the target BER (such as either 10^{-12} or 10^{-15}). The corresponding coding gain (CG) and net coding gain (NCG) are respectively defined as [4]

$$\text{CG} = 20\log_{10}\left[\text{erfc}^{-1}(2\text{BER}_{\text{ref}})\right] - 20\log_{10}\left[\text{erfc}^{-1}(2\text{BER}_{\text{in}})\right] \quad (\text{dB}) \tag{6.31}$$

$$\text{NCG} = 20\log_{10}\left[\text{erfc}^{-1}(2\text{BER}_{\text{ref}})\right] - 20\log_{10}\left[\text{erfc}^{-1}(2\text{BER}_{\text{in}})\right] + 10\log_{10}R \quad (\text{dB}). \tag{6.32}$$

All coding gains reported in this chapter are in fact NCG, although they are sometimes called coding gains only, because this is common practice in the coding theory literature [15,17–19,26].

6.2.5 Syndrome Decoding and Standard Array

The received vector $\boldsymbol{r} = \boldsymbol{x} + \boldsymbol{e}$ ($\boldsymbol{x}$ is the codeword and $\boldsymbol{e}$ is the error pattern introduced above) is a codeword if the following *syndrome equation* is satisfied: $\boldsymbol{s} = \boldsymbol{rH}^{\mathrm{T}} = \boldsymbol{0}$. The syndrome has the following important properties:

1. The syndrome is a function of the error pattern only. This property can easily be proved from the definition of a syndrome as follows: $\boldsymbol{s} = \boldsymbol{rH}^{\mathrm{T}} = (\boldsymbol{x} + \boldsymbol{e})\boldsymbol{H}^{\mathrm{T}} = \boldsymbol{xH}^{\mathrm{T}} + \boldsymbol{eH}^{\mathrm{T}} = \boldsymbol{eH}^{\mathrm{T}}$.
2. All error patterns that differ by a codeword have the same syndrome. This property can also be proved from the syndrome definition. Let $\boldsymbol{x}_i$ be the ith ($i = 0,1,\ldots,2^{k-1}$) codeword. The set of error patterns that differ by a codeword is known as a coset: $\{\boldsymbol{e}_i = \boldsymbol{e} + \boldsymbol{x}_i;\ i = 0,1,\ldots,2^{k-1}\}$. The syndrome corresponding to the ith error pattern from this set, $\boldsymbol{s}_i = \boldsymbol{r}_i\boldsymbol{H}^{\mathrm{T}} = (\boldsymbol{x}_i + \boldsymbol{e})\boldsymbol{H}^{\mathrm{T}} = \boldsymbol{x}_i\boldsymbol{H}^{\mathrm{T}} + \boldsymbol{eH}^{\mathrm{T}} = \boldsymbol{eH}^{\mathrm{T}}$, is a function of the error pattern only, and therefore all error patterns from the coset have the same syndrome.
3. The syndrome is a function of only those columns of a parity-check matrix corresponding to the error locations. The parity-check matrix can be written in the following form: $\boldsymbol{H} = [\boldsymbol{h}_1 \ldots \boldsymbol{h}_n]$, where the ith element $\boldsymbol{h}_i$ denotes the ith column of $\boldsymbol{H}$. Based on the syndrome definition for an error pattern $\boldsymbol{e} = [e_1 \ \ldots \ e_n]$, the following is valid:

$$\boldsymbol{s} = \boldsymbol{eH}^{\mathrm{T}} = \begin{bmatrix} e_1 & e_2 & \ldots & e_n \end{bmatrix} \begin{bmatrix} \boldsymbol{h}_1^{\mathrm{T}} \\ \boldsymbol{h}_2^{\mathrm{T}} \\ \ldots \\ \boldsymbol{h}_n^{\mathrm{T}} \end{bmatrix} = \sum_{i=1}^{n} e_i \boldsymbol{h}_i^{\mathrm{T}}, \tag{6.33}$$

   which proves the claim of property 3.
4. A syndrome decoding procedure for an (n,k) linear block code can correct up to t errors, providing that Hamming bound (6.21) is satisfied (this property will be proved in the next subsection).

By using property 2, 2^k codewords partition the space of all received words into 2^k disjoint subsets. Any received word within the subset will be decoded as the unique codeword. A *standard array* is a technique by which this partition can be achieved, and can be constructed using the following two steps [15,17–19,22,24–26]:

1. Write down 2^k codewords as elements of the first row, with the all-zero codeword as the leading element.

$x_1 = 0$	x_2	x_3	...	x_i	...	x_{2^k}
e_2	$x_2 + e_2$	$x_3 + e_2$	...	$x_i + e_2$	...	$x_{2^k} + e_2$
e_3	$x_2 + e_3$	$x_3 + e_3$	...	$x_i + e_2$	...	$x_{2^k} + e_3$
...	...			...		... ...
e_j	$x_2 + e_j$	$x_3 + e_j$	...	$x_i + e_j$	...	$x_{2^k} + e_j$
...	...			...		...
$e_{2^{n-k}}$	$x_2 + e_{2^{n-k}}$	$x_3 + e_{2^{n-k}}$	...	$x_i + e_{2^{n-k}}$	...	$x_{2^k} + e_{2^{n-k}}$

FIGURE 6.8

Standard array architecture.

2. Repeat the steps 2(a) and 2(b) until all 2^n words are exhausted.
   - **(a)** Of the remaining unused n-tuples, select one with the least weight for the leading element of the next row.
   - **(b)** Complete the current row by adding the leading element to each nonzero codeword appearing in the first row and writing down the resulting sum in the corresponding column.

The standard array for an (n,k) block code obtained by this algorithm is illustrated in Figure 6.8. The columns represent 2^k disjoint sets, and every row represents the coset of the code, with leading elements being called coset leaders.

Example. The standard array of (5,2) code $C = \{(00000),(11010),(10101),(01111)\}$ is given in Table 6.2. The parity-check matrix of this code is given by:

$$H = \begin{bmatrix} 1 & 0 & 0 & 1 & 1 \\ 0 & 1 & 0 & 1 & 0 \\ 0 & 0 & 1 & 0 & 1 \end{bmatrix}.$$

Because the minimum distance of this code is 3 (first, second, and fourth columns add to zero), this code is able to correct all single errors. For example, if the word 01010 is received it will be decoded to the topmost codeword 11010 of the column in which it lies. In the same table corresponding syndromes are provided as well.

The syndrome decoding procedure has three steps [15,17–19,22,24–26]:

1. For the received vector r, compute the syndrome $s = rH^{\mathrm{T}}$. From property 3 we can establish one-to-one correspondence between the syndromes and error

Table 6.2 Standard array of (5,2) code and corresponding decoding table

	Codewords				Syndrome s	Error pattern
Coset leader	00000	11010	10101	01111	000	00000
	00001	11011	10100	01110	101	00001
	00010	11000	10111	01101	110	00010
	00100	11110	10001	01011	001	00100
	01000	10010	11101	00111	010	01000
	10000	01010	00101	11111	100	10000
	00011	11001	10110	01100	011	00011
	00110	11100	10011	01001	111	00110

patterns (see Table 6.2), leading to a look-up table (LUT) containing the syndrome and corresponding error pattern (the coset leader).

2. Within the coset characterized by the syndrome $\boldsymbol{s}$, identify the coset leader, say $\boldsymbol{e}_0$. The coset leader corresponds to the error pattern with the largest probability of occurrence.
3. Decode the received vector as $\boldsymbol{x}=\boldsymbol{r}+\boldsymbol{e}_0$.

Example. Let the received vector for the (5,2) code example above be $\boldsymbol{r}=(01010)$. The syndrome can be computed as $\boldsymbol{s}=\boldsymbol{r}\boldsymbol{H}^{\mathrm{T}}=(100)$, and the corresponding error pattern from the LUT is found to be $\boldsymbol{e}_0=(10000)$. The decoded word is obtained by adding the error pattern to the received word $\boldsymbol{x}=\boldsymbol{r}+\boldsymbol{e}_0=(11010)$, and the error on the first bit position is corrected.

The standard array can be used to determine the probability of word error as follows:

$$P_{\mathrm{w}}(e)=1-\sum_{i=0}^{n}\alpha_i\, p^i(1-p)^{n-i}, \tag{6.34}$$

where α_i is the number of coset leaders of weight i (the distribution of weights is also known as *weight distribution* of coset leaders) and p is the crossover probability of BSC. Any error pattern that is not a coset leader will result in decoding error. For example, the weight distribution of coset leaders in (5,2) code is $\alpha_0=1$, $\alpha_1=5$, $\alpha_2=2$, $\alpha_i=0$, $i=3,4,5$, which leads to the following word error probability:

$$P_{\mathrm{w}}(e)=1-(1-p)^5-5p(1-p)^4-2p^2(1-p)^3\big|_{p=10^{-3}}=7.986\cdot10^{-6}.$$

We can use Eq. (6.34) to estimate the coding gain of a given linear block code. For example, the word error probability for Hamming (7,4) code is

$$P_{\mathrm{w}}(e)=1-(1-p)^7-7p(1-p)^6=\sum_{i=2}^{7}\binom{7}{i}p^i(1-p)^{7-i}\approx21p^2.$$

In the previous section, we established the following relationship between bit and word error probabilities: $P_b \approx P_w(e)(2t+1)/n = (3/7)P_w(e) \approx (3/7)21p^2 = 9p^2$. Therefore, the crossover probability can be evaluated as

$$p = \sqrt{P_b}/3 = (1/2)\operatorname{erfc}\left(\sqrt{\frac{RE_b}{N_0}}\right).$$

From this expression we can easily calculate the required SNR to achieve target P_b. By comparing this obtained SNR with the corresponding SNR for uncoded BPSK we can evaluate the corresponding coding gain.

In order to evaluate the probability of undetected error, we have to determine the other codeword weights as well. Because the undetected errors are caused by error patterns being identical to the nonzero codewords, the undetected error probability can be evaluated by

$$P_u(e) = \sum_{i=1}^{n} A_i p^i (1-p)^{n-i} = (1-p)^n \sum_{i=1}^{n} A_i \left(\frac{p}{1-p}\right), \tag{6.35}$$

where p is the crossover probability and A_i denotes the number of codewords of weight i. The codeword weight can be determined by the *MacWilliams identity* that establishes the connection between codeword weights A_i and the codeword weights of corresponding dual code B_i by [15]

$$A(z) = 2^{-(n-k)}(1+z)^n B\left(\frac{1-z}{1+z}\right), \; A(z) = \sum_{i=0}^{n} A_i z^i, \; B(z) = \sum_{i=0}^{n} B_i z^i, \tag{6.36}$$

where $A(z)$ ($B(z)$) represents the polynomial representation of codeword weights (dual codeword weights). By substituting $z = p/(1-p)$ in (6.36) and knowing that $A_0 = 1$ we obtain:

$$A\left(\frac{p}{1-p}\right) - 1 = \sum_{i=1}^{n} A\left(\frac{p}{1-p}\right)^i. \tag{6.37}$$

Substituting (6.37) into (6.35) we obtain:

$$P_u(e) = (1-p)^n \left[A\left(\frac{p}{1-p}\right) - 1\right]. \tag{6.38}$$

An alternative expression for $P_u(e)$ in terms of $B(z)$ can be obtained from (6.36), which is more suitable for use when $n-k<k$, as follows:

$$P_u(e) = 2^{-(n-k)} B(1-2p) - (1-p)^n. \tag{6.39}$$

For large n, k, and $n-k$ the use of the MacWilliams identity is impractical; in this case an upper bound on the average probability of undetected error of an (n,k) systematic code should be used instead:

$$\overline{P_u}(e) \le 2^{-(n-k)}[1 - (1-p)^n]. \tag{6.40}$$

For a q-ary maximum-distance separable code, which satisfies the Singleton bound introduced in the next section with equality, we can determine a closed formula for weight distribution [15]:

$$A_i = \binom{n}{i}(q-1)\sum_{j=0}^{i-d_{\min}}(-1)^j\binom{i-1}{j}q^{i-d_{\min}-j}, \tag{6.41}$$

where $d_{\min}$ is the minimum distance of the code, $A_0 = 1$, and $A_i = 0$ for $i \in [1, d_{\min} - 1]$.

6.2.6 Important Coding Bounds

In this section we describe several important coding bounds, including the Hamming, Plotkin, Gilbert–Varshamov, and Singleton bounds [15,17–19, 22,24–26]. The *Hamming* bound has already been introduced for binary LBCs by (6.21). The Hamming bound for a q-ary (n,k) LBC is given by

$$\begin{aligned}&\left[1 + (q-1)\binom{n}{1} + (q-1)^2\binom{n}{2} + \ldots + (q-1)^i\binom{n}{i} + \ldots\right.\\&\left.+ (q-1)^t\binom{n}{t}\right]q^k \le q^n,\end{aligned} \tag{6.42}$$

where t is the error correction capability and $(q-1)^i\binom{n}{i}$ is the number of received words that differ from a given code word in i symbols. Namely, there are n chooses i ways in which symbols can be chosen out of n and there are $(q-1)^i$ possible choices for symbols. The codes satisfying the Hamming bound with equality are known as *perfect codes*. Hamming codes are perfect codes because $n = 2^{n-k} - 1$, which is equivalent to $(1+n)2^k = 2^n$, so that the inequality above is satisfied with equality. The $(n,1)$ repetition code is also a perfect code. The three-error correcting (23,12) Golay code is another example of a perfect code because

$$\left[1 + \binom{23}{1} + \binom{23}{2} + \binom{23}{3}\right]2^{12} = 2^{23}.$$

The *Plotkin* bound is the bound on the minimum distance of a code:

$$d_{\min} \le \frac{n2^{k-1}}{2^k - 1}. \tag{6.43}$$

Namely, if all codewords are written as the rows of a $2^k \times n$ matrix, each column will contain 2^{k-1} zeros and 2^{k-1} ones, with the total weight of all codewords being $n2^{k-1}$.

The *Gilbert–Varshamov* bound is based on the property that the minimum distance d_{min} of a linear (n,k) block code can be determined as the minimum number of columns in matrix $\boldsymbol{H}$ that sum to zero:

$$\binom{n-1}{1} + \binom{n-1}{2} + \ldots + \binom{n-1}{d_{min}-2} < 2^{n-k} - 1. \tag{6.44}$$

Another important bound is the *Singleton bound*:

$$d_{min} \leq n - k + 1. \tag{6.45}$$

This bound is straightforward to prove. Let only one bit of value 1 be present in the information vector. If it is involved in $n-k$ parity checks, then the total number of ones in the codeword cannot be larger than $n-k+1$. The codes satisfying the Singleton bound with equality are known as the *maximum-distance separable* (MDS) codes (e.g. RS codes are MDS codes).

6.3 CYCLIC CODES

The most commonly used class of linear block codes is the class of cyclic codes. Examples of cyclic codes include BCH codes, Hamming codes, and Golay codes. RS codes are also cyclic but nonbinary codes. Even LDPC codes can be designed in cyclic or quasi-cyclic fashion.

Let us observe a vector space of dimension n. The subspace of this space is a *cyclic code* if for any codeword $\boldsymbol{c}(c_0,c_1,\ldots,c_{n-1})$ the arbitrary cyclic shift $\boldsymbol{c}_j(c_{n-j},c_{n-j+1},\ldots,c_{n-1},c_0,c_1,\ldots,c_{n-j-1})$ is another codeword. With every codeword $\boldsymbol{c}(c_0,c_1,\ldots,c_{n-1})$ from a cyclic code, we associate the *codeword polynomial*:

$$c(x) = c_0 + c_1x + c_2x^2 + \ldots + c_{n-1}x^{n-1}. \tag{6.46}$$

The jth cyclic shift, observed mod(x^n-1), is also a codeword polynomial:

$$c^{(j)}(x) = x^j c(x) \bmod (x^n - 1). \tag{6.47}$$

It is straightforward to show that the observed subspace is cyclic if composed of polynomials divisible by a polynomial $g(x) = g_0 + g_1x + \ldots + g_{n-k}x^{n-k}$ that divides x^n-1 at the same time. The polynomial $g(x)$, of degree $n-k$, is called the *generator polynomial* of the code. If $x^n - 1 = g(x)h(x)$, then a polynomial of degree k is called a *parity-check polynomial.* The generator polynomial has the following three important properties [15,17–19,22,24–26]:

1. The generator polynomial of an (n,k) cyclic code is unique (usually proved by contradiction).
2. Any multiple of the generator polynomial is a codeword polynomial.
3. The generator polynomial and parity-check polynomial are factors of x^n-1.

The generator polynomial $g(x)$ and the parity-check polynomial $h(x)$ serve the same role as the generator matrix $\boldsymbol{G}$ and parity-check matrix $\boldsymbol{H}$ of a linear block code. n-tuples

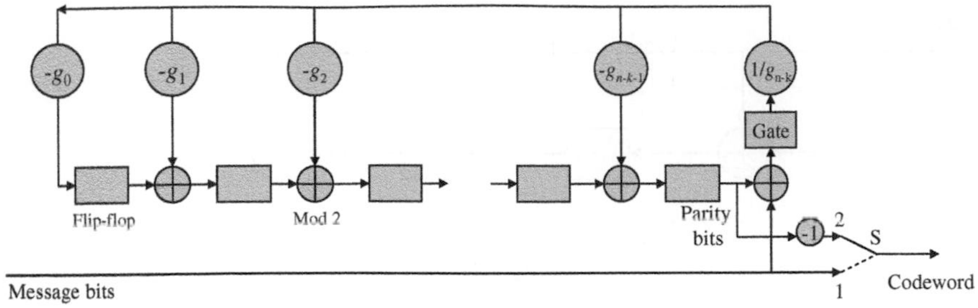

FIGURE 6.9

Systematic cyclic encoder.

related to the k polynomials $g(x)$, $xg(x)$,...,$x^{k-1}g(x)$ may be used in rows of the $k \times n$ generator matrix $\boldsymbol{G}$, while n-tuples related to the $(n-k)$ polynomials $x^k h(x^{-1})$, $x^{k+1}h(x^{-1})$,...,$x^{n-1}h(x^{-1})$ may be used in rows of the $(n-k) \times n$ parity-check matrix $\boldsymbol{H}$.

To encode we have simply to multiply the message polynomial $m(x) = m_0 + m_1 x + \ldots + m_{k-1}x^{k-1}$ by the generator polynomial $g(x)$, i.e. $c(x) = m(x)g(x)$ mod$(x^n - 1)$, where $c(x)$ is the codeword polynomial. To encode in *systematic* form we have to find the remainder of $x^{n-k}m(x)/g(x)$ and add it to the shifted version of message polynomial $x^{n-k}m(x)$, i.e. $c(x) = x^{n-k}m(x) + \text{rem}[x^{n-k}m(x)/g(x)]$, where rem[ ] denotes the remainder of a given entity. The general circuit for generating the codeword polynomial in systematic form is given in Figure 6.9. The encoder operates as follows. When the switch S is in position 1 and the gate is closed (on), the information bits are shifted into the shift register and at the same time transmitted onto the channel. Once all information bits are shifted into the register in k shifts, with the gate being open (off), the switch S is moved to position 2, and the content of the $(n-k)$ shift register is transmitted onto the channel.

To check if the received word polynomial is the codeword polynomial $r(x) = r_0 + r_1 x + \ldots + r_{n-1}x^{n-1}$ we have simply to determine the *syndrome polynomial* $s(x) = \text{rem}[r(x)/g(x)]$. If $s(x)$ is zero then there is no error introduced during transmission. The corresponding circuit is shown in Figure 6.10.

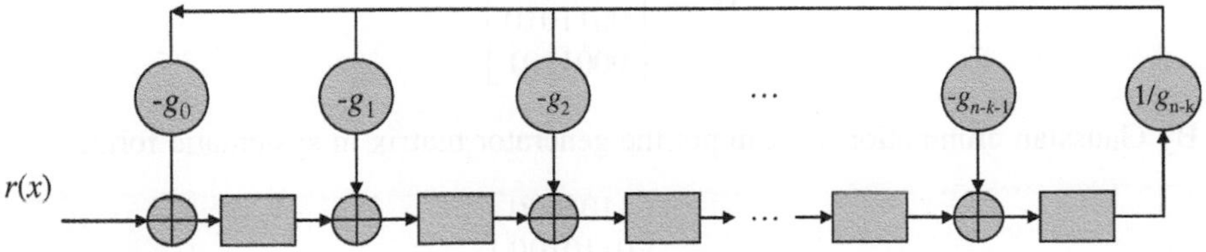

FIGURE 6.10

Syndrome calculator.

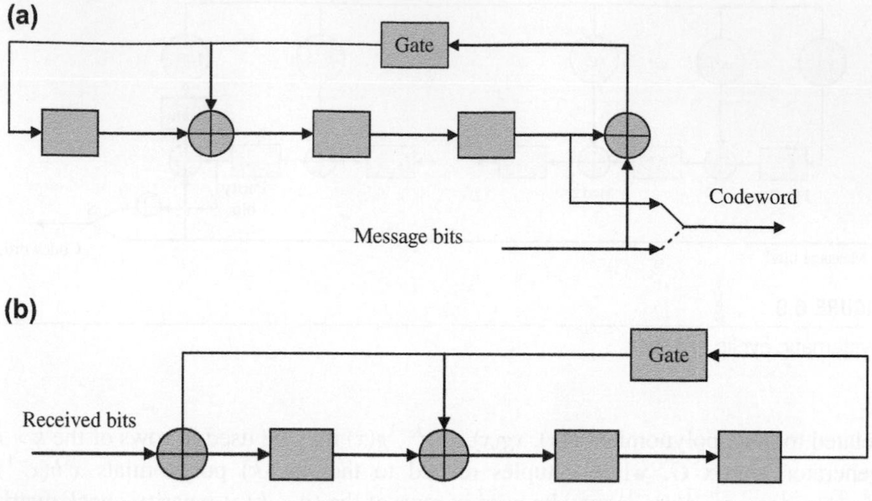

FIGURE 6.11

Hamming (7,4) encoder (a) and syndrome calculator (b).

For example, the encoder and syndrome calculator for (7,4) Hamming code are given in Figure 6.11a and b respectively. The generating polynomial is given by $g(x) = 1 + x + x^3$. The polynomial $x^7 + 1$ can be factorized as follows: $x^7 + 1 = (1 + x)(1 + x^2 + x^3)(1 + x + x^3)$. If we select $g(x) = 1 + x + x^3$ as the generator polynomial, based on property 3 of the generator polynomial, the corresponding parity-check polynomial will be $h(x) = (1 + x)(1 + x^2 + x^3) = 1 + x + x^2 + x^4$. The message sequence 1001 can be represented in polynomial form by $m(x) = 1 + x^3$. For the representation in systematic form, we have to multiply $m(x)$ by x^{n-k} to obtain $x^{n-k}\mathrm{m}(x) = x^3\mathrm{m}(x) = x^3 + x^6$. The codeword polynomial is obtained by $c(x) = x^{n-k}m(x) + \mathrm{rem}[x^{n-k}m(x)/g(x)] = x + x^3 + \mathrm{rem}[(x + x^3)/(1 + x + x^3)] = x + x^2 + x^3 + x^6$. The corresponding codeword is 0111001. To obtain the generator matrix of this code we can use the polynomials $g(x) = 1 + x + x^3$, $xg(x) = x + x^2 + x^4$, $x^2g(x) = x^2 + x^3 + x^5$, and $x^3g(x) = x^3 + x^4 + x^6$, and write the corresponding n-tuples in the form of a matrix as follows:

$$G' = \begin{bmatrix} 1101000 \\ 0110100 \\ 0011010 \\ 0001101 \end{bmatrix}.$$

By Gaussian elimination we can put the generator matrix in systematic form:

$$G = \begin{bmatrix} 1101000 \\ 0110100 \\ 1110010 \\ 1010001 \end{bmatrix}.$$

The parity-check matrix can be obtained from the following polynomials:

$$x^4h(x^{-1}) = 1 + x^2 + x^3 + x^4,\ x^5h(x^{-1}) = x + x^3 + x^4 + x^5,\ x^6h$$
$$(x^{-1}) = x^2 + x^3 + x^5 + x^6,$$

by writing down the corresponding n-tuples in the form of a matrix:

$$\boldsymbol{H}' = \begin{bmatrix} 1011100 \\ 0101110 \\ 0010111 \end{bmatrix}.$$

The matrix $\boldsymbol{H}$ can be put in systematic form by Gaussian elimination:

$$\boldsymbol{H} = \begin{bmatrix} 1001011 \\ 0101110 \\ 0010111 \end{bmatrix}.$$

The syndrome polynomial $s(x)$ has the following three important properties, which can be used to simplify the implementation of decoders [15,17–19,22,24–26]:

1. The syndrome of the received word polynomial $r(x)$ is also the syndrome of the corresponding error polynomial $e(x)$.
2. The syndrome of a cyclic shift of $r(x)$, $xr(x)$, is determined by $xs(x)$.
3. The syndrome polynomial $s(x)$ is identical to the error polynomial $e(x)$, if the errors are confined to the $(n - k)$ parity-check bits of the received word polynomial $r(x)$.

Maximal-length codes $(n = 2^m - 1, m)$ $(m \geq 3)$ are dual Hamming codes and have minimum distance $d_{\min} = 2^m - 1$. The parity-check polynomial for (7,3) maximal-length codes is therefore $h(x) = 1 + x + x^3$. The encoder for (7,3) maximum-length code is given in Figure 6.12. The generator polynomial gives one period of maximum-length code, providing that encoder is initialized to 0...01. For example, the generator polynomial for the (7,3) maximum-length code above is $g(x) = 1 + x + x^2 + x^4$, and the output sequence is given by:

$$\underbrace{100}_{\text{initial state}} \qquad \underbrace{1110100}_{g(x)=1+x+x^2+x^4}.$$

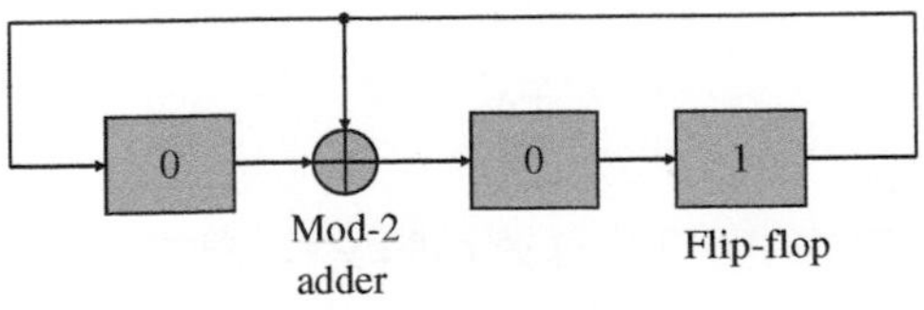

FIGURE 6.12

Encoder for the (7,3) maximal-length code.

Cyclic redundancy check (CRC) codes are very popular codes for error detection. Any (n,k) CRC codes are capable of detecting [15,17–19,22,24–26]:

- All error bursts of length $n-k$, with an error burst of length $n-k$ being defined as a contiguous sequence of $n-k$ bits in which the first and last bits or any other intermediate bits are received in error.
- A fraction of error bursts of length equal to $n-k+1$; the fraction equals $1-2^{-(n-k-1)}$.
- A fraction of error of length greater than $n-k+1$; the fraction equals $1-2^{-(n-k-1)}$.
- All combinations of $d_{\min}-1$ (or fewer) errors.
- All error patterns with an odd number of errors if the generator polynomial $g(x)$ for the code has an even number of nonzero coefficients.

In Table 6.3 are listed the generator polynomials of several CRC codes, which are currently used in various communication systems.

Decoding of cyclic codes is composed of the same three steps used in the decoding of linear block codes, namely syndrome computation, error pattern identification, and error correction [15,17–19,22,24–26]. The *Meggit decoder* configuration, which is implemented based on syndrome property 2 (also known as the Meggit theorem), is shown in Figure 6.13. The syndrome is calculated by dividing the received word by the generating polynomial $g(x)$, and at the same time the received word is shifted into the buffer register. Once the last bit of the received word enters the decoder, the gate is turned off. The syndrome is further read into the error pattern detection circuit, implemented as the combinational logic circuit, which generates 1 if and only if the content of the syndrome register corresponds to a correctible error pattern at the highest-order position x^{n-1}. By adding the output of the error pattern detector to the bit in error, the error can be corrected and, at the

Table 6.3 Generator polynomials of several CRC codes

CRC codes	Generator polynomial	$n-k$
CRC-8 code (IEEE 802.16, WiMax)	$1+x^2+x^8$	8
CRC-16 code (IBM CRC-16, ANSI, USB, SDLC)	$1+x^2+x^{15}+x^{16}$	16
CRC-ITU (X25, V41, CDMA, Bluetooth, HDLC, PPP)	$1+x^5+x^{12}+x^{16}$	16
CRC-24 (WLAN, UMTS)	$1+x+x^5+x^6+x^{23}+x^{24}$	24
CRC-32 (Ethernet)	$1+x+x^2+x^4+x^5+x^7+x^8+x^{10}+x^{11}+x^{12}+x^{16}+x^{22}+x^{23}+x^{26}+x^{32}$	32

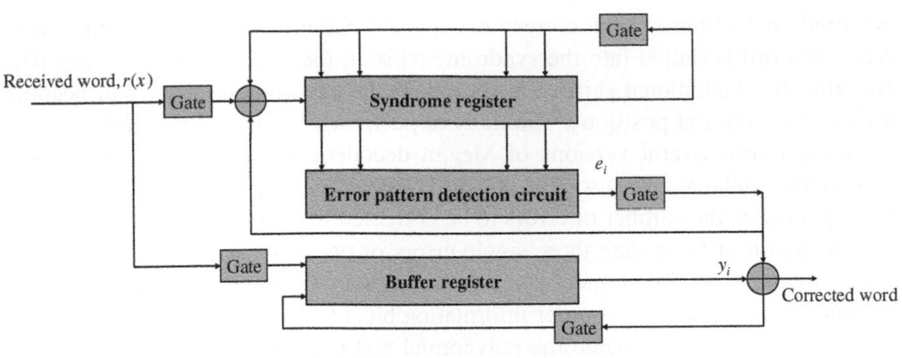

FIGURE 6.13

A Meggit decoder configuration.

same time, the syndrome has to be modified. If an error occurred on position x^l, by cyclically shifting the received word $n-l-1$ times, the erroneous bit will appear in position x^{n-1}, and can be corrected. The decoder therefore corrects the errors in a bit-by-bit fashion until the entire received word is read out from the buffer register.

The Hamming (7,4) cyclic decoder configuration, for generating polynomial $g(x)=1+x+x^3$, is shown in Figure 6.14. Because we expect only single errors, the error polynomial corresponding to the highest order position is $e(x)=x^6$ and the corresponding syndrome is $s(x)=1+x^2$. Once this syndrome is detected, the erroneous bit at position x^6 is to be corrected. Let us now assume that an error

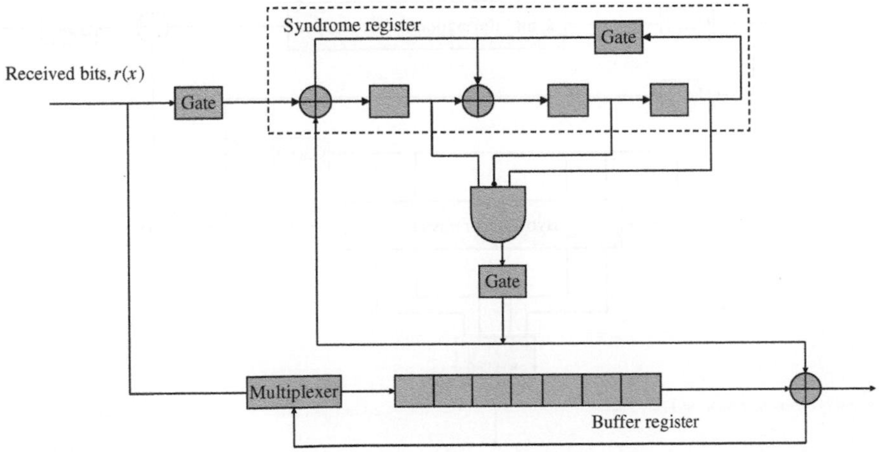

FIGURE 6.14

Hamming (7,4) decoder configuration.

occurred on position x^i, with corresponding error pattern $e(x) = x^i$. Once the entire received word is shifted into the syndrome register, the error syndrome is not 101. But after $6 - i$ additional shifts, the content of the syndrome register will become 101, and the error at position x^6 (initially at position x^i) will be corrected.

There exists several versions of Meggit decoders; however, the basic idea is essentially similar to that described above. The complexity of this decoder increases very quickly as the number of errors to be corrected increases, and it is rarely used for correction of more than three single errors or one burst of errors. The Meggit decoder can be simplified under certain assumptions. Let us assume that errors are confined to only the highest order information bits of the received polynomial $r(x)$: $x^k, x^{k+1}, \ldots, x^{n-1}$, so that syndrome polynomial $s(x)$ is given by:

$$s(x) = r(x) \bmod g(x) = [c(x) + e(x)] \bmod g(x) = e(x) \bmod g(x), \quad (6.48)$$

where $r(x)$ is the received polynomial, $c(x)$ is the codeword polynomial, and $g(x)$ is the generator polynomial. The corresponding error polynomial can be estimated by

$$\begin{aligned} e'(x) &= e'(x) \bmod g(x) = s'(x) = x^{n-k} s(x) \bmod g(x) \\ &= e_k + e_{k+1}x + \ldots + e_{n-2}x^{n-k-2} + e_{n-1}x^{n-k-1}. \end{aligned} \quad (6.49)$$

The error polynomial will be at most of degree $n - k - 1$ because $\deg[g(x)] = n - k$. Therefore, the syndrome register content is identical to the error pattern, and we say that the error pattern is *trapped* in the syndrome register. The corresponding decoder, known as the *error-trapping decoder*, is shown in Figure 6.15

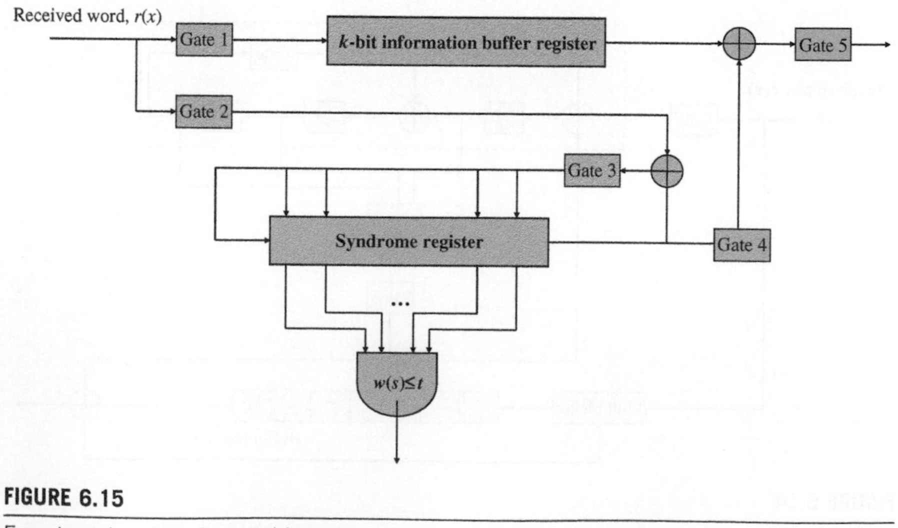

FIGURE 6.15

Error-trapping decoding architecture.

[15,17]. If t or fewer errors occur in $n-k$ consecutive locations, it can be shown that the error pattern is trapped in the syndrome register only when the weight of the syndrome $w(s)$ is less than or equal to t [15,17]. Therefore, the test for the error-trapping condition is to check if the weight of the syndrome is t or less. With gates 1, 2, and 3 closed (4 and 5 open), the received word is shifted into the syndrome register from the right end, which is equivalent to premultiplying the received polynomial $r(x)$ by x^{n-k}. Once the highest order k bits (corresponding to information bits) are shifted into the information buffer, gate 1 is opened. Gate 2 is opened once all n bits from the received word are shifted into the syndrome register. The syndrome register at this point contains the syndrome corresponding to $x^{n-k}r(x)$. If its weight is t or less, gates 4 and 5 are closed (all others are opened) and the corrected information is shifted out. If $w(s) > t$, the errors are not confined to the $n-k$ higher-order positions of $s(x)$; we keep shifting with gate 3 being on (other gates are switched off), until $w(s) \leq t$. If $w(s)$ never goes $\leq t$ and the syndrome register is shifted k times, either an error pattern with errors confined to $n-k$ consecutive end-around locations has occurred or an uncorrectable error pattern has occurred. For an additional explanation of this decoder, and other types of cyclic decoders, the interested reader is referred to Ref. [15].

6.4 BOSE–CHAUDHURI–HOCQUENGHEM (BCH) CODES

The BCH codes, the most famous cyclic codes, were discovered by Hocquenghem in 1959 and by Bose and Chaudhuri in 1960 [15,17–19]. Among many different decoding algorithms, the most important are the Massey–Berlekamp algorithm and Chien's search algorithm. An important subclass of BCH is a class of Reed–Solomon codes proposed in 1960. Before we continue further with a study of BCH codes, we have to introduce some properties of finite fields [15,19,40,41].

6.4.1 Galois Fields

Before we can proceed further with finite fields, we introduce the concepts of ring, field, and congruencies from abstract algebra [40,41]. A *ring* is defined as a set of elements R with two operations, addition "+" and multiplication "•", satisfying the following three properties: (i) R is an Abelian (commutative) group under addition; (ii) the multiplication operation is associative; and (iii) multiplication is associative over addition. A *field* is defined as a set of elements F with two operations, addition "+" and multiplication "•", satisfying the following properties: (i) F is an Abelian group under addition operation, with 0 being the identity element; (ii) the nonzero elements of F form an Abelian group under the multiplication, with 1 being the identity element; and (iii) the multiplication operation is distributive over the addition operation.

The quantity a is said to be *congruent* to quantity b observed per modulus n, denoted as $a \equiv b \pmod{n}$, if $a - b$ is divisible by n. If $x \equiv a \pmod{n}$, then a is called a *residue* to x to modulus n. A *class of residues* to modulus n is the class of all integers congruent to a given residue (mod n), and every member of the class is called a representative of the class. There are n classes, represented by (0),(1),(2),…,(n – 1), and the representatives of these classes are called a *complete system* of incongruent residues to modulus n. If i and j are two members of a complete system of incongruent residues to modulus n, then addition and multiplication between i and j can be introduced by

1. $i + j = (i + j) \pmod{n}$
2. $i \cdot j = (i \cdot j) \pmod{n}$.

A complete system of residues (mod n) forms a *commutative ring* with unity element. Let s be a nonzero element of these residues. Then s possesses an inverse element if and only if n is a prime, p. When p is a prime, a complete system of residues (mod p) forms a *Galois* (finite) *field*, and is commonly denoted by GF(p).

Let P(x) be any given polynomial in x of degree m with coefficients belonging to GF(p), and let $F(x)$ be any polynomial in x with integral coefficients. Then $F(x)$ may be expressed as [40]

$$F(x) = f(x) + p \cdot q(x) + P(x) \cdot Q(x),$$

where $f(x) = a_0 + a_1x + a_2x^2 + \ldots + a_{m-1}x^{m-1}$, $a_i \in \mathrm{GF}(p)$. This relationship may be written as $F(x) \equiv f(x) \bmod\{p,P(x)\}$, and we say that $f(x)$ is the *residue* of $F(x)$ modulus p and $P(x)$. If p and $P(x)$ are kept fixed but $f(x)$ is varied, p^m classes can be formed (because each coefficient of $f(x)$ may take p values of GF(p)). The classes defined by $f(x)$ form a commutative (Abelian) ring, which will be a field if and only if $P(x)$ is *irreducible* over GF(p) (not divisible with any other polynomial of degree $m - 1$ or less) [15,17,19]. The finite field formed by p^m classes of residues is called a *Galois field of order* p^m and is denoted by GF(p^m). Two important *properties* of GF(q), $q = p^m$, are given below [15,17,19]:

1. The roots of polynomial $x^{q-1} - 1$ are all nonzero elements of GF(q).
2. Let $P(x)$ be an irreducible polynomial of degree m with coefficients from GF(p) and β be a root from the extended field GF($q = p^m$). Then all the m roots of $P(x)$ are $\beta, \beta^p, \beta^{p^2}, \ldots, \beta^{p^{m-1}}$.

The nonzero elements of GF(p^m) can be represented as polynomials of degree at most $m - 1$ or as powers of a *primitive root* α such that [15,17,19]

$$\alpha^{p^m-1} = 1, \quad \alpha^d \neq 1 \text{ (for } d \text{ dividing } p^m - 1).$$

Therefore, the primitive element (root) is a field element that generates all nonzero field elements as its successive powers. An irreducible polynomial that has a primitive element as its root is called a *primitive polynomial.*

The function $P(x)$ is said to be a *minimum polynomial* for generating the elements of $GF(p^m)$ and represents the smallest degree polynomial over $GF(p)$ having a field element $\beta \in GF(p^m)$ as a root. To obtain a minimum polynomial we have to divide $x^q - 1$ $(q = p^m)$ by the least common multiple (LCM) of all factors of the form $x^d - 1$, where d is a divisor of $p^m - 1$, and obtain the so-called *cyclotomic equation* (i.e. the equation having as its roots all primitive roots of the equation $x^{q-1} - 1 = 0$). The order of this equation is $O(p^m - 1)$, where $O(k)$ is the number of all positive integers less than k and relatively prime to it. By substituting each coefficient in this equation by the least nonzero residue to modulus p, we get the cyclotomic polynomial of order $O(p^m - 1)$. Let $P(x)$ be an irreducible factor of this polynomial, then $P(x)$ is a minimum polynomial, which is in general not unique.

Example. Let us determine the minimum polynomial for generating the elements of $GF(2^3)$. The cyclotomic polynomial is $(x^7 - 1)/(x - 1) = x^6 + x^5 + x^4 + x^3 + x^2 + x + 1 = (x^3 + x^2 + 1)(x^3 + x + 1)$. Hence, $P(x)$ can be either $x^3 + x^2 + 1$ or $x^3 + x + 1$. Let us choose $P(x) = x^3 + x^2 + 1$. The degree of this polynomial is $\deg[P(x)] = 3$. Now we explain how we can construct $GF(2^3)$ using $P(x)$. The construction always starts with elements from the basic field (in this case $GF(2) = \{0,1\}$). All nonzero elements can be obtained as successive powers of primitive root α, until no new element is generated, which is given in the first column of Table 6.4. The second column is obtained by exploiting the primitive polynomial $P(x) = x^3 + x^2 + 1$. α is the root of $P(x)$ and therefore $P(\alpha) = \alpha^3 + \alpha^2 + 1 = 0$, and α^3 can be expressed as $\alpha^2 + 1$. α^4 can be expressed as $\alpha\alpha^3 = \alpha(\alpha^2 + 1) = \alpha^3 + \alpha = \alpha^2 + 1 + \alpha$. The third column in Table 6.4 is obtained by reading off coefficients in the second column, with the leading coefficient multiplying α^2.

Table 6.4 Three different representations of $GF(2^3)$ generated by $x^3 + x^2 + 1$

Power of α	Polynomial	Three-tuple
0	0	000
α^0	1	001
α^1	α	010
α^2	α^2	100
α^3	$\alpha^2 + 1$	101
α^4	$\alpha^2 + \alpha + 1$	111
α^5	$\alpha + 1$	011
α^6	$\alpha^2 + \alpha$	110
α^7	1	001

6.4.2 The Structure and Decoding of BCH Codes

Equipped with this knowledge of Galois fields, we can continue our description of the structure of BCH codes. Let the finite field GF(q) (*symbol field*) and extension field GF(q^m) (*locator field*), $m \geq 1$, be given. For every m_0 ($m_0 \geq 1$) and Hamming distance d there exists a BCH code with the generating polynomial $g(x)$, if and only if it is of smallest degree with coefficients from GF(q) and with roots from the extension field GF(q^m) as follows [15,22]:

$$\alpha^{m_0}, \alpha^{m_0+1}, \ldots, \alpha^{m_0+d-2}, \tag{6.50}$$

where α is from GF(q^m). The codeword length is determined as the least common multiple of orders of roots. (The order of an element β from a finite field is the smallest positive integer j such that $\beta^j = 1$.)

It can be shown that for any positive integers m ($m \geq 3$) and t ($t < 2^{m-1}$) there exists a *binary* BCH code having the following properties [15,17,22]:

- Codeword length: $n = 2^m - 1$
- Number of parity bits: $n - k \leq mt$
- Minimum Hamming distance: $d \geq 2t + 1$.

This code is able to correct up to t errors. The generator polynomial can be found as the LCM of the minimal polynomials of α^i [15,17,22]:

$$g(x) = \text{LCM}[P_{\alpha^1}(x), P_{\alpha^3}(x), \ldots, P_{\alpha^{2t-1}}(x)], \tag{6.51}$$

where α is a primitive element in GF(2^m) and $P_{\alpha^i}(x)$ is the minimal polynomial of α^i.

Let $c(x) = c_0 + c_1 x + c_2 x^2 + \ldots + c_{n-1} x^{n-1}$ be the codeword polynomial, and let the roots of generator polynomial be $\alpha, \alpha^2, \ldots, \alpha^{2t}$, where t is the error correction capability of the BCH code. Because the generator polynomial $g(x)$ is the factor of codeword polynomial $c(x)$, the roots of $g(x)$ must also be the roots of $c(x)$:

$$c(\alpha^i) = c_0 + c_1 \alpha^i + \ldots + c_{n-1} \alpha^{(n-1)i} = 0; \quad 1 \leq i \leq 2t. \tag{6.52}$$

This equation can also be written as an inner (scalar) product of codeword vector $c = [c_0 \; c_1 \; \ldots \; c_{n-1}]$ and the following vector $[1 \; \alpha^i \; \alpha^{2i} \; \ldots \; \alpha^{2(n-1)i}]$:

$$[c_0 \; c_1 \ldots c_{n-1}] \begin{bmatrix} 1 \\ \alpha^i \\ \ldots \\ \alpha^{(n-1)i} \end{bmatrix} = 0; \quad 1 \leq i \leq 2t. \tag{6.53}$$

Equation (6.53) can also be written as the following matrix:

$$[c_0\ c_1 \ldots c_{n-1}]\begin{bmatrix} \alpha^{n-1} & \alpha^{n-2} & \ldots & \alpha & 1 \\ (\alpha^2)^{n-1} & (\alpha^2)^{n-2} & \ldots & \alpha^2 & 1 \\ (\alpha^3)^{n-1} & (\alpha^3)^{n-2} & \ldots & \alpha^3 & 1 \\ \ldots & \ldots & \ldots & \ldots & \ldots \\ (\alpha^{2t})^{n-1} & (\alpha^{2t})^{n-2} & \ldots & \alpha^{2t} & 1 \end{bmatrix}^T = \boldsymbol{cH}^{\mathrm{T}} = \mathbf{0}, \tag{6.54}$$

$$\boldsymbol{H} = \begin{bmatrix} \alpha^{n-1} & \alpha^{n-2} & \ldots & \alpha & 1 \\ (\alpha^2)^{n-1} & (\alpha^2)^{n-2} & \ldots & \alpha^2 & 1 \\ (\alpha^3)^{n-1} & (\alpha^3)^{n-2} & \ldots & \alpha^3 & 1 \\ \ldots & \ldots & \ldots & \ldots & \ldots \\ (\alpha^{2t})^{n-1} & (\alpha^{2t})^{n-2} & \ldots & \alpha^{2t} & 1 \end{bmatrix},$$

where $\boldsymbol{H}$ is the parity-check matrix of the BCH code. Using property 2 of GF(q) from the previous section, we conclude that α^i and α^{2i} are the roots of the same minimum polynomial, so that the even rows in $\boldsymbol{H}$ can be omitted to get the final version of the parity-check matrix of BCH codes:

$$\boldsymbol{H} = \begin{bmatrix} \alpha^{n-1} & \alpha^{n-2} & \ldots & \alpha & 1 \\ \alpha^{3(n-1)} & \alpha^{3(n-2)} & \ldots & \alpha^3 & 1 \\ \alpha^{5(n-1)} & \alpha^{5(n-2)} & \ldots & \alpha^5 & 1 \\ \ldots & \ldots & \ldots & \ldots & \ldots \\ \alpha^{(2t-1)(n-1)} & \alpha^{(2t-1)(n-2)} & \ldots & \alpha^{2t-1} & 1 \end{bmatrix}. \tag{6.55}$$

For example, the (15,7) two-error correcting BCH code has the generator polynomial [17]:

$$\begin{aligned} g(x) &= \mathrm{LCM}[\phi_\alpha(x), \phi_{\alpha^3}(x)] \\ &= \mathrm{LCM}\left[x^4 + x + 1, (x+\alpha^3)(x+\alpha^6)(x+\alpha^9)(x+\alpha^{12})\right] \\ &= x^8 + x^7 + x^6 + x^4 + 1 \end{aligned}$$

and the parity check matrix [17]

$$\begin{aligned} \boldsymbol{H} &= \begin{bmatrix} \alpha^{14} & \alpha^{13} & \alpha^{12} & \alpha^{11} & \ldots & \alpha & 1 \\ \alpha^{42} & \alpha^{39} & \alpha^{36} & \alpha^{33} & \ldots & \alpha^3 & 1 \end{bmatrix} \\ &= \begin{bmatrix} \alpha^{14} & \alpha^{13} & \alpha^{12} & \alpha^{11} & \alpha^{10} & \alpha^9 & \alpha^8 & \alpha^7 & \alpha^6 & \alpha^5 & \alpha^4 & \alpha^3 & \alpha^2 & \alpha & 1 \\ \alpha^{12} & \alpha^9 & \alpha^6 & \alpha^3 & 1 & \alpha^{12} & \alpha^9 & \alpha^6 & \alpha^3 & 1 & \alpha^{12} & \alpha^9 & \alpha^6 & \alpha^3 & 1 \end{bmatrix}. \end{aligned}$$

In the previous expression we have used the fact that in GF(2^4), $\alpha^{15} = 1$. The primitive polynomial used to design this code was $p(x) = x^4 + x + 1$. Every element in GF(2^4) can be represented as four-tuple, as shown in Table 6.5.

To create the second column we have used the relation $\alpha^4 = \alpha + 1$, and the four-tuples are obtained by reading off the coefficients in the second column. By replacing the powers of α in the parity-check matrix above by corresponding four-tuples, the parity-check matrix can be written in the following binary form:

$$\boldsymbol{H} = \begin{bmatrix} 1 & 1 & 1 & 1 & 0 & 1 & 0 & 1 & 1 & 0 & 0 & 1 & 0 & 0 & 0 \\ 0 & 1 & 1 & 1 & 1 & 0 & 1 & 0 & 1 & 1 & 0 & 0 & 1 & 0 & 0 \\ 0 & 0 & 1 & 1 & 1 & 1 & 0 & 1 & 0 & 1 & 1 & 0 & 0 & 1 & 0 \\ 1 & 1 & 1 & 0 & 1 & 0 & 1 & 1 & 0 & 0 & 1 & 0 & 0 & 0 & 1 \\ 1 & 1 & 1 & 1 & 0 & 1 & 1 & 1 & 1 & 0 & 1 & 1 & 1 & 1 & 0 \\ 1 & 0 & 1 & 0 & 0 & 1 & 0 & 1 & 0 & 0 & 1 & 0 & 1 & 0 & 0 \\ 1 & 1 & 0 & 0 & 0 & 1 & 1 & 0 & 0 & 0 & 1 & 1 & 0 & 0 & 0 \\ 1 & 0 & 0 & 0 & 1 & 1 & 0 & 0 & 0 & 1 & 1 & 0 & 0 & 0 & 1 \end{bmatrix}.$$

Table 6.6 lists the parameters of several BCH codes generated by primitive elements of order less than $2^5 - 1$ that are of interest for high-speed communications. The complete list can be found in Appendix C of Ref. [15].

Generally there is no need for q to be a prime, it could be a prime power. However, the symbols must be taken from GF(q) and the roots from GF(q^m).

Table 6.5 GF(2^4) generated by $x^4 + x + 1$

Power of α	Polynomial of α	Four-tuple
0	0	0000
α^0	1	0001
α^1	α	0010
α^2	α^2	0100
α^3	α^3	1000
α^4	$\alpha + 1$	0011
α^5	$\alpha^2 + \alpha$	0110
α^6	$\alpha^3 + \alpha^2$	1100
α^7	$\alpha^3 + \alpha + 1$	1011
α^8	$\alpha^2 + 1$	0101
α^9	$\alpha^3 + \alpha$	1010
α^{10}	$\alpha^2 + \alpha + 1$	0111
α^{11}	$\alpha^3 + \alpha^2 + \alpha$	1110
α^{12}	$\alpha^3 + \alpha^2 + \alpha + 1$	1111
α^{13}	$\alpha^3 + \alpha^2 + 1$	1101
α^{14}	$\alpha^3 + 1$	1001

Table 6.6 A set of primitive binary BCH codes

n	*k*	*t*	Generator polynomial (in octal form)
15	11	1	23
63	57	1	103
63	51	2	12471
63	45	3	1701317
127	120	1	211
127	113	2	41567
127	106	3	11554743
127	99	4	3447023271
255	247	1	435
255	239	2	267543
255	231	3	156720665
255	223	4	75626641375
255	215	5	23157564726421
255	207	6	16176560567636227
255	199	7	7633031270420722341
255	191	8	2663470176115333714567
255	187	9	52755313540001322236351
255	179	10	22624710717340432416300455

Among different classes of the nonbinary BCH codes, the Reed–Solomon codes are the best known; and these codes are explained briefly in the next section.

The BCH codes can be decoded in the same way as any other cyclic code class. For example, Figure 6.16 shows the error-trapping decoder for the BCH (15,7) double-error correcting code. The operating principle of this circuit has been explained in the previous section. Here we explain the decoding process by employing algorithms especially developed for decoding BCH codes. Let $g(x)$ be the generator polynomial with corresponding roots $\alpha, \alpha^2, \ldots, \alpha^{2t}$. Let $c(x) = c_0 + c_1x + c_2x^2 + \ldots + c_{n-1}x^{n-1}$ be the codeword polynomial, $r(x) = r_0 + r_1x + r_2x^2 + \ldots + r_{n-1}x^{n-1}$ be the received word polynomial, and $e(x) = e_0 + e_1x + e_2x^2 + \ldots + e_{n-1}x^{n-1}$ be the error polynomial. The roots of the generator polynomial are also the roots of the codeword polynomial, i.e.

$$c(\alpha^i) = 0, \ i = 0, 1, \ldots, 2t. \tag{6.56}$$

For binary BCH codes the only nonzero element is 1, therefore the indices i of coefficients $e_i \neq 0$ (or $e_i = 1$) determine the error locations. For nonbinary BCH

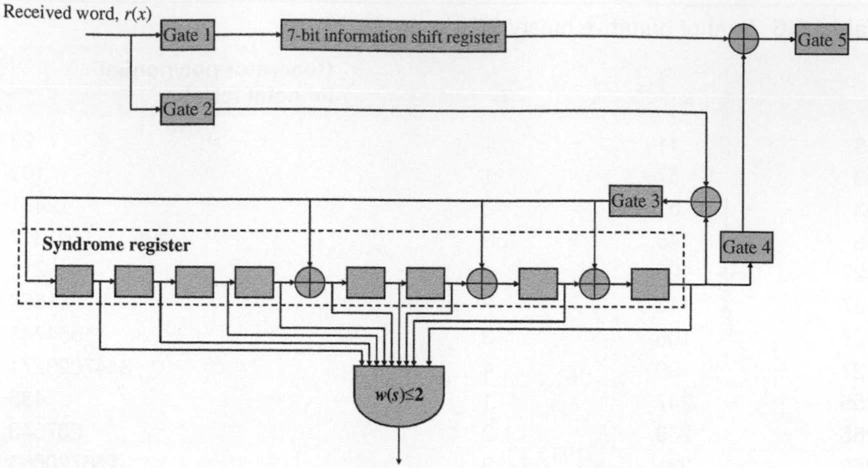

FIGURE 6.16

Error-trapping decoder for (15,7) BCH code generated by $g(x) = 1 + x^4 + x^6 + x^7 + x^8$.

codes the error magnitudes are also important in addition to error locations. By evaluating the received word polynomial $r(x)$ for α^i we obtain:

$$r(\alpha^i) = c(\alpha^i) + e(\alpha^i) = e(\alpha^i) = S_i, \tag{6.57}$$

where S_i is the ith component of the syndrome vector, defined by

$$\boldsymbol{S} = [S_1 \quad S_2 \quad \ldots \quad S_{2t}] = \boldsymbol{r}\boldsymbol{H}^{\mathrm{T}}. \tag{6.58}$$

The BCH code is able to correct up to t errors. Let us assume that the error polynomial $e(x)$ has no more than t errors. It can then be written as

$$e(x) = e_{j_1}x^{j_1} + e_{j_2}x^{j_2} + \ldots + e_{j_l}x^{j_l} + \ldots + e_{j_v}x^{j_v}; \quad 0 \le v \le t. \tag{6.59}$$

e_{j_v} is the error magnitude and j_l is the error location. The corresponding syndrome components can be obtained from (6.56) and (6.59) as follows:

$$S_i = e_{j_1}(\alpha^i)^{j_1} + e_{j_2}(\alpha^i)^{j_2} + \ldots + e_{j_l}(\alpha^i)^{j_l} + \ldots + e_{j_v}(\alpha^i)^{j_v}; \quad 0 \le v \le t, \tag{6.60}$$

where α^{j_l} is the error location number. Notice that the error magnitudes are from the symbol field, while the error location numbers are from the extension field. In order to avoid double indexing we introduce the notation $X_l = \alpha^{j_l}$, $Y_l = e_{j_l}$. The pairs (X_l, Y_l) completely identify the errors ($l \in [1,v]$). We then have to solve the following set of equations:

$$\begin{aligned} S_1 &= Y_1X_1 + Y_2X_2 + \ldots + Y_vX_v \\ S_2 &= Y_1X_1^2 + Y_2X_2^2 + \ldots + Y_vX_v^2 \\ &\ldots \\ S_{2t} &= Y_1X_1^{2t} + Y_2X_2^{2t} + \ldots + Y_vX_v^{2t}. \end{aligned} \tag{6.61}$$

The procedure to solve this system of equations represents the corresponding decoding algorithm. Direct solution of this system of equations is impractical. There exist many different algorithms to solve the system of equations (6.61), ranging from iterative to Euclidean algorithms [15,17,19]. A very popular decoding algorithm for BCH codes is the *Massey–Berlekamp algorithm* [15,17,19]. In this algorithm BCH decoding is observed as a shift register synthesis problem: Given the syndromes S_i, we have to find the minimal length shift register that generates the syndromes. Once we determine the coefficients of this shift register, we construct the *error locator* polynomial [17,19]:

$$\sigma(x) = \prod_{i=1}^{v}(1 + X_i x) = \sigma_v x^v + \sigma_{v-1}x^{v-1} + \ldots + \sigma_1 x + 1, \tag{6.62}$$

where the σ_i values, also known as elementary symmetric functions, are given by Viète's formulas:

$$\begin{aligned} \sigma_1 &= X_1 + X_2 + \ldots + X_v \\ \sigma_2 &= \sum_{i<j} X_i X_j \\ \sigma_3 &= \sum_{i<j<k} X_i X_j X_k \\ &\ldots \\ \sigma_v &= X_1 X_2 \ldots X_v. \end{aligned} \tag{6.63}$$

Because $\{X_l\}$ are the inverses of the roots of $\sigma(x)$, $\sigma(1/X_l) = 0 \ \forall \ l$, we can write [17,19]:

$$X_l^v \sigma(X_l^{-1}) = X_l^v + \sigma_1 X_l^{v-1} + \ldots + \sigma_v. \tag{6.64}$$

By multiplying the previous equation by X_l^j and performing summation over l for fixed j, we obtain:

$$S_{v+j} + \sigma_1 S_{v+j-1} + \ldots + \sigma_v S_j = 0; \quad j = 1, 2, \ldots, v. \tag{6.65}$$

This equation can be rewritten as follows:

$$S_{v+j} = -\sum_{i=1}^{v} \sigma_i S_{v+j-i}; \quad j = 1, 2, \ldots, v. \tag{6.66}$$

S_{v+j} represents the output of the shift register, shown in Figure 6.17. The Massey–Berlekamp algorithm is summarized by the flow chart shown in Figure 6.18, which is self-explanatory. Once the locator polynomial is determined, we have to find the roots and invert them to obtain the error locators.

To determine the error magnitudes, we have to define another polynomial, known as the *error evaluator* polynomial [17]:

$$\varsigma(x) = 1 + (S_1 + \sigma_1)x + (S_2 + \sigma_1 S_1 + \sigma_2)x^2 + \ldots + (S_v + \sigma_1 S_{v-1} + \ldots + \sigma_v)x^v. \tag{6.67}$$

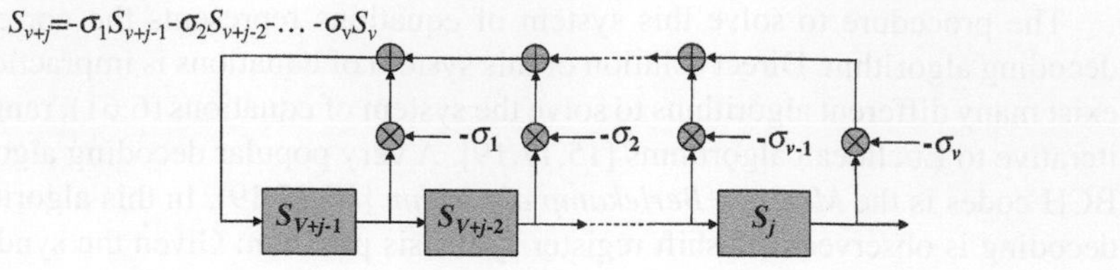

FIGURE 6.17

A shift register that generates the syndromes S_j.

The error magnitudes are then obtained from [17]

$$Y_l = \frac{\varsigma(X_l^{-1})}{\prod_{i=1,\ i\neq l}^{v}(1+X_iX_l^{-1})}. \tag{6.68}$$

For binary BCH codes $Y_l = 1$, so we do not need to evaluate the error magnitudes.

6.5 REED–SOLOMON (RS) CODES, CONCATENATED CODES, AND PRODUCT CODES

The Reed–Solomon (RS) codes were discovered in 1960 and represent a special class of nonbinary BCH codes [29,30]. RS codes represent the most commonly used nonbinary codes. Both the code symbols and the roots of the generating polynomial are from the locator field. In other words, the symbol field and the locator field are the same ($m = 1$) for RS codes. The codeword length of RS codes is determined by $n = q^m - 1 = q - 1$, so that RS codes are relatively short codes. The minimum polynomial for some element β is $P_\beta(x) = x - \beta$. If α is the primitive element of GF(q) (q is a prime or prime power), the generator polynomial for a t-error-correcting Reed–Solomon code is given by [15,17,19]

$$g(x) = (x - a)(x - \alpha^2)\ldots(x - \alpha^{2t}). \tag{6.69}$$

The generator polynomial degree is $2t$ and it is the same as the number of parity symbols $n - k = 2t$, while the block length of the code is $n = q - 1$. Since the minimum distance of BCH codes is $2t + 1$, the minimum distance of RS codes is $d_{\min} = n - k + 1$, therefore satisfying the Singleton bound ($d_{\min} \leq n - k + 1$) with equality and belonging to the class of *maximum-distance separable* (MDS) *codes*. When $q = 2^m$, the RS code parameters are $n = m(2^m - 1)$, $n - k = 2mt$, and $d_{\min} = 2mt + 1$. Therefore, the minimum distance of RS codes, when observed as binary codes, is large. The RS codes may be considered as burst error correcting codes, and as such are suitable for bursty-error-prone channels. This binary code is able to correct up to t bursts of length m. Equivalently, this binary code is able to correct a single burst of length $(t - 1)m + 1$.

The weight distribution of RS codes can be determined by [15]

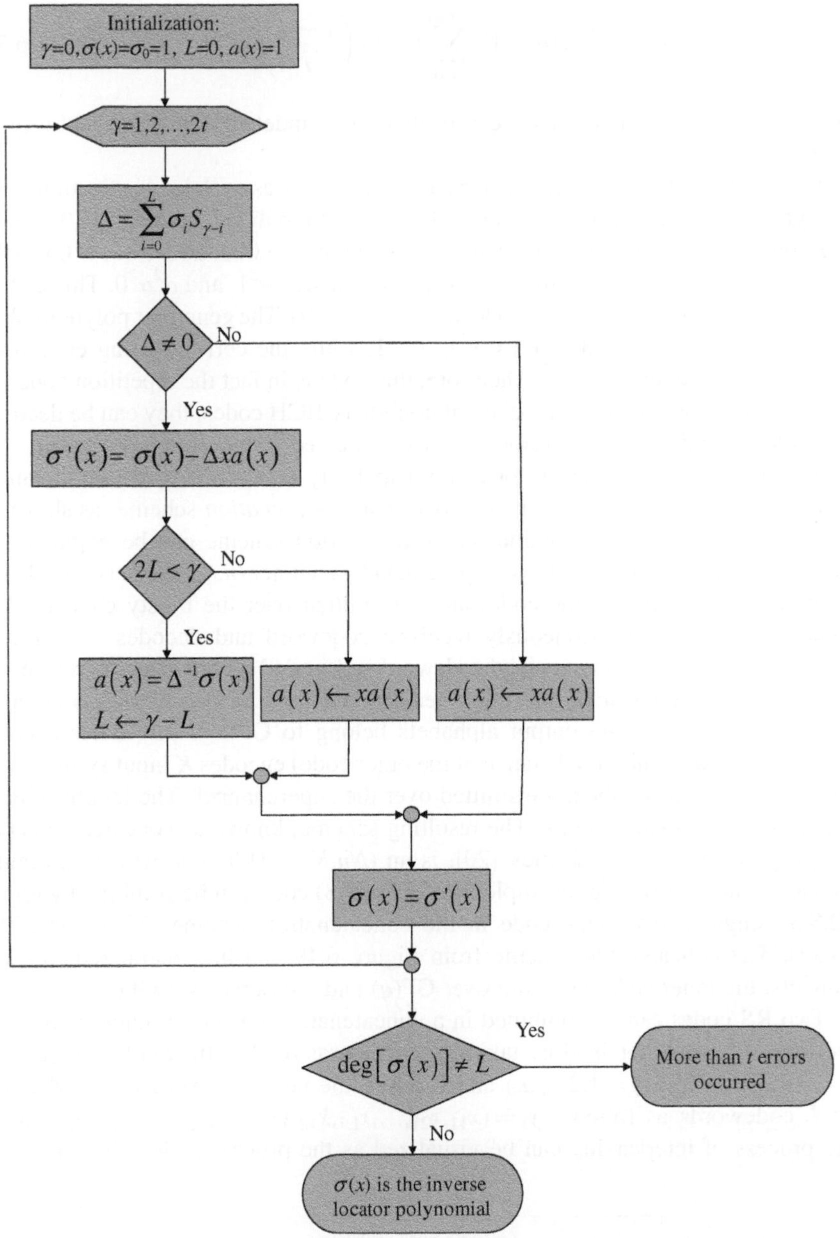

FIGURE 6.18

Flow chart of the Massey–Berlekamp algorithm. Δ denotes the error (discrepancy) between the syndrome and the shift register output, $a(x)$ stores the content of the shift register (normalized by Δ^{-1}) prior to lengthening.

$$A_i = \binom{n}{i}(q-1)\sum_{j=0}^{i-d_{\min}}(-1)^j\binom{i-1}{j}q^{i-d_{\min}-j}, \tag{6.70}$$

and by using this expression we can evaluate the undetected error probability by Eq. (6.38).

Example. Let GF(4) be generated by $1+x+x^2$, as was explained in Section 6.4.1. The symbols of GF(4) are 0, 1, α, and α^2. The generator polynomial for RS(3,2) code is given by $g(x)=x-\alpha$. The corresponding codewords are 000, 101, $\alpha 0\alpha$, $\alpha^2 0\alpha^2$, 011, 110, $\alpha 1\alpha^2$, $\alpha^2 1\alpha$, $0\alpha\alpha$, $1\alpha\alpha^2$, $\alpha\alpha 0$, $\alpha^2\alpha 1$, $0\alpha^2\alpha^2$, $1\alpha^2\alpha$, $\alpha\alpha^2 1$, and $\alpha^2\alpha^2 0$. This code is essentially the even parity-check code ($\alpha^2+\alpha+1=0$). The generator polynomial for RS(3,1) is $g(x)=(x-\alpha)(x-\alpha^2)=x^2+x+1$, while the corresponding codewords are 000, 111, $\alpha\alpha\alpha$, and $\alpha^2\alpha^2\alpha^2$. Therefore, this code is in fact the repetition code.

Since RS codes are a special class of nonbinary BCH codes, they can be decoded using the same decoding algorithm already explained in the previous section.

To improve the burst error correction capability of RS codes, RS code can be combined with an inner binary block code in a *concatenation* scheme, as shown in Figure 6.19. The key idea behind the concatenation scheme can be explained as follows [17]. Consider the codeword generated by an inner (n,k,d) code (with d being the minimum distance of the code) and transmitted over the bursty channel. The decoder processes the erroneously received codeword and decodes it correctly. However, occasionally the received codeword is decoded incorrectly. Therefore, the inner encoder, the channel, and the inner decoder may be considered as a superchannel whose input and output alphabets belong to GF(2^k). The outer encoder (N,K,D) (D is the minimum distance of the outer code) encodes K input symbols and generates N output symbols transmitted over the superchannel. The length of each symbol is k information digits. The resulting scheme, known as concatenated code and proposed initially by Forney [20], is an $(Nn,Kk \geq Dd)$ code with a minimum distance of at least Dd. For example, RS(255,239,8) code can be combined with the (12,8,3) single parity-check code in the concatenation scheme $(12 \cdot 255,239 \cdot 8,\geq 24)$. The concatenated scheme from Figure 6.19 can be generalized to q-ary channels, the inner code operating over GF(q) and the outer over GF(q^k).

Two RS codes can be combined in a concatenated scheme by interleaving. An *interleaved code* is obtained by taking L codewords (of length N) of a given code $\boldsymbol{x}_j=(x_{j1},x_{j2},\ldots,x_{jN})$ $(j=1,2,\ldots,L)$ and forming the new codeword by interleaving the L codewords as follows: $\boldsymbol{y}_i=(x_{11},x_{21},\ldots,x_{L1},x_{12},x_{22},\ldots,x_{L2},\ldots,x_{1N},x_{2N},\ldots,x_{LN})$. The process of interleaving can be visualized as the process of forming an $L \times N$

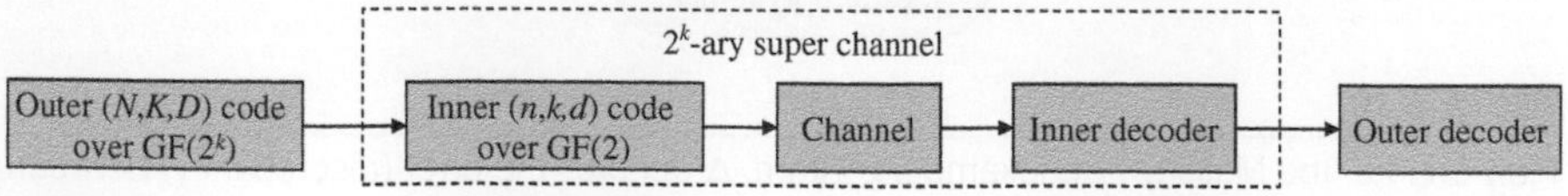

FIGURE 6.19

Concatenated $(Nn,Kk,\geq Dd)$ code.

matrix of L codewords written row by row and transmitting the matrix column by column, as given below:

$$\begin{array}{l} x_{11}x_{12}\ldots x_{1N} \\ x_{21}x_{22}\ldots x_{2N} \\ \ldots \\ x_{L1}x_{L2}\ldots x_{LN} \end{array}$$

The parameter L is known as the interleaving *degree*. The transmission must be postponed until L codewords are collected. To be able to transmit a column whenever a new codeword becomes available, the codewords should be arranged down diagonals as given below, and the interleaving scheme is known as *delayed interleaving* (one-frame delayed interleaving):

$$\begin{array}{cccccc} x_{i-(N-1),1} & \cdots & x_{i-2,1} & x_{i-1,1} & x_{i,1} & \\ & x_{i-(N-1),2} & \cdots & x_{i-2,2} & x_{i-1,2} & x_{i,2} \\ & & \cdots\cdots & \cdots\cdots & \cdots\cdots & \\ & & & x_{i-(N-1),N-1} & x_{i-(N-2),N-1} & \\ & & & & x_{i-(N-1),N} & x_{i-(N-2),N} \end{array}$$

Each new codeword completes one column of this array. In the example above the codeword $\boldsymbol{x}_i$ completes the column (frame) $x_{i,1},x_{i-1,2},\ldots,x_{i-(N-1),N}$. A generalization of this scheme, in which the components of the ith codeword $\boldsymbol{x}_i$, say $x_{i,j}$ and $x_{i,j+1}$, are spaced λ frames apart is known as λ-frame delayed interleaved.

Another way to deal with burst errors is to arrange two RS codes in *turbo product* manner, as shown in Figure 6.20. A product code [3–6] is an (n_1n_2,k_1k_2,d_1d_2) code in which codewords form an $n_1 \times n_2$ array such that each row is a codeword from an (n_1,k_1,d_1) code C_1, and each column is a codeword from an (n_2,k_2,d_2) code C_2, with n_i, k_i and d_i $(i = 1,2)$ being the codeword length, dimension, and minimum distance respectively of the ith component code. Turbo product codes were proposed by

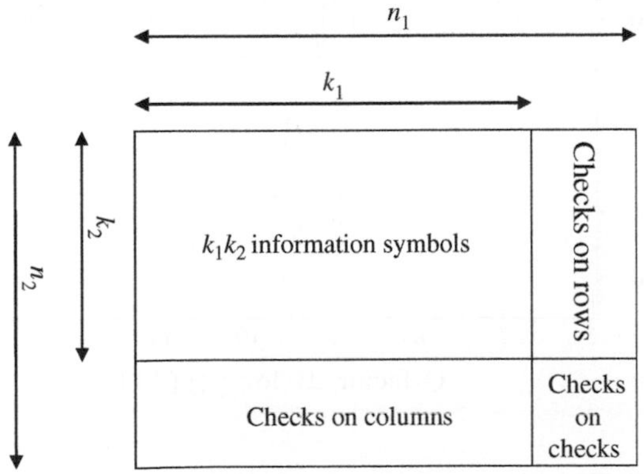

FIGURE 6.20

Structure of a codeword of a turbo product code.

Elias [16]. Both binary (such as binary BCH codes) and nonbinary codes (such as RS codes) may be arranged in product code manner. It is possible to show [17] that the minimum distance of a product code is the product of the minimum distances of the component codes. It is straightforward to show that the product code is able to correct a burst error of length $b = \max(n_1 b_2, n_2 b_1)$, where b_i is the burst error capability of component code $i = 1,2$.

The results of Monte Carlo simulations for different RS concatenation schemes on an optical on–off keying AWGN channel are shown in Figure 6.21. Interestingly, the concatenation scheme RS(255,239) + RS(255,223) of code rate $R = 0.82$ outperforms the concatenation scheme RS(255,223) + RS(255,223) of lower code rate $R = 0.76$, as well as the concatenation scheme RS(255,223) + RS(255,239) of the same code rate.

6.6 CONCLUDING REMARKS

The standard FEC schemes that belong to the class of hard decision codes have been described in this chapter. More powerful FEC schemes belong to the class of soft

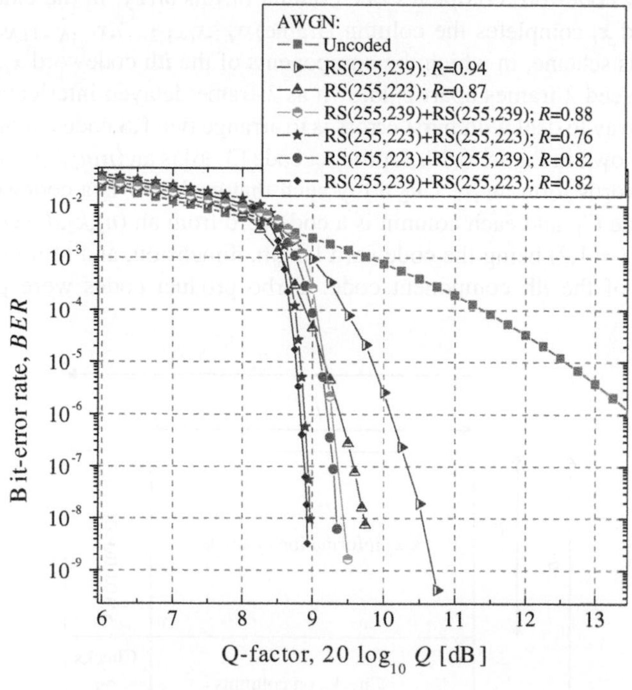

FIGURE 6.21

BER performance of concatenated RS codes on an optical on–off keying channel. The Q-factor is defined as $Q = (\mu_1 - \mu_0)/(\sigma_1 + \sigma_0)$, where μ_i and σ_i are the mean value and standard deviation corresponding to symbol i ($i = 0,1$).

iteratively decodable codes, but their description is beyond the scope of this chapter. In recent books [37,38], the author and colleagues have described several classes of iteratively decodable codes, such as turbo codes, turbo-product codes, LDPC codes, GLDPC codes, and nonbinary LDPC codes. An FPGA implementation of decoders for binary LDPC codes has also been discussed. It was then explained how to combine multilevel modulation and channel coding optimally by using coded modulation. An LDPC-coded turbo equalizer was considered as a candidate for dealing with various channel impairments simultaneously.

In Section 6.1 classical channel coding preliminaries were introduced, namely basic definitions, channel models, the concept of channel capacity, and statement of the channel coding theorem. Section 6.2 covered the basics of linear block codes, such as definitions of generator and parity-check matrices, syndrome decoding, distance properties of LBCs, and some important coding bounds. In Section 6.3 cyclic codes were introduced. The BCH codes were described in Section 6.4. The RS, concatenated, and product codes were described in Section 6.5. After this short summary section, a set of problems is provided for readers to gain a deeper understanding of classical error correction concepts.

6.7 PROBLEMS

1. Prove the following properties of mutual information:
   (a) Mutual information is *symmetric*: $I(X,Y)=I(Y,X)$.
   (b) Mutual information of a channel is *non-negative*: $I(X,Y)\geq 0$.
   (c) Mutual information of a channel is related to the joint entropy $H(X,Y)$ as follows:

$$I(X;Y)=H(X)+H(Y)-H(X,Y),$$

$$H(X,Y)=\sum_j\sum_k p(x_j;y_k)\log_2\left[\frac{1}{p(x_j;y_k)}\right].$$

2. Let us observe an M-ary input M-ary output symmetric channel, such as M-ary PPM, for which $p(y_j|x_i)=P_s/(M-1)$ and $p(y_j|x_j)=1-P_s$, where P_s is the symbol error probability. Derive the following expression for channel capacity, in bits/symbol:

$$C=\log_2 M+(1-P_s)\log_2(1-P_s)+P_s\log_2\left(\frac{P_s}{M-1}\right).$$

   Plot the channel capacity as a function of P_s by using M as a parameter.
3. By using the information capacity theorem and Fano's inequality, generate the minimum bit error rate versus signal-to-noise ratio plots for different code rates. Observe the binary-phase shift keying transmission over a zero-mean additive Gaussian noise channel.
4. The binary erasure channel (BEC) has two inputs and three outputs, as depicted in Figure 6.P4. The inputs are labeled as x_0 and x_1, and the outputs as y_0, y_1,

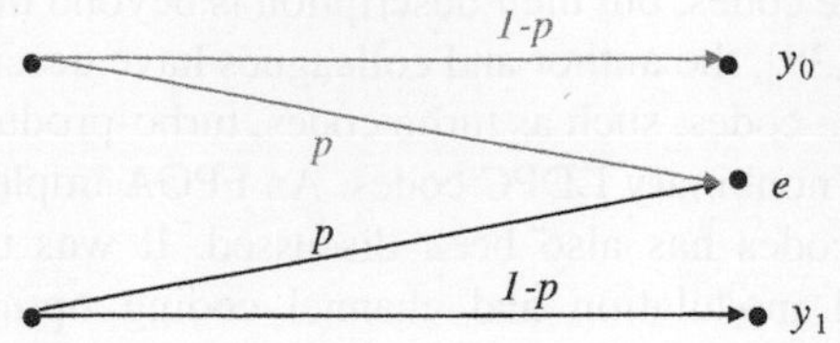

FIGURE 6.P4

Binary erasure channel (BEC) model.

and e. A fraction p of the incoming bits are erased by the channel. Determine the capacity of this channel.

5. The binary symmetric channel (BSC) is described by the following channel matrix:

$$P_{\mathrm{BSC}} = \begin{bmatrix} q & p \\ p & q \end{bmatrix}, \quad p + q = 1.$$

Determine the channel capacity of this channel and plot its dependence against the crossover probability p.

Observe now the channel that is obtained by cascading the BSC channels. Determine the channel matrix after cascading two and three BSC channels. Plot the channel capacity of cascaded channels against p. Discuss the results.

6. The binary asymmetric channel (BAC) is described by the following channel matrix:

$$P_{\mathrm{BAC}} = \begin{bmatrix} q_1 & p_1 \\ p_2 & q_2 \end{bmatrix}, \quad p_i + q_i = 1 \quad (i = 1, 2).$$

Determine the channel capacity of this channel and plot its dependence against p_1 by using p_2 as the parameter. Discuss the BAC and BSC results.

7. An (n,k) linear block code is described by the following parity-check matrix:

$$H = \begin{bmatrix} 0 & 0 & 0 & 0 & 0 & 0 & 0 & 1 & 1 & 1 & 1 & 1 & 1 & 1 & 1 \\ 0 & 0 & 0 & 1 & 1 & 1 & 1 & 0 & 0 & 0 & 0 & 1 & 1 & 1 & 1 \\ 0 & 1 & 1 & 0 & 0 & 1 & 1 & 0 & 0 & 1 & 1 & 0 & 0 & 1 & 1 \\ 1 & 0 & 1 & 0 & 1 & 0 & 1 & 0 & 1 & 0 & 1 & 0 & 1 & 0 & 1 \end{bmatrix}.$$

(a) Determine the code parameters: codeword length, number of information bits, code rate, overhead, minimum distance, and error correction capability.

(b) Represent the matrix H in systematic form and determine the generator matrix G of the corresponding systematic code.

8. Consider a (7,4) code with generator matrix:

$$G = \begin{bmatrix} 0 & 1 & 0 & 1 & 1 & 0 & 0 \\ 1 & 0 & 1 & 0 & 1 & 0 & 0 \\ 0 & 1 & 1 & 0 & 0 & 1 & 0 \\ 1 & 1 & 0 & 0 & 0 & 0 & 1 \end{bmatrix}.$$

(a) Find all the codewords of the code.
(b) What is the minimum distance of the code?
(c) Determine the parity-check matrix of the code.
(d) Determine the syndrome for the received vector [1101011].
(e) Assuming that an information bit sequence of all zeros is transmitted, find all minimum weight error patterns $\boldsymbol{e}$ that result in a valid codeword that is not the all-zero codeword.
(f) Use row and column operations to transform $\boldsymbol{G}$ to systematic form and find its corresponding parity-check matrix. Sketch a shift register implementation of this systematic code.
(g) All Hamming codes have a minimum distance of 3. What is the error correction and error detection capability of a Hamming code?

9. Consider the following set:

$$C = \{(000000), (001011), (010101), (011110), (100111), \\ (101100), (110010), (111001)\}.$$

(a) Show that C is a vector space over GF(2).
(b) Determine the dimension of this set.
(c) Determine the set of basis vectors.

If the set C is considered as the codebook of an LBC:

(d) Determine the parameters of this code.
(e) Determine the generator and parity-check matrices in systematic form.
(f) Determine the minimum distance and error correction capability of this code.

10. Let C be an (n,k) LBC of minimum Hamming distance d. Define a new code C_e by adding an additional overall parity-check equation. The code obtained is known as *extended* code.
(a) If the minimum distance of the original code is d, determine the minimum distance of the extended code.
(b) Show that the extended code is an $(n+1,k)$ code.

11. The maximum-length C^m_{dual} is the dual of $(2^m - 1, 2^m - 1 - m, 3)$ of the corresponding Hamming code. Let $m = 3$.
(a) List all codewords of this code.
(b) Show that C^3_{dual} is an (7,3,4) LBC.

12. Determine an (8,4) code obtained by extending the (7,4) Hamming code. Determine its minimum distance and error correction capability. Show that the

(8,4) code and its dual are identical. Such codes are called self-dual and are very important in quantum error correction.

13. Let $\boldsymbol{G} = [\boldsymbol{I}_k|\boldsymbol{P}]$ be the generator matrix of a systematic (n,k) LBC. Show that this code is self-dual if and only if $\boldsymbol{P}$ is a square and $\boldsymbol{P}\boldsymbol{P}^{\mathrm{T}} = \boldsymbol{I}$. Show that for dual codes $n = 2k$. Finally, design self-dual codes for $n = 4$, 6, and 8.

14. Prove the following properties of a standard array:

(a) All n-tuples of a row are distinct.

(b) Each n-tuple appears exactly once in the standard array.

(c) There are exactly 2^{n-k} rows in the standard array.

(d) For perfect codes (satisfying the Hamming bound with equality) all n-tuples of weight $t = \mathrm{int}[(d_{\min} - 1)/2]$ or less appear as coset leaders (int $[x]$ is the integer part of x).

(e) For quasi-perfect codes, in addition to all n-tuples of weight t or less, some but not all n-tuples of weight $t + 1$ appear as coset leaders.

(f) All elements in the same row (coset) have the same syndrome.

(g) Elements in different rows have different syndromes.

(h) There are 2^{n-k} different syndromes corresponding to 2^{n-k} rows.

15. For the extended code from Problem 12 determine the standard array and syndrome decoding table. If the received word was (10101010), what would be the result of standard array decoding? Determine the weight distribution of coset leaders and the word error probability for a crossover probability of BSC of 10^{-4}. Finally, determine the coding gain at target BER of 10^{-6}.

16. The (n,k) LBC can be *shortened* by deleting the information symbols, which is equivalent to deleting the parity columns in the parity-check matrix $\boldsymbol{H} = [\boldsymbol{P}^{\mathrm{T}}|\boldsymbol{I}_{n-k}]$. Let us remove all columns of even weight in $\boldsymbol{P}^{\mathrm{T}}$ of the $(2^m - 1, 2^m - 1 - m)$ Hamming code. Derive the new code parameters in terms of the original code and determine the minimum distance of the shortened code. What is the error correction capability of this shortened code?

17. The *Reed–Muller (RM)* codes have been used in deep-space communications such as Mariner 9, Voyager 1, and Voyager 2. The rth order RM code has the following generator matrix:

$$\boldsymbol{G} = \begin{bmatrix} \boldsymbol{G}_0 \\ \boldsymbol{G}_1 \\ \vdots \\ \boldsymbol{G}_r \end{bmatrix},$$

where $\boldsymbol{G}_0$ is an all-one row vector of length $n = 2^m$. The submatrix $\boldsymbol{G}_1$ has dimensions $m \times 2^m$ in which columns represent all possible m-tuples. The submatrix $\boldsymbol{G}_2$ is obtained from $\boldsymbol{G}_1$ by taking two row-products of $\boldsymbol{G}_1$, with only indices in increasing order taken into account. The product of two rows $\boldsymbol{a} = (a_1,\ldots,a_n)$ and $\boldsymbol{b} = (b_1,\ldots,b_n)$ is defined by $\boldsymbol{ab} = (a_1b_1,\ldots,a_nb_n)$. The submatrix $\boldsymbol{G}_3$ is obtained from $\boldsymbol{G}_2$ by taking three row-products of $\boldsymbol{G}_1$, and so on.

(a) For $m=4$ and $r=3$ provide the corresponding generator matrix and determine the code parameters.
(b) Show that the zero-order RM code is in fact the $(n,1)$ repetition code.
(c) Determine the minimum distance and parameters of the first-order RM codes.
(d) Prove that for any RM code the following is valid:

$$n=\sum_{i=0}^{r}\binom{m}{i}, \quad n-k=\sum_{i=0}^{m-1-i}\binom{m}{i}.$$

(e) Determine the first-order RM code for $m=3$ and the second-order RM code for $m=4$. How are these related to the Hamming codes?

18. Prove that an LBC can correct all error patterns having e_c or fewer errors and simultaneously detect all error patterns containing e_d ($e_d \geq e_c$) or fewer errors if the minimum distance d_{min} satisfies the following inequality: $d_{min} \geq e_d + e_c + 1$.

19. Let the polynomial x^7+1 over GF(2) be given. Determine the generator and parity-check polynomials of cyclic (7,4) code. By using these polynomials create the corresponding generator and parity-check matrices. Encode the information sequence (1101) using this cyclic code. Provide the encoder and syndrome circuits.

20. Design a (15,11) cyclic code that is capable of correcting single errors. Provide the generator and parity-check polynomials. Provide the corresponding Meggit decoder. Finally, provide the corresponding error-trapping decoder.

21. Consider the three-error-correcting (15,5) BCH code with generator polynomial $g(x)=1+x+x^2+x^4+x^5+x^8+x^{10}$.
(a) Decode $r(x)=1+x^5+x^6+x^7+x^9$ using the error-trapping decoder.
(b) Decode $r(x)=1+x+x^3+x^7$ using the Massey–Berlekamp algorithm.

22. In GSM systems Fire codes are used to correct the burst of errors. An error burst of length B is defined as a contiguous sequence of B bits in which the first and last bits or any other intermediate bits are received in error. The (n,k) codes for burst error correction should satisfy the *Reiger bound*: $n-k \geq 2B$. The *Fire code* over GF(q) is constructed using the following generator polynomial:

$$g(x)=(x^{2B-1}-1)p(x), \quad \deg(p(x))=m \geq B, \quad \mathrm{rem}\left[(x^{2B-1}-1)/p(x)\right] \neq 0.$$

Determine the block length of this code. For $n-k=40$ determine the parameter B and parameters corresponding to the Fire code so that the Reiger bound is satisfied.

23. This problem is also related to the Fire codes. Let us consider the primitive polynomial $p(x)=1+x+x^6$ over GF(2). Use this primitive polynomial to determine the Fire code that can correct bursts of length 7 or less.

Table 6.P25 Weight distribution of the dual of two-error-corrective primitive binary BCH codes of length $2^m - 1$

Weight w	Number of vectors with weight w, B_w
Odd **m** $\geq$ **3**	
0	1
$2^{m-1} - 2^{(m+1)/2-1}$	$(2^{m-2} + 2^{(m-1)/2-1})(2^m - 1)$
2^{m-1}	$(2^m - 2^{m-1} + 1)(2^m - 1)$
$2^{m-1} + 2^{(m+1)/2-1}$	$(2^{m-2} - 2^{(m-1)/2-1})(2^m - 1)$
Even **m** $\geq$ **4**	
0	1
$2^{m-1} - 2^{(m+2)/2-1}$	$2^{(m-2)/2-1}(2^{(m-2)/2} + 1)(2^m - 1)/3$
$2^{m-1} - 2^{m/2-1}$	$2^{(m+2)/2-1}(2^{m/2} + 1)(2^m - 1)/3$
2^{m-1}	$(2^{m-2} + 1)(2^m - 1)$
$2^{m-1} + 2^{m/2-1}$	$2^{(m+2)/2-1}(2^{m/2} - 1)(2^m - 1)/3$
$2^{m-1} + 2^{(m+2)/2-1}$	$2^{(m-2)/2-1}(2^{(m-2)/2} - 1)(2^m - 1)/3$

(modified from Ref. [15])

24. Let α be the primitive element in GF(2^4). Determine the generator polynomial and parity-check matrix of binary BCH code with error correction capability of two and codeword length of 15.

25. Consider the two-error-correcting binary BCH (1023,1003) code. By using Table 6.P25 and the MacWilliams identities determine the weight enumerator of this code. Plot $\log_{10}(A_i)$ as a function of Hamming weight.

26. Consider the RS (31,15) code.

(a) Determine the number of bits per symbol in the code and determine the block length in bits.

(b) Determine the minimum distance of the code and the error correction capability.

27. Consider the RS (n,k) code:

(a) Prove that the minimum distance is given by $d_{\min} = n - k + 1$.

(b) Prove that the weight distribution of RS codes can be determined by

$$A_i = \binom{n}{i}(q-1)\sum_{j=0}^{i-d_{\min}}(-1)^j\binom{i-1}{j}q^{i-d_{\min}-j}.$$

28. Let GF(2^3) be obtained as shown in Table 6.4.

(a) Determine the generator and parity-check polynomials of RS code capable of correcting two errors. If the binary message (011000011) is to be transmitted determine the codeword polynomial.

(b) Determine the generator polynomial of *systematic* RS code capable of correcting two errors. If the binary message (011000011) is to be transmitted determine the codeword polynomial.

29. Design the RS code over GF(2^4), given in Table 6.5, capable of correcting three errors. If the received word $\boldsymbol{r} = (000\alpha^7 00\alpha^3 00000\alpha^4 00)$ was obtained, using the Massey–Berlekamp algorithm determine the error polynomial.

30. Design a concatenated code using (15,9) three-symbol-error-correcting RS code as the outer code and (7,4) binary Hamming code as the inner code.

(a) Determine the overall codeword length and the number of binary information bits contained in the codeword.

(b) Determine the error correction capability of this code.

References

[1] C. Berrou, A. Glavieux, P. Thitimajshima, Near Shannon limit error-correcting coding and decoding: Turbo codes, in: Proc. 1993 Int. Conf. Commun. (ICC 1993), pp. 1064–1070.

[2] C. Berrou, A. Glavieux, Near optimum error correcting coding and decoding: Turbo codes, IEEE Trans. Commun. 10 (October 1996) 1261–1271.

[3] R.M. Pyndiah, Near optimum decoding of product codes, IEEE Trans. Commun. 46 (1998) 1003–1010.

[4] T. Mizuochi, Recent progress in forward error correction and its interplay with transmission impairments, IEEE J. Sel. Top. Quantum Electron. 12 (4) (July/August 2006) 544–554.

[5] R.G. Gallager, Low Density Parity Check Codes, MIT Press, Cambridge, MA, 1963.

[6] I.B. Djordjevic, S. Sankaranarayanan, S.K. Chilappagari, B. Vasic, Low-density parity-check codes for 40 Gb/s optical transmission systems, IEEE/LEOS J. Sel. Top. Quantum Electron 12 (4) (July/August 2006) 555–562.

[7] I.B. Djordjevic, O. Milenkovic, B. Vasic, Generalized low-density parity-check codes for optical communication systems, IEEE/OSA J. Lightwave Technol 23 (May 2005) 1939–1946.

[8] B. Vasic, I.B. Djordjevic, R. Kostuk, Low-density parity check codes and iterative decoding for long haul optical communication systems, IEEE/OSA J. Lightwave Technol. 21 (February 2003) 438–446.

[9] I.B. Djordjevic, et al., Projective plane iteratively decodable block codes for WDM high-speed long-haul transmission systems, IEEE/OSA J. Lightwave Technol. 22 (March 2004) 695–702.

[10] O. Milenkovic, I.B. Djordjevic, B. Vasic, Block-circulant low-density parity-check codes for optical communication systems, IEEE/LEOS J. Sel. Top. Quantum Electron 10 (March/April 2004) 294–299.

[11] B. Vasic, I.B. Djordjevic, Low-density parity check codes for long haul optical communications systems, IEEE Photon. Technol. Lett. 14 (August 2002) 1208–1210.

[12] I.B. Djordjevic, M. Arabaci, L. Minkov, Next generation FEC for high-capacity communication in optical transport networks, IEEE/OSA J. Lightwave Technol. Vol. 27, no. 16, pp. 3518–3530, August 15, 2009. (Invited paper).

[13] S. Chung, et al., On the design of low-density parity-check codes within 0.0045 dB of the Shannon limit, IEEE Commun. Lett. 5 (February 2001) 58–60.

[14] I.B. Djordjevic, L.L. Minkov, H.G. Batshon, Mitigation of linear and nonlinear impairments in high-speed optical networks by using LDPC-coded turbo equalization, IEEE J. Sel. Areas Commun., Optical Commun. Netw. 26 (6) (August 2008) 73–83.

[15] S. Lin, D.J. Costello, Error Control Coding: Fundamentals and Applications, Prentice-Hall, 1983.

[16] P. Elias, Error-free coding. IRE Trans. Inform, Theory IT-4 (September 1954) 29–37.

[17] J.B. Anderson, S. Mohan, Source and Channel Coding: An Algorithmic Approach, Kluwer Academic, Boston, MA, 1991.

[18] F.J. MacWilliams, N.J.A. Sloane, The Theory of Error-Correcting Codes, North Holland, Amsterdam, 1977.

[19] S.B. Wicker, Error Control Systems for Digital Communication and Storage, Prentice-Hall, Englewood Cliffs, NJ, 1995.

[20] G.D. Forney Jr., Concatenated Codes, MIT Press, Cambridge, MA, 1966.

[21] L.R. Bahl, J. Cocke, F. Jelinek, J. Raviv, Optimal decoding of linear codes for minimizing symbol error rate, IEEE Trans. Inform. Theory IT-20 (2) (March 1974) 284–287.

[22] D.B. Drajic, An Introduction to Information Theory and Coding, second ed., Akademska Misao, Belgrade, 2004 (in Serbian).

[23] W.E. Ryan, Concatenated convolutional codes and iterative decoding, in: J.G. Proakis (Ed.), Wiley Encyclopedia in Telecommunications, John Wiley, 2003.

[24] S. Haykin, Communication Systems, John Wiley, 2004.

[25] J.G. Proakis, Digital Communications, McGraw-Hill, Boston, MA, 2001.

[26] R.H. Morelos-Zaragoza, The Art of Error Correction Coding, John Wiley, Boston, MA, 2002.

[27] T.M. Cover, J.A. Thomas, Elements of Information Theory, John Wiley, New York, 1991.

[28] F.M. Ingels, Information and Coding Theory, Intext Educational, Scranton, 1971.

[29] I.S. Reed, G. Solomon, Polynomial codes over certain finite fields, SIAM J. Appl. Math. 8 (1960) 300–304.

[30] S.B. Wicker, V.K. Bhargva (Eds.), Reed–Solomon Codes and Their Applications, IEEE Press, New York, 1994.

[31] J.K. Wolf, Efficient maximum likelihood decoding of linear block codes using a trellis, IEEE Trans. Inform. Theory IT-24 (1) (January 1978) 76–80.

[32] B. Vucetic, J. Yuan, Turbo Codes – Principles and Applications, Kluwer Academic, Boston, MA, 2000.

[33] M. Ivkovic, I.B. Djordjevic, B. Vasic, Calculation of achievable information rates of long-haul optical transmission systems using instanton approach, IEEE/OSA J. Lightwave Technol. 25 (May 2007) 1163–1168.

[34] D. Divsalar, F. Pollara, Turbo codes for deep-space communications, TDA Progress Report 42-120 15 (February 1995) 29–39.

[35] M.E. van Valkenburg, Network Analysis, third ed., Prentice-Hall, Englewood Cliffs, NJ, 1974.

[36] I.B. Djordjevic, M. Arabaci, L. Minkov, Next generation FEC for high-capacity communication in optical transport networks, IEEE/OSA J. Lightwave Technol. 27 (16) (15 August 2009) 3518–3530 (invited paper).

[37] W. Shieh, I. Djordjevic, OFDM for Optical Communications, Elsevier, October 2009.

[38] I.B. Djordjevic, W. Ryan, B. Vasic, Coding for Optical Channels, Springer, March 2010.

[39] M.F. Barsoum, B. Moision, M. Fitz, D. Divsalar, J. Hamkins, Iterative coded pulse-position-modulation for deep-space optical communications, in: Proc. ITW 2007, Lake Tahoe, CA, 2–6 September 2007, pp. 66–71.

[40] D. Raghavarao, Constructions and Combinatorial Problems in Design of Experiments, Dover, New York, 1988 (reprint).

[41] C.C. Pinter, A Book of Abstract Algebra, Dover, New York, 2010 (reprint).

CHAPTER

Quantum Error Correction

7

CHAPTER OUTLINE

Quantum Information Processing and Quantum Error Correction. DOI: 10.1016/B978-0-12-385491-9.00007-1

This chapter considers quantum error correction codes (QECCs) [1–25]. The concept of this chapter is to gradually introduce the reader to the quantum error correction coding principles, moving from an intuitive description to a rigorous mathematical description. The chapter starts with Pauli operators, basic definitions, and representation of quantum errors. Although the Pauli operators were introduced in Chapter 2, they are re-introduced here in the context of quantum errors and quantum error correction. Next, basic quantum codes, such as three-qubit flip code, three-qubit phase flip code, and Shor's nine-qubit code, are presented. Projection measurements are used to determine the error syndrome and perform corresponding error correction actions. Further, the stabilizer formalism and the stabilizer group are introduced. The basic stabilizer codes are described as well. The whole of the next chapter is devoted to the stabilizer codes; here only the basic concepts are introduced. An important class of codes, the class of Calderbank–Shor–Steane (CSS) codes [1,2], is described next. The connection between classical and quantum codes is then established, and two classes of CSS codes, dual-containing and quantum codes derived from classical codes over GF(4), are described. The concept of quantum error correction is then formally introduced, followed by the necessary and sufficient conditions for quantum code to correct a given set of errors. Then, the minimum distance of a quantum code is defined and used to relate it to the error correction capability of a quantum code. The CSS codes are then revisited by using this mathematical framework. Then, important quantum coding bounds, such as the Hamming quantum bound, quantum Gilbert–Varshamov bound, and quantum Singleton bound (also known as the Knill–Laflamme bound [17]), are discussed. Next, the concept of operator-sum representation is introduced, and used to provide a physical interpretation and to describe the measurement of the environment. Finally, several important quantum channel models are introduced; such as the depolarizing channel, amplitude damping channel, and generalized amplitude damping channel. After the summary section, a set of problems for self-study is provided, which enables the reader to better understand the underlying concepts of quantum error correction.

7.1 PAULI OPERATORS (REVISITED)

The *Pauli operators* $G = \{I,X,Y,Z\}$, as already introduced in Chapter 2, can be represented in matrix form as follows:

$$X = \sigma_x \doteq \begin{bmatrix} 0 & 1 \\ 1 & 0 \end{bmatrix}, \quad Y = \sigma_y \doteq \begin{bmatrix} 0 & -j \\ j & 0 \end{bmatrix}, \quad Z = \sigma_z \doteq \begin{bmatrix} 1 & 0 \\ 0 & -1 \end{bmatrix}. \tag{7.1}$$

Their action on qubit $|\psi\rangle = a|0\rangle + b|1\rangle$ can be described as:

$$\begin{aligned} X(a|0\rangle + b|1\rangle) &= a|1\rangle + b|0\rangle \quad Y(a|0\rangle + b|1\rangle) = j(a|1\rangle - b|0\rangle) \\ Z(a|0\rangle + b|1\rangle) &= a|0\rangle - b|1\rangle. \end{aligned} \tag{7.2}$$

Therefore, the action of the X operator is to introduce the bit flip, the action of the Z operator is to introduce the phase flip, and the action of the Yoperator is to introduce simultaneously the bit and phase flips. The properties of Pauli operators can be summarized as:

$$X^2 = I,\quad Y^2 = I,\quad Z^2 = I$$
$$XY = jZ \quad YX = -jZ \quad YZ = jX \quad ZY = -jX \quad ZX = jY \quad XZ = -jY. \tag{7.3}$$

The properties of Pauli operators given by (7.3) can also be summarized by the following multiplication table:

×	I	X	Y	Z
I	I	X	Y	Z
X	X	I	jZ	−jY
Y	Y	−jZ	Y	jX
Z	Z	jY	−jX	I

It is interesting to notice that two Pauli operators commute only if they are identical or one of them is the identity operator, otherwise they anticommute. Because the set G is not closed under multiplication, it is not a multiplicative group. However, from Chapter 2 we know that two states $|\psi\rangle$ and $e^{j\theta}|\psi\rangle$ that differ only in global phase shift cannot be distinguished in quantum mechanics because the results of a measurement are the same. Therefore, often in quantum error correction we can simply omit the imaginary unit j. The corresponding multiplication table in which the imaginary unit is ignored is as follows:

×	I	X	Y	Z
I	I	X	Y	Z
X	X	I	Z	Y
Y	Y	Z	Y	X
Z	Z	Y	X	I

Clearly, all operators now commute and the corresponding set with such defined multiplication is a commutative (Abelian) group. Such a group can be called a "projective" Pauli group [15], and it is isomorphic to the group of binary two-tuples $(Z_2)^2 = \{00,01,10,11\}$ with the addition table given by:

+	00	01	11	10
00	00	01	11	10
01	01	00	10	11
11	11	10	00	01
10	10	11	01	00

The projective Pauli group is also isomorphic to the quartenary group $F_4 = \{0, 1, \omega, \varpi\}$, $\varpi = 1 + \omega$, with corresponding addition table given by:

+	0	ϖ	1	ω
0	0	ϖ	1	ω
ϖ	ϖ	0	ω	1
1	1	ω	0	ϖ
ω	ω	1	ϖ	0

Based on these tables we can establish the following correspondence among the projective Pauli group G', $(Z_2)^2$, and F_4:

G'	$(Z_2)^2$	F_4
I	00	0
X	01	ϖ
Y	11	1
Z	10	ω

This correspondence can be used to relate the quantum codes to classical codes over $(Z_2)^2$ and F_4.

In addition to Pauli operators, the Hadamard operator (gate) will also be used a lot in this chapter. Its action can be described as:

$$\begin{aligned} H(a|0\rangle + b|1\rangle) &= a\frac{1}{\sqrt{2}}(|0\rangle + |1\rangle) + b\frac{1}{\sqrt{2}}(|0\rangle - |1\rangle) \\ &= \frac{1}{\sqrt{2}}[(a+b)|0\rangle + (a-b)|1\rangle]. \end{aligned} \tag{7.4}$$

The quantum error correction code can be defined as mapping from a K-qubit space to an N-qubit space. To facilitate its definition, we introduce the concept of Pauli operators. A *Pauli operator on N qubits* has the form [12] $cO_1O_2\ldots O_N$, where each $O_i \in \{I,X,Y,Z\}$ and $c = j^l$ $(l = 1,2,3,4)$. This operator takes $|i_1i_2\ldots i_N\rangle$ to $cO_1|i_1\rangle \otimes O_2|i_2\rangle \ldots \otimes O_N|i_N\rangle$. For example, the action of $IXZ(|000\rangle + |111\rangle) = |010\rangle - |101\rangle$ is to bit-flip the second qubit and phase-flip the third qubit if it was $|1\rangle$. For convenience, we will also often use a shorthand notation for representing Pauli operators, in which only the non-identity operators O_i are written, while the identity operators are assumed. For example, $IXIZI$ can be denoted as X_2Z_4, meaning that operator X acts on the second qubit and operator Z on the fourth qubit (the action of identity operators I to other qubits is simply omitted since it does not cause any change). Two Pauli operators *commute* if and only if there is an even number of places where they have different Pauli matrices, neither of which is the identity I. For example, XXI and IYZ do not commute, whereas XXI and ZYX do commute. If two

Pauli operators do not commute they anticommute, since their individual Pauli matrices either commute or anticommute.

The set of Pauli operators on N qubits form the *multiplicative Pauli group* G_N. For a multiplicative group we can define the *Clifford operator* [8] U as the operator that preserves the elements of the Pauli group under conjugation, namely $\forall\ O \in G_N : UOU^{\dagger} \in G_N$. The encoded operator for quantum error correction typically belongs to the Clifford group. To implement any unitary operator from the Clifford group, the use of CNOT U_{CNOT}, Hadamard H, and phase gate P is sufficient. The operation of the H gate is already given by (7.4), the action of the P gate can be described as $P(a|0\rangle + b|1\rangle) = a|0\rangle + jb|1\rangle$, and the action of the CNOT gate can be described as $U_{\text{CNOT}}|i\rangle|j\rangle = |i\rangle|i \oplus j\rangle$.

Every quantum channel error E (be it either discrete or continuous) can be described as a superposition of elements from the discrete set $\{I,X,Y,Z\}$, given by $E = e_1 I + e_2 X + e_3 Y + e_4 Z$. For example, the error $E = \begin{bmatrix} 1 & 0 \\ 0 & 0 \end{bmatrix}$ can be represented in terms of Pauli operators as $E = (I + Z)/2$. An error operator that affects several qubits can be written as a weighted sum of Pauli operators $\Sigma c_i P_i$ acting on the ith qubits. An error may act not only on the code qubits but also on the environment. Given an initial state $|\psi\rangle|\phi\rangle^{\text{e}}$, which is a tensor product of code qubits $|\psi\rangle$ and environmental states $|\phi\rangle^{\text{e}}$, any error acting on both the code and the environment can be written as a weighted sum $\Sigma c_{i,j} P_i P_j^{\text{e}}$ of Pauli operators that act on both code and environment qubits. If S_i are syndrome operators that identify the error term $P_{\alpha'}|\psi\rangle$, then the operators $S_i I^{\text{e}}$ will pick up terms of the form $\Sigma_{\beta'} c_{\alpha',\beta'} P_{\alpha'}|\psi\rangle P_{\beta'}^{\text{e}}|\phi\rangle^{\text{e}}$ from the quantum noise affected state, and these terms can be written as $P_{\alpha'}|\psi\rangle|\mu\rangle^{\text{e}}$ for some new environmental state $|\mu\rangle^{\text{e}}$. Therefore, the measurement of the syndrome restores a tensor product state of qubits and environment, suggesting that the code and the environment evolve independently of each other.

7.2 QUANTUM ERROR CORRECTION CONCEPTS

Quantum error correction is essentially more complicated than classical error correction. Difficulties of quantum error correction can be summarized as follows. (i) The no-cloning theorem indicates that it is impossible to make a copy of an arbitrary quantum state $|\psi\rangle = \alpha|0\rangle + \beta|1\rangle$. For this to be possible it is clear from $|\psi\rangle|\psi\rangle = \alpha^2|00\rangle + \alpha\beta|01\rangle + \alpha\beta|10\rangle + \beta^2|11\rangle$, that $\alpha\beta$ must be equal to 0, so that the state $|\psi\rangle$ is no longer arbitrary. (ii) Quantum errors are continuous and a qubit can be in any superposition of the two basis states. (iii) Measurements destroy quantum information. Quantum error correction consists of four major steps: encoding, error detection, error recovery, and decoding. The elements of quantum error correction codes are shown in Figure 7.1. Based on a discussion of quantum errors and Pauli operators on N qubits, the quantum error correction code can be defined as follows. The $[N,K]$ *quantum error correction code* performs encoding of the quantum state of K qubits,

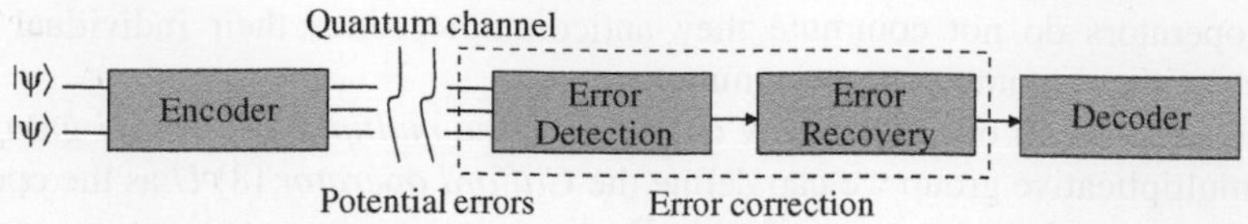

FIGURE 7.1

Quantum error correction principle.

specified by 2^K complex coefficients α_s ($s = 0,1, \ldots, 2^K - 1$), into a quantum state of N qubits, in such a way that errors can be detected and corrected, and all 2^K complex coefficients can be perfectly restored, up to the global phase shift. The sender (Alice) encodes quantum information in state $|\psi\rangle$ with the help of local ancillary qubits $|0\rangle$, and then sends the encoded qubits over a noisy quantum channel (say a free-space optical channel or optical fiber). The receiver (Bob) performs multi-qubit measurement on all qubits to diagnose the channel error and performs a recovery unitary operation R to reverse the action of the channel. The principles of quantum error correction will be more evident after several simple quantum codes provided below.

7.2.1 Three-Qubit Flip Code

Assume we want to send a single qubit $|\Psi\rangle = \alpha|0\rangle + \beta|1\rangle$ through the quantum channel in which, during transmission, the transmitted qubit can be flipped to $X|\Psi\rangle = \beta|0\rangle + \alpha|1\rangle$ with probability p. Such a quantum channel is called a *bit-flip channel* and it can be described as shown in Figure 7.2a.

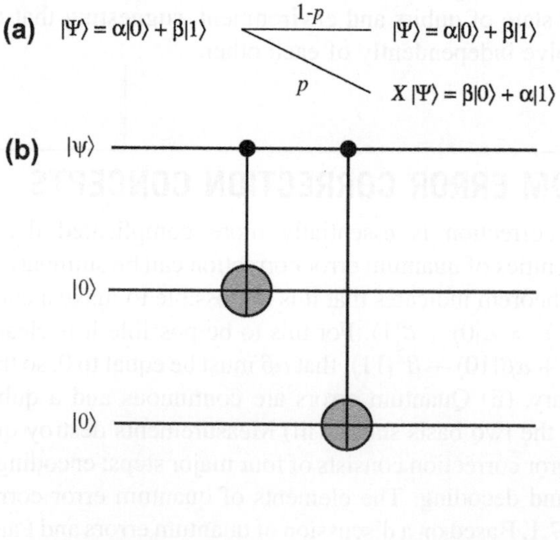

FIGURE 7.2

(a) The bit-flipping channel model. (b) The three-qubit flip-code encoder.

Three-qubit flip code sends the same qubit three times, and therefore represents the repetition code equivalent. The corresponding codewords in this code are $|\bar{0}\rangle = |000\rangle$ and $|\bar{1}\rangle = |111\rangle$. The three-qubit flip-code encoder is shown in Figure 7.2b. One input qubit and two ancillaries are used at the input encoder, which can be represented by $|\psi_{123}\rangle = \alpha|000\rangle + \beta|100\rangle$. The first ancillary qubit (the second qubit at the encoder input) is controlled by the information qubit (the first qubit at the encoder input) so that its output can be represented by $\text{CNOT}_{12}(\alpha|000\rangle + \beta|100\rangle) = \alpha|000\rangle + \beta|110\rangle$ (if the control qubit is $|1\rangle$ the target qubit is flipped, otherwise it stays unchanged). The output of the first CNOT gate is used as input to the second CNOT gate in which the second ancillary qubit (the third qubit) is controlled by the information qubit (the first qubit) so that the corresponding encoder output is obtained as $\text{CNOT}_{13}(\alpha|000\rangle + \beta|110\rangle) = \alpha|000\rangle + \beta|111\rangle$, which indicates that the basis codewords are indeed $|\bar{0}\rangle$ and $|\bar{1}\rangle$.

With this code, we are able to correct a single qubit flip, which occurs with probability $(1-p)^3 + 3p(1-p)^2 = 1 - 3p^2 + 2p^3$. Therefore, the probability of an error remaining uncorrected or wrongly corrected with this code is $3p^2 - 2p^3$. It is clear from Figure 7.3 that the three-qubit bit-flip encoder is a *systematic encoder* in which the information qubit is unchanged, and the ancillary qubits are used to impose the encoding operation and create the parity qubits (the output qubits 2 and 3).

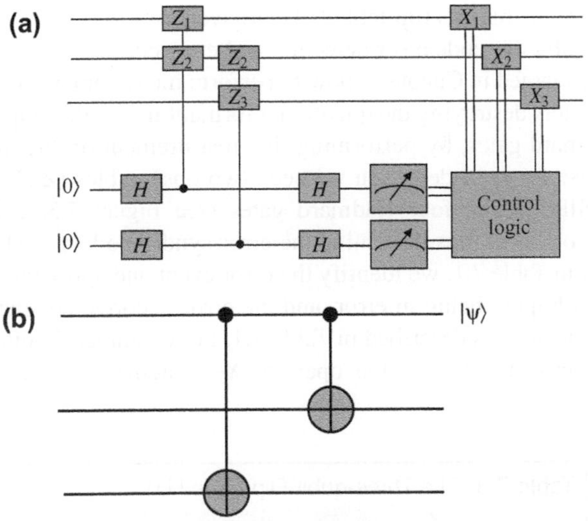

FIGURE 7.3

(a) Three-qubit flip code error detection and error correction circuit. (b) Decoder circuit configuration.

Example. Let us assume that a qubit flip occurred on the first qubit leading to received quantum word $|\psi_r\rangle = \alpha|100\rangle + \beta|011\rangle$. The error correction, as indicated above, consists of two steps. The first step is error detection, in which measurement is made with projection operators P_i ($i = 0,1,2,3$), defined as

$$P_0 = |000\rangle\langle 000| + |111\rangle\langle 111| \qquad P_1 = |100\rangle\langle 100| + |011\rangle\langle 011|$$
$$P_2 = |010\rangle\langle 010| + |101\rangle\langle 101| \qquad P_3 = |001\rangle\langle 001| + |110\rangle\langle 110|.$$

The projection operator P_i determines whether bit-flip error occurs on the ith qubit location, and P_0 means there is no bit-flip error at all. The syndrome measurements give the result $\langle\psi_r|P_0|\psi_r\rangle = 0$, $\langle\psi_r|P_1|\psi_r\rangle = 1$, $\langle\psi_r|P_2|\psi_r\rangle = 0$, $\langle\psi_r|P_3|\psi_r\rangle = 0$, which can be represented as syndrome vector $S = [0\ 1\ 0\ 0]$, indicating that the error occurred in the first qubit. The second step is error recovery, in which we flip the first qubit back to the original one by applying the X_1 operator to it. Another approach would be to perform measurements on the observables Z_1Z_2 and Z_2Z_3. The result of measurement is the eigenvalue ± 1, and corresponding eigenvectors are two valid codewords, namely $|000\rangle$ and $|111\rangle$. The observables can be represented as follows:

$$Z_1Z_2 = (|00\rangle\langle 11| + |11\rangle\langle 11|) \otimes I - (|01\rangle\langle 01| + |10\rangle\langle 10|) \otimes I$$
$$Z_2Z_3 = I \otimes (|00\rangle\langle 11| + |11\rangle\langle 11|) - I \otimes (|01\rangle\langle 01| + |10\rangle\langle 10|).$$

It can be shown that $\langle\psi_r|Z_1Z_2|\psi_r\rangle = -1$, $\langle\psi_r|Z_2Z_3|\psi_r\rangle = +1$, indicating that an error occurred on either the first or second qubit, but not on the second or third qubit. The intersection reveals that the first qubit was in error. By using this approach we can create a three-qubit look-up table (LUT), given as Table 7.1.

The three-qubit flip-code error detection and error correction circuit is shown in Figure 7.3. We learned in Chapter 3 how to perform measurements of an observable without completely destroying the quantum information with the help of an ancillary and two Hadamard gates, by performing the measurement on the ancillary qubits instead. Because for error detection we need two observables (Z_1Z_2 and Z_2Z_3), we need two ancillaries and four Hadmard gates (see Figure 7.3a). The results of measurements on ancillaries determine the error syndrome $[\pm 1\ \pm 1]$ and, based on the LUT given in Table 7.1, we identify the error event and apply the corresponding X_i gate on the ith qubit being in error, and the error is corrected since $X^2 = I$. The control logic operation is described in Table 7.1. For example, if both outputs at the measurements circuits are -1, the operator X_2 is activated. The last step is to

Table 7.1 The Three-qubit Flip-Code LUT

Z_1Z_2	Z_2Z_3	Error
+1	+1	I
+1	−1	X_3
−1	+1	X_1
−1	−1	X_2

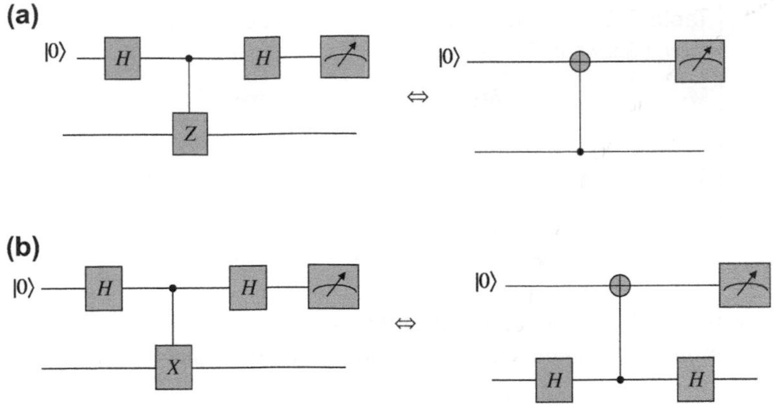

FIGURE 7.4

The equivalent circuits for measuring: (a) the *Z* operator; (b) the *X* operator.

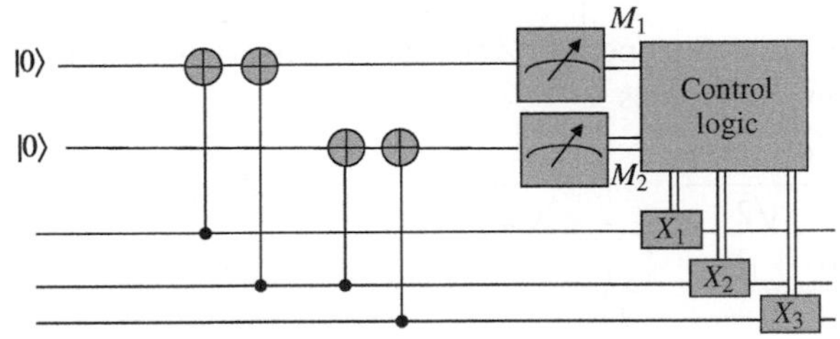

FIGURE 7.5

The equivalent three-qubit flip-code error detection and error correction circuit.

perform decoding, as shown in Figure 7.3b, by simply reversing the order of elements in the corresponding encoder.

From the universal quantum gates section we know that CNOT and Hadamard gates are sufficient for many quantum error correction codes. Therefore, we can use the equivalent circuits for measuring the Z and X operators shown in Figure 7.4a and b respectively to implement the three-qubit flip-code error detection and correction circuits, as shown in Figure 7.5. The corresponding LUT for this representation is given as Table 7.2. The control logic operates as described in this table. For example, if $M_1 = M_2 = 1$, the operator X_2 is activated.

7.2.2 Three-Qubit Phase-Flip Code

Assume we want to send a single qubit $|\Psi\rangle = \alpha|0\rangle + \beta|1\rangle$ through the quantum channel in which, during transmission, the qubit can be phase-flipped as follows,

Table 7.2 The Three-qubit Code Syndrome LUT for Equivalent Circuit Shown in Figure 7.5

M_1	M_2	Error
0	0	I
1	0	X_1
1	1	X_2
0	1	X_3

$Z|\Psi\rangle = \alpha|0\rangle - \beta|1\rangle$, with certain probability p. This quantum channel is known as the quantum *phase-flip channel*, and the corresponding channel model is shown in Figure 7.6.

In order to protect against phase-flip errors, we work in a diagonal basis (DB) instead:

$$|+\rangle = \frac{|0\rangle + |1\rangle}{\sqrt{2}} \qquad |-\rangle = \frac{|0\rangle - |1\rangle}{\sqrt{2}}. \tag{7.5}$$

In this basis, the phase-flip operator Z acts as an ordinary qubit-flip operator because the action on new base states is

$$\begin{aligned} Z|+\rangle &= \frac{Z|0\rangle + Z|1\rangle}{\sqrt{2}} = \frac{|0\rangle - |1\rangle}{\sqrt{2}} = |-\rangle \qquad Z|-\rangle = \frac{Z|0\rangle - Z|1\rangle}{\sqrt{2}} \\ &= \frac{|0\rangle + |1\rangle}{\sqrt{2}} = |+\rangle. \end{aligned} \tag{7.6}$$

The alternative syndrome measurement for this code can therefore be described as $H^{\otimes 3}Z_1Z_2H^{\otimes 3} = X_1X_2$ and $H^{\otimes 3}Z_2Z_3H^{\otimes 3} = X_2X_3$. For base conversion we can use the Hadamard gate, whose action is described by Eq. (7.4). The three-qubit phase-flip encoding circuit, shown in Figure 7.7, can be implemented using the three-qubit flip encoder shown in Figure 7.3 followed by three Hadamard gates to perform the basis conversion.

7.2.3 Shor's Nine-Qubit Code

The key idea of Shor's nine-qubit code is to first encode the qubit using the phase-flip encoder and then to encode each of the three resulting qubits again with the

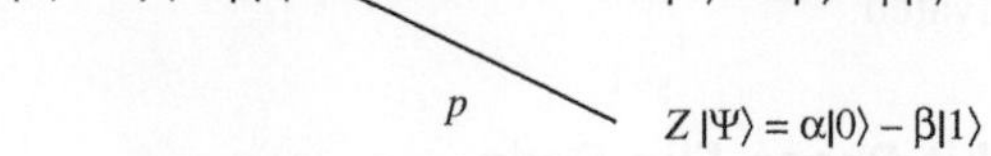

FIGURE 7.6

The phase-flipping channel model.

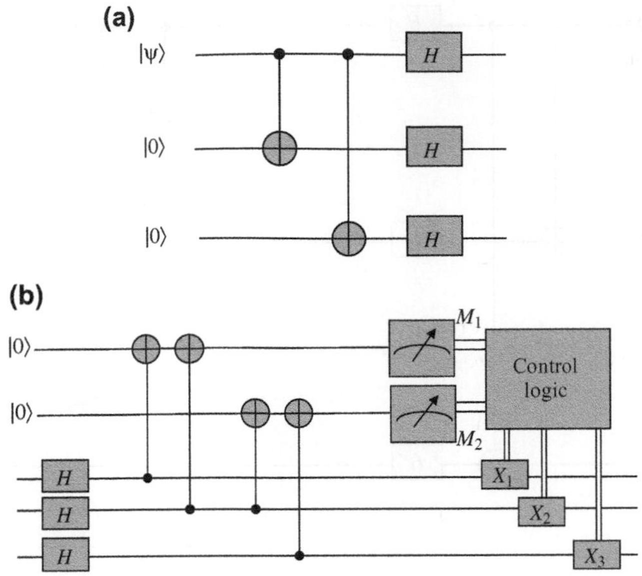

FIGURE 7.7

Three-qubit phase-flip: (a) encoder circuit; (b) error detection and recovery circuits.

bit-flip encoder, therefore providing protection against both qubit-flip and phase-flip errors. The two base codewords can be obtained, based on this description, as follows:

$$
\begin{aligned}
|\bar{0}\rangle &= \frac{1}{2\sqrt{2}}(|000\rangle + |111\rangle)(|000\rangle + |111\rangle)(|000\rangle + |111\rangle) \\
&= \frac{1}{2\sqrt{2}}(|000000000\rangle + |000000111\rangle + |000111000\rangle + |111000000\rangle \\
&\quad + |000111111\rangle + |111111000\rangle + |111000111\rangle + |111111111\rangle) \\
|\bar{1}\rangle &= \frac{1}{2\sqrt{2}}(|000\rangle - |111\rangle)(|000\rangle - |111\rangle)(|000\rangle - |111\rangle) \\
&= \frac{1}{2\sqrt{2}}(|000000000\rangle - |000000111\rangle - |000111000\rangle - |111000000\rangle \\
&\quad + |000111111\rangle + |111111000\rangle + |111000111\rangle + |111111111\rangle).
\end{aligned} \tag{7.7}
$$

The Shor encoder, shown in Figure 7.8, is composed of two stages: stage (i) is the phase-flip encoder and in stage (ii) every qubit from the phase-flip encoder is further

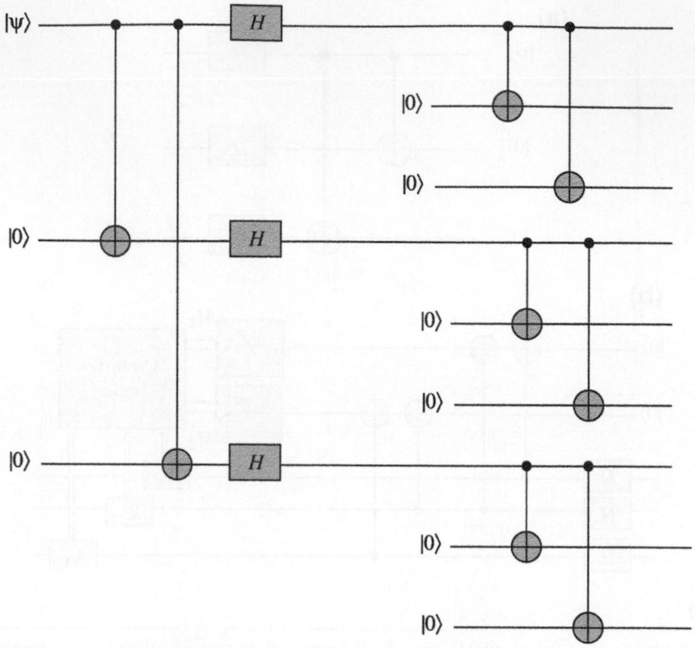

FIGURE 7.8

Nine-qubit Shor code encoding circuit.

encoded using the qubit-flip encoder. Therefore, this code can be considered as a *concatenated code*. Shor's nine-qubit error correction and decoder circuit, shown in Figure 7.9, is composed of two stages: (i) the first stage comprises three three-qubit bit-flip detector, recovery, and decoder circuits followed by three Hadamard gates to perform CB conversion; and (ii) an additional three-qubit bit-flip error detector, recovery, and decoder circuit.

Because the component codes in Shor's code are systematic, while the overall code is not systematic, the first, fourth, and seventh qubits are used, after CB conversion, as inputs to the second stage three-qubit bit-flip code detector, recovery, and detector circuit.

After this introductory treatment of quantum error correction, in the next section we describe a very important class of quantum error correction codes, namely stabilizer codes. The stabilizer codes are described in this section on a conceptual level and in Section 7.3 in a more formal way. Because this is an important class of quantum codes, the whole of the next chapter is devoted to the stabilizer codes.

7.2.4 Stabilizer Code Concepts

To facilitate the description of stabilizer codes we introduce several definitions. The Pauli group over one qubit is defined as $G_1 = \{\pm I, \pm jI, \pm X, \pm jX, \pm Y, \pm jY, \pm Z, \pm jZ\}$.

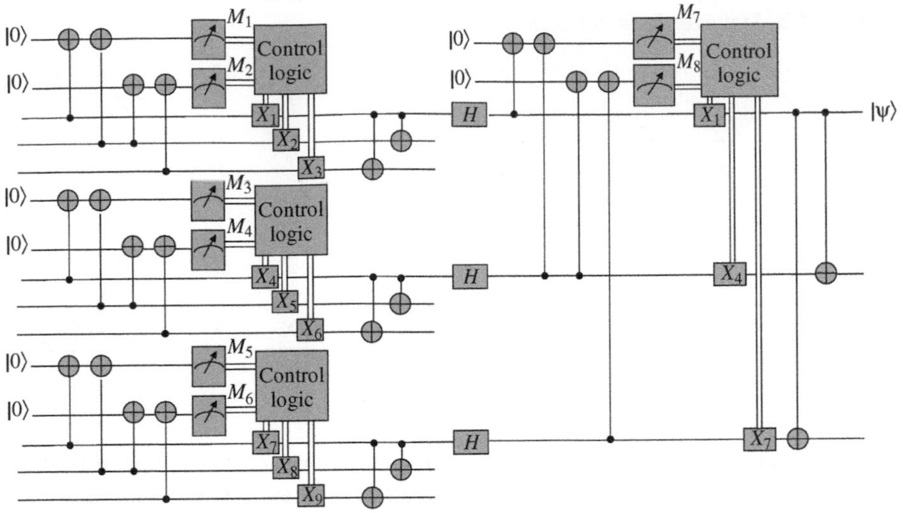

FIGURE 7.9

Error correction and decoding circuit for nine-qubit Shor code.

The Pauli group G_N on N qubits consists of all Pauli operators of the form $cO_1O_2\ldots O_N$, where $O_i \in \{I, X, Y, \text{or } Z\}$, and $c \in \{1,-1,j,-j\}$. Therefore, any Pauli vector on N qubits can be written uniquely as the product of X- and Z-containing operators together with the phase factor (± 1 or $\pm j$). In the stabilizer framework, a codeword is defined to be a ket $|\psi\rangle$, which is a $+1$ eigenket of all stabilizers s_i, so that $s_i|\psi\rangle = |\psi\rangle$ for all i. In other words, the set of operators s_i that "stabilizes" ("fixes") the codeword states forms the group S that is called the stabilizer group. A *stabilizer group* S consists of a set of Pauli matrices (X, Y, Z together with multiplicative factors ± 1, $\pm j$), with a property that any two operators in group S commute so that they can be measured simultaneously. We say that an operator fixes the state if the state is an eigenket with eigenvalue $+1$ for this operator.

Example. An alternative syndrome measurement basis for Shor's code is given by Z_1Z_2; Z_2Z_3; Z_3Z_4; Z_4Z_5; Z_5Z_6; Z_7Z_8; Z_8Z_9; $X_1X_2X_3X_4X_5X_6$; and $X_4X_5X_6X_7X_8X_9$. The Z syndromes are used for bit-flip error detection and the X syndromes for phase-flip error detection. The key idea of the stabilizer formalism is that quantum states can be more efficiently described by working on the operators that stabilize them than by working explicitly on the state. All operators that fix the base codewords of Shor's code $|\bar{0}\rangle$ and $|\bar{1}\rangle$, introduced earlier can be written as a product of eight operators M_i, shown in Table 7.3.

The stabilizer S is an Abelian subgroup of G_N. Let V_S denote the vector space stabilized by S, i.e. the set of states on N qubits that are fixed by every element from S. The stabilizer S is described by its generators g_i. Any subgroup of G_N can be used as stabilizer, providing that the following two conditions are satisfied: (i) the elements of S commute with each other and (ii) $-I$ is not an element of S.

Table 7.3 Stabilizer Table for Shor's Code

	1	2	3	4	5	6	7	8	9	
M_1	Z	Z	I	I	I	I	I	I	I	Z_1Z_2
M_2	I	Z	Z	I	I	I	I	I	I	Z_2Z_3
M_3	I	I	I	Z	Z	I	I	I	I	Z_4Z_5
M_4	I	I	I	I	Z	Z	I	I	I	Z_5Z_6
M_5	I	I	I	I	I	I	Z	Z	I	Z_7Z_8
M_6	I	I	I	I	I	I	I	Z	Z	Z_8Z_9
M_7	X	X	X	X	X	X	I	I	I	$X_1X_2X_3X_4X_5X_6$
M_8	I	I	I	X	X	X	X	X	X	$X_4X_5X_6X_7X_8X_9$

An $[N,K]$ *stabilizer code* is defined as a vector subspace V_S stabilized by $S = \langle g_1, g_2, \ldots, g_{N-K}\rangle$. Any element s from S can be written as a unique product of powers of the generators, as described in Section 7.3. This interpretation is very similar to classical error correction described in the previous chapter, in which the basis vectors g_i in code space have a similar role as stabilizer generators in quantum error correction.

Example. Let us observe the Bell (EPR) state $|\psi\rangle = (|00\rangle + |11\rangle)/\sqrt{2}$. It can be easily shown that $X_1X_2|\psi\rangle = |\psi\rangle$ and $Z_1Z_2|\psi\rangle = |\psi\rangle$. Therefore, the EPR state is stabilized by X_1X_2 and Z_1Z_2. It can be shown that $|\psi\rangle$ is the unique state that is stabilized by the two operators.

Let us analyze what is going to happen when a unitary gate U is applied to a vector subspace V_S stabilized by S:

$$s \in S: \; U|\psi\rangle = Us|\psi\rangle = UsU^{+}U|\psi\rangle = (UsU^{+})U|\psi\rangle. \quad (7.8)$$

Therefore, the state $U|\psi\rangle$ is stabilized by UsU^{+}. We say the set of U such that $UG_NU^{+} = G_N$ is the *normalizer* of G_N, denoted by $N(G_N)$. With this formalism we can define the *distance* of a stabilizer code to be the minimum weight of an element of $N(S) - S$. More details on stabilizer groups and stabilizer codes can be found in Chapter 8 (see also Section 7.3). In the next section, we turn our attention to establishing the relationship between quantum and classical codes.

7.2.5 Relationship Between Quantum and Classical Codes

As was indicated above, given any Pauli operator on N qubits, we can write it uniquely as a product of an X-containing operator and a Z-containing operator and a phase factor (± 1, $\pm j$). For instance, $XIYZYI = -(XIXIXI) \cdot (IIZZZI)$. We can now express the X operator as a binary string of length N, with "1" standing for X and "0" for I, and do the same for the Z operator. Thus, each stabilizer can be written as the X string followed by the Z string, giving a matrix of width $2N$. We mark the boundary between the two types of strings with vertical bars, so, for instance, the set of generators of Shor's code appears as the *quantum-check matrix A*. The quantum-check matrix of Shor's code (see Table 7.3) is given by

$$\mathbf{A} = \begin{bmatrix} X \qquad Z \\ 111111000|000000000 \\ 000111111|000000000 \\ 000000000|110000000 \\ 000000000|011000000 \\ 000000000|000110000 \\ 000000000|000011000 \\ 000000000|000000110 \\ 000000000|000000011 \end{bmatrix}.$$

The commutativity of stabilizers now appears as *orthogonality of rows* with respect to a *twisted* (*sympletic*) *product*, formulated as follows. If the kth row is $r_k = (x_k;z_k)$, where x_k is the X binary string and z_k the Z string, then the twisted product of rows k and l is defined by

$$r_k \odot r_l = x_k \cdot z_l + x_l \cdot z_k \text{mod} 2, \tag{7.9}$$

where $x_k \cdot z_l$ is the dot (scalar) product defined by $x_k \cdot z_l = \sum_j x_{kj} z_{lj}$. The twisted product is zero if and only if there is an even number of places where the operators corresponding to rows k and l differ (and neither are the identity), i.e. if the operators commute. If we write the quantum check $\mathbf{A}$ as $\mathbf{A} = (\mathbf{A}_1|\mathbf{A}_2)$, then the condition that the twisted product is zero for all k and l can be written compactly as

$$\mathbf{A}_1\mathbf{A}_2^{\mathrm{T}} + \mathbf{A}_2\mathbf{A}_1^{\mathrm{T}} = \mathbf{0}. \tag{7.10}$$

A *Pauli error operator E* can be interpreted as a binary string $\boldsymbol{e}$ of length $2N$, in which we reverse the order of X and Z strings. For example, $E = Z_1X_2Y_9$ can be written as $\boldsymbol{e} = [100000001|010000001]$. Thus, the quantum syndrome for the noise is exactly the classical syndrome $\boldsymbol{e}\mathbf{A}^{\mathrm{T}}$, considering $\mathbf{A}$ as a parity-check matrix and $\boldsymbol{e}$ as binary noise vector. In this formalism, we perform the ordinary dot product (mod 2) of error vector $\boldsymbol{e}$ with the corresponding row of the quantum-check matrix; if the result is 0 the error operator and corresponding stabilizer of the row commute, otherwise the result is 1 and they do not commute.

In conclusion, the properties of stabilizer codes can be inferred from those of a special class of classical codes. Given any binary matrix of size $M \times 2N$ that has the property that the twisted product of any two rows is zero, an equivalent quantum code can be constructed that encodes $N - M$ qubits in N qubits. Several examples are provided in the next three subsections to illustrate this representation.

7.2.6 Quantum Cyclic Codes

An $(N;K) = (5;1)$ quantum code is generated by the following four stabilizers: *XZZXI*, *IXZZX*, *XIXZZ*, and *ZXIXZ*. The corresponding quantum-check matrix is

$$A = \begin{bmatrix} X \quad Z \\ 10010|01100 \\ 01001|00110 \\ 10100|00011 \\ 01010|10001 \end{bmatrix}.$$

A correctable set of errors consists of all operators with one non-identity term, e.g. *XIIII*, *IIIYI*, or *IZIII*. These correspond to binary strings such as 00000|10000 for *XIIII*, 00010|00010 for *IIIYI*, and so on. There are 15 of these, and each has a distinct syndrome, thereby using $2^4 - 1$ possible nonzero syndromes. Because this code satisfies the quantum Hamming inequality (see Section 7.4) with equality, it belongs to the class of *perfect quantum codes.*

7.2.7 Calderbank–Shor–Steane (CSS) Codes

An important class of codes, invented by Calderbank, Shor and Steane, and known as the class of CSS codes [1,2], has the form:

$$A = \begin{bmatrix} H & | & 0 \\ 0 & | & G \end{bmatrix}, \qquad HG^{\mathrm{T}} = \mathbf{0}, \tag{7.11}$$

where H and G are $M \times N$ matrices. The condition $HG^{\mathrm{T}} = \mathbf{0}$ ensures that the twisted product condition is satisfied. As there are $2M$ stabilizer conditions applying to N qubit states, $N - 2M$ qubits are encoded in N qubits. Special cases of CSS codes are dual-containing codes, also known as weakly self-dual codes, in which $H = G$, so that the quantum-check matrix A has the following form:

$$A = \begin{bmatrix} H & | & 0 \\ 0 & | & H \end{bmatrix}, \qquad HH^{\mathrm{T}} = \mathbf{0}. \tag{7.12}$$

The condition $HH^{\mathrm{T}} = 0$ is equivalent to $C^{\perp}(H) \subset C(H)$, where $C(H)$ is the code having H as its parity-check matrix and $C^{\perp}(H)$ as its dual code. An example of a dual-containing code is *Steane's* seven-qubit code, defined by the Hamming (7,4) code:

$$H = \begin{bmatrix} 0001111 \\ 0110011 \\ 1010101 \end{bmatrix}.$$

The rows have an even number of ones, and any two of them overlap by an even number of ones, so that $C^{\perp}(H) \subset C(H)$. Here $M = 3$, $N = 7$, and so $N - 2M = 1$, and thus one qubit is encoded in seven qubits, representing the [7,1] code.

7.2.8 Quantum Codes Over GF(4)

Let the elements of GF(4) be 0, 1, w, and $w^2 = w' = 1 + w$. One can write a row $r_1 = (a_1a_2...a_n|b_1b_2...b_n)$ of quantum-check matrix in the binary string representation as a vector over GF(4) as $\rho_1 = (a_1 + b_1w; a_2 + b_2w;...; a_n + b_nw)$. Given a second row $r_2 = (c_1c_2 \ldots c_n|d_1d_2 \ldots d_n)$, with $\rho_2 = (c_1 + d_1w; c_2 + d_2w;...; c_n + d_nw)$, the *Hermitian inner product* is defined by

$$\rho_1 \cdot \rho_2 = \sum_i (a_i + b_iw')(c_i + d_iw) = \sum_i [(a_ic_i + b_id_i + b_ic_i) + (a_id_i + b_ic_i)w]. \tag{7.13}$$

Because $a_id_i + b_ic_i = r_1 \odot r_2$, the orthogonality in the Hermitian sense ($\rho_1 \cdot \rho_2 = 0$) leads to orthogonality in a sympletic sense too ($r_1 \odot r_2 = 0$). The opposite is not true because the term $a_ic_i + b_id_i + b_ic_i$ is not necessary zero when $a_id_i + b_ic_i = 0$. Therefore, the codes over GF(4) satisfying the property that any two rows are orthogonal in Hemitian inner product sense can be used as quantum codes.

7.3 QUANTUM ERROR CORRECTION

In the previous section, quantum error correction was introduced on a conceptual level. Here, we describe the quantum codes more formally. The section starts with redundancy, followed by stabilizer groups, and quantum syndrome decoding. We also formally establish the connection between classical and quantum codes. The necessary and sufficient conditions for quantum error correction are discussed and a detailed quantum error correction example is given. Further, the distance properties of quantum codes are discussed and distance is related to error correction capability. Finally, quantum encoder and decoder implementations are described.

7.3.1 Redundancy and Quantum Error Correction

A quantum error correction code (QECC) that encodes K qubits into N qubits, denoted as $[N,K]$, is defined by an encoding mapping U from the K-qubit Hilbert space H_2^K onto a 2^K-dimensional subspace C_q of the N-qubit Hilbert space H_2^N. The subspace C_q is called the *code space*, the states belonging to C_q are called the *codewords*, and the encoded computational basis kets are called the *basis codewords*. The single-qubit computational basis (CB) states are typically chosen to be the eigenkets of Z_j:

$$Z_j|b_j\rangle = (-1)^{b_j}|b_j\rangle; \quad b_j = 0, 1; \quad j = 1, 2, ..., K. \tag{7.14}$$

The CB states of H_2^K are given by

$$|\boldsymbol{b}\rangle \equiv |b_1 \ldots b_K\rangle = |b_1\rangle \otimes \ldots \otimes |b_K\rangle. \tag{7.15}$$

The encoded CB states are obtained by applying the encoding mapping to the CB states as follows:

$$|\overline{\boldsymbol{b}}\rangle \equiv |\overline{b_1 \ldots b_K}\rangle = U|b_1 \ldots b_K\rangle, \tag{7.16}$$

where the action of encoding mapping is given by $Z_j \xrightarrow{U} \overline{Z}_j = UZ_jU^{\dagger}$. The encoded CB states are simultaneous eigenkets of $\{Z_j\text{: } j=1,\ldots,K\}$ because

$$\begin{aligned}\overline{Z}_j|\overline{\boldsymbol{b}}\rangle &= UZ_jU^{\dagger}|\overline{b_1 \ldots b_K}\rangle = UZ_jU^{\dagger}(U|b_1 \ldots b_K\rangle) = UZ_j|\boldsymbol{b}\rangle = U(-1)^{b_j}|\boldsymbol{b}\rangle \\ &= (-1)^{b_j}|\overline{\boldsymbol{b}}\rangle.\end{aligned} \tag{7.17}$$

Therefore, the encoding mapping preserves the eigenvalues.

The simplest quantum error correction codes are *canonical codes* [16,22], which can be introduced by the following trivial encoding mapping U_c:

$$U_c: \quad |\psi\rangle \to |0\rangle|\psi\rangle. \tag{7.18}$$

In canonical codes (see Figure 7.10), the quantum register containing $N-K$ ancillaries $|\boldsymbol{0}\rangle_{N-K} = \underbrace{|0\rangle \otimes \ldots \otimes |0\rangle}_{N-K}$ is appended to the information quantum register containing K qubits. The basis for single-qubit errors is given by $\{I,X,Y,Z\}$. The basis for N-qubit errors is obtained by forming all possible direct products:

$$E = j^l O_1 \otimes \ldots \otimes O_N; \quad O_i \in \{I,X,Y,Z\}, \ l = 0,1,2,3. \tag{7.19}$$

The N-qubit error basis can be transformed into multiplicative *Pauli group* G_N if we allow E to be premultiplied by -1 and $\pm$j. By noticing that $Y = -jXZ$, any error in the Pauli group of N-qubit errors can be represented by

$$\begin{aligned}&E = j^{l'}X(\boldsymbol{a})Z(\boldsymbol{b}); \ \boldsymbol{a} = a_1 \ldots a_N; \ \boldsymbol{b} = b_1 \ldots b_N; \ a_i, b_i = 0,1; \ l' = 0,1,2,3 \\ &X(\boldsymbol{a}) \equiv X_1^{a_1} \otimes \ldots \otimes X_N^{a_N}; \ Z(\boldsymbol{b}) \equiv Z_1^{b_1} \otimes \ldots \otimes Z_N^{b_N}.\end{aligned} \tag{7.20}$$

In (7.20), the subscript i ($i = 0,1, \ldots, N$) denotes the location of the qubit to which operator X_i (Z_i) is applied; a_i (b_i) takes the value 1 if the ith X (Z) operator is to be included and the value 0 if the same operator is to be excluded. Therefore, to uniquely determine the error operator (up to the phase constant $j^{l'} = 1,\text{j},-1,-\text{j}$ for $l' = 0,1,2,3$) it is sufficient to specify vectors $\boldsymbol{a}$ and $\boldsymbol{b}$.

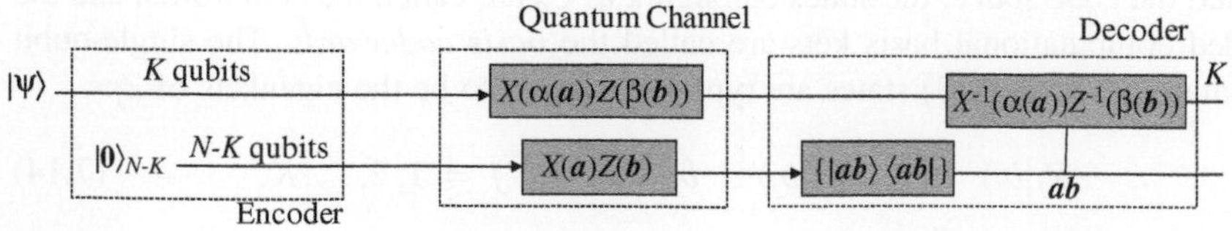

FIGURE 7.10

Canonical quantum error correction coding.

For example, the error $E = Z_1X_2Y_9 = -jZ_1X_2X_9Z_9$ can be identified (up to the phase constant) by specifying $\boldsymbol{a} = (010000001)$ and $\boldsymbol{b} = (100000001)$. Notice that this representation is equivalent to the classical representation of error E from Section 7.2 as $\boldsymbol{e} = [100000001|010000001]$, by reversing the positions of $\boldsymbol{a}$ and $\boldsymbol{b}$.

In Figure 7.10, we employ this simplified notation and represent the quantum error as a product of X- and Z-containing operators. We observe separately possible errors introduced on information qubits, denoted as $X(\alpha(\boldsymbol{a}))$ and $Z(\beta(\boldsymbol{b}))$, and errors introduced on ancillaries, denoted as $X(\boldsymbol{a})$ and $Z(\boldsymbol{b})$, where α and β are functions of $\boldsymbol{a}$ and $\boldsymbol{b}$ respectively. The action of a correctable quantum error E

$$E \in \boldsymbol{E}_c = \{X(\boldsymbol{a})Z(\boldsymbol{b}) \otimes X(\alpha(\boldsymbol{a}))Z(\beta(\boldsymbol{b})) : \boldsymbol{a},\boldsymbol{b} \in F_2^{N-K};\ F_2 = \{0,1\}\};$$
$$\alpha, \beta : F_2^{N-K} \rightarrow F_2^{N-K} \tag{7.21}$$

can be described as follows:

$$E(|\mathbf{0}\rangle|\psi\rangle) = X(\boldsymbol{a})Z(\boldsymbol{b})|0\rangle \otimes X(\alpha(\boldsymbol{a}))Z(\beta(\boldsymbol{b}))|\psi\rangle. \tag{7.22}$$

Since $X(\boldsymbol{a})Z(\boldsymbol{b})|\mathbf{0}\rangle = X(\boldsymbol{a})|\mathbf{0}\rangle = |\boldsymbol{a}\rangle$, we obtain:

$$E(|\mathbf{0}\rangle|\psi\rangle) = |\boldsymbol{a}\rangle \otimes X(\alpha(\boldsymbol{a}))Z(\beta(\boldsymbol{b}))|\psi\rangle = |\boldsymbol{a}\rangle|\psi'\rangle. \tag{7.23}$$

On the receiver side, we perform the measurements on ancillaries to determine the syndrome $\boldsymbol{S} = (\boldsymbol{a},\boldsymbol{b})$, without affecting the information qubits. Once the syndromes are determined, we perform reverse recovery operator actions, denoted in Figure 7.10, as $X^{-1}(\alpha(\boldsymbol{a}))$ and $Z^{-1}(\beta(\boldsymbol{b}))$, and the proper information state is recovered. Notice that by employing entanglement between the source and destination we can simplify the decoding process, which is the subject of the chapter on entanglement-assisted quantum error correction (see also Refs [16,22]).

An arbitrary quantum error-correcting code can be observed as a generalization of canonical code, as shown in Figure 7.11, where U is the corresponding encoding operator. The set of errors introduced by a channel, which can be corrected by this code, can now be represented by

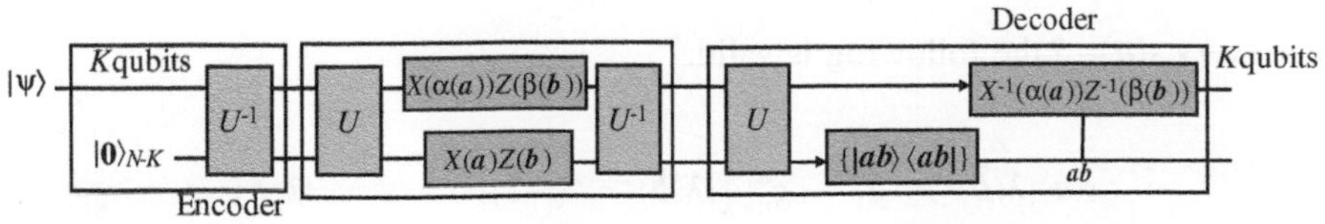

FIGURE 7.11

An arbitrary quantum error-correcting code represented as a generalization of canonical code.

$$E = \{U[X(\boldsymbol{a})Z(\boldsymbol{b})\otimes X(\alpha(\boldsymbol{a}))Z(\beta(\boldsymbol{b}))]U^{-1}: \\ \boldsymbol{a},\boldsymbol{b}\in F_2^{N-K};\ F_2 = \{0,1\}\};\quad \alpha,\beta : F_2^{N-K}\rightarrow F_2^{N-K}. \tag{7.24}$$

7.3.2 Stabilizer Group S

The quantum stabilizer code C_q, with parameters $[N,K]$, can be defined as the unique subspace of H_2^N that is fixed by the elements from stabilizer S of C_q as given by

$$\forall\ s\in S,\quad |c\rangle\in C_q:\quad s|c\rangle = |c\rangle. \tag{7.25}$$

The stabilizer group S is constructed from a set of $N-K$ operators $g_1, \ldots, g_{N-K}$, also known as *generators* of S. Generators have the following properties: (i) they commute among each other; (ii) they are unitary and Hermitian; and (iii) they have an order of 2 because $g_i^2 = I$. Any element $s\in S$ can be written as a unique product of powers of the generators:

$$s = g_1^{a_1}\ldots g_{N-K}^{a_{N-K}},\quad a_i\in\{0,1\};\quad i = 1,\ldots,N-K. \tag{7.26}$$

When a_i equals 1 (0) the ith generator is included (excluded) from the product of generators. Therefore, the elements of S can be labeled by simple strings of length $N-K$, namely $\boldsymbol{a} = a_1\ldots a_{N-K}$. Here we can draw a parallel with classical error correction, in which any codeword in a block code can be represented as a linear combination of basis codewords. The eigenvalues of generators g_i, $\{\text{eig}(g_i)\}$, can be found starting from property (iii):

$$|\text{eig}(g_i)\rangle = I|\text{eig}(g_i)\rangle = g_i^2|\text{eig}(g_i)\rangle = \text{eig}^2(g_i)|\text{eig}(g_i)\rangle. \tag{7.27}$$

From (7.27) it is clear that $\text{eig}^2(g_i) = 1$, meaning that eigenvalues are $\text{eig}(g_i) = \pm 1 = (-1)^{\lambda_i}$; $\lambda_i = 0,1$. Because the "parent" space for C_q is 2^N-dimensional space, we need to determine N commuting operators that specify a unique state $|\psi\rangle\in H_2^N$. The following 2^N simultaneous eigenstates of $\{g_1, g_2, \ldots, g_{N-K}; \overline{Z}_1, \overline{Z}_2, \ldots, \overline{Z}_K\}$ can be used as a basis for H_2^N. Namely, we have chosen the subset $\{\overline{Z}_i\}$ in such a way that each element of this subset commutes with all generators $\{g_i\}$ as well as among all elements of the subset itself. The eigenkets can be labeled by $\boldsymbol{\lambda} = \lambda_1\ldots\lambda_{N-K}$ and encoded CBs satisfy:

$$g_i|\boldsymbol{\lambda};\overline{\boldsymbol{b}}\rangle = (-1)^{\lambda_i}|\boldsymbol{\lambda};\overline{\boldsymbol{b}}\rangle\qquad \overline{Z}_j|\boldsymbol{\lambda};\overline{\boldsymbol{b}}\rangle = (-1)^{b_j}|\boldsymbol{\lambda};\overline{\boldsymbol{b}}\rangle;\quad i = 1,\ldots,N-K; \\ j = 1,\ldots,K;\quad \lambda_i, b_j = 0,1. \tag{7.28}$$

For every $s(\boldsymbol{a})\in S$ the following is valid:

$$s(\boldsymbol{a})|\boldsymbol{\lambda};\overline{\boldsymbol{b}}\rangle = g_1^{a_1}\ldots g_{n-k}^{a_{n-k}}|\boldsymbol{\lambda};\overline{\boldsymbol{b}}\rangle = (-1)^{\sum_{i=1}^{N-K}\lambda_i a_i \bmod 2}|\boldsymbol{\lambda};\overline{\boldsymbol{b}}\rangle \\ = (-1)^{\boldsymbol{\lambda}\cdot\boldsymbol{a}}|\boldsymbol{\lambda};\overline{\boldsymbol{b}}\rangle;\quad \boldsymbol{\lambda}\cdot\boldsymbol{a} = \sum_{i=1}^{N-K}\lambda_i a_i \bmod 2. \tag{7.29}$$

From Eq. (7.25) it is clear that

$$s(\boldsymbol{a})|\boldsymbol{\lambda};\bar{\boldsymbol{b}}\rangle = |\boldsymbol{\lambda};\bar{\boldsymbol{b}}\rangle \tag{7.30}$$

when $\boldsymbol{\lambda} \cdot \boldsymbol{a} = 0 \bmod 2$. The simplest way to satisfy (7.30) is for $\lambda = 0 \ldots 0$, meaning that eigenkets $\{|\boldsymbol{\lambda} = 0 \ldots 0;\bar{\boldsymbol{b}}\rangle\}$ can be used as encoded CB states.

The following interpretation of the same result is due to Gaitan [14]. The set $C(\boldsymbol{\lambda}) = \{|\boldsymbol{\lambda};\bar{\boldsymbol{b}}\rangle : \bar{\boldsymbol{b}} \in F_2^K\}$ (where $F_2 = \{0,1\}$) is clearly the subspace of H_N^2 whose elements are simultaneous eigenvectors of the generators $g_1, \ldots, g_{N-K}$ with corresponding eigenvalues $(-1)^{\lambda_1}, \ldots, (-1)^{\lambda_{N-K}}$. It can be shown that the set of 2^{N-K} subspaces $\{C(\boldsymbol{\lambda})$: $\boldsymbol{\lambda} \in F_2^{N-K}\}$ partitions H_N^2. Because the quantum stabilizer code C_q is fixed by stabilizer S, the states $|c\rangle \in C_q$ are fixed by $g_1, \ldots, g_{N-K}$ with corresponding eigenvalues $\text{eig}(g_i) = 1$ $(i = 1,2, \ldots, N - K)$, meaning that $C_q \subset C(\boldsymbol{\lambda} = 0 \ldots 0)$. Since both C_q and $C(\boldsymbol{\lambda} = 0 \ldots 0)$ are 2^K-dimensional, it must be $C_q = C(\boldsymbol{\lambda} = 0 \ldots 0)$. Therefore, the encoded CB states are the eigenkets $\{|\boldsymbol{\lambda} = 0 \ldots 0;\bar{\boldsymbol{b}}\rangle\}$.

7.3.3 Quantum-Check Matrix and Syndrome Equation

In the previous section, we have already established the connection between quantum and classical codes, but we did not explain how this connection was actually derived, which is the topic of this subsection. The following two theorems are important in establishing the connection between classical and quantum codes.

Theorem 7.1. Let E be an error and S the stabilizer group for a quantum stabilizer code C_q. If S contains an element that anticommutes with E, then for all $|c\rangle, |c'\rangle \in C_q$, $E|c\rangle$ is orthogonal to $|c'\rangle$:

$$\langle c'|E|c\rangle = 0. \tag{7.31}$$

This theorem is quite straightforward to prove. Since $|c\rangle, |c'\rangle \in C_q$ and the error E anticommutes with $s \in S$ $(\{E, s\} = 0)$, we can write

$$E|c\rangle = Es|c\rangle = -sE|c\rangle. \tag{7.32}$$

By multiplying by $\langle c'|$ from the left side we obtain:

$$\langle c'|E|c\rangle = -\langle c'|sE|c\rangle = -\langle c'|E|c\rangle. \tag{7.33}$$

By solving (7.33) for $\langle c'|E|c\rangle$ we obtain $\langle c'|E|c\rangle = 0$.

Theorem 7.2. Let E be an error and C_q be a quantum stabilizer code with generators $g_1, \ldots, g_{N-K}$. The image $E(C_q)$ under E is $C(\boldsymbol{\lambda})$, where $\boldsymbol{\lambda} = \lambda_1 \ldots \lambda_{N-K}$, such that

$$\lambda_i = \begin{cases} 0, & [E, g_i] = 0 \\ 1, & \{E, g_i\} = 0 \end{cases} \quad (i = 1, \ldots, N - K). \tag{7.34}$$

This theorem can be proved by observing the simultaneous action of generator g_i and an error operator E on $|c\rangle \in C_q$:

$$g_i E|c\rangle = (-1)^{\lambda_i} E g_i|c\rangle = (-1)^{\lambda_i} E|c\rangle. \tag{7.35}$$

It is clear from (7.35) that for $\lambda_i = 0$, E and g_i commute since $g_iE = Eg_i$. On the other hand, for $\lambda_i = 1$, from (7.35) we obtain $g_iE = -Eg_i$, which means that E and g_i anticommute ($\{E, g_i\} = 0$). These two cases can be written in one equation as follows:

$$\lambda_i = \begin{cases} 0, & [E, g_i] = 0 \\ 1, & \{E, g_i\} = 0 \end{cases} \quad (i = 1, \ldots, N-K),$$

therefore proving Eq. (7.34). What remains is to prove that the image $E(C_q)$ equals $C(\boldsymbol{\lambda}) = \{|\boldsymbol{\lambda}; \bar{\boldsymbol{\delta}}\rangle : \bar{\boldsymbol{\delta}} \in F_2^K\}$. The ket $E|c\rangle$ can be represented in terms of eigenkets $|\boldsymbol{\lambda}; \bar{\boldsymbol{a}}\rangle$ as follows:

$$E|c\rangle = \sum_{\lambda'} \sum_{\bar{\boldsymbol{a}}} \alpha(\boldsymbol{\lambda}'; \bar{\boldsymbol{\delta}}) |\boldsymbol{\lambda}'; \bar{\boldsymbol{a}}\rangle, \tag{7.36}$$

where $\alpha(\boldsymbol{\lambda}'; \bar{\boldsymbol{a}})$ are projections along basis kets $|\boldsymbol{\lambda}'; \bar{\boldsymbol{a}}\rangle$. Let us now employ the property we just derived and observe the action of g_i and E on $|c\rangle$:

$$g_iE|c\rangle = \sum_{\lambda'} \sum_{\bar{\boldsymbol{a}}} \alpha(\boldsymbol{\lambda}'; \bar{\boldsymbol{a}}) g_i |\boldsymbol{\lambda}'; \bar{\boldsymbol{a}}\rangle = \sum_{\lambda} \sum_{\bar{\boldsymbol{a}}} (-1)^{\lambda_i} \alpha(\boldsymbol{\lambda}; \bar{\boldsymbol{a}}) |\boldsymbol{\lambda}; \bar{\boldsymbol{a}}\rangle. \tag{7.37}$$

It is clear from (7.37) and (7.35), in which the right-hand side of the equation is a function only of eigenvalues $(-1)^{\lambda_i}$, that

$$E|c\rangle = \sum_{\bar{\boldsymbol{a}}} \alpha(\boldsymbol{\lambda}; \bar{\boldsymbol{a}}) |\boldsymbol{\lambda}; \bar{\boldsymbol{a}}\rangle, \tag{7.38}$$

and therefore $E(C_q) = C(\boldsymbol{\lambda})$. Because $I(C_q) = C(0\ldots0)$, where I is the identity operator, it is clear that $\boldsymbol{\lambda} = \lambda_1 \ldots \lambda_{N-K}$ can be used as the error syndrome of E, denoted as $S(E)$. Therefore, the *error syndrome* is defined as $S(E) = \lambda_1 \ldots \lambda_{n-k}$, where λ_i is defined by (7.34). Based on the discussion from Section 7.2, we can represent the generator g_i as a binary vector of length $2N$: $\boldsymbol{g}_i = (\boldsymbol{a}_i | \boldsymbol{b}_i)$, where the row vector $\boldsymbol{a}_i$ ($\boldsymbol{b}_i$) is obtained from the operator representation by replacing the X operators (Z operators) by 1 and identity operators by 0. The binary representations of all generators can be used to create the *quantum-check matrix* $\boldsymbol{A}$ obtained by writing the binary representation of g_i as the ith row of $(\boldsymbol{A})_i$ as follows:

$$\boldsymbol{A} = \begin{pmatrix} (\boldsymbol{A})_1 \\ \vdots \\ (\boldsymbol{A})_{n-k} \end{pmatrix}, \quad (\boldsymbol{A})_i = \boldsymbol{g}_i = (\boldsymbol{a}_i | \boldsymbol{b}_i). \tag{7.39}$$

By representing the error operator E as binary error row vector $\boldsymbol{e} = (\boldsymbol{c}|\boldsymbol{d})$, but now with $\boldsymbol{c}$ ($\boldsymbol{d}$) obtained from error operator E by replacing the Z operators (X operators) with ones and identity operators by zeros, the syndrome equation can be written as

$$S(E) = \boldsymbol{e}\boldsymbol{A}^{\mathrm{T}}, \tag{7.40}$$

which is very similar to classical error correction.

7.3.4 Necessary and Sufficient Conditions for Quantum Error Correction Coding

Quantum error correction determines the error syndrome with the help of ancilla qubits. The measured value of syndrome S determines a set of errors $\boldsymbol{E}_S = (E_1, E_2, \ldots)$ with syndromes being equal to the measured value: $S(E_i) = S$, $\forall\ E_i \in \boldsymbol{E}_S$. In analogy to classical error correction, this set of errors $\boldsymbol{E}_S$ having the same syndrome S can be called the *coset*. Among different candidate errors we chose for the expected error the most probable one, say E_S. We then perform recovery operation by simply applying E_S^{+}. If the actual error differs from true error E_S, the resulting state $E_S^{+}E\,|s\rangle$ will be from C_q but different from uncorrupted state $|s\rangle$. The following two conditions should be satisfied during design of a quantum code: (i) the encoded CB states must be chosen carefully so that the environment is not able to distinguish among different CBs; and (ii) the corrupted images of codewords must be orthogonal among each other. These two conditions can be formulated as a theorem that provides the *necessary and sufficient conditions* to be satisfied so that a quantum code C_q to be able to correct a given set of errors $\boldsymbol{E}$.

Theorem 7.3: Necessary and sufficient conditions for QECCs [14]. The code C_q is an $\boldsymbol{E}$ error correcting code if and only if $\forall\ |\bar{i}\rangle, |\bar{j}\rangle\ (\bar{i} \neq \bar{j})$ and $\forall\ E_a,\ E_b \in \boldsymbol{E}$, the following two conditions are valid:

$$\langle \bar{i}|E_a^{\dagger}E_b|\bar{i}\rangle = \langle \bar{j}|E_a^{\dagger}E_b|\bar{j}\rangle \qquad \langle \bar{i}|E_a^{\dagger}E_b|\bar{j}\rangle = 0. \tag{7.41}$$

The first condition indicates that the action of the environment is similar for all CBs, so that the channel is not able to distinguish among different CBs. The second condition indicates that the images of codewords are orthogonal to each other. The proof of this theorem is left to the reader.

7.3.5 A Quantum Stabilizer Code for a Phase-Flip Channel (Revisited)

The different quantum error models considered so far assume: (i) errors on different qubits are independent; (ii) single qubit errors are equally likely; and (iii) the single-qubit error probability is the same for all qubits. Several such models, including the phase-flip channel model, have already been introduced in the previous section. The phase-flip channel produces eight possible errors on the three qubits, as shown in Table 7.4.

At least one of the two generators must anticommute with each of the single-qubit errors $\{E_1, E_2, E_3\}$, justifying the following selection of generators:

$$g_1 = X_1X_2 \qquad g_2 = X_1X_3.$$

The error syndrome can easily be calculated from Theorem 7.2 by

$$S(E) = \lambda_1 \ldots \lambda_{N-K}; \qquad \lambda_i = \begin{cases} 0, & [E, g_i] = 0 \\ 1, & \{E, g_i\} = 0, \end{cases}$$

Table 7.4 Syndrome LUT for the Phase-flip Channel

Error E	Error probability $P_e(E)$	Error syndrome $S(E)$
$E_0 = I$	$(1-p)^3$	00
$E_1 = Z_1$	$p(1-p)^2$	11
$E_2 = Z_2$	$p(1-p)^2$	10
$E_3 = Z_3$	$p(1-p)^2$	01
$E_4 = Z_1Z_2$	$p^2(1-p)$	01
$E_5 = Z_1Z_3$	$p^2(1-p)$	10
$E_6 = Z_2Z_3$	$p^2(1-p)$	11
$E_7 = Z_1Z_2Z_3$	p^3	00

and is given in the third column of Table 7.4. The expected error determined from the syndrome and corresponding recovery operator are provided in Table 7.5, in which, of several candidate error operators having the same syndrome, the error with lowest weight is selected.

The stabilizer group is the set of elements of the form $s(p) = g_1^{a1} g_2^{p2}$, where $a_1, a_2 =$ 0,1. By simply varying a_i (0 or 1) we obtain the following stabilizer:

$$S = \{I, X_1X_2, X_1X_3, X_2X_3\}.$$

The single-qubit CB states can be chosen to be the eigenkets of X:

$$|0\rangle \equiv |\mathrm{eig}(X) = +1\rangle \qquad |1\rangle \equiv |\mathrm{eig}(X) = -1\rangle.$$

The stabilizer S must fix the code space C_q, including the encoded CB states:

$$|0\rangle \rightarrow |\overline{0}\rangle = |000\rangle \qquad |1\rangle \rightarrow |\overline{1}\rangle = |111\rangle.$$

This mapping is consistent with no-cloning theorem which claims that arbitrary kets cannot be cloned, and the cloning of orthogonal kets is allowed.

The *failure probability* for error correction P_f is the probability that one of the unexpected errors $\{E_4, E_5, E_6, E_7\}$ occurs:

$$P_f = 3p^2(1-p) + p^3.$$

Table 7.5 Syndrome LUT and Recovery Operators for the Phase-flip Channel

Most probable error E_s	Error syndrome $S(E)$	Recovery operator R_s
$E_{00} = I$	00	$R_{00} = I$
$E_{01} = Z_3$	01	$R_{01} = Z_3$
$E_{10} = Z_2$	10	$R_{10} = Z_2$
$E_{11} = Z_1$	11	$R_{11} = Z_1$

Example. Say the error syndrome was $S = 10$, while the actual error was $E_5 = Z_1Z_3$. Based on the LUT (see Table 7.5) we will choose $E_{10} = Z_2$ so that the corresponding recovery operator will be $R_{10} = E_{10}^\dagger = Z_2$. The transmitted codeword can be represented as $|c\rangle = a|\overline{0}\rangle + b|\overline{1}\rangle$. The action of the channel leads to the state $E_5|c\rangle$, while the simultaneous action of the E_5 recovery operator leads to $|\psi\rangle = E_{10}^\dagger E_5|c\rangle = Z_1Z_2Z_3|c\rangle$. Because the action of the resulting operator is as follows:

$$Z\left\{\begin{array}{l}|+1\rangle\\|-1\rangle\end{array}\right. = \left\{\begin{array}{l}|-1\rangle\\|+1\rangle\end{array}\right. \Rightarrow E_{10}^\dagger E_5\left\{\begin{array}{l}|\overline{0}\rangle\\|\overline{1}\rangle\end{array}\right. = Z_1Z_2Z_3\left\{\begin{array}{l}|\overline{0}\rangle\\|\overline{1}\rangle\end{array}\right. = \left\{\begin{array}{l}|\overline{1}\rangle\\|\overline{0}\rangle\end{array}\right.,$$

and the final state $|\psi_f\rangle = a|\overline{1}\rangle + b|\overline{0}\rangle$ is different from the original state, and so the error correction has failed to correct the error, but it was able to return the received state back to the code space.

7.3.6 Distance Properties of Quantum Error Correction Codes

A QECC is defined by the encoding operation of mapping K qubits into N qubits. The QECC $[N,K]$ can be interpreted as a 2^K-dimensional subspace C_q of N-qubit Hilbert space H_N^2, together with corresponding recovery operation R. As mentioned above, this subspace (C_q) is called the code space, the kets belonging to C_q are known as codewords, and the encoded computational basis kets are called the basis codewords. The basis for single-qubit errors is given by $\{I,X,Y,Z\}$, as described above. Because any error, be either discrete or continuous, can be represented as a linear combination of base errors, the linear combination of correctable errors will also be a correctable error. The basis for N-qubit quantum errors is obtained by forming all possible direct products:

$$\begin{aligned}
E &= j^l O_1 \otimes \ldots \otimes O_N = j^{l'} X(\boldsymbol{a}) Z(\boldsymbol{b}); \quad \boldsymbol{a} = a_1 \ldots a_N; \quad \boldsymbol{b} = b_1 \cdots b_N; \\
&\quad a_i, b_i = 0,1; \quad l, l' = 0,1,2,3 \\
X(\boldsymbol{a}) &\equiv X_1^{a_1} \otimes \ldots \otimes X_N^{a_N}; \quad Z(\boldsymbol{b}) \equiv Z_1^{b_1} \otimes \ldots \otimes Z_N^{b_N}; \quad O_i \in \{I, X, Y, Z\}.
\end{aligned} \tag{7.42}$$

The *weight* of an error operator $E(\boldsymbol{a},\boldsymbol{b})$ is defined to be the number of qubits different from the identity operator. Necessary and sufficient conditions for a quantum error correction code to correct a set of errors $E = \{E_p\}$, given by Eq. (7.41), can be consolidated as follows:

$$\langle \overline{i}|E_p^\dagger E_q|\overline{j}\rangle = C_{pq}\delta_{\overline{i}\overline{j}}, \tag{7.43}$$

where the matrix elements C_{pq} satisfy the condition $C_{pq} = C_{qp}^*$, so that the square matrix $\boldsymbol{C} = (C_{pq})$ is Hermitian. A QECC for which matrix $\boldsymbol{C}$ is singular is said to be *degenerate*. If we interpret $E_p^\dagger E_q$ as a new error operator E, Eq. (7.43) can be rewritten as

$$\langle \overline{i}|E|\overline{j}\rangle = C_E\delta_{\overline{i}\overline{j}}. \tag{7.44}$$

We say that a QECC has a *distance D* if all errors of weight less than D satisfy Eq. (7.44), and there exists at least one error of weight D to violate it. In other words, the distance of QECC is the weight of the smallest weight D of error E that cannot be detected by the code. Similarly to classical codes, we can relate the distance D to the error correction capability t as follows: $D \geq 2t + 1$. Namely, since $E = E_p^{\dagger} E_q$ the weight of error operator E will be $\mathrm{wt}(E_p^{\dagger} E_q) = 2t$. If we are only interested in detecting errors but not correcting them, the error detection capability d is related to the distance D by $D \geq d + 1$. Since we are interested only in detection of errors we can set $E_p = I$ to obtain $\mathrm{wt}(E_p^{\dagger} E_q) = \mathrm{wt}(E_q) = d$. If we are interested in simultaneously detecting d errors and correcting t errors, the distance of the code must be $D \geq d + t + 1$. The following theorem can be used to determine if a given quantum code C_q of quantum distance D is degenerate or not.

Theorem 7.4. The quantum code C_q of distance D is a *degenerate code* if and only if its stabilizer S contains an element with weight less than D (excluding the identity element).

The theorem can be proved as follows. If the code C_q is a degenerate code of distance D, there will exist two correctable errors E_1, E_2 such that their action on a CB codeword is the same: $E_1|\bar{i}\rangle = E_2|\bar{i}\rangle$. By multiplying by $E_2^{\dagger}$ from the left we obtain $E_2^{\dagger} E_1|\bar{i}\rangle = |\bar{i}\rangle$, which indicates that the error $E_2^{\dagger} E_1 \in S$. Since any correctable error satisfies $\langle \bar{i}|E|\bar{j}\rangle = C_E \delta_{\bar{i}\bar{j}}$, we obtain $\langle \bar{i}|E_2^{\dagger} E_1|\bar{j}\rangle = C_{12}\delta_{\bar{i}\bar{j}}$. Because the C_q code has a distance D it is clear that $\mathrm{wt}(E_2^{\dagger} E_1) < D$. On the other hand, if there exists an $s \in S$ with $\mathrm{wt}(s) < D$, we can find another $s_a \in S$ so that $s_a s = s_b \in S$. By multiplying both sides by $s_a^{\dagger}$, we obtain $s = s_a^{\dagger} s_b$. From the definition of stabilizer codes we know that $s|\bar{i}\rangle = |\bar{i}\rangle$, $\forall\, s \in S$. By expressing $s = s_a^{\dagger} s_b$, we obtain $s_a^{\dagger} s_b|\bar{i}\rangle = |\bar{i}\rangle$, which by multiplying by s_a from the left becomes $s_b|\bar{i}\rangle = s_a|\bar{i}\rangle$, which is equivalent to $(s_a - s_b)|\bar{i}\rangle = 0$. Since the matrix S_{ab} is singular, the code C_q is degenerate.

7.3.7 Calderbank–Shor–Steane (CSS) Codes (Revisited)

The CSS codes are constructed from two classical binary codes C and C' satisfying the following three properties:

1. C and C' are (n,k,d) and (n',k',d') codes respectively
2. $C' \subset C$
3. C and $C'^{\perp}$ are both t-error-correcting codes.

The code construction partitions C into the cosets of C': $C = C' \cup (c_1 + C') \cup \ldots \cup (c_N + C')$, where $c_1, c_2, \ldots, c_N \in C$, and N is the number of cosets determined by (from Lagrange's theorem) $N = 2^k/2^{k'} = 2^{k-k'}$. The *basis codewords* are obtained by identifying each one with the corresponding coset $|\bar{c}_i\rangle \Leftrightarrow c_i + C'$, so that

$$|\bar{c}_i\rangle = \frac{1}{\sqrt{2^{k'}}} \sum_{c' \in C'} |c_i + c'\rangle; \quad i = 1, \ldots, N. \tag{7.45}$$

If $v, w \in C$ so that $v - w = d \in C'$ (belong to the same coset of C'), then $|\overline{v}\rangle = |\overline{w}\rangle$ because

$$|\overline{v}\rangle = \frac{1}{\sqrt{2^{k'}}} \sum_{c' \in C'} |v + c'\rangle \overset{v-w=d}{=} \frac{1}{\sqrt{2^{k'}}} \sum_{c' \in C'} |w + d + c'\rangle$$
$$\overset{d+c'=d'}{=} \frac{1}{\sqrt{2^{k'}}} \sum_{d' \in C'} |w + d'\rangle = |\overline{w}\rangle. \quad (7.46)$$

The code space C_q is a subspace of H_N^2 spanned by the basis codewords and is therefore $N = 2^{k-k'}$-dimensional.

Example: CSS [7,3,1] code. For Steane [7,1,3] code, C is the Hamming (7,4,3) and C' is (7,3,4) maximum-length code:

$$H(C) = H(C'_\perp) = \begin{bmatrix} 0 & 0 & 0 & 1 & 1 & 1 & 1 \\ 0 & 1 & 1 & 0 & 0 & 1 & 1 \\ 1 & 0 & 1 & 0 & 1 & 0 & 1 \end{bmatrix}.$$

Based on (7.12), (7.39), and matrix H above, we obtain the following generators:

$$g_1 = X_4X_5X_6X_7 \qquad g_2 = X_2X_3X_6X_7 \qquad g_3 = X_1X_3X_5X_7$$
$$g_4 = Z_4Z_5Z_6Z_7 \qquad g_5 = Z_2Z_3Z_6Z_7 \qquad g_6 = Z_1Z_3Z_5Z_7.$$

The number of cosets is $N = 2^{k-k'} = 2$, and the corresponding cosets are

$$(0000000 + C') = \{000\,0\,000,\ 011\,0\,011,\ 101\,0\,101,\ 110\,0\,110,\ 000\,1\,111,\ 011\,1\,100,\ 101\,1\,010,\ 110\,1\,001\}$$
$$(1111111 + C') = \{111\ 1\,111,\ 100\,1\,100,\ 010\,1\,010,\ 001\,1\,001,\ 111\,0\,000,\ 100\,0\,011,\ 010\,0\,101,\ 001\,0\,110\}.$$

Therefore, based on (7.45), we can represent the basis codewords by

$$|\overline{0}\rangle = \frac{1}{\sqrt{2^3}}[|0000000\rangle + |0110011\rangle + |1010101\rangle + |1100110\rangle + |0001111\rangle + |0111100\rangle + |1011010\rangle + |1101001\rangle]$$

$$|\overline{1}\rangle = \frac{1}{\sqrt{2^3}}[|1111111\rangle + |1001100\rangle + |0101010\rangle + |0011001\rangle + |1110000\rangle + |1000011\rangle + |0100101\rangle + |0010110\rangle].$$

It is straightforward to show that the generators $g_1, \ldots, g_6$ fix the basis codewords.

By definition, the distance D of a QECC is the weight of the smallest error not satisfying Eq. (7.44). We can easily find that $\langle\overline{0}|X_1X_2X_3|\overline{1}\rangle = 1$. Since the weight of error $X_1X_2X_3$ is 3, the distance of the code is 3 and the error correction capability is 1. It can be shown that Steane [7,3,1] code is nondegenerate since the stabilizer S does not contain any element of weight smaller than 3.

7.3.8 Encoding and Decoding Circuits of Quantum Stabilizer Codes

In this section we briefly describe the implementation of encoders and decoders for quantum stabilizer codes. Efficient encoding and decoding of quantum stabilizer codes are described in the next chapter. Here we provide very simple descriptions of encoders and decoders, which is provided for completeness of presentation. In particular, decoders are quite easy to implement by using the stabilizer formalism. The error detector can be obtained by concatenation of circuits corresponding to different stabilizers. The transmission error can be identified as an intersection of corresponding syndrome measurements. Let us observe the implementation of stabilizer $S_a = X_1X_4X_5X_6$. The syndrome quantum circuit for measurement of stabilizer S_a is shown in Figure 7.12a, b. In Figure 7.12a the quantum syndrome implementation circuit is based on Hadamard (H) and controlled-X gates, while in Figure 7.12b the corresponding implementation is based on H and CNOT gates only. The initial and final Hadamard gates are applied on the ancillary. We perform the measurement on the ancillary qubit only in the $\{|0\rangle,|1\rangle\}$ basis, which gives the outcome 0 if S_a has the outcome +1, and the outcome 1 when S_a has the outcome −1. This can be explained as follows [12]:

$$\begin{aligned}(I\otimes H)S_a^c(I\otimes H)|\psi\rangle|0\rangle &= (I\otimes H)S_a^c|\psi\rangle(|0\rangle+|1\rangle)/\sqrt{2}\\ &= (I\otimes H)(|\psi\rangle|0\rangle + S_a|\psi\rangle|1\rangle)/\sqrt{2}\\ &= [|\psi\rangle(|0\rangle+|1\rangle) + S_a|\psi\rangle(|0\rangle-|1\rangle)]/2\\ &= [(I+S_aI)|\psi\rangle|0\rangle + (I-S_aI)|\psi\rangle|1\rangle]/2. \end{aligned} \tag{7.47}$$

Because measuring the ancillary projects $|\psi\rangle$ on the eigenkets of S_a, we will get outcome 0 when S_a has outcome +1 and 1 when S_a has outcome −1. (In (7.47) we use the superscript c to denote the control operation; for $c = 0$ the action S_a is excluded.)

We now describe the implementation of encoders for a dual-containing code defined by a full-rank matrix $\boldsymbol{H}$ with $N > 2M$, based on the proposal of MacKay et al. [12]. We first need to transform the matrix $\boldsymbol{H}$ by Gauss elimination into the form $\tilde{\boldsymbol{H}} = [\boldsymbol{I}_M|\boldsymbol{P}_{M\times(N-M)}]$, where $\boldsymbol{I}$ is the identity matrix of size $M \times M$ and $\boldsymbol{P}$ is a binary matrix of size $M \times (N-M)$ of full rank. We further transform the matrix $\boldsymbol{P}$ also by Gaussian elimination into the form $\tilde{\boldsymbol{P}} = [\boldsymbol{I}_M|\boldsymbol{Q}_{M\times(N-2M)}]$, where $\boldsymbol{Q}$ is a binary matrix of size $M \times (N-2M)$. Clearly, for arbitrary string $\boldsymbol{f}$ of length $K = N-2M$, $[0|\boldsymbol{Qf}|\boldsymbol{f}]$ is a codeword of $C(\boldsymbol{H})$. The encoder can now be implemented in two stages, as shown in Figure 7.13, which is modified from Ref. [12].

The first stage performs the following mapping:

$$|\boldsymbol{0}\rangle_M|\boldsymbol{0}\rangle_M|\boldsymbol{s}\rangle_K \rightarrow |\boldsymbol{0}\rangle_M|\boldsymbol{Qs}\rangle_M|\boldsymbol{s}\rangle_K, \tag{7.48}$$

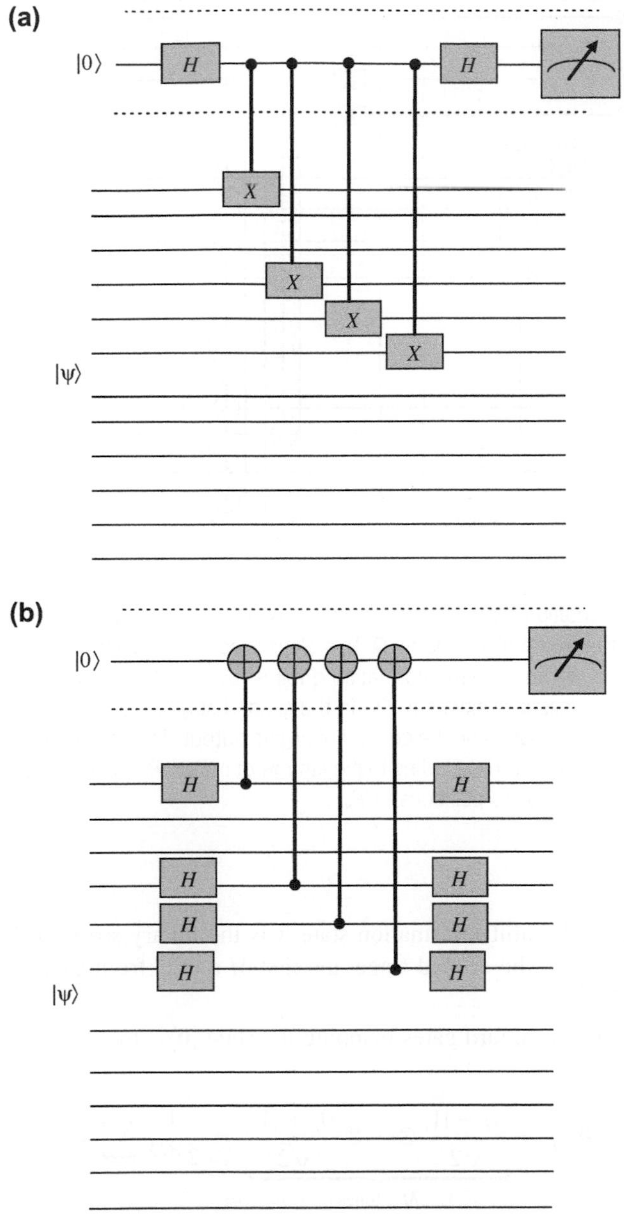

FIGURE 7.12

The syndrome quantum circuit for stabilizer $S_a = X_1X_4X_5X_6$: (a) based on H and controlled-X gates; (b) based on H and CNOT gates.

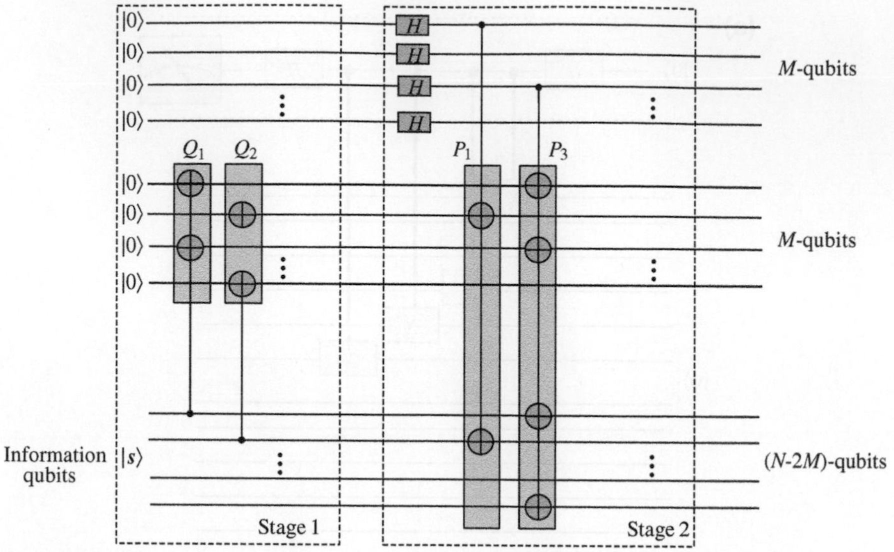

FIGURE 7.13

Encoder implementation for dual-containing codes. The block Q_k in the first stage corresponds to the kth column of submatrix $\boldsymbol{Q}$. This block is controlled by the kth information qubit and executed only when the kth information qubit is $|1\rangle$. The CNOT gates are placed according to the nonzero positions in Q_k. In the second stage, the rows of matrix $\boldsymbol{P}$ are conditionally executed based on the content of H gate output. The kth Hadamard gate output controls the set of gates corresponding to the kth row of matrix $\boldsymbol{P}$. The CNOT gates are placed in accordance with nonzero positions in P_k.

where $|\boldsymbol{s}\rangle_K$ is the K-qubit information state, $\boldsymbol{s}$ is the binary string of length K, and $|\boldsymbol{0}\rangle_M = \underbrace{|0\rangle \ldots |0\rangle}_{M \text{ times}}$. In the second stage, the first M qubits from stage 1 (ancillaries) are processed by Hadamard gates mapping the state $|\boldsymbol{0}\rangle_M$ to

$$|\boldsymbol{0}\rangle_M \rightarrow \underbrace{\frac{|0\rangle + |1\rangle}{\sqrt{2}} \otimes \cdots \otimes \frac{|0\rangle + |1\rangle}{\sqrt{2}}}_{M \text{ times}} = \frac{1}{2^{M/2}} \sum_{\boldsymbol{v}} |\boldsymbol{v}\rangle_M, \tag{7.49}$$

where vector $\boldsymbol{v}$ covers all possible binary M-tuples. Let us now introduce the binary vector $\boldsymbol{y}$ to denote the concatenation of binary strings $\boldsymbol{Qs}$ and $\boldsymbol{s}$ in (7.48). From Figure 7.13 it is clear that the purpose of stage 2 is to conditionally execute the row operators of $\tilde{\boldsymbol{H}} = [\boldsymbol{I}_M | \boldsymbol{P}_{M\times(N-M)}]$ on target qubits $|\boldsymbol{y}\rangle_{N-M}$. The qubits $|\boldsymbol{v}\rangle_M$ serve as control qubits. Notice that this stage is very similar to classical encoders of systematic linear block codes. Therefore, the operation of the final stage is to perform the following mapping:

$$\frac{1}{2^{M/2}}\sum_{v}|v\rangle_M|y\rangle_{N-M} \rightarrow \frac{1}{2^{M/2}}\sum_{v}\left(\prod_{m=1}^{M}P_m^c\right)|v\rangle_M|y\rangle_{N-M}$$
$$= \frac{1}{2^{M/2}}\sum_{x\in C^{\perp}(H)}|x+y\rangle_N,\ y\in C(\boldsymbol{H}). \tag{7.50}$$

The most general codeword of the form

$$|\psi\rangle = \sum_{y\in C(\boldsymbol{H})}\alpha_y\sum_{x\in C^{\perp}(H)}|x+y\rangle_N \tag{7.51}$$

can be obtained if we start from $\sum_s \alpha_s|s\rangle_K$ instead of $|s\rangle_K$.

7.4 IMPORTANT QUANTUM CODING BOUNDS

In this section we describe several important quantum coding bounds, including the Hamming, Gilbert–Varshamov, and Singleton bounds. We also discuss quantum weight enumerators and the quantum MacWilliams identity.

7.4.1 Quantum Hamming Bound

The quantum Hamming bound for an $[N,K]$ QECC of error correction capability t is given by

$$\sum_{j=0}^{t}\binom{N}{j}3^j2^K \leq 2^N, \tag{7.52}$$

where K is the information word length, and N is the codeword length. This inequality is quite straightforward to prove. The number of information words can be found as 2^K, the number of error locations is N chooses j, and the number of possible errors $\{X,Y,Z\}$ at every location is 3^j. The total number of errors for all codewords that can be corrected cannot be larger than the code space, which is 2^N-dimensional. Analogously to classical codes, the QECCs satisfying the Hamming inequality with equality can be called *perfect* codes. For $t = 1$, the quantum Hamming bound becomes $(1 + 3N)2^K \leq 2^N$. For an $[N,1]$ quantum code with $t = 1$, the quantum Hamming bound is simply $2(1 + 3N) \leq 2^N$. The smallest possible N to satisfy the Hamming bound is $N = 5$, which represents the perfect code. This cyclic code has already been introduced in Section 7.2. It is interesting to note that the quantum Hamming bound is identical to the classical Hamming bound for q-ary linear block codes (LBCs) (see Section 6.7.2) by setting $q = 4$, corresponding to the cardinality of set of errors $\{I,X,Y,Z\}$. This is consistent with the connection we established between QECCs and classical codes over GF(4).

The asymptotic quantum Hamming bound can be obtained by letting $N \to \infty$. For very large N, the last term in summation (7.52) dominates and we can write $\binom{N}{t} 3^t 2^K \leq 2^N$. By taking the $\log_2(\,)$ from both sides of the inequality we obtain: $\log_2 \binom{N}{t} + t\log_2 3 + K \leq N$. By using the approximation $\log_2 \binom{N}{t} \simeq NH(t/N)$, where $H(p)$ is the binary entropy function $H(p) = -p\log p - (1-p)\log(1-p)$, we obtain the following *asymptotic quantum Hamming bound*:

$$\frac{K}{N} \leq 1 - H(t/N) - \frac{t}{N}\log_2 3. \tag{7.53}$$

7.4.2 Quantum Gilbert–Varshamov Bound

Let us consider an $[N,K]$ QECC with distance D. Following the analogy with classical q-ary LBC we established in the previous subsection, we expect that the quantum Gilbert–Varshamov bound is going to be the same as the classical one for $q = 4$:

$$\sum_{j=0}^{D-1} \binom{N}{j} 3^j 2^K \geq 2^N. \tag{7.54}$$

Because all basis codewords and all errors E from Pauli group G_N with wt(E) $< D$ must satisfy the equation

$$\langle \bar{i}|E|\bar{j}\rangle = C_E \delta_{\bar{j}\bar{j}}, \tag{7.55}$$

the number of such errors is $N_E = \sum_{j=0}^{D-1} \binom{N}{j} 3^j$. The following derivation is due to Gottesman [8] (see also Ref. [14]). Consider the state $|\psi_1\rangle$ satisfying (7.55) for all possible errors E, which can be used as a basis codeword. Let us observe now the-space orthogonal to both $|\psi_1\rangle$ and $E|\psi_1\rangle$. The dimensionality of this subspace is $2^N - N_E$. Consider now the state $|\psi_2\rangle$ from this subspace satisfying (7.55) and determine the subspace orthogonal to both $E|\psi_1\rangle$ and $E|\psi_2\rangle$. The dimensionality of this subspace is $2^N - 2N_E$. We can iterate this procedure i ($<2^K$) times. In the ith step the corresponding subspace will have dimensionality $2^N - iN_E > 0$, where $i < 2^K$. The inequality $2^N - iN_E > 0$ can be rewritten by substituting the expression N_E as follows:

$$\sum_{j=0}^{D-1} \binom{N}{j} 3^j i < 2^N; \quad i < 2^K. \tag{7.56}$$

Since i is strictly smaller than 2^K for $i = 2^K$ we obtain the inequality (7.54). Similarly to the quantum Hamming bound, the quantum Gilbert–Varshamov bound is identical to the classical Gilbert–Varshamov bound for q-ary LBCs by setting $q = 4$ (corresponding to the cardinality of the set of single-qubit errors $\{I,X,Y,Z\}$).

The asymptotic quantum Gilbert–Varshamov bound can be determined by following a similar procedure as for the asymptotic quantum Hamming bound to obtain:

$$\frac{K}{N} \geq 1 - H(d/N) - \frac{d}{N}\log_2 3. \tag{7.57}$$

By combining the asymptotic quantum Hamming and Gilbert–Varshamov bounds we obtain the following upper and lower bounds for quantum code rate $R = K/N$:

$$1 - H(D/N) - \frac{D}{N}\log_2 3 \leq \frac{K}{N} \leq 1 - H(t/N) - \frac{t}{N}\log_2 3. \tag{7.58}$$

7.4.3 Quantum Singleton Bound (Knill–Laflamme Bound)

In Chapter 6, we derived the classical Singleton bound for (n,k,d) LBC as follows:

$$d \leq n - k + 1 \Leftrightarrow n - k \geq d - 1 = 2t + 1 - 1 = 2t. \tag{7.59}$$

For the quantum $[N,K]$ codes obtained by CSS construction, we have $2M = 2(n - k)$ stabilizer conditions applied to N qubit states so that the corresponding Singleton bound will be $2M = 2(n - k) \geq 4t$. Because in CSS construction $N - 2M$ qubits are encoded into N qubits, the quantum Singleton bound becomes

$$N - K \geq 4t \Leftrightarrow N - K \geq 2(D - 1) \Leftrightarrow \frac{K}{N} \leq 1 - \frac{2}{N}(D - 1). \tag{7.60}$$

The quantum Singleton bound is also known as Knill-Laflamme bound [17]. Analogously to classical codes, the quantum codes satisfying the quantum Singleton bound with equality can be called *maximum distance separable* (MDS) quantum codes. By combining the asymptotic Gilbert–Varshamov and Singleton bounds we obtain the following bounds for quantum code rate:

$$1 - H(D/N) - \frac{D}{N}\log_2 3 \leq \frac{K}{N} \leq 1 - \frac{2}{N}(D - 1). \tag{7.61}$$

For entanglement-assisted-like QECCs [16,22], the Singleton bound can be derived as follows. Consider the $[N,K,D]$ QECC and assume that K Bell states are shared between the source and destination. We further encode the transmitter portion of Bell states only. The corresponding codeword can be represented in the following form:

$$\overbrace{\boldsymbol{d}}^{K}\ \overbrace{\boldsymbol{a}}^{D-1}\ \overbrace{\boldsymbol{b}}^{D-1}\ \overbrace{\boldsymbol{c}}^{N-2(D-1)}, \quad K = 2(D - 1), \tag{7.62}$$

where $\boldsymbol{d}$ is the receiver portion of Bell states, which stays unencoded. The von Neumann entropy of density operator ρ, introduced in Chapter 2, is defined as

$S(\rho) = -\mathrm{Tr}\,\rho\log\rho$. We know from the quantum information theory chapter that $S(\rho)$ vanishes for a non-entangled pure state, and it can be as large as $\log N_D$ for a maximally entangled state for two N_D-state systems $|\psi\rangle = \frac{1}{\sqrt{N_D}}\sum_i |ii\rangle$, where $|ij\rangle$ is the computational basis of a composite system. An important property of the von Neumann entropy is the *subadditivity property.* Let a composite system S be a bipartite system composed of component subsystems S_1 and S_2. The subadditivity property can be stated as $S(\rho(S_1S_2)) \leq S(\rho(S_1)) + S(\rho(S_2))$, with equality being satisfied if the component susbsystems are independent of each other: $\rho(S_1S_2) = \rho(S_1)\otimes\rho(S_2)$. It is clear from (7.62) that

$$S(\rho(\boldsymbol{da})) = S(\rho(\boldsymbol{bc}))\text{ and }S(\rho(\boldsymbol{db})) = S(\rho(\boldsymbol{ac})). \tag{7.63}$$

From the subadditivity property it follows that

$$S(\rho(\boldsymbol{bc})) \leq S(\rho(\boldsymbol{b})) + S(\rho(\boldsymbol{c}))\text{ and }S(\rho(\boldsymbol{ac})) \leq S(\rho(\boldsymbol{a})) + S(\rho(\boldsymbol{c})). \tag{7.64}$$

Since $\boldsymbol{d}$ and $\boldsymbol{a}$ ($\boldsymbol{d}$ and $\boldsymbol{b}$) blocks are independent of each other we obtain:

$$S(\rho(\boldsymbol{da})) = S(\rho(\boldsymbol{d})) + S(\rho(\boldsymbol{a}))\text{ and }S(\rho(\boldsymbol{db})) = S(\rho(\boldsymbol{d})) + S(\rho(\boldsymbol{b})). \tag{7.65}$$

By properly combining (7.63)—(7.65) we obtain:

$$\begin{aligned} &S(\rho(\boldsymbol{d})) \leq S(\rho(\boldsymbol{b})) + S(\rho(\boldsymbol{c})) - S(\rho(\boldsymbol{a})) \\ \text{and } &S(\rho(\boldsymbol{d})) \leq S(\rho(\boldsymbol{a})) + S(\rho(\boldsymbol{c})) - S(\rho(\boldsymbol{b})) \end{aligned} \tag{7.66}$$

By summing the two inequalities in (7.66) and dividing by 2 we obtain the Singleton bound:

$$\log_2 2^K \leq \log_2 2^{N-2(D-1)} \Leftrightarrow K \leq N - 2(D-1) \Leftrightarrow N - K \geq 2(D-1) \tag{7.67}$$

7.4.4 Quantum Weight Enumerators and Quantum MacWilliams Identity

We have shown in Chapter 6 that in classical error correction the codeword weight distribution can be used in the determination of word error probability and some other properties of the code. Similar ideas can be adopted in quantum error correction. Let A_w be the number of stabilizer S elements of weight w, and B_w be the number of elements of the same weight in the centralizer of S, denoted as $C(S)$. The polynomials $A(z) = \Sigma_w A_w z^w$ and $B(z) = \Sigma_w B_w z^w$ ($A_0 = B_0 = 1$) can be used as the *weight enumerators* of S and $C(S)$ respectively.

Let F be a finite field GF(q), where q is a prime power, and let F^N be a vector space of dimension N over F. A *linear code* C_F of length N over F is a subspace of F^N. Let $C_F{}^\perp$ denote a dual code of C_F and let the elements of F be denoted by $\omega_0 = 0, \omega_1, \ldots, \omega_{q-1}$. The *composition* of a vector $\boldsymbol{v} \in F^N$ is defined to be $\mathrm{comp}(\boldsymbol{v}) = \boldsymbol{s} =$

$(s_0, s_1, \ldots, s_{q-1})$, where s_i is the number of coordinates of $\boldsymbol{v}$ equal to ω_i. Clearly $\sum_{i=0}^{q-1} s_i = N$. The *Hamming weight* of vector $\boldsymbol{v}$, denoted as wt($\boldsymbol{v}$), is the number of nonzero coordinates, namely $\mathrm{wt}(\boldsymbol{v}) = \sum_{i=1}^{q-1} s_i(\boldsymbol{v})$. Let A_i be the number of vectors in C_F having $\mathrm{wt}(\boldsymbol{v}) = i$. Then $\{A_i\}$ will be the weight enumerator of C_F. Similarly, let $\{B_i\}$ denote the weight numerator of dual code. The weight enumerators are related by the following MacWilliams identity [19,20]:

$$\begin{aligned}\sum_{i=0}^{n} B_i z^i &= \frac{1}{|C_F|}\sum_{i=0}^{n} A_i (1+(q-1)z)^{n-i}(1-z)^i \\ &= \frac{1}{|C_F|}(1+(q-1)z)^n \sum_{i=0}^{n} A_i \left(\frac{1-z}{1+(q-1)z}\right)^i. \end{aligned} \tag{7.68}$$

The expression (7.68) can be rewritten in terms of polynomials introduced above as follows:

$$B(z) = \frac{1}{|C_F|}(1+(q-1)z)^n A\left(\frac{1-z}{1+(q-1)z}\right). \tag{7.69}$$

We have already shown above that both quantum Hamming and Gilbert–Varshamov bounds can be obtained as special cases of classical nonbinary codes for $q = 4$ (corresponding to the cardinality of set $\{I,X,Y,Z\}$). Therefore, the quantum MacWilliams identity for $[N,K]$ quantum code can be obtained from (7.69) by setting $q = 4$:

$$B(z) = \frac{1}{2^{N-K}}(1+3z)^n A\left(\frac{1-z}{1+3z}\right). \tag{7.70}$$

By matching the coefficients of z^i in (7.68) and by setting $q = 4$ we establish the relationship between B_i and A_i:

$$B_i = \frac{1}{2^{N-K}}\sum_{w=0}^{N}\left[\sum_{j=0}^{i}(-1)^j 3^{i-j}\binom{w}{j}\binom{N-w}{i-j}\right]A_w. \tag{7.71}$$

The weight distribution of classical RS (n,k,d) code with symbols from GF(q) is given by

$$A_w = \binom{n}{w}(q-1)\sum_{j=0}^{w-d}(-1)^j\binom{w-1}{j}q^{w-d-j}. \tag{7.72}$$

If the quantum code is derived from RS code then the weight distribution A_w can be found in closed form by setting $q = 4$ in (7.72).

Readers interested in rigorous derivation of (7.70) and (7.71) and some other quantum bounds are referred to Refs [8,14]. In the next section, we introduce the operator-sum representation that is a powerful tool in the study of different quantum channels. Namely we can describe the action of a quantum channel by so-called superoperator or quantum operation, expressed in terms of channel operation elements. Several important quantum channel models will then be discussed using this framework. This study is useful in quantum information theory to determine the quantum channel capacity and for quantum code design.

7.5 QUANTUM OPERATIONS (SUPEROPERATORS) AND QUANTUM CHANNEL MODELS

In this section we are concerned with the operator-sum representation of a quantum operation (also known as superoperator), and with various quantum channel models, including depolarizing channel, amplitude damping channel, and generalized amplitude damping channel. We will follow the interpretation of superoperators due to Gaitan [14].

7.5.1 Operator-Sum Representation

Let the composite system C be composed of quantum register Q and environment E. This kind of system can be modeled as a closed quantum system. Because the composite system is closed, its dynamic is unitary, and the final state is specified by a unitary operator U as $U(\rho\otimes\varepsilon_0)U^\dagger$, where ρ is a density operator of the initial state of quantum register Q, and ε_0 is the initial density operator of the environment E. The reduced density operator of Q upon interaction ρ_f can be obtained by tracing out the environment:

$$\rho_f = \mathrm{Tr}_E[U(\rho\otimes\varepsilon_0)U^\dagger]\equiv\xi(\rho). \tag{7.73}$$

The transformation (mapping) of the initial density operator ρ to the final density operator ρ_f, denoted as $\xi : \rho\rightarrow\rho_f$, given by Eq. (7.73), is often called the *superoperator* or *quantum operation*. The final density operator can be expressed in so-called *operator-sum representation* as follows:

$$\rho_f = \sum_k E_k\rho E_k^\dagger, \tag{7.74}$$

where E_k are the operation elements for the superoperator. The operator-sum representation is extremely suitable to represent the action of a channel and to study the quantum channel capacity, which is the subject of the quantum information theory chapter. To derive (7.74) we perform spectral decomposition of the environment initial state:

$$\varepsilon_0 = \sum_l \lambda_l|\phi_l\rangle\langle\phi_l|; \quad |\phi_l\rangle = \text{eigenkets } (\varepsilon_0) \tag{7.75}$$

and use the definition of the trace operation from Chapter 2, namely $\mathrm{Tr}(X) = \sum_i\langle a^{(i)}|X|a^{(i)}\rangle$. From the properties of trace operation we know that it is independent of the basis, and we choose for the basis the orthonormal basis of environmental Hilbert space H_E: $\{|e_m\rangle\}$. Therefore, by using spectral decomposition and trace definition, the final density operator becomes

$$\begin{aligned}\rho_f &= \sum_m \langle e_m|U\{\rho\otimes\varepsilon_0\}U^\dagger|e_m\rangle = \sum_{l,m}\lambda_l\langle e_m|U\{\rho\otimes|\phi_l\rangle\langle\phi_l|\}U^\dagger|e_m\rangle \\ &= \sum_{l,m} E_{lm}\rho E_{lm}^\dagger, \quad E_{lm} = \sqrt{\lambda_l}\langle e_m|U|\phi_l\rangle,\end{aligned} \tag{7.76}$$

where E_{lm} are the operation elements of the superoperator. By using a single subscript k instead of the double subscript lm, we obtain the operator-sum representation (7.74). The operator-sum representation can be used to classify quantum operations into two categories: (i) *trace preserving*, when $\mathrm{Tr}\ \xi(\rho) = \mathrm{Tr}\ \rho = 1$; and (ii) *nontrace preserving*, when $\mathrm{Tr}\ \xi(\rho) < 1$.

Starting from the trace-preserving condition:

$$\mathrm{Tr}\ \rho = \mathrm{Tr}\ \xi(\rho) = \mathrm{Tr}\Big[\sum_k E_k\rho E_k^\dagger\Big] = \mathrm{Tr}\Big[\rho\sum_k E_kE_k^\dagger\Big] = 1,$$

we obtain:

$$\sum_k E_kE_k^\dagger = I. \tag{7.77}$$

For nontrace-preserving quantum operation, Eq. (7.77) is not satisfied, and informally we can write $\sum_k E_kE_k^\dagger < I$.

Notice that in the derivation above we assumed that quantum register and environment are initially non-entangled and that the environment was in a pure state, which does not reduce the generality of the discussion. Observe again the situation in which the quantum register Q and its environment E are initially non-entangled, with corresponding density operators ρ and $\varepsilon_0 = |e_0\rangle\langle e_0|$ respectively. Upon interaction we perform *measurement* on observable $M = \Sigma_m\mu_mP_m$ of E, where $\{\mu_m\}$ are the eigenvalues of M and P_m is the projection operator onto the subspace of states of E with eigenvalues μ_m. For the outcome μ_m, the final (normalized) state of the composite system is given by

$$\rho_{\mathrm{tot}}^k = \frac{P_kU(\rho\otimes|e_0\rangle\langle e_0|)U^\dagger P_k}{\mathrm{Tr}(P_kU(\rho\otimes|e_0\rangle\langle e_0|)U^\dagger P_k)}. \tag{7.78}$$

The denominator can be simplified by applying the tracing per environment first:

$$\begin{aligned}\mathrm{Tr}(P_kU(\rho\otimes|e_0\rangle\langle e_0|)U^\dagger P_k) &= \mathrm{Tr}_Q\mathrm{Tr}_E(P_kU(\rho\otimes|e_0\rangle\langle e_0|)U^\dagger P_k) \\ &= \mathrm{Tr}_Q(E_k\rho E_k^\dagger), \quad E_k = \langle\mu_k|U|e_0\rangle,\end{aligned} \tag{7.79}$$

and by substituting this result in (7.78) we obtain:

$$\rho_{\text{tot}}^k = \frac{P_k U(\rho \otimes |e_0\rangle\langle e_0|) U^\dagger P_k}{\text{Tr}_Q(E_k \rho E_k^\dagger)}. \tag{7.80}$$

In practice, we are interested in the (normalized) reduced density operator for the quantum register Q, which can be obtained by tracing out the environment:

$$\rho^k = \frac{\text{Tr}_E(P_k U(\rho \otimes |e_0\rangle\langle e_0|) U^\dagger P_k)}{\text{Tr}_Q(E_k \rho E_k^\dagger)} = \frac{E_k \rho E_k^\dagger}{\text{Tr}_Q(E_k \rho E_k^\dagger)}. \tag{7.81}$$

In Chapter 2 we learned that the probability of obtaining μ_k from the measurement of M can be determined by $\Pr(\mu_k) = \text{Tr}(\rho_f P_k)$. Therefore, the *probability of measurement* being μ_k can be obtained from:

$$\begin{aligned} P(k) &= \Pr(\mu_k) = \text{Tr}[P_k U(\rho \otimes |e_0\rangle\langle e_0|) U^\dagger] \underset{\text{Tr } AB = \text{Tr } BA}{\overset{P_k^2 = P_k}{=}} \text{Tr}[P_k U(\rho \otimes |e_0\rangle\langle e_0|) U^\dagger P_k] \\ &= \text{Tr}_Q(E_k \rho E_k^\dagger). \end{aligned} \tag{7.82}$$

If the measurement outcome is not observed, the composite system will be in a mixed state:

$$\rho_{\text{tot}} = \sum_k P(k) \rho_{\text{tot}}^k. \tag{7.83}$$

The reduced density operator upon measurement will then be

$$\rho_Q = \sum_k P(k) \text{Tr}_E(\rho_{\text{tot}}^k) = \sum_k P(k) \rho_k = \sum_k E_k \rho E_k^\dagger. \tag{7.84}$$

Therefore, the representation of ρ_Q is the same as that of the operator-sum representation. In the derivation of (7.74) we did not perform any measurement; nevertheless, we obtained the same result. This conclusion can be formulated as the *principle of implicit measurement*: Once a subsystem E of composite system C has finished interacting with the rest of the composite system $Q = C - E$, the reduced density operator for Q will not be affected by any operations that are carried out solely on E. As a consequence of this principle, we can interpret the quantum operation of the environment to the quantum register as performing the mapping $\rho \to \rho_k$ with probability $P(k)$.

Without loss of generality let us assume that quantum register Q and the environment E are initially in product state $\rho \otimes \varepsilon_0$, and ε_0 is in the mixed state $\varepsilon_0 = \Sigma_l \lambda_l |\phi_l\rangle\langle\phi_l| (\text{Tr}(\varepsilon_0) = \Sigma_l \lambda_l = 1)$. Further, Q and E undergo the interaction described by U, and upon interaction we perform measurement, which leaves the state of E in subspace $S \subset H_E$, where H_E is the Hilbert space of the environment. The projection operator P_E associated with S and the superoperator can be represented in terms of the orthonormal basis for S $\{|g_m\rangle\}$ as follows:

$$P_E = \sum_m |g_m\rangle\langle g_m|. \tag{7.85}$$

Let $S^{\perp}$ denote the dual space with orthonormal basis $\{|h_n\rangle\}$. Since the Hilbert space of the environment H_E can be written as $S \oplus S^{\perp} = H_E$, the combined basis sets $\{|g_m\rangle\}$ and $\{|h_n\rangle\}$ can be used to create the basis for H_E. The reduced density operator for the quantum register after the measurement will be:

$$\xi(\rho) = \mathrm{Tr}_E(P_E U(\rho \otimes \varepsilon_0)U^{\dagger} P_E) = \sum_{l,m} E_{l,m}\rho E_{l,m}^{\dagger}, \quad E_{lm} = \sqrt{\lambda_l}\langle g_m|U|\phi_l\rangle. \tag{7.86}$$

The probability that after the measurement the state of E is found in the subspace S is then as follows:

$$P(S) = \mathrm{Tr}_Q(\xi(\rho)), \tag{7.87}$$

because $P(S)$ is the probability $0 \le \mathrm{Tr}_Q(\xi(\rho)) \le 1$, and the quantum operation is nontrace preserving. If we do not observe the outcome of the measurement the corresponding setting is called *nonselective dynamics*, otherwise it is called *selective dynamics*. For nonselective dynamics it is clear that $P_E = I_E$ and the set $\{|g_m\rangle\}$ spans H_E (meaning that $H_E = S$). Because of this fact we expect that the completeness relationship involving the operation elements E_{lm} from (7.84) is satisfied:

$$\begin{aligned}\sum_{lm} E_{lm}^{\dagger}E_{lm} &= \sum_{lm} \lambda_l \langle\phi_l|U^{\dagger}|g_m\rangle\langle g_m|U|\phi_l\rangle \\ &= \sum_{l} \lambda_l \langle\phi_l|U^{\dagger}U|\phi_l\rangle \overset{U^{\dagger}U = I_Q \otimes I_E}{=} \sum_{l} \lambda_l \langle\phi_l|\phi_l\rangle I_Q \\ &= I_Q \sum_{l} \lambda_l \overset{\sum_l \lambda_l = \mathrm{Tr}\,\varepsilon_0 = 1}{=} I_Q.\end{aligned} \tag{7.88}$$

The quantum operation now is clearly trace preserving and nothing has been learned about the final state of the environment. Therefore, nonselective dynamics corresponds to trace-preserving quantum operations.

In selective dynamics we observe the outcome of the measurement of E so that now $P_E \neq I_E$ and $S \subset H_E$. Since $P(S) + P(S^{\perp}) = 1$, the trace of quantum operation is $\mathrm{Tr}(\xi(\rho)) = P(S) = 1 - P(S^{\perp}) < 1$ and the quantum operation is not trace preserving. Therefore, since the selective dynamics of quantum operations is non-trace preserving we can get some knowledge about the final state of the environment. We have already mentioned above that the Hilbert space of the environment H_E can be written as $S \oplus S^{\perp} = H_E$, and the combined basis set $\{|g_m\rangle\}$ of S and basis set $\{|h_n\rangle\}$ of $S^{\perp}$ can be used as the basis for H_E. We also defined earlier the element operations corresponding to S as $E_{lm} = \sqrt{\lambda_l}\langle g_m|U|\phi_l\rangle$. In similar fashion we define the element operations corresponding to $S^{\perp}$ as $H_{ln} = \sqrt{\lambda_l}\langle h_n|U|\phi_l\rangle$. We can show that the completeness relation is now satisfied when both element operations E_{lm} and H_{ln} are involved:

$$\sum_{lm} E_{lm}^{\dagger}E_{lm} + \sum_{\text{ln}} H_{\text{ln}}^{\dagger}H_{ln} = \sum_{lm} \lambda_l \langle\phi_l|U^{\dagger}|g_m\rangle\langle g_m|U|\phi_l\rangle + \sum_{ln} \lambda_l \langle\phi_l|U^{\dagger}|h_n\rangle\langle h_n|U|\phi_l\rangle$$
$$= \sum_{l} \lambda_l \langle\phi_l|U^{\dagger}\left\{\sum_{m}|g_m\rangle\langle g_m| + \sum_{n}|h_n\rangle\langle h_n|\right\}U|\phi_l\rangle$$
$$= \sum_{l} \lambda_l \langle\phi_l|U^{\dagger}U|\phi_l\rangle = I_Q. \tag{7.89}$$

This derivation justifies our previous claim that for nonpreserving operations we can write informally that $\Sigma_{lm}E_{lm}E_{lm}^{\dagger} < I$.

It is clear from the discussion above that the operator-sum representation is base dependent. By choosing a different basis for the environment and a different basis for S from those discussed above, we will obtain a different operator-sum representation. Therefore, the operator-sum representation is not unique. We also expect that we should be able to transform one set of elementary operations used in the operator-sum representation to another set of elementary operations by properly choosing the unitary transformation. In the remainder of this section we describe several important quantum channels by employing the operator-sum representation.

7.5.2 Depolarizing Channel

The depolarizing channel (see Figure 7.14) with probability $1-p$ leaves the qubit as it is, while with probability p it moves the initial state into $\rho_f = I/2$ that maximizes the von Neumann entropy $S(\rho) = -\text{Tr}\ \rho \log \rho = 1$. The properties describing the model can be summarized as follows:

1. Qubit errors are independent,
2. Single-qubit errors (X, Y, Z) are equally likely, and
3. All qubits have the same single-error probability $p/4$.

Operation elements E_i of the channel model should be selected as follows:

$$E_0 = \sqrt{1-3p/4I};\ E_1 = \sqrt{p/4X};\ E_1 = \sqrt{p/4X};\ E_1 = \sqrt{p/4Z}. \tag{7.90}$$

The action of a depolarizing channel is to perform the mapping $\rho \to \xi(\rho) = \sum_i E_i \rho E_i^{\dagger}$, where ρ is the initial density operator. Without loss of generality we will assume that the initial state was pure $|\psi\rangle = a|0\rangle + b|1\rangle$, so that

$$\rho = |\psi\rangle\langle\psi| = (a|0\rangle + b|1\rangle)(\langle 0|a^* + \langle 1|b^*)$$
$$= |a|^2\begin{bmatrix}1 & 0\\ 0 & 0\end{bmatrix} + ab^*\begin{bmatrix}0 & 1\\ 0 & 0\end{bmatrix} + a^*b\begin{bmatrix}0 & 0\\ 1 & 0\end{bmatrix} + |b|^2\begin{bmatrix}0 & 0\\ 0 & 1\end{bmatrix} = \begin{bmatrix}|a|^2 & ab^*\\ a^*b & |b|^2\end{bmatrix}. \tag{7.91}$$

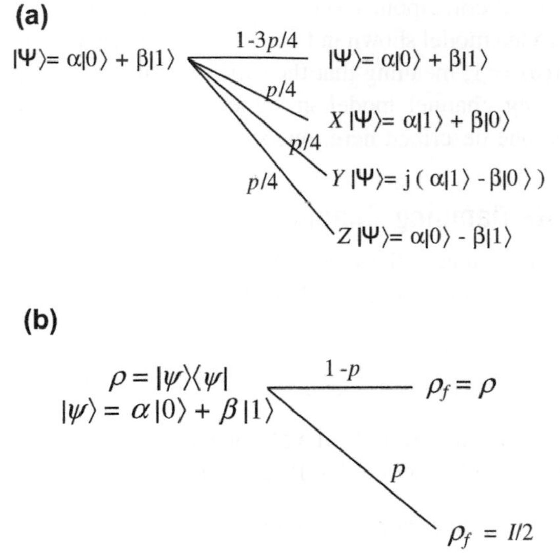

FIGURE 7.14

Depolarizing channel model: (a) operator description; (b) density operator description.

The resulting quantum operation can be represented using the operator-sum representation as follows:

$$\begin{aligned}
\xi(\rho) &= \sum_i E_i \rho E_i^{\dagger} = \left(1-\frac{3p}{4}\right)\rho + \frac{p}{4}(X\rho X + Y\rho Y + Z\rho Z) \\
&= \left(1-\frac{3p}{4}\right)\begin{bmatrix} |a|^2 & ab^* \\ a^*b & |b|^2 \end{bmatrix} + \frac{p}{4}\begin{bmatrix} 0 & 1 \\ 1 & 0 \end{bmatrix}\begin{bmatrix} |a|^2 & ab^* \\ a^*b & |b|^2 \end{bmatrix}\begin{bmatrix} 0 & 1 \\ 1 & 0 \end{bmatrix} \\
&\quad + \frac{p}{4}\begin{bmatrix} 1 & 0 \\ 0 & -1 \end{bmatrix}\begin{bmatrix} |a|^2 & ab^* \\ a^*b & |b|^2 \end{bmatrix}\begin{bmatrix} 1 & 0 \\ 0 & -1 \end{bmatrix} + \frac{p}{4}\begin{bmatrix} 0 & -j \\ j & 0 \end{bmatrix}\begin{bmatrix} |a|^2 & ab^* \\ a^*b & |b|^2 \end{bmatrix}\begin{bmatrix} 0 & -j \\ j & 0 \end{bmatrix} \\
&= \left(1-\frac{3p}{4}\right)\begin{bmatrix} |a|^2 & ab^* \\ a^*b & |b|^2 \end{bmatrix} + \frac{p}{4}\begin{bmatrix} |b|^2 & a^*b \\ ab^* & |a|^2 \end{bmatrix} + \frac{p}{4}\begin{bmatrix} |a|^2 & -ab^* \\ -a^*b & |b|^2 \end{bmatrix} \\
&\quad + \frac{p}{4}\begin{bmatrix} |b|^2 & -a^*b \\ -ab^* & |a|^2 \end{bmatrix} = \begin{bmatrix} \left(1-\frac{p}{2}\right)|a|^2 + \frac{p}{2}|b|^2 & (1-p)ab^* \\ (1-p)a^*b & \frac{p}{2}|a|^2 + (1-\frac{p}{2})|b|^2 \end{bmatrix} \\
&= \begin{bmatrix} (1-p)|a|^2 + \frac{p}{2}(|a|^2+|b|^2) & (1-p)ab^* \\ (1-p)a^*b & \frac{p}{2}(|a|^2+|b|^2) + (1-p)|b|^2 \end{bmatrix} \\
&\overset{|a|^2+|b|^2=1}{=} (1-p)\begin{bmatrix} |a|^2 & ab^* \\ a^*b & |b|^2 \end{bmatrix} + \frac{p}{2}I = (1-p)\rho + \frac{p}{2}I.
\end{aligned} \tag{7.92}$$

The first line in (7.92) corresponds to the model shown in Figure 7.14a and the last line corresponds to the model shown in Figure 7.14b. It is clear from (7.91) and (7.92) that Tr $\xi(\rho) = \text{Tr}(\rho) = 1$, meaning that the superoperator is trace preserving. Notice that the depolarizing channel model in some other books/papers can be slightly different from the one described here; one such model is given in Problem 15.

7.5.3 Amplitude Damping Channel

In certain quantum channels the errors X, Y, and Z do not occur with the same probability. In an amplitude damping channel, the operation elements are given by

$$E_0 = \begin{pmatrix} 1 & 0 \\ 0 & \sqrt{1-\varepsilon^2} \end{pmatrix}, \qquad E_1 = \begin{pmatrix} 0 & \varepsilon \\ 0 & 0 \end{pmatrix}. \tag{7.93}$$

Spontaneous emission is an example of a physical process that can be modeled using the amplitude damping channel model. If $|\psi\rangle = a\,|0\rangle + b|1\rangle$ is the initial qubit state $\left(\rho = \begin{bmatrix} |a|^2 & ab^* \\ a^*b & |b|^2 \end{bmatrix}\right)$, the effect of the amplitude damping channel is to perform the following mapping:

$$\begin{aligned} \rho \to \xi(\rho) &= E_0 \rho E_0^\dagger + E_1 \rho E_1^\dagger = \begin{pmatrix} 1 & 0 \\ 0 & \sqrt{1-\varepsilon^2} \end{pmatrix} \begin{bmatrix} |a|^2 & ab^* \\ a^*b & |b|^2 \end{bmatrix} \begin{pmatrix} 1 & 0 \\ 0 & \sqrt{1-\varepsilon^2} \end{pmatrix} \\ &+ \begin{pmatrix} 0 & \varepsilon \\ 0 & 0 \end{pmatrix} \begin{bmatrix} |a|^2 & ab^* \\ a^*b & |b|^2 \end{bmatrix} \begin{pmatrix} 0 & 0 \\ \varepsilon & 0 \end{pmatrix} \\ &= \begin{pmatrix} |a|^2 & ab^*\sqrt{1-\varepsilon^2} \\ a^*b\sqrt{1-\varepsilon^2} & |b|^2(1-\varepsilon^2) \end{pmatrix} + \begin{pmatrix} |b|^2\varepsilon^2 & 0 \\ 0 & 0 \end{pmatrix} \\ &= \begin{pmatrix} |a|^2+\varepsilon^2|b|^2 & ab^*\sqrt{1-\varepsilon^2} \\ a^*b\sqrt{1-\varepsilon^2} & |b|^2(1-\varepsilon^2) \end{pmatrix}. \end{aligned} \tag{7.94}$$

The probabilities $P(0)$ and $P(1)$ that E_0 and E_1 occur are given by

$$\begin{aligned} P(0) &= \text{Tr}(E_0 \rho E_0^\dagger) = \text{Tr}\begin{pmatrix} |a|^2 & ab^*\sqrt{1-\varepsilon^2} \\ a^*b\sqrt{1-\varepsilon^2} & |b|^2(1-\varepsilon^2) \end{pmatrix} = 1 - \varepsilon^2|b|^2 \\ P(1) &= \text{Tr}\left(E_1 \rho E_1^\dagger\right) = \text{Tr}\begin{pmatrix} |b|^2\varepsilon^2 & 0 \\ 0 & 0 \end{pmatrix} = \varepsilon^2|b|^2. \end{aligned} \tag{7.95}$$

The corresponding amplitude damping channel model is shown in Figure 7.15.

We have shown in Chapter 2 that for a two-level system the density operator can be written as $\rho = (I + \boldsymbol{R}\cdot\boldsymbol{\sigma})/2$, where $\boldsymbol{\sigma} = [X\ \ Y\ \ Z]$ and $\boldsymbol{R} = [R_x\ R_y\ R_z]$ is the Bloch vector, whose components can be determined by $R_x = \text{Tr}(\rho X)$,

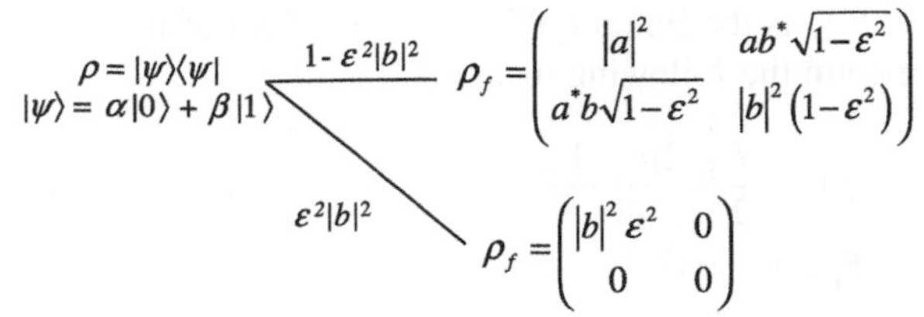

FIGURE 7.15

Amplitude damping channel model.

$R_y = \mathrm{Tr}(\rho Y)$, $R_z = \mathrm{Tr}(\rho Z)$. Using this result we can determine the Bloch vector $\boldsymbol{R}_\rho$ for ρ by

$$R_{\rho,x} = \mathrm{Tr}(\rho X) = \mathrm{Tr}\left(\begin{bmatrix} |a|^2 & ab^* \\ a^*b & |b|^2 \end{bmatrix}\begin{bmatrix} 0 & 1 \\ 1 & 0 \end{bmatrix}\right) = ab^* + a^*b$$
$$= 2\mathrm{Re}(a^*b) \quad R_{\rho,y} = 2\mathrm{Im}(a^*b) \quad R_{\rho,z} = |a|^2 - |b|^2 \tag{7.96}$$

and $\boldsymbol{R}_\xi$ for $\xi(\rho)$ by

$$R_{\xi,x} = \mathrm{Tr}(\rho_f X) = \mathrm{Tr}\left(\begin{pmatrix} |a|^2 + \varepsilon^2|b|^2 & ab^*\sqrt{1-\varepsilon^2} \\ a^*b\sqrt{1-\varepsilon^2} & |b|^2(1-\varepsilon^2) \end{pmatrix}\begin{bmatrix} 0 & 1 \\ 1 & 0 \end{bmatrix}\right)$$
$$= (ab^* + a^*b)\sqrt{1-\varepsilon^2} = R_{\rho,x}\sqrt{1-\varepsilon^2} \tag{7.97}$$
$$R_{\xi,y} = R_{\rho,y}\sqrt{1-\varepsilon^2} \quad R_{\xi,z} = R_{\rho,z}(1-\varepsilon^2) + \varepsilon^2.$$

Example If the initial state was $|1\rangle$ the Bloch vectors $\boldsymbol{R}_\rho$ and $\boldsymbol{R}\xi$ for ρ and $\xi(\rho)$ are given respectively by

$$\boldsymbol{R}_\rho = -\hat{z} \qquad \boldsymbol{R}_\xi = (-1 + 2\varepsilon^2)\hat{z}.$$

Clearly, the initial amplitude of the Bloch vector is reduced. The error introduced by the channel can be interpreted as $X + jY$.

Example On the other hand, if the initial state was $|0\rangle$, $\xi(\rho)$ is given by $\xi(\rho) = |0\rangle\langle 0|$, suggesting that the initial state was not changed at all. Therefore, $\Pr(|1\rangle \rightarrow |0\rangle) = \varepsilon^2$.

7.5.4 Generalized Amplitude Damping Channel

In a generalized amplitude damping channel, the operation elements are given by

$$E_0 = \sqrt{p}\begin{pmatrix} 1 & 0 \\ 0 & \sqrt{1-\gamma} \end{pmatrix}, \qquad E_1 = \sqrt{p}\begin{pmatrix} 0 & \sqrt{\gamma} \\ 0 & 0 \end{pmatrix},$$
$$E_2 = \sqrt{1-p}\begin{pmatrix} \sqrt{1-\gamma} & 0 \\ 0 & 1 \end{pmatrix}, \qquad E_3 = \sqrt{1-p}\begin{pmatrix} 0 & 0 \\ \sqrt{\gamma} & 0 \end{pmatrix}. \tag{7.98}$$

If $|\psi\rangle = a\,|0\rangle + b|1\rangle$ is the initial qubit state, the effect of the amplitude damping channel is to perform the following mapping:

$$\rho \to \xi(\rho) = \xi(\rho) = \frac{I}{2} + \frac{2p-1}{2}Z + \frac{1}{2}[R_{\rho,x}\sqrt{1-\gamma}X + R_{\rho,y}\sqrt{1-\gamma}Y + R_{\rho,z}(1-\gamma)Z], \tag{7.99}$$

where the initial Bloch vector is given by $R_\rho = (R_{\rho,x}, R_{\rho,y}, R_{\rho,z}) = (2\mathrm{Re}(a^*b), 2\mathrm{Im}(a^*b), |a|^2 - |b|^2)$. The operation elements E_i ($i = 0,1,2,3$) occur with probabilities:

$$\begin{aligned} P(0) &= \mathrm{Tr}(E_0\rho E_0^\dagger) &&= p\left[1 - \frac{\gamma}{2}(1 - R_{\rho,z})\right] \\ P(1) &= \mathrm{Tr}(E_1\rho E_1^\dagger) &&= p\frac{\gamma}{2}(1 - R_{\rho,z}) \\ P(2) &= \mathrm{Tr}(E_2\rho E_2^\dagger) &&= (1-p)\left[1 - \frac{\gamma}{2}(1 + R_{\rho,z})\right] \\ P(3) &= \mathrm{Tr}(E_3\rho E_3^\dagger) &&= (1-p)\frac{\gamma}{2}(1 + R_{\rho,z}). \end{aligned} \tag{7.100}$$

The quantum errors considered so far are uncorrelated. If the error probability of correlated errors drops sufficiently rapidly with the number of errors so that correlated errors with more than t errors are unlikely, we can use quantum error correction that can fix up to t errors. Errors that "knock" the qubit state out of this two-dimensional Hilbert space are known as *leakage errors*.

7.6 SUMMARY

This chapter has considered quantum error correction concepts, ranging from an intuitive description to a rigorous mathematical framework. After the Pauli operators were introduced, we described basic quantum codes, such as three-qubit flip code, three-qubit phase-flip code, Shor's nine-qubit code, stabilizer codes, and CSS codes. We then formally introduced quantum error correction, including quantum error correction mapping, quantum error representation, stabilizer group definition, quantum-check matrix representation, and the quantum syndrome equation. We further provided the necessary and sufficient conditions for quantum error correction, discussed the distance properties and error correction capability, and revisited the CSS codes. Section 7.4 analyzed important quantum coding bounds, including the quantum Hamming bound, Gilbert—Varshamov bound, and Singleton bound. In the same section we also discussed quantum weight enumerators and quantum MacWilliams identities. We further discussed the quantum superoperators and various quantum channels, including quantum depolarization, amplitude damping, and generalized amplitude damping channels. For a deeper understanding of quantum error correction, in Section 7.7 we provide a set of problems.

7.7 PROBLEMS

1. The set of elements from the Pauli group G_N that commute with all elements from stabilizer S is called the *centralizer* of S, denoted by $C(S)$. On the other hand, the set of elements from the Pauli group G_N that fix S under conjugation is called the *normalizer* of S, denoted as $N(S)$. In other words, $N(S) = \{e \in G_N | eSe^\dagger = S\}$.
   (a) Show that the centralizer of S is a subgroup of G_N.
   (b) Show that $N(S) = C(S)$.
2. Let C_q be a quantum error correcting code with distance D. Prove that this code is degenerate code if and only if its stabilizer S has an element with weight less than D (excluding the identity element).
3. In order to ensure that the set of N-qubit errors G_N, defined by (7.20), is a group, commonly we need to work with the *quotient group* G_N/C, where $C = \{\pm I, \pm jI\}$; so that the coset $eC = \{\pm e, \pm je\}$ is considered as a single error since all errors from this coset have the same syndrome.
   (a) Show that the orders of G_N and G_N/C are 2^{2N+2} and 2^{2N} respectively.
   (b) Show that $\forall\ e_1, e_2 \in G_N$: $[e_1,e_2] = 0$ or $\{e_1,e_2\} = 0$. Therefore, any two errors from Pauli group G_N either commute or anticommute.
   (c) Show that $\forall\ e \in G_N$ the following is valid: (i) $e^2 = \pm I$; (ii) $e^+ = \pm e$; (iii) $e^{-1} = e^+$.
4. Let us observe Shor's [9,1,3] code, whose basis codewords are given by

$$\begin{aligned}|\bar{0}\rangle &= \frac{1}{2\sqrt{2}}(|000\rangle + |111\rangle)(|000\rangle + |111\rangle)(|000\rangle + |111\rangle)\\ &= \frac{1}{2\sqrt{2}}(|000000000\rangle + |000000111\rangle + |000111000\rangle + |111000000\rangle +\\ &\qquad |000111111\rangle + |111111000\rangle + |111000111\rangle + |111111111\rangle)\end{aligned}$$

$$\begin{aligned}|\bar{1}\rangle &= \frac{1}{2\sqrt{2}}(|000\rangle - |111\rangle)(|000\rangle - |111\rangle)(|000\rangle - |111\rangle)\\ &= \frac{1}{2\sqrt{2}}(|000000000\rangle - |000000111\rangle - |000111000\rangle - |111000000\rangle +\\ &\qquad |000111111\rangle + |111111000\rangle + |111000111\rangle - |111111111\rangle).\end{aligned}$$

   (a) By using the fact that this code can be obtained by concatenating two [3,1,1] codes, its generators can be written as:

$$\begin{aligned}g_1 &= Z_1Z_2 \quad g_2 = Z_1Z_3 \quad g_3 = Z_4Z_5\\ g_4 &= Z_4Z_6 \quad g_5 = Z_7Z_8 \quad g_6 = Z_7Z_9\\ g_7 &= X_1X_2X_3X_4X_5X_6 \qquad g_8 = X_1X_2X_3X_7X_8X_9.\end{aligned}$$

Show that these generators fix the basis codewords.

(b) Find the distance of this code.

(c) Determine if this code is degenerate or nondegenerate.

5. The quantum-check matrix is given by

$$A = \left[\begin{array}{c|c} 1001 & 0110 \\ 1111 & 1001 \end{array}\right]$$

(a) Determine the generators of the corresponding stabilizer code.

(b) Find the parameters $[N,K]$ of the code.

(c) Determine the stabilizer of this code.

(d) Find the distance of the code.

(e) Determine if this code is degenerate or nondegenerate.

(f) Provide the syndrome decoder implementation.

6. The quantum-check matrix is given by

$$A = \left[\begin{array}{c|c} 11111111 & 00000000 \\ 00111100 & 00010111 \\ 01011010 & 00101011 \\ 00101011 & 01001101 \\ 01001101 & 10001110 \end{array}\right].$$

(a) Show that the twisted product between any two rows is zero.

(b) Determine the generators of the corresponding stabilizer code.

(c) Find the parameters $[N,K,D]$ of the code.

(d) Provide the syndrome decoder implementation.

7. The quantum check matrix $\boldsymbol{A}$ of a CSS code is given by

$$\boldsymbol{A} = \left[\begin{array}{c|c} \boldsymbol{H} & \boldsymbol{0} \\ \boldsymbol{0} & \boldsymbol{H} \end{array}\right].$$

Can the classical code with matrix $\boldsymbol{H}$ below be used in CSS code? If not, why? Can you modify the matrix $\boldsymbol{H}$ so that it can be used to construct the corresponding CSS code? Determine the parameters of the obtained code.

$$\boldsymbol{H} = \begin{bmatrix} 0 & 0 & 1 & 0 & 1 & 0 & 1 \\ 1 & 0 & 0 & 1 & 1 & 0 & 0 \\ 0 & 1 & 0 & 1 & 0 & 0 & 1 \\ 1 & 0 & 0 & 0 & 0 & 1 & 1 \\ 0 & 0 & 1 & 1 & 0 & 1 & 0 \\ 1 & 1 & 1 & 0 & 0 & 0 & 0 \\ 0 & 1 & 0 & 0 & 1 & 1 & 0 \end{bmatrix}.$$

8. Prove that the quantum code C_q is an $\boldsymbol{E}$ error-correcting code if and only if $\forall\ |\bar{i}\rangle, |\bar{j}\rangle\ (\bar{i} \neq \bar{j})$ and $\forall\ E_a,\ E_b \in \boldsymbol{E}$, the following two conditions are valid:

$$\langle \bar{i}|E_a^\dagger E_b|\bar{i}\rangle = \langle \bar{j}|E_a^\dagger E_b|\bar{j}\rangle \qquad \langle \bar{i}|E_a^\dagger E_b|\bar{j}\rangle = 0.$$

9. For the decoder shown in Figure 7.P9 provide the corresponding generators and determine the code parameters. Also determine the quantum-check matrix. Can this code be related to CSS codes? If yes, how?

10. For the encoder shown in Figure 7.P10 provide the corresponding decoder. For the obtained decoder provide the corresponding generators and determine the code parameters. Finally, provide the quantum-check matrix.

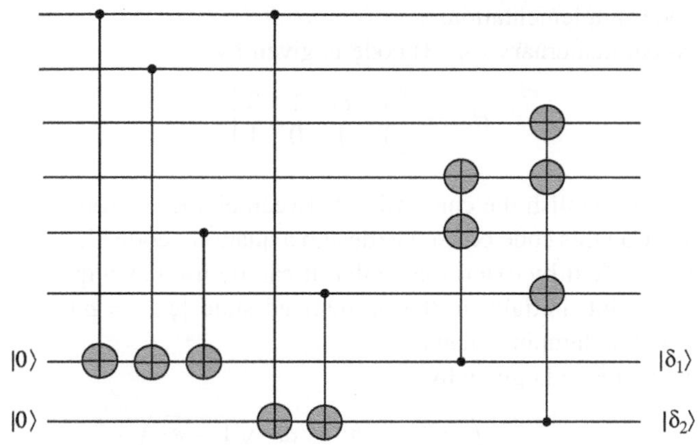

FIGURE 7.P9

Decoder circuit under study. The information qubits are denoted by $|\delta_1\rangle$ and $|\delta_2\rangle$.

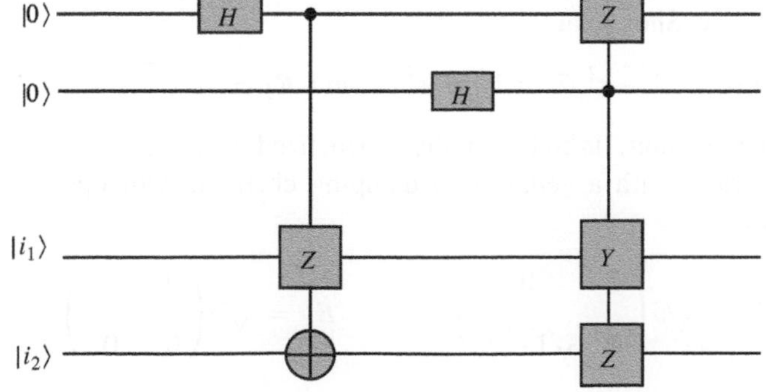

FIGURE 7.P10

Encoder circuit under study. The information qubits are denoted by $|i_1\rangle$ and $|i_2\rangle$.

11. Let the dual-containing CSS code be given by the following quantum-check matrix:

$$A = \begin{bmatrix} H & | & 0 \\ 0 & | & H \end{bmatrix} \qquad H = \begin{bmatrix} 1 & 0 & 0 & 1 & 1 & 1 \\ 1 & 1 & 1 & 0 & 0 & 1 \\ 0 & 1 & 1 & 1 & 1 & 0 \end{bmatrix}.$$

(a) By using MacKay's encoder implementation shown in Figure 7.13, provide the corresponding encoder implementation.

(b) Based on the quantum-check matrix above and the syndrome decoder approach implementation shown in Figure 7.12, provide the corresponding decoder implementation.

12. The classical quaternary (4,2,3) code is given by

$$H_4 = \begin{bmatrix} 1 & \omega & 1 & 0 \\ 1 & 1 & 0 & 1 \end{bmatrix}.$$

How can we establish the connection between elements from GF(4) and Pauli operators? Can this code be used to design a quantum code? If not, explain why. Can you modify it by extension so that it can be used as a quantum code?

13 Consider a qubit, initially in the normalized state $|\psi> = a|0> + b|1>$, that interacts with a damping channel.

(a) Show that $\xi(\rho)$ is given by

$$\xi(\rho) = \begin{pmatrix} |a|^2 + \varepsilon^2 |b|^2 & ab^* \sqrt{1-\varepsilon^2} \\ ab^* \sqrt{1-\varepsilon^2} & |b|^2(1-\varepsilon^2) \end{pmatrix}.$$

(b) Derive the probabilities $P(0)$ and $P(1)$ that errors E_0 and E_1 will occur. Also derive the initial and final Bloch vectors $\boldsymbol{R}_\rho$ and $\boldsymbol{R}_\xi$.

(c) Let $\rho = (1/2)(|0> + |1>)(<0| + <1|)$, corresponding to Bloch vector $\boldsymbol{R}_\rho = \hat{\boldsymbol{x}}$. Show that

$$p(0) = 1 - \varepsilon^2/2,\ p(1) = \varepsilon^2/2 \text{ and } \boldsymbol{R}_\xi = \sqrt{1-\varepsilon^2}\hat{\boldsymbol{x}} + \varepsilon^2 \hat{\boldsymbol{z}}.$$

14. Consider a qubit, initially in the normalized state $|\psi> = a|0> + b|1>$, that interacts with a *generalized* damping channel, with operation elements given by

$$E_0 = \sqrt{p}\begin{pmatrix} 1 & 0 \\ 0 & \sqrt{1-\gamma} \end{pmatrix}, \qquad E_1 = \sqrt{p}\begin{pmatrix} 0 & \sqrt{\gamma} \\ 0 & 0 \end{pmatrix}$$

$$E_2 = \sqrt{1-p}\begin{pmatrix} \sqrt{1-\gamma} & 0 \\ 0 & 1 \end{pmatrix}, \qquad E_3 = \sqrt{1-p}\begin{pmatrix} 0 & 0 \\ \sqrt{\gamma} & 0 \end{pmatrix}.$$

The Bloch vector $\boldsymbol{R}_\rho$ for initial $\rho = |\psi\rangle\langle\psi|$ is given by

$$R_{\rho,x} = 2\mathrm{Re}(a^*b) \qquad R_{\rho,y} = 2\mathrm{Im}(a^*b) \qquad R_{\rho,z} = |a|^2 - |b|^2.$$

(a) Show that the quantum operation associated with this channel is trace preserving.
(b) Determine $\xi(\rho)$ and the final Bloch vector $\boldsymbol{R}_\xi$.
(c) Determine the probabilities $P(i)$ that processes E_i $(i = 0,1,2,3)$ occur.
(d) Show that density matrix $\rho_{fp} = (1/2)(I + \boldsymbol{R}_{fp}\cdot\boldsymbol{\sigma})$, $\boldsymbol{R}_{fp} = (2p-1)\hat{z}$ is a fixed point for the channel $\xi(\rho_{fp}) = \rho_{fp}$.

15. Consider a qubit, initially in the normalized state $|\psi\rangle = a|0\rangle + b|1\rangle$, that interacts with a *depolarizing* channel, with operation elements now given by

$$E_1 = \sqrt{1-p}I, \qquad E_2 = \sqrt{\frac{p}{3}}X$$

$$E_3 = \sqrt{\frac{p}{3}}Z, \qquad E_4 = -j\sqrt{\frac{p}{3}}Y.$$

where X, Y, and Z are the Pauli matrices. The depolarizing channel leaves the qubit intact with probability $1-p$; with probability $p/3$ it performs either E_2 or E_3 or E_4.

(a) Is the quantum operation associated with this channel trace preserving or nontrace preserving?
(b) Determine the superoperator $\xi(\rho)$, where ρ is the initial state density operator.
(c) Determine the fidelity F, which represents the "quality of transmitted qubit", defined as

$$F = \sum_i (\mathrm{Tr}\,\rho E_i)(\mathrm{Tr}\,\rho E_i^\dagger)$$

References

[1] A.R. Calderbank, P.W. Shor, Good quantum error-correcting codes exist, Phys. Rev. A 54 (1996) 1098–1105.

[2] A.M. Steane, Error correcting codes in quantum theory, Phys. Rev. Lett. 77 (1996) 793.

[3] A.M. Steane, Simple quantum error-correcting codes, Phys. Rev. A 54 (6) (December 1996) 4741–4751.

[4] A.R. Calderbank, E.M. Rains, P.W. Shor, N.J.A. Sloane, Quantum error correction and orthogonal geometry, Phy. Rev. Lett. 78 (3) (January 1997) 405–408.

[5] E. Knill, R. Laflamme, Concatenated Quantum Codes (1996). Available at: http://arxiv.org/abs/quant-ph/9608012.

[6] R. Laflamme, C. Miquel, J.P. Paz, W.H. Zurek, Perfect quantum error correcting code, Phys. Rev. Lett. 77 (1) (July 1996) 198–201.

[7] D. Gottesman, Class of quantum error correcting codes saturating the quantum Hamming bound, Phys. Rev. A 54 (September 1996) 1862–1868.
[8] D. Gottesman, Stabilizer Codes and Quantum Error Correction. Ph.D. dissertation, California Institute of Technology, Pasadena, CA, 1997.
[9] R. Cleve, D. Gottesman, Efficient computations of encoding for quantum error correction, Phys. Rev. A 56 (July 1997) 76–82.
[10] A.Y. Kitaev, Quantum error correction with imperfect gates, in: Quantum Communication, Computing, and Measurement, Plenum Press, New York, 1997, pp. 181–188.
[11] C.H. Bennett, D.P. DiVincenzo, J.A. Smolin, W.K. Wootters, Mixed-state entanglement and quantum error correction, Phys. Rev. A 54 (6) (November 1996) 3824–3851.
[12] D.J.C. MacKay, G. Mitchison, P.L. McFadden, Sparse-graph codes for quantum error correction, IEEE Trans. Inform. Theory 50 (2004) 2315–2330.
[13] M.A. Neilsen, I.L. Chuang, Quantum Computation and Quantum Information, Cambridge University Press, Cambridge, 2000.
[14] F. Gaitan, Quantum Error Correction and Fault Tolerant Quantum Computing, CRC Press, 2008.
[15] G.D. Forney Jr., M. Grassl, S. Guha, Convolutional and tail-biting quantum error-correcting codes, IEEE Trans. Inform. Theory 53 (March 2007) 865–880.
[16] M.-H. Hsieh, Entanglement-Assisted Coding Theory. Ph.D. dissertation, University of Southern California, August 2008.
[17] E. Knill, R. Laflamme, Theory of quantum error-correcting codes, Phys. Rev. A 55 (1997) 900.
[18] J. Preskill, Ph219/CS219 Quantum Computing (Lecture Notes), Caltech (2009). Available at: http://theory.caltech.edu/people/preskill/ph229/.
[19] F.J. MacWilliams, N.J.A. Sloane, J.-M. Goethals, The MacWilliams identities for nonlinear codes, Bell Syst. Technol. J. 51 (4) (April 1972) 803–819.
[20] F.J. MacWilliams, N.J.A. Sloane, The Theory of Error Correcting Codes, North-Holland Mathematical Library, New York, 1977.
[21] A.R. Calderbank, E.M. Rains, P.W. Shor, N.J.A. Sloane, Quantum error correction via codes over GF(4), IEEE Trans. Inform. Theory 44 (1998) 1369–1387.
[22] I. Devetak, T.A. Brun, M.-H. Hsieh, Entanglement-assisted quantum error-correcting codes, in: V. Sidoravičius (Ed.), New Trends in Mathematical Physics, Selected Contributions of the XVth International Congress on Mathematical Physics, Springer, 2009, pp. 161–172.
[23] I.B. Djordjevic, Quantum LDPC codes from balanced incomplete block designs, IEEE Commun. Lett. 12 (May 2008) 389–391.
[24] I.B. Djordjevic, Photonic quantum dual-containing LDPC encoders and decoders, IEEE Photon. Technol. Lett. 21 (13) (1 July 2009) 842–844.
[25] I.B. Djordjevic, Photonic entanglement-assisted quantum low-density parity-check encoders and decoders, Opt. Lett. 35 (9) (1 May 2010) 1464–1466.

CHAPTER

Quantum Stabilizer Codes and Beyond

8

CHAPTER OUTLINE

In this chapter, a well-studied class of quantum codes is described, namely stabilizer codes [1–52] and related quantum codes. The basic concept of stabilizer codes was introduced in Chapter 7. Here the stabilizer codes are described in a more rigorous way. After rigorous introduction of stabilizer codes (Section 8.1), their properties and encoded operations are introduced in Section 8.2. Further, the stabilizer codes are described by using the finite geometry interpretation in Section 8.3. The standard form of stabilizer codes is introduced in Section 8.4, which is further used to provide the efficient encoder and decoder implementations in Section 8.5. Section 8.6 concerns nonbinary stabilizer codes. Next, subsystem codes are described in Section 8.7, entanglement-assisted codes in Section 8.8, and topological codes in Section 8.9. For a better understanding of this material, a set of problems is provided at the end of the chapter.

8.1 STABILIZER CODES

Quantum stabilizer codes are based on the stabilizer concept, which is introduced in the Appendix (see also Chapter 7). Before the stabilizer code is formally introduced, the Pauli group definition from the previous chapter is revisited. A *Pauli operator on N qubits* has the form $cO_1O_2 \ldots O_N$, where each $O_i \in \{I, X, Y, Z\}$ and $c = j^l$

Quantum Information Processing and Quantum Error Correction. DOI: 10.1016/B978-0-12-385491-9.00008-3

($l = 1,2,3,4$). The set of Pauli operators on N qubits forms the *multiplicative Pauli group* G_N. It has been shown earlier that the basis for single-qubit errors is given by $\{I,X,Y,Z\}$, while the basis for N-qubit errors is obtained by forming all possible direct products as follows:

$$E = O_1 \otimes \ldots \otimes O_N; \quad O_i \in \{I, X, Y, Z\}. \tag{8.1}$$

The N-qubit error basis can be transformed into multiplicative Pauli group G_N if it is allowed to premultiply E by j^l ($l = 0,1,2,3$). By noticing that $Y = -jXZ$, any error in the Pauli group of N-qubit errors can be represented by

$$\begin{aligned} &E = j^{\lambda} X(\boldsymbol{a}) Z(\boldsymbol{b}); \quad \boldsymbol{a} = a_1 \ldots a_N; \quad \boldsymbol{b} = b_1 \ldots b_N; \quad a_i, b_i = 0, 1; \quad \lambda = 0, 1, 2, 3 \\ &X(\boldsymbol{a}) \equiv X_1^{a_1} \otimes \ldots \otimes X_N^{a_N}; \quad Z(\boldsymbol{b}) \equiv Z_1^{b_1} \otimes \ldots \otimes Z_N^{b_N} \end{aligned} \tag{8.2}$$

In (8.2), the subscript i ($i = 0,1,\ldots,N$) is used to denote the location of the qubit to which operator X_i (Z_i) is applied; a_i (b_i) takes the value 1 (0) if the ith X (Z) operator is to be included (excluded). Therefore, to uniquely determine the error operator (up to the phase constant $j^{\lambda} = 1, j, -1, -j$ for $l' = 0,1,2,3$), it is sufficient to specify the binary vectors $\boldsymbol{a}$ and $\boldsymbol{b}$.

The action of the Pauli group G_N on a set W is further studied. The *stabilizer* of an element $w \in W$, denoted as S_w, is the set of elements from G_N that *fix* w: $S_w = \{g \in G_N | gw = w\}$. It is straightforward to show that S_w forms the group. Let S be the largest Abelian subgroup of G_N that fixes all elements from quantum code C_Q, which is commonly called the stabilizer group. The *stabilizer group* $S \subseteq G_N$ is constructed from a set of $N-K$ operators $g_1,\ldots,g_{N-K}$, also known as *generators* of S. Generators have the following properties: (i) they commute among each other; (ii) they are unitary and Hermitian; and (iii) they have the order of 2 because $g_i^2 = I$. Any element $s \in S$ can be written as a unique product of powers of the generators:

$$s = g_1^{c_1} \ldots g_{N-K}^{c_{N-K}}, \quad c_i \in \{0, 1\}; \quad i = 1, \ldots, N-K. \tag{8.3}$$

When c_i equals 1 (0), the ith generator is included (excluded) from the product of generators above. Therefore, the elements from S can be labeled by simple binary strings of length $N-K$, namely $\boldsymbol{c} = c_1 \ldots c_{N-K}$. The eigenvalues of generators g_i, denoted as $\{\text{eig}(g_i)\}$, can be found starting from property (iii) as follows:

$$|\text{eig}\ (g_i)\rangle = I|\text{eig}\ (g_i)\rangle = g_i^2|\text{eig}\ (g_i)\rangle = \text{eig}^2\ (g_i)|\text{eig}\ (g_i)\rangle. \tag{8.4}$$

From (8.4) it is clear that $\text{eig}^2\ (g_i) = 1$, meaning that eigenvalues are given by $\text{eig}(g_i) = \pm 1 = (-1)^{\lambda_i}; \quad \lambda_i = 0, 1$.

The quantum stabilizer code C_Q, with parameters $[N,K]$, can now be defined as the unique subspace of Hilbert space H_2^N that is fixed by the elements from stabilizer S of C_Q as follows:

$$C_Q = \bigcap_{s \in S} \{|c\rangle \in H_2^N |\ s|c\rangle = |c\rangle\}. \tag{8.5}$$

Notice that S cannot include $-I$ and jI because the corresponding code space C_Q will be trivial. Namely, if $-I$ is included in S, then it must fix arbitrary codeword $-I|c\rangle = |c\rangle$, which can be satisfied only when $|c\rangle = 0$. Since $(jI)^2 = -I$, the same conclusion can be made about jI. Because the original space for C_Q is 2^N-dimensional space, it is necessary to determine N commuting operators that specify a unique state ket $|\psi\rangle \in H_2^N$. The following simultaneous eigenkets of $\{g_1, g_2, \ldots, g_{N-K}; \overline{Z}_1, \overline{Z}_2, \ldots, \overline{Z}_K\}$ can be used as a basis for H_2^N. Namely, if the subset $\{\overline{Z}_i\}$ is chosen in such a way that each element from this subset commutes with all generators $\{g_i\}$, as well as among all elements of the subset itself, the elements from the set above can be used as basis kets. This consideration will be the subject of interest in the next section. In the rest of this section, instead, some important properties of the stabilizer group and stabilizer codes are studied.

In most cases for QECC it is sufficient to observe the *quotient group* G_N/C, $C = \{\pm I, \pm jI\}$ (see Appendix for the definition of a quotient group). Since in quantum mechanics it is not possible to distinguish unambiguously two states that differ only in a global phase constant (see Chapter 2), the coset $EC = \{\pm E, \pm jE\}$ can be considered as a single error. From Eq. (8.2) it is clear that any error from Pauli group G_N can be uniquely described by two binary vectors $\boldsymbol{a}$ and $\boldsymbol{b}$ each of length N. The number of such errors is therefore $2^{2N} \times 4$, where the factor 4 originates from the global phase constant j^l $(l = 0,1,2,3)$. Therefore, the order of G_N is 2^{2N+2}, while the order of the quotient group G_N/C is $2^{2N+2}/4 = 2^{2N}$.

Two important *properties* of N-qubit errors are:

(i) $\forall\ E_1, E_2 \in G_N$ the following is valid: $[E_1,E_2] = 0$ or $\{E_1,E_2\} = 0$. In other words, the elements from the Pauli group either commute or anticommute.

(ii) For $\forall\ E \in G_N$ the following is valid: (1) $E^2 = \pm I$, (2) $E^\dagger = \pm E$, and (3) $E^{-1} = E^\dagger$.

In order to prove property (i), the fact that operators X_i and Z_j commute if $i \neq j$ and anticommute when $i = j$ can be used. By using the error representation given in Eq. (8.2), the error E_1E_2 can be represented as

$$E_1E_2 = j^{l_1+l_2}X(\boldsymbol{a}_1)Z(\boldsymbol{b}_1)X(\boldsymbol{a}_2)Z(\boldsymbol{b}_2) = j^{l_1+l_2}(-1)^{\boldsymbol{a}_1\boldsymbol{b}_2+\boldsymbol{b}_1\boldsymbol{a}_2}X(\boldsymbol{a}_2)Z(\boldsymbol{b}_2)X(\boldsymbol{a}_1)Z(\boldsymbol{b}_1)$$
$$= (-1)^{\boldsymbol{a}_1\boldsymbol{b}_2+\boldsymbol{b}_1\boldsymbol{a}_2}E_2E_1 = \begin{cases} E_2E_1, & \boldsymbol{a}_1\boldsymbol{b}_2 + \boldsymbol{b}_1\boldsymbol{a}_2 = 0 \bmod 2 \\ -E_2E_1, & \boldsymbol{a}_1\boldsymbol{b}_2 + \boldsymbol{b}_1\boldsymbol{a}_2 = 1 \bmod 2, \end{cases} \tag{8.6}$$

where the notation $\boldsymbol{ab} = \Sigma_i a_i b_i$ is used to denote the dot product. The first claim of property (ii) can easily be proved from Eq. (8.2) as follows:

$$E^2 = j^{2l'}X(\boldsymbol{a})Z(\boldsymbol{b})X(\boldsymbol{a})Z(\boldsymbol{b}) = (-1)^{l'+\boldsymbol{ab}}X(\boldsymbol{a})X(\boldsymbol{a})Z(\boldsymbol{b})Z(\boldsymbol{b}) = (-1)^{l'+\boldsymbol{ab}}I = \pm I, \tag{8.7}$$

where the following property of Pauli operators $Z^2 = X^2 = I$ is used. The second claim of property (ii) can also be proved by again using the definition in Eq. (8.2) by

$$E^{\dagger} = (j^{l'}X(\boldsymbol{a})Z(\boldsymbol{b}))^{\dagger} = (-j)^{l'}Z^{\dagger}(\boldsymbol{b})X^{\dagger}(\boldsymbol{a}) = (-j)^{l'}Z(\boldsymbol{b})X(\boldsymbol{a})$$
$$= (-j)^{l'}(-1)^{\boldsymbol{ab}}X(\boldsymbol{a})Z(\boldsymbol{b}) = \pm E, \tag{8.8}$$

where the Hermitian property of Pauli matrices $X^{\dagger}=X$, $Z^{\dagger}=Z$ is used. Finally, claim 3 of property (ii) can be proved by using Eq. (8.8) as follows:

$$EE^{\dagger} = E(\pm E) = \pm I, \tag{8.9}$$

indicating that $E^{-1}=E^{\dagger}$.

In the previous chapter, the concept of the syndrome of an error was introduced as follows. Let C_Q be a quantum stabilizer code with generators $g_1, g_2, \ldots, g_{N-K}$ and let $E \in G_N$ be an error. The *error syndrome* for error E is defined by the bit string $S(E)=[\lambda_1\lambda_2\ldots\lambda_{N-K}]^{\mathrm{T}}$ with component bits being determined by

$$\lambda_i = \begin{cases} 0, & [E, g_i]=0 \\ 1, & \{E, g_i\}=0 \end{cases} \quad (i=1,\ldots,N-K). \tag{8.10}$$

Similarly as in syndrome decoding for classical error correction, it is expected that errors with nonvanishing syndrome are correctable. In Chapter 7 (Section 7.3, Theorem 7.1), it was proved that if an error E anticommutes with some element from stabilizer group S, then the image of the computational basis (CB) codeword is orthogonal to another CB codeword so that the following can be written:

$$\langle \bar{i}|E|\bar{j}\rangle = 0. \tag{8.11}$$

It is also known from Chapter 7 (see Eq. (7.44)) that

$$\langle \bar{i}|E|\bar{j}\rangle = C_E\delta_{\bar{i}\bar{j}}. \tag{8.12}$$

Therefore, from (8.11) and (8.12) it is clear that $C_E=0$, which indicates that errors with nonvanishing syndrome are correctable.

Let the set of errors $E=\{E_m\}$ from G_N for which $S(E^{+}{}_mE_n) \neq 0$ be observed. From Theorem 7.1 from Section 7.3 it is clear that

$$\langle \bar{i}|E_m^{\dagger}E_n|\bar{j}\rangle = 0. \tag{8.13}$$

From Eq. (8.12) it can concluded that $C_{mn}=0$, indicating that this set of errors can be corrected.

The third case of interest is errors E with vanishing syndrome $S(E)=0$. Clearly these errors commute with all generators from S. Because the stabilizer is Abelian, $S \subseteq C(S)$, where $C(S)$ is the *centralizer* of S introduced in the Appendix, i.e. the set of errors $E \in G_N$ that commute with all elements from S. There are two options for error E: (i) $E \in S$, in which case there is no need to correct the error since it fixes all codewords already; and (ii) $E \in C(S)-S$, in which case the error is not detectable. This observation can formally be proved as follows. Given that $E \in C(S)-S$, $s \in S$, $|c\rangle \in C_Q$, the following is valid:

$$sE|c\rangle = Es|c\rangle = E|c\rangle. \tag{8.14}$$

Because $E \notin S$, $E|c\rangle = |c'\rangle \neq |c\rangle$, indicating that the error E resulted in a codeword different from the original codeword. Without loss of generality it is assumed that the erroness codeword is a basis codeword $|c'\rangle = |\bar{i}\rangle$, which can be expanded in terms of basis codewords, as can any other codeword, as follows:

$$E|\bar{i}\rangle = \sum_j p_j|\bar{j}\rangle \neq |\bar{i}\rangle. \tag{8.15}$$

Multiplying by the dual of the basis codeword $\langle\bar{k}|$ from the left, the following is obtained:

$$\langle\bar{k}|E|\bar{i}\rangle = p_k \neq 0 \quad (k \neq i). \tag{8.16}$$

For the error E to be detectable, Eq. (8.12) requires p_k to be 0 when k is different from i, while Eq. (8.16) claims that $p_k \neq 0$, which indicates that the error E is not detectable.

Because $C(S)$ is the subgroup of G_N, from the Appendix (see Theorem T5) it is known that its cosets can be used to partition G_N as follows:

$$EC\,(S) = \{Ec|c \in C(S)\}. \tag{8.17}$$

From the Lagrange theorem, the following is true:

$$|G_N| = |C(S)|[G_N : C(S)] = 2^{2N+2}. \tag{8.18}$$

Based on the syndrome definition in Eq. (8.10), it is clear that with an error syndrome binary vector of length $N-K$ the 2^{N-K} different errors can be identified, which means that the index of $C(S)$ in G_N is $[G_N{:}C(S)] = 2^{N-K}$. From (8.18) the following is obtained:

$$|C(S)| = |G_N|/[G_N : C(S)] = 2^{N+K+2}. \tag{8.19}$$

Finally, since $C = \{\pm I, \pm jI\}$ the quotient group $C(S)/C$ order is

$$|C(S)/C| = 2^{N+K}. \tag{8.20}$$

Similarly as in classical error correction, where the set of errors that differ in a codeword have the same syndrome, it is expected that errors $E_1, E_2 \in G_N$, belonging to the same coset $EC(S)$, will have the same syndrome $S(E) = \lambda = [\lambda_1 \ldots \lambda_{N-K}]^T$. This observation can be expressed as the following theorem.

Theorem 8.1. The set of errors $\{E_i|E_i \in G_N\}$ have the same syndrome $\lambda = [\lambda_1 \ldots \lambda_{N-K}]^T$ if and only if they belong to the same coset $EC(S)$ given by (8.17).

Proof. In order to prove the theorem, without loss of generality, two arbitrary errors E_1 and E_2 will be observed. First, by assuming that these two errors have the same syndrome, namely $S(E_1) = S(E_2) = \lambda$, it will be proved that they belong to the same coset. For each generator g_i the following is valid:

$$E_1E_2g_i = (-1)^{\lambda_i} E_1g_iE_2 = (-1)^{2\lambda_i} g_iE_1E_2 = g_iE_1E_2, \tag{8.21}$$

which means that E_1E_2 commutes with g_i and therefore $E_1E_2 \in C(S)$. From claim (2) of property (ii) we know that $E_1^\dagger = \pm E_1$, and since $E_1E_2 \in C(S)$ then

$E_1^\dagger E_2 = c \in C(S)$. By multiplying $E_1^\dagger E_2 = c$ by E_1 from the left $E_2 = E_1 c$ is obtained, which means that $E_2 \in E_1 C(S)$. Since E_2 belongs to both $E_2 C(S)$ and $E_1 C(S)$ (and there is no intersection between equivalence classes), it can be concluded that $E_1 C(S) = E_2 C(S)$, meaning that errors E_1 and E_2 belong to the same coset. To complete the proof of Theorem 8.1, the opposite part of the claim will be proved. Namely, by assuming that two errors E_1 and E_2 belong to the same coset, it will be proved that they have the same syndrome. Since $E_1, E_2 \in E_1 C(S)$ from the definition of coset (8.17), it is clear that $E_2 = E_1 c$ and therefore $E_1^\dagger E_2 = c \in C(S)$, indicating that $E_1^\dagger E_2$ commutes with all generators from S. From claim (2) of property (ii) it is clear that $E_1^\dagger = \pm E_1$, meaning that $E_1 E_2$ must commute with all generators from S, which can be written as

$$[E_1 E_2, g_i] = 0, \quad \forall g_i \in S. \tag{8.22}$$

Let the syndromes corresponding to E_1 and E_2 be denoted by $S(E_1) = [\lambda_1 \ldots \lambda_{N-K}]^{\mathrm{T}}$ and $S(E_2) = [\lambda'_1 \ldots \lambda'_{N-K}]^{\mathrm{T}}$ respectively. Further, the commutation relations between $E_1 E_2$ and generator g_i are established as

$$E_1 E_2 g_i = (-1)^{\lambda'_i} E_1 g_i E_2 = (-1)^{\lambda_i + \lambda'_i} g_i E_1 E_2. \tag{8.23}$$

Because of commutation relation (8.22) clearly $\lambda_i + \lambda'_i = 0 \bmod 2$, which is equivalent to $\lambda'_i = \lambda_i \bmod 2$, therefore proving that errors E_1 and E_2 have the same syndrome.

From the previous discussion at the beginning of the section, it is clear that nondetectable errors belong to $C(S) - S$, and the *distance D* of a QECC can be defined as the lowest weight among weights of all elements from $C(S) - S$. The distance of a QECC is related to the error correction capability of the codes, as discussed in the section on distance properties of QECCs in the previous chapter. In the same section, the concept of *degenerate* codes is introduced as codes for which the matrix $\boldsymbol{C}$ with elements from (8.12) is singular. In the same chapter a useful theorem is proved, which can be used as a relatively simple check on degeneracy of a QECC. This theorem claims that a quantum stabilizer code with distance D is a degenerate code if and only if its stabilizer S contains at least one element (not counting the identity) of weight less than D. Therefore, two errors E_1 and E_2 are degenerate if $E_1 E_2 \in S$. It can be shown that for nondegenerate quantum stabilizer codes, linearly independent correctable errors have different syndromes. Let E_1 and E_2 be two arbitrary linearly independent correctable errors with corresponding syndromes $S(E_1)$ and $S(E_2)$ respectively. Because these two errors are correctable their weight in $E_1^\dagger E_2$ must be smaller than the distance of the code and since the code is nondegenerate clearly $E_1^\dagger E_2 \notin S$, $E_1^\dagger E_2 \notin C(S) - S$. Since $E_1^\dagger E_2$ does not belong to either S or $C(S) - S$ it must belong to $G_N - C(S)$ and anticommute with at least one element, say g, from S. If E_1 anticommutes with g, E_2 must commute with it, and vice versa, and so $S(E_1) \neq S(E_2)$. From Eq. (8.11) it is clear that $\langle \bar{i} | E_1^\dagger E_2 | \bar{j} \rangle = 0$ and, therefore, because of (8.12), $C_{12} = 0$. In conclusion, only degenerate codes can have $\langle \bar{i} | E_1^\dagger E_2 | \bar{j} \rangle \neq 0$.

In some books and papers, the description of stabilizer quantum codes is based on the normalizer concept introduced in the Appendix instead of the centralizer concept employed in this section. Namely, the *normalizer* $N_G(S)$ of the stabilizer S is defined as the set of errors $E \in G_N$ that fix S under conjugation, in other words $N_G(S) = \{E \in G_N | ESE^\dagger = S\}$. These two concepts are equivalent to each other since it can be proved that $N_G(S) = C(S)$. In order to prove this claim let us observe an element c from $C(S)$. Then for every element $s \in S$ the following is valid: $csc^\dagger = scc^\dagger = s$, which indicates that $C(S) \subseteq N(S)$. Let us now observe an element n from $N(S)$. By using the property (i) that the elements n and s either commute or anticommute we have $nsn^\dagger = \pm snn^\dagger = \pm s$. Since $-I$ cannot be an element from S, clearly $nsn^\dagger = s$ or equivalently $ns = sn$, meaning that s and n commute and therefore $N(S) \subseteq C(S)$. Since the only way to satisfy both the $C(S) \subseteq N(S)$ and $N(S) \subseteq C(S)$ requirements is by setting $N(S) = C(S)$, this proves that these two representations of quantum stabilizer codes are equivalent.

8.2 ENCODED OPERATORS

An $[N,K]$ QECC introduces an encoding map U from 2^K-dimensional Hilbert space, denoted by H_2^K, to 2^N-dimensional Hilbert space, denoted by H_2^N, as follows:

$$|\delta\rangle \in H_2^K \rightarrow |c\rangle = U|\delta\rangle \qquad O \in G_k \rightarrow \overline{O} = UOU^\dagger. \tag{8.24}$$

In particular, Pauli operators X and Z are mapped to

$$X_i \rightarrow \overline{X}_i = UX_iU^\dagger \quad \text{and} \quad Z_i \rightarrow \overline{Z}_i = UZ_iU^\dagger \tag{8.25}$$

respectively. We have shown in the previous section that arbitrary N-qubit error can be written in terms of X- and Z-containing operators as given by Eq. (8.2). In similar fashion, the unencoded operator O can also be represented in terms of X- and Z-containing operators as follows:

$$O = j^\lambda X(\boldsymbol{a})Z(\boldsymbol{b}); \quad \boldsymbol{a} = a_1 \ldots a_K; \quad \boldsymbol{b} = b_1 \ldots b_K; \quad a_i, b_i \in \{0,1\}. \tag{8.26}$$

The encoding given by Eq. (8.24) maps the operator O to encoded operator $\overline{O}$ as follows:

$$\begin{aligned}
\overline{O} &= U\left[j^\lambda X(\boldsymbol{a})Z(\boldsymbol{b})\right]U^\dagger = j^\lambda UX_1^{a_1}U^\dagger UX_2^{a_2} \ldots U^\dagger UX_N^{a_N}U^\dagger UZ_1^{b_1}U^\dagger UZ_2^{b_2} \ldots U^\dagger UZ_N^{b_N}U^\dagger \\
&= j^\lambda (UX_1^{a_1}U^\dagger)(UX_2^{a_2}U^\dagger) \ldots U^\dagger (UX_N^{a_N}U^\dagger)(UZ_1^{b_1}U^\dagger)UZ_2^{b_2} \ldots U^\dagger (UZ_N^{b_N}U^\dagger) = \\
&= j^\lambda \underbrace{(\overline{X}_1)^{a_1}(\overline{X}_2)^{a_2} \ldots (\overline{X}_N)^{a_N}}_{\overline{X}(\boldsymbol{a})} \underbrace{(\overline{Z}_1)^{b_1} \ldots (\overline{Z}_N)^{b_N}}_{\overline{Z}(\boldsymbol{b})} = j^\lambda \overline{X}(\boldsymbol{a})\overline{Z}(\boldsymbol{b}).
\end{aligned} \tag{8.27}$$

By using Eq. (8.27), we can easily prove the following commutativity properties:

$$[\overline{X}_i, \overline{X}_j] = [\overline{Z}_i, \overline{Z}_j] = 0$$
$$[\overline{X}_i, \overline{Z}_j] = 0 \quad (i \neq j)$$
$$\{\overline{X}_i, \overline{Z}_i\} = 0. \tag{8.28}$$

It can also be proved that $\overline{X}_i, \overline{Z}_i$ commute with the generators g_i from S, indicating that *Pauli encoded operators belong to C(S)*.

The *decoding map* $U^\dagger$returns back the codewords to unencoded kets, i.e. $U^\dagger|c\rangle = |\delta\rangle$; $|c\rangle \in H_2^N$, $|\delta\rangle \in H_2^K$. To demonstrate this, let us start with the definition of a stabilizer:

$$s|c\rangle = |c\rangle, \quad \forall s \in S \tag{8.29}$$

and apply $U^\dagger$ on both sides of (8.29) to obtain:

$$U^\dagger s|c\rangle = U^\dagger|c\rangle. \tag{8.30}$$

By inserting $UU^\dagger = I$ between s and $|c\rangle$ we obtain:

$$U^\dagger s(UU^\dagger)|c\rangle = U^\dagger|c\rangle \Leftrightarrow (U^\dagger sU)|\delta\rangle = |\delta\rangle. \tag{8.31}$$

Namely, since $U^\dagger: |c\rangle \in H_2^N \rightarrow |\delta\rangle \in H_2^K$, it is clear that stabilizer elements must be mapped to identity operators $I_K \in G_K$, meaning that $U^\dagger sU = I_k$. In order to be able to better characterize the decoding mapping, we will prove that *centralizer C(S) is generated by encoded Pauli operators* $\{\overline{X}_i, \overline{Z}_i | i = 1, \ldots, K\}$ *and generators of S*, $\{g_i$: $i = 1,\ldots,N - K\}$. In order to prove this claim, let us consider the set W of 2^{N+K+2} operators generated by all possible powers of g_k, X_l, and Z_m as follows:

$$\begin{aligned} W = &\left\{\widehat{O}(\boldsymbol{a},\boldsymbol{b},\boldsymbol{c}) = j^\lambda(\overline{X}_1)^{a_1}\ldots(\overline{X}_K)^{a_K}(\overline{Z}_1)^{b_1}\ldots(\overline{Z}_K)^{b_K}(g_1)^{c_1}\cdots(g_{N-K})^{c_{N-K}}\right\} \\ &= \{j^\lambda\overline{X}(\boldsymbol{a})\overline{Z}(\boldsymbol{b})g(\boldsymbol{c})\}; \\ &\boldsymbol{a} = a_1\ldots a_K; \ \ \boldsymbol{b} = b_1\ldots b_K; \\ &\boldsymbol{c} = c_1\ldots c_{N-K}; \ \ g(\boldsymbol{c}) = (g_1)^{c_1}\ldots(g_{N-K})^{c_{N-K}}. \end{aligned} \tag{8.32}$$

From commutation relations and from (8.27) we can conclude that encoded operators $\overline{O}$ commute with generators of stabilizer S, and therefore $\overline{O} \in C(S)$. From (8.32) we see that an operator $\widehat{O}(\boldsymbol{a},\boldsymbol{b},\boldsymbol{c})$ can be represented as the product of operators $\overline{O}$ and $g(s)$. Since $\overline{O} \in C(S)$ and $g(s)$ contains generators of stabilizer S, clearly $\widehat{O}(\boldsymbol{a},\boldsymbol{b},\boldsymbol{c}) \in C(S)$. The cardinality of set W is $|W| = 2^{K+K+N-K+2} = 2^{N+K+2}$ and it is the same as $|C(S)|$ (see Eq. (8.20)), which leads us to the conclusion that

$C(S) = W$. Let us now observe an error E from $C(S)$. The decoding mapping will perform the following action:

$$\begin{aligned}\overline{E} \rightarrow U^{\dagger}\overline{E}U &= U^{\dagger}\left[j^{\lambda}\overline{X}(\boldsymbol{a})\overline{Z}(\boldsymbol{b})\overline{g}(\boldsymbol{c})\right]U \\ &= j^{\lambda}U^{\dagger}(\overline{X}_1)^{a_1}UU^{\dagger}\ldots(\overline{X}_K)^{a_K}UU^{\dagger}(\overline{Z}_1)^{b_1}UU^{\dagger}\ldots(\overline{Z}_K)^{b_K}UU^{\dagger} \\ &\quad\times(g_1)^{c_1}UU^{\dagger}\ldots UU^{\dagger}(g_{N-K})^{c_{N-K}}U^{\dagger} \\ &= j^{\lambda}(U^{\dagger}UX_1^{a_1}U^{\dagger}U)\ldots(U^{\dagger}UX_K^{a_K}U^{\dagger}U)(U^{\dagger}UZ_1^{b_1}U^{\dagger}U)\ldots(U^{\dagger}UZ_K^{b_K}U^{\dagger}U) \\ &\quad\times(U^{\dagger}g_1^{c_1}U)\ldots(U^{\dagger}g_{N-k}^{c_{N-K}}U).\end{aligned} \tag{8.33}$$

By using the fact that stabilizer elements are mapped to identity during demapping, i.e. $U^{\dagger}sU = I_k$, Eq. (8.33) becomes

$$U^{\dagger}\overline{E}U = j^{\lambda}(X_1^{a_1})\ldots(X_K^{a_K})(Z_1^{b_1})\ldots(Z_K^{b_K}) = j^{\lambda}X(\boldsymbol{a})Z(\boldsymbol{b}). \tag{8.34}$$

Therefore, decoding mapping is essentially mapping from $C(S)$ to G_K. The image of this mapping is G_K, and based on Eq. (8.31) the kernel of mapping is S. In other words:

$$\mathrm{Im}(U^{\dagger}) = G_K \quad \text{and} \quad \mathrm{Ker}(U^{\dagger}) = S. \tag{8.35}$$

It can also be shown that decoding mapping U^{+} is a *homomorphism* from $C(S) \rightarrow G_K$. That is, from (8.34) it is clear that decoding mapping is onto (surjective). Let us now observe two operators from $C(S)$, denoted as $\overline{O}_1$ and $\overline{O}_2$. Decoding will lead to

$$U^{\dagger}:\ \overline{O}_1 \rightarrow U^{\dagger}\overline{O}_1U = U^{\dagger}UO_1U^{\dagger}U = O_1, \quad U^{\dagger}:\ \overline{O}_2 \rightarrow O_2. \tag{8.36}$$

The decoding mapping of product of $\overline{O}_1$ and $\overline{O}_2$ is clearly

$$U^{\dagger}:\ \overline{O}_1\overline{O}_2 \rightarrow U^{\dagger}\overline{O}_1\overline{O}_2U = U^{\dagger}UO_1U^{\dagger}UO_2U^{\dagger}U = O_1O_2. \tag{8.37}$$

Since the image of the product of operators equals the product of the images of corresponding operators, the decoding mapping is clearly a homomorphism. Let us now apply the fundamental homomorphism theorem from the Appendix on the decoding homomorphism $U^{\dagger}: C(S) \rightarrow G_K$, which claims that the quotient group $C(S)/\mathrm{Ker}(U^{\dagger}) = C(S)/S$ is isomorphic to G_K, so that we can write

$$C(S)/S \cong G_K. \tag{8.38}$$

Because two isomorphic groups have the same order and from the previous section, we know that the order of an unencoded group is $|G_K| = 2^{2K+2}$. We conclude that

$$|C(S)/S| = |G_K| = 2^{2K+2}. \tag{8.39}$$

In the previous section, we have shown that an error $E \in C(S) - S$ is not detectable by the QECC scheme because these errors perform maping from one codeword to another codeword and can therefore be used to implement encoded operations. The errors E and Es ($s \in S$) essentially perform the same encoded operation, which means that encoded operations can be identified with cosets of S in $C(S)$. This observation is consistent with the isomorphism given by Eq. (8.38). Since the same vectors $\boldsymbol{a}$ and $\boldsymbol{b}$ uniquely determine both O and $\overline{O}$ (see Eqs (8.26) and (8.27)), and because of isomorphism (8.38), we can identify encoded operators $\overline{O}$ with cosets of quotient group $C(S)/S$. Therefore, we have established bijection between the encoded operations and the cosets of $C(S)/S$. In conclusion, the Pauli encoded operators $\{\overline{X}_i, \overline{Z}_j | i,j = 1, \ldots, K\}$ must belong to different cosets. Finally, we are going to show that the quotient group can be represented by

$$C(S)/S = \cup\{\overline{X}_m S, \overline{Z}_n S: m, n = 1, \ldots, K\} \wedge \{j^l: l = 0, 1, 2, 3\}. \qquad (8.40)$$

This representation can be established by labeling the cosets with the help of (8.27) as follows:

$$\overline{O}S = j^l \overline{X}(\boldsymbol{a})\overline{Z}(\boldsymbol{b})S = j^l (\overline{X}_1 S)^{a_1} \ldots (\overline{X}_k S)^{a_K} (\overline{Z}_1 S)^{b_1} \ldots (\overline{Z}_1 S)^{b_K}. \qquad (8.41)$$

Indeed, the cosets have similar forms as given by Eq. (8.40).

[5,1,3] code example (cyclic and quantum perfect code). If the generators are given by

$$g_1 = X_1 Z_2 Z_3 X_4, \quad g_2 = X_2 Z_3 Z_4 X_5, \quad g_3 = X_1 X_3 Z_4 Z_5, \quad g_4 = Z_1 X_2 X_4 Z_5,$$

the stabilizer elements can be generated as $s = g_1^{c_1} \ldots g_4^{c_4}, \quad c_i \in \{0, 1\}$. The encoded Pauli operators are given by $\overline{X} = X_1 X_2 X_3 X_4 X_5$, $\overline{Z} = Z_1 Z_2 Z_3 Z_4 Z_5$. Since $g_1 \overline{X} = Y_2 Y_3 X_5 \in C(S) - S$, $\text{wt}(g_1 \overline{X}) = 3$ and the distance is $D = 3$. Since the smallest weight of stabilizer elements is 4, which is larger than D, clearly the code is nondegenerate. Because the Hamming bound is satisfied with equality, the code is perfect. The basis codewords are given by

$$|\overline{0}\rangle = \sum_{s \in S} s|0000\rangle$$

$$|\overline{1}\rangle = \overline{X}|\overline{0}\rangle.$$

8.3 FINITE GEOMETRY INTERPRETATION

The quotient group G_N/C, where $C = \{\pm I, \pm jI\}$, as discussed above, is the normal subgroup of G_N, and therefore it can be constructed from its cosets with coset multiplication (see Appendix) being defined as $(E_1 C)(E_2 C) = (E_1 E_2)C$, where E_1 and E_2 are two error operators. We have shown in the previous chapter and Section 8.2 that bijection between G_N/C and $2N$-dimensional Hilbert space F_2^{2N} can be established by Eq. (8.2). We have also shown (see Eq. (8.6)) that two error operators either commute or anticommute. We can modify the commutativity relation (8.6) as follows:

$$E_1E_2 = \begin{cases} E_2E_1, & \boldsymbol{a}_1\boldsymbol{b}_2 + \boldsymbol{b}_1\boldsymbol{a}_2 = 0 \bmod 2 \\ -E_2E_1, & \boldsymbol{a}_1\boldsymbol{b}_2 + \boldsymbol{b}_1\boldsymbol{a}_2 = 1 \bmod 2 \end{cases} = \begin{cases} E_2E_1, & \boldsymbol{a}_1\boldsymbol{b}_2 + \boldsymbol{b}_1\boldsymbol{a}_2 = 0 \\ -E_2E_1, & \boldsymbol{a}_1\boldsymbol{b}_2 + \boldsymbol{b}_1\boldsymbol{a}_2 = 1, \end{cases} \tag{8.42}$$

where the dot product is now observed per mod 2: $\boldsymbol{a}_1\boldsymbol{b}_2 = a_{11}b_{21} + \ldots a_{1N}b_{2N}$ mod 2. The inner product of the type $\boldsymbol{a}_1\boldsymbol{b}_2 + \boldsymbol{b}_1\boldsymbol{a}_2$ is known as the *sympletic (twisted) inner product* and can be denoted by $v(E_1)\odot v(E_2)$, where $v(E_i) = [\boldsymbol{a}_i|\boldsymbol{b}_i]$ $(i = 1,2)$ is the vector representation of E_i. The addition operation in the sympletic inner product is also per mod 2, meaning that it performs mapping of two vectors from F_2^{2N} into $F_2 = \{0,1\}$. Based on (8.42) and sympletic inner product notation, the commutativity relation can be written as

$$v(E_1)\odot v(E_2) = \begin{cases} 0, & [E_1, E_2] = 0 \\ 1, & \{E_1, E_2\} = 0. \end{cases} \tag{8.43}$$

Important properties of the sympletic inner product are:

(i) Self-orthogonality: $v\odot v = 0$
(ii) Symmetry: $u\odot v = v\odot u$
(iii) Bilinearity property:

$$(u+v)\odot w = u\odot w + v\odot w \qquad u\odot(v+w) = u\odot v + u\odot w, \quad \forall u,v,w \in F_2^{2N}.$$

Earlier we defined the weight of an N-dimensional operator as the number of non-identity operators. Given the fact that an N-dimensional operator E can be represented in a vector (finite geometry) representation by $v(E) = [a_1 \ldots a_N|b_1b_2 \ldots b_N]$, the *weight* of a vector $v(E)$, denoted as wt($v(E)$), sometimes called *sympletic weight* and denoted by swt($v(E)$), is equal to the number of components i for which $a_i = 1$ and/or $b_i = 1$. In other words, $\text{wt}([\boldsymbol{a}|\boldsymbol{b}]) = |\{i|(a_i,b_i) \neq (0,0)\}|$. The weight in a finite geometry representation is equal to the weight in an operator form representation, defined as the number of non-identity tensor product components in $E = cO_1 \ldots O_N$ (where $O_i \in \{I,X,Y,Z\}$ and $c = j^\lambda$, $\lambda = 0,1,2,3$), i.e. $\text{wt}(E) = |\{O_i \neq I\}|$. In similar fashion, the (sympletic) *distance* between two vectors $v(E_1) = (\boldsymbol{a}_1|\boldsymbol{b}_1)$ and $v(E_2) = (\boldsymbol{a}_2|\boldsymbol{b}_2)$ can be defined as the weight of their sum, i.e. $D(v(E_1),v(E_2)) = \text{wt}(v(E_1)+v(E_2))$. For example, let us observe the operators $E_1 = X_1Z_2Z_3X_4$ and $E_2 = X_2Z_3Z_4$. The corresponding finite geometry representations are $v(E_1) = (1001|0110)$ and $v(E_2) = (0100|0011)$, the weights are 4 and 3 respectively, and the distance between them is

$$D(v(E_1), v(E_2)) = \text{wt}(v(E_1) + v(E_2)) = \text{wt}(1101|0101) = 3.$$

The *dual* quantum error correction code can be introduced in similar fashion as was done in classical error correction in Chapter 6. The QECC code, as described in

Section 8.1, can be described in terms of stabilizer S. Any element s from S can be represented as a vector $v(s) \in F_2^{2N}$. The stabilizer generators

$$g_i = j^l X(\boldsymbol{a}_i) Z(\boldsymbol{b}_i); \quad i = 1, \ldots N - K \tag{8.44}$$

can be represented, using a finite geometry formalism, by

$$\boldsymbol{g}_i \doteq v(g_i) = (\boldsymbol{a}_i | \boldsymbol{b}_i). \tag{8.45}$$

The *dual* (null) space $S_\perp$ can then be defined as the set of vectors orthogonal to stabilizer generator vectors, in other words $\{v_d \in F_2^{2n}: \quad v_d \odot v(s) = 0\}$. Because of (8.43), $S_\perp$ is in fact the image of $C(S)$.

We turn our attention now to the *syndrome* representation using a finite geometry representation. An error operator E from G_N can be represented using a finite geometry interpretation as follows:

$$E \in G_N \rightarrow \boldsymbol{v}_E = v(E) = (\boldsymbol{a}_E | \boldsymbol{b}_E) \in F_2^{2N}. \tag{8.46}$$

Based on (8.10) and (8.43), the syndrome components can be determined by

$$S(E) = [\lambda_1 \lambda_2 \ldots \lambda_{N-K}]^{\mathrm{T}}; \quad \lambda_i = \boldsymbol{v}_E \odot \boldsymbol{g}_i, \quad i = 1, \ldots, N - K. \tag{8.47}$$

Notice that the ith component of syndrome λ_i can be calculated by matrix multiplication, instead of sympletic product, as follows:

$$\begin{aligned} &\lambda_i = \boldsymbol{v}_E \odot \boldsymbol{g}_i = \boldsymbol{b}_E \boldsymbol{a}_i + \boldsymbol{a}_E \boldsymbol{b}_i = (\boldsymbol{b}_E | \boldsymbol{a}_E) \begin{pmatrix} \boldsymbol{a}_i \\ \hline \boldsymbol{b}_i \end{pmatrix} = \boldsymbol{v}_E \boldsymbol{J} \boldsymbol{g}_i^{\mathrm{T}} = \boldsymbol{v}_E \boldsymbol{J} (\boldsymbol{A})_i^{\mathrm{T}}; \\ &(\boldsymbol{A})_i = \boldsymbol{g}_i = (\boldsymbol{a}_i | \boldsymbol{b}_i); \quad i = 1, \cdots, N - K, \end{aligned} \tag{8.48}$$

where $\boldsymbol{J}$ is the matrix used to permute the location of X- and Z-containing operators in $\boldsymbol{v}_{\mathrm{E}}$, i.e.

$$\boldsymbol{v}_E \boldsymbol{J} = (\boldsymbol{a}_E | \boldsymbol{b}_E) \begin{pmatrix} \boldsymbol{0}_N & \boldsymbol{I}_N \\ \boldsymbol{I}_N & \boldsymbol{0}_N \end{pmatrix} = (\boldsymbol{b}_E | \boldsymbol{a}_E), \quad \boldsymbol{J} = \begin{pmatrix} \boldsymbol{0}_N & \boldsymbol{I}_N \\ \boldsymbol{I}_N & \boldsymbol{0}_N \end{pmatrix},$$

with $\boldsymbol{0}_N$ being an $N \times N$ all-zero matrix and $\boldsymbol{I}_N$ being the $N \times N$ identity matrix. Equation (8.48) allows us to represent the sympletic inner product as matrix multiplication. By defining the (quantum) check matrix as follows:

$$\boldsymbol{A} = \begin{pmatrix} (\boldsymbol{A})_1 \\ (\boldsymbol{A})_2 \\ \vdots \\ (\boldsymbol{A})_{N-K} \end{pmatrix}, \tag{8.49}$$

where each row is the finite geometry representation of generator g_i, we can represent (8.47) and (8.48) as follows:

$$S(E) = [\lambda_1 \lambda_2 \ldots \lambda_{N-K}]^{\mathrm{T}} = \boldsymbol{v}_E \boldsymbol{J} \boldsymbol{A}^{\mathrm{T}}. \tag{8.50}$$

The syndrome equation (8.50) is very similar to the classical error correction syndrome equation (see Chapter 6).

The *encoded Pauli operators* can be represented using finite geometry formalism as follows:

$$v(\overline{X}_i) = (\boldsymbol{a}(\overline{X}_i)|\boldsymbol{b}(\overline{X}_i)) \qquad v(\overline{Z}_i) = (\boldsymbol{a}(\overline{Z}_i)|\boldsymbol{b}(\overline{Z}_i)). \tag{8.51}$$

The commutativity properties (8.28) can now be written as

$$\begin{aligned} &v(\overline{X}_m) \odot \boldsymbol{g}_n = 0; \quad n = 1, \dots, N-K \\ &v(\overline{X}_m) \odot v(\overline{X}_n) = 0; \quad n = 1, \dots, K \\ &v(\overline{X}_m) \odot v(\overline{Z}_n) = 0; \quad n \neq m \\ &v(\overline{X}_m) \odot v(\overline{Z}_m) = 1. \end{aligned} \tag{8.52}$$

From (8.52) it is clear that in order to completely characterize the encoded Pauli operators in finite geometry representation, we have to solve the system of $N-K$ equations (given by (8.52)). Since there are $2N$ unknowns, we have $N-K$ degrees of freedom, which can be used to simplify encoder and decoder implementations as described in the following sections. Since there are $N-K$ degrees of freedom, there are 2^{N-K} ways in which the Pauli encoded operators $\overline{X}_i$ can be chosen, which is the same as the number of cosets in $C(S)$.

Example: [4,2,2] code. The generators and encoded Pauli operators are given by

$$\begin{aligned} g_1 &= X_1Z_2Z_3X_4 & g_2 &= Y_1X_2X_3Y_4 \\ \overline{X}_1 &= X_1Y_3Y_4 & \overline{Z}_1 &= Y_1Z_2Y_3 \\ \overline{X}_2 &= X_1X_3Z_4 & \overline{Z}_2 &= X_2Z_3Z_4. \end{aligned}$$

The corresponding finite geometry representations are given by

$$\begin{aligned} v(g_1) &= (1001|0110) & v(g_2) &= (1111|1001) \\ v(\overline{X}_1) &= (1011|0011) & v(\overline{X}_2) &= (1010|0001) \\ v(\overline{Z}_1) &= (1010|1110) & v(\overline{Z}_2) &= (0100|0011). \end{aligned}$$

Finally, the quantum-check matrix is given as

$$\boldsymbol{A} = \begin{pmatrix} 1001|0110 \\ 1111|1001 \end{pmatrix}.$$

Before concluding this section, we show how to represent an arbitrary $s \in S$ by using the finite geometry formalism. Based on generator representation (8.44), we can express an arbitrary element s from S as

$$\begin{aligned} s &= (g_1)^{c_1} \dots (g_{N-K})^{c_{N-K}} = \prod_{i=1}^{N-K} g_i^{c_i} = \prod_{i=1}^{N-K} (j^{\lambda} X(\boldsymbol{a}_i) Z(\boldsymbol{b}_i))^{c_i} \\ &= j^{\lambda'} \prod_{i=1}^{N-K} \prod_k (X_k)^{c_i a_k} (Z_k)^{c_i b_k} = j^{\lambda'} X\left(\sum_{i=1}^{N-K} c_i \boldsymbol{a}_i\right) Z\left(\sum_{i=1}^{N-K} c_i \boldsymbol{b}_i\right). \end{aligned} \tag{8.53}$$

The corresponding finite geometry representation of a stabilizer element s, based on (8.45), can be written as follows:

$$v_s = \sum_{i=1}^{N-K} c_i \boldsymbol{g}_i. \tag{8.54}$$

Therefore, the set of vectors $\{\boldsymbol{g}_i: i = 1,\ldots,N-K\}$ spans a linear subspace of F_2^{2N}, which represents the image of the stabilizer S. Equation (8.54) is very similar to classical linear block encoding from Chapter 6.

8.4 STANDARD FORM OF STABILIZER CODES

The quantum-check matrix of QECC C_q, denoted by A_q, can be written, based on (8.49), as follows:

$$\boldsymbol{A}_q = \begin{pmatrix} \boldsymbol{g}_1 \\ \vdots \\ \boldsymbol{g}_{N-K} \end{pmatrix} = \left(\begin{array}{ccc|ccc} a_{1,1} & \cdots & a_{1,N} & b_{1,1} & \cdots & b_{1,N} \\ & & \vdots & \vdots & & \\ a_{N-K,1} & \cdots & a_{N-K,N} & b_{N-K,1} & \cdots & b_{N-K,N} \end{array}\right) = (\boldsymbol{A}'|\boldsymbol{B}'), \tag{8.55}$$

where the ith row corresponds to the finite geometry representation of the generator g_i (see Eq. (8.45)). Based on linear algebra we can make the following two observations:

- **Observation 1**: Swapping the mth and nth qubits in the quantum register corresponds to the mth and nth columns being swapped within both submatrices $\boldsymbol{A}'$ and $\boldsymbol{B}'$.
- **Observation 2**: Adding the nth row of $\boldsymbol{A}_q$ to the mth row ($m \neq n$) maps $\boldsymbol{g}_m \rightarrow \boldsymbol{g}_m + \boldsymbol{g}_n$ and generator $g_m \rightarrow g_m g_n$, so that the codewords and stabilizer are invariant to this change.

Therefore, we can perform *Gauss–Jordan elimination* to put the quantum-check matrix into the following form:

$$\boldsymbol{A}_q = \left(\begin{array}{cc|cc} \overbrace{\boldsymbol{I}}^{r} & \overbrace{\boldsymbol{A}}^{N-r} & \overbrace{\boldsymbol{B}}^{r} & \overbrace{\boldsymbol{C}}^{N-r} \\ \boldsymbol{0} & \boldsymbol{0} & \boldsymbol{D} & \boldsymbol{E} \end{array}\right) \begin{array}{l} \}r \\ \}N-K-r \end{array}; \quad r = \mathrm{rank}(\boldsymbol{A}_q). \tag{8.56}$$

In the second step, we perform further Gauss–Jordan elimination on submatrix $\boldsymbol{E}$ to obtain:

$$\boldsymbol{A}_q = \left(\begin{array}{ccc|ccc} \overbrace{\boldsymbol{I}}^{r} & \overbrace{\boldsymbol{A}_1}^{N-K-r-s} & \overbrace{\boldsymbol{A}_2}^{K+s} & \overbrace{\boldsymbol{B}}^{r} & \overbrace{\boldsymbol{C}_1}^{N-K-r} & \overbrace{\boldsymbol{C}_2}^{K+s} \\ \boldsymbol{0} & \boldsymbol{0} & \boldsymbol{0} & \boldsymbol{D}_1 & \boldsymbol{I} & \boldsymbol{E}_2 \\ \boldsymbol{0} & \boldsymbol{0} & \boldsymbol{0} & \boldsymbol{D}_2 & \boldsymbol{0} & \boldsymbol{0} \end{array}\right) \begin{array}{l} \}r \\ \}N-K-r-s. \\ \}s \end{array} \tag{8.57}$$

Since the last s generators do not commute with the first r operators we have to set $s=0$, which leads to the standard form of quantum-check matrix:

$$A_q = \left(\begin{array}{ccc|ccc} \overbrace{I_r}^{r} & \overbrace{A_1}^{N-K-r} & \overbrace{A_2}^{K} & \overbrace{B}^{r} & \overbrace{C_1}^{N-K-r} & \overbrace{C_2}^{K} \\ \mathbf{0} & \mathbf{0} & \mathbf{0} & D & I_{N-K-r} & E \end{array} \right) \begin{array}{l} \}r \\ \}N-K-r \end{array}. \tag{8.58}$$

Example. The standard form for Steane's code can be obtained as follows:

$$A_q = \left(\begin{array}{c|c} 0001111 & 0000000 \\ 0110011 & 0000000 \\ 1010101 & 0000000 \\ 0000000 & 0001111 \\ 0000000 & 0110011 \\ 0000000 & 1010101 \end{array} \right)$$

$$\sim \left[\begin{array}{ccccccc|ccccccc} 1 & 0 & 0 & 0 & 1 & 1 & 1 & 0 & 0 & 0 & 0 & 0 & 0 & 0 \\ 0 & 1 & 0 & 1 & 0 & 1 & 1 & 0 & 0 & 0 & 0 & 0 & 0 & 0 \\ 0 & 0 & 1 & 1 & 1 & 1 & 0 & 0 & 0 & 0 & 0 & 0 & 0 & 0 \\ 0 & 0 & 0 & 0 & 0 & 0 & 0 & 1 & 0 & 1 & 1 & 0 & 0 & 1 \\ 0 & 0 & 0 & 0 & 0 & 0 & 0 & 0 & 1 & 1 & 0 & 1 & 0 & 1 \\ 0 & 0 & 0 & 0 & 0 & 0 & 0 & 1 & 1 & 1 & 0 & 0 & 1 & 0 \end{array} \right].$$

We turn our attention now to the *standard form* representation for the *encoded Pauli operators*. Based on the previous section, Pauli encoded operators $\overline{X}_i$ can be represented using a finite geometry representation as follows:

$$v(\overline{X}_i) = \left(\begin{array}{ccc|ccc} \overbrace{u_1(i)}^{r} & \overbrace{u_2(i)}^{N-K-r} & \overbrace{u_3(i)}^{K} & \overbrace{v_1(i)}^{r} & \overbrace{v_2(i)}^{N-K-r} & \overbrace{v_3(i)}^{K} \end{array} \right). \tag{8.59}$$

Since encoded Pauli operators have $N-K$ degrees of freedom, as shown in the previous section, we can set arbitrary $N-K$ components of (8.59) to zero:

$$v(\overline{X}_i) = \left(\begin{array}{ccc|ccc} \mathbf{0} & \overbrace{u_2(i)}^{N-K-r} & \overbrace{u_3(i)}^{K} & \overbrace{v_1(i)}^{r} & \mathbf{0} & \overbrace{v_3(i)}^{K} \end{array} \right). \tag{8.60}$$

Because $\overline{X}_i \in C(S)$, it must commute with all generators of S. Based on Eq. (8.50), for matrix multiplication $v(\overline{X}_i)A_q^{\mathrm{T}}$ we have to change the positions of X- and Z-containing operators so that the commutativity can be expressed as

$$\begin{pmatrix} \mathbf{0} \\ \hline \mathbf{0} \end{pmatrix} = (v_1(i)\ \mathbf{0}\ v_3(i) | \mathbf{0}\ u_2(i)\ u_3(i)) \left(\begin{array}{ccc|ccc} I & A_1 & A_2 & B & C_1 & C_2 \\ \mathbf{0} & \mathbf{0} & \mathbf{0} & D & I & E \end{array} \right)^{\mathrm{T}}. \tag{8.61}$$

After matrix multiplication in (8.61), we obtain the following two sets of linear equations:

$$\begin{aligned} v_1(i) + A_2 v_3(i) + C_1 u_2(i) + C_2 u_3(i) &= 0 \\ u_2(i) + E\ u_3(i) &= 0. \end{aligned} \tag{8.62}$$

For convenience, let us write the Pauli encoded operators as the following matrix:

$$\begin{aligned} \overline{X} &= \begin{pmatrix} v(\overline{X}_1) \\ \vdots \\ v(\overline{X}_K) \end{pmatrix} = (\mathbf{0}\quad u_2\quad u_3 \,|\, v_1\quad \mathbf{0}\quad v_3); \\ u_2 &= \begin{pmatrix} u_{2,1}(1) & \dots & u_{2,N-K-r}(1) \\ \vdots & & \vdots \\ u_{2,1}(K) & \dots & u_{2,N-K-r}(K) \end{pmatrix} \\ u_3 &= \begin{pmatrix} u_{3,1}(1) & \dots & u_{3,K}(1) \\ \vdots & & \vdots \\ u_{3,1}(K) & \dots & u_{3,K}(K) \end{pmatrix}; \quad v_1 = \begin{pmatrix} v_{1,1}(1) & \dots & v_{1,r}(1) \\ \vdots & & \vdots \\ v_{1,1}(K) & \dots & v_{1,r}(K) \end{pmatrix}, \\ v_3 &= \begin{pmatrix} v_{3,1}(1) & \dots & v_{3,K}(1) \\ \vdots & & \vdots \\ v_{3,1}(K) & \dots & v_{3,K}(K) \end{pmatrix}, \end{aligned} \tag{8.63}$$

where each row represents one of the encoded Pauli operators. Based on (8.28), we conclude that the Pauli encoded operators $\overline{X}_i$ must commute among each other, which in matrix representation can be written as

$$\mathbf{0} = (v_1\quad \mathbf{0}\quad v_3 \,|\, \mathbf{0}\quad u_2\quad u_3)(\mathbf{0}\quad u_2\quad u_3 \,|\, v_1\quad \mathbf{0}\quad v_3)^{\mathrm{T}} = v_3 u_3^{\mathrm{T}} + u_3 v_3^{\mathrm{T}}. \tag{8.64}$$

By solving the system of equations (8.62) and (8.64) we obtain the following solution:

$$u_3 = I,\ v_3 = \mathbf{0},\ u_2 = E,\ v_1 = C_1 E + C_2. \tag{8.65}$$

Based on (8.63) we derive the standard form of encoded Pauli operators $\overline{X}_i$:

$$\overline{X} = (\mathbf{0}\quad E^{\mathrm{T}}\quad I \,|\, (E^{\mathrm{T}} C_1^{\mathrm{T}} + C_2^{\mathrm{T}})\quad \mathbf{0}\quad \mathbf{0}). \tag{8.66}$$

The standard form for the encoded Pauli operators $\overline{Z}_i$ can be derived in similar fashion. Let us first introduce the matrix representation of encoded Pauli operators $\overline{Z}_i$, similarly to (8.63):

$$\overline{Z} = \begin{pmatrix} v(\overline{Z}_1) \\ \vdots \\ v(\overline{Z}_K) \end{pmatrix} = (\mathbf{0}\quad u_2'\quad u_3' \,|\, v_1'\quad \mathbf{0}\quad v_3'). \tag{8.67}$$

The encoded Pauli operators $\overline{Z}_i$ must commute with generators g_i, which in matrix form can be represented by

$$\begin{pmatrix} \mathbf{0} \\ \hline \mathbf{0} \end{pmatrix} = (\mathbf{v}_1' \;\; \mathbf{0} \;\; \mathbf{v}_3' | \mathbf{0} \;\; \mathbf{u}_2' \;\; \mathbf{u}_3') \left(\begin{array}{ccc|ccc} \boldsymbol{I} & \boldsymbol{A}_1 & \boldsymbol{A}_2 & \boldsymbol{B} & \boldsymbol{C}_1 & \boldsymbol{C}_2 \\ \mathbf{0} & \mathbf{0} & \mathbf{0} & \boldsymbol{D} & \boldsymbol{I} & \boldsymbol{E} \end{array} \right)^{\mathrm{T}}, \tag{8.68}$$

which leads to the following set of linear equations:

$$\begin{aligned} \mathbf{v}_1' + \boldsymbol{A}_2\mathbf{v}_3' + \boldsymbol{C}_1\mathbf{u}_2' + \boldsymbol{C}_2\mathbf{u}_3' &= \mathbf{0} \\ \mathbf{u}_2' + \boldsymbol{E}\;\mathbf{u}_3' &= \mathbf{0}. \end{aligned} \tag{8.69}$$

Further, the encoded Pauli operators $\overline{Z}_i$ commute with $\overline{X}_i$ ($j \neq i$), but anticommute for $j = i$ (see Eq. (8.28)), which in matrix form can be represented as

$$\boldsymbol{I} = (\mathbf{0} \;\; \mathbf{u}_2' \;\; \mathbf{u}_3' \,|\, \mathbf{v}_1' \;\; \mathbf{0} \;\; \mathbf{v}_3')(\mathbf{v}_1 \;\; \mathbf{0} \;\; \mathbf{0} \,|\, \mathbf{0} \;\; \mathbf{u}_2 \;\; \boldsymbol{I})^{\mathrm{T}} = \mathbf{v}_3', \tag{8.70}$$

and the corresponding solution is $\mathbf{v}_3' = \boldsymbol{I}$. Finally, the $\overline{Z}_i$ operators commute among themselves, which in matrix form can be written as

$$\mathbf{0} = (\mathbf{0} \;\; \mathbf{u}_2' \;\; \mathbf{u}_3' \,|\, \mathbf{v}_1' \;\; \mathbf{0} \;\; \boldsymbol{I})(\mathbf{v}_1' \;\; \mathbf{0} \;\; \boldsymbol{I} \,|\, \mathbf{0} \;\; \mathbf{u}_2' \;\; \mathbf{u}_3')^{\mathrm{T}} = \mathbf{u}_3' + (\mathbf{u}_3')^{\mathrm{T}}, \tag{8.71}$$

whose solution is $\mathbf{u}_3' = \mathbf{0}$. By substituting solutions for (8.70) and (8.71) we obtain:

$$\mathbf{v}_1' = \boldsymbol{A}_2, \;\; \mathbf{u}_2' = \mathbf{0}. \tag{8.72}$$

Based on solutions of (8.70), (8.71), and (8.72), from (8.67) we derive the standard form of the encoded Pauli operators $\overline{Z}_i$ as

$$\overline{\mathbf{Z}} = (\mathbf{0} \;\; \mathbf{0} \;\; \mathbf{0} \,|\, \boldsymbol{A}_2^{\mathrm{T}} \;\; \mathbf{0} \;\; \boldsymbol{I}). \tag{8.73}$$

Example: Standard form of [5,1,3] code. The generators of this code are

$g_1 = X_1Z_2Z_3X_4$, $g_2 = X_2Z_3Z_4X_5$, $g_3 = X_1X_3Z_4Z_5$, and $g_4 = Z_1X_2X_4Z_5$.

The standard form of quantum-check matrix can be obtained by Gauss elimination:

$$\boldsymbol{A}_q = \left(\begin{array}{ccccc|ccccc} 1 & 0 & 0 & 1 & 0 & 0 & 1 & 1 & 0 & 0 \\ 0 & 1 & 0 & 0 & 1 & 0 & 0 & 1 & 1 & 0 \\ 1 & 0 & 1 & 0 & 0 & 0 & 0 & 0 & 1 & 1 \\ 0 & 1 & 0 & 1 & 0 & 1 & 0 & 0 & 0 & 1 \end{array} \right)$$

$$\sim \left(\begin{array}{ccccc|ccccc} 1 & 0 & 0 & 0 & 1 & 1 & 1 & 0 & 1 & 1 \\ 0 & 1 & 0 & 0 & 1 & 0 & 0 & 1 & 1 & 0 \\ 0 & 0 & 1 & 0 & 1 & 1 & 1 & 0 & 0 & 0 \\ 0 & 0 & 0 & 1 & 1 & 1 & 0 & 1 & 1 & 1 \end{array} \right).$$

From Eq. (8.58) we conclude that

$$r=4,\ \mathbf{A}_1=0,\ \mathbf{C}_1=0,\ \mathbf{E}=0,\ \mathbf{B}=\begin{pmatrix}1&1&0&1\\0&0&1&1\\1&1&0&0\\1&0&1&1\end{pmatrix},$$

$$\mathbf{A}_2=\begin{pmatrix}1\\1\\1\\1\end{pmatrix},\quad \mathbf{C}_2=\begin{pmatrix}1\\0\\0\\1\end{pmatrix}.$$

From (8.66), the standard form for encoded Pauli X operators is

$$\begin{aligned}\overline{\mathbf{X}}&=\begin{pmatrix}\mathbf{0} & \mathbf{E}^{\mathrm{T}} & \mathbf{I}\,|\,(\mathbf{E}^{\mathrm{T}}\mathbf{C}_1^{\mathrm{T}}+\mathbf{C}_2^{\mathrm{T}}) & \mathbf{0} & \mathbf{0}\end{pmatrix}\\ &=\begin{pmatrix}0 & 0 & 0 & 0 & 1\,|\,1 & 0 & 0 & 1 & 0\end{pmatrix}.\end{aligned}$$

From (8.73), the standard form for encoded Pauli Z operators is

$$\overline{\mathbf{Z}}=\begin{pmatrix}\mathbf{0} & \mathbf{0} & \mathbf{0}\,|\,\mathbf{A}_2^{\mathrm{T}} & \mathbf{0} & \mathbf{I}\end{pmatrix}=\begin{pmatrix}0 & 0 & 0 & 0 & 0\,|\,1 & 1 & 1 & 1 & 1\end{pmatrix}.$$

By using $XZ=-jY$, we obtain from the standard form of $\mathbf{A}_q$ the following generators (by ignoring the global phase constant):

$$g_1=Y_1Z_2Z_4Y_5,\quad g_2=X_2Z_3Z_4X_5,\ g_3=Z_1Z_2X_3X_5\ \text{ and }\ g_4=Z_1Z_3Y_4Y_5.$$

The encoded Pauli operators, based on the above, can be written in operator form as

$$\overline{X}=Z_1Z_4X_5,\quad \overline{Z}=Z_1Z_2Z_3Z_4Z_5.$$

8.5 EFFICIENT ENCODING AND DECODING

This section considers efficient encoder and decoder implementations, initially introduced by Gottesman and Cleve [7–9] (see also Ref. [14]).

8.5.1 Efficient Encoding

Here we study an efficient implementation of an encoder for [N,K] quantum stabilizer code. The unencoded K-qubit CB states can be obtained as simultaneous eigenkets of Pauli Z_i operators from

$$\begin{aligned}&Z_i|\delta_1\ldots\delta_K\rangle=(-1)^{\delta_i}|\delta_1\ldots\delta_k\rangle;\ \delta_i=0,1;\ i=1,\ldots,K\\ &|\delta_1\ldots\delta_k\rangle=X_1^{\delta_1}\ldots X_K^{\delta_K}|0\ldots0\rangle_K.\end{aligned}\tag{8.74}$$

The encoding operator U maps:

- The unencoded K-qubit CB ket $|\delta_1\ldots\delta_K\rangle$ to N-qubit basis codeword $\overline{|\delta_1\ldots\delta_K\rangle}=U|\delta_1\ldots\delta_K\rangle$.
- The single-qubit Pauli operator X_i to the N-qubit encoded operator $\overline{X}_i$, i.e. $X_i\to\overline{X}_i=UX_iU^{\dagger}$.

It follows from the above and Eq. (8.74) that

$$
\begin{aligned}
|\overline{\delta_1 \ldots \delta_k}\rangle &= U|\delta_1 \ldots \delta_k\rangle = U(X_1^{\delta_1} \ldots X_K^{\delta_K})(U^\dagger U)|0 \ldots 0\rangle_K \\
&= \left[U(X_1^{\delta_1} \ldots X_K^{\delta_K})U^\dagger\right] \underbrace{U|0 \ldots 0\rangle_K}_{|\overline{0 \ldots 0}\rangle} = (\overline{X}_1)^{\delta_1} \ldots (\overline{X}_K)^{\delta_K}|\overline{0 \ldots 0}\rangle.
\end{aligned} \tag{8.75}
$$

By defining the basis codeword as

$$
|\overline{0 \ldots 0}\rangle = \sum_{s \in S} s|0 \ldots 0\rangle_N, \tag{8.76}
$$

we can simplify the implementation of the encoder.

It can be shown by mathematical induction that

$$
\sum_{s \in S} s = \prod_{i=1}^{N-K} (I_N + g_i). \tag{8.77}
$$

For $N - K = 1$, we have only one generator g so that the left-hand side of Eq. (8.77) becomes

$$
\sum_{s \in S} s = \sum_{i=0}^{1} g^i = I + g. \tag{8.78}
$$

Let us assume that Eq. (8.77) is correct for $N - K = M$, and we can prove that the same equation is also valid for $N - K = M + 1$ as follows:

$$
\begin{aligned}
\sum_{s \in S} s &= \sum_{c_1=0;\ldots;c_M=0;c_{M+1}=0}^{1} g_1^{c_1} \ldots g_M^{c_M} g_{M+1}^{c_{M+1}} = \prod_{i=1}^{M}(I_N + g_i) \sum_{c_{M+1}=0}^{1} g_{M+1}^{c_{M+1}} \\
&= \prod_{i=1}^{M}(I_N + g_i) \cdot (I_N + g_{M+1}) = \prod_{i=1}^{M+1}(I_N + g_i).
\end{aligned} \tag{8.79}
$$

By using (8.77), Eq. (8.76) can be written as

$$
|\overline{0 \ldots 0}\rangle = \prod_{i=1}^{N-K} (I_N + g_i)|0 \ldots 0\rangle_N. \tag{8.80}
$$

Since $|\overline{\delta_1 \ldots \delta_K}\rangle = (\overline{X}_1)^{\delta_1} \ldots (X_K)^{\delta_K}|\overline{0 \ldots 0}\rangle$, by using (8.80) the N-qubit basis codeword can be obtained as

$$
|\overline{\delta_1 \ldots \delta_K}\rangle = \prod_{i=1}^{N-K} (I_N + g_i)\overline{X}_1^{\delta_1} \ldots \overline{X}_K^{\delta_K}|0 \ldots 0\rangle_N. \tag{8.81}
$$

We will use Eq. (8.81) to efficiently implement the encoder with the help of ancillary qubits. The encoding operation can thus be represented as the following mapping:

$$|0\ldots0\delta_1\ldots\delta_K\rangle \equiv |0\ldots0\rangle_{N-K}\otimes|\delta_1\ldots\delta_K\rangle \rightarrow |\overline{\delta_1\ldots\delta_K}\rangle. \tag{8.82}$$

Based on Eq. (8.81), we first have to determine the *action of encoded Pauli operators* $\overline{X}_i$:

$$\overline{X}_1^{\delta_1}\ldots\overline{X}_K^{\delta_K}|0\ldots0\rangle_N = \breve{U}_1\ldots\breve{U}_K|0\ldots0\delta_1\ldots\delta_K\rangle, \tag{8.83}$$

where $\breve{U}_i$ are controlled operations, whose action will be determined below, based on the standard form of $\overline{X}_i$:

$$v(\overline{X}_i) = \left(\overbrace{\mathbf{0}}^{r}\quad \overbrace{\mathbf{u}_2(i)}^{N-K-r}\quad \overbrace{\mathbf{u}_3(i)}^{K}\;\middle|\;\overbrace{\mathbf{v}_1(i)}^{r}\quad \mathbf{0}\quad \mathbf{0}\right);\quad \mathbf{u}_3(i) = (0\ldots1_i\ldots0). \tag{8.84}$$

By introducing the following two definitions:

$$S_x[\mathbf{u}_2(i)] \equiv X_{r+1}^{u_{2,1}(i)}\ldots X_{N-K}^{u_{2,N-K-r}(i)};\; S_z[\mathbf{v}_1(i)] \equiv Z_1^{v_{1,1}(i)}\ldots Z_r^{v_{1,r}(i)}, \tag{8.85}$$

we can express the action of encoded Pauli operator $\overline{X}_i$, based on (8.84) and (8.85), as follows:

$$\overline{X}_i = S_x[\mathbf{u}_2(i)]S_z[\mathbf{v}_1(i)]X_{N-K+i}. \tag{8.86}$$

In a special case, for $i=K$, Eq. (8.86) becomes

$$\overline{X}_k = S_x[\mathbf{u}_2(K)]S_z[\mathbf{v}_1(K)]X_N, \tag{8.87}$$

so that the action of $\overline{X}_K^{\delta_K}$ on $|0\ldots0\rangle_N$ is as follows:

$$\begin{aligned}\overline{X}_K^{\delta_k}|0\ldots0\rangle_N &\underset{S_z[\mathbf{v}_1(K)]|0\ldots1\rangle_N=|0\ldots1\rangle_N}{\overset{X_N|0\ldots0\rangle_N=|0\ldots1\rangle_N}{=}} \begin{cases}|0\ldots0\rangle_N, & \delta_K=0\\ S_x[\mathbf{u}_2(K)]|0\ldots1\rangle_N, & \delta_K=1\end{cases}\\ &= \{S_x[\mathbf{u}_2(K)]\}^{\delta_K}|0\ldots\delta_K\rangle_N.\end{aligned} \tag{8.88}$$

The overall action of $\overline{X}_1^{\delta_1}\ldots\overline{X}_K^{\delta_K}$ on $|0\ldots0\rangle_N$ can be obtained in iterative fashion:

$$\overline{X}_1^{\delta_1}\ldots\overline{X}_K^{\delta_K}|0\ldots0\rangle_N = \prod_{i=1}^{K}\{S_x[\mathbf{u}_2(i)]\}^{\delta_i}|0\ldots0\delta_1\ldots\delta_K\rangle_N. \tag{8.89}$$

Therefore, the action of operators in (8.83) is

$$\breve{U}_i = \{S_x[\mathbf{u}_2(i)]\}^{\delta_i}. \tag{8.90}$$

Therefore, the action of $\breve{U}_i$ is a controlled-$S_x[\mathbf{u}_2(i)]$ operation, controlled by information qubit δ_i, associated with the $(N-K+i)$th qubit. The qubits affected by $\breve{U}_i$

operation are the target qubits at positions $r+1$ to $N-K$. The number of required two-qubit gates is $\leq K(N-K-r)$.

In order to determine N-qubit basis codewords, based on Eq. (8.81), we now have to apply the operator $G=\prod_{i=1}^{N-K}(I_N+g_i)$ on $\overline{X}_1^{\delta_1}\ldots\overline{X}_K^{\delta_K}|0\ldots0\rangle_N$. The action of G can simply be determined by the following factorization:

$$G=\prod_{i=1}^{N-K}(I_N+g_i)=G_1G_2;\ G_1=\prod_{i=1}^{r}(I_N+g_i),\quad G_2=\prod_{i=r+1}^{N-K}(I_N+g_i). \tag{8.91}$$

Since the qubits at positions $r+1$ to $N-K$ are affected by $\breve{U}_i$, the G_1 and $\breve{U}_i$ operations commute (it does not matter in which order we apply them) and we can write

$$|\overline{\delta_1\ldots\delta_K}\rangle=G_1G_2\overline{X}_1^{\delta_1}\ldots\overline{X}_K^{\delta_K}|0\ldots0\rangle_N=G_1\overline{X}_1^{\delta_1}\ldots\overline{X}_K^{\delta_K}G_2|0\ldots0\rangle_N. \tag{8.92}$$

From the standard form of the quantum-check matrix (8.58) it is clear that G_2 is composed of Z operators only and since $Z_i|0\ldots0\rangle_N=|0\ldots0\rangle_N$, the state $|0\ldots0\rangle_N$ is invariant with respect to operation G_2 and we can write

$$|\overline{\delta_1\ldots\delta_K}\rangle=G_1\overline{X}_1^{\delta_1}\ldots X_K^{\delta_K}|0\ldots0\rangle_N=G_1\breve{U}_T|0\ldots0\delta_1\ldots\delta_K\rangle,\quad \breve{U}_T=\breve{U}_1\ldots\breve{U}_K. \tag{8.93}$$

To complete the encoder implementation, we need to determine the *action of* G_1, which contains factors like $(I+g_i)$. From the standard form (8.58) we know that the finite geometry representation of generator g_i is given by

$$g_i=(0\ldots01_i0\ldots0A_1(i)A_2(i)|B(i)C_1(i)C_2(i)). \tag{8.94}$$

From (8.94) it is clear that the ith position operator X is always present, while the presence of operator Z is dependent on the content of $B_i(i)$, so that we can express g_i as follows:

$$\underline{g_i=T_iX_iZ_i^{B_i(i)}};\quad i=1,\ldots,r, \tag{8.95}$$

where T_i is the operator derived from g_i by removing all operators associated with the ith qubit.

Example. The generator g_1 for [5,1,3] code is given by $g_1=Y_1Z_2Z_4Y_5$. By removing all operators associated with the ith qubit we obtain $T_1=Z_2Z_4Y_5$.

The action of $(I+g_i)$ on $|\psi\rangle=\overline{X}_1^{\delta_1}\ldots\overline{X}_K^{\delta_K}|0\ldots0\rangle_N$, based on (8.95), is given by

$$(I+g_i)|\psi\rangle=\breve{U}_T|0\ldots0\delta_1\ldots\delta_K\rangle+T_iX_iZ_i^{B_i(i)}\breve{U}_T|0\ldots0\delta_1\ldots\delta_K\rangle. \tag{8.96}$$

From (8.95) it is clear that the action of g_i affects the qubits at positions 1 to r, while $\breve{U}_T$ affects qubits from $r+1$ to $N-K$ (see Eqs (8.50) and (8.85)), which indicates that these operators commute, $Z_i^{B_i(i)}\breve{U}_T=\breve{U}_TZ_i^{B_i(i)}$. Because the state $|0\rangle$ is invariant on action of Z, the state $|0...0\delta_1...\delta_K\rangle$ is also invariant on action of $Z_i^{B_i(i)}$, i.e. $Z_i^{B_i(i)}|0...0\delta_1...\delta_K\rangle = |0...0\delta_1...\delta_K\rangle$. On the other hand, the action of the X operator on state $|0\rangle$ is to perform qubit flipping to $|1\rangle$, so that we can write $X_i|0...0\delta_1...\delta_K\rangle = |0...01_i0...0\delta_1...\delta_K\rangle$. The overall action of $(I+g_i)$ on $|\psi\rangle$ is therefore

$$(I+g_i)|\psi\rangle = \breve{U}_T|0...0\delta_1...\delta_K\rangle + T_i\breve{U}_T|0...1_i...0\delta_1...\delta_K\rangle. \tag{8.97}$$

Equation (8.97) can be expressed in terms of Hadamard gate action on the ith qubit:

$$\begin{aligned} H_i|\psi\rangle &= \breve{U}_T H_i|0...0_i...0\delta_1...\delta_K\rangle^{H_i|\delta_i\rangle=\frac{1}{\sqrt{2}}\left[|0\rangle+(-1)^{\delta_i}|1\rangle\right],\ \delta_i=0,1} \\ &= \breve{U}_T\{|0...0_i...0\delta_1...\delta_K\rangle + |0...1_i...0\delta_1...\delta_K\rangle\}. \end{aligned} \tag{8.98}$$

Let us further apply the operator controlled T_i, i.e. $T_i^{\alpha_i}$ $(\alpha_i = 0, 1)$, in Eq. (8.98) from the left to obtain:

$$T_i^{\alpha_i}H_i|\psi\rangle = \breve{U}_T|0...0_i...0\delta_1...\delta_K\rangle + T_i\breve{U}_T|0...1_i...0\delta_1...\delta_k\rangle, \tag{8.99}$$

which is the same as the action of $I+g_i$ given by (8.97). We conclude, therefore, that

$$(I+g_i)|\psi\rangle = T_i^{\alpha_i}H_i|\psi\rangle, \tag{8.100}$$

meaning that *the action of* $(I+g_i)$ *on* $|\psi>$ *is to first apply Hadamard gate* H_i *on* $|\psi>$ *followed by the controlled-*T_j *gate* (the T_i gate is a controlled qubit at position i). Since T_i acts on no more than $N-1$ qubits, and controlled-T_i gates on no more than $N-1$ two-qubit gates, plus one-qubit Hadamard gate H_i is applied, the number of gates needed to perform operation (8.100) is $(N-1)+1$ gates.

Based on the definition of G_1 operation (see Eq. (8.91)), we have to iterate the action $(I+g_i)$ per $i=1,...,r$ to obtain the *basis codewords*:

$$\overline{|\delta_1...\delta_k\rangle} = \underbrace{\prod_{i=1}^{N-K}(I+g_i)}_{\prod_{i=1}^{r}T_i^{\alpha_i}H_i}\underbrace{\overline{X}_1^{\delta_1}...\overline{X}_k^{\delta_k}}_{\prod_{m=1}^{K}\breve{U}_m}|0...0\rangle_N = \left(\prod_{i=1}^{r}T_i^{\alpha_i}H_i\right)\left(\prod_{m=1}^{K}\breve{U}_m\right)|0...0\delta_1...\delta_K\rangle. \tag{8.101}$$

From Eq. (8.101) it is clear that the number of required Hadmard gates is r, the number of two-qubit gates implementing controlled-T_i gates is $r(N-1)$, and the number of controlled-$S_x[u_2(i)]$ gates (operating on qubits $r+1$ to $N-K$) is $K(N-K-r)$, so that the number of total gates is not larger than

$$N_{\text{gates}} = r[(N-1)+1] + K(N-K-r) = (K+r)(N-K). \tag{8.102}$$

Since the number of gates is a linear function of codeword length N, this implementation can be called "efficient". Based on Eqs (8.90) and (8.85) we can rewrite Eq. (8.101) in a form suitable for implementation:

$$|\overline{\delta_1 \ldots \delta_k}\rangle = \left(\prod_{i=1}^{r} T_i^{\alpha_i} H_i\right)\left(\prod_{m=1}^{K}\{S_x[\boldsymbol{u}_2(m)]\}^{\delta_m}\right)|0\ldots 0\delta_1\ldots\delta_K\rangle,$$
$$S_x[\boldsymbol{u}_2(m)] = X_{r+1}^{u_{2,1}(m)} \ldots X_{N-K}^{u_{2,N-K-r}(m)}. \quad (8.103)$$

Example. Let us now study the implementation of the encoding circuit for [5,1,3] quantum stabilizer code; the same code was observed in the previous section. The generators derived from the standard form representation were found to be

$$g_1 = Y_1 Z_2 Z_4 Y_5, \quad g_2 = X_2 Z_3 Z_4 X_5, \quad g_3 = Z_1 Z_2 X_3 X_5 \quad \text{and} \quad g_4 = Z_1 Z_3 Y_4 Y_5.$$

The encoded Pauli X operator was found as

$$\overline{X} = Z_1 Z_4 X_5,$$

and the corresponding finite geometry representation is

$$v(\overline{X}) = (0 \quad 0 \quad 0 \quad 0 \quad 1 | 1 \quad 0 \quad 0 \quad 1 \quad 0).$$

Since $r = 4$, $N = 5$, and $K = 1$, from (8.84) it is clear that $N - K - r = 0$ and so $\boldsymbol{u}_2 = 0$, meaning that $S_x[u_2] = I$. From Eq. (8.103) we see that we need to determine the T_i operators, which can be simply determined from generators g_i by removing the ith qubit term:

$$T_1 = Z_2 Z_4 Y_5, \quad T_2 = Z_3 Z_4 X_5, \quad T_3 = Z_1 Z_2 X_5 \text{ and } T_4 = Z_1 Z_3 Y_5.$$

Based on Eq. (8.103), the basis codewords can be obtained as:

$$|\overline{\delta}\rangle = \left[\prod_{i=1}^{4} (T_i)^{\delta} H_i\right] |0000\delta\rangle,$$

and the corresponding encoder implementation is shown in Figure 8.1.

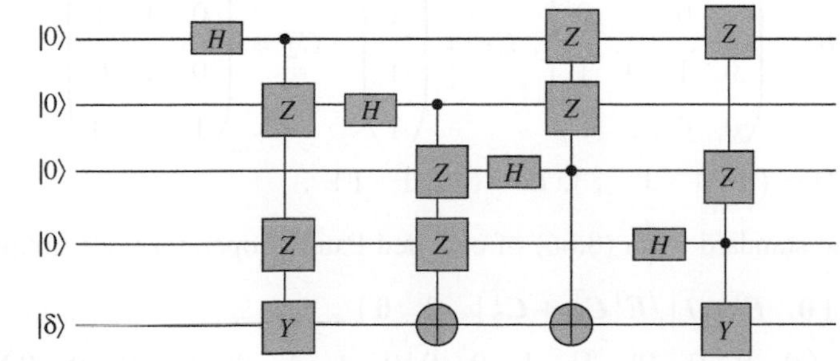

FIGURE 8.1

Efficient implementation of encoding circuit for [5,1,3] quantum stabilizer code. Since the action of Z in T_i^{δ} $(i = 1,2)$ is trivial, it can be omitted.

Example. The generators for an [8,3,3] quantum stabilizer code are given by

$$g_1 = X_1X_2X_3X_4X_5X_6X_7X_8, \quad g_2 = Z_1Z_2Z_3Z_4Z_5Z_6Z_7Z_8, \quad g_3 = X_2X_4Y_5Z_6Y_7Z_8,$$
$$g_4 = X_2Z_3Y_4X_6Z_7Y_8 \text{ and } g_5 = Y_2X_3Z_4X_5Z_6Y_8.$$

The corresponding quantum-check matrix, based on the generators above, is given by

$$A = \left(\begin{array}{cccccccc|cccccccc} 1 & 1 & 1 & 1 & 1 & 1 & 1 & 1 & 0 & 0 & 0 & 0 & 0 & 0 & 0 & 0 \\ 0 & 0 & 0 & 0 & 0 & 0 & 0 & 0 & 1 & 1 & 1 & 1 & 1 & 1 & 1 & 1 \\ 0 & 1 & 0 & 1 & 1 & 0 & 1 & 0 & 0 & 0 & 0 & 0 & 1 & 1 & 1 & 1 \\ 0 & 1 & 0 & 1 & 0 & 1 & 0 & 1 & 0 & 0 & 1 & 1 & 0 & 0 & 1 & 1 \\ 0 & 1 & 1 & 0 & 1 & 0 & 0 & 1 & 0 & 1 & 0 & 1 & 0 & 1 & 0 & 1 \end{array}\right)$$

By Gauss elimination we can put the quantum-check matrix above in standard form as in (8.58) as follows:

$$A = \left(\begin{array}{cccc|c|ccc|cccc|c|ccc} 1 & 0 & 0 & 0 & 1 & 1 & 1 & 0 & 0 & 1 & 0 & 0 & 1 & 1 & 0 & 1 \\ 0 & 1 & 0 & 0 & 1 & 1 & 0 & 1 & 0 & 0 & 1 & 0 & 1 & 0 & 1 & 1 \\ 0 & 0 & 1 & 0 & 1 & 0 & 1 & 1 & 0 & 1 & 0 & 1 & 1 & 0 & 1 & 0 \\ 0 & 0 & 0 & 1 & 0 & 1 & 1 & 1 & 0 & 0 & 1 & 1 & 1 & 1 & 0 & 0 \\ 0 & 0 & 0 & 0 & 0 & 0 & 0 & 0 & 1 & 1 & 1 & 1 & 1 & 1 & 1 & 1 \end{array}\right).$$

From code parameters and Eq. (8.58) it is clear that $r = 4, \;\; K = 3, \;\; N - K - r = 1$, and

$$A_1 = \begin{pmatrix} 1 \\ 1 \\ 1 \\ 0 \end{pmatrix}, \quad A_2 = \begin{pmatrix} 1 & 1 & 0 \\ 1 & 0 & 1 \\ 0 & 1 & 1 \\ 1 & 1 & 1 \end{pmatrix}$$

$$B = \begin{pmatrix} 0 & 1 & 0 & 0 \\ 0 & 0 & 1 & 0 \\ 0 & 1 & 0 & 1 \\ 0 & 0 & 1 & 1 \end{pmatrix}, \quad C_1 = \begin{pmatrix} 1 \\ 1 \\ 1 \\ 1 \end{pmatrix}, \quad C_2 = \begin{pmatrix} 1 & 0 & 1 \\ 0 & 1 & 1 \\ 0 & 1 & 0 \\ 1 & 0 & 0 \end{pmatrix},$$

$$D = (1 \;\; 1 \;\; 1 \;\; 1), \; E = (1 \;\; 1 \;\; 1).$$

Based on the standard form (8.66) of encoded Pauli X operators, we obtain:

$$\overline{X} = \left(\mathbf{0} \;\; E^{\mathrm{T}} \;\; I \,\middle|\, (E^{\mathrm{T}}C_1^{\mathrm{T}} + C_2^{\mathrm{T}}) \;\; \mathbf{0} \;\; \mathbf{0}\right)$$
$$= \left(\begin{array}{cccc|c|ccc|cccc|c|ccc} 0 & 0 & 0 & 0 & 1 & 1 & 0 & 0 & 0 & 1 & 1 & 0 & 0 & 0 & 0 & 0 \\ 0 & 0 & 0 & 0 & 1 & 0 & 1 & 0 & 1 & 0 & 0 & 1 & 0 & 0 & 0 & 0 \\ 0 & 0 & 0 & 0 & 1 & 0 & 0 & 1 & 0 & 0 & 1 & 1 & 0 & 0 & 0 & 0 \end{array}\right).$$

Based on the standard form (8.73) of encoded Pauli Z operators, we obtain:

$$\overline{Z} = (\mathbf{0} \quad \mathbf{0} \quad \mathbf{0} \,|\, A_2^{\mathrm{T}} \quad \mathbf{0} \quad I)$$
$$= \left(\begin{array}{cccccccc|cccccccc} 0 & 0 & 0 & 0 & 0 & 0 & 0 & 0 & 1 & 1 & 0 & 1 & 0 & 1 & 0 & 0 \\ 0 & 0 & 0 & 0 & 0 & 0 & 0 & 0 & 1 & 0 & 1 & 1 & 0 & 0 & 1 & 0 \\ 0 & 0 & 0 & 0 & 0 & 0 & 0 & 0 & 0 & 1 & 1 & 1 & 0 & 0 & 0 & 1 \end{array}\right).$$

From the encoded Pauli X matrix above, (8.84), and (8.103), it is clear that

$$\boldsymbol{u}_2 = \begin{pmatrix} 1 \\ 1 \\ 1 \end{pmatrix}, \quad S_x[\boldsymbol{u}_2(1)] = X_5, \quad S_x[\boldsymbol{u}_2(2)] = X_5, \quad S_x[\boldsymbol{u}_2(3)] = X_5.$$

The generators derived from the standard form quantum-check matrix are obtained as

$$g_1 = X_1Z_2Y_5Y_6X_7Z_8, \quad g_2 = X_2Z_3Y_5X_6Z_7Y_8, \quad g_3 = Z_2X_3Z_4Y_5Y_7X_8,$$
$$g_4 = Z_3Y_4Z_5Y_6X_7X_8, \text{ and } g_5 = Z_1Z_2Z_3Z_4Z_5Z_6Z_7Z_8.$$

The corresponding T_i operators are obtained from generators g_i by omitting the ith term:

$$T_1 = Z_2Y_5Y_6X_7Z_8, \; T_2 = Z_3Y_5X_6Z_7Y_8, \; T_3 = Z_2Z_4Y_5Y_7X_8, \; T_4 = Z_3Z_5Y_6X_7X_8.$$

Based on Eq. (8.103), we obtain the efficient implementation of the encoder shown in Figure 8.2.

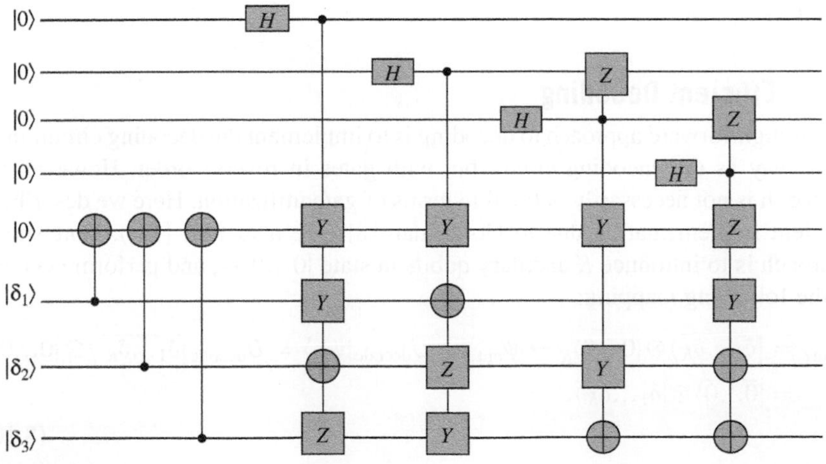

FIGURE 8.2

Efficient implementation of encoding circuit for [8,3,3] quantum stabilizer code.

To further simplify the encoder implementation, we can modify the standard form (8.58) by Gauss elimination to obtain:

$$A_q = \left(\begin{array}{ccc|ccc} \overbrace{I_r}^{r} & \overbrace{A_1}^{N-K-r} & \overbrace{A_2}^{K} & \overbrace{B}^{r} & \overbrace{0}^{N-K-r} & \overbrace{C}^{K} \\ 0 & 0 & 0 & D & I_{N-K-r} & E \end{array}\right) \begin{array}{l} \}r \\ \}N-K-r \end{array}. \tag{8.104}$$

The corresponding encoded Pauli operators can be represented by

$$\overline{X} = (0 \quad E^{\mathrm{T}} \quad I_K \,|\, C^{\mathrm{T}} \quad 0 \quad 0), \quad \overline{Z} = (0 \quad 0 \quad 0 \,|\, A_2^{\mathrm{T}} \quad 0 \quad I_K). \tag{8.105}$$

From (8.104) it is clear that the generators can be represented, using finite geometry representation, as

$$g_i = (0 \;\; 0 \;\; \ldots \;\; 1_i \;\; 0 \;\; \ldots \;\; 0 \;\; a_{r+1} \;\; \ldots \;\; a_N| \;\; b_1 \;\; \ldots \;\; b_r \;\; 0 \;\; \ldots \;\; 0 \;\; b_{N-K+1} \;\; \ldots \;\; b_N). \tag{8.106}$$

Based on (8.105), the ith encoded Pauli X operator can be represented as

$$v(\overline{X}_i) = (0 \;\ldots\; 0 \;\; a'_{r+1} \;\ldots\; a'_{N-K} \;\; 0 \;\ldots\; 0 \;\; 1_{N-K+i} \;\; 0 \;\ldots\; 0| \;\; b'_1 \;\ldots\; b'_r \;\; 0 \;\ldots\; 0). \tag{8.107}$$

Based on (8.103), (8.106), and (8.107), we show in Figure 8.3 the two-stage encoder configuration corresponding to the standard form (8.104). The first stage is related to controlled-X operations in which the ith information qubit (or equivalently the $(N-K+i)$th codeword qubit) is used to control ancillary qubits at positions $r+1$ to $N-K$, based on (8.107). In the second stage, the ith ancillary qubit upon application of the Hadamard gate is used to control qubits at codeword locations $i+1$ to N, based on (8.106).

8.5.2 Efficient Decoding

The straightforward approach to decoding is to implement the decoding circuit in the same way as the encoding circuit but with gates in reverse order. However, this approach is not necessarily optimal in terms of gate utilization. Here we describe an efficient implementation due to Gottesman [8] (see also Ref. [14]). Gottesman's approach is to introduce K ancillary qubits in state $|0\ldots0>_K$, and perform decoding as the following mapping:

$$\begin{aligned} |\psi_{\text{in}}\rangle &= |\overline{\delta_1\ldots\delta_K}\rangle \otimes |0\cdots0\rangle_K \to |\psi_{\text{out}}\rangle = U_{\text{decode}}|\psi_{\text{in}}\rangle = U_{\text{decode}}|\overline{\delta_1\ldots\delta_K}\,\rangle \otimes |0\ldots0\rangle_K \\ &= |\overline{0\ldots0}\rangle \otimes |\delta_1\ldots\delta_K\rangle. \end{aligned} \tag{8.108}$$

With respect to Eq. (8.108) the following two remarks are important:

- **Remark 1.** U_{decode} is a linear operator. It is, therefore, sufficient to concentrate on the decoding of basis codewords.

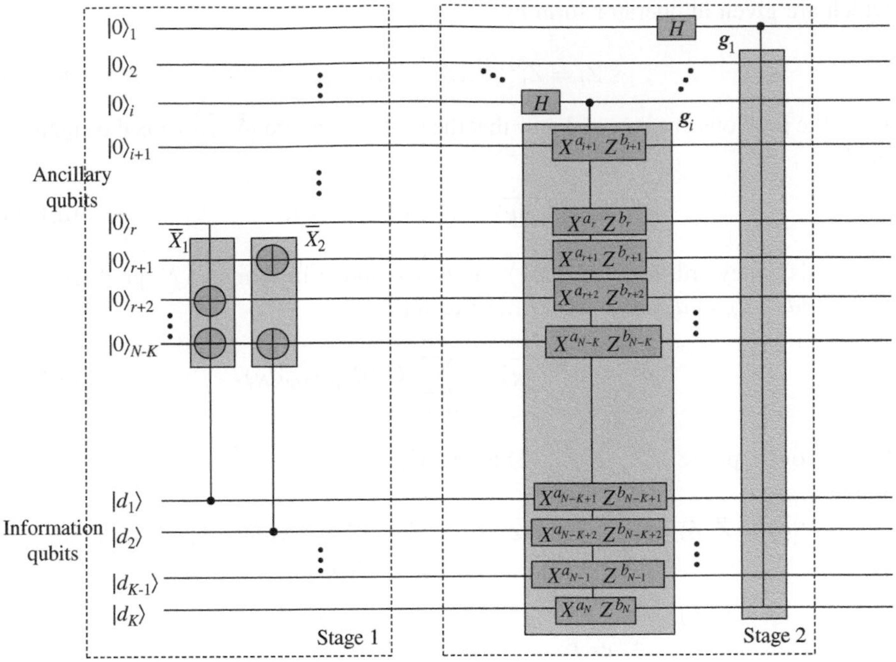

FIGURE 8.3

Efficient encoder implementation based on standard form (8.104).

- **Remark 2.** The decoding procedure in addition to decoded state $|\delta_1 \ldots \delta_K>$ also returns N qubits in state $|\overline{0 \ldots 0}\rangle$, which is needed in the encoding process. This occurs as a result of applying the operator G on ket $|0 \ldots 0>_N$. Therefore, by implementing the encoder and decoder in the transceiver by applying Gottesman's approach, we can save Nr gates in the encoding process.

Decoding can be performed in two stages:

1. By applying properly the CNOT gates to the ancillary qubits we put the ancillaries into decoded state $|\delta_1 \ldots \delta_K>$, i.e. we perform the following mapping:

$$|\psi_{\text{in}}\rangle = |\overline{\delta_1 \ldots \delta_K}\rangle \otimes |0 \ldots 0\rangle_K \rightarrow |\overline{\delta_1 \ldots \delta_K}\rangle \otimes |\delta_1 \ldots \delta_K\rangle.$$

2. By applying a controlled-X_i operation to each encoded qubit i conditioned on the ith ancillary qubit, we perform the following mapping: $|\overline{\delta_1 \ldots \delta_K}\rangle \otimes |\delta_1 \ldots \delta_K\rangle \rightarrow |\overline{0 \ldots 0}\rangle \otimes |\delta_1 \ldots \delta_K\rangle$.

The starting point for the first stage is the standard form of encoded Pauli operators $\overline{Z}_i$:

$$\overline{\mathbf{Z}} = (\mathbf{0} \quad \mathbf{0} \quad \mathbf{0} \,|\, \mathbf{A}_2^{\mathrm{T}} \quad \mathbf{0} \quad I), \;\; v(\overline{Z}_i) = (v_l),$$
$$v_l = \begin{cases} A_{2,l}(i); & l = 1, \ldots, r \\ \delta_{l,N-K+i}; & l = r+1, \ldots, N, \end{cases} \tag{8.109}$$

which are given in operator form by

$$\overline{Z}_i = Z_1^{A_{2,1}(i)} \ldots Z_r^{A_{2,r}(i)} Z_{N-K+i}. \tag{8.110}$$

From the previous chapter we know that the basis codeword $|\overline{\delta_1 \ldots \delta_K}\rangle$ is the eigenket of $\overline{Z}_i$:

$$\overline{Z}_i |\overline{\delta_1 \ldots \delta_K}\rangle = (-1)^{\delta_i} |\overline{\delta_1 \ldots \delta_K}\rangle. \tag{8.111}$$

The basis codeword $|\overline{\boldsymbol{\delta}}\rangle = |\overline{\delta_1 \ldots \delta_K}\rangle$ can be expanded in terms of N-qubit CB kets $\{|\boldsymbol{d}\rangle = |d_1 \ldots d_N\rangle |\ d_i = 0,1;\ i = 1,\ldots,N\}$ as follows:

$$|\overline{\boldsymbol{\delta}}\rangle = |\overline{\delta_1 \ldots \delta_K}\rangle = \sum_{\boldsymbol{d} \in F_2^N} C_{\overline{\boldsymbol{\delta}}}(\boldsymbol{d}) |d_1 \ldots d_N\rangle. \tag{8.112}$$

Let us now apply (8.111) in (8.112) to obtain:

$$\begin{aligned}(-1)^{\delta_i} |\overline{\boldsymbol{\delta}}\rangle = \overline{Z}_i |\overline{\boldsymbol{\delta}}\rangle &= \sum_{\boldsymbol{d} \in F_2^n} C_{\overline{\boldsymbol{\delta}}}(\boldsymbol{d}) \overline{Z}_i |d_1 \ldots d_N\rangle = \sum_{\boldsymbol{d} \in F_2^N} C_{\overline{\boldsymbol{\delta}}}(\boldsymbol{d}) (-1)^{v(\overline{Z}_i)\cdot \boldsymbol{d}} |d_1 \ldots d_N\rangle \\ &= (-1)^{v(\overline{Z}_i)\cdot \boldsymbol{d}} \sum_{\boldsymbol{d} \in F_2^N} C_{\overline{\boldsymbol{\delta}}}(\boldsymbol{d}) |d_1 \ldots d_N\rangle = (-1)^{v(\overline{Z}_i)\cdot \boldsymbol{d}} |\overline{\boldsymbol{\delta}}\rangle.\end{aligned} \tag{8.113}$$

From Eq. (8.113) it is obvious that

$$\delta_i = v(\overline{Z}_i)\cdot \boldsymbol{d}. \tag{8.114}$$

We indicated above that in the *first stage* of the decoding procedure we need to properly apply CNOT gates on ancillaries, and it is therefore important to determine the action of the X_{a_i} gate on $|\overline{\delta_1 \ldots \delta_K}\rangle \otimes |0\ldots 0\rangle_K$. By applying the expansion (8.112) we obtain:

$$\begin{aligned}X_{a_1}^{\delta_1} |\overline{\delta_1 \ldots \delta_K}\rangle \otimes |0\ldots 0\rangle_K &= \sum_{\boldsymbol{d} \in F_2^N} C_{\overline{\boldsymbol{\delta}}}(\boldsymbol{d}) |d_1 \ldots d_N\rangle \otimes \left[X_{a_1}^{\delta_1} |0\ldots 0\rangle_K \right] \overset{\delta_1 = v(\overline{Z}_1)\cdot \boldsymbol{d}}{=} \\ &\sum_{\boldsymbol{d} \in F_2^N} C_{\overline{\boldsymbol{\delta}}}(\boldsymbol{d}) |d_1 \ldots d_N\rangle \otimes X_{a_1}^{v(\overline{Z}_1)\cdot \boldsymbol{d}} |0\ldots 0\rangle_K.\end{aligned} \tag{8.115}$$

From (8.115) it is clear that we need to determine the action of $X_{a_1}^{v(\overline{Z}_1)\cdot \boldsymbol{d}}$ to $|0\rangle$:

$$X_{a_1}^{v(\overline{Z}_1)\cdot \boldsymbol{d}} |0\rangle = X_{a_1}^{v_1(\overline{Z}_1) d_1} \ldots X_{a_1}^{v_N(\overline{Z}_1) d_N} |0\rangle = |v(\overline{Z}_1)\cdot \boldsymbol{d}\rangle = |\delta_1\rangle. \tag{8.116}$$

Therefore, the action of the $X_{a_1}^{\delta_1}$ gate on $|\overline{\delta_1 \ldots \delta_K}\rangle \otimes |0\ldots 0\rangle_K$ is

$$X_{a_1}^{\delta_1} |\overline{\delta_1 \ldots \delta_k}\rangle \otimes |0\ldots 0\rangle_K = |\overline{\delta_1 \ldots \delta_K}\rangle \otimes |\delta_1 \ldots 0\rangle_K. \tag{8.117}$$

By applying a similar procedure on the remained ancillary qubits we obtain:

$$\prod_{i=1}^{K} X_{a_i}^{\delta_i} |\overline{\delta_1 \dots \delta_K}\rangle \otimes |0\dots 0\rangle_K = |\overline{\delta_1 \dots \delta_K}\rangle \otimes |\delta_1 \dots \delta_K\rangle. \tag{8.118}$$

In the *second stage* of decoding, as indicated above, we have to perform the mapping $|\overline{\delta_1 \dots \delta_K}\rangle \otimes |\delta_1 \dots \delta_K\rangle \rightarrow |\overline{0 \dots 0}\rangle \otimes |\delta_1 \dots \delta_K\rangle$. This mapping can be implemented by applying a controlled-X_i operation to the ith encoded qubit, which serves as target qubit. The ith ancillary qubit serves as the control qubit. This action can be described as follows:

$$(\overline{X}_1)^{\delta_1} |\overline{\delta_1 \dots \delta_K}\rangle \otimes |\delta_1 \dots \delta_K\rangle = |\overline{(\delta_1 \oplus \delta_1) \dots \delta_k}\rangle \otimes |\delta_1 \dots \delta_K\rangle = |\overline{0 \dots \delta_K}\rangle \otimes |\delta_1 \dots \delta_K\rangle. \tag{8.119}$$

Further, we perform a similar procedure on the remaining encoded qubits:

$$\prod_{i=1}^{K} (\overline{X}_i)^{\delta_i} |\overline{\delta_1 \dots \delta_K}\rangle \otimes |\delta_1 \dots \delta_K\rangle = |\overline{0 \dots 0}\rangle \otimes |\delta_1 \dots \delta_K\rangle. \tag{8.120}$$

The *overall decoding process*, by applying (8.118) and (8.120), can be represented by

$$\prod_{i=1}^{K} (\overline{X}_i)^{\delta_i} \prod_{l=1}^{K} X_{a_l}^{\delta_l} |\overline{\delta_1 \dots \delta_K}\rangle \otimes |0\dots 0\rangle_K = |\overline{0 \dots 0}\rangle \otimes |\delta_1 \dots \delta_K\rangle. \tag{8.121}$$

The *total number of gates for decoding*, based on Eq. (8.121), can be found as

$$K(r+1) + K(N-K+1) = K(N-K+r+2) \leq (K+r)(N-K). \tag{8.122}$$

Because $\overline{Z}_i$ acts on $r+1$ qubits (see Eq. (8.105)) and there are K such actions based on (8.121), controlled-X_i operations require no more than $N-K+1$ two-qubit gates and there are K such actions (see (8.121)). On the other hand, the number of gates required in reverse encoding is $(K+r)(N-K)$ (see Eq. (8.102)). Therefore, the decoder implemented based on Eq. (8.121) is more efficient than decoding based on reverse efficient encoding.

Example. The decoding circuit for [4,2,2] quantum stabilizer code can be obtained based on (8.121). The starting point is the generators of the code:

$$g_1 = X_1 Z_2 Z_3 X_4 \text{ and } g_2 = Y_1 X_2 X_3 Y_4.$$

The corresponding quantum-check matrix and its standard from (obtained by Gaussian elimination) are given by

$$\mathbf{A} = \left(\begin{array}{cccc|cccc} 1 & 0 & 0 & 1 & 0 & 1 & 1 & 0 \\ 1 & 1 & 1 & 1 & 1 & 0 & 0 & 1 \end{array}\right) \sim \left(\begin{array}{cccc|cccc} 1 & 0 & 0 & 1 & 0 & 1 & 1 & 0 \\ 0 & 1 & 1 & 0 & 1 & 1 & 1 & 1 \end{array}\right).$$

The encoded Pauli operators in finite geometry form are given by

$$\overline{\mathbf{X}} = (\mathbf{0} \quad \mathbf{E}^{\mathrm{T}} \quad \mathbf{I} \,|\, (\mathbf{E}^{\mathrm{T}}\mathbf{C}_1^{\mathrm{T}} + \mathbf{C}_2^{\mathrm{T}}) \quad \mathbf{0} \quad \mathbf{0}) = \left(\begin{array}{cccc|cccc} 0 & 0 & 1 & 0 & 1 & 1 & 0 & 0 \\ 0 & 0 & 0 & 1 & 0 & 1 & 0 & 0 \end{array}\right)$$

$$\overline{\mathbf{Z}} = (\mathbf{0} \quad \mathbf{0} \quad \mathbf{0} \,|\, \mathbf{A}_2^{\mathrm{T}} \quad \mathbf{0} \quad \mathbf{I}) = \left(\begin{array}{cccc|cccc} 0 & 0 & 0 & 0 & 0 & 1 & 1 & 0 \\ 0 & 0 & 0 & 0 & 1 & 0 & 0 & 1 \end{array}\right)$$

The corresponding operator representation of Pauli encoded operators is

$$\begin{aligned} \overline{X}_1 &= Z_1 Z_2 X_3, \quad & \overline{X}_2 &= Z_2 X_4, \\ \overline{Z}_1 &= Z_2 Z_3, \quad & \overline{Z}_2 &= Z_1 Z_4. \end{aligned}$$

The decoding circuit obtained by using Eq. (8.121) is shown in Figure 8.4. The decoding circuit has two stages. In the first stage we apply CNOT gates on ancillaries, with control states being the encoded states. The Pauli encoded state $\overline{Z}_1 = Z_2 Z_3$ indicates that the first ancillary qubit is controlled by encoded kets at positions 2 and 3. The Pauli encoded state $\overline{Z}_2 = Z_1 Z_4$ indicates that the second ancillary qubit is controlled by encoded kets at positions 1 and 4. The second stage is implemented based on Pauli encoded operators $\overline{X}_i$. The ancillary qubits now serve as control qubits, and encoded qubits as target qubits. The Pauli encoded state $\overline{X}_1 = Z_1 Z_2 X_3$ indicates that the first ancillary qubit controls the encoded kets at positions 1, 2 and 3, and corresponding controlled gates are Z at positions 1 and 2, and X at position 3. The Pauli encoded state $\overline{X}_2 = Z_2 X_4$ indicates that the second ancillary qubit controls the encoded kets at positions 2 and 4, while the corresponding controlled gates are Z at position 2 and X at position 4.

Example. The decoding circuit for [5,1,3] quantum stabilizer code can be obtained in similar fashion to the previous example, and the starting point is the generators of the code:

$$g_1 = Y_1 Z_2 Z_4 Y_5, \quad g_2 = X_2 Z_3 Z_4 Z_5, \quad g_3 = Z_1 Z_2 X_3 X_5 \text{ and } g_4 = Z_1 Z_3 Y_4 Y_5.$$

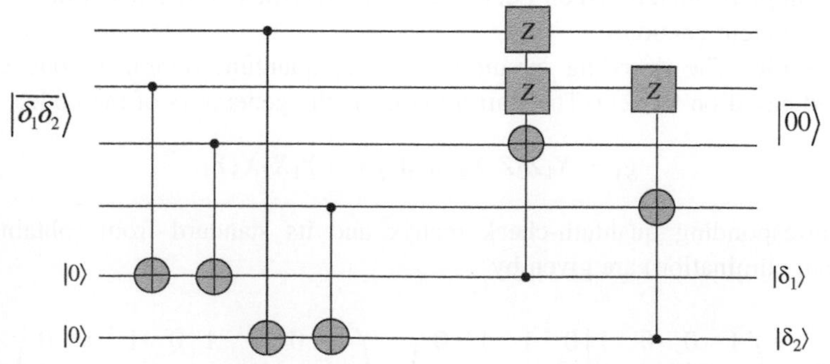

FIGURE 8.4

Efficient implementation of decoding circuit for [4,2,2] quantum stabilizer code.

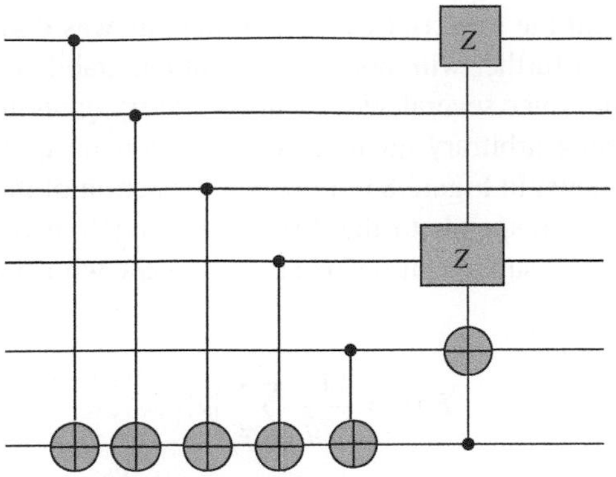

FIGURE 8.5

Efficient implementation of decoding circuit for [5,1,3] quantum stabilizer code.

Following a similar procedure as in the example above, the standard form Pauli encoded operators are given by

$$\overline{Z} = Z_1 Z_2 Z_3 Z_4 Z_5 \text{ and } \overline{X} = Z_1 Z_4 X_5.$$

The corresponding decoding circuit is shown in Figure 8.5.

8.6 NONBINARY STABILIZER CODES

In the previous sections, we considered quantum stabilizer codes with finite geometry representation over F_2^{2N}, $F_2 = \{0,1\}$. Because these codes are defined over F_2^{2N} they can be called "binary" stabilizer codes. In this section, we are concerned with stabilizer codes defined over F_q^{2N}, where $q = p^m$ is a prime power (p is a prime and $m \geq 1$ is an integer) [22–27]. This class of codes can be called the "nonbinary" stabilizer codes. Although many definitions and properties from previous sections are applicable here, certain modifications are needed as described below. First of all we operate on q-ary quantum digits, which in analogy with qubits can be called "qudits". Secondly, we need to extend the definitions of quantum gates to qudits. In the previous chapter, we saw that arbitrary qubit error can be represented in terms of Pauli operators $\{I,X,Y,Z\}$. We have also seen that the Y operator can be expressed in terms of X and Z operators. A similar strategy can be applied here. We need to extend the definitions of X and Z operators to qudits as follows:

$$X(a)|x\rangle = |x+a\rangle, \qquad Z(b) = \omega^{\mathrm{tr}(bx)}|x\rangle; \quad x,a,b \in F_q, \tag{8.123}$$

where $\mathrm{tr}(\cdot)$ denotes the trace operation from F_q to F_p and ω is a pth rooth of unity, namely $\omega = \exp(j2\pi/p)$. The trace operation from F_{q^m} to F_q is defined as

$$\mathrm{tr}_{q^m}(x) = \sum_{i=0}^{m-1} x^{q^i}. \tag{8.124}$$

If F_q is the prime field the subscript can be omitted, as was done in Eq. (8.123).

Before we proceed further with nonbinary quantum stabilizer codes, we believe it is convenient to introduce several additional *nonbinary quantum gates*, which will be useful in describing arbitrary quantum computation on qudits. The nonbinary quantum gates are shown in Figure 8.6, in which the action of the gates is described as well. The F gate corresponds to the discrete Fourier transform (DFT) gate. Its action on ket $|0\rangle$ is the superposition of all basis kets with the same probability amplitude:

$$F|0\rangle = \frac{1}{\sqrt{q}} \sum_{u \in F_q} |u\rangle. \tag{8.125}$$

From Chapter 2 we know that an operator A can be represented in the form $A = \Sigma_{i,j} A_{ij} |i\rangle\langle j|$. Based on Figure 8.6, we determine the action of operator $FX(b)F^{\dagger}$ as follows:

$$\begin{aligned} FX(a)F^{\dagger} &= \frac{1}{\sqrt{q}} \sum_{i,j \in F_q} \omega^{\mathrm{tr}(ij)} |i\rangle\langle j| \sum_{x \in F_q} \omega |x+a\rangle\langle x| \frac{1}{\sqrt{q}} \sum_{l,k \in F_q} \omega^{-\mathrm{tr}(lk)} |l\rangle\langle k| \\ &= \frac{1}{q} \sum_{i,j \in F_q} \omega^{\mathrm{tr}(ia)} \underbrace{\sum_{x \in F_q} \omega^{\mathrm{tr}(ix - xl)}}_{1,\, i=l \quad 0,\, i \neq l} |i\rangle\langle l| \\ &= \sum_{i \in F_q} \omega^{\mathrm{tr}(ia)} |i\rangle\langle i| = Z(a). \end{aligned} \tag{8.126}$$

By using the basic nonbinary gates shown in Figure 8.6 and Eq. (8.123), we can perform more complicated operations, as illustrated in Figure 8.7. These circuits will be used later in the section for efficient implementation of quantum nonbinary encoders and decoders. Notice that the circuit at the bottom of Figure 8.7 is obtained by concatenation of the circuits at the top.

It can be shown that the set of errors $\varepsilon = \{X(a)X(b) | a, b \in F_q\}$ satisfies the following properties: (i) it contains the identity operator; (ii) $\mathrm{tr}(E_1^{\dagger} E_2) = 0 \;\; \forall E_1, E_2 \in \varepsilon$; and (iii) $\forall E_1, E_2 \in \varepsilon: \; E_1 E_2 = cE_3, \; E_3 \in \varepsilon, \; c \in F_q$.

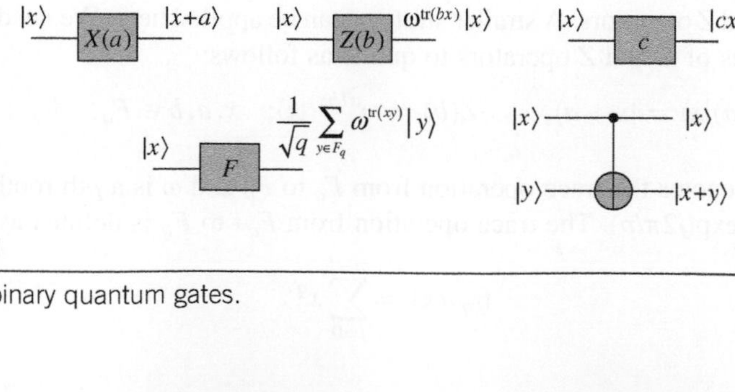

FIGURE 8.6

Basic nonbinary quantum gates.

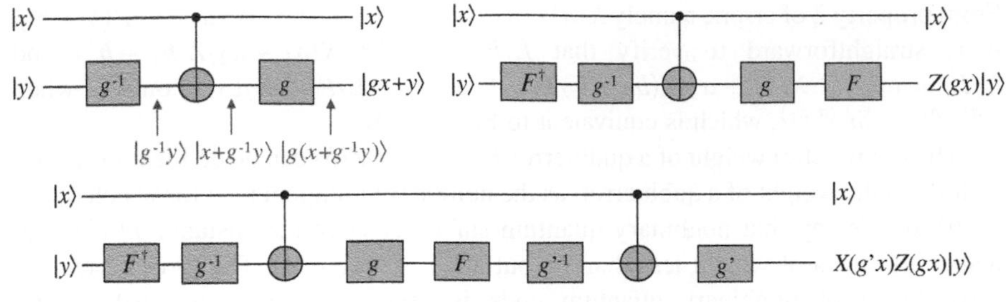

FIGURE 8.7

Nonbinary quantum circuits derived from basic nonbinary gates.

This set forms an error basis for the set of $q \times q$ matrices, and sometimes is called "nice error basis". It can also be shown that: (1) $X(a)Z(b) = \omega^{-tr(ab)}Z(b)X(a)$ and (2) $X(a + a')Z(b + b') = \omega^{-tr(a'b)}X(a)Z(b)X(a')Z(b')$.

Example. The 4-ary nice error basis over $F_4 = \{0, 1, \alpha, \overline{\alpha}\}$ can be obtained as the tensor product of the Pauli basis:

$$X(0) = II, \;\; X(1) = IX, \;\; X(\alpha) = XI, \;\; X(\overline{\alpha}) = XX,$$
$$Z(0) = II, \; Z(1) = ZI, \;\; Z(\alpha) = ZZ, \; Z(\overline{\alpha}) = IZ.$$

In order to determine the nice basis error on N qudits we introduce the following notation: $X(\boldsymbol{a}) = X(a_1) \otimes \ldots \otimes X(a_N)$ and $Z(\boldsymbol{b}) = Z(b_1) \otimes \ldots \otimes Z(b_N)$, where $\boldsymbol{a} = (a_1,\ldots,a_N)$, $\boldsymbol{b} = (b_1,\ldots,b_N)$, and $a_i, b_i \in F_q$. The set $\varepsilon_N = \{X(\boldsymbol{a})X(\boldsymbol{b})|\boldsymbol{a},\boldsymbol{b} \in F_q^N\}$ is the nice error basis defined over F_q^{2N}. Similarly to the Pauli multiplicative group, we can define the *error group* by

$$G_N = \left\{\omega^c X(\boldsymbol{a})Z(\boldsymbol{b}) \middle| \boldsymbol{a},\boldsymbol{b} \in F_q^N, c \in F_p\right\}. \tag{8.127}$$

Let S be the largest Abelian subgroup of G_N that fixes all elements from quantum code C_Q, called the stabilizer group. The $[N,K]$ nonbinary stabilizer code C_Q is defined as the K-dimensional subspace of the N-qudit Hilbert space H_q^N as follows:

$$C_Q = \bigcap_{s \in S} \left\{|c\rangle \in H_q^N \middle| s|c\rangle = |c\rangle\right\}. \tag{8.128}$$

Clearly, the definition of nonbinary stabilizer codes is a straightforward generalization of quantum stabilizer code over H_2^N. Therefore, similar properties, definitions, and theorems as introduced in previous sections are applicable here. For example, two errors $E_1 = \omega^{c_1}X(\boldsymbol{a}_1)Z(\boldsymbol{b}_1)$, $E_2 = \omega^{c_2}X(\boldsymbol{a}_2)Z(\boldsymbol{b}_2) \in G_N$ commute if and only if their trace sympletic product vanishes, i.e.

$$\mathrm{tr}(\boldsymbol{a}_1\boldsymbol{b}_2 - \boldsymbol{a}_2\boldsymbol{b}_1) = 0. \tag{8.129}$$

From property 2 of errors, namely $X(a)Z(b)X(a')Z(b') = \omega^{\mathrm{tr}(a'b)}X(a+a')Z(b+b')$, it is straightforward to verify that $E_1E_2 = \omega^{\mathrm{tr}(\boldsymbol{a}_2\boldsymbol{b}_1)}X(\boldsymbol{a}_1+\boldsymbol{a}_2)Z(\boldsymbol{b}_1+\boldsymbol{b}_2)$ and $E_2E_1 = \omega^{\mathrm{tr}(\boldsymbol{a}_1\boldsymbol{b}_2)}X(\boldsymbol{a}_1+\boldsymbol{a}_2)Z(\boldsymbol{b}_1+\boldsymbol{b}_2)$. Clearly $E_1E_2 = E_2E_1$ only when $\omega^{\mathrm{tr}(\boldsymbol{a}_2\boldsymbol{b}_1)} = \omega^{\mathrm{tr}(\boldsymbol{a}_1\boldsymbol{b}_2)}$, which is equivalent to Eq. (8.129).

The (sympletic) weight of a qudit error $E = \omega^c X(\boldsymbol{a})Z(\boldsymbol{b})$ can be defined, in similar fashion to the weight of a qubit error, as the number of components i for which $(a_i,b_i) \neq (0,0)$. We say that nonbinary quantum stabilizer code has distance D if it can detect all errors of weight less than D, but none of weight D. The error correction capability t of nonbinary quantum code is related to minimum distance by $t = \lfloor (D-1)/2 \rfloor$. We say that nonbinary quantum stabilizer code is nondegenerate if its stabilizer group S does not contain an element of weight smaller than t. From the definitions above, we can see that nonbinary quantum stabilizer codes are a straightforward generalization of the corresponding qubit stabilizer codes. The key difference is that instead of the sympletic product we need to use the trace-sympletic product. Similar theorems can be proved and properties can be derived by following similar procedures, by taking into account the differences outlined above. On the other hand, the quantum hardware implementation is more challenging. For example, instead of using Pauli X and Z gates we have to use the $X(a)$ and $Z(b)$ gates shown in Figure 8.6. We can also define the standard form, but instead of using F_2 during Gauss elimination, we have to perform Gauss elimination in F_q instead. For example, we can determine the syndrome based on syndrome measurements, as illustrated in previous sections. Let the generator $\boldsymbol{g}_i$ be given as follows:

$$\boldsymbol{g}_i = [\boldsymbol{a}_i|\boldsymbol{b}_i] = [0 \ldots 0 \; a_i \ldots a_N \mid 0 \ldots \; 0 \; b_i \ldots b_N] \in F_q^{2N}, \quad a_i, b_i \in F_q. \qquad (8.130)$$

The quantum circuit shown in Figure 8.8 will provide the nonzero measurement if a detectable error does not commute with a multiple of $\boldsymbol{g}_i$. By comparison

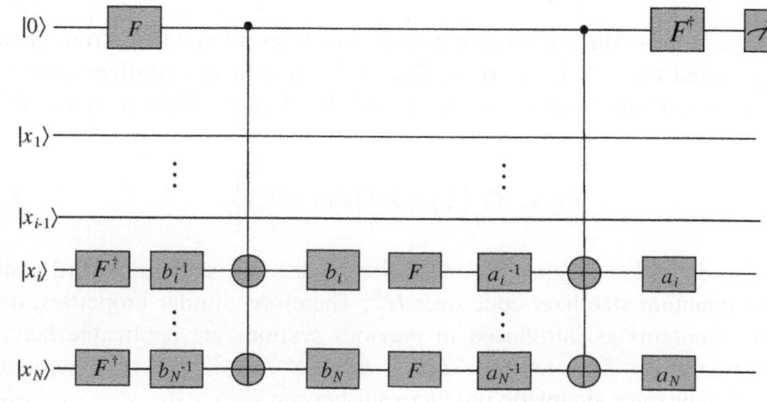

FIGURE 8.8

Nonbinary syndrome decoding circuit corresponding to generator g_i.

with the corresponding qubit syndrome decoding circuits from the previous chapter we conclude that the F gate for nonbinary quantum codes has a similar role to the Hadamard gate for codes over F_2. Notice that the syndrome circuit is a generalization of the bottom circuit in Figure 8.7. From Figure 8.8 it is clear that if the input to the stabilizer circuit is $E|\psi\rangle$, where E is the qudit error and $|\psi\rangle$ is the codeword, then the stabilizer circuit performs the following mapping:

$$|0\rangle E|\psi\rangle \to \sum_{u\in F_q} F^{\dagger}|u\rangle X(u\boldsymbol{a}_i)Z(u\boldsymbol{b}_i)E|\psi\rangle. \tag{8.131}$$

The qudit error E, by ignoring the complex phase constant, can be represented as $E = X(\boldsymbol{a})Z(\boldsymbol{b})$. Based on the text below Eq. (8.129), it is clear that $X(\boldsymbol{a}_i)Z(\boldsymbol{b}_i)E = \omega^{\mathrm{tr}(\boldsymbol{a}\boldsymbol{b}_i-\boldsymbol{a}_i\boldsymbol{b})}EX(\boldsymbol{a}_i)Z(\boldsymbol{b}_i)$. Based on Figure 8.6 we can show that $X(u\boldsymbol{a}_i)Z(u\boldsymbol{b}_i)E = \omega^{\mathrm{tr}(u(\boldsymbol{a}\boldsymbol{b}_i-\boldsymbol{a}_i\boldsymbol{b}))}EX(\boldsymbol{a}_i)Z(\boldsymbol{b}_i)$, and by substituting in (8.131) we obtain:

$$|0\rangle E|\psi\rangle \to \sum_{u\in F_q} F^{\dagger}|u\rangle X(u\boldsymbol{a}_i)Z(u\boldsymbol{b}_i)E|\psi\rangle = \sum_{u\in F_q} F^{\dagger}|u\rangle \omega^{\mathrm{tr}(u(\boldsymbol{a}\boldsymbol{b}_i-\boldsymbol{a}_i\boldsymbol{b}))}EX(\boldsymbol{a}_i)Z(\boldsymbol{b}_i)|\psi\rangle. \tag{8.132}$$

Since $X(\boldsymbol{a}_i)Z(\boldsymbol{b}_i)|\psi\rangle = |\psi\rangle$, because $X(\boldsymbol{a}_i)Z(\boldsymbol{b}_i) \in S$, Eq. (8.132) becomes

$$\begin{aligned}
|0\rangle E|\psi\rangle &\to \left[\sum_{u\in F_q} F^{\dagger}|u\rangle \omega^{\mathrm{tr}(u(\boldsymbol{a}\boldsymbol{b}_i-\boldsymbol{a}_i\boldsymbol{b}))}\right] E|\psi\rangle \\
&= \left[\sum_{u\in F_q}\sum_{v\in F_q} \omega^{-\mathrm{tr}(uv)}\omega^{\mathrm{tr}(u(\boldsymbol{a}\boldsymbol{b}_i-\boldsymbol{a}_i\boldsymbol{b}))}|v\rangle\right] E|\psi\rangle \\
&= \sum_{v\in F_q}|v\rangle \underbrace{\sum_{u\in F_q}\omega^{\mathrm{tr}(u(\boldsymbol{a}\boldsymbol{b}_i-\boldsymbol{a}_i\boldsymbol{b})-uv)}}_{1,\ \text{for } v=\boldsymbol{a}\boldsymbol{b}_i-\boldsymbol{a}_i\boldsymbol{b}} E|\psi\rangle = |\boldsymbol{a}\boldsymbol{b}_i-\boldsymbol{a}_i\boldsymbol{b}\rangle E|\psi\rangle.
\end{aligned} \tag{8.133}$$

From (8.133) it is evident that the result of the measurement gives us the ith syndrome component $\lambda_i = \mathrm{tr}(\boldsymbol{a}\boldsymbol{b}_i - \boldsymbol{a}_i\boldsymbol{b})$. By $N-K$ measurements on corresponding generators $\boldsymbol{g}_i$ $(i=1,\ldots,N-K)$ we obtain the following syndrome $S(E)$:

$$S(E) = [\lambda_1\lambda_2\ldots\lambda_{N-K}]^{\mathrm{T}};\ \lambda_i = \mathrm{tr}(\boldsymbol{a}\boldsymbol{b}_i - \boldsymbol{a}_i\boldsymbol{b}),\ i = 1,\ldots,N-K. \tag{8.134}$$

Clearly, the syndrome equation (8.134) is very similar to (8.47), and similar conclusions can be drawn. The correctable qudit error maps the code space to q^K-dimensional subspace of q^N-dimensional Hilbert space. Since there are $N-K$ generators, or equivalently syndrome positions, there are q^{N-K} different cosets. All qudit errors belonging to the same coset have the same syndrome. By selecting the most probable qudit error for the coset representative, typically lowest weight error, we can uniquely identify the qudit error and consequently perform the error correction.

We can define the projector P onto the code space C_Q as follows:

$$P = \frac{1}{|S|}\sum_{s \in S} s. \tag{8.135}$$

Clearly, $Ps = s$ for every $s \in S$. Further,

$$P^2 = \frac{1}{|S|}\sum_{s \in S} Ps = \frac{1}{|S|}\sum_{s \in S} s = P. \tag{8.136}$$

Since $s^\dagger \in S$, $P^\dagger = P$ is valid, which indicates that P is an orthogonal projector. The dimensionality of C_Q is $\dim C_Q = q^N/|S|$.

Quantum bounds similar to those described in the previous chapter can be derived for nonbinary stabilizer codes as well. For example, the Hamming bound for nonbinary quantum stabilizer code $[N,K,D]$ (where K is the number of information qudits and N is the codeword length) over F_q is given by

$$\sum_{i=0}^{\lfloor (D-1)/2 \rfloor} \binom{N}{i} (q^2 - 1)^i q^K \leq q^N. \tag{8.137}$$

This inequality is obtained from inequality (8.52) in Chapter 7 by substituting the term 3^i with $(q^2 - 1)^i$, and it can be proved using similar methodology. The number of possible information qudit words is now q^K, and the number of possible codewords is q^N.

8.7 SUBSYSTEM CODES

The subsystem codes [26–31] represent a generalization of nonbinary stabilizer codes from the previous section. They can also be considered as a generalization of decoherence-free subspaces (DFSs) [32,33] and noiseless subsystems (NSs) [34]. The key idea behind subsystem codes is to decompose the quantum code C_Q as the tensor product of two subsystems A and B as $C_Q = A \otimes B$. The information qudits belong to subsystem A and non-information qudits, also known as gauge qudits, belong to subsystem B. We are only concerned with errors introduced in subsystem A.

It was seen in the previous chapter that both noise $\mathscr{E}$ and recovery $\mathscr{R}$ processes can be described as quantum operations $\mathscr{E},\mathscr{R}$: $L(H) \rightarrow L(H)$, where $L(H)$ is the space of linear operators on Hilbert space H. These mappings can be represented in terms of the operator-sum representation,

$\mathscr{E}(\rho)=\sum_i E_i \rho E_i^\dagger$, $E_i \in L(H)$, $\mathscr{E}=\{E_i\}$. Given the quantum code C_Q that is subspace of H, we say that the set of errors $\mathscr{E}$ are correctable if there exists a recovery operation $\mathscr{R}$ such that $\mathscr{R}\mathscr{E}(\rho)=\rho$ from any state ρ from $L(C_Q)$. In terms of subsystem codes, we say that there exists a recovery operation such that for any ρ^A from $L(A)$ and ρ^B from $L(B)$ the following is valid: $\mathscr{R}\mathscr{E}(\rho^A \otimes \rho^B)=\rho^A \otimes \rho'^B$, where the state ρ'^B is not relevant. The necessary and sufficient condition for the set of errors $\mathscr{E}=\{E_m\}$ to be correctable is that $PE_m^\dagger E_n P = I^A \otimes g_{mn}^B$, $\forall m,n$, where P is the projector on code subspace. This set of errors is called the correctable set of errors. Clearly, the linear combination of errors from $\mathscr{E}$ is also correctable so that it makes sense to observe the correctable set of errors as linear space with a properly chosen operator basis. If we are concerned with quantum registers/system composed of N qubits, the corresponding error operators are Pauli operators. The Pauli group on N qubits, G_N, has already been introduced in Section 8.1. The stabilizer formalism can also be used in describing the stabilizer subsystem codes. Stabilizer subsystem codes are determined by a subgroup of G_N that contains the element jI, called the *gauge group* $\mathscr{G}$, and by the stabilizer group S that is properly chosen such that $S'=j^\lambda S$ is the center of $\mathscr{G}$, denoted as $Z(\mathscr{G})$ (i.e. the set of elements from $\mathscr{G}$ that commute with all elements from $\mathscr{G}$). We are concerned with the following decomposition of H: $H=C\oplus C^\perp=(H_A \otimes H_B)\oplus C^\perp$, where $C^\perp$ is the dual of $C=H_A\otimes H_B$. The gauge operators are chosen in such a way that they act trivially on subsystem A, but generate full algebra of the subsystem B. The information is encoded on subsystem A, while the subsystem B is used to absorb the effects of gauge operations. The Pauli operators for K logical qubits are obtained from the isomorphism $N(\mathscr{G})/S \simeq G_K$, where $N(\mathscr{G})$ is the normalizer of $\mathscr{G}$. On the other hand, subsystem B consists of R gauge qubits recovered from isomorphism $\mathscr{G}/S \simeq G_N$, where $N=K+R+s$, $s \geq 0$. If $\tilde{X}_1$ and $\tilde{Z}_1$ represent the images of X_i and Z_i under automorphism U of G_N, the stabilizer can be described by $S=\langle \tilde{Z}_1,\ldots,\tilde{Z}_s\rangle$; $R+s\leq N$; $R,s\geq 0$. The gauge group can be specified by $\mathscr{G}=\langle jI,\tilde{Z}_1,\cdots,\tilde{Z}_{s+R},\tilde{X}_{s+1},\cdots,\tilde{X}_R\rangle$. The images must satisfy the commutative relations (8.28). The logical (encoded) Pauli operators will then be $\overline{X}_1=\tilde{X}_{s+R+1},\overline{Z}_1=\tilde{Z}_{s+R+1},\ldots,\overline{X}_K=\tilde{X}_N,\overline{Z}_K=\tilde{Z}_N$. The detectable errors are the elements of $G_N-N(S)$ and undetectable errors are in $N(S)-\mathscr{G}$. Undetectable errors are related to the logical Pauli operators since $N(S)/\mathscr{G}\simeq N(\mathscr{G})/S'$. Thus, if $n\in N(S)$ there exists $g\in\mathscr{G}$ such that $ng\in N(\mathscr{G})$ and if $g'\in\mathscr{G}$ such that $ng'\in N(\mathscr{G})$, $gg'\in\mathscr{G}\cap N(\mathscr{G})=S'$. The distance D of this code is defined as the minimum weight among undetectable errors. The subsystem code encodes K qubits into N-qubit codewords and has R gauge qubits, and can therefore be denoted as $[N,K,R,D]$ code.

We turn our attention now to subsystem codes defined as subspace of q^N-dimensional Hilbert space H_q^N. For a and b from F_q, we define unitary operators (qudit errors) $X(a)$ and $Z(b)$ as follows:

$$X(a)|x\rangle=|x+a\rangle, \qquad Z(b)=\omega^{\mathrm{tr}(bx)}|x\rangle;\ \ x,a,b\in F_q;\ \ \omega=e^{j2\pi/p}. \qquad (8.138)$$

As expected, since nonbinary stabilizer subsystem codes are a generalization of nonbinary stabilizer codes, similar qudit errors occur. The trace operation is defined by (8.124). The qudit *error group* is defined by

$$G_N = \left\{ \omega^c X(\boldsymbol{a})Z(\boldsymbol{b}) \middle| \boldsymbol{a},\boldsymbol{b} \in F_q^N, c \in F_p \right\};$$
$$\boldsymbol{a} = (a_1 \ldots a_N), \boldsymbol{b} = (b_1 \ldots b_N); \quad a_i, b_i \in F_q. \tag{8.139}$$

Let C_Q be a quantum code such that $H = C_Q \oplus C_Q^{\perp}$. The [N,K,R,D] subsystem code over F_q is defined as the decomposition of code space C_Q into a tensor product of two subsystems A and B such that $C_Q = A \otimes B$, where dimensionality of A equals $\dim A = q^K$, $\dim B = q^R$, and all errors of weight less than $D_{\min}$ on subsystem A can be detected. What is interesting about this class of codes is that when constructed from classical codes, the corresponding classical code does not need to be dual containing (self-orthogonal). Notice that subsystem codes can also be defined over F_{q^2}.

For [N,K,R,$D_{\min}$] stabilizer subsystem codes over F_2 it can be shown that the centralizer of S is given by (see Problem 13)

$$C_{G_N}(S) = \langle \mathcal{G}, \overline{X}_1, \overline{Z}_1, \ldots, \overline{X}_K, \overline{Z}_K \rangle. \tag{8.140}$$

Since $C_Q = A \otimes B$, then $\dim A = 2^K$ and $\dim B = 2^R$. From Eq. (8.135) we know that stabilizer S can be used as projector to C_Q. The dimensionality of quantum code defined by stabilizer S is 2^{K+R}. The stabilizer S therefore defines an [N, K+R,D] stabilizer code. Based on (8.140) we conclude that image operators $\tilde{Z}_i, \tilde{X}_i$; $i = s+1, \ldots, R$ behave as encoded operators on gauge qubits, while $\overline{Z}_i, \overline{X}_i$ act on information qubits. In total we have a set of $2(K+R)$ encoded operators of [N, K+R,D] stabilizer codes given by

$$\left\{ \overline{X}_1, \overline{Z}_1, \ldots, \overline{X}_K, \overline{Z}_K, \tilde{X}_{s+1}, \tilde{Z}_{s+1}, \ldots, \tilde{X}_{s+R}, \tilde{Z}_{s+R} \right\}. \tag{8.141}$$

Therefore, with (8.141), we have just established the connection between stabilizer codes and subsystem codes. Since subsystem codes are more flexible for design, they can be used to design new classes of stabilizer codes. More importantly, Eq. (8.141) give us an opportunity to implement encoders and decoders for subsystem codes using similar approaches to those already developed for stabilizer codes, such as standard form formalism. For example, subsystem code [4,1,1,2] is described by stabilizer group $S = \langle \tilde{Z}_1, \tilde{Z}_2 \rangle$ and gauge group $\mathcal{G} = \langle S, \tilde{X}_3, \tilde{Z}_3, jI \rangle$, where $\tilde{Z}_1 = X_1X_2X_3X_4$, $\tilde{Z}_2 = Z_1Z_2Z_3Z_4$, $\tilde{Z}_3 = Z_1Z_2$, and $\tilde{X}_3 = X_2X_4$. The encoded operators are given by $\overline{X}_1 = X_2X_3$ and $\overline{Z}_1 = Z_2Z_4$. Corresponding [4,2] stabilizer code, based on (8.141), is described by generators $g_1 = \tilde{X}_3$, $g_2 = \tilde{Z}_3$, $g_3 = \overline{X}_1$, and $g_4 = \overline{Z}_1$. In Figure 8.9, we show two possible encoder versions of the same [4,1,1,2] subsystem code. Notice that the state of gauge qubits is irrelevant as far as information qubits are concerned. They can be randomly chosen. In Figure 8.9b, we show that by proper setting of gauge qubits we can simplify encoder implementation.

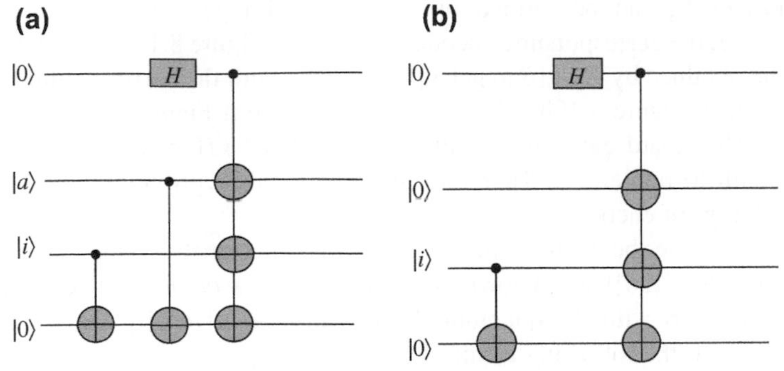

FIGURE 8.9

Encoder implementation of [4,1,1;2] subsystem code: (a) gauge qubit $|a\rangle$ has arbitrary value; (b) gauge qubit is set to $|0\rangle$.

Based on Eq. (8.141) and the text around it, we can conclude that with the help of stabilizer S and gauge $\mathcal{G}$ groups of subsystem code, we are able to create a stabilizer S_A whose encoder can be used as an encoder for subsystem code providing that gauge qubits are initialized to the $|0\rangle$ qubit. Clearly $S_A \supseteq S$, which indicates that the space stabilized by S_A is the subspace of $C_Q = A \otimes B$, which on the other hand is stabilized by S. Since $|S_A|/|S| = 2^R$, the dimension of the subspace stabilized by S_A is $2^{K+R-R} = 2^K$. If the subsystem stabilizer and gauge groups are given by $S = \langle \tilde{Z}_1, \ldots, \tilde{Z}_s \rangle$, $s = N - K - R$, and $\mathcal{G} = \langle jI, S, \tilde{Z}_{s+1}, \ldots, \tilde{Z}_{s+R}, \tilde{X}_{s+1}, \ldots, \tilde{X}_{s+R} \rangle$ respectively, then the stabilizer group of S_A will be given by $S_A = \langle S, \tilde{Z}_{s+1}, \ldots, \tilde{Z}_{s+R} \rangle$. The quantum-check matrix for S_A can be put in standard form (8.104) as follows:

$$A(S_A) = \left(\begin{array}{ccc|ccc} \overbrace{I_r}^{r} & \overbrace{A_1}^{N-K-r} & \overbrace{A_2}^{K} & \overbrace{B}^{r} & \overbrace{0}^{N-K-r} & \overbrace{C}^{K} \\ 0 & 0 & 0 & D & I_{N-K-r} & E \end{array} \right) \begin{array}{l} \}r \\ \}N-K-r \end{array}, \quad r = \text{rank}(A). \tag{8.142}$$

The encoded Pauli operators of S_A are given by

$$\overline{X} = (0 \;\; E^{\mathrm{T}} \;\; I_K \,|\, C^{\mathrm{T}} \;\; 0 \;\; 0), \quad \overline{Z} = (0 \;\; 0 \;\; 0 \,|\, A_2^{\mathrm{T}} \;\; 0 \;\; I_K). \tag{8.143}$$

The encoder circuit for $[N,K,R,D]$ subsystem code can be implemented as shown in Figure 8.3, wherein all ancillary qubits are initialized to the $|0\rangle$ qubit.

As an illustration, let us study the encoder implementation of [9,1,4,3] Bacon–Shor code [35,36] (see also Ref. [26]). The stabilizer S and gauge $\mathcal{G}$ groups of this code are given by

$$S = \langle \tilde{Z}_1 = X_1X_2X_3X_7X_8X_9, \tilde{Z}_2 = X_4X_5X_6X_7X_8X_9, \tilde{Z}_3 = Z_1Z_3Z_4Z_6Z_7Z_9, \\ \tilde{Z}_4 = Z_2Z_3Z_5Z_6Z_8Z_9 \rangle$$

$$\mathcal{G} = \langle S, \mathcal{G}_X, \mathcal{G}_Z \rangle, \quad \mathcal{G}_X = \langle \tilde{X}_5 = X_2X_5, \tilde{X}_6 = X_3X_6, \tilde{X}_7 = X_6X_9, \\ \tilde{X}_8 = X_1X_2X_3X_4X_5X_6 \rangle,$$

$$\mathcal{G}_Z = \langle \tilde{Z}_5 = Z_1Z_3, \tilde{Z}_6 = Z_4Z_6, \tilde{Z}_7 = Z_2Z_3, \tilde{Z}_8 = Z_5Z_6 \rangle.$$

The stabilizer S_A can be formed by adding the images from $\mathcal{G}_Z$ to S to get $S_A = \langle S, \mathcal{G}_Z \rangle$; the corresponding encoder is shown in Figure 8.10a. The stabilizer S_A can also be obtained by $\mathcal{G}_X$ to S to get $S_A = \langle S, \mathcal{G}_X \rangle$, with the corresponding encoder being shown in Figure 8.10b. Clearly, the encoder from Figure 8.10b has a larger number of Hadamard gates, but smaller number of CNOT gates, which are much more difficult to implement. Therefore, the gauge qubits provide flexibility in the implementation of encoders.

We further describe another encoder implementation method due to Grassl et al. [37] (see also Ref. [26]), also known as the *conjugation method*. The key idea of this method is to start with the quantum-check matrix $\boldsymbol{A}$, given by Eq. (8.142), and transform it into the following form:

$$\boldsymbol{A}' = [\boldsymbol{0}\ \boldsymbol{0} \mid \boldsymbol{I}_{N-K}\ \ \boldsymbol{0}]. \tag{8.144}$$

During this transformation we have to memorize the actions we performed, which is equivalent to the gates being employed on information and ancillary qubits in order to perform this transformation. Notice that the quantum-check matrix in (8.144) corresponds to canonical quantum codes described in the previous chapter, which simply perform the following mapping:

$$|\psi\rangle \rightarrow \underbrace{|0\rangle \otimes \ldots \otimes |0\rangle}_{N-K \text{ times}} |\psi\rangle = |\boldsymbol{0}\rangle_{N-K}|\psi\rangle. \tag{8.145}$$

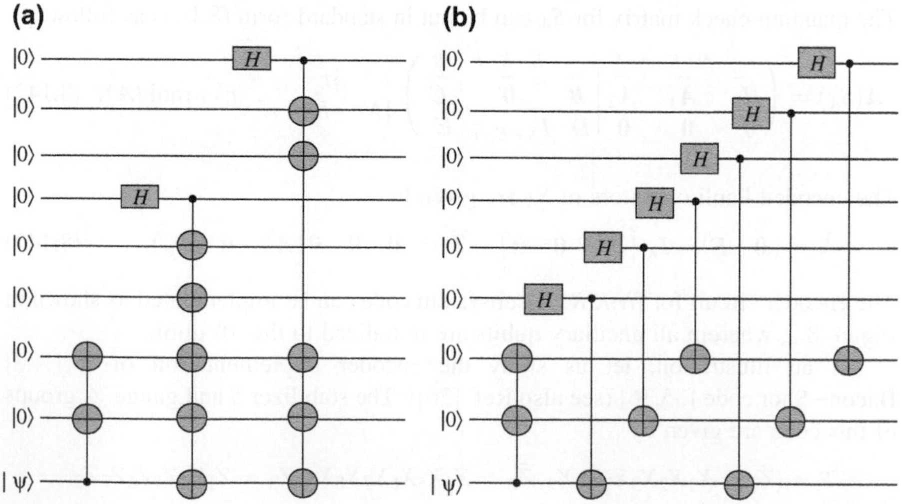

FIGURE 8.10

Encoder for [9,1,4;3] subsystem code (or equivalently [9,1,3] stabilizer code) for S_A given by: (a) $S_A = \langle S, \mathcal{G}_Z \rangle$; (b) $S_A = \langle S, \mathcal{G}_X \rangle$.

In this encoding mapping, $N-K$ ancillary qubits are appended to K information qubits. The corresponding encoded Pauli operators are given by

$$\overline{X} = (\mathbf{0}\ I_K \mid \mathbf{0}\ \mathbf{0}),\quad \overline{Z} = (\mathbf{0}\ \mathbf{0} \mid \mathbf{0}\ I_K). \tag{8.146}$$

Let us observe one particular generator of quantum-check matrix, say $\boldsymbol{g}=(a_1,\ldots,a_N|b_1,\ldots,b_N)$. The action of the Hadamard gate on the ith qubit is to swap positions of a_i and b_i as follows, where we underline the affected qubit for clarity:

$$(a_1 \ldots \underline{a_i} \ldots a_N | b_1 \ldots \underline{b_i} \ldots b_N) \xrightarrow{H} (a_1 \ldots \underline{b_i} \ldots a_N | b_1 \ldots \underline{a_i} \ldots b_N). \tag{8.147}$$

The action of the CNOT gate on $\boldsymbol{g}$, where the ith qubit is the control qubit and the jth qubit is the target qubit, whose action is denoted as CNOT^{i,j}, can be described as

$$(a_1 \ldots \underline{a_j} \ldots a_N | b_1 \ldots \underline{b_i} \ldots b_N) \xrightarrow{\text{CNOT}^{i,j}}$$
$$(a_1 \ldots a_{j-1}\, \underline{a_j + a_i}\, a_{j+1} \ldots a_N | b_1 \ldots b_{i-1}\, \underline{b_i + b_j}\, b_{i+1} \ldots b_N). \tag{8.148}$$

Therefore, the jth entry in the X portion and the ith entry in the Z portion are affected. Finally, the action of phase gate P on the ith qubit is to perform the following mapping:

$$(a_1 \ldots \underline{a_i} \ldots a_N | b_1 \ldots \underline{b_i} \ldots b_N) \xrightarrow{P} (a_1 \ldots \underline{a_i} \ldots a_N | b_1 \ldots \underline{a_i + b_i} \ldots b_N). \tag{8.149}$$

Based on Eqs (8.147)–(8.149), we can create the look-up table (LUT) in Table 8.1.

Based on Table 8.1, the transformation of $\boldsymbol{g}=(a_1,\ldots,a_N|b_1,\ldots,b_N)$ to $\boldsymbol{g}'=(a'_1,\ldots,a'_N|0,\ldots,0)$ can be achieved by application of Hadamard H and phase P gates as follows:

$$\bigotimes_{i=1}^{N} H^{\overline{a}_i b_i} P^{a_i b_i},\quad \overline{a}_i = a_i + 1 \bmod 2. \tag{8.150}$$

For example, $\boldsymbol{g}=(10010|01110)$ can be transformed to $\boldsymbol{g}'=(11110|00000)$ by application of the sequence of gates $H_2 H_3 P_4$, based on Table 8.1 or Eq. (8.150).

Table 8.1 Look-Up-Table of Actions of Gates I, H, and P on the ith Qubit

(a_i,b_i)	Gate	(c,d)
(0,0)	I	(0,0)
(1,0)	I	(1,0)
(0,1)	H	(1,0)
(1,1)	P	(1,0)

The ith qubit (a_i, b_i), through the action of the corresponding gate, is transformed into (c,d).

Based on Eq. (8.148) above, the error operator $\boldsymbol{e} = (a_1,\ldots,a_i = 1,\ldots,a_N|\ 0,\ldots,0)$ can be converted to $\boldsymbol{e}' = (0,\ldots,\ a_i' = 1,0\ldots0|0\ldots0)$ by the application of the following sequence of gates:

$$\prod_{m=1,m\neq n}^{N} (\text{CNOT}^{m,n})^{a_n}. \tag{8.151}$$

For example, the error operator $\boldsymbol{e} = (11110|00000)$ can be transformed to $e' = (0100|0000)$ by the application of a sequence of CNOT gates: $\text{CNOT}^{2,1}$ $\text{CNOT}^{2,3}\text{CNOT}^{2,4}$.

Therefore, the first stage is to convert the Z portion of the stabilizer matrix to an all-zero submatrix by application of H and P gates by means of Eq. (8.150). In the second stage, we convert every row of the stabilizer matrix in the first stage $(\boldsymbol{a}'|\boldsymbol{0})$ to $(0\ldots a_i'= 10\ldots0|0\ldots0)$ by means of Eq. (8.151). After the second stage, the stabilizer matrix has the form $(\boldsymbol{I}_{N-K}0|\boldsymbol{00})$. In the third stage, we transform the obtained stabilizer matrix to $(\boldsymbol{00}|\boldsymbol{I}_{N-K}\boldsymbol{0})$ by applying the H gates to the first $N-K$ qubits. The final form of the stabilizer matrix is that of canonical code, which simply appends $N-K$ ancillary qubits in the $|0\rangle$ state to K information qubits, i.e. it performs the mapping (8.145). During this three-stage transformation encoded Pauli operators are transformed into the form given by Eq. (8.146). Let us denote the sequence of gates being applied during this three-stage transformation as $\{U\}$. By applying this sequence of gates in opposite order we obtain the encoding circuit of the corresponding stabilizer code.

Let us now apply the conjugation method on subsystem codes. The key difference with respect to stabilizer codes is that we need to observe the whole gauge group instead. One option is to create the stabilizer group S_A based on the gauge group $\mathcal{G} = <S, \mathcal{G}_Z, \mathcal{G}_X>$ as either $S_A = <S, \mathcal{G}_Z>$ or $S_A = <S, \mathcal{G}_X>$ and then apply the conjugation method. The second option is to transform the gauge group, represented using finite geometry formalism, as follows:

$$\mathcal{G} = \begin{bmatrix} S \\ \mathcal{G}_Z \\ \mathcal{G}_X \end{bmatrix} \overset{\{U\}}{\rightarrow} \mathcal{G}' = \begin{bmatrix} \mathbf{0} & \mathbf{0} & \mathbf{0}| & \boldsymbol{I}_s & \mathbf{0} & \mathbf{0} \\ \mathbf{0} & \mathbf{0} & \mathbf{0}| & \mathbf{0} & \boldsymbol{I}_R & \mathbf{0} \\ \mathbf{0} & \boldsymbol{I}_R & \mathbf{0}| & \mathbf{0} & \mathbf{0} & \mathbf{0} \end{bmatrix}. \tag{8.152}$$

With transformation (8.152) we perform the following mapping:

$$|\phi\rangle|\psi\rangle \rightarrow \underbrace{|0\rangle \otimes \ldots \otimes |0\rangle}_{s=N-K-R\ \text{times}} |\phi\rangle|\psi\rangle = |\mathbf{0}\rangle_s|\phi\rangle|\psi\rangle, \tag{8.153}$$

where quantum state $|\psi\rangle$ corresponds to information qubits, while the quantum state $|\phi\rangle$ represents the gauge qubits. Therefore, with transformation (8.152) we obtain the *canonical subsystem code*, i.e. the subsystem code in which $s = N-K-R$

ancillary qubits in the $|0\rangle$ state are appended to information and gauge qubits. By applying the sequence of gates $\{U\}$ in (8.152) in the opposite order we obtain the encoder of the subsystem code.

As an illustrative example, let us observe again the subsystem [4,1,1,2] code, whose gauge group can be represented using the finite geometry representation as follows:

$$\mathcal{G} = \left[\begin{array}{cccc|cccc} 1 & 1 & 1 & 1 & 0 & 0 & 0 & 0 \\ 0 & 0 & 0 & 0 & 1 & 1 & 1 & 1 \\ \hline 0 & 0 & 0 & 0 & 0 & 0 & 1 & 1 \\ 0 & 1 & 0 & 1 & 0 & 0 & 0 & 0 \end{array}\right].$$

By applying a sequence of transformations, we can represent the [4,1,1,2] subsystem code in canonical form:

$$\mathcal{G} \xrightarrow{U_1=\mathrm{CNOT}^{1,2}\mathrm{CNOT}^{1,3}\mathrm{CNOT}^{1,4}} \left[\begin{array}{cccc|cccc} 1 & 0 & 0 & 0 & 0 & 0 & 0 & 0 \\ 0 & 0 & 0 & 0 & 0 & 1 & 1 & 1 \\ \hline 0 & 0 & 0 & 0 & 0 & 0 & 1 & 1 \\ 0 & 1 & 0 & 1 & 0 & 0 & 0 & 0 \end{array}\right] \xrightarrow{U_2=H_2H_3H_4} \left[\begin{array}{cccc|cccc} 1 & 0 & 0 & 0 & 0 & 0 & 0 & 0 \\ 0 & 1 & 1 & 1 & 0 & 0 & 0 & 0 \\ \hline 0 & 0 & 1 & 1 & 0 & 0 & 0 & 0 \\ 0 & 0 & 0 & 0 & 0 & 1 & 0 & 1 \end{array}\right]$$

$$\xrightarrow{U_3=\mathrm{CNOT}^{2,3}\mathrm{CNOT}^{2,4}} \left[\begin{array}{cccc|cccc} 1 & 0 & 0 & 0 & 0 & 0 & 0 & 0 \\ 0 & 1 & 0 & 0 & 0 & 0 & 0 & 0 \\ \hline 0 & 0 & 1 & 1 & 0 & 0 & 0 & 0 \\ 0 & 0 & 0 & 0 & 0 & 0 & 0 & 1 \end{array}\right] \xrightarrow{U_4=\mathrm{CNOT}^{4,3}} \left[\begin{array}{cccc|cccc} 1 & 0 & 0 & 0 & 0 & 0 & 0 & 0 \\ 0 & 1 & 0 & 0 & 0 & 0 & 0 & 0 \\ \hline 0 & 0 & 0 & 1 & 0 & 0 & 0 & 0 \\ 0 & 0 & 0 & 0 & 0 & 0 & 0 & 1 \end{array}\right]$$

$$\xrightarrow{U_5=H_1H_2} \left[\begin{array}{cccc|cccc} 0 & 0 & 0 & 0 & 1 & 0 & 0 & 0 \\ 0 & 0 & 0 & 0 & 0 & 1 & 0 & 0 \\ \hline 0 & 0 & 0 & 1 & 0 & 0 & 0 & 0 \\ 0 & 0 & 0 & 0 & 0 & 0 & 0 & 1 \end{array}\right].$$

By starting from canonical codeword $|0\rangle|0\rangle|\psi\rangle|\phi\rangle$ and applying the sequence of transformations $\{U_i\}$ in the opposite order we obtain the encoder shown in Figure 8.11a. An alternative encoder can be obtained by swapping the control and target qubits of U_3 and U_4 to obtain the encoder shown in Figure 8.11b. By setting the gauge qubit to $|0\rangle$ and by several simple modifications we obtain the optimized encoder circuit shown in Figure 8.11c.

Before concluding this section, we provide the following theorem, whose proof is left to the reader, which allows us to establish the connection between classical codes and quantum subsystem codes.

Theorem 8.2. Let C be a classical linear subcode of F_2^{2N} and let D denote its subcode $D = C \cap C^{\perp}$. If $x = |C|$ and $y = |D|$, then there exits subsystem code

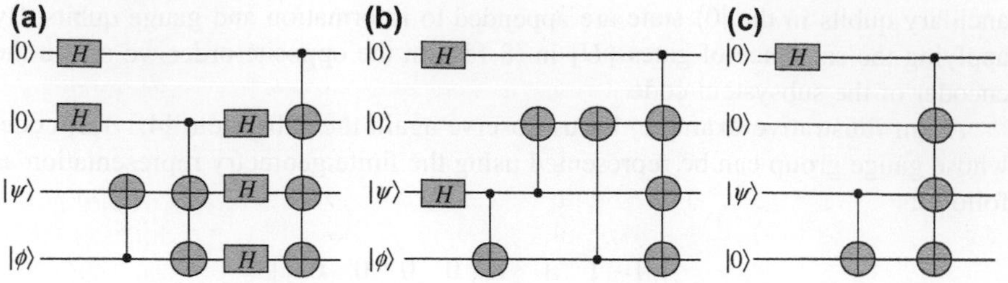

FIGURE 8.11

Encoder for [4,1,1,2] subsystem code by conjugation method: (a) the first version; (b) the second version; (c) optimized version.

$C_Q = A \otimes B$, where $\dim A = q^N/(xy)^{1/2}$ and $\dim B = (x/y)^{1/2}$. The minimum distance of subsystem A is given by

$$D_{\min} = \begin{cases} \mathrm{wt}((C + C^{\perp}) - C) = \mathrm{wt}(D^{\perp} - C), & D^{\perp} \neq C \\ \mathrm{wt}(D^{\perp}), & D^{\perp} = C \end{cases},$$

$$C^{\perp} = \left\{ u \in F_q^{2N} \middle| u \odot v = 0, \ \forall v \in C \right\}. \tag{8.154}$$

The minimum distance of $[N,K,R]$ subsystem code $D_{\min}$, given by (8.154), is more challenging to determine than that of stabilizer code because of the set difference. The definition can be relaxed and we can define minimum distance as $D_m = \mathrm{wt}(C)$. Further, we say that subsystem code is nondegenerate (pure) if $D_m \geq D_{\min}$. The Hamming and Singleton bounds can now be simply determined by observing the corresponding bounds of nonbinary stabilizer codes. Since $\dim A = q^K$, $\dim B = q^R$, the Singleton bound of $[N,K,R,D_{\min}]$ subsystem code is given by

$$K + R \leq N - 2D_{\min} + 2. \tag{8.155}$$

The corresponding Hamming quantum bound of subsystem code is given, based on (8.137), by

$$\sum_{i=0}^{\lfloor (D_{\min}-1)/2 \rfloor} \binom{N}{i} (q^2 - 1)^i q^{K+R} \leq q^N. \tag{8.156}$$

8.8 ENTANGLEMENT-ASSISTED (EA) QUANTUM CODES

The EA-QECCs make use of pre-existing entanglement between transmitter and receiver to improve the reliability of transmission [19,39,40]. In CSS codes only dual-containing classical codes can be used. In the EA-QECC concept, arbitrary classical code can be used as quantum code, providing that entanglement exists between sender and receiver. That is, if classical codes are not dual containing they correspond to a set of stabilizer generators that do not commute. By providing entanglement between source and destination, the corresponding generators can be embedded into a larger set of commuting generators, which gives a well-defined code space. Notice that the whole of Chapter 9 is devoted to EA codes; in this section, we provide some basic facts about them, which are provided for completeness of presentation in this chapter. We start our discussion with *superdense coding* [38]. In this scheme, the transmitter (Alice) and receiver (Bob) share the entangled state, also known as the ebit state [19,39]:

$$|\Phi\rangle = \frac{1}{\sqrt{2}}(|0\rangle \otimes |0\rangle + |1\rangle \otimes |1\rangle).$$

The first half of the entangled pair belongs to Alice and the second half to Bob. The state $|\Phi\rangle$ is the simultaneous (+1,+1) eigenstate of the commuting operators $Z \otimes Z$ and $X \otimes X$. Alice sends a two-bit message $(a_1,a_2) \in F_2^2$ to Bob as follows. Alice performs the encoding operation on half of the ebit, based on the message (a_1,a_2):

$$(Z^{a_1}X^{a_2} \otimes I^B)|\Phi\rangle = |a_1, a_2\rangle. \tag{8.157}$$

Alice sends the encoded state to Bob over the "perfect" qubit channel. Bob performs decoding based on measurement in the

$$\{(Z^{a_1}X^{a_2} \otimes I^B)|\Phi\rangle : (a_1, a_2) \in F_2^2\} \tag{8.158}$$

basis, i.e. by simultaneously measuring the $Z \otimes Z$ and $X \otimes X$ observables. This scheme can be generalized as follows. Alice and Bob share the state $|\Phi\rangle^{\otimes m}$, which is the simultaneous +1 eigenstate of

$$\begin{aligned} &Z(e_1)\otimes Z(e_1)\ldots Z(e_m)\otimes Z(e_m) \ \wedge \ X(e_1)\otimes X(e_1)\ldots X(e_m)\otimes X(e_m); \\ &e_i = (0\ldots01_i0\ldots0). \end{aligned} \tag{8.159}$$

Alice encodes the message $(a_1,a_2) \in F_2^{2m}$, producing the encoded state:

$$(Z(a_1)X(a_2) \otimes I^B)|\Phi\rangle = |a_1, a_2\rangle. \tag{8.160}$$

Bob performs decoding by simultaneously measuring the following observables:

$$Z(e_1)\otimes Z(e_1)\ldots Z(e_m)\otimes Z(e_m) \ \wedge \ X(e_1)\otimes X(e_1)\ldots X(e_m)\otimes X(e_m). \tag{8.161}$$

Let G_N be the Pauli group on N qubits. For any non-Abelian subgroup $S \subset G_N$ of size 2^{N-K}, there exists a set of generators for S $\{\underline{Z}_1,\ldots,\underline{Z}_{s+c}, \underline{X}_{s+1},\ldots,\underline{X}_{s+c}\}$ $(s + 2c = N - K)$ with the following commutation properties:

$$\begin{aligned} &[\underline{Z}_i, \underline{Z}_j] = 0, \quad \forall i,j; \quad [\underline{X}_i, \underline{X}_j] = 0, \quad \forall i,j; \quad [\underline{X}_i, \underline{Z}_j] = 0, \\ &\forall i \neq j; \ \{\underline{Z}_i, \underline{X}_i\} = 0, \quad \forall i. \end{aligned} \tag{8.162}$$

The non-Abelian group can be partitioned into:

1. A commuting subgroup, the *isotropic subgroup*, $S_I = \{Z_1,\ldots,Z_s\}$.
2. An *entanglement subgroup* $S_E = \{Z_{s+1}, Z_{s+c}, X_{s+1},\ldots,X_{s+c}\}$ with anticommuting pairs. The anticommuting pairs (Z_i, X_i) are shared between source A and destination B.

For example, the state shared between A and B, the ebit (entanglement qubit),

$$|\phi^{AB}\rangle = \frac{1}{\sqrt{2}}(|00\rangle^{AB} + |11\rangle^{AB}),$$

has two operators to fix it, X^AX^B and Z^AZ^B. These two operators commute $[X^AX^B, Z^AZ^B] = 0$, while local operators anticommute $\{X^A, X^B\} = \{Z^A, Z^B\} = 0$. We can therefore use entanglement to resolve anticommutativity of two generators in the non-Abelian group S. From the previous chapter we know that two basic properties of Pauli operators are: (i) two Pauli operators *commute* if and only if there is an even number of places where they have different Pauli matrices, neither of which is the identity I; and (ii) if two Pauli operators do not commute, they anticommute, since their individual Pauli matrices either commute or anticommute. Let us observe the following four-qubit operators:

$$\begin{array}{cccc|c} Z & X & Z & I & X \\ Z & Z & I & Z & Z \\ Y & X & X & Z & I \\ Z & Y & Y & X & I \end{array}.$$

Clearly, they do not commute. However, if we extend them by adding an additional qubit as given above, they can be embedded in a larger space in which they commute.

The operating principle of EA quantum error correction is illustrated in Figure 8.12. The sender A (Alice) encodes quantum information in state $|\psi\rangle$ with the help of local ancillary qubits $|0\rangle$ and her half of shared ebits $|\phi^{AB}\rangle$, and then sends the encoded qubits over a noisy quantum channel (say a free space optical channel or optical fiber). The receiver B (Bob) performs multi-qubit measurement on all qubits to diagnose the channel error and performs a recovery unitary operation R to reverse the action of the channel. Notice that the channel does not affect at all the receiver's half of shared ebits. The *operation steps* of an entanglement-assisted quantum error correction scheme are:

1. The sender and receiver share c ebits before quantum communication starts, and the sender employs a ancillary qubits. The *unencoded state* is a simultaneous $+1$ eigenstate of the following operators:

$$\{Z_{s+1}|Z_1, \ldots, Z_{s+c}|Z_c, X_{s+1}|X_1, \ldots, X_{s+c}|X_c, Z_1, \ldots, Z_s\}, \tag{8.163}$$

where the first half of the ebits correspond to the sender and the second half to the receiver. The sender encodes K information qubits with the help of a ancillary

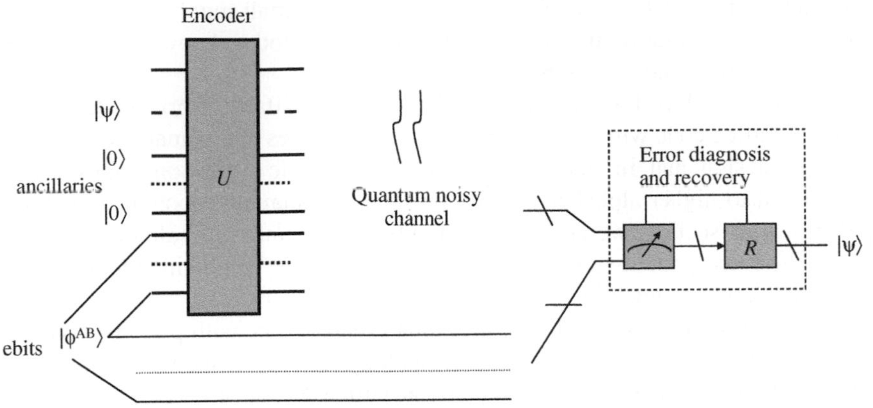

FIGURE 8.12

EA quantum error correction operating principle.

qubits and the sender's half of the c ebits. The encoding unitary transforms the unencoded operators to the following encoded operators:

$$\{\underline{Z}_{s+1}|Z_1, \ldots, \underline{Z}_{s+c}|Z_c, \underline{X}_{s+1}|X_1, \ldots, \underline{X}_{s+c}|X_c, \underline{Z}_1, \ldots, \underline{Z}_s\}. \tag{8.164}$$

2. The sender transmits N qubits over a noisy quantum channel, which affects only these N qubits, not the receiver's half of the c ebits.
3. The receiver combines the received qubits with c ebits and performs the measurement on all $N+c$ qubits.
4. Upon identification of error, the receiver performs the recovery operation to reverse the quantum channel error.

8.9 TOPOLOGICAL CODES

Topological quantum error correction codes [42–50] are typically defined on a two-dimensional lattice, with quantum parity check being geometrically local. The locality of parity check is of crucial importance since the syndrome measurements are easy to implement when the qubits involved in syndrome verification are in close proximity to each other. In addition, the possibility of implementing the universal quantum gates topologically increases the interest in topological quantum codes. On the other hand, it may be argued that quantum LDPC codes can be designed in such a way that quantum parity checks are only local. In addition, through the concept of subsystem codes, the portion of qubits that does not carry any encoded information can be interpreted as the portion of gauge qubits. The gauge qubits can be used to "absorb" the effect of errors. Moreover, the subsystem codes allow syndrome measurements with a smaller number of qubit interactions.

The combination of locality of toplogical codes and small number of interactions lead to a new generation of quantum error correction codes, known as topological subsystem codes, due to Bombin [46].

The topogical codes on square lattice such as Kitaev's toric code [43,44], quantum lattice code with a boundary [45], surface codes and planar codes [45] are easier to design and implement [46–48]. These basic codes on qubits can be generalized to higher alphabets [49,50] resulting in quantum topological codes on qudits. In this section, we are concerned only with quantum topological codes on qubits, the corresponding topological codes on qudits can be obtained following the similar analogy provided in Section 8.6 (see also Refs [49,50]).

Kitaev's toric code [43,44] is defined on a square lattice with periodic boundary conditions, meaning that the lattice has a topology of torus as shown in Figure 8.13. The qubits are associated with the edges of the lattice. For an $m \times m$ square lattice on the torus there are $N = 2m^2$ qubits. For each vertex v and plaquette (or face) p (see Figure 8.13, right), we associate the stabilizer operators as follows:

$$A_v = \underset{i \in n(v)}{\otimes} X_i \quad \text{and} \quad B_p = \underset{i \in n(p)}{\otimes} Z_i, \tag{8.165}$$

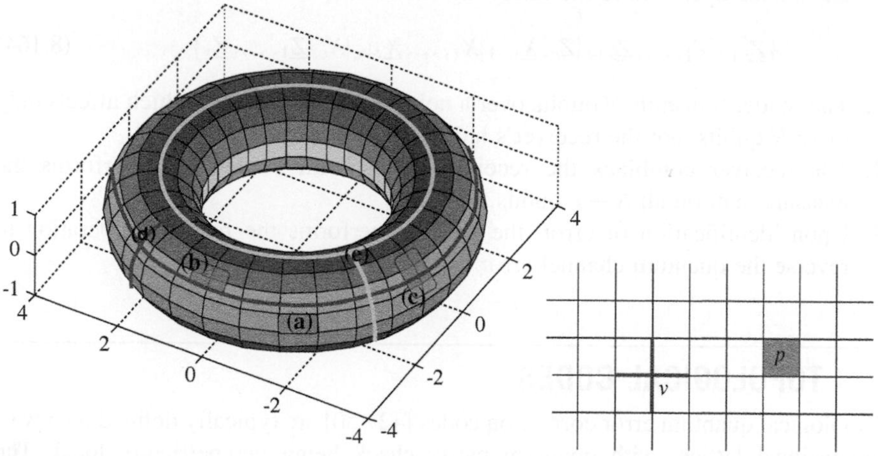

FIGURE 8.13

The square lattice on the torus. The qubits are associated with the edges. The figure on the right shows an enlarged portion of the square lattice on the torus. The Z-containing operators represent the string on the lattice, while the X-containing operators represent the strings on the dual lattice. The elementary trivial loop (cycle) denoted by (a) corresponds to a certain B_p operator, while the elementary trivial cycle denoted by (b) corresponds to a certain A_v operator on the lattice dual. The trivial loops can be obtained as a product of individual elementary loops – see, for example, the cycle denoted by (c). The nontrivial loops on the dual of the lattice, such as (d), or on the lattice, such as (e), correspond to the generators from the set of encoded (logical) Pauli operators.

where X_i and Z_i denote the corresponding Pauli X and Z operators on position i, $n(v)$ denotes the neighborhood of vertex v, i.e. the set of edges incident to vertex v, while $n(p)$ denotes the neighborhood of plaquette p, i.e. the set of edges encircling the face p. (We use the symbol $\otimes$ to denote the tensor product, as we have done earlier.) Based on the properties of Pauli operators (see Chapter 2) it is clear that operators A_v and B_p mutually commute. The commutation of operators A_v among themselves is trivial to prove. Since operators A_v and B_p have either no or two edges in common (see Figure 8.13, right), they commute (an even number of anticommuting Pauli operators results in commuting stabilizer operators). The operators A_v and B_p are Hermitian and have eigenvalues 1 and -1. Let H_2^N denote the Hilbert space, where $N=2m^2$. The toric code space C_Q can then be defined as follows:

$$C_Q = \{|c\rangle \in H_2^N |\ A_v|c\rangle = |c\rangle,\ B_p|c\rangle = |c\rangle;\quad \forall\ v,p\}. \tag{8.166}$$

Equation (8.166) represents the eigenvalue equation, and since A_v and B_p mutually commute, they have common eigenkets. The stabilizer group is defined as $S=\langle A_v, B_p\rangle$.

Since $\Pi_v A_v = I$, $\Pi_p B_p = I$ there are $M=2m^2-2$ independent stabilizer operators. The number of information qubits can be determined by $K=N-M=2$ and the code space dimensionality is dim $C_Q=2^{N-K}=4$. Clearly, this toric code has a low quantum code rate. Another interesting property to notice is that stabilizers A_v contain only X operators, while stabilizers B_p contain only Z operators. Clearly, this toric code represents a particular instance of CSS codes, described in Chapter 7. The encoder and decoder can therefore be implemented as described in Section 8.5. We have seen above that commutation of A_v and B_p arises from the fact that these two operators have either no or two common edges, which is equivalent to the classical code with the property that any two rows overlap in an even number of positions, representing a dual-containing code.

Let us now observe the error E acting on codeword $|c\rangle$, which results in state $E|c\rangle$ that is not necessarily the eigenket with eigenvalue $+1$ of vertex and plaquette operators. If E and A_p commute then the ket $E|c\rangle$ will be the $+1$ eigenket, otherwise it will be the -1 eigenket. A similar conclusion applies to B_p. By performing measurements on each vertex A_v and plaquette B_p operators we obtain the syndrome pair $s=(s_v,s_p)$, where $s_i=\pm 1$, $i\in\{v,p\}$. The intersection of syndrome pairs will give us a set of errors having the same syndrome, namely the coset (the set of errors that differ in an element from stabilizer S). Of all the possible errors from the coset we choose the most probable one, typically the lowest weight one. We can also consider stabilizer elements individually instead of in pairs, and the corresponding error syndrome can be obtained as explained in the previous chapter or in Sections 8.1 and 8.3. Another approach would be to select the most likely error compatible with the syndrome, i.e.

$$E_{\text{ML}} = \arg\max_{E\in\mathscr{E}} P(E),\quad \mathscr{E} = \{E|EA_v = s_vA_vE,\ EB_p = s_pB_pE\}, \tag{8.167}$$

where $P(E)$ is the probability of occurrence of a given error E, and the subscript ML is used to denote maximum-likelihood decision. For example, for the depolarizing channel from the previous chapter, $P(E)$ is given by

$$P(E) = (1-p)^{N-\mathrm{wt}(E)} \frac{p^{\mathrm{wt}(E)}}{3}, \tag{8.168}$$

where with probability $1-p$ we leave the qubit unaffected and apply the Pauli operators X, Y, and Z each with probability $p/3$. The error correction action is to apply the error E_{ML} selected according to (8.167). If the proper error was selected, the overall action of the quantum channel and error correction would be $E^2 = \pm I$. Notice that this approach is not necessarily the optimum approach as will become apparent soon. From Figure 8.13 it is clear that only Z-containing operators correspond to the strings on the lattice, while only X-containing operators correspond to the strings on the dual lattice. The plaquette operators are represented as elementary cycles (loops) on the lattice, while the vertex operators are represented by elementary loops on the dual lattice. Trivial loops can be obtained as the product of elementary loops. Therefore, the plaquette operators generate a group of homologically trivial loops on the torus, while the vertex operators generate homologically trivial loops on the dual lattice. Two error operators E_1 and E_2 have the same effect on a codeword $|c\rangle$ if E_1E_2 contains only homologically trivial loops, and we say that these two errors are homologically equivalent. This concept is very similar to the error coset concept introduced in the previous chapter. There exist four independent operators that perform the mapping of the stabilizer S to itself. These operators commute with all operators from S (both A_v and B_p values), but do not belong to S and are known as encoded (logical) Pauli operators. The encoded Pauli operators correspond to the homologically nontrivial loops of Z and X type, as shown in Figure 8.13. Let the set of logical Pauli operators be denoted by L. Because the operators $l \in L$ and ls for every $s \in S$ are equivalent only homology classes (cosets) are important. The corresponding maximum-likelihood error operator selection can be performed as follows:

$$l_{\mathrm{ML}} = \arg\max_{l \in L} \sum_{s \in S} P(E = lsR(s)), \tag{8.169}$$

where $R(s)$ is the representative error from the coset corresponding to syndrome s. The error correction is performed by applying the operator $l_{\mathrm{ML}}R(s)$. Notice that the algorithm above can be computationally extensive. An interesting algorithm with complexity of $\mathcal{O}(m^2 \log m)$ can be found in Refs [47,48]. Since we have already established the connection between quantum LDPC codes and torus codes we can use the LDPC decoding algorithms described in Chapter 10 instead. We now determine encoded (logical) Pauli operators following the description of Bravyi and Kitaev [45]. Let us consider an operator of the form:

$$\overline{Y}(c, c') = \prod_{i \in c} Z_i \prod_{j \in c'} X_j, \tag{8.170}$$

where c is an I cycle (loop) on the lattice and c' is an I cycle on the dual lattice. It is straightforward to show that $\overline{Y}(c,c')$ commutes with all stabilizers and thus it performs mapping of code subspace C_Q to itself. Since this mapping depends on homology classes (cosets) of c and c', we can denote the encoded Pauli operators by $\overline{Y}([c],[c'])$. It can be shown that $\overline{Y}([c],[c'])$ forms a linear basis for $L(C_Q)$. The logical operators $\overline{X}_1(0,[c_1']),\ \overline{X}_2(0,[c_2']),\ \overline{Z}_1([c_1],0)$, and $\overline{Z}_2([c_2],0)$, where c_1,c_2 are the cycles on the original lattice and c_1', c_2' are the cycles on the dual lattice, can be used as generators of $L(C_Q)$.

We turn our attention now to the quantum topological codes on a lattice with boundary, introduced by Bravyi and Kitaev [45]. Instead of dealing with lattices on the torus, here we consider the finite square lattice on the plane. Two types of boundary can be introduced: the z-type boundary shown in Figure 8.14a and the x-type boundary shown in Figure 8.14b. In Figure 8.14c, we provide an illustrative example [45] for a 2×3 lattice.

The simplest form of quantum boundary code can be obtained by simply alternating x-type and z-type boundaries, as illustrated in Figure 8.14c. Since an $n \times m$ lattice has $(n+1)(m+1)$ horizontal edges and nm vertical edges, by associating the edges with qubits, the corresponding codeword length will be $N = 2nm + n + m + 1$. The stabilizers can be formed in very similar fashion to toric codes. The vertex and plaquette operators can be defined in a fashion similar to (8.165), and the code subspace can be introduced by (8.166). The free ends of edges do not contribute to stabilizers. Notice that inner vertex and plaquette stabilizers are of weight 4, while outer stabilizers are of weight 3. This quantum code is an example of irregular quantum stabilizer code. The number of plaquette and vertex stabilizers is determined by $n(m+1)$ and $(n+1)m$ respectively. For example, as shown in Figure 8.14c, the vertex stabilizers are given by

$$\begin{aligned}
&A_{v_1} = X_{B_1v_1}X_{v_1v_2}X_{v_1v_4}, &&A_{v_2} = X_{v_1v_2}X_{v_2v_3}X_{v_2v_5}, &&A_{v_3} = X_{v_2v_3}X_{v_3B_2}X_{v_3v_6},\\
&A_{v_4} = X_{B_1v_4}X_{v_4v_5}X_{v_1v_4}X_{v_4v_7}, &&A_{v_5} = X_{v_4v_5}X_{v_5v_6}X_{v_2v_5}X_{v_5v_8}, &&A_{v_6} = X_{v_5v_6}X_{v_6B_2}X_{v_3v_6}X_{v_6v_9},\\
&A_{v_7} = X_{B_1v_7}X_{v_7v_8}X_{v_4v_7}, &&A_{v_8} = X_{v_7v_8}X_{v_8v_9}X_{v_5v_8}, &&A_{v_9} = X_{v_8v_9}X_{v_9B_2}X_{v_6v_9}.
\end{aligned}$$

On the other hand, the plaquette stabilizers are given by

$$\begin{aligned}
&B_{p_1} = Z_{B_1'p_1}Z_{p_1p_5}Z_{p_1p_2}, \quad B_{p_2} = Z_{B_1'p_2}Z_{p_2p_6}Z_{p_1p_2}Z_{p_2p_3}, \quad B_{p_3} = Z_{B_1'p_3}Z_{p_3p_7}Z_{p_2p_3}Z_{p_3p_4},\\
&B_{p_4} = Z_{B_1'p_4}Z_{p_4p_8}Z_{p_3p_4}, \quad B_{p_5} = Z_{p_1p_5}Z_{p_5B_2'}Z_{p_5p_6}, \quad B_{p_6} = Z_{p_2p_6}Z_{p_6B_2'}Z_{p_5p_6}Z_{p_6p_7},\\
&B_{p_7} = Z_{p_3p_7}Z_{p_7B_2'}Z_{p_6p_7}Z_{p_7p_8}, \quad B_{p_8} = Z_{p_4p_8}Z_{p_8B_2'}Z_{p_7p_8}.
\end{aligned}$$

From this example it is clear that the number of independent stabilizers is given by $N - K = 2nm + n + m$, so that the number of information qubits of this code is only $K = 1$. The encoded (logical) Pauli operators can also be obtained by (8.170), but now with I-cycle c (c') being defined as a string that ends on the boundary of the lattice (dual lattice). An error E is undetectable if it commutes with stabilizers, i.e. it can be represented as a linear combination of operators given by (8.170). Since the distance is defined as the minimum weight of error that cannot be detected, I-cycle c (c') has length $n+1$ ($m+1$) and the minimum distance is determined by $D = \min(n+1, m+1)$.

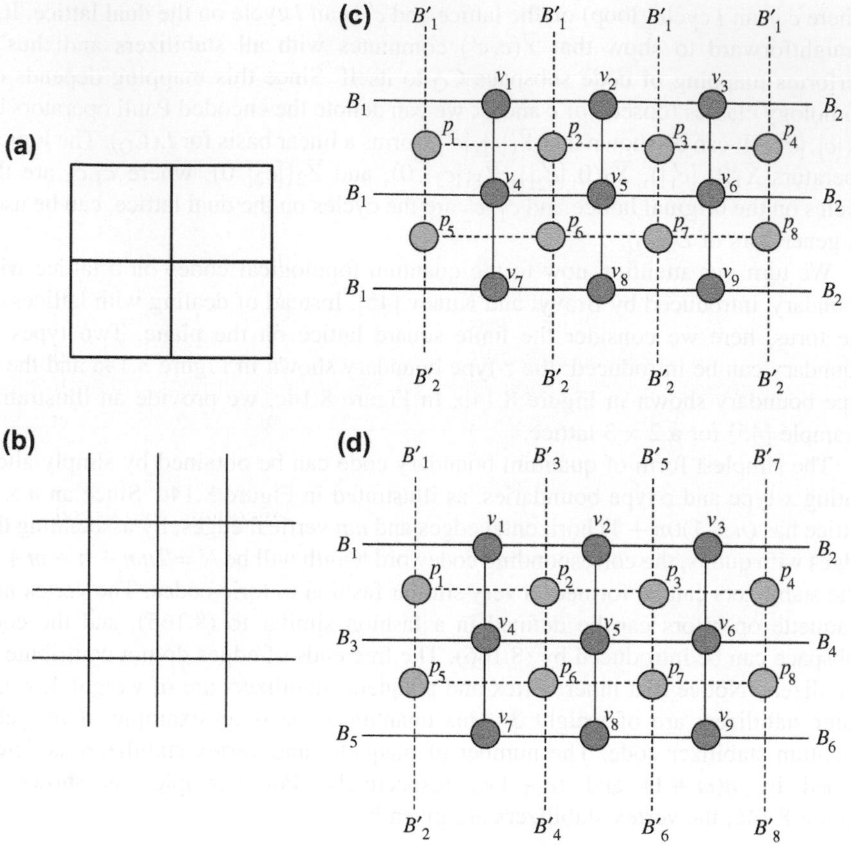

FIGURE 8.14

Quantum codes derived from the square lattice on the plane: (a) *z*-type boundary; (b) *x*-type boundary; (c) 2 × 3 lattice based code; (d) a generalization of topological code (c). (Modified from Ref. [45].) B_1 and B_2 denote the boundaries of the original lattice, while B_1' and B_2' denote the boundaries of the dual lattice.

The parameters of this quantum code are $[2nm+n+m+1, 1, \min(n+1, m+1)]$. Clearly, by letting one dimension go to infinity the minimum distance will be infinitely large, but quantum code rate will tend to zero.

This class of codes is generalized in many different ways: (i) by setting $B_1 = B_2 = B$ we obtain the square lattice on the *cylinder*; (ii) by setting $B_1 = B_2 = B_1' = B_2' = B$ we obtain the square lattice on the *sphere*; and (iii) by assuming that free edges end at different boundaries we obtain a stronger code, since the number of cycles of length 4 in the corresponding bipartite graph representation (see Chapter 10 for definition) is smaller. The 2 × 3 lattice based code belonging to class (iii) is shown in Figure 8.13c, and its properties are studied in Problem 28. Notice that the lattice does not need to be square/rectangular, which can further be used as another generalization. Several examples of these codes are provided in Refs [44,45]. By employing the good properties of topological codes and subsystem codes, we can design the so-called subsystem topological codes [46].

8.10 SUMMARY

This chapter has been concerned with stabilizer codes and their relatives. Stabilizer codes were introduced in Section 8.1. Their basic properties were discussed in Section 8.2. Encoded operations were introduced in the same section. In Section 8.3, the finite geometry representation of stabilizer codes was introduced. This representation was used in Section 8.4 to introduce the so-called standard form of the stabilizer code. The standard form is a basic representation of efficient encoder and decoder implementations (Section 8.5). Section 8.6 covered nonbinary stabilizer codes, which generalizes the previous sections. Subsystem codes were introduced in Section 8.7. In the same section efficient encoding and decoding of subsystem codes was discussed as well. Entanglement-assisted codes were introduced in Section 8.8. An important class of quantum codes, topological codes, was discussed in Section 8.9. Finally, for a better understanding of the material presented in this chapter, a set of problems is provided in the next section.

8.11 PROBLEMS

1. By using Eq. (8.28), prove the following commutative properties:

$$\left[\overline{X}_i, \overline{X}_j\right] = \left[\overline{Z}_i, \overline{Z}_j\right] = 0$$
$$\left[\overline{X}_i, \overline{Z}_j\right] = 0 \quad (i \neq j)$$
$$\{\overline{X}_i, \overline{Z}_i\} = 0$$

   Prove also that $\overline{X}_i, \overline{Z}_i$ commute with the generators g_i from S.
2. The [4,2] QECC is given by the generators $g_1 = X_1Z_2Z_3X_4$, $g_2 = Y_1X_2X_3Y_4$. Determine appropriately the Pauli encoded operators so that the commutation relations from Problem 1 are satisfied. What is the distance of this code? Is this code degenerate? Determine the basis codewords.
3. The generators for [8,3,3] code are given by

$$g_1 = X_1X_2X_3X_4X_5X_6X_7X_8, \quad g_2 = Z_1Z_2Z_3Z_4Z_5Z_6Z_7Z_8, \quad g_3 = X_2X_4Y_5Z_6Y_7Z_8,$$
$$g_4 = X_2Z_3Y_4X_6Z_7Y_8, \text{ and } \quad g_5 = Y_2X_3Z_4X_5Z_6Y_8.$$

   (a) Determine the standard form of quantum-check matrix and determine the corresponding generators in standard form.
   (b) Determine encoded Pauli operators in standard form.
   (c) Determine the efficient encoder implementation.

(d) Determine the efficient decoder implementation.

(e) Prove that the distance of this code is $D = 3$.

(f) Determine whether this code is nondegenerate.

4. The quantum-check matrix A of a CSS code is given by

$$A = \left[\begin{array}{c|c} H & 0 \\ 0 & H \end{array}\right] \qquad H = \begin{bmatrix} 1 & 0 & 0 & 1 & 1 & 1 \\ 1 & 1 & 1 & 0 & 0 & 1 \\ 0 & 1 & 1 & 1 & 1 & 0 \end{bmatrix}.$$

(a) Determine the standard form of quantum-check matrix and determine the corresponding generators in standard form.

(b) Determine encoded Pauli operators in standard form.

(c) Given the results of (a) and (b), determine the efficient encoding circuit for the corresponding quantum stabilizer code.

(d) Given the results of (a) and (b), determine the efficient decoding circuit for the corresponding quantum stabilizer code.

5. The quantum-check matrix of a stabilizer code is given below as matrix A.

(a) Determine the standard form of quantum-check matrix and determine the corresponding generators in standard form.

(b) Determine encoded Pauli operators in standard form.

(c) Determine the code parameters.

(d) Given the results of (a) and (b), determine the efficient encoding circuit for the corresponding quantum stabilizer code.

(e) Given the results of (a) and (b), determine the efficient decoding circuit for the corresponding quantum stabilizer code.

$$A = \begin{bmatrix} 1 & 1 & 1 & 1 & 1 & 1 & 1 & 1 & 1 & 0 & 0 & 0 & 0 & 0 & 0 & 0 & 0 & 0 \\ 0 & 0 & 0 & 1 & 1 & 1 & 1 & 1 & 1 & 0 & 0 & 0 & 0 & 0 & 0 & 0 & 0 & 0 \\ 0 & 0 & 0 & 0 & 0 & 0 & 0 & 0 & 0 & 1 & 0 & 1 & 1 & 0 & 1 & 1 & 0 & 1 \\ 0 & 0 & 0 & 0 & 0 & 0 & 0 & 0 & 0 & 0 & 1 & 1 & 0 & 1 & 1 & 0 & 1 & 1 \\ 0 & 0 & 0 & 0 & 0 & 0 & 0 & 0 & 0 & 1 & 0 & 1 & 0 & 0 & 0 & 0 & 0 & 0 \\ 0 & 0 & 0 & 0 & 0 & 0 & 0 & 0 & 0 & 0 & 0 & 0 & 1 & 0 & 1 & 0 & 0 & 0 \\ 0 & 0 & 0 & 0 & 0 & 0 & 0 & 0 & 0 & 0 & 1 & 1 & 0 & 0 & 0 & 0 & 0 & 0 \\ 0 & 0 & 0 & 0 & 0 & 0 & 0 & 0 & 0 & 0 & 0 & 0 & 0 & 1 & 1 & 0 & 0 & 0 \end{bmatrix}.$$

6. We would like to design a CSS code based on the following parity-check matrix of classical code:

$$H = \begin{bmatrix} 0 & 0 & 1 & 0 & 1 & 0 & 1 \\ 1 & 0 & 0 & 1 & 1 & 0 & 0 \\ 0 & 1 & 0 & 1 & 0 & 0 & 1 \\ 1 & 0 & 0 & 0 & 0 & 1 & 1 \\ 0 & 0 & 1 & 1 & 0 & 1 & 0 \\ 1 & 1 & 1 & 0 & 0 & 0 & 0 \\ 0 & 1 & 0 & 0 & 1 & 1 & 0 \end{bmatrix}.$$

(a) Can this matrix $\boldsymbol{H}$ be used to design a CSS code? If not, how can it be modified so that it can be used to design a corresponding CSS code?
(b) Determine the standard form of quantum-check matrix and determine the corresponding generators in standard form.
(c) Determine encoded Pauli operators in standard form.
(d) Determine the code parameters.
(e) Given the results of (a) and (b), determine the efficient encoding circuit for the corresponding quantum stabilizer code.
(f) Given the results of (a) and (b), determine the efficient decoding circuit for the corresponding quantum stabilizer code.

7. For nonbinary quantum gates from Figure 8.6, show that: (1) $F^{-1}Z(b)F = X(b)$ and (2) $F^{-1}X(a)F = Z(-a)$.

8. Show that the set of errors from Section 8.6, $\varepsilon = \{X(a)X(b)|a, b \in F_q\}$, satisfies the following properties: (i) it contains the identity operator; (ii) $\mathrm{tr}(E_1^\dagger E_2) = 0$ $\forall E_1, E_2 \in \varepsilon$; and (iii) $\forall E_1, E_2 \in \varepsilon: \ E_1E_2 = cE_3, \ E_3 \in \varepsilon, \ c \in F_q$.

9. If the operators $X(a)$ and $Z(b)$ are defined as $X(a)|x\rangle = |x + a\rangle, \ \ Z(b) = \omega^{\mathrm{tr}(bx)}|x\rangle; \ \ x, a, b \in F_q$, show that: (1) $X(a)Z(b) = \omega^{-\mathrm{tr}(ab)}Z(b)X(a)$ and (2) $X(a + a')Z(b + b') = \omega^{-\mathrm{tr}(a'b)}X(a)\,Z(b)X(a')Z(b')$.

10. Show that the syndrome circuit from Figure 8.8, for input $E|\psi\rangle$, where E is the qudit error and $|\psi\rangle$ is the codeword, performs the following mapping: $|0\rangle E|\psi\rangle \rightarrow \sum_{u\in F_q} F^\dagger|u\rangle X(u\boldsymbol{a}_i)Z(u\boldsymbol{b}_i)E|\psi\rangle$.

11. Prove that the Hamming bound for nonbinary quantum stabilizer code [N,K,D] (where K is the number of information qudits and N is the codeword length) over F_q is given by

$$\sum_{i=0}^{\lfloor (D-1)/2 \rfloor} \binom{N}{i} (q^2 - 1)^i q^K \leq q^N.$$

12. Prove that the Singleton bound for nonbinary quantum stabilizer code [N,K,D] (where K is the number of information qudits and N is the codeword length) over F_q is given by $K \leq N - 2D + 2$.

13. We say that the nonbinary quantum stabilizer code [N,K,D] (where K is the number of information qudits and N is the codeword length) is nondegenerate (or pure) if and only if its stabilizer S does not contain any nonscalar error operator of weight less than D. Prove that there does not exist any *perfect* nondegenerate code of distance greater than 3.

14. A quantum stabilizer code [N,K,D] (where K is the number of information qudits and N is the codeword length) exists only if there exists a classical linear code $C \subseteq F_q^{2N}$ of size $|C| = q^{N-K}$ such that $C \subseteq C^\perp$ and $\mathrm{wt}(C^\perp - C) = D$. Prove the claim.

15. CSS code construction. Let C_1 and C_2 denote two classical linear codes with parameters (n_1,k_1,d_1) and (n_2,k_2,d_2) such that $C_2^\perp \subseteq C_1$. Then there exists an [n,$k_1 + k_2 - n$,d] quantum stabilizer code over F_q such that

$d = \min\{\mathrm{wt}(c) | c \in (C_1 - C_2^{\perp}) \cup (C_2 - C_1^{\perp})\}$ is nondegenerate to $\min\{d_1, d_2\}$. Prove the claim.

16. Let C be a classical linear code over F_q with parameters (n,k,d) that contains its dual, $C^{\perp} \subseteq C$. Then there exists an $[n, 2k - n, \geq d]$ quantum stabilizer code over F_q that is nondegenerate to d. Prove the claim.

17. Let C_Q be an $[N,K,R,D]$ stabilizer subsystem code over F_2 with stabilizer S and gauge group $\mathcal{G}$. Denote the encode operators $\overline{X}_i, \overline{Z}_i$; $i = 1, \ldots, K$ satisfying commutative relations (8.28). Then there exist operators $\tilde{X}_i, \tilde{Z}_i \in G_N$; $i = 1, \ldots, N$ such that:

(i) $S = \langle \tilde{Z}_1, \ldots, \tilde{Z}_s \rangle$; $R + s \leq N$; $R, s \geq 0$.
(ii) $\mathcal{G} = \langle jI, \tilde{Z}_1, \ldots, \tilde{Z}_{s+R}, \tilde{X}_{s+1}, \ldots, \tilde{X}_R \rangle$.
(iii) $\overline{X}_1 = \tilde{X}_{s+R+1}, \overline{Z}_1 = \tilde{Z}_{s+R+1}, \ldots, \overline{X}_K = \tilde{X}_N, \overline{Z}_K = \tilde{Z}_N$.
(iv) $C_{G_N}(S) = \langle \mathcal{G}, \overline{X}_1, \overline{Z}_1, \ldots, \overline{X}_K, \overline{Z}_K \rangle$.

Prove the claims (i)–(iv).

18. For [9,1,4,3] Bacon–Shor code described in Section 8.7, describe how the encoder architectures shown in Figure 8.9 are obtained.

19. Let us observe one particular generator of quantum-check matrix, say $\mathbf{g} = (a_1, \ldots, a_N | b_1, \ldots, b_N)$. Show that the action of H and P gates on the ith qubit is given by Eqs (8.147) and (8.149) respectively. Show that the action of the CNOT gate on $\mathbf{g}$, where the ith qubit is the control qubit and the jth qubit is the target qubit, the action denoted as $\mathrm{CNOT}^{i,j}$, is given by Eq. (8.148).

20. Based on Table 8.1, show that the transformation of $\mathbf{g} = (a_1, \ldots, a_N | b_1, \ldots, b_N)$ to $\mathbf{g}' = (a'_1, \ldots, a'_N | 0, \ldots, 0)$ can be achieved by application of the Hadamard and phase gates as follows:

$$\bigotimes_{i=1}^{N} H^{\overline{a}_i b_i} P^{a_i b_i}, \quad \overline{a}_i = a_i + 1 \bmod 2.$$

21. Based on Eq. (8.148), show that the error operator $\mathbf{e} = (a_1, \ldots, a_i = 1, \ldots, a_N | 0, \ldots, 0)$ can be converted to $\mathbf{e}' = (0, \ldots, a'_i = 1, 0 \ldots 0 | 0 \ldots 0)$ by the application of the following sequence of gates:

$$\prod_{m=1, m \neq n}^{N} (\mathrm{CNOT}^{m,n})^{a_n}.$$

22. For quantum stabilizer codes from Problems 3–6 determine the corresponding encoding circuits by using the conjugation method.

23. For [9,1,4,3] Bacon–Shor code described in Section 8.7, determine the encoding circuit by using the conjugation method.

24. Show that by transforming the gauge group, represented using finite geometry formalism, as follows:

$$\mathcal{G} = \begin{bmatrix} S \\ \mathcal{G}_Z \\ \mathcal{G}_X \end{bmatrix} \xrightarrow{\{U\}} \mathcal{G}' = \left[\begin{array}{ccc|ccc} \mathbf{0} & \mathbf{0} & \mathbf{0} & I_s & \mathbf{0} & \mathbf{0} \\ \mathbf{0} & \mathbf{0} & \mathbf{0} & \mathbf{0} & I_R & \mathbf{0} \\ \mathbf{0} & I_R & \mathbf{0} & \mathbf{0} & \mathbf{0} & \mathbf{0} \end{array}\right],$$

we essentially implement the following mapping:

$$|\phi\rangle|\psi\rangle \rightarrow \underbrace{|0\rangle \otimes \ldots \otimes |0\rangle}_{s=N-K-R \text{ times}} |\phi\rangle|\psi\rangle = |\mathbf{0}\rangle_s|\phi\rangle|\psi\rangle.$$

Which sequence of gates should be applied on the gauge group to perform the transformation above?

25. Explain how the encoder circuit shown in Figure 8.11c has been derived from Figure 8.11b.

26. Prove Theorem 8.2.

27. Let us observe the torus code with an $m \times m$ lattice. Let us further consider the operator of the form:

$$\overline{Y}(c, c') = \prod_{i \in c} Z_i \prod_{j \in c'} X_j,$$

where c is an I cycle (loop) on the lattice and c' is an I cycle on the dual lattice.

(a) Show that $\overline{Y}(c, c')$ operators commute with all stabilizers and thus perform mapping of code subspace C_Q to itself.

(b) Show that $\overline{Y}([c], [c'])$ forms a linear basis for $L(C_Q)$.

(c) Show that the logical operators $\overline{X}_1(0, [c_1']), \overline{X}_2(0, [c_2']), \overline{Z}_1([c_1], 0)$, and $\overline{Z}_2([c_2], 0)$, where c_1, c_2 are the cycles on the original lattice and c_1', c_2' are the cycles on the dual lattice, can be used as generators of $L(C_Q)$.

28. Let us observe the plane code with $n \times m$ lattice with boundaries as illustrated in Figure 8.14d. Determine the quantum code parameters [N,K,D]. For the 2×3 example shown in Figure 8.14d determine the stabilizer and encoded Pauli operators. Determine the corresponding quantum-check matrix and represent it in standard form. Finally, provide efficient encoder and decoder implementations.

References

[1] A.R. Calderbank, P.W. Shor, Good quantum error-correcting codes exist, Phys. Rev. A 54 (1996) 1098–1105.

[2] A.M. Steane, Error correcting codes in quantum theory, Phys. Rev. Lett. 77 (1996) 793.

[3] A.M. Steane, Simple quantum error-correcting codes, Phys. Rev. A 54 (6) (December 1996) 4741–4751.

[4] A.R. Calderbank, E.M. Rains, P.W. Shor, N.J.A. Sloane, Quantum error correction and orthogonal geometry, Phy. Rev. Lett. 78 (3) (January 1997) 405–408.

[5] E. Knill, R. Laflamme, Concatenated quantum codes. Available at: http://arxiv.org/abs/quant-ph/9608012, 1996.

[6] R. Laflamme, C. Miquel, J.P. Paz, W.H. Zurek, Perfect quantum error correcting code, Phys. Rev. Lett. 77 (1) (July 1996) 198–201.

[7] D. Gottesman, Class of quantum error correcting codes saturating the quantum Hamming bound, Phys. Rev. A 54 (September 1996) 1862–1868.

[8] D. Gottesman, Stabilizer Codes and Quantum Error Correction. Ph.D. dissertation, California Institute of Technology, Pasadena, CA, 1997.

[9] R. Cleve, D. Gottesman, Efficient computations of encoding for quantum error correction, Phys. Rev. A 56 (July 1997) 76–82.

[10] A.Y. Kitaev, Quantum error correction with imperfect gates, in: Quantum Communication, Computing, and Measurement, Plenum Press, New York, 1997, pp. 181–188.

[11] C.H. Bennett, D.P. DiVincenzo, J.A. Smolin, W.K. Wootters, Mixed-state entanglement and quantum error correction, Phys. Rev. A 54 (6) (November 1996) 3824–3851.

[12] D.J.C. MacKay, G. Mitchison, P.L. McFadden, Sparse-graph codes for quantum error correction, IEEE Trans. Inform. Theory 50 (2004) 2315–2330.

[13] M.A. Neilsen, I.L. Chuang, Quantum Computation and Quantum Information, Cambridge University Press, Cambridge, 2000.

[14] F. Gaitan, Quantum Error Correction and Fault Tolerant Quantum Computing, CRC Press, 2008.

[15] G.D. Forney Jr., M. Grassl, S. Guha, Convolutional and tail-biting quantum error-correcting codes, IEEE Trans. Inform. Theory 53 (March 2007) 865–880.

[16] A.R. Calderbank, E.M. Rains, P.W. Shor, N.J.A. Sloane, Quantum error correction via codes over GF(4), IEEE Trans. Inform. Theory 44 (1998) 1369–1387.

[17] I.B. Djordjevic, Quantum LDPC codes from balanced incomplete block designs, IEEE Commun. Lett. 12 (May 2008) 389–391.

[18] I.B. Djordjevic, Photonic quantum dual-containing LDPC encoders and decoders, IEEE Photon. Technol. Lett. 21 (13) (1 July 2009) 842–844.

[19] I.B. Djordjevic, Photonic entanglement-assisted quantum low-density parity-check encoders and decoders, Opt. Lett. 35 (9) (1 May 2010) 1464–1466.

[20] S.A. Aly, A. Klappenecker, P.K. Sarvepalli, On quantum and classical BCH codes, IEEE Trans. Inform. Theory 53 (3) (2007) 1183–1188.

[21] S.A. Aly, A. Klappenecker, P.K. Sarvepalli, Primitive quantum BCH codes over finite fields, Proc. Int. Symp. Inform. Theory 2006 (ISIT 2006) (2006) 1114–1118.

[22] A. Ashikhmin, E. Knill, Nonbinary quantum stabilizer codes, IEEE Trans. Inform. Theory 47 (7) (November 2001) 3065–3072.

[23] A. Ketkar, A. Klappenecker, S. Kumar, P.K. Sarvepalli, Nonbinary stabilizer codes over finite fields, IEEE Trans. Inform. Theory 52 (11) (November 2006) 4892–4914.

[24] J.-L. Kim, J. Walker, Nonbinary quantum error-correcting codes from algebraic curves, Discrete Math 308 (14) (2008) 3115–3124.

[25] P.K. Sarvepalli, A. Klappenecker, Nonbinary quantum Reed–Muller codes. Proc. Int. Symp. Inform, Theory 2005 (ISIT 2005) (2005) 1023–1027.

[26] P.K. Sarvepalli, Quantum Stabilizer Codes and Beyond. Ph.D. dissertation, Texas A&M University, August 2008.

[27] S.A.A. Aly Ahmed, Quantum Error Control Codes. Ph.D. dissertation, Texas A&M University, May 2008.

[28] S.A. Aly, A. Klappenecker, Subsystem code constructions, Proc. ISIT 2008, Toronto, Canada, 6–11 July 2008, pp. 369–373.

[29] S. Bravyi, Subsystem codes with spatially local generators, Phys. Rev. A 83 (2011). 012320-1–012320-9.

[30] A. Klappenecker, P.K. Sarvepalli, On subsystem codes beating the quantum Hamming or Singleton bound, Proc. Math. Phys. Eng. Sci. 463 (2087) (8 November 2007) 2887–2905.
[31] A. Klappenecker, P.K. Sarvepalli, Clifford code construction of operator quantum error correcting codes (April 2006), quant-ph/0604161.
[32] D.A. Lidar, I.L. Chuang, K.B. Whaley, Decoherence-free subspaces for quantum computation, Phys. Rev. Lett. 81 (12) (1998) 2594–2597.
[33] P.G. Kwiat, A.J. Berglund, J.B. Altepeter, A.G. White, Experimental verification of decoherence-free subspaces, Science 290 (20 October 2000) 498–501.
[34] L. Viola, E.M. Fortunato, M.A. Pravia, E. Knill, R. Laflamme, D.G. Cory, Experimental realization of noiseless subsystems for quantum information processing, Science 293 (14 September 2001) 2059–2063.
[35] D. Bacon, Operator quantum error correcting subsystems for self-correcting quantum memories, Phys. Rev. A 73 (1) (2006). 012340-1–012340-13.
[36] D. Bacon, A. Casaccino, Quantum error correcting subsystem codes from two classical linear codes, Proc. 44th Annual Allerton Conf. on Communication, Control, and Computing, Monticello, IL, 2006, pp. 520–527.
[37] M. Grassl, M. Rötteler, T. Beth, Efficient quantum circuits for non-qubit quantum error-correcting codes, Int. J. Found. Comput. Sci. 14 (5) (2003) 757–775.
[38] A. Harrow, Superdense coding of quantum states, Phys. Rev. Lett. 92 (18) (7 May 2004). 187901-1–187901-4.
[39] M.-H. Hsieh, Entanglement-Assisted Coding Theory. Ph.D. dissertation, University of Southern California, August 2008.
[40] I. Devetak, T.A. Brun, M.-H. Hsieh, Entanglement-assisted quantum error-correcting codes, in: V. Sidoravičius (Ed.), New Trends in Mathematical Physics, Selected Contributions of the XVth International Congress on Mathematical Physics, Springer, 2009, pp. 161–172.
[41] E. Knill, R. Laflamme, Theory of quantum error-correcting codes, Phys. Rev. A 55 (1997) 900.
[42] J. Preskill, Ph219/CS219 Quantum Computing (Lecture Notes), Caltech (2009). Available at: http://theory.caltech.edu/people/preskill/ph229/.
[43] A. Kitaev, Topological quantum codes and anyons. In Quantum Computation: A Grand Mathematical Challenge for the Twenty-First Century and the Millennium, Proc. Symp. Appl. Math., Washington, DC, 2000, Vol. 58, Providence, RI: Am. Math. Soc., 2002, pp. 267–272.
[44] A.Y. Kitaev, Fault-tolerant quantum computation by anyons, Ann. Phys. 303 (1) (2003) 2–30. Available at: http://arxiv.org/abs/quant-ph/9707021.
[45] S.B. Bravyi, A.Y. Kitaev, Quantum Codes on a Lattice with Boundary. Available at: http://arxiv.org/abs/quant-ph/9811052.
[46] H. Bombin, Topological subsystem codes, Phys. Rev. A 81 (3 March 2010). 032301-1–032301-15.
[47] G. Duclos-Cianci, D. Poulin, Fast decoders for topological subsystem codes, Phys. Rev. Lett. 104 (2010). 050504-1–050504-4.
[48] G. Duclos-Cianci, D. Poulin, A renormalization group decoding algorithm for topological quantum codes, Proc. 2010 IEEE Information Theory Workshop-ITW 2010, Dublin, Ireland, 30 August–3 September 2010.

[49] M. Suchara, S. Bravyi, B. Terhal, Constructions and noise threshold of topological subsystem codes, J. Phys. A: Math. Theory 44 (2011), paper no. 155301.

[50] P. Sarvepalli, Topological color codes over higher alphabet, Proc. 2010 IEEE Information Theory Workshop-ITW 2010, Dublin, Ireland, 2010.

[51] F. Hernando, M.E. O'Sullivan, E. Popovici, S. Srivastava, Subfield-subcodes of generalized toric codes, Proc. ISIT 2010, Austin, TX, 13–18 June 2010, pp. 1125–1129.

[52] H. Bombin, Topological order with a twist: Ising anyons from an Abelian model, Phys. Rev. Lett. 105 (2010). 030403-1–030403-4.

CHAPTER

Entanglement-Assisted Quantum Error Correction

9

CHAPTER OUTLINE

In this chapter, the entanglement-assisted (EA) quantum error correction codes (QECCs) [1−13] are described, which make use of pre-existing entanglement between transmitter and receiver to improve the reliability of transmission. In the previous chapter, the concept of EA quantum codes was briefly introduced. EA quantum codes can be considered as a generalization of superdense coding [14,15]. They can also be considered as particular instances of subsystem codes [16−18], described in Chapter 8. A key advantage of EA quantum codes compared to CSS codes is that EA quantum codes do not require the corresponding classical codes, from which they are derived, to be dual containing. That is, arbitrary classical code can be used to design EA quantum codes, providing that there exists a pre-entanglement between source and destination. A general description of entanglement-assisted quantum error correction is provided in Section 9.1. In Section 9.2, the entanglement-assisted canonical code is studied, along with its error correction capability. Further, in Section 9.3 the concept of EA canonical code is generalized to arbitrary EA code. Also in this section, the design of EA codes from classical codes, in particular classical quaternary codes, is described. In Section 9.4, encoding and decoding for EA quantum codes are discussed. In Section 9.5, the concept of operator quantum error correction is introduced. Entanglement-assisted operator quantum error correction is considered in Section 9.6. In this chapter, the notation due to Brun, Devetak, Hsieh, and Wilde [1−6], who invented this class of quantum codes, is used.

Quantum Information Processing and Quantum Error Correction. DOI: 10.1016/B978-0-12-385491-9.00009-5

9.1 ENTANGLEMENT-ASSISTED QUANTUM ERROR CORRECTION PRINCIPLES

The block scheme of entanglement-assisted quantum code, which requires a certain number of entangled qubits (ebits) to be shared between the source and destination, is shown in Figure 9.1. The source encodes quantum information in K-qubit state $|\psi\rangle$ with the help of local ancillary qubits $|0\rangle$ and source-half of shared ebits (e ebits) into N qubits, and then sends the encoded qubits over a noisy quantum channel (such as a free-space or fiber-optic channel). The receiver performs decoding on all qubits ($N+e$ qubits) to diagnose the channel error and performs a recovery unitary operation to reverse the action of the channel. Notice that the channel does not affect the receiver's half of shared ebits at all. By omitting the ebits, the conventional quantum coding scheme is obtained.

It was discussed in previous chapters that both noise N and recovery $\mathscr{R}$ processes can be described as quantum operations $\mathscr{N},\mathscr{R}$: $L(H) \rightarrow L(H)$, where $L(H)$ is the space of linear operators on Hilbert space H. These mappings can be represented in terms of the operator-sum representation:

$$\mathscr{N}(\rho) = \Sigma_i E_i \rho E_i^{\dagger}, \quad E_i \in L(H), \quad \mathscr{N} = \{E_i\}. \tag{9.1}$$

Given the quantum code C_Q that is subspace of H, we say that the set of errors $\mathscr{E}$ are correctable if there exists a recovery operation $\mathscr{R}$ such that $\mathscr{R}\mathscr{N}(\rho) = \rho$ for any state ρ from $L(C_Q)$. Further, each error operator E_i can be expanded in terms of Pauli operators:

$$\begin{gathered} E_i = \sum_{[\boldsymbol{a}|\boldsymbol{b}] \in F_2^{2N}} \alpha_{i,[\boldsymbol{a}|\boldsymbol{b}]} E_{[\boldsymbol{a}|\boldsymbol{b}]}, \quad E_{[\boldsymbol{a}|\boldsymbol{b}]} = j^{\lambda} X(\boldsymbol{a}) Z(\boldsymbol{b}); \quad \boldsymbol{a} = a_1 \ldots a_N; \ \boldsymbol{b} = b_1 \ldots b_N; \\ a_i, b_i = 0,1; \ \lambda = 0,1,2,3 \quad X(\boldsymbol{a}) \equiv X_1^{a_1} \otimes \ldots \otimes X_N^{a_N}; \ Z(\boldsymbol{b}) \equiv Z_1^{b_1} \otimes \ldots \otimes Z_N^{b_N}. \end{gathered} \tag{9.2}$$

An $[N,K,e;D]$ EA-QECC consists of: (i) an encoding map,

$$\hat{U}_{\mathrm{enc}}: \ L^{\otimes K} \otimes L^{\otimes e} \rightarrow L^{\otimes N}, \tag{9.3}$$

and (ii) decoding trace-preserving mapping,

$$\mathscr{D}: \ L^{\otimes N} \otimes L^{\otimes e} \rightarrow L^{\otimes K}, \tag{9.4}$$

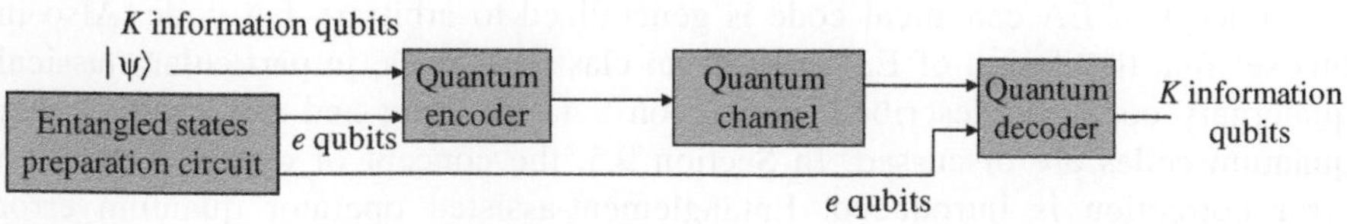

FIGURE 9.1

A generic entanglement-assisted quantum error correction scheme.

such that the composition of encoding mapping, channel transformation, and decoding mapping is the identity mapping:

$$\mathscr{D} \circ \mathscr{N} \circ \breve{U}_{\text{enc}} \circ \breve{U}_{\text{app}} = I^{\otimes K}; \quad I: \quad L \rightarrow L; \quad U_{\text{app}}|\psi\rangle = |\psi\rangle|\Phi^{\otimes e}\rangle, \tag{9.5}$$

where U_{app} is an operation that simply appends the transmitter half of the maximum entangled state $|\Phi^{\otimes e}\rangle$.

9.2 ENTANGLEMENT-ASSISTED CANONICAL QUANTUM CODES

The simplest EA quantum code, the EA canonical code, performs the following trivial encoding operation:

$$\hat{U}_c: \quad |\psi\rangle \rightarrow |\mathbf{0}\rangle_l \otimes |\Phi^{\otimes e}\rangle \otimes |\psi\rangle_K; \; l = N - K - e. \tag{9.6}$$

The canonical EA code, shown in Figure 9.2, is therefore an extension of the canonical code presented in Chapter 7, in which e ancillary qubits are replaced by e maximally entangled kets shared between transmitter and receiver. Since the number of information qubits is K and codeword length is N, the number of remaining ancillary qubits in the $|0\rangle$ state is $l = N - K - e$.

The EA canonical code can correct the following set of errors:

$$\mathscr{N}_{\text{canonical}} = \{X(\boldsymbol{a})Z(\boldsymbol{b}) \otimes Z(\boldsymbol{b}_{\text{eb}})X(\boldsymbol{a}_{\text{eb}}) \otimes X(\alpha(\boldsymbol{a}, \boldsymbol{b}_{\text{eb}}, \boldsymbol{a}_{\text{eb}}))Z(\beta(\boldsymbol{a}, \boldsymbol{b}_{\text{eb}}, \boldsymbol{a}_{\text{eb}})):$$

$$\boldsymbol{a}, \boldsymbol{b} \in F_2^l; \boldsymbol{b}_{\text{eb}}, \boldsymbol{a}_{\text{eb}} \in F_2^e\} \quad \alpha, \beta: \quad F_2^l \times F_2^e \times F_2^e \rightarrow F_2^K, \tag{9.7}$$

where the Pauli operators $X(\boldsymbol{c})$ ($\boldsymbol{c} \in \{\boldsymbol{a}, \boldsymbol{a}_{\text{eb}}, \alpha(\boldsymbol{a}, \boldsymbol{b}_{\text{eb}}, \boldsymbol{a}_{\text{eb}})\}$) and $Z(\boldsymbol{d})$ ($\boldsymbol{d} \in \{\boldsymbol{b}, \boldsymbol{b}_{\text{eb}}, \beta(\boldsymbol{a}, \boldsymbol{b}_{\text{eb}}, \boldsymbol{a}_{\text{eb}})\}$) are introduced in (9.2).

In order to prove claim (9.7), let us observe one particular error E_c from the set of errors (9.7) and study its action on the transmitted codeword $|\mathbf{0}\rangle_l \otimes |\Phi^{\otimes e}\rangle \otimes |\psi\rangle_K$:

$$X(\boldsymbol{a})Z(\boldsymbol{b})|\mathbf{0}\rangle_l \otimes (Z(\boldsymbol{b}_{\text{eb}})X(\boldsymbol{a}_{\text{eb}}) \otimes I^{R_x})|\Phi^{\otimes e}\rangle \otimes X(\alpha(\boldsymbol{a}, \boldsymbol{b}_{\text{eb}}, \boldsymbol{a}_{\text{eb}}))Z(\beta(\boldsymbol{a}, \boldsymbol{b}_{\text{eb}}, \boldsymbol{a}_{\text{eb}}))|\psi\rangle. \tag{9.8}$$

The action of operator $X(\boldsymbol{a})Z(\boldsymbol{b})$ on state $|\mathbf{0}\rangle_l$ is given by $X(\boldsymbol{a})Z(\boldsymbol{b})|\mathbf{0}\rangle_l \overset{Z(\boldsymbol{b})|\mathbf{0}\rangle_l = |\mathbf{0}\rangle_l}{=} X(\boldsymbol{a})|\mathbf{0}\rangle_l = |\boldsymbol{a}\rangle$. On the other hand, the action of $Z(\boldsymbol{b}_{\text{eb}})X(\boldsymbol{a}_{\text{eb}})$ on the maximum entangled state is $(Z(\boldsymbol{b}_{\text{eb}})X(\boldsymbol{a}_{\text{eb}}) \otimes I^{R_x})|\Phi^{\otimes e}\rangle = |\boldsymbol{b}_{\text{eb}}, \boldsymbol{a}_{\text{eb}}\rangle$, where I^{R_x} denotes the identity operators applied on the receiver half of the maximum entangled state. Finally, the action of channel on information portion of the codeword is given by $X(\alpha(\boldsymbol{a}, \boldsymbol{b}_{\text{eb}}, \boldsymbol{a}_{\text{eb}}))Z(\beta(\boldsymbol{a}, \boldsymbol{b}_{\text{eb}}, \boldsymbol{a}_{\text{eb}}))|\psi\rangle = |\psi'\rangle$. Clearly, the vector $(\boldsymbol{a} \; \boldsymbol{b} \; \boldsymbol{b}_{\text{eb}} \; \boldsymbol{a}_{\text{eb}})$ uniquely specifies the error operator E_c, and can be called the syndrome vector. Since the state $|\mathbf{0}\rangle_l$ is invariant on the action of $Z(\boldsymbol{b})$, the vector $\boldsymbol{b}$ can be omitted from

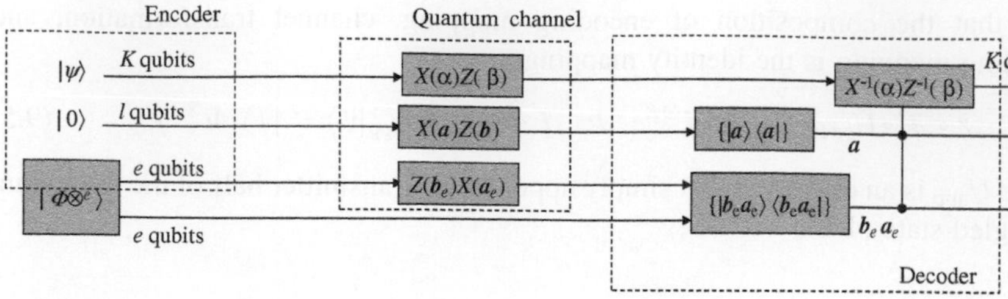

FIGURE 9.2

An entanglement-assisted canonical code.

the syndrome vector. To obtain the $\boldsymbol{a}$ portion of the syndrome, we have to perform simultaneous measurements on $Z(\boldsymbol{d}_i)$, where $\boldsymbol{d}_i = (0\ldots01_i0\ldots0)$. On the other hand, in order to obtain the $(\boldsymbol{b}_{\text{eb}}\, \boldsymbol{a}_{\text{eb}})$ portion of the syndrome vector, we have to perform measurements on $Z(\boldsymbol{d}_1)Z(\boldsymbol{d}_1),\ldots,Z(\boldsymbol{d}_e)Z(\boldsymbol{d}_e)$ for $\boldsymbol{b}_{\text{eb}}$ and $X(\boldsymbol{d}_1)X(\boldsymbol{d}_1),\ldots,X(\boldsymbol{d}_e)X(\boldsymbol{d}_e)$ for $\boldsymbol{a}_{\text{eb}}$. Once the syndrome vector is determined, we apply $X^{-1}(\alpha(\boldsymbol{a},\boldsymbol{b}_e,\boldsymbol{a}_e))$ $Z^{-1}(\beta(\boldsymbol{a},\boldsymbol{b}_e,\boldsymbol{a}_e))$ to undo the action of the channel. From syndrome measurements it is clear that we applied exactly the same measurements as in superdense coding, indicating that EA codes can indeed be interpreted as a generalization of superdense coding (see Section 8.8 or Refs [4,15]). In order to avoid the need for measurement we can perform the following controlled unitary decoding:

$$U_{\text{canocnical, dec}} = \sum_{\boldsymbol{a},\boldsymbol{b}_{\text{eb}},\boldsymbol{a}_{\text{eb}}} |\boldsymbol{a}\rangle\langle\boldsymbol{a}| \otimes |\boldsymbol{b}_{\text{eb}},\boldsymbol{a}_{\text{eb}}\rangle\langle\boldsymbol{b}_{\text{eb}},\boldsymbol{a}_{\text{eb}}| \otimes X^{-1} \times (\alpha(\boldsymbol{a},\boldsymbol{b}_{\text{eb}},\boldsymbol{a}_{\text{eb}}))Z^{-1}(\beta(\boldsymbol{a},\boldsymbol{b}_{\text{eb}},\boldsymbol{a}_{\text{eb}})). \tag{9.9}$$

EA error correction can also be described using the stabilizer formalism interpretation. Let $S_{\text{canonical}}$ be the non-Abelian group generated by

$$S_{\text{canonical}} = \langle S_{\text{canonical, I}}, S_{\text{canonical, E}}\rangle;\ S_{\text{canonical, I}} = \langle Z_1,\ldots,Z_l\rangle,\ S_{\text{canonical, E}} = \langle Z_{l+1},\ldots,Z_{l+e},X_{l+1},\ldots,X_{l+e}\rangle, \tag{9.10}$$

where $S_{\text{canonical,I}}$ denotes the isotropic subgroup and $S_{\text{canonical,E}}$ the sympletic subgroup of anticommuting operators. The representation (9.10) has certain similarities with subsystem codes from Section 8.7 of the previous chapter. Namely, by interpreting $S_{\text{canonical,I}}$ as the stabilizer group and $S_{\text{canonical,E}}$ as the gauge group, it turns out that EA codes represent a generalization of the subsystem codes from Section 8.7. Moreover, the subsystem codes do not require entanglement to be established between transmitter and receiver. On the other hand, EA codes with few ebits have reasonable complexity of encoder and decoder, compared to dual-containing codes, and as such represent promising candidates for fault-tolerant quantum computing and quantum teleportation applications.

We perform an Abelian extension of the non-Abelian group $S_{\text{canonical}}$, denoted as $S_{\text{canonical,ext}}$, which now acts on $N+e$ qubits, as follows:

$$\begin{aligned}&Z_1 \otimes I, \ \ldots, Z_l \otimes I \\ &Z_{l+1} \otimes Z_1, \ X_{l+1} \otimes X_1, \ \ldots, Z_{l+e} \otimes Z_e, \ X_{l+e} \otimes X_e.\end{aligned} \tag{9.11}$$

Out of $N+e$ qubits, the first N correspond to the transmitter and an additional e qubits correspond to the receiver. The operators on the right-hand side of (9.11) act on receiver qubits. This extended group is Abelian and can be used as the stabilizer group that fixes the code subspace $C_{\text{canonical,EA}}$. The obtained stabilizer group can be called the entanglement-assisted stabilizer group. The number of information qubits of this stabilizer can be determined by $K=N-l-e$, so that the code can be denoted as $[N,K,e;D]$, where D is the distance of the code.

Before concluding this section, it will be useful to identify the correctable set of errors. In analogy with the stabilizer codes described in Chapter 8, we expect the correctable set of errors of EA-QECC, defined by $S_{\text{canonical}} = \langle S_{\text{canonical,I}}, S_{\text{canonical,E}} \rangle$, to be given by

$$\mathcal{N}_{\text{canonical}} = \left\{ E_m \middle| \ \forall E_1, E_2 \Rightarrow E_2^\dagger E_1 \in S_{\text{canonical, I}} \cup (G_N - C(S_{\text{canonical}})) \right\}, \tag{9.12}$$

where $C(S_{\text{canonical}})$ is the centralizer of $S_{\text{canonical}}$ (i.e. the set of errors that commute with all elements from $S_{\text{canonical}}$). We have already shown above (see Eq. (9.7) and corresponding text) that every error can be specified by the syndrome ($\boldsymbol{a}$ $\boldsymbol{b}_{\text{eb}}$ $\boldsymbol{a}_{\text{eb}}$). If two errors E_1 and E_2 have the same syndrome ($\boldsymbol{a}$ $\boldsymbol{b}_{\text{eb}}$ $\boldsymbol{a}_{\text{eb}}$), then $E_2^\dagger E_1$ will have all-zero syndrome and clearly such an error belongs to $S_{\text{canonical,I}}$. Such an error $E_2^\dagger E_1$ is trivial, as it fixes all codewords. If two errors have different syndromes, then the syndrome for $E_2^\dagger E_1$, which has the form ($\boldsymbol{a}$ $\boldsymbol{b}_{\text{eb}}$ $\boldsymbol{a}_{\text{eb}}$), indicates that such an error can be corrected provided that it does not belong to $C(S_{\text{canonical}})$. That is, when an error belongs to $C(S_{\text{canonical}})$, then its action results in another codeword different from the original codeword, so that such an error is undetectable.

9.3 GENERAL ENTANGLEMENT-ASSISTED QUANTUM CODES

From group theory it is known that if V is an arbitrary subgroup of G_N of order 2^M then there exists a set of generators $\{\overline{X}_{p+1},\ldots,\overline{X}_{p+q};\overline{Z}_1,\ldots,\overline{Z}_{p+q}\}$ satisfying similar commutation properties to Eq. (8.28) of the previous chapter, namely:

$$[\overline{X}_m, \overline{X}_n] = [\overline{Z}_m, \overline{Z}_n] = 0 \ \ \forall m,n; \quad [\overline{X}_m, \overline{Z}_n] = 0 \ \ \forall m \neq n; \ \ \{\overline{X}_m, \overline{Z}_n\} = 0 \ \ \forall m = n. \tag{9.13}$$

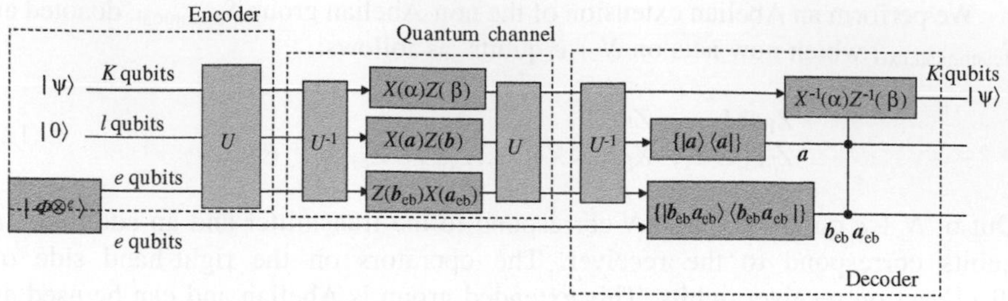

FIGURE 9.3

An entanglement-assisted code obtained as a generalization of the canonical EA code.

We also know from the theorem in Section 2.7 that the unitary equivalent observables, A and UAU^{-1}, have identical spectra. This means that we can establish bijection between two groups V and S that will preserve their commutation relations, so that $\forall\ v \in V$ there exists corresponding $s \in S$ such that $v = UsU^{-1}$ (up to a general phase constant). By using these results, we can establish bijection between canonical entanglement-assisted quantum code described by $S_{\text{canonical}} = \langle S_{\text{canonical, I}}, S_{\text{canonical, E}}\rangle$ and a general EA code characterized by $S = \langle S_{\text{I}}, S_{\text{E}}\rangle$ with $S = US_{\text{canonical}}U^{-1}$ (or equivalently $S_{\text{canonical}} = U^{-1}SU$), where U is a corresponding unitary operator.

Based on the discussion above, we can generalize the results from the previous section as follows. Given a general group $S = <S_{\text{I}}, S_{\text{E}}>$ ($|S_{\text{I}}| = 2^{N-K-e}$ and $|S_{\text{E}}| = 2^{2e}$), there exists an $[N,K,e;D]$ EA-QECC code, illustrated in Figure 9.3, denoted as C_{EA}, defined by the encoding–decoding pair (E,D), with the following properties:

1. The EA code C_{EA} can correct any error from the following set of errors $\mathcal{N}$:

$$\mathcal{N} = \left\{ E_m \middle| \forall E_1, E_2 \Rightarrow E_2^{\dagger} E_1 \in S_{\text{I}} \cup (G_N - C(S)) \right\}. \tag{9.14}$$

2. The code space C_{EA} is a simultaneous eigenspace of the Abelian extension of S, denoted as S_{ext}.
3. To decode, the error syndrome is obtained by simultaneously measuring the observables from S_{ext}.

Because the commutation relations of S are the same as that of EA canonical code $C_{\text{canonical,EA}}$, we can find a unitary operator U such that $S_{\text{canonical}} = USU^{-1}$. Based on Figure 9.3 it is clear that the encoding mapping can be represented as a composition of unitary operator U and encoding mapping of canonical code $\mathscr{E}_{\text{canonical}}$, i.e. $\mathscr{E} = U \circ \mathscr{E}_{\text{canonical}}$. In similar fashion, the decoding mapping can be represented by $D = D_{\text{canonical}} \circ \hat{U}^{-1}$, $\hat{U} = \text{extension}(U)$. Further, the composition of encoding, error, and decoding mapping would result in the identity operator:

$$D \circ \mathcal{N} \circ E = D_{\text{canonical}} \circ \hat{U}^{-1} \circ \hat{U}\mathcal{N}_{\text{canonical}}\hat{U}^{-1} \circ \hat{U}E = I^{\otimes K}. \tag{9.15}$$

The set of errors given by (9.14) is the correctable set of errors because $C_{\text{canonical,EA}}$ is simultaneous eigen-space of $S_{\text{canonical,ext}}$, and $S_{\text{ext}} = US_{\text{canonical,ext}}U^{-1}$, and by definition $C_{\text{EA}} = U(C_{\text{canonical,EA}})$, so that we conclude that C_{ext} is a simultaneous eigenspace of S_{ext}. We learned in the previous section that the decoding operation of $D_{\text{canonical}}$ involves: (i) measuring the set of generators of $S_{\text{canonical,ext}}$, yielding the error syndrome according to the error E_c; and (ii) performing the recovery operation to undo the action of the channel error E_c. The decoding action of C_{EA} is equivalent to the measurement of $S_{\text{ext}} = US_{\text{canonical,ext}}U^{-1}$, followed by the recovery operation UE_cU^{-1} and U^{-1} to undo the encoding.

The distance D of EA code can be introduced in similar fashion to that of a stabilizer code. The distance D of EA code is defined as the minimum weight among undetectable errors. In other words, we say that EA quantum code has distance D if it can detect all errors of weight less than D, but none of weight D. The error correction capability t of EA quantum code is related to minimum distance D by $t = \lfloor (D-1)/2 \rfloor$. The weight of a qubit error $[\mathbf{u}|\mathbf{v}]$ ($\mathbf{u},\mathbf{v} \in F_2^N$) can be defined, in similar fashion to that for stabilizer codes, as the number of components i for which $(u_i,b_i) \neq (0,0)$ or equivalently $\text{wt}([\mathbf{u}|\mathbf{v}]) = \text{wt}(\mathbf{u}+\mathbf{v})$, where "+" denotes bit-wise addition mod 2.

EA quantum bounds are similar to the quantum coding bounds presented in Chapter 7. For example, the EA quantum Singleton bound is given by

$$N + e - K \geq 2(D-1).$$

9.3.1 EA-QECCs Derived from Classical Quaternary and Binary Codes

We turn our attention now to establishing the connection between EA codes and quaternary classical codes, which are defined over $F_4 = \text{GF}(4) = \{0,1,\omega,\omega^2 = \varpi\}$. As a reminder, the addition and multiplication tables for F_4 are given below:

+	0	ϖ	1	ω
0	0	ϖ	1	ω
ϖ	ϖ	0	ω	1
1	1	ω	0	ϖ
ω	ω	1	ϖ	0

×	0	ϖ	1	ω
0	0	0	0	0
ϖ	0	ω	ϖ	1
1	0	ϖ	1	ω
ω	0	1	ω	ϖ

On the other hand, the addition table for binary two-tuples, $F_2^2 = \{00,01,10,11\}$, is:

+	00	01	11	10
00	00	01	11	10
01	01	00	10	11
11	11	10	00	01
10	10	11	01	00

Further, from Chapter 7 we know that the multiplication table of Pauli matrices is given by:

×	I	X	Y	Z
I	I	X	Y	Z
X	X	I	jZ	$-jY$
Y	Y	$-jZ$	I	jX
Z	Z	jY	$-jX$	I

If we ignore the phase factor in Pauli matrix multiplication, we can establish the following correspondences between G_1, F_2^2, and F_4:

G	F_2^2	F_4
I	00	0
X	10	ω
Y	11	1
Z	01	ϖ

It can easily be shown that the mapping f: $F_4 \rightarrow F_2^2$ is an isomorphism. Therefore, if a classical code (n,k,d) C_4 over GF(4) exists, then an $[N,K,e,D] = [n,2k-n+e,e;D]$ EA-QECC exists for some non-negative integer e, which can be described by the following quantum-check matrix:

$$A_{\text{EA}} = f(\tilde{\boldsymbol{H}}_4); \quad \tilde{\boldsymbol{H}}_4 = \begin{pmatrix} \omega \boldsymbol{H}_4 \\ \overline{\omega} \boldsymbol{H}_4 \end{pmatrix}, \tag{9.16}$$

where $\boldsymbol{H}_4$ denotes the parity-check matrix of classical code (n,k,d) over F_4. The sympletic product of binary vectors from (9.16) is equal to the trace of the product of their GF(4) representations as follows:

$$\boldsymbol{h}_m \odot \boldsymbol{h}_n = \text{tr}\left(f^{-1}(\boldsymbol{h}_m)\overline{f^{-1}(\boldsymbol{h}_n)}\right), \quad \text{tr}(w) = w + \overline{w}, \tag{9.17}$$

where $\boldsymbol{h}_m$ ($\boldsymbol{h}_n$) is the mth (nth) row of $\boldsymbol{A}_{\mathrm{EA}}$ and the overbar denotes the conjugation. Based on Chapter 7, the sympletic product of $\boldsymbol{A}_{\mathrm{EA}}$ is given by

$$\begin{aligned}\mathrm{tr}\left\{\begin{pmatrix}\omega\boldsymbol{H}_4\\ \overline{\omega}\boldsymbol{H}_4\end{pmatrix}\begin{pmatrix}\omega\boldsymbol{H}_4\\ \overline{\omega}\boldsymbol{H}_4\end{pmatrix}^\dagger\right\} &= \mathrm{tr}\left\{\begin{pmatrix}\omega\boldsymbol{H}_4\\ \overline{\omega}\boldsymbol{H}_4\end{pmatrix}\begin{pmatrix}\overline{\omega}\boldsymbol{H}_4^\dagger & \omega\boldsymbol{H}_4^\dagger\end{pmatrix}\right\}\\ &= \mathrm{tr}\left\{\begin{pmatrix}\boldsymbol{H}_4\boldsymbol{H}_4^\dagger & \overline{\omega}\boldsymbol{H}_4\boldsymbol{H}_4^\dagger\\ \omega\boldsymbol{H}_4\boldsymbol{H}_4^\dagger & \boldsymbol{H}_4\boldsymbol{H}_4^\dagger\end{pmatrix}\right\} = \mathrm{tr}\left\{\begin{pmatrix}1 & \overline{\omega}\\ \omega & 1\end{pmatrix}\otimes\boldsymbol{H}_4\boldsymbol{H}_4^\dagger\right\}\\ &= \begin{pmatrix}1 & \overline{\omega}\\ \omega & 1\end{pmatrix}\otimes\boldsymbol{H}_4\boldsymbol{H}_4^\dagger + \begin{pmatrix}1 & \omega\\ \overline{\omega} & 1\end{pmatrix}\otimes\overline{\boldsymbol{H}}_4\boldsymbol{H}_4^{\mathrm{T}}\end{aligned} \tag{9.18}$$

The rank of trace of $\tilde{\boldsymbol{H}}_4\tilde{\boldsymbol{H}}_4^\dagger$ will not change if we perform the similarity transformation $\boldsymbol{B}\boldsymbol{A}\mathrm{tr}(\tilde{\boldsymbol{H}}_4\tilde{\boldsymbol{H}}_4^\dagger)\boldsymbol{A}^\dagger\boldsymbol{B}^\dagger$, where $\boldsymbol{A}=\begin{bmatrix}1 & \overline{\omega}\\ \omega & 1\end{bmatrix}\otimes\boldsymbol{I}$, $\boldsymbol{B}=\begin{bmatrix}1 & 0\\ 1 & 1\end{bmatrix}\otimes\boldsymbol{I}$, to obtain:

$$\boldsymbol{B}\boldsymbol{A}\tilde{\boldsymbol{H}}_4\tilde{\boldsymbol{H}}_4^\dagger\boldsymbol{A}^\dagger\boldsymbol{B}^\dagger = \begin{bmatrix}\overline{\boldsymbol{H}}_4\boldsymbol{H}_4^{\mathrm{T}} & \boldsymbol{0}\\ \boldsymbol{0} & \boldsymbol{H}_4\boldsymbol{H}_4^\dagger\end{bmatrix} = \overline{\boldsymbol{H}}_4\boldsymbol{H}_4^{\mathrm{T}} + \boldsymbol{H}_4\boldsymbol{H}_4^\dagger. \tag{9.19}$$

The rank of $\boldsymbol{B}\boldsymbol{A}\mathrm{tr}(\tilde{\boldsymbol{H}}_4\tilde{\boldsymbol{H}}_4^\dagger)\boldsymbol{A}^\dagger\boldsymbol{B}^\dagger$ is then obtained as

$$\begin{aligned}\mathrm{rank}\,(\boldsymbol{B}\boldsymbol{A}\mathrm{tr}(\tilde{\boldsymbol{H}}_4\tilde{\boldsymbol{H}}_4^\dagger)\boldsymbol{A}^\dagger\boldsymbol{B}^\dagger) &= \mathrm{rank}\,(\overline{\boldsymbol{H}}_4\boldsymbol{H}_4^{\mathrm{T}}+\boldsymbol{H}_4\boldsymbol{H}_4^\dagger) = \mathrm{rank}\,(\overline{\boldsymbol{H}}_4\boldsymbol{H}_4^{\mathrm{T}}) + \mathrm{rank}\,(\boldsymbol{H}_4\boldsymbol{H}_4^\dagger)\\ &= 2\mathrm{rank}\,(\boldsymbol{H}_4\boldsymbol{H}_4^\dagger) = 2e,\quad e=\mathrm{rank}\,(\boldsymbol{H}_4\boldsymbol{H}_4^\dagger).\end{aligned} \tag{9.20}$$

Equation (9.20) indicates that the number of required ebits is $e=\mathrm{rank}(\boldsymbol{H}_4\boldsymbol{H}_4^\dagger)$, so that the parameters of this EA code, based on Eq. (9.16), are $[N,K,e]=[n,n+e-2(n-k),e]=[n,2k-n+e,e]$.

Example. The EA code derived from (4,2,3) 4-ary classical code described by parity-check matrix

$$\boldsymbol{H}_4=\begin{pmatrix}1 & 1 & 0 & 1\\ 1 & \omega & 1 & 0\end{pmatrix}$$

has parameters [4,1,1,3], and based on (9.16) the corresponding quantum-check matrix is given by

$$\boldsymbol{A}_{\mathrm{EA}}=f\left(\tilde{\boldsymbol{H}}_4\right)=f\left(\begin{bmatrix}\omega\boldsymbol{H}_4\\ \overline{\omega}\boldsymbol{H}_4\end{bmatrix}\right)=f\left(\begin{bmatrix}\omega & \omega & 0 & \omega\\ \omega & \overline{\omega} & \omega & 0\\ \overline{\omega} & \overline{\omega} & 0 & \overline{\omega}\\ \overline{\omega} & 1 & \overline{\omega} & 0\end{bmatrix}\right)=\begin{pmatrix}X & X & I & X\\ X & Z & X & I\\ Z & Z & I & Z\\ Z & Y & Z & I\end{pmatrix}.$$

In the remainder of this section, we describe how to relate EA quantum codes to binary classical codes. Let $\boldsymbol{H}$ be a binary parity-check matrix of dimensions $(n-k)\times n$. We will show that the corresponding EA-QECC has the parameters

$[n, 2k-n+e; e]$, where $e = \text{rank}(\boldsymbol{H}\boldsymbol{H}^{\text{T}})$. The quantum-check matrix of CSS-like EA quantum code can be represented as follows:

$$\boldsymbol{A}_{\text{EA}} = \left(\begin{array}{c|c} \boldsymbol{H} & \boldsymbol{0} \\ \boldsymbol{0} & \boldsymbol{H} \end{array}\right). \tag{9.21}$$

The rank of the sympletic product of $\boldsymbol{A}_{\text{AE}}$, based on (9.17), is given by

$$\begin{aligned}
\text{rank}\{\boldsymbol{A}_{\text{EA}} \odot \boldsymbol{A}_{EA}^{\text{T}}\} &= \text{rank}\left\{\begin{pmatrix} \boldsymbol{H} \\ \boldsymbol{0} \end{pmatrix}\begin{pmatrix} \boldsymbol{0} & \boldsymbol{H}^{\text{T}} \end{pmatrix} + \begin{pmatrix} \boldsymbol{0} \\ \boldsymbol{H} \end{pmatrix}\begin{pmatrix} \boldsymbol{H}^{\text{T}} & \boldsymbol{0} \end{pmatrix}\right\} \\
&= \text{rank}\left\{\begin{pmatrix} \boldsymbol{0} & \boldsymbol{H}\boldsymbol{H}^{\text{T}} \\ \boldsymbol{0} & \boldsymbol{0} \end{pmatrix} + \begin{pmatrix} \boldsymbol{0} & \boldsymbol{0} \\ \boldsymbol{H}\boldsymbol{H}^{\text{T}} & \boldsymbol{0} \end{pmatrix}\right\} \\
&= \text{rank}\{\boldsymbol{H}\boldsymbol{H}^{\text{T}} + \boldsymbol{H}\boldsymbol{H}^{\text{T}}\} = 2\text{rank}\{\boldsymbol{H}\boldsymbol{H}^{\text{T}}\} \\
&= 2e, \quad e = \text{rank}\{\boldsymbol{H}\boldsymbol{H}^{\text{T}}\}.
\end{aligned} \tag{9.22}$$

From (9.22) it is clear that the number of required qubits is $e = \text{rank}(\boldsymbol{H}\boldsymbol{H}^{\text{T}})$, so that the parameters of this EA code are $[N, K, e] = [n, n+e-2(n-k), e] = [n, 2k-n+e, e]$. Dual-containing codes can therefore be considered as EA codes for which $e = 0$.

The EA code given by (9.21) can be generalized as follows:

$$\boldsymbol{A}_{\text{EA}} = (\boldsymbol{A}_x | \boldsymbol{A}_z), \tag{9.23}$$

where the $(n-k) \times n$ submatrix $\boldsymbol{A}_{\text{x}}$ ($\boldsymbol{A}_{\text{z}}$) contains only X operators (Z operators). The rank of the sympletic product of $\boldsymbol{A}_{\text{AE}}$ can be obtained as

$$\text{rank}\{\boldsymbol{A}_{\text{EA}} \odot \boldsymbol{A}_{EA}^{\text{T}}\} = \text{rank}\{\boldsymbol{A}_x\boldsymbol{A}_z^{\text{T}} + \boldsymbol{A}_z\boldsymbol{A}_x^{\text{T}}\} = 2e, \; e = \text{rank}\{\boldsymbol{A}_x\boldsymbol{A}_z^{\text{T}} + \boldsymbol{A}_z\boldsymbol{A}_x^{\text{T}}\}/2. \tag{9.24}$$

The parameters of this EA code are $[N, K, e] = [n, n+e-(n-k), e] = [n, k+e, e]$.

Another generalization of (9.21) is given by

$$\boldsymbol{A}_{\text{EA}} = \left(\begin{array}{c|c} \boldsymbol{H}_x & \boldsymbol{0} \\ \boldsymbol{0} & \boldsymbol{H}_z \end{array}\right), \tag{9.25}$$

where the $(n-k_x) \times n$ submatrix $\boldsymbol{H}_x$ contains only X operators, while the $(n-k_z) \times n$ submatrix $\boldsymbol{H}_z$ contains only Z operators. The rank of the sympletic product of $\boldsymbol{A}_{\text{AE}}$ is given by

$$\begin{aligned}
\text{rank}\{\boldsymbol{A}_{\text{EA}} \odot \boldsymbol{A}_{\text{EA}}^{\text{T}}\} &= \text{rank}\left\{\begin{pmatrix} \boldsymbol{H}_x \\ \boldsymbol{0} \end{pmatrix}\begin{pmatrix} \boldsymbol{0} & \boldsymbol{H}_z^{\text{T}} \end{pmatrix} + \begin{pmatrix} \boldsymbol{0} \\ \boldsymbol{H}_z \end{pmatrix}\begin{pmatrix} \boldsymbol{H}_x^{\text{T}} & \boldsymbol{0} \end{pmatrix}\right\} \\
&= \text{rank}\left\{\begin{pmatrix} \boldsymbol{0} & \boldsymbol{H}_x\boldsymbol{H}_z^{\text{T}} \\ \boldsymbol{0} & \boldsymbol{0} \end{pmatrix} + \begin{pmatrix} \boldsymbol{0} & \boldsymbol{0} \\ \boldsymbol{H}_z\boldsymbol{H}_x^{\text{T}} & \boldsymbol{0} \end{pmatrix}\right\} \\
&= \text{rank}\{\boldsymbol{H}_x\boldsymbol{H}_z^{\text{T}} + \boldsymbol{H}_z\boldsymbol{H}_x^{\text{T}}\} = 2e, \quad e = \text{rank}\{\boldsymbol{H}_x\boldsymbol{H}_z^{\text{T}}\}.
\end{aligned} \tag{9.26}$$

The EA code parameters are $[N,K,e;D]=[n,n+e-(n-k_x)-(n-k_z),e;\min(d_x,d_z)]=[n,k_x+k_z-n+e,e;\min(d_x,d_z)]$.

9.4 ENCODING AND DECODING FOR ENTANGLEMENT-ASSISTED QUANTUM CODES

Encoders and decoders for entanglement-assisted codes can be implemented in similar fashion as for subsystem codes, described in the previous chapter: (i) using standard format formalism [21,22] and (ii) by using the conjugation method due to Grassl et al. [23] (see also Refs [7,24]).

In the previous section, we defined an EA $[N,K,e]$ code by using a non-Abelian subgroup S of G_N (multiplicative Pauli group on N qubits) with $2e+l$ generators, where $l=N-K-e$. The subgroup S can be described by the set of independent generators $\{\overline{X}_{e+1},\ldots,\overline{X}_{e+l};\overline{Z}_1,\ldots,\overline{Z}_{e+l}\}$, which satisfy the commutation relations (9.13). The non-Abelian subgroup S can be decomposed into two subgroups: the commuting isotropic group $S_I=\{\overline{Z}_1,\ldots,\overline{Z}_l\}$ and the entanglement subgroup with anticommuting pairs $\{\overline{X}_{l+1},\ldots,\overline{X}_{l+e};\overline{Z}_{l+1},\ldots,\overline{Z}_{l+e}\}$. This decomposition allows us to determine the EA code parameters. Because the isotropic subgroup is a commuting group, it corresponds to ancillary qubits. On the other hand, since the entanglement subgroup is composed of anticommuting pairs, its elements correspond to the transmitter and receiver halves of ebits. This decomposition can be performed using the sympletic Gram–Schmidt procedure [7]. Let us assume that the following generators create the subgroup S: $g_1,\ldots,g_M$. We start the procedure with generator g_1 and check if it commutes with all other generators. If it commutes, we remove it from further consideration. If g_1 anticommutes with a generator g_i, we relabel g_2 as g_i and vice versa. For the remaining generators we perform the following manipulation:

$$g_m = g_m {g_1}^{f(g_2,g_m)} {g_2}^{f(g_1,g_m)},\quad f(a,b)=\begin{cases}0, & [a,b]=0\\ 1, & \{a,b\}=0\end{cases};\quad m=3,4,\ldots,M. \tag{9.27}$$

The generators g_1 and g_2 are then removed from further discussion, and the same algorithm is applied on the remaining generators. At the end of this procedure, the generators removed from consideration create the code generators satisfying the commutation relations (9.13).

We can perform an extension of the non-Abelian subgroup S into Abelian and then apply the standard form procedure for encoding and decoding, which was fully explained in the previous chapter. Here, instead, we will concentrate to the conjugation method for EA encoder implementation. The key idea is very similar to that of subsystem codes, described in Section 8.7. That is, we represent the non-Abelian group S using finite geometry interpretation. We then perform Gauss elimination to transform the EA code into canonical EA code:

$$|c\rangle \rightarrow |\Phi^{\otimes e}\rangle \underbrace{|0\rangle\otimes\ldots\otimes|0\rangle}_{l=N-K-e\ \text{times}} |\Phi^{\otimes e}\rangle|\psi\rangle_K = |\Phi^{\otimes e}\rangle|\mathbf{0}\rangle_l|\psi\rangle_K, \tag{9.28}$$

where $|c\rangle$ is the EA codeword, $|\Phi^{\otimes e}\rangle$ is the ebit state, and $|\mathbf{0}\rangle_l$ are l ancillary states prepared into the $|0\rangle$ state. In other words, we have to apply the sequence of gates $\{U\}$ to perform the following transformation of the quantum-check matrix of EA code:

$$\mathbf{A}_{\mathrm{EA}} = [\mathbf{A}_x|\mathbf{A}_z] \xrightarrow{\{U\}} \mathbf{A}_{\mathrm{EA,\,canonical}} = \left[\begin{array}{ccc|ccc} \overbrace{\mathbf{0}}^{e} & \overbrace{\mathbf{0}}^{l} & \overbrace{\mathbf{0}}^{K} & \boldsymbol{I}_e & \mathbf{0} & \mathbf{0} \\ \mathbf{0} & \mathbf{0} & \mathbf{0} & \mathbf{0} & \mathbf{0} & \mathbf{0} \\ \boldsymbol{I}_e & \mathbf{0} & \mathbf{0} & \mathbf{0} & \boldsymbol{I}_l & \mathbf{0} \\ \mathbf{0} & \mathbf{0} & \mathbf{0} & \mathbf{0} & \mathbf{0} & \mathbf{0} \end{array}\right]. \tag{9.29}$$

The right-hand side of Eq. (9.29) is obtained from Eq. (9.11). We apply the same set of rules as for subsystem codes. The action of the Hadamard gate on the ith qubit is to change the position of a_i and b_i as follows, where we underline the affected qubits for convenience:

$$(a_1 \ldots \underline{a_i} \ldots a_N | b_1 \ldots \underline{b_i} \ldots b_N) \xrightarrow{H_i} (a_1 \ldots \underline{b_i} \ldots a_N | b_1 \ldots \underline{a_i} \ldots b_N). \tag{9.30}$$

The action of the CNOT gate on generator $\boldsymbol{g}$, where the ith qubit is the control qubit and the jth qubit is the target qubit, the action denoted as $\mathrm{CNOT}_{i,j}$, can be described as

$$(a_1 \ldots \underline{a_j} \ldots a_N | b_1 \ldots \underline{b_i} \ldots b_N) \xrightarrow{\mathrm{CNOT}_{i,j}} (a_1 \ldots a_{j-1}\, \underline{a_j + a_i}\, a_{j+1} \ldots a_N | b_1 \ldots b_{i-1}\, \underline{b_i + b_j}\, b_{i+1} \ldots b_N). \tag{9.31}$$

Therefore, the jth entry in the X portion and the ith entry in the Z portion are affected. The action of phase gate P on the ith qubit is to perform the following mapping:

$$(a_1 \ldots \underline{a_i} \ldots a_N | b_1 \ldots \underline{b_i} \ldots b_N) \xrightarrow{P} (a_1 \ldots \underline{a_i} \ldots a_N | b_1 \ldots \underline{a_i + b_i} \ldots b_N). \tag{9.32}$$

Finally, the application of a CNOT gate between the ith and jth qubits three times in an alternative fashion leads to swapping of the ith and jth columns in both the X and Z portions of the quantum-check matrix, $\mathrm{SWAP}_{i,j} = \mathrm{CNOT}_{i,j}\mathrm{CNOT}_{j,i}\mathrm{CNOT}_{i,j}$, as shown in Chapter 3:

$$(a_1 \ldots \underline{a_i} \ldots \underline{a_j} \ldots a_N | b_1 \ldots \underline{b_i} \ldots \underline{b_j} \ldots b_N) \xrightarrow{\mathrm{SWAP}_{i,j}} (a_1 \ldots \underline{a_j} \ldots \underline{a_i} \ldots a_N | b_1 \ldots \underline{b_j} \ldots \underline{b_i} \ldots b_N). \tag{9.33}$$

Let us denote the sequence of gates being applied during this transformation to canonical EA code as $\{U\}$. By applying this sequence of gates in the opposite order we obtain the encoding circuit of the corresponding EA code. Notice that adding the nth row to the mth row maps $g_m \to g_m g_n$ so that codewords are invariant to this change.

Since this method is very similar to that used for subsystem codes in the previous chapter, we turn our attention now to the sympletic Gram–Schmidt method due to Wilde [7], which has already been discussed above at the operator level. Here we describe this method using a finite geometry interpretation, based on Eqs (9.30)–(9.33). Let us observe the following example due to Wilde [7]:

$$
A_{\mathrm{EA}} = \left[\begin{array}{cccc|cccc} 0&1&0&0&1&0&1&0\\ 0&0&0&0&1&1&0&1\\ 1&1&1&0&0&1&0&0\\ 1&1&0&1&0&0&0&0 \end{array}\right]
\xrightarrow{\mathrm{SWAP}_{1,2}}
\left[\begin{array}{cccc|cccc} 1&0&0&0&0&1&1&0\\ 0&0&0&0&1&1&0&1\\ 1&1&1&0&1&0&0&0\\ 1&1&0&1&0&0&0&0 \end{array}\right]
\xrightarrow{H_2,H_3}
\left[\begin{array}{cccc|cccc} 1&1&1&0&0&0&0&0\\ 0&1&0&0&1&0&0&1\\ 1&0&0&0&1&1&1&0\\ 1&0&0&1&0&1&0&0 \end{array}\right]
$$

$$
\xrightarrow{\mathrm{CNOT}_{1,2},\mathrm{CNOT}_{1,3}}
\left[\begin{array}{cccc|cccc} 1&0&0&0&0&0&0&0\\ 0&1&0&0&1&0&0&1\\ 1&1&1&0&1&1&1&0\\ 1&1&1&1&1&1&0&0 \end{array}\right]
\xrightarrow{H_1,H_4}
\left[\begin{array}{cccc|cccc} 0&0&0&0&1&0&0&0\\ 1&1&0&1&0&0&0&0\\ 1&1&1&0&1&1&1&0\\ 1&1&1&0&1&1&0&1 \end{array}\right]
\xrightarrow{\mathrm{CNOT}_{1,2},\mathrm{CNOT}_{1,4}}
\left[\begin{array}{cccc|cccc} 0&0&0&0&1&0&0&0\\ 1&0&0&0&0&0&0&0\\ 1&0&1&1&0&1&1&0\\ 1&0&1&1&1&1&0&1 \end{array}\right]
$$

$$
\xrightarrow{\substack{\mathrm{row}_4\leftarrow\mathrm{row}_4+\mathrm{row}_1\\ \mathrm{row}_3\leftarrow\mathrm{row}_3+\mathrm{row}_2\\ \mathrm{row}_4\leftarrow\mathrm{row}_4+\mathrm{row}_2}}
\left[\begin{array}{cccc|cccc} 0&0&0&0&1&0&0&0\\ 1&0&0&0&0&0&0&0\\ 0&0&1&1&0&1&1&0\\ 0&0&1&1&0&1&0&1 \end{array}\right]
\xrightarrow{H_2}
\left[\begin{array}{cccc|cccc} 0&0&0&0&1&0&0&0\\ 1&0&0&0&0&0&0&0\\ 0&1&1&1&0&0&1&0\\ 0&1&1&1&0&0&0&1 \end{array}\right]
\xrightarrow{\mathrm{CNOT}_{2,3},\mathrm{CNOT}_{2,4}}
\left[\begin{array}{cccc|cccc} 0&0&0&0&1&0&0&0\\ 1&0&0&0&0&0&0&0\\ 0&1&0&0&0&1&1&0\\ 0&1&0&0&0&1&0&1 \end{array}\right]
$$

$$
\xrightarrow{P_2,H_3}
\left[\begin{array}{cccc|cccc} 0&0&0&0&1&0&0&0\\ 1&0&0&0&0&0&0&0\\ 0&1&1&0&0&0&0&0\\ 0&1&0&0&0&0&0&1 \end{array}\right]
\xrightarrow{\mathrm{CNOT}_{2,3}}
\left[\begin{array}{cccc|cccc} 0&0&0&0&1&0&0&0\\ 1&0&0&0&0&0&0&0\\ 0&1&0&0&0&0&0&0\\ 0&1&1&0&0&0&0&1 \end{array}\right]
\xrightarrow{\mathrm{row}_4\leftarrow\mathrm{row}_4+\mathrm{row}_3}
\left[\begin{array}{cccc|cccc} 0&0&0&0&1&0&0&0\\ 1&0&0&0&0&0&0&0\\ 0&1&0&0&0&0&0&0\\ 0&0&1&0&0&0&0&1 \end{array}\right]
$$

$$
\xrightarrow{H_2,H_4}
\left[\begin{array}{cccc|cccc} 0&0&0&0&1&0&0&0\\ 1&0&0&0&0&0&0&0\\ 0&0&0&0&0&1&0&0\\ 0&0&1&1&0&0&0&0 \end{array}\right]
\xrightarrow{\mathrm{CNOT}_{3,4}}
\left[\begin{array}{cccc|cccc} 0&0&0&0&1&0&0&0\\ 1&0&0&0&0&0&0&0\\ 0&0&0&0&0&1&0&0\\ 0&0&1&0&0&0&0&0 \end{array}\right]
\xrightarrow{H_3}
\left[\begin{array}{cccc|cccc} 0&0&0&0&1&0&0&0\\ 1&0&0&0&0&0&0&0\\ 0&0&0&0&0&1&0&0\\ 0&0&0&0&0&0&1&0 \end{array}\right].
$$

Clearly, we have transformed the initial EA into canonical form (9.29). The corresponding encoding circuits can be obtained by applying the set of gates to the canonical codeword, used to transform the quantum-check matrix into canonical form, but now in the opposite order. The encoding circuit for this EA code is shown in Figure 9.4.

9.5 OPERATOR QUANTUM ERROR CORRECTION CODES (SUBSYSTEM CODES)

The operator quantum error correction codes (OQECCs) [2,4,16–18,24–28] represent a very important class of quantum codes, which are also known as subsystem codes. This class of codes has already been discussed in the previous chapter. Nevertheless, the following interpretation due to Hsieh, Devetak, and Brun [2,4] is useful for a deeper understanding of the concept.

We start our description with canonical OQECCs. An $[N,K,R]$ canonical OQECC can be represented by the following mapping:

$$
|\phi\rangle|\psi\rangle \rightarrow \underbrace{|0\rangle\otimes\ldots\otimes|0\rangle}_{s=N-K-R\ \text{times}}|\phi\rangle_R|\psi\rangle_K = |\mathbf{0}\rangle_s|\phi\rangle_R|\psi\rangle_K, \tag{9.34}
$$

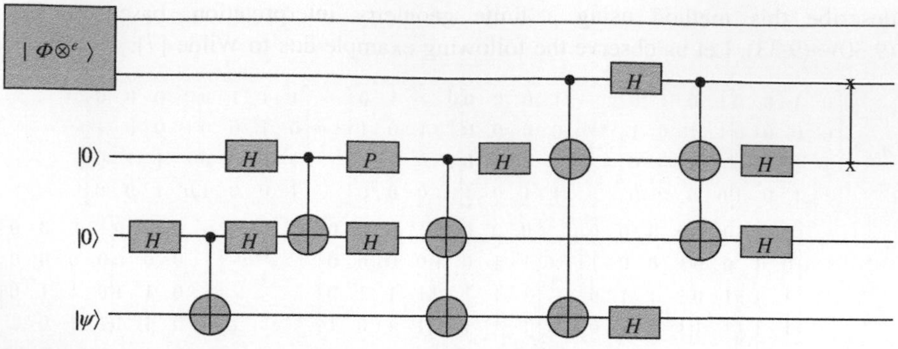

FIGURE 9.4

The encoding circuit for the EA code from the example above.

where $|\psi\rangle$ represents a K-qubit information state, $|\phi\rangle$ represents the R-qubit gauge state, and $|\mathbf{0}\rangle_s$ represent s ancillary qubits. So the key difference with respect to canonical stabilizer code is that R ancillary $|0\rangle$ qubits are converted into gauge qubits. This canonical OQECC code can correct the following set of errors:

$$\mathcal{N}_{\text{canonical,OQECC}} = \{X(\boldsymbol{a})Z(\boldsymbol{b})\otimes X(\boldsymbol{c})Z(\boldsymbol{d})\otimes X(\alpha(\boldsymbol{a}))Z(\beta(\boldsymbol{a})): \\ \boldsymbol{a},\boldsymbol{b}\in F_2^s;\ \boldsymbol{c},\boldsymbol{d}\in F_2^R\};\quad \alpha,\beta\colon F_2^s\to F_2^K. \tag{9.35}$$

In order to prove claim (9.35), let us observe one particular error E_c from the set of errors (9.35) and study its action on the transmitted codeword $|\mathbf{0}\rangle_s|\phi\rangle_R|\psi\rangle_K$:

$$X(\boldsymbol{a})Z(\boldsymbol{b})|\mathbf{0}\rangle_s\otimes(X(\boldsymbol{c})Z(\boldsymbol{d}))|\phi\rangle\otimes X(\alpha(\boldsymbol{a}))Z(\beta(\boldsymbol{a}))|\psi\rangle. \tag{9.36}$$

The action of operator $X(\boldsymbol{a})Z(\boldsymbol{b})$ on state $|\mathbf{0}\rangle_s$ is given by $X(\boldsymbol{a})Z(\boldsymbol{b})|\mathbf{0}\rangle_s \overset{Z(b)|0\rangle_s=|0\rangle_s}{=} X(\boldsymbol{a})|\mathbf{0}\rangle_s = |\boldsymbol{a}\rangle$. On the other hand, the action of $X(\boldsymbol{c})Z(\boldsymbol{d})$ on the gauge state is $(X(\boldsymbol{c})Z(\boldsymbol{d}))|\phi\rangle = |\phi'\rangle$. Finally, the action of a channel on the information portion of the codeword is given by $X(\alpha(\boldsymbol{a}))Z(\beta(\boldsymbol{a}))|\psi\rangle = |\psi'\rangle$. Clearly, the vector $(\boldsymbol{a}\ \boldsymbol{b}\ \boldsymbol{c}\ \boldsymbol{d})$ uniquely specifies the error operator E_c, and can be called the syndrome vector. Since the state $|\mathbf{0}\rangle_s$ is invariant on action of $Z(\boldsymbol{b})$, the vector $\boldsymbol{b}$ can be omitted from the syndrome vector. Also, since the final state of the gauge subsystem is irrelevant to the information qubit states, the vectors $\boldsymbol{c}$ and $\boldsymbol{d}$ can also be omitted from consideration. To obtain the $\boldsymbol{a}$ portion of the syndrome, we have to perform simultaneous measurements on $Z(\boldsymbol{d}_i)$, where $\boldsymbol{d}_i=(0\ldots01_i0\ldots0)$ on ancillary qubits. Once the syndrome vector $\boldsymbol{a}$ is determined, we apply the operator $X^{-1}(\alpha(\boldsymbol{a}))Z^{-1}(\beta(\boldsymbol{a}))$ to the information portion of the codeword to undo the action of the channel, as illustrated

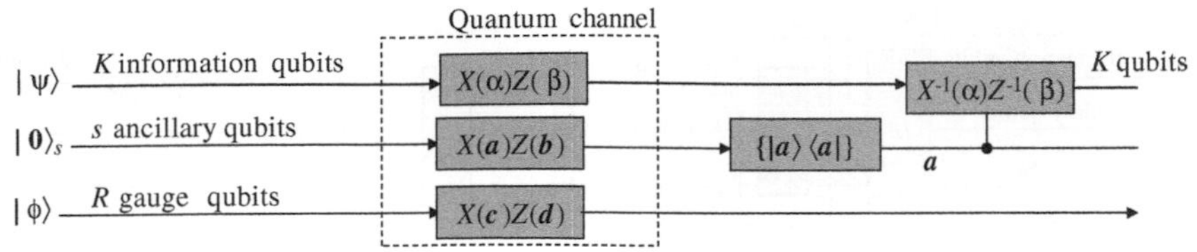

FIGURE 9.5

Canonical OQECC encoder and decoder principles.

in Figure 9.5. In order to avoid the need for measurement we can perform the following controlled unitary decoding operation:

$$U_{\text{canocnical, OECC, dec}} = \sum_{a} |\boldsymbol{a}\rangle\langle\boldsymbol{a}| \otimes \boldsymbol{I} \otimes X^{-1}(\alpha(\boldsymbol{a}))Z^{-1}(\beta(\boldsymbol{a})), \tag{9.37}$$

and discard the unimportant portion of the received quantum word.

The OQECCs can also be described by using the stabilizer formalism interpretation as follows:

$$S_{\text{canonical, OQECC}} = \langle S_{\text{canonical, OQECC, I}}, S_{\text{canonical, OQECC, E}} \rangle$$
$$S_{\text{canonical, OQECC, I}} = \langle Z_1, \ldots, Z_s \rangle, \quad S_{\text{canonical, OQECC, E}} = \langle Z_{s+1}, \ldots, Z_{s+R}, X_{s+1}, \ldots, X_{s+R} \rangle, \tag{9.38}$$

where $S_{\text{canonical, OQECC,I}}$ denotes the isotropic subgroup of size 2^s and $S_{\text{canonical, OQECC,E}}$ denotes the sympletic subgroup of size 2^{2R}. The isotropic subgroup defines a 2^{K+R}-dimensional code space, while the sympletic subgroup defines all possible operations on gauge qubits. The OQECC can correct the following set of errors:

$$\mathcal{N} = \left\{ E_m \middle| \forall E_1, E_2 \Rightarrow E_2^\dagger E_1 \in S_{\text{canonical, OQECC}} \cup (G_N - C(S_{\text{canonical, OQECC, I}})) \right\}. \tag{9.39}$$

Using a similar approach as in Section 9.3, we can establish bijection between canonical OQECC described by $S_{\text{canonical, OQECC}} = \langle S_{\text{canonical, OQECC, I}}, S_{\text{canonical, OQECC, E}} \rangle$ and a general OQECC characterized by $S_{\text{OQECC}} = \langle S_{\text{OQECC,I}}, S_{\text{OQECC,E}} \rangle$ with $S_{\text{OQECC}} = U S_{\text{canonical,OQECC}} U^{-1}$ (or equivalently $S_{\text{canonical,OQECC}} = U^{-1} S_{\text{OQECC}} U$), where U is a corresponding unitary operator. This $[N,K,R;D]$ OQECC code, illustrated in Figure 9.6, denoted as C_{OQECC}, is defined by the encoding–decoding pair (E,D) with the following properties:

(i) The OQECC code C_{OQECC} can correct any error for the following set of errors:

$$\mathcal{N} = \left\{ E_m \middle| \forall E_1, E_2 \Rightarrow E_2^\dagger E_1 \in S_{\text{OQECC}} \cup (G_N - C(S_{\text{OQECC, I}})) \right\}. \tag{9.40}$$

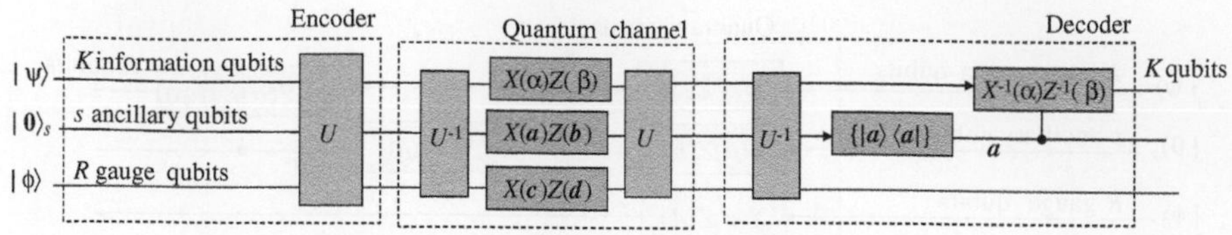

FIGURE 9.6

EA-OQECC encoder and decoder principles.

(ii) The code space C_{OQECC} is a simultaneous eigenspace of $S_{\text{OQECC,I}}$.

(iii) To decode, the error syndrome is obtained by simultaneously measuring the observables from $S_{\text{OQECC,I}}$.

The encoding and decoding of OQECC (subsystem) codes has already been described in the previous chapter.

9.6 ENTANGLEMENT-ASSISTED OPERATOR QUANTUM ERROR CORRECTION CODING (EA-OQECC)

The entanglement-assisted OQECCs represent the generalization of EA codes and OQECCs (also known as subsystem codes). That is, in addition to information and gauge qubits used in OQECCs, several entanglement qubits (ebits) are shared

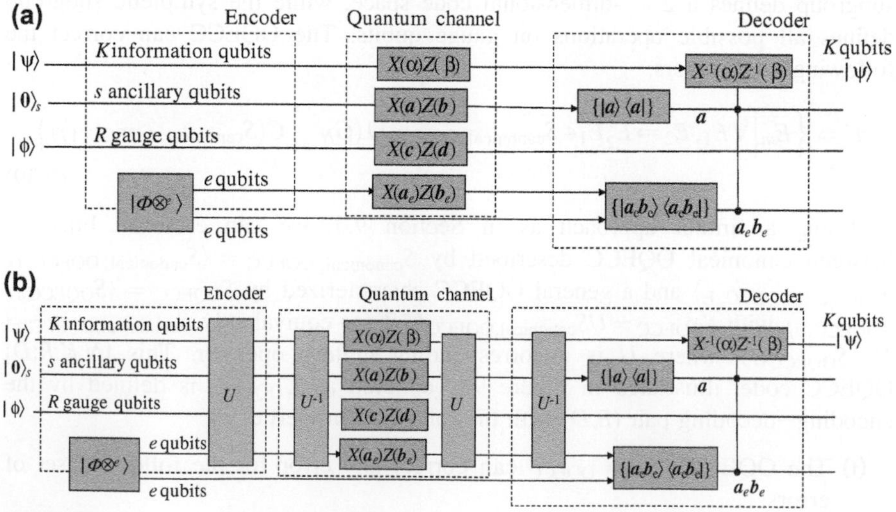

FIGURE 9.7

EA-OQECCs: (a) canonical code; (b) general code.

between the source and destination. The simplest EA-QECC, the canonical code, performs the following trivial encoding operation:

$$\hat{U}_c:\ |\psi\rangle_K \to |\psi\rangle_K \otimes |\mathbf{0}\rangle_s \otimes |\phi\rangle_R \otimes |\Phi^{\otimes e}\rangle;\ s = N - K - R - e. \tag{9.41}$$

The codeword in the canonical [*N*,*K*,*R*,*e*] EA-OQECC, shown in Figure 9.7a, is composed of *K* information qubits, *R* gauge qubits, and *e* maximally entangled kets shared between transmitter and receiver. Since the number of information qubits is *K* and codeword length is *N*, the number of ancillary qubits remaining in the $|0\rangle$ state is $s = N - K - R - e$.

The canonical EA-OQECC can correct the following set of errors:

$$\mathcal{N}_{\text{canonical}} = \{X(\alpha(\mathbf{a},\mathbf{a}_{\text{eb}},\mathbf{b}_{\text{eb}}))Z(\beta(\mathbf{a},\mathbf{a}_{\text{eb}},\mathbf{b}_{\text{eb}}))\otimes X(\mathbf{a})Z(\mathbf{b})\otimes X(\mathbf{c})Z(\mathbf{d})\otimes X(\mathbf{a}_{\text{eb}})Z(\mathbf{b}_{\text{eb}}):$$
$$\mathbf{a},\mathbf{b}\in F_2^s;\ \mathbf{c},\mathbf{d}\in F_2^R;\ \mathbf{b}_{\text{eb}},\mathbf{a}_{\text{eb}}\in F_2^e\};\quad \alpha,\beta:\ F_2^s\times F_2^e\times F_2^e\to F_2^K, \tag{9.42}$$

where the Pauli operators *X*(***e***) (***e*** ∈ $\{\mathbf{a},\mathbf{a}_{\text{eb}},\mathbf{c},\alpha(\mathbf{a},\mathbf{b}_{\text{eb}},\mathbf{a}_{\text{eb}})\}$) and *Z*(***e***) (***e*** ∈ $\{\mathbf{b},\mathbf{b}_{\text{eb}},\mathbf{d},\beta(\mathbf{a},\mathbf{b}_{\text{eb}},\mathbf{a}_{\text{eb}})\}$) are introduced by (9.2).

In order to prove claim (9.42), let us observe one particular error E_c from the set of errors (9.42) and study its action on the transmitted codeword $|\psi\rangle_K \otimes |\mathbf{0}\rangle_s \otimes |\phi\rangle_R \otimes |\Phi^{\otimes e}\rangle$:

$$X(\alpha(\mathbf{a},\mathbf{a}_{\text{eb}},\mathbf{b}_{\text{eb}}))Z(\beta(\mathbf{a},\mathbf{a}_{\text{eb}},\mathbf{b}_{\text{eb}}))|\psi\rangle \otimes X(\mathbf{a})Z(\mathbf{b})|\mathbf{0}\rangle_s \otimes X(\mathbf{c})Z(\mathbf{d})|\phi\rangle_R \otimes$$
$$(X(\mathbf{a}_{\text{eb}})Z(\mathbf{b}_{\text{eb}})\otimes I^{R_x})|\Phi^{\otimes e}\rangle. \tag{9.43}$$

The action of operator *X*(***a***)*Z*(***b***) on state $|\mathbf{0}\rangle_s$ is given by $X(\mathbf{a})Z(\mathbf{b})|\mathbf{0}\rangle_s \overset{Z(\mathbf{b})|\mathbf{0}\rangle_s=|\mathbf{0}\rangle_s}{=} X(\mathbf{a})|\mathbf{0}\rangle_s = |\mathbf{a}\rangle$. On the other hand, the action of $X(\mathbf{a}_{\text{eb}})Z(\mathbf{b}_{\text{eb}})$ on the maximum entangled state is $(X(\mathbf{a}_{\text{eb}})Z(\mathbf{b}_{\text{eb}})\otimes I^{R_x})|\Phi^{\otimes e}\rangle = |\mathbf{a}_{\text{eb}},\mathbf{b}_{\text{eb}}\rangle$, where I^{R_x} denotes the identity operators applied on the receiver half of the maximum entangled state. The action of a channel on gauge qubits is described by $(X(\mathbf{c})Z(\mathbf{d}))|\phi\rangle = |\phi'\rangle$. Finally, the action of a channel on the information portion of the codeword is given by $X(\alpha(\mathbf{a},\mathbf{a}_{\text{eb}},\mathbf{b}_{\text{eb}}))$ $Z(\beta(\mathbf{a},\mathbf{a}_{\text{eb}},\mathbf{b}_{\text{eb}}))|\psi\rangle = |\psi'\rangle$. Clearly, the vector ($\mathbf{a}\ \mathbf{b}\ \mathbf{c}\ \mathbf{d}\ \mathbf{a}_{\text{eb}}\ \mathbf{b}_{\text{eb}}$) uniquely specifies the error operator E_c, and can be called the syndrome vector. Since the state $|\mathbf{0}\rangle_s$ is invariant on the action of *Z*(***b***), the vector ***b*** can be omitted from the syndrome vector. Similarly, since the state of gauge qubits is irrelevant to the information qubits state, the vectors ***c*** and ***d*** can also be dropped from the syndrome vector. To obtain the ***a*** portion of the syndrome, we have to perform simultaneous measurements on $Z(\mathbf{d}_i)$, where $\mathbf{d}_i = (0\ldots01_i0\ldots0)$. On the other hand, in order to obtain the ($\mathbf{a}_{\text{eb}}\ \mathbf{b}_{\text{eb}}$) portion of the syndrome vector, we have to perform measurements on $Z(\mathbf{d}_1)Z(\mathbf{d}_1)$, ..., $Z(\mathbf{d}_e)Z(\mathbf{d}_e)$ for $\mathbf{b}_{\text{eb}}$ and $X(\mathbf{d}_1)X(\mathbf{d}_1)$, ..., $X(\mathbf{d}_e)$ $X(\mathbf{d}_e)$ for $\mathbf{a}_{\text{eb}}$. Once the syndrome vector is determined, we apply $X^{-1}(\alpha(\mathbf{a},\mathbf{a}_e,\mathbf{b}_e))$ $Z^{-1}(\beta(\mathbf{a},\mathbf{b}_e,\mathbf{a}_e))$ to undo the action of the channel. In order to avoid the need for measurement, we can perform the following controlled unitary decoding:

$$U_{\text{canocnical, dec}} = \sum_{\mathbf{a},\mathbf{b}_{\text{eb}},\mathbf{a}_{\text{eb}}} \{X^{-1}(\alpha(\mathbf{a},\mathbf{a}_{\text{eb}},\mathbf{b}_{\text{eb}}))Z^{-1}(\beta(\mathbf{a},\mathbf{a}_{\text{eb}},\mathbf{b}_{\text{eb}}))$$
$$\otimes |\mathbf{a}\rangle\langle\mathbf{a}| \otimes I \otimes |\mathbf{a}_{\text{eb}},\mathbf{b}_{\text{eb}}\rangle\langle\mathbf{a}_{\text{eb}},\mathbf{b}_{\text{eb}}|\}. \tag{9.44}$$

The canonical EA-OQECC can also be described by using the stabilizer formalism interpretation. Let $S_{\text{canonical}}$ be the non-Abelian group generated by

$$S_{\text{canonical}} = \langle S_{\text{canonical, I}}, S_{\text{canonical, E}}, S_{\text{canonical, G}} \rangle, \tag{9.45}$$

where $S_{\text{canonical,I}}$ denotes the isotropic subgroup, $S_{\text{canonical,E}}$ the entanglement subgroup, and $S_{\text{canonical,G}}$ the gauge subgroup, given respectively by

$$\begin{aligned} S_{\text{canonical, I}} &= \langle Z_1, \ldots, Z_s \rangle, \\ S_{\text{canonical, E}} &= \langle Z_{s+1}, \ldots, Z_{s+e}, X_{s+1}, \ldots, X_{s+e} \rangle, \\ S_{\text{canonical, G}} &= \langle Z_{s+e+1}, \ldots, Z_{s+e+R}, X_{s+e+1}, \ldots, X_{s+e+R} \rangle. \end{aligned} \tag{9.46}$$

By using the finite geometry representation, the generators of (9.45) and (9.46) can be arranged as follows:

$$A_{\text{EA-OQECC,canonical}} = \left[\begin{array}{cccc|cccc} \overbrace{}^{K} & \overbrace{}^{s} & \overbrace{}^{R} & \overbrace{}^{e} & \overbrace{}^{K} & \overbrace{}^{s} & \overbrace{}^{R} & \overbrace{}^{e} \\ \mathbf{0} & \mathbf{0} & \mathbf{0} & \mathbf{0} & \mathbf{0} & I_s & \mathbf{0} & \mathbf{0} \\ \mathbf{0} & \mathbf{0} & \mathbf{0} & \mathbf{0} & \mathbf{0} & \mathbf{0} & I_R & \mathbf{0} \\ \mathbf{0} & \mathbf{0} & I_R & \mathbf{0} & \mathbf{0} & \mathbf{0} & \mathbf{0} & \mathbf{0} \\ \mathbf{0} & \mathbf{0} & \mathbf{0} & \mathbf{0} & \mathbf{0} & \mathbf{0} & \mathbf{0} & I_e \\ \mathbf{0} & \mathbf{0} & \mathbf{0} & I_e & \mathbf{0} & \mathbf{0} & \mathbf{0} & \mathbf{0} \end{array}\right]. \tag{9.47}$$

The EA-OQECC can correct the following set of errors:

$$\begin{aligned} \mathcal{N} = \Big\{ E_m \Big| \, \forall E_1, E_2 \Rightarrow E_2^\dagger E_1 \in \langle S_{\text{canonical, I}}, S_{\text{canonical, G}} \rangle \\ \cup (G_N - C(\langle S_{\text{canonical, I}}, S_{\text{canonical, G}} \rangle)) \Big\}. \end{aligned} \tag{9.48}$$

Using a similar approach as in Section 9.3, we can establish the following bijection between canonical EA-OQECC described by $S_{\text{canonical}} = \langle S_{\text{canonical, I}}, S_{\text{canonical, E}}, S_{\text{canonical, G}} \rangle$, and a general EA-OQECC characterized by $S = \langle S_{\text{I}}, S_{\text{E}}, S_{\text{G}} \rangle$, with $S = U S_{\text{canonical}} U^{-1}$ (or equivalently $S_{\text{canonical}} = U^{-1} S U$), where U is a corresponding unitary operator. This [N,K,R,e] EA-OQECC, illustrated in Figure 9.7b, is defined by the encoding–decoding pair (E,D) with the property that it can correct any error from the following set of errors:

$$\mathcal{N} = \Big\{ E_m \Big| \, \forall E_1, E_2 \Rightarrow E_2^\dagger E_1 \in \langle S_{\text{I}}, S_{\text{G}} \rangle \cup (G_N - C(\langle S_{\text{I}}, S_{\text{G}} \rangle)) \Big\}. \tag{9.49}$$

The EA-OQECC is very flexible, as it can be transformed into another code by adding or removing certain stabilizer elements. It can be shown that an [$N,K,R,e;D_1$] can be transformed into an [$N,K+R,0,e;D_2$] or an [$N,K,0,e;D_3$] code. The minimum distance of EQ-OQECCs is defined in fashion similar to that in previous sections. For encoder implementation, we can use the conjugation method, in which we employ the gates described in Section 9.4 to transform the EA-OQECC into the corresponding canonical form (9.47). By applying this sequence of gates in the opposite order we obtain the encoding circuit of the corresponding EA-OQECC.

9.6.1 EA-OQECCs Derived from Classical Binary and Quaternary Codes

The EA-OQECCs can be derived from the EA codes described in Section 9.3, by properly rearranging and combining the stabilizers.

Example. Let us observe the [8,1,1;3] EA code, described by the following group:

$$S_{\mathrm{EA}} = \langle S_{\mathrm{I}}, S_{\mathrm{E}} \rangle,\quad S_{\mathrm{I}} = \langle Z_1Z_2, Z_1Z_3, Z_4Z_5, Z_4Z_6, Z_7Z_8, X_1X_2X_3X_4X_5X_6 \rangle,$$
$$S_{\mathrm{E}} = \langle Z_8, X_1X_2X_3X_7X_8 \rangle.$$

The corresponding Pauli encoded operators are given by $\{Z_1Z_4Z_8, X_4X_5X_6\}$. By using this EA code we can derive the following $[8,1,R=2,e=1;3]$ EA-OQECC:

$$S_{\mathrm{EA}} = \langle S_{\mathrm{I}}, S_{\mathrm{E}}, S_{\mathrm{G}} \rangle,\quad S_{\mathrm{I}} = \langle Z_1Z_2Z_4Z_5, Z_1Z_3Z_4Z_6, Z_7Z_8, X_1X_2X_3X_4X_5X_6 \rangle,$$
$$S_{\mathrm{E}} = \langle Z_8, X_1X_2X_3X_7X_8 \rangle, S_{\mathrm{G}} = \langle Z_1Z_2, X_2X_5, Z_4Z_6, X_3X_6 \rangle.$$

For CSS-like codes, the number of ebits can be determined by $e = \mathrm{rank}(\boldsymbol{H}\boldsymbol{H}^{\mathrm{T}})$, where $\boldsymbol{H}$ is the parity-check matrix of classical code. For example, the BCH codes described in Chapter 6 can be used in the design of EA codes. The BCH code over GF(2^m) has codeword length $n = 2^m - 1$, minimum distance $d \geq 2t+1$, number of parity bits $n-k \leq mt$, and its parity-check matrix is given by

$$\boldsymbol{H} = \begin{bmatrix} \alpha^{n-1} & \alpha^{n-2} & \cdots & \alpha & 1 \\ \alpha^{3(n-1)} & \alpha^{3(n-2)} & \cdots & \alpha^3 & 1 \\ \alpha^{5(n-1)} & \alpha^{5(n-2)} & \cdots & \alpha^5 & 1 \\ \cdots & \cdots & \cdots & \cdots & \cdots \\ \alpha^{(2t-1)(n-1)} & \alpha^{(2t-1)(n-2)} & \cdots & \alpha^{2t-1} & 1 \end{bmatrix}. \tag{9.50}$$

In Chapter 6, we described how to obtain a binary representation of this parity-check matrix. For example, for $m = 6$, the number of ebits required is $e = \mathrm{rank}(\boldsymbol{H}\boldsymbol{H}^{\mathrm{T}}) = 6$. This BCH (63,39;9) code, therefore, can be used to design $[n = 63, 2k - n + e = 21, e = 6]$ EA code. By inspection we can find that the last six rows of the binary matrix $\boldsymbol{H}$ are sympletic pairs that form an entanglement group. The gauge group can be formed by removing one sympletic pair at a time from the entanglement group and adding it to the gauge group. Clearly, the following set of quantum codes can be obtained using this approach: $\{[63,21,R=0,e=6]$, $[63,21,R=1,e=5]$, $[63,21,R=2,e=4]$, $[63,21,R=3,e=3]$, $[63,21,R=4,e=2]$, $[63,21,R=5,e=1]$, $[63,21,R=6,e=0]\}$. The first code from the set is a pure EA code, and the last code from the set is a pure subsystem code, while the quantum codes in between are EA-OQECCs.

The EA-OQECC codes can also be derived from classical low-density parity-check (LDPC) codes. Chapter 10 considers the different classes of quantum codes derived from classical LDPC codes [3,29–33].

Quaternary classical codes can also be used in the design of EA-QEECCs. The first step is to start with the corresponding EA code, using the concepts outlined in

Section 9.3. The second step is to carefully move some of the sympletic pairs from the entanglement group to the gauge group so that the minimum distance is not affected that much. Let us consider the following (15,10,4) quaternary code [4]:

$$\boldsymbol{H}_4 = \begin{bmatrix} 1 & 0 & 0 & 0 & 1 & 1 & \alpha^2 & 0 & 1 & \alpha^2 & 0 & \alpha & \alpha^2 & 1 & 0 \\ 0 & 1 & 0 & 0 & 1 & 0 & \alpha & \alpha^2 & 1 & \alpha & 0 & 0 & 1 & \alpha & 1 \\ 0 & 0 & 1 & 0 & \alpha & \alpha^2 & 1 & \alpha & 1 & 0 & 0 & \alpha & 1 & \alpha^2 & \alpha \\ 0 & 0 & 0 & 1 & 1 & \alpha^2 & 0 & 1 & \alpha^2 & \alpha & 0 & \alpha^2 & 1 & 0 & \alpha^2 \\ 0 & 0 & 0 & 0 & 0 & 0 & 0 & 0 & 0 & 0 & 1 & 0 & 0 & 0 & 0 \end{bmatrix}, \tag{9.51}$$

where the elements of GF(4) are $\{0,1,\alpha,\alpha^2\}$, generated by the primitive polynomial $p(x)=x^2+x+1$. By using Eq. (9.16) we obtain the following quantum-check matrix of $[15,9,e=4;4]$ EA code:

$$\begin{aligned} \boldsymbol{A}_{\mathrm{EA}} &= f\left(\tilde{\boldsymbol{H}}_4\right) = f\left(\begin{bmatrix} \alpha \boldsymbol{H}_4 \\ \alpha^2 \boldsymbol{H}_4 \end{bmatrix}\right) \\ &= \begin{bmatrix} X & I & I & I & X & X & Y & I & Z & Y & I & Z & Y & X & I \\ I & X & I & I & X & I & Z & Y & X & Z & I & I & X & Z & X \\ I & I & X & I & Z & Y & X & Z & X & I & I & Z & X & Y & Z \\ I & I & I & X & X & Y & I & X & Y & Z & I & Y & X & I & Y \\ I & I & I & I & I & I & I & I & I & I & X & I & I & I & I \\ Z & I & I & I & Z & Z & X & I & Z & X & I & Y & X & Z & I \\ I & Z & I & I & Z & I & Y & X & Z & Y & I & I & Z & Y & Z \\ I & I & Z & I & Y & X & Z & Y & Z & I & I & Y & Z & X & Y \\ I & I & I & Z & Z & X & I & Z & X & Y & I & X & Z & I & X \\ I & I & I & I & I & I & I & I & I & I & Z & I & I & I & I \end{bmatrix}. \end{aligned} \tag{9.52}$$

From Eq. (9.52) it is clear that the stabilizers in the third and fourth rows commute, and therefore they represent the isotropic subgroup. The remaining eight anti-commute and represent the entanglement subgroup. By removing any two anti-commuting pairs from the entanglement subgroup, say the first two, and moving them to the gauge subgroup, we obtain the $[15,9,e=3,R=1]$ EA-OQECC. By using the MAGMA approach [34], we can identify an EA-QECC derived from (9.52) having the largest possible minimum distance.

9.7 SUMMARY

This chapter has considered entanglement-assisted quantum error correction codes, which use pre-existing entanglement between transmitter and receiver to improve the reliability of transmission. A key advantage of EA quantum codes compared to CSS codes is that EA quantum codes do not impose the dual-containing constraint. Therefore, arbitrary classical codes can be used to design EA quantum codes. The number of required ebits has been determined by $e=\mathrm{rank}(\boldsymbol{H}\boldsymbol{H}^{\mathrm{T}})$, where $\boldsymbol{H}$ is the parity-check matrix of the corresponding classical code. A general description of entanglement-assisted quantum error correction was provided in Section 9.1. In Section 9.2, we studied entanglement-assisted canonical code and its error correction capability. In Section 9.3, the concept of EA canonical code to arbitrary EA code was generalized. We also described how to design EA codes from classical codes, in particular classical quaternary codes. In Section 9.4, encoding for EA quantum codes was discussed. In Section 9.5, the concept of operator quantum error correction, also known as subsystem codes, was introduced. Entanglement-assisted operator quantum error correction was discussed in Section 9.6.

In the following section, we provide a set of problems that will help the reader gain better understanding of the material in this chapter.

9.8 PROBLEMS

1. If V is an arbitrary subgroup of G_N of order 2^M, then there exists a set of generators $\{\overline{X}_{p+1},\ldots,\overline{X}_{p+q};\overline{Z}_1,\ldots,\overline{Z}_{p+q}\}$ satisfying similar commutation properties to Eq. (8.28) of the previous chapter:

$$[\overline{X}_m,\overline{X}_n]=[\overline{Z}_m,\overline{Z}_n]=0\ \ \forall m,n;\quad [\overline{X}_m,\overline{Z}_n]=0\ \ \forall m\neq n;$$
$$\{\overline{X}_m,\overline{Z}_n\}=0\ \ \forall m=n.$$

   Prove this claim.

2. If there exists bijection between two groups V and S that preserves their commutation relations, prove that $\forall\ v\in V$ there exists corresponding $s\in S$ such that $v=UsU^{-1}$ (up to a general phase constant).

3. As a generalization of the EA quantum codes represented above, we can design a *continuous-variable* EA quantum code given by

$$A_{\text{EA}}=(A_x|A_z),$$

   where A_{EA} is an $(n-k)\times 2n$ real matrix, and both A_x and A_z are $(n-k)\times n$ real matrices. The number of required entangled states is

$$e=\text{rank}\{A_zA_x^{\text{T}}-A_xA_z^{\text{T}}\}/2.$$

   Prove this claim.

4. A formula similar to that from the previous example can be derived for EA qudit codes, by replacing the subtraction operation by subtraction mod p (where p is a prime) as follows:

$$e=\text{rank}\{(A_zA_x^{\text{T}}-A_xA_z^{\text{T}})\text{mod}\,p\}/2.$$

   The quantum-check matrix of this qudit code is given by $A_{\text{EA}}=(A_x|A_z)$, with matrix elements being from finite field $F_p=\{0,1,\ldots,p-1\}$. Notice that error operators need to be redefined as it was done in the previous chapter. The corresponding entanglement qudits (edits) have the form $\sum_{m=0}^{p-1}|m\rangle|m\rangle/\sqrt{p}$. Prove the claims. Can you generalize the EA qudit code design over GF(q) ($q=p^m$, $m>1$)?

5. This problem concerns extended EA codes. If an EA $[N,K,e;D]$ code exists prove that an extended $[N+1,K-1,e';D']$ also exists for some e' and $D'\geq D$. Provide the quantum-check matrix of the extended EA code, assuming that the quantum-check matrix of the original EA code is known.

6. In certain quantum technologies, the CNOT gate is challenging to implement. For EA code given by the quantum-check matrix below, determine the encoding circuit without using any SWAP operators because they require the use of three CNOT gates.

$$\mathbf{A}_{\mathrm{EA}}=\left[\begin{array}{cccc|cccc} 0&1&0&0&1&0&1&0\\ 0&0&0&0&1&1&0&1\\ 1&1&1&0&0&1&0&0\\ 1&1&0&1&0&0&0&0 \end{array}\right].$$

7. By using the standard form method, conjugation method, and sympletic Gram–Schmidt method for EA code encoder implementation, described by the quantum-check matrix below, provide the corresponding encoding circuits. Discuss the complexity of the corresponding realizations.

$$\mathbf{A}_{\mathrm{EA}}=\begin{bmatrix} 0&0&1&0&1&0&1\\ 1&0&0&1&1&0&0\\ 0&1&0&1&0&0&1\\ 1&0&0&0&0&1&1\\ 0&0&1&1&0&1&0\\ 1&1&1&0&0&0&0\\ 0&1&0&0&1&1&0 \end{bmatrix}.$$

8. Prove that the canonical OQECC can correct the following set of errors:

$$\mathcal{N}=\left\{E_m\,\middle|\,\forall E_1,E_2\Rightarrow E_2^{\dagger}E_1\in S_{\text{canonical, OQECC}}\cup\left(G_N-C\left(S_{\text{canonical, OQECC, I}}\right)\right)\right\}.$$

9. Prove that $[N,K,R,D]$ OQECC code has the following properties:

(i) It can correct any error for the following set of errors:

$$\mathcal{N}=\left\{E_m\,\middle|\,\forall E_1,E_2\Rightarrow E_2^{\dagger}E_1\in S_{\text{OQECC}}\cup\left(G_N-C\left(S_{\text{OQECC, I}}\right)\right)\right\}.$$

(ii) The code space C_{OQECC} is a simultaneous eigenspace of $S_{\text{OQECC,I}}$.

(iii) To decode, the error syndrome is obtained by simultaneously measuring the observables from $S_{\text{OQECC,I}}$.

10. Prove that canonical $[N,K,R,e]$ EA-OQECC can correct the following set of errors:

$$\mathcal{N}=\left\{E_m \middle| \forall E_1,E_2 \Rightarrow E_2^{\dagger}E_1 \in \langle S_{\text{canonical, I}}, S_{\text{canonical, G}}\rangle \right.$$
$$\left. \cup (G_N - C(\langle S_{\text{canonical, I}}, S_{\text{canonical, G}}\rangle))\right\}.$$

Prove also that $[N,K,R,e]$ EA-OQECC can correct the following set of errors:

$$\mathcal{N}=\left\{E_m \middle| \forall E_1,E_2 \Rightarrow E_2^{\dagger}E_1 \in \langle S_{\text{I}}, S_{\text{G}}\rangle \cup (G_N - C(\langle S_{\text{I}}, S_{\text{G}}\rangle))\right\}.$$

11. Prove that an $[N,K,R,e;D_1]$ EA-OQECC can be transformed into an $[N,K+R,0,e;D_2]$ or an $[N,K,0,e;D_3]$ code. Show that the minimum distances satisfy the inequalities $D_2 \le D_1 \le D_3$.

12. The (15,7) BCH code has the following parity-check matrix:

$$\boldsymbol{H}=\begin{bmatrix}
1&1&1&1&0&1&0&1&1&0&0&1&0&0&0\\
0&1&1&1&1&0&1&0&1&1&0&0&1&0&0\\
0&0&1&1&1&1&0&1&0&1&1&0&0&1&0\\
1&1&1&0&1&0&1&1&0&0&1&0&0&0&1\\
1&1&1&1&0&1&1&1&1&0&1&1&1&1&0\\
1&0&1&0&0&1&0&1&0&0&1&0&1&0&0\\
1&1&0&0&0&1&1&0&0&0&1&1&0&0&0\\
1&0&0&0&1&1&0&0&0&1&1&0&0&0&1
\end{bmatrix}.$$

Determine the parameters of the corresponding EA code. Describe all possible EA-OQECCs that can be derived from this code.

13. Let us observe an EA code described by the following quantum-check matrix:

$$\boldsymbol{A}_{\text{EA}}=\begin{bmatrix}
X&I&I&I&X&X&Y&I&Z&Y&I&Z&Y&X&I\\
I&X&I&I&X&I&Z&Y&X&Z&I&I&X&Z&X\\
I&I&X&I&Z&Y&X&Z&X&I&I&Z&X&Y&Z\\
I&I&I&X&X&Y&I&X&Y&Z&I&Y&X&I&Y\\
I&I&I&I&I&I&I&I&I&I&X&I&I&I&I\\
Z&I&I&I&Z&Z&X&I&Z&X&I&Y&X&Z&I\\
I&Z&I&I&Z&I&Y&X&Z&Y&I&I&Z&Y&Z\\
I&I&Z&I&Y&X&Z&Y&Z&I&I&Y&Z&X&Y\\
I&I&I&Z&Z&X&I&Z&X&Y&I&X&Z&I&X\\
I&I&I&I&I&I&I&I&I&I&Z&I&I&I&I
\end{bmatrix}.$$

By using this quantum-check matrix determine an EA-OQECC of largest possible minimum distance. What are the parameters of this code? By using the conjugation method provide the corresponding encoding circuit.

References

[1] T. Brun, I. Devetak, M.H. Hsieh, Correcting quantum errors with entanglement, Science 314 (20 October 2006) 436–439.

[2] M.-H. Hsieh, I. Devetak, T. Brun, General entanglement-assisted quantum error correcting codes, Phys. Rev. A 76 (19 December 2007), 062313-1–062313-7.

[3] I.B. Djordjevic, Photonic entanglement-assisted quantum low-density parity-check encoders and decoders, Opt. Lett. 35 (9) (1 May 2010) 1464–1466.

[4] M.-H. Hsieh, Entanglement-Assisted Coding Theory. Ph.D. dissertation, University of Southern California, August 2008.

[5] I. Devetak, T.A. Brun, M.-H. Hsieh, Entanglement-assisted quantum error-correcting codes, in: V. Sidoravičius (Ed.), New Trends in Mathematical Physics, Selected Contributions of the XVth International Congress on Mathematical Physics, Springer, 2009, pp. 161–172.

[6] M.-H. Hsieh, W.-T. Yen, L.-Y. Hsu, High performance entanglement-assisted quantum LDPC codes need little entanglement, IEEE Trans. Inform. Theory 57 (3) (March 2011) 1761–1769.

[7] M.M. Wilde, Quantum Coding with Entanglement. Ph.D. dissertation, University of Southern California, August 2008.

[8] M.M. Wilde, T.A. Brun, Optimal entanglement formulas for entanglement-assisted quantum coding, Phys. Rev. A 77 (2008), paper 064302.

[9] M.M. Wilde, H. Krovi, T.A. Brun, Entanglement-assisted quantum error correction with linear optics, Phys. Rev. A 76 (2007), paper 052308.

[10] M.M. Wilde, T.A. Brun, Protecting quantum information with entanglement and noisy optical modes, Quantum Inform. Process 8 (2009) 401–413.

[11] M.M. Wilde, T.A. Brun, Entanglement-assisted quantum convolutional coding, Phys. Rev. A 81 (2010), paper 042333.

[12] M.M. Wilde, D. Fattal, Nonlocal quantum information in bipartite quantum error correction, Quantum Inform. Process 9 (2010) 591–610.

[13] Y. Fujiwara, D. Clark, P. Vandendriessche, M. De Boeck, V.D. Tonchev, Entanglement-assisted quantum low-density parity-check codes, Phys. Rev. A 82 (4) (2010), paper 042338.

[14] C.H. Bennett, S.J. Wiesner, Communication via one- and two-particle operators on Einstein–Podolsky–Rosen states, Phys. Rev. Lett. 69 (1992) 2881–2884.

[15] A. Harrow, Superdense coding of quantum states, Phys. Rev. Lett. 92 (18) (7 May 2004). 187901-1–187901-4.

[16] S.A. Aly, A. Klappenecker, Subsystem code constructions, Proc. ISIT 2008, Toronto, Canada, (6–11 July 2008), pp. 369–373.

[17] S. Bravyi, Subsystem codes with spatially local generators, Phys. Rev. A 83 (2011). 012320-1–012320-9.

[18] A. Klappenecker, P.K. Sarvepalli, On subsystem codes beating the quantum Hamming or Singleton bound, Proc. Math. Phys. Eng. Sci. 463 (2087) (8 November 2007) 2887–2905.

[19] M.M. Wilde, Quantum-shift-register circuits, Phys. Rev. A 79 (6) (2009), paper 062325.

[20] S. Taghavi, T.A. Brun, D.A. Lidar, Optimized entanglement-assisted quantum error correction, Phys. Rev. A 82 (4) (2010), paper 042321.

[21] D. Gottesman, Stabilizer Codes and Quantum Error Correction. Ph.D. dissertation, California Institute of Technology, Pasadena, CA, 1997.

[22] R. Cleve, D. Gottesman, Efficient computations of encoding for quantum error correction, Phys. Rev. A 56 (July 1997) 76−82.

[23] M. Grassl, M. Rötteler, T. Beth, Efficient quantum circuits for non-qubit quantum error-correcting codes, Int. J. Found. Comput. Sci. 14 (5) (2003) 757−775.

[24] P.K. Sarvepalli, Quantum Stabilizer Codes and Beyond. Ph.D. dissertation, Texas A&M University, August 2008.

[25] D. Bacon, Operator quantum error correcting subsystems for self correcting quantum memories, Phys. Rev. A 73 (1) (30 January 2006). 012340-1−012340-13.

[26] D. Bacon, A. Casaccino, Quantum error correcting subsystem codes from two classical linear codes, Proc. 44th Annual Allerton Conf., Allerton House, UIUC, IL (27−29 September 2006), pp. 520−527.

[27] D. Kribs, R. Laflamme, D. Poulin, Unified and generalized approach to quantum error correction, Phys. Rev. Lett. 94 (18) (9 May 2005), paper 180501.

[28] M.A. Nielsen, D. Poulin, Algebraic and information-theoretic conditions for operator quantum error correction, Phys. Rev. A 75 (21 June 2007), paper 064304.

[29] I.B. Djordjevic, Quantum LDPC codes from balanced incomplete block designs, IEEE Commun. Lett. 12 (May 2008) 389−391.

[30] I.B. Djordjevic, Photonic quantum dual-containing LDPC encoders and decoders, IEEE Photon. Technol. Lett. 21 (13) (1 July 2009) 842−844.

[31] T. Camara, H. Ollivier, J.-P. Tillich, A class of quantum LDPC codes: Construction and performances under iterative decoding, Proc. ISIT 2007 (24−29 June 2007) 811−815.

[32] S. Lin, S. Zhao, A class of quantum irregular LDPC codes constructed from difference family, Proc. ICSP (24−28 October 2010) 1585−1588.

[33] Y. Fujiwara1, D. Clark, P. Vandendriessche, M. De Boeck, V.D. Tonchev, Entanglement-assisted quantum low-density parity-check codes, Phys. Rev. A 82 (4) (2010), paper 042338.

[34] W. Bosma, J.J. Cannon, C. Playoust, The MAGMA algebra system I: The user language, J. Symb. Comput. 4 (1997) 235−266.

This page intentionally left blank

CHAPTER

Quantum Low-Density Parity-Check Codes

10

CHAPTER OUTLINE

This chapter is concerned with quantum low-density parity-check (LDPC) codes, which have many advantages compared to other classes of quantum codes due to the sparseness of their quantum-check matrices. Both semi-random and structured quantum LDPC codes are described. Key advantages of structured quantum LDPC codes compared to other codes include: (i) regular structure in the corresponding parity-check (***H***) matrices leads to low complexity encoders/decoders; and (ii) their sparse ***H*** matrices require a small number of interactions per qubit to determine the error location. The chapter begins with the introduction of classical LDPC codes in Section 10.1, their design and decoding algorithms. Dual-containing quantum LDPC codes are then described in Section 10.2. Section 10.3 covers entanglement-assisted quantum LDPC codes. Section 10.4 describes the probabilistic sum-product algorithm based on the quantum-check matrix instead of the classical parity-check matrix. Notice that encoders for quantum dual-containing LDPC codes can be implemented based on either the standard form method or conjugation method (described in Chapter 8). On the other hand, encoders for entanglement-assisted LDPC codes can be implemented as described in Chapter 9. Since there is no difference in encoder implementation of quantum LDPC codes compared to the other classes of quantum block codes described in the previous two chapters, we do not discuss encoder implementation here but concentrate instead on design and decoding algorithms for quantum LDPC codes.

Quantum Information Processing and Quantum Error Correction. DOI: 10.1016/B978-0-12-385491-9.00010-1

10.1 CLASSICAL LDPC CODES

Codes on graphs, such as turbo codes [1] and LDPC codes [2–27], have revolutionized communications, and are becoming standard in many applications. LDPC codes, invented by Gallager in the 1960s, are linear block codes for which the parity-check matrix has a low density of ones [4]. LDPC codes have generated great interests in the coding community recently [2–21], and this has resulted in a great deal of understanding of the different aspects of LDPC codes and their decoding process. An iterative LDPC decoder based on the sum-product algorithm (SPA) has been shown to achieve a performance as close as 0.0045dB to the Shannon limit [20]. The inherent low complexity of this decoder opens up avenues for use in different high-speed applications. Because of their low decoding complexity, LDPC codes are intensively studied for quantum applications as well.

If the parity-check matrix has a low density of ones and the number of ones per row and per column are both constant, the code is said to be a *regular LDPC* code. To facilitate implementation at high speed, we prefer the use of regular rather than irregular LDPC codes. The graphical representation of LDPC codes, known as the bipartite (Tanner) graph representation, is helpful in efficient description of LDPC decoding algorithms. A *bipartite (Tanner) graph* is a graph whose nodes may be separated into two classes (*variable* and *check* nodes), and where *undirected edges* may only connect two nodes not residing in the same class. The Tanner graph of a code is drawn according to the following rule: A check (function) node c is connected to a variable (bit) node v whenever element h_{cv} in a parity-check matrix $\boldsymbol{H}$ is a one. In an $m \times n$ parity-check matrix, there are $m = n - k$ check nodes and n variable nodes.

Example 10.1. As an illustrative example, consider the $\boldsymbol{H}$ matrix of the following code:

$$\boldsymbol{H} = \begin{bmatrix} 1 & 0 & 1 & 0 & 1 & 0 \\ 1 & 0 & 0 & 1 & 0 & 1 \\ 0 & 1 & 1 & 0 & 0 & 1 \\ 0 & 1 & 0 & 1 & 1 & 0 \end{bmatrix}.$$

For any valid codeword $\boldsymbol{x} = [x_0 x_1 \ldots x_{n-1}]$, the checks used to decode the codeword are written as:

- Equation (c_0): $x_0 + x_2 + x_4 = 0 \pmod 2$
- Equation (c_1): $x_0 + x_3 + x_5 = 0 \pmod 2$
- Equation (c_2): $x_1 + x_2 + x_5 = 0 \pmod 2$
- Equation (c_3): $x_1 + x_3 + x_4 = 0 \pmod 2$.

The bipartite graph (Tanner graph) representation of this code is given in Figure 10.1a. The circles represent the bit (variable) nodes while squares represent the check (function) nodes. For example, the variable nodes x_0, x_2, and x_4 are involved in equation (c_0), and are therefore connected to the check node c_0. A closed path in a bipartite graph comprising l edges that close back on themselves is called

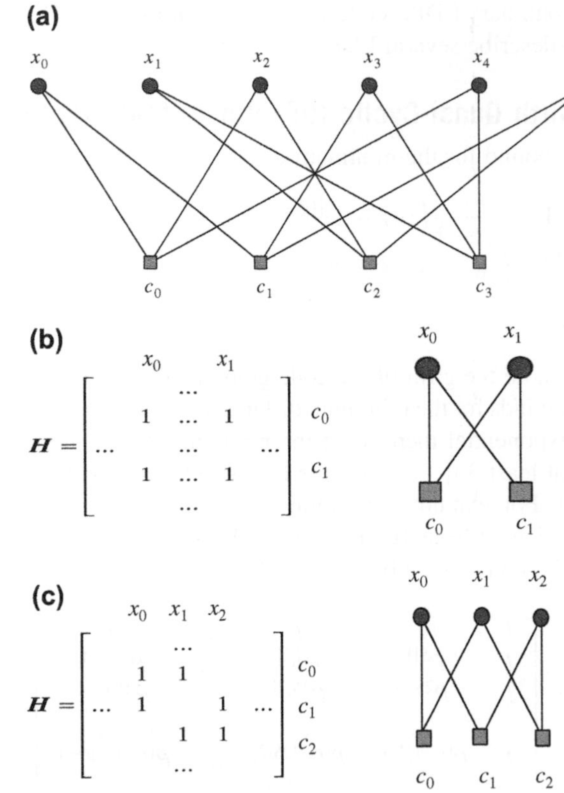

FIGURE 10.1

(a) Bipartite graph of (6, 2) code described by the $\boldsymbol{H}$ matrix. (b, c) Cycles in a Tanner graph: (b) cycle of length 4; (c) cycle of length 6.

a *cycle* of length l. The shortest cycle in the bipartite graph is called the *girth*. The girth influences the minimum distance of LDPC codes, correlates the extrinsic log-likelihood ratios, and therefore affects the decoding performance. The use of large-girth LDPC codes is preferable because the large girth increases the minimum distance and decorrelates the extrinsic info in the decoding process. To improve the iterative decoding performance, we have to avoid cycles of length 4, and preferably 6 as well. To check for the existence of short cycles, one has to search over the $\boldsymbol{H}$ matrix for the patterns shown in Figure 10.1b, c.

In the rest of this section, we describe a method for designing large-girth quasi-cyclic (QC) LDPC codes (subsection 10.1.1), and an efficient and simple variant of the sum-product algorithm (SPA) suitable for use in high-speed communications, namely the min-sum-with-correction-term algorithm (subsection 10.1.2). We also evaluate bit error rate (BER) performance of QC-LDPC codes against concatenated codes and turbo-product codes (TPCs) (subsection 10.1.3). Further, in Section 10.1.4

we introduce the nonbinary LDPC codes. Finally, in addition to QC-LDPC codes, in Section 10.1.5 we describe several LDPC code designs.

10.1.1 Large-Girth Quasi-Cyclic (QC) Binary LDPC Codes

Based on Tanner's bound for the minimum distance of an LDPC code [22]:

$$d \geq \begin{cases} 1 + \dfrac{w_c}{w_c - 2}\left((w_c - 1)^{\lfloor (g-2)/4 \rfloor} - 1\right), & g/2 = 2m+1 \\ 1 + \dfrac{w_c}{w_c - 2}\left((w_c - 1)^{\lfloor (g-2)/4 \rfloor} - 1\right) + (w_c - 1)^{\lfloor (g-2)/4 \rfloor}, & g/2 = 2m \end{cases} \tag{10.1}$$

(where g and w_c denote the girth of the code graph and the column weight respectively, and where d stands for the minimum distance of the code), it follows that large girth leads to an exponential increase in the minimum distance, provided that the column weight is at least 3 ($\lfloor\ \rfloor$ denotes the largest integer less than or equal to the enclosed quantity). For example, the minimum distance of girth-10 codes with column weight $r = 3$ is at least 10. The parity-check matrix of regular QC LDPC codes [23,24] can be represented by

$$\boldsymbol{H} = \begin{bmatrix} I & I & I & \dots & I \\ I & P^{S[1]} & P^{S[2]} & \dots & P^{S[c-1]} \\ I & P^{2S[1]} & P^{2S[2]} & \dots & P^{2S[c-1]} \\ \dots & \dots & \dots & \dots & \dots \\ I & P^{(r-1)S[1]} & P^{(r-1)S[2]} & \dots & P^{(r-1)S[c-1]} \end{bmatrix}, \tag{10.2}$$

where I is a $B \times B$ (B is a prime number) identity matrix, P is a $B \times B$ permutation matrix given by $P = (p_{ij})_{B\times B}$, $p_{i,i+1} = p_{B,1} = 1$ (zero otherwise), and where r and c represent the number of block rows and block columns respectively in (10.2). The set of integers S is to be carefully chosen from the set $\{0,1,\dots,B-1\}$ so that cycles of short length, in the corresponding Tanner (bipartite) graph representation of (10.2), are avoided. According to Theorem 2.1 in Ref. [24], we have to avoid cycles of length $2k$ ($k = 3$ or 4) defined by the following equation:

$$S[i_1]j_1 + S[i_2]j_2 + \dots + S[i_k]j_k = S[i_1]j_2 + S[i_2]j_3 + \dots + S[i_k]j_1 \bmod p, \tag{10.3}$$

where the closed path is defined by (i_1,j_1), (i_1,j_2), (i_2,j_2), (i_2,j_3), ..., (i_k,j_k), (i_k,j_1) with the pair of indices denoting row–column indices of permutation blocks in (10.2) such that $l_m \neq l_{m+1}$, $l_k \neq l_1$ ($m = 1,2,\dots,k$; $l \in \{i,j\}$). Therefore, we have to identify the sequence of integers $S[i] \in \{0,1,\dots,B-1\}$ ($i = 0,1,\dots,r-1$; $r < B$) not satisfying Eq. (10.3), which can be done either by computer search or in combinatorial fashion. For example, to design the QC LDPC codes in Ref. [18], we introduced the concept of the cyclic-invariant difference set (CIDS). The CIDS-based codes come naturally as girth-6 codes, and to increase the girth we had to selectively remove certain elements from a CIDS. The design of LDPC codes of rate above 0.8, column weight 3, and girth-10 using the CIDS approach is a very challenging and still an open problem. Instead, in our recent paper [23], we solved this problem by developing an efficient computer search algorithm. We add an integer at a time from the set $\{0,1,\ldots,B-1\}$ (not used before) to the initial set S and check if Eq. (10.3) is satisfied. If Eq. (10.3) is satisfied, we remove that integer from the set S and continue our search with another integer from set $\{0,1,\ldots,B-1\}$ until we exploit all the elements from $\{0,1,\ldots,B-1\}$. The code rate of these QC codes, R, is lower bounded by

$$R \geq \frac{|S|B - rB}{|S|B} = 1 - r/|S|, \tag{10.4}$$

and the codeword length is $|S|B$, where $|S|$ denotes the cardinality of set S. For a given code rate R_0, the number of elements from S to be used is $\lfloor r/(1-R_0) \rfloor$. With this algorithm, LDPC codes of arbitrary rate can be designed.

Example 10.2. By setting $B = 2311$, the set of integers to be used in (10.2) is obtained as $S = \{1, 2, 7, 14, 30, 51, 78, 104, 129, 212, 223, 318, 427, 600, 808\}$. The corresponding LDPC code has rate $R_0 = 1 - 3/15 = 0.8$, column weight 3, girth-10, and length $|S|B = 15 \cdot 2311 = 34{,}665$. In the example above, the initial set of integers was $S = \{1, 2, 7\}$, and the set of rows to be used in (10.2) is $\{1, 3, 6\}$. The use of a different initial set will result in a different set from that obtained above.

Example 10.3. By setting $B = 269$, the set S is obtained as $S = \{0, 2, 3, 5, 9, 11, 12, 14, 27, 29, 30, 32, 36, 38, 39, 41, 81, 83, 84, 86, 90, 92, 93, 95, 108, 110, 111, 113, 117, 119, 120, 122\}$. If 30 integers are used, the corresponding LDPC code has rate $R_0 = 1 - 3/30 = 0.9$, column weight 3, girth-8, and length $30 \bullet 269 = 8070$.

10.1.2 Decoding of Binary LDPC Codes

In this subsection, we describe the min-sum-with-correction-term decoding algorithm [25,26]. It is a simplified version of the original algorithm proposed by Gallager [4]. Gallager proposed a near-optimal iterative decoding algorithm for LDPC codes that computes the distributions of the variables in order to calculate the *a posteriori probability* (APP) of a bit v_i of a codeword $\boldsymbol{v} = [v_0\ v_1\ \ldots\ v_{n-1}]$ being equal to 1, given a received vector $\boldsymbol{y} = [y_0\ y_1\ \ldots\ y_{n-1}]$. This iterative decoding scheme involves passing the extrinsic info back and forth among the c nodes and the v nodes over the edges to update the distribution estimation. Each iteration in this scheme is composed of two half-iterations. In Figure 10.2, we illustrate both the first and second halves of an iteration of the algorithm. As an example, in Figure 10.2a, we show the message sent from v-node v_i to c-node c_j. The v_i node collects the information from the channel (y_i sample), in addition to extrinsic info from other c nodes connected to the v_i node, processes them and sends the extrinsic info (not already available info) to c_j. This extrinsic info contains the information about the probability $\Pr(c_i = b|y_0)$, where $b \in \{0,1\}$. This is performed on all c nodes connected to the v_i node. On the other hand, Figure 10.2b shows the extrinsic info sent from c-node c_i to v-node v_j, which contains the information about $\Pr(c_i$ equation is satisfied$|\boldsymbol{y})$. This is done repeatedly to all the c nodes connected to the v_i node.

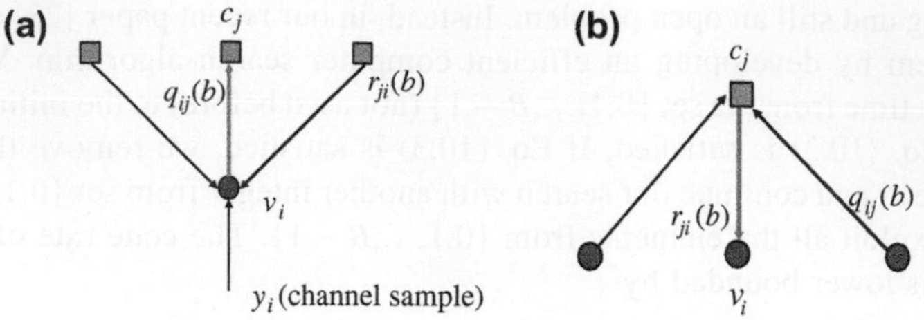

FIGURE 10.2

The half-iterations of the sum-product algorithm: (a) first half-iteration – extrinsic info sent from *v* nodes to *c* nodes; (b) second half-iteration – extrinsic info sent from *c* nodes to *v* nodes.

After this intuitive description, we describe the min-sum-with-correction-term algorithm in more detail [25] because of its simplicity and suitability for high-speed implementation. Generally, we can either compute the APP $\Pr(v_i|\mathbf{y})$ or the APP ratio $l(v_i) = \Pr(v_i = 0|\mathbf{y})/\Pr(v_i = 1|\mathbf{y})$, which is also referred to as the likelihood ratio. In the log-domain version of the sum-product algorithm, we replace these likelihood ratios with log-likelihood ratios (LLRs) due to the fact that the probability domain includes many multiplications that lead to numerical instabilities, whereas the computation using LLRs involves addition only. Moreover, the log-domain representation is more suitable for finite precision representation. Thus, we compute the LLRs using $L(v_i) = \log[\Pr(v_i = 0|\mathbf{y})/\ \Pr(v_i = 1|\mathbf{y})]$. For the final decision, if $L(v_i) > 0$ we decide in favor of 0 and if $L(v_i) < 0$ we decide in favor of 1. To further explain the algorithm, we introduce the following notation due to MacKay [27] (see also Ref. [25]):

$V_j = \{v \text{ nodes connected to } c\text{-node } c_j\}$
$V_j \backslash i = \{v \text{ nodes connected to } c\text{-node } c_j\} \backslash \{v\text{-node } v_i\}$
$C_i = \{c \text{ nodes connected to } v\text{-node } v_i\}$
$C_i \backslash j = \{c \text{ nodes connected to } v\text{-node } v_i\} \backslash \{c\text{-node } c_j\}$
$M_v(\sim i) = \{\text{messages from all } v \text{ nodes except node } v_i\}$
$M_c(\sim j) = \{\text{messages from all } c \text{ nodes except node } c_j\}$
$P_i = \Pr(v_i = 1|y_i)$
S_i = event that the check equations involving c_i are satisfied
$q_{ij}(b) = \Pr(v_i = b|S_i, y_i, M_c(\sim j))$
$r_{ji}(b) = \Pr(\text{check equation } c_j \text{ is satisfied}|v_i = b, M_v(\sim i))$.

In the log-domain version of the sum-product algorithm, all the calculations are performed in the log-domain as follows:

$$L(v_i) = \log\left[\frac{\Pr(v_i = 0|y_i)}{\Pr(v_i = 1|y_i)}\right],\quad L(r_{ji}) = \log\left[\frac{r_{ji}(0)}{r_{ji}(1)}\right],\quad L(q_{ji}) = \log\left[\frac{q_{ji}(0)}{q_{ji}(1)}\right]. \tag{10.5}$$

The algorithm starts with the initialization step, where we set $L(v_i)$ as follows:

$$
\begin{aligned}
L(v_i) &= (-1)^{y_i} \log\left(\frac{1-\varepsilon}{\varepsilon}\right), \quad \text{for BSC} \\
L(v_i) &= \frac{2y_i}{\sigma^2}, \quad \text{for binary input AWGN} \\
L(v_i) &= \log\left(\frac{\sigma_1}{\sigma_0}\right) - \frac{(y_i-\mu_0)^2}{2\sigma_0^2} + \frac{(y_i-\mu_1)^2}{2\sigma_1^2}, \quad \text{for BA-AWGN} \\
L(v_i) &= \log\left(\frac{\Pr(v_i=0|y_i)}{\Pr(v_i=1|y_i)}\right), \quad \text{for an abritrary channel,}
\end{aligned}
\tag{10.6}
$$

where ε is the probability of error in the binary symmetric channel (BSC), σ^2 is the variance of the distribution of the additive white Gaussian noise (AWGN), and μ_j and σ_j^2 ($j=0,1$) represent the mean and the variance of the Gaussian process corresponding to the bits $j=0,1$ of a binary asymmetric (BA)-AWGN channel. After initialization of $L(q_{ij})$, we calculate $L(r_{ji})$ as follows:

$$
\begin{aligned}
L(r_{ji}) &= L\left(\sum_{i' \in V_j \backslash i} b_{i'}\right) = L(\ldots \oplus b_k \oplus b_l \oplus b_m \oplus b_n \ldots) \\
&= \ldots L_k \boxplus L_l \boxplus L_m \boxplus L_n \boxplus \ldots,
\end{aligned}
\tag{10.7}
$$

where $\oplus$ denotes modulo-2 addition and $\boxplus$ denotes a pairwise computation defined by

$$
L_1 \boxplus L_2 = \prod_{k=1}^{2} \operatorname{sign}(L_k) \cdot \phi\left(\sum_{k=1}^{2} \phi(|L_k|)\right), \quad \phi(x) = -\log\tanh(x/2). \tag{10.8}
$$

Upon calculation of $L(r_{ji})$, we update

$$
L(q_{ij}) = L(v_i) + \sum_{j' \in C_i \backslash j} L(r_{j'i}), \quad L(Q_i) = L(v_i) + \sum_{j \in C_i} L(r_{ji}). \tag{10.9}
$$

Finally, the decision step is as follows:

$$
\hat{v}_i = \begin{cases} 1, & L(Q_i) < 0 \\ 0, & \text{otherwise.} \end{cases} \tag{10.10}
$$

If the syndrome equation $\hat{\boldsymbol{v}}\boldsymbol{H}^{\mathrm{T}} = \mathbf{0}$ is satisfied or the maximum number of iterations is reached, we stop; otherwise, we recalculate $L(r_{ji})$, update $L(q_{ij})$ and $L(Q_i)$, and check again. It is important to set the number of iterations high enough to ensure that most of the codewords are decoded correctly and low enough not to affect the processing time. It is important to mention that a decoder for good LDPC codes requires fewer number of iterations to guarantee successful decoding. The *Gallager log-domain SPA* can be formulated as follows:

0. *Initialization*: For $j = 0,1,\ldots,n-1$, initialize the messages to be sent from v-node i to c-node j to channel LLRs, namely $L(q_{ij}) = L(v_i)$.
1. *c-node update rule*: For $j = 0,1,\ldots,n-k-1$, compute $L(r_{ji}) = \left(\prod_{i' \in C_j \backslash i} \alpha_{i'j}\right) \phi\left[\sum_{i' \in C_j \backslash i} \phi(\beta_{i'j})\right]$, where $\alpha_{ij} = \text{sign}[L(q_{ij})], \beta_{ij} = |L(q_{ij})|$, and $\phi(x) = -\log\tanh(x/2) = \log[(e^x+1)/(e^x-1)]$.
2. *v-node update rule*: For $i = 0,1,\ldots,n-1$, set $L(q_{ij}) = L(v_i) + \sum_{j' \in C_i \backslash j} L(r_{j'i})$ for all c nodes for which $h_{ji} = 1$.
3. *Bit decisions*: Update $L(Q_i)$ $(i = 0,\ldots,n-1)$ using $L(Q_i) = L(v_i) + \sum_{j \in C_i} L(r_{ji})$ and set $\hat{v}_i = 1$ when $L(Q_i) < 0$ (otherwise $\hat{v}_i = 0$). If $\hat{\boldsymbol{v}}\boldsymbol{H}^{\mathrm{T}} = \boldsymbol{0}$ or a predetermined number of iterations has been reached then stop, otherwise go to step 1.

Because the c-node update rule involves log and tanh functions, it is computationally intensive, and there exist many approximations. Very popular is the *min-sum-plus-correction-term approximation* [6]. It can be shown that the "box-plus" operator $\boxplus$ can also be calculated by

$$L_1 \boxplus L_2 = \prod_{k=1}^{2} \text{sign}(L_k) \cdot \min(|L_1|, |L_2|) + c(x,y), \tag{10.11}$$

where $c(x,y)$ denotes the correction factor defined by

$$c(x,y) = \log[1+\exp(-|x+y|)] - \log[1+\exp(-|x-y|)], \tag{10.12}$$

commonly implemented as a look-up table (LUT).

10.1.3 BER Performance of Binary LDPC Codes

The results of simulations for an AWGN channel model are given in Figure 10.3, where we compare large-girth LDPC codes (Figure 10.3a) against RS codes, concatenated RS codes, TPCs, and other classes of LDPC codes. In all simulation results in this section, we maintain double precision. For the LDPC(16935,13550) code, we also provide three- and four-bit fixed-point simulation results (see Figure 10.3a). Our results indicate that the four-bit representation performs comparably to the double-precision representation whereas the three-bit representation performs 0.27 dB worse than the double-precision representation at a BER of $2 \cdot 10^{-8}$. The girth-10 LDPC(24015,19212) code of rate 0.8 outperforms the concatenation RS(255,239) + RS(255,223) (of rate 0.82) by 3.35 dB and RS(255,239) by 4.75 dB, both at BER of 10^{-7}. The same LDPC code outperforms the projective geometry (PG) $(2,2^6)$-based LDPC(4161,3431) (of rate 0.825) of girth-6 by 1.49 dB at BER of 10^{-7}, and outperforms CIDS-based LDPC(4320,3242) of rate 0.75 and girth-8 LDPC codes by 0.25 dB. At BER of 10^{-10}, it outperforms

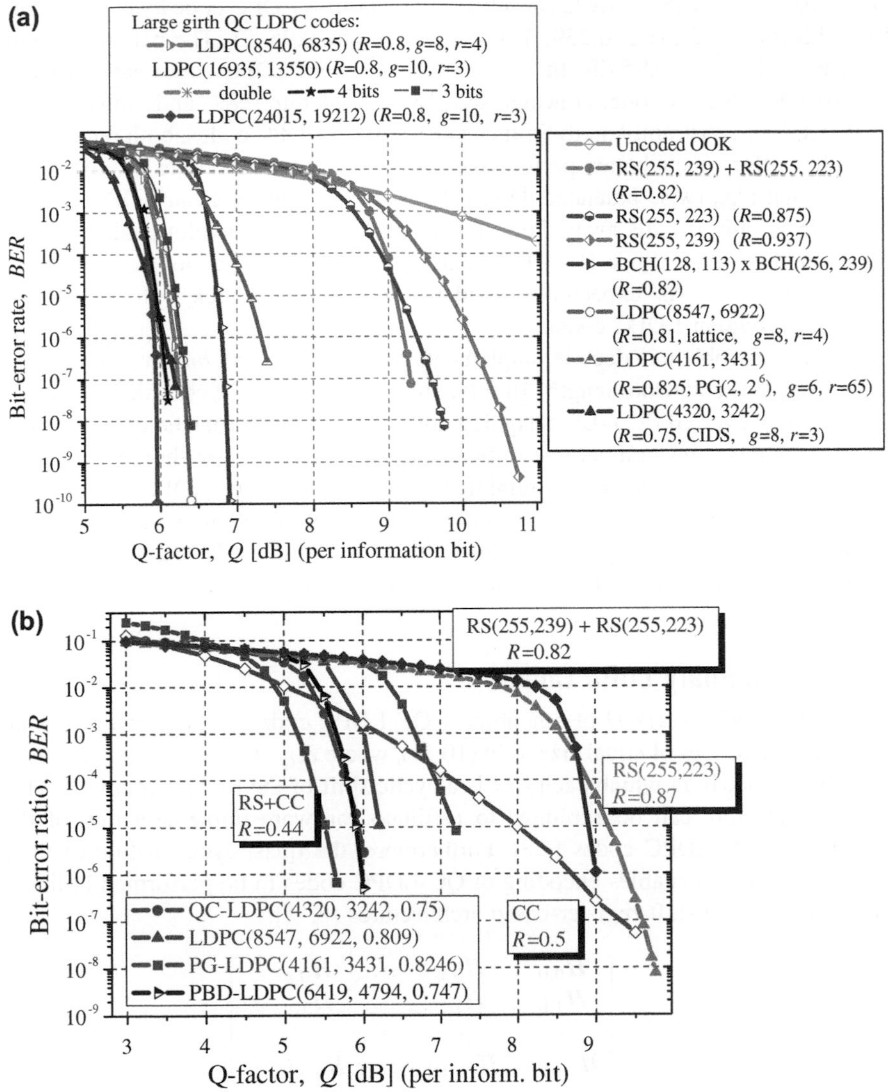

FIGURE 10.3

(a) Large-girth QC LDPC codes against RS codes, concatenated RS codes, TPCs, and girth-6 LDPC codes on an AWGN channel model. (b) LDPC codes versus convolutional, concatenated RS, and concatenation of convolutional and RS codes on an AWGN channel. Number of iterations in sum-product-with-correction-term algorithm was set to 25. The Q factor is defined as $Q = (\mu_1 - \mu_0)/(\sigma_1 + \sigma_0)$, where μ_i is the mean value corresponding to bit $i\,(i = 0,1)$ and σ_i is the corresponding standard deviation.

(After Ref. [2]; IEEE 2009; reprinted with permission.)

lattice-based LDPC(8547,6922) of rate 0.81 and girth-8 LDPC code by 0.44 dB, and BCH(128,113) × BCH(256,239) TPC of rate 0.82 by 0.95 dB. The net coding gain at BER of 10^{-12} is 10.95 dB. In Figure 10.3b, different LDPC codes are compared against RS(255,223) code, concatenated RS code of rate 0.82, and convolutional code (CC) (of constraint length 5). It can be seen that LDPC codes, both regular and irregular, offer much better performance than hard-decision codes. It should be noticed that a pairwise balanced design (PBD) [28]-based irregular LDPC code of rate 0.75 is only 0.4 dB away from the concatenation of convolutional RS codes (denoted in Figure 10.3b as RS + CC) with significantly lower code rate $R = 0.44$ at a BER of 10^{-6}. As expected, irregular LDPC codes (black colored curves) outperform regular LDPC codes.

The main problem in decoder implementation for large-girth *binary* LDPC codes is the excessive codeword length. To solve this problem, we will consider *nonbinary* LDPC codes over GF(2^m). By designing codes over higher-order fields, we aim to achieve coding gains comparable to binary LDPC codes but for shorter codeword lengths. Notice also that it is straightforward to relate 4-ary LDPC codes (over GF(4)) to entanglement-assisted quantum LDPC codes given by Eq. (9.16) in Chapter 9. Moreover, properly designed classical nonbinary LDPC codes can be used as nonbinary stabilizer codes, described in Section 8.6.

10.1.4 Nonbinary LDPC Codes

The parity-check matrix $\boldsymbol{H}$ of a nonbinary QC-LDPC code can be organized as an array of submatrices of equal size as in (10.13), where $\boldsymbol{H}_{i,j}$, $0 \le i < \gamma$, $0 \le j < \rho$, is a $B \times B$ submatrix in which each row is a cyclic shift of the row preceding it. This modular structure can be exploited to facilitate hardware implementation of the decoders of QC-LDPC codes [7,8]. Furthermore, the quasi-cyclic nature of their generator matrices enables encoding of QC-LDPC codes to be performed in linear time using simple shift-register-based architectures [8,11]:

$$\boldsymbol{H} = \begin{bmatrix} \boldsymbol{H}_{0,0} & \boldsymbol{H}_{0,1} & \cdots & \boldsymbol{H}_{0,\rho-1} \\ \boldsymbol{H}_{1,0} & \boldsymbol{H}_{1,1} & \cdots & \boldsymbol{H}_{1,\rho-1} \\ \vdots & \vdots & \ddots & \vdots \\ \boldsymbol{H}_{\gamma-1,0} & \boldsymbol{H}_{\gamma-1,1} & \cdots & \boldsymbol{H}_{\gamma-1,\rho-1} \end{bmatrix}. \tag{10.13}$$

If we select the entries of $\boldsymbol{H}$ from the binary field GF(2), as was done in the previous section, then the resulting QC-LDPC code is a binary LDPC code. On the other hand, if the selection is made from the Galois field of q elements denoted by GF(q), then we obtain a q-ary QC-LDPC code. In order to decode binary LDPC codes, as described above, an iterative message-passing algorithm is referred to as SPA. For nonbinary LDPC codes, a variant of the SPA known as the q-ary SPA (QSPA) is used [9]. When the field order is a power of 2, i.e. $q = 2^m$, where m is an integer and $m \ge 2$, a fast Fourier transform (FFT)-based implementation of QSPA, referred to as FFT-QSPA, significantly reduces the computational complexity of QSPA.

FFT-QSPA is further analyzed and improved in Ref. [10]. A mixed-domain FFT-QSPA implementation (MD-FFT-QSPA) aims to reduce the hardware implementation complexity by transforming multiplications in the probability domain into additions in the log domain whenever possible. It also avoids instability issues commonly faced in probability-domain implementations.

Following the code design discussed above, we generated (3,15)-regular, girth-8 LDPC codes over the fields $GF(2^p)$, where $0 \leq p \leq 7$. All the codes had a code rate (R) of at least 0.8 and hence an overhead $OH = (1/R - 1)$ of 25% or less. We compared the BER performances of these codes against each other and against some other well-known codes, such as RS(255,239), RS(255,223) and concatenated RS(255,239) + RS(255,223) codes, and BCH(128,113) × BCH(256,239) TPC. We used the binary AWGN (BI-AWGN) channel model in our simulations and set the maximum number of iterations to 50. In Figure 10.4a, we present the BER performances of the set of nonbinary LDPC codes discussed above. Using the figure, we can conclude that when we fix the girth of a nonbinary regular, rate-0.8 LDPC code at eight, increasing the field order above eight exacerbates the BER performance. In

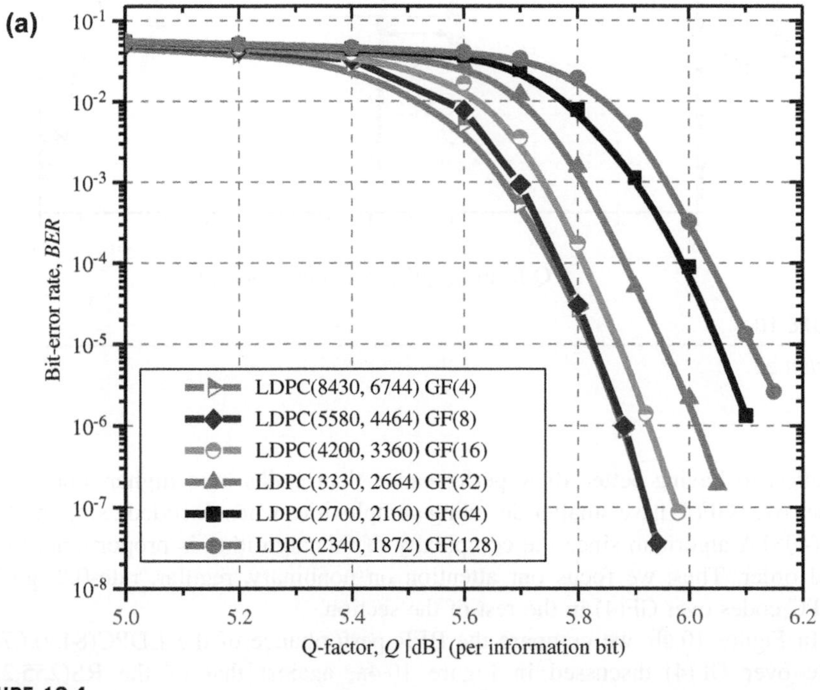

FIGURE 10.4

(a) Comparison of nonbinary, (3,15)-regular, girth-8 LDPC codes over a BI-AWGN channel. (b) Comparison of four-ary (3,15)-regular, girth-8 LDPC codes: a binary, girth-10 LDPC code, three RS codes, and a TPC code.

(After Ref. [2]; IEEE 2009; reprinted with permission.)

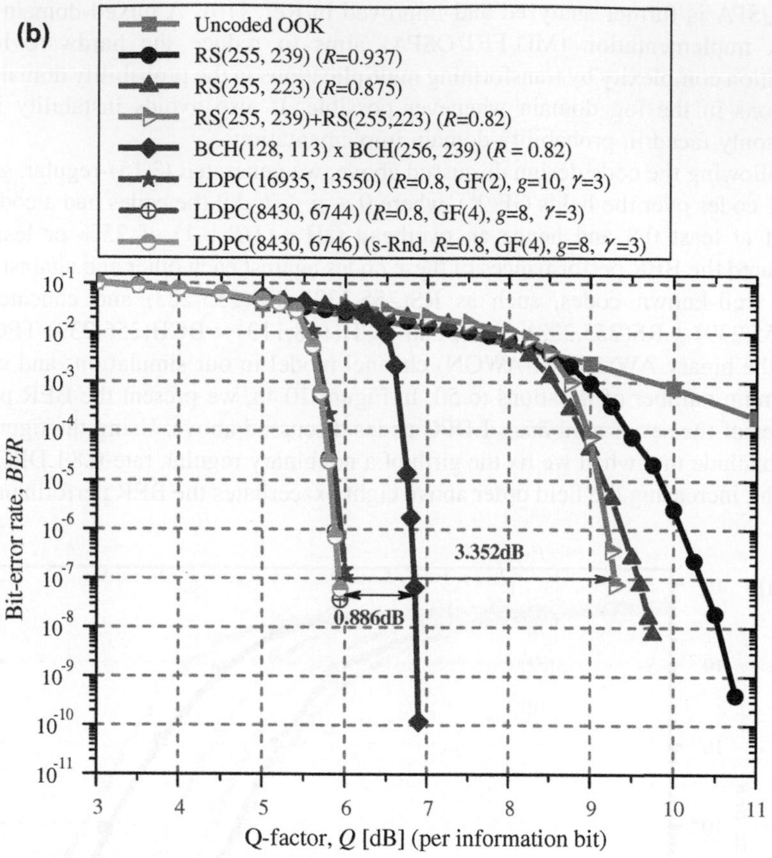

FIGURE 10.4

Continued

addition to having better BER performance than codes over higher-order fields, codes over GF(4) have smaller decoding complexities when decoded using the MD-FFT-QSPA algorithm since the complexity of this algorithm is proportional to the field order. Thus, we focus our attention on nonbinary, regular, rate-0.8, girth-8 LDPC codes over GF(4) in the rest of the section.

In Figure 10.4b, we compare the BER performance of the LDPC(8430,6744) code over GF(4) discussed in Figure 10.4a, against that of the RS(255,239) code, RS(255,223) code, RS(255,239) + RS(255,223) concatenation code, and BCH(128,113) × BCH(256,239) TPC. We observe that the LDPC code over GF(4) outperforms all of these codes by a significant margin. In particular, it provides an additional coding gain of 3.363 and 4.401 dB at a BER of 10^{-7} when compared to the concatenation code RS(255,239) + RS(255,223) and the RS(255,239) code respectively. Its coding gain improvement over BCH(128,113) × BCH(256,239) TPC is 0.886 dB at a BER of 4×10^{-8}. We also present in Figure 10.4b a competitive, binary, (3,15)-regular, LDPC(16935,13550) code proposed in Ref. [23]. We can see that the four-ary, (3,15)-regular, girth-8 LDPC(8430,6744) code beats the bit-length-matched binary LDPC code by a margin of 0.089 dB at a BER of 10^{-7}. More importantly, the complexity of the MD-FFT-QSPA used for decoding the nonbinary LDPC code is lower than the min-sum-with-correction-term algorithm [6] used for decoding the corresponding binary LDPC code. When the MD-FFT-QSPA is used for decoding a (γ,ρ)-regular q-ary LDPC (N/log q, K/log q) code, which is bit-length-matched to a (γ,ρ)-regular binary LDPC(N,K) code, the complexity is given by $(M/\log q)2\rho q(\log q + 1 - 1/(2\rho))$ additions, where $M = N - K$ is the number of check nodes in the binary code. On the other hand, to decode the bit-length-matched binary counterpart using the min-sum-with-correction-term algorithm [6], one needs $15M(\rho - 2)$ additions. Thus, a (3,15)-regular four-ary LDPC code requires 91.28% of the computational resources required for decoding a (3,15)-regular binary LDPC code of the same rate and bit length.

10.1.5 LDPC Code Design

The most obvious way to design LDPC codes is to construct a low-density parity-check matrix with prescribed properties. Some important designs, among others, include: (i) Gallager codes (semi-random construction) [4]; (ii) MacKay codes (semi-random construction) [26]; (iii) finite-geometry-based LDPC codes [29–31]; (iv) combinatorial design-based LDPC codes [32], (v) quasi-cyclic (array, block-circulant) LDPC codes [2,3,18,23,24] (and references therein); (vi) irregular LDPC codes [31]; and (vii) Tanner codes [22]. Quasi-cyclic LDPC codes of large girth have already been discussed in Section 10.1.1. In the rest of this section we describe briefly several classes of LDPC codes, namely Gallager, Tanner, and MacKay codes. Other classes of codes, commonly referred to as structured (because they have regular structure in their parity-check matrices, which can be exploited to facilitate hardware implementation of encoders and decoders), will be described in the corresponding sections relating to quantum LDPC codes. Notice that the basic design principles for both classical and quantum LDPC codes are very similar, and the same designs already used for classical LDPC code design can be used to design quantum LDPC codes. Therefore, various code designs from Chapter 17 of Ref. [21], with certain modifications, can be used for quantum LDPC code design as well.

10.1.5.1 Gallager Codes

The $\boldsymbol{H}$ matrix for Gallager code has the following general form:

$$\boldsymbol{H} = [\boldsymbol{H}_1 \quad \boldsymbol{H}_2 \quad \ldots \quad \boldsymbol{H}_{w_c}]^{\mathrm{T}}, \tag{10.14}$$

where $\boldsymbol{H}_1$ is a $p \times p \cdot w_r$ matrix of row weight w_r, and $\boldsymbol{H}_i$ are column-permuted versions of the $\boldsymbol{H}_1$ submatrix. The row weight of $\boldsymbol{H}$ is w_r, and column weight is w_c.

The permutations are carefully chosen to avoid cycles of length 4. The $\boldsymbol{H}$ matrix is obtained by computer search.

10.1.5.2 Tanner Codes and Generalized LDPC (GLDPC) Codes

In Tanner codes [22], each bit node is associated with a code bit and each check node is associated with a subcode whose length is equal to the degree of the node, which is illustrated in Figure 10.5. Notice that so-called generalized LDPC codes [33–37] were inspired by Tanner codes.

Example 10.4. Let us consider the following example:

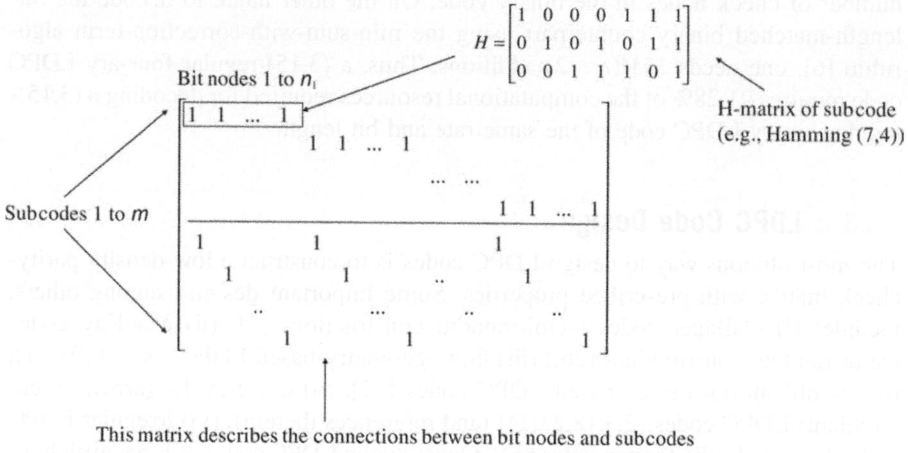

The Tanner code design in this example is performed in two stages. We first design the global code by starting from an identity matrix $I_{m/2}$ and replace every nonzero element with n_1 ones, and every zero element by n_1 zeros. The lower submatrix is obtained by concatenating the identity matrices I_{n_1}. In the second stage, we substitute all-one row vectors (of length n_1) with the parity-check matrix of local

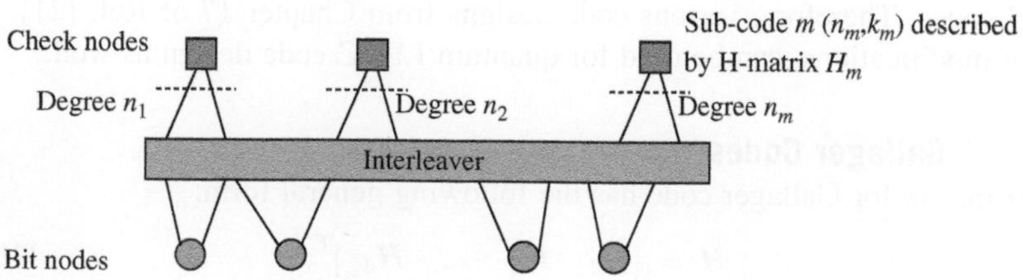

FIGURE 10.5

Tanner codes.

linear block code (n_1,k_1), such as Hamming, BCH, or RM code. The resulting parity-check matrix is used as the parity-check matrix of an LDPC code.

The GLDPC codes can be constructed in similar fashion. To construct a GLDPC code, one can replace each single parity-check equation of a global LDPC code by the parity-check matrix of a simple linear block code, known as the constituent (local) code. This construction was proposed by Lentmaier and Zigangirov [35], and we will refer to it as the LZ-GLDPC code construction. One illustrative example is shown in Figure 10.6. In another construction proposed by Boutros et al. [34], referred to here as B-GLDPC code construction, the parity-check matrix $\boldsymbol{H}$ is a sparse matrix partitioned into W submatrices, $\boldsymbol{H}_1,\ldots,\boldsymbol{H}_W$. The $\boldsymbol{H}_1$ submatrix is a block-diagonal matrix generated from an identity matrix by replacing the ones with a parity-check matrix $\boldsymbol{H}_0$ of a local code of codeword length n and dimension k:

$$\boldsymbol{H} = [\boldsymbol{H}_1^{\mathrm{T}} \quad \boldsymbol{H}_2^{\mathrm{T}} \quad \ldots \quad \boldsymbol{H}_W^{\mathrm{T}}]^{\mathrm{T}}; \; \boldsymbol{H}_1 = \begin{bmatrix} \boldsymbol{H}_0 & & 0 \\ & \boldsymbol{H}_0 & \\ & & \ldots \\ 0 & & \boldsymbol{H}_0 \end{bmatrix}. \qquad (10.15)$$

Each submatrix $\boldsymbol{H}_j$ in (10.15) is derived from $\boldsymbol{H}_1$ by random column permutations. The code rate of a GLDPC code is lower bounded by

$$R = K/N \geq 1 - W(1 - k/n), \qquad (10.16)$$

where K and N denote the dimension and the codeword length of a GLDPC code, W is the column weight of a global LDPC code, and k/n is the code rate of a local code (k and n denote the dimension and the codeword length of a local code). The GLDPC codes can be classified as follows [33]: (i) GLDPC codes with algebraic local codes of short length, such as Hamming codes, BCH codes, RS codes, or Reed–Muller codes; (ii) GLDPC codes for which the local codes are high-rate regular or irregular LDPC codes with large minimum distance; and (iii) *fractal* GLDPC codes in which

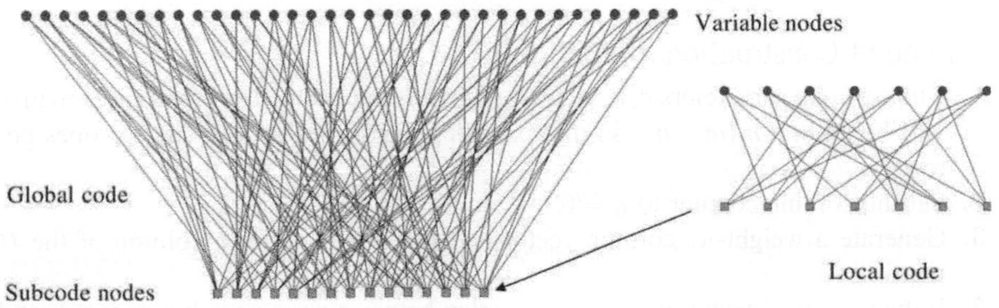

FIGURE 10.6

Construction of LZ-GLDPC codes.

the local code is in fact another GLDPC code. For more details on LZ-GLDPC- and B-GLDPC-like codes, and their generalization-fractal GLDPC codes (a local code is another GLDPC code), the interested reader is referred to Refs [33,37] (and references therein).

10.1.5.3 MacKay Codes

After MacKay [27], below are listed several ways to generate sparse matrices in order of increasing algorithm complexity (not necessarily improved performance):

1. The $\boldsymbol{H}$ matrix is generated by starting from an all-zero matrix and randomly inverting w_c (not necessarily distinct bits) in each column. The resulting LDPC codes is an irregular code.
2. The $\boldsymbol{H}$ matrix is generated by randomly creating weight-w_c columns.
3. The $\boldsymbol{H}$ matrix is generated with weight-w_c columns and uniform row weight (as near as possible).
4. The $\boldsymbol{H}$ matrix is generated with weight-w_c columns, weight-w_r rows, and no two columns having overlap larger than 1.
5. The $\boldsymbol{H}$ matrix is generated as in (4), and short cycles are avoided.
6. The $\boldsymbol{H}$ matrix is generated as in (5), and can be represented as $\boldsymbol{H}=[\boldsymbol{H}_1|\boldsymbol{H}_2]$, where $\boldsymbol{H}_2$ is invertible or at least has a full rank.

The construction via (5) may lead to an $\boldsymbol{H}$ matrix that is not of full rank. Nevertheless, it can be put in the following form by column swapping and Gauss–Jordan elimination:

$$\boldsymbol{H} = \begin{bmatrix} \boldsymbol{P}^{\mathrm{T}} & \boldsymbol{I} \\ \boldsymbol{0} & \boldsymbol{0} \end{bmatrix}, \tag{10.17}$$

and by eliminating the all-zero submatrix we obtain the parity-check matrix of systematic LDPC code:

$$\tilde{\boldsymbol{H}} = \begin{bmatrix} \boldsymbol{P}^{\mathrm{T}} & \boldsymbol{I} \end{bmatrix}. \tag{10.18}$$

Of the various construction algorithms listed above, construction (5) will be described in more detail.

Outline of Construction Algorithm (5)

1. Choose code parameters n, k, w_c, w_r, and g (g is the girth). The resulting $\boldsymbol{H}$ matrix will be an $m \times n$ ($m = n - k$) matrix with w_c ones per column and w_r ones per row.
2. Set the column counter to $i_c = 0$.
3. Generate a weight-w_c column vector and place it in the i_cth column of the $\boldsymbol{H}$ matrix.
4. If the weight of each row $\le w_r$, the overlap between any two columns is ≤ 1, and if all cycle lengths are $\ge g$, then increment the counter $i_c = i_c + 1$.
5. If $i_c = n$ stop, else go to step 3.

This algorithm could take hours to run with no guarantee of regularity of the $\boldsymbol{H}$ matrix. Moreover, it may not finish at all, and we need to restart the search with another set of parameters. Richardson and Urbanke proposed a linear complexity in length technique based on the $\boldsymbol{H}$ matrix [38].

An alternative approach to simplify encoding is to design the codes via algebraic, geometric, or combinatorial methods [24,29,30,39]. The $\boldsymbol{H}$ matrix of those designs can be put in cyclic or quasi-cyclic form, leading to encoder implementations based on shift registers and mod-2 adders. Since these classes of codes can also be used for design of quantum LDPC codes, we postpone their description until later sections.

10.2 DUAL-CONTAINING QUANTUM LDPC CODES

It has been shown by Calderbank, Shor, and Steane [40,41] that quantum codes known as CSS codes can be designed using a pair of conventional linear codes satisfying the *twisted property*, i.e. one of the codes includes the dual of another code. This class of quantum codes has already been studied in Chapter 7. Among CSS codes, particularly simple are those based on dual-containing codes [42], whose (quantum) check matrix can be represented by [42]

$$\boldsymbol{A} = \left[\begin{array}{c|c} \boldsymbol{H} & \boldsymbol{0} \\ \boldsymbol{0} & \boldsymbol{H} \end{array}\right], \tag{10.19}$$

where $\boldsymbol{H}\boldsymbol{H}^{\mathrm{T}} = \boldsymbol{0}$, which is equivalent to $C^{\perp}(\boldsymbol{H}) \subset C(\boldsymbol{H})$, where $C(\boldsymbol{H})$ is the code having $\boldsymbol{H}$ as the parity-check matrix and $C^{\perp}(\boldsymbol{H})$ is its corresponding dual code. It has been shown in Ref. [42] that the requirement $\boldsymbol{H}\boldsymbol{H}^{\mathrm{T}} = \boldsymbol{0}$ is satisfied when rows of $\boldsymbol{H}$ have an even number of ones, and any two of them overlap by an even number of ones. The LDPC codes satisfying these two requirements in Ref. [42] were designed by exhaustive computer search, in Ref. [43] they were designed as codes over GF(4) by identifying the Pauli operators $\{I,X,Y,Z\}$ with elements from GF(4), while in Ref. [44] they were designed in quasi-cyclic fashion. In what follows we will show how to design dual-containing LDPC codes using the combinatorial objects known as balanced incomplete block designs (BIBDs) [45–49]. Notice that the theory behind BIBDs is well known (see Ref. [50]), and BIBDs of unity index have already been used to design LDPC codes of girth-6 [51]. Notice, however, that dual-containing LDPCs are girth-4 LDPC codes, and they can be designed based on BIBDs with even index [45,46].

A balanced incomplete block design, denoted as BIBD(v,b,r,k,λ), is a collection of subsets (also known as *blocks*) of a set V of size v, with the size of each subset being k, so that: (i) each pair of elements (also known as *points*) occurs in *exactly* λ of the subsets; and (ii) every element occurs in exactly r subsets. The BIBD parameters satisfy the following two conditions [50]: (a) $vr = bk$ and (b) $\lambda(v-1) = r(k-1)$. Because the BIBD parameters are related (conditions (a) and (b)) it is sufficient to identify only three of them: v, k, and λ. It can easily be verified [51] that a point-block incident matrix represents a parity-check matrix H of an LDPC code of code rate R lower bounded by $R \geq [b - \text{rank}(\boldsymbol{H})]/b$, where b is the codeword length, and with rank( ) denoting the rank of the parity-check matrix. The parameter k corresponds to the column weight, r to the row weight, and v to the number of parity checks. The corresponding quantum code rate is lower bounded by $R_Q \geq [b - 2\text{rank}(\boldsymbol{H})]/b$. By selecting the index of BIBD $\lambda = 1$, the parity-check matrix has a girth of at least 6. For a classical LDPC code to be applicable in quantum error correction the following two conditions have to be satisfied [42]: (1) the LDPC code must contain its dual or equivalently any two rows of the parity-check matrix must have even overlap and the row weight must be even ($\boldsymbol{H}\boldsymbol{H}^{\mathrm{T}} = \boldsymbol{0}$), and (2) the code must have rate greater than 1/2. The BIBDs with even index λ satisfy condition (1). The parameter λ corresponds to the number of ones in which two rows overlap.

Example 10.5. The parity-check matrix from BIBD(7,7,4,4,2) = {{1,2,4,6}, {2,6,3,7}, {3,5,6,1}, {4,3,2,5}, {5,1,7,2}, {6,7,5,4}, {7,4,1,3}}, given as the $\boldsymbol{H}_1$ matrix below, has rank 3, even overlap between any two rows, and row weight even as well (or equivalently $\boldsymbol{H}_1\boldsymbol{H}_1^{\mathrm{T}} = 0$). Figure 10.7 shows the equivalence between this BIBD, parity-check matrix $\boldsymbol{H}$, and the Tanner graph. The blocks correspond to the bit nodes and provide the position of nonzero elements in corresponding columns. For example, block $B_1 = \{1,2,4,6\}$ corresponds to the bit node v_1 and it is involved in parity checks c_1, c_2, c_4, and c_6. Therefore, we establish edges between variable node v_1 and set check nodes c_1, c_2, c_4, and c_6. The B_1 block also gives the positions of nonzero elements in the first column of the parity-check matrix.

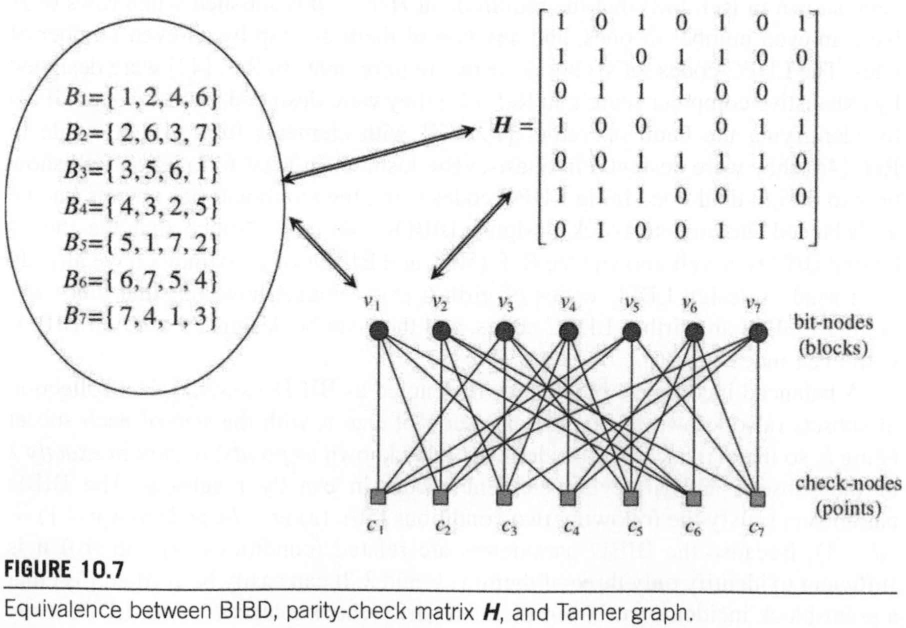

FIGURE 10.7

Equivalence between BIBD, parity-check matrix $\boldsymbol{H}$, and Tanner graph.

Notice that so-called λ configurations (in the definition of BIBD the word "exactly" from condition (i) is replaced by "*at most*") are not applicable here. However, the $\boldsymbol{H}$ matrix from BIBD of unity index can be converted to an H matrix satisfying condition (1) by adding a column with all ones, which is equivalent to adding an additional block to unity index BIBD having all elements from V. For example, the parity-check matrix from BIBD(7,7,3,3,1), after the addition of a column with all ones, is given as the matrix $\boldsymbol{H}_2$ below, and satisfies the condition $\boldsymbol{H}_2\boldsymbol{H}_2^{\mathrm{T}} = 0$. The following method, due to Bose [52], is a powerful method to design many different BIBDs with desired index λ. Let S be a set of elements. Associate with each element u from S n symbols u_1, u_2, …, u_n. Let sets S_1, …, S_t satisfy the following three conditions: (i) every set S_i $(i = 1,\ldots,t)$ contains k symbols (the symbols from the same set are different from one another); (ii) among kt symbols in t sets exactly r symbols belong to each of n classes ($nr = kt$); and (iii) the differences between t sets are symmetrically repeated so that each repeats λ times. If s is an element from S, from each set S_i we are able to form another set $S_{i,s}$ by adding s to S_i, keeping the class number (subscript) unchanged; then sets $S_{i,s}$ $(i = 1,\ldots,t;\ s \in S)$ represent an (mn,nt,r,k,λ) BIBD. By observing the elements from BIBD blocks as positions of ones in corresponding columns of a parity-check matrix, the code rate of an LDPC code thus obtained is lower bounded by $R \geq 1 - m/t$ ($R_Q \geq 1 - 2m/t$), and the codeword length is determined by $b = nt$.

$$\boldsymbol{H}_1 = \begin{bmatrix} 1 & 0 & 1 & 0 & 1 & 0 & 1 \\ 1 & 1 & 0 & 1 & 1 & 0 & 0 \\ 0 & 1 & 1 & 1 & 0 & 0 & 1 \\ 1 & 0 & 0 & 1 & 0 & 1 & 1 \\ 0 & 0 & 1 & 1 & 1 & 1 & 0 \\ 1 & 1 & 1 & 0 & 0 & 1 & 0 \\ 0 & 1 & 0 & 0 & 1 & 1 & 1 \end{bmatrix}, \quad \boldsymbol{H}_2 = \begin{bmatrix} 1 & 1 & 1 & 0 & 0 & 0 & 0 & 1 \\ 1 & 0 & 0 & 1 & 1 & 0 & 0 & 1 \\ 1 & 0 & 0 & 0 & 0 & 1 & 1 & 1 \\ 0 & 1 & 0 & 1 & 0 & 1 & 0 & 1 \\ 0 & 1 & 0 & 0 & 1 & 0 & 1 & 1 \\ 0 & 0 & 1 & 1 & 0 & 0 & 1 & 1 \\ 0 & 0 & 1 & 0 & 1 & 1 & 0 & 1 \end{bmatrix}.$$

In the rest of this section we introduce several constructions employing the method due to Bose [45,46].

Construction 1. If $6t + 1$ is a prime or prime power and θ is a primitive root of GF($6t + 1$), then the t initial sets $S_i = (0,\ \theta^i,\theta^{2t+i},\theta^{4t+i})$ $(i = 0,1,\ldots,t - 1)$ form a BIBD($6t + 1$,t($6t + 1$),$4t$,4,2). The BIBD is formed by adding the elements from GF($6t + 1$) to the initial blocks S_i. Because the index of BIBD is even ($\lambda = 2$) and row weight $r = 4t$ is even, the corresponding LDPC code is a dual-containing code ($\boldsymbol{H}\boldsymbol{H}^{\mathrm{T}} = \boldsymbol{0}$). The quantum code rate for this construction, and Constructions 2 and 3 as well, is lower bounded by $R_Q \geq (1 - 2/t)$. The BIBD(7,7,4,4,2) given above is obtained using this construction method. For $t = 30$ a dual-containing LDPC(5430,5249) code is obtained. The corresponding quantum LDPC code has rate 0.934, which is significantly higher than any of the codes introduced in Ref. [42] ($R_Q = 1/4$).

Construction 2. If $10t + 1$ is a prime or prime power and θ is a primitive root of GF($10t + 1$), then the t initial sets $S_i = (\theta^i,\theta^{2t+i},\theta^{4t+i},\theta^{6t+i})$ form a BIBD

$(10t+1, t(10t+1), 4t, 4, 2)$. For example, for $t=24$ dual-containing LDPC (5784,5543) code is obtained, and the corresponding CSS code has rate 0.917.

Construction 3. If $5t+1$ is a prime or prime power and θ is a primitive root of GF($5t+1$), then the t initial sets $(\theta^i, \theta^{2t+i}, \theta^{4t+i}, \theta^{6t+i}, \theta^{8t+i})$ form a BIBD($5t+1, t(5t+1), 5t, 5, 4$). Notice that parameter t has to be even for LDPC code to satisfy condition (1). For example, for $t=30$ the dual-containing LDPC(4530,4379) code is obtained, and the corresponding CSS LDPC code has rate 0.934.

Construction 4. If $2t+1$ is a prime or prime power and θ is a primitive root of GF($2t+1$), then the $5t+2$ initial sets

$$\begin{array}{ll}
(\theta_1^i, \theta_1^{t+i}, \theta_3^{i+a}, \theta_3^{t+i+a}, 0_2) & (i=0,1,\ldots,t-1) \\
(\theta_2^i, \theta_2^{t+i}, \theta_4^{i+a}, \theta_4^{t+i+a}, 0_3) & (i=0,1,\ldots,t-1) \\
(\theta_3^i, \theta_3^{t+i}, \theta_5^{i+a}, \theta_5^{t+i+a}, 0_4) & (i=0,1,\ldots,t-1) \\
(\theta_4^i, \theta_4^{t+i}, \theta_1^{i+a}, \theta_1^{t+i+a}, 0_5) & (i=0,1,\ldots,t-1) \\
(\theta_5^i, \theta_5^{t+i}, \theta_2^{i+a}, \theta_2^{t+i+a}, 0_1) & (i=0,1,\ldots,t-1) \\
(0_1, 0_2, 0_3, 0_4, 0_5), (0_1, 0_2, 0_3, 0_4, 0_5) &
\end{array}$$

form a BIBD($10t+5, (5t+2)(2t+1), 5t+2, 5, 2$). Similarly as in the previous construction, the parameter t has to be even. The quantum code rate is lower bounded by $R_Q \geq [1 - 2 \cdot 5/(5t+2)]$, and the codeword length is determined by $(5t+2)(2t+1)$. For $t=30$ the dual-containing LDPC(5490,5307) is obtained, and corresponding quantum LDPC codes has rate 0.934. The following two constructions are obtained by converting unity index BIBD into $\lambda=2$BIBD.

Construction 5. If $12t+1$ is a prime or prime power and θ is a primitive root of GF($12t+1$), then the t initial sets $(0, \theta^i, \theta^{4t+i}, \theta^{8t+i})$ $(i=0,2,\ldots,2t-2)$ form a BIBD($12t+1, t(12t+1), 4t, 4, 1$). To convert this unity index BIBD into $\lambda=2$BIBD we have to add an additional block $(1,\ldots,12t+1)$.

Construction 6. If $20t+1$ is a prime or prime power and θ is a primitive root of GF($20t+1$), then the t initial sets $(\theta^i, \theta^{4t+i}, \theta^{8t+i}, \theta^{12t+i}, \theta^{16t+i})$ $(i=0,2,\ldots,2t-2)$ form a BIBD($20t+1, t(20t+1), 5t, 5, 1$). Similarly as in the previous construction, in order to convert this unity index BIBD into $\lambda=2$ BIBD we have to add an additional block $(1,\ldots,20t+1)$.

Construction 7. If $2(2\lambda+1)t+1$ is a prime power and θ is a primitive root of GF[$2(2\lambda+1)t+1$], then the t initial sets $S_i=(\theta^i, \theta^{2t+i}, \theta^{4t+i}, \ldots, \theta^{4\lambda t+i})$ $(i=0,1,\ldots,t-1)$ form a BIBD($2(2\lambda+1)t+1, t[2(2\lambda+1)t+1], (2\lambda+1)t, 2\lambda+1, \lambda$). The BIBD is formed by adding the elements from GF[$2(2\lambda+1)t+1$] to the initial blocks S_i. For any even index λ and even parameter t (the row weight is even), the corresponding LDPC code is a dual-containing code ($\boldsymbol{H}\boldsymbol{H}^{\mathrm{T}}=0$). The quantum code rate for this construction is lower bounded by $R_Q \geq (1-2/t)$, and the minimum distance is lower bounded by $d_{\min} \geq 2\lambda+2$. For any odd index λ design we have to add an additional block $(1,2,\ldots,2(2\lambda+1)t+1)$, so that the row weight of $\boldsymbol{H}$ becomes even.

Construction 8. If $2(2\lambda-1)t+1$ is a prime power and θ is a primitive root of GF[$2(2\lambda-1)t+1$], then the t initial sets $S_i=(0, \theta^i, \theta^{2t+i}, \ldots, \theta^{(4\lambda-1)t+i})$ form a BIBD($2(2\lambda-1)t+1, [2(2\lambda-1)t+1]t, 2\lambda t, 2\lambda, \lambda$). For any even index λ, the corresponding LDPC code is a dual-containing code. The quantum LPDC code rate is lower bounded by $R_Q \geq (1-2/t)$, and the minimum distance is lower bounded by $d_{min} \geq 2\lambda + 1$.

Construction 9. If $(\lambda-1)t$ is a prime power and θ is a primitive root of GF$[(\lambda-1)t+1]$, then the t initial sets $(0,\theta^i,\theta^{t+i},\ldots,\theta^{(\lambda-2)t+i})$ form a BIBD$[(\lambda-1)t+1,((\lambda-1)t+1)t,\lambda t,\lambda,\ \lambda]$. Again for an even index λ the quantum LDPC code of rate $R_Q \geq (1-2/t)$ is obtained, whose minimum distance is lower bounded by $d_{min} \geq \lambda + 1$. Similarly as in the constructions 5 and 6, for any odd index λ design we have to add an additional block $(1,2,\ldots,(\lambda-1)t)$.

Construction 10. If $2k-1$ is a prime power and θ is a primitive root of GF$(2k-1)$, then the initial sets

$$\left(0,\theta^i,\theta^{i+2},\ldots,0^{i+2k-4}\right),\quad \left(\infty,\theta^{i+1},\theta^{i+3},\ldots,0^{i+2k-3}\right);\quad (i=0,1)$$

form a BIBD$(2k,4(2k-1),2(2k-1),k,2(k-1))$. For even k the quantum code rate is lower bounded by $R_Q \geq [1-1/(2\ k-1)]$, the codeword length is determined by $4(2k-1)$, and the minimum distance is lower bounded by $d_{min} \geq k+1$.

We performed simulations for error-correcting performance of BIBD-based dual-containing codes described above as the function of noise level by Monte Carlo simulations. We simulated the classical binary symmetric channel (BSC) to be compatible with current literature [42–44]. The results of simulations are shown in Figure 10.8 for 30 iterations in the sum-product-with-correction-term algorithm described in the previous section. We simulated dual-containing LDPC codes of high rate and moderate length, so that corresponding quantum LDPC code has a rate around 0.9. BER curves correspond to the $C/C^{\perp}$ case, and are obtained by counting the errors only on those codewords from C not belonging to $C^{\perp}$. The codes from BIBD with index $\lambda = 2$ outperform the codes with index $\lambda = 4$. The codes derived from unity index BIBDs by adding an all ones column outperform the codes derived from BIBDs of even index. In the simulations presented in Figure 10.8, codes with parity-check matrices with column weight $k = 4$ or 5 are observed. The code from projective geometry (PG) is an exception. It is based on BIBD$(s^2+s+1,s+1,1)$, where s is a prime power; in our example parameter s was set to 64. The LDPC(4162,3432) code based on this BIBD outperforms other codes; however, the code rate is lower ($R = 0.825$) and the column weight is large (65). For more details on PG codes and secondary structures developed from them, the interested reader is referred to the next section.

To improve the BER performance of proposed high-rate sparse dual-containing codes we employed an efficient algorithm due to Sankaranarayanan and Vasic [53], for removing cycles of length 4 in the corresponding bipartite graph. As shown in Figure 10.8, this algorithm can significantly improve the BER performance, especially for weak (index-4 BIBD-based) codes. For example, with LDPC(3406,3275) code, a BER of 10^{-5} can be achieved at a crossover probability of $1.14 \cdot 10^{-4}$ if the sum-product-with-correction-term algorithm is employed ($g = 4$ curve), while the same BER can be achieved at a crossover probability of $6.765 \cdot 10^{-4}$ when the algorithm proposed in Ref. [53] is employed ($g = 6$ curve). Notice that this

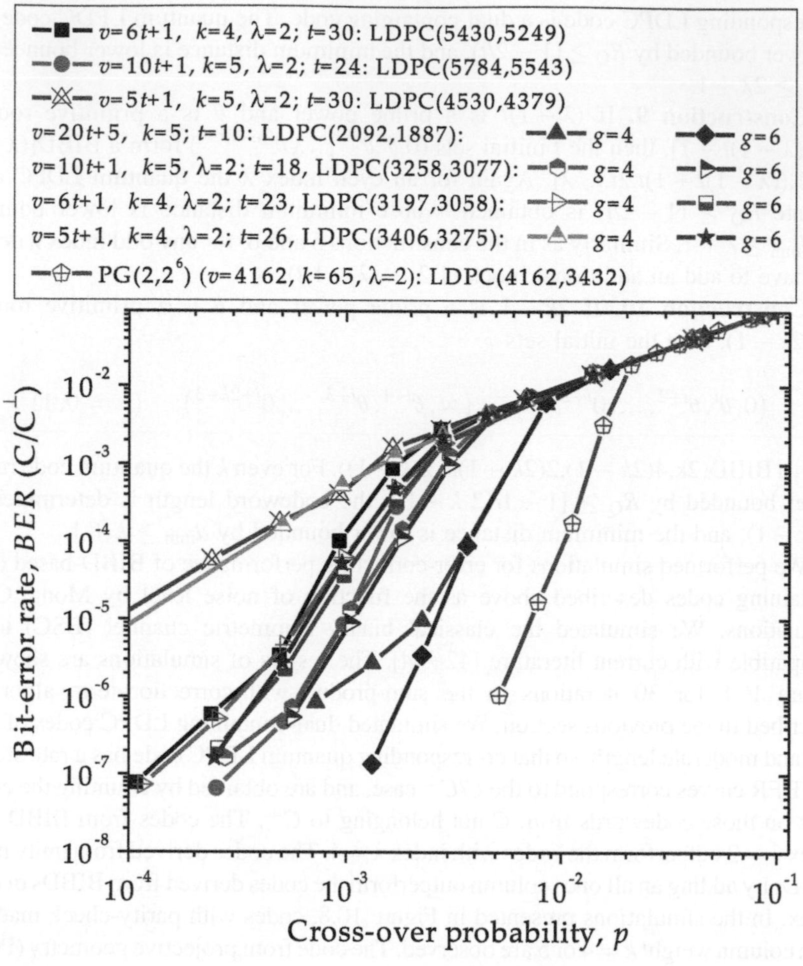

FIGURE 10.8

BERs against crossover probability on a binary symmetric channel.

(After Ref. [45]; IEEE 2008; reprinted with permission.)

algorithm modifies the parity-check matrix by adding the auxiliary variables and checks, so that the four-cycles are removed in the modified parity-check matrix. The algorithm attempts to minimize the required number of auxiliary variable/check nodes while removing the four-cycles. The modified parity-check matrix is used only in the decoding phase, while the encoder remains the same.

In Figure 10.9 we further compare BER performance of three quantum LDPC codes of quantum rate above 0.9, which are designed by employing Constructions 7 and 8: (i) quantum LDPC(7868,7306,0.9285,$\geq$6) code from Construction 7 by setting $t=28$ and $\lambda=2$; (ii) quantum LDPC(8702,8246,0.946,$\geq$5) code from

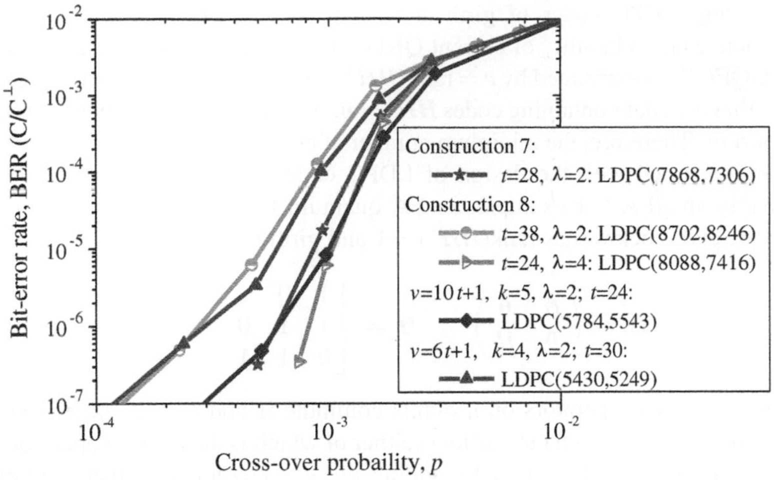

FIGURE 10.9

BERs against crossover probability on a binary symmetric channel.

(After Ref. [46]; IEEE 2009; reprinted with permission.)

Construction 8 by setting $t=38$ and $\lambda=2$; and (iii) quantum LDPC(8088, 7416,0.917,$\geq$9) from Construction 8 by setting $t=24$ and $\lambda=4$. For comparison purposes two curves for quantum LDPC codes from Figure 10.8 are plotted as well. The BIBD codes from Constructions 7 and 8 outperform the codes from previous constructions for BERs around 10^{-7}. The code from BIBD with index $\lambda=4$ and Construction 8 outperforms the codes with index $\lambda=2$ from both constructions, because it has larger minimum distance. The code with index $\lambda=2$ from Construction 7 outperforms the corresponding code from Construction 8.

10.3 ENTANGLEMENT-ASSISTED QUANTUM LDPC CODES

The EA-QECCs [54–60] make use of pre-existing entanglement between transmitter and receiver to improve the reliability of transmission. Dual-containing LDPC codes themselves have a number of advantages compared to other classes of quantum codes, because of the sparseness of their quantum parity-check matrices, which leads to a small number of interactions being required to perform decoding. However, the dual-containing LDPC codes are in fact girth-4 codes, and the existence of short cycles worsens the error performance by correlating the extrinsic information in the decoding process. By using the EA-QECC concept, arbitrary classical code can be used as quantum code. That is, if classical codes are not dual containing they correspond to a set of stabilizer generators that do not commute. By providing entanglement between source and destination, the corresponding generators can be embedded into larger set of commuting generators, which gives a well-defined code space.

By using LDPC codes of girth at least 6 we can dramatically improve the performance and reliability of current QKD schemes. The number of ebits required for EA-QECC is determined by $e = \text{rank}(\boldsymbol{H}\boldsymbol{H}^{\text{T}})$, as described in the previous chapter. Notice that for dual-containing codes $\boldsymbol{H}\boldsymbol{H}^{\text{T}} = \boldsymbol{0}$, meaning that the number of required ebits is zero. Therefore, the minimum number of ebits in an EA-QECC is 1, and here we are concerned with the design of LDPC codes of girth $g \geq 6$ with $e = 1$ or reasonably small e. For example, an EA quantum LDPC code of quantum check matrix $\boldsymbol{A}$, given below, has $\text{rank}(\boldsymbol{H}\boldsymbol{H}^{\text{T}}) = 1$ and girth 6:

$$\boldsymbol{A} = \begin{pmatrix} \boldsymbol{H} & \boldsymbol{0} \\ \boldsymbol{0} & \boldsymbol{H} \end{pmatrix} \quad \boldsymbol{H} = \begin{bmatrix} 1 & 0 & 1 & 1 & 1 \\ 1 & 1 & 0 & 0 & 1 \\ 0 & 1 & 1 & 1 & 0 \end{bmatrix}. \tag{10.20}$$

Because two Pauli operators on n qubits commute if and only if there is an even number of places in which they differ (neither of which is the identity operator I), we can extend the generators in $\boldsymbol{A}$ by adding the $e = 1$ column so that they can be embedded into a larger Abelian group; this procedure is known as Abelianization in abstract algebra. For example, by adding the $(0\ 1\ 1)^{\text{T}}$ column to the $\boldsymbol{H}$ matrix above, newly obtained code is dual-containing quantum code, and the corresponding quantum-check matrix is

$$\boldsymbol{A}' = \begin{pmatrix} \boldsymbol{H}' & \boldsymbol{0} \\ \boldsymbol{0} & \boldsymbol{H}' \end{pmatrix} \quad \boldsymbol{H}' = \left(\begin{array}{ccccc|c} 1 & 0 & 1 & 1 & 1 & 0 \\ 1 & 1 & 0 & 0 & 1 & 1 \\ 0 & 1 & 1 & 1 & 0 & 1 \end{array}\right). \tag{10.21}$$

The last column in $\mathbf{H}'$ corresponds to the ebit, the qubit that is shared between the source and destination, which is illustrated in Figure 10.10. The source encodes quantum information in state $|\psi\rangle$ with the help of local ancillary qubits $|0\rangle$ and source's half of the shared ebits, and then sends the encoded qubits over a noisy quantum channel (say a free-space optical channel or optical fiber). The receiver performs decoding on all qubits to diagnose the channel error and performs a recovery unitary operation to reverse the action of the channel. Notice that the channel does not affect the receiver's half of the shared ebits at all. The encoding of quantum LDPC codes is very similar to the different classes of EA block codes described in the previous chapter. Since the encoding of EA LDPC is not different from that of any other EA block code, in the rest of this section we deal with the

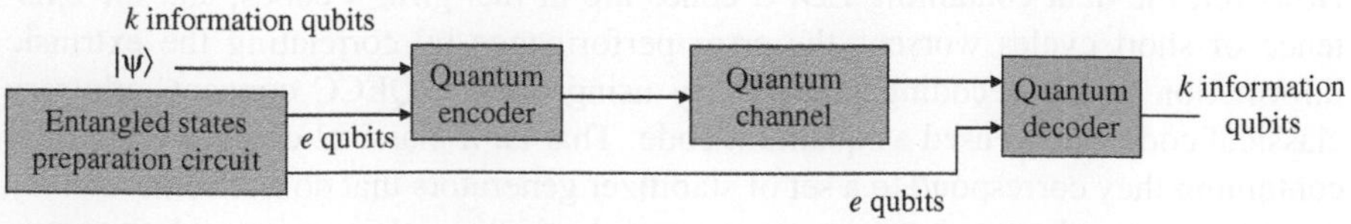

FIGURE 10.10

The operating principle of entanglement-assisted quantum codes.

design of EA LDPC codes instead. The following theorem can be used for design of EA codes from BIBDs that require only one ebit to be shared between source and destination.

Theorem 10.1. Let $\boldsymbol{H}$ be a $b \times v$ parity-check matrix of an LDPC code derived from a BIBD(v,k,1) of odd r. The rank of $\boldsymbol{HH}^{\mathrm{T}}$ is equal to 1, while the corresponding EA LDPC code of CSS type has the parameters $[v,v-2b+1]$ and requires one ebit to be shared between source and destination.

Proof. Because any two rows or columns in $\boldsymbol{H}$ of size $b \times v$, derived from BIBD(v,k,1), overlap in *exactly* one position, by providing that row weight r is odd, the matrix $\boldsymbol{HH}^{\mathrm{T}}$ is an all-one matrix. The rank of the all-one matrix $\boldsymbol{HH}^{\mathrm{T}}$ is 1. Therefore, LDPC codes from BIBDs of unity index ($\lambda = 1$) and odd r have $e = \mathrm{rank}(\boldsymbol{HH}^{\mathrm{T}}) = 1$, while the number of required ebits to be shared between source and destination is $e = 1$. If the EA code is put in CSS form, then the parameters of EA codes will be $[v,v-2b+1]$.

Because the girth of codes derived by employing the Theorem 10.1 is 6, they can significantly outperform quantum dual-containing LDPC codes. Notice that certain blocks in so-called λ configurations [50] have overlap of zero and therefore cannot be used for designing EA codes with $e = 1$. Shrikhande [61] has shown that the generalized Hadamard matrices, affine resolvable BIBDs, group-divisible designs and orthogonal arrays of strength 2 are all related, and because for these combinatorial objects $\lambda = 0$ or 1, they are not suitable in the design of EA codes of $e = 1$. Notice that if the quantum-check matrix is put in the form $\boldsymbol{A} = [\boldsymbol{H}|\boldsymbol{H}]$ then the corresponding EA code has parameters $[v,v-b+1]$. This design can be generalized by employing two BIBDs having the same parameter v, as explained in the theorem below.

Theorem 10.2. Let $\boldsymbol{H}_1$ be a $b_1 \times v$ parity-check matrix of an LDPC code derived from a BIBD(v,k_1,1) of odd r_1 and $\boldsymbol{H}_2$ be a $b_2 \times v$ parity-check matrix of an LDPC code derived from a BIBD(v,k_2,1) of odd r_2. The number of ebits required in the corresponding EA code with quantum-check matrix $\boldsymbol{A} = [\boldsymbol{H}_1|\boldsymbol{H}_2]$ is $e = \mathrm{rank}(\boldsymbol{H}_1\boldsymbol{H}_2^{\mathrm{T}} + \boldsymbol{H}_2\boldsymbol{H}_1^{\mathrm{T}})/2 = 1$, while the corresponding EA LDPC code has the parameters $[v,v-b+1]$.

The proof of this theorem is a straightforward generalization of the proof of Theorem 10.1. Below we describe several designs belonging to the class of codes from Theorem 10.1.

Design 1: Steiner triple system. If $6t+1$ is a prime power and θ is a primitive root of GF($6t+1$), then the following t initial blocks $(\theta^i,\theta^{2t+i},\theta^{4t+i})$ $(i = 0,1,\ldots,t-1)$ form BIBD($6t+1$,3,1) with $r = 3t$. The BIBD is formed by adding the elements from GF($6t+1$) to the initial blocks. The corresponding LDPC code has $\mathrm{rank}(\boldsymbol{HH}^{\mathrm{T}}) = 1$ and girth of 6.

Design 2: Projective planes. A finite projective plane of order q, say PG(2,q), is a $(q^2+q+1,q+1,1)$ BIBD, $q \geq 2$ and q is a power of prime [17,21,29,30,56]. The finite projective plane-based geometries are summarized in Table 10.1. The point set of the design consists of all the *points* on PG(2,q), and the block set of the design consist of all *lines* on PG(2,q). The points and lines of a finite projective plane satisfy the following set of axioms: (i) every line consists of the same number of

Table 10.1 Finite Projective Plane $2(v,k,1)$-Based Geometries

k	v	Parameter	Name
$q+1$	q^2+q+1	q – a prime power	Projective planes
q	q^2	q – a prime power	Affine planes
$q/2$	$q(q-1)/2$	q – a prime power and even	Oval designs
$\sqrt{q}+1$	$q\sqrt{q}+1$	q – a prime power and square	Unitals

points; (ii) any two points on the plane are connected by a unique line; (iii) any two lines on the plane intersect at a unique point; and (iv) a fixed number of lines pass through any point on the plane. Since any two lines on a projective plane intersect at a unique point, there are no parallel lines on the plane ($\lambda = 1$). The incidence matrix (parity-check matrix of corresponding LDPC codes) of such a design, for $q = 2^s$, is cyclic and hence any row of the matrix can be obtained by cyclic shifting (right or left) another row of the matrix. Since the row weight $q+1 = 2^s+1$ is odd, the corresponding LDPC code has rank($\boldsymbol{H}\boldsymbol{H}^{\mathrm{T}}$) = 1. The minimum distance of codes from projective planes, affine planes, oval designs, and unitals [62] is at least 2^s+2, 2^s+1, $2^{s-1}+1$ and $2^{s/2}+2$ respectively. Corresponding ranks of $\boldsymbol{H}$ matrices are given respectively by [62] 3^s+1, 3^s, 3^s-2^s, and $2^{3s/2}$. Notice, however, that affine plane-based codes have rank($\boldsymbol{H}\boldsymbol{H}^{\mathrm{T}}$) > 1, because they contain parallel lines. The code rate of classical projective plane codes is given by $R = (q^2+q+1-\mathrm{rank}(\boldsymbol{H}))/(q^2+q+1) = (q^2+q-3^s)/(q^2+q+1)$, while the quantum code rate of the corresponding quantum code is given by

$$\begin{aligned} R_{\mathrm{Q}} &= \left(q^2+q+1-2\,\mathrm{rank}(\boldsymbol{H})+\,\mathrm{rank}\left(\boldsymbol{H}\boldsymbol{H}^{\mathrm{T}}\right)\right)/\left(q^2+q+1\right) \\ &= \left(q^2+q-2\cdot 3^s\right)/\left(q^2+q+1\right). \end{aligned} \tag{10.22}$$

Design 3: *m*-dimensional projective geometries. The finite projective geometries [28,50] PG(m, p^s) are constructed using $(m+1)$-tuples $\boldsymbol{x} = (x_0, x_1, \ldots, x_m)$ of elements x_i from GF(p^s) (p is a prime, s is a positive integer), not all simultaneously equal to zero, called points. Two $(m+1)$-tuples $\boldsymbol{x}$ and $\boldsymbol{y} = (y_0, y_1, \ldots, y_m)$ represent the same point if $\boldsymbol{y} = \lambda\boldsymbol{x}$, where λ is a nonzero element from GF(p^s). Therefore, each point can be represented in p^s-1 ways (an equivalence class). The number of points in PG(m, p^s) is given by

$$v = \left[p^{(m+1)s}-1\right]/\left(p^s-1\right). \tag{10.23}$$

The points (equivalence classes) can be represented by $[\alpha^i] = \{\alpha^i, \beta\alpha^i, \ldots, \beta^{p^s-2}\alpha^i\}$ $(0 \le i \le v)$, where $\beta = \alpha^v$. Let $[\alpha^i]$ and $[\alpha^j]$ be two distinct points on PG(m,p^s), then the line passing through them consists of points of the form $[\lambda_1\alpha^i + \lambda_2\alpha^j]$, where $\lambda_1, \lambda_2 \in$ GF(p^s). Because $[\lambda_1\alpha^i + \lambda_2\alpha^j]$ and $[\beta^k\lambda_1\alpha^i + \beta^k\lambda_2\alpha^j]$ represent the same point, each line in PG(m, p^s) consists of $k = (p^{ms}-1)/(p^s-1)$

points. The number of lines intersecting at a given point is given by k, and the number of lines in m-dimensional PG is given by

$$b = \left[p^{s(m+1)} - 1\right](p^{sm} - 1)/\left[\left(p^{2s} - 1\right)\left(p^s - 1\right)\right]. \qquad (10.24)$$

The parity-check matrix is obtained as the incidence matrix $\boldsymbol{H} = (h_{ij})_{b \times v}$ with rows corresponding to the lines and columns to the points, and columns being arranged in the following order: $[\alpha^0]$, …, $[\alpha^v]$. This class of codes is sometimes called type I projective geometry codes [21]. The $h_{ij} = 1$ if the jth point belongs to the ith line, and zero otherwise. Each row has weight $p^s + 1$ and each column has weight k. Any two rows or columns have exactly one "1" in common, and providing that p^s is even then the row weight $(p^s + 1)$ will be odd, and the corresponding LDPC code will have $\text{rank}(\boldsymbol{HH}^{\text{T}}) = 1$. The quantum code rate of corresponding codes will be given by

$$R_{\text{Q}} = \left(v - 2\text{rank}(\boldsymbol{H}) + \text{rank}\left(\boldsymbol{HH}^{\text{T}}\right)\right)/v. \qquad (10.25)$$

By defining a parity-check matrix as an incidence matrix with lines corresponding to columns and points to rows, the corresponding LDPC code will have $\text{rank}(\boldsymbol{HH}^{\text{T}}) = 1$ providing that row weight k is odd. This type of PG codes is sometimes called type II PG codes [21].

Design 4: Finite geometry codes on *m*-flats [62]. In projective geometric terminology, an $(m + 1)$-dimensional element of the set is referred to as an *m-flat*. For examples, *points* are 0-flats, *lines* are 1-flats, *planes* are 2-flats, and *m-dimensional* spaces are $(m - 1)$-flats. An l-dimension flat is said to be contained in an m-dimensional flat if it satisfies a set-theoretic containment. Since any m-flat is a finite vector subspace, it is justified to talk about a basis set composed of basis elements independent in GF(p) (p is a prime). In the following discussion, the basis element of a 0-flat is referred to as a point and, generally, this should cause no confusion because this basis element is representative of the 0-flat. Using the above definitions and linear algebraic concepts, it is straightforward to count the number of m-flats, defined by a basis with $m + 1$ points, in PG(n, q). The number of such flats, say $N_{\text{PG}}(m,n,q)$, is the number of ways of choosing $m + 1$ independent points in PG(n,q) divided by the number of ways of choosing $m + 1$ independent points in any m-flat:

$$\begin{aligned} N_{\text{PG}}(m,n,q) &= \frac{\left(q^{n+1} - 1\right)\left(q^{n+1} - q\right)\left(q^{n+1} - q^2\right)\ldots\left(q^{n+1} - q^m\right)}{(q^{m+1} - 1)(q^{m+1} - q)(q^{m+1} - q^2)\ldots(q^{m+1} - q^m)} \\ &= \prod_{i=0}^{m} \frac{\left(q^{n+1-i} - 1\right)}{\left(q^{m+1-i} - 1\right)}. \end{aligned} \qquad (10.26)$$

Hence, the number of 0-flats in PG(n, q) is $N_{\text{PG}}(0,n,q) = \dfrac{q^{n+1} - 1}{q - 1}$, and the number of 1-flats is

$$N_{\text{PG}}(1,n,q) = \frac{(q^{n+1} - 1)(q^n - 1)}{(q^2 - 1)(q - 1)}. \qquad (10.27)$$

When $n=2$, the number of points is equal to the number of lines, and this agrees with the dimensions of point-line incidence matrices of projective plane codes, introduced above.

An algorithm to construct an m-flat in PG(n,q) begins by recognizing that the elements of GF(q^{n+1}) can be used to represent points of PG(n,q) in the following manner [62,63]. If α is a primitive element of GF(q^{n+1}), then α^{v}, where $v=(q^{n+1}-1)/(q-1)$, is a primitive element of GF(q). Each one of the first v powers of α can be taken as the basis of one of the 0-flats in PG(n, q). In other words, if α^{i} is the basis of a 0-flat in PG(n, q), then every α^{k}, such that $i \equiv k \bmod v$, is contained in the subspace. In a similar fashion, a set of $m+1$ powers of α that are independent in GF(q) forms a basis of an m-flat. If $\alpha^{s_1}, \alpha^{s_2}, \ldots, \alpha^{s_{m+1}}$ is a set of $m+1$ basis elements of an m-flat, then any point in the flat can be written as

$$\zeta_i = \sum_{j=1}^{m+1} \varepsilon_{ij} \alpha^{s_j}, \tag{10.28}$$

where ε_{ij} are chosen such that no two vectors $<\varepsilon_{i1}, \varepsilon_{i2}, \ldots, \varepsilon_{i(m+1)}>$ are linear multiples over GF(q). Now every ζ_i can be equivalently written as a power of the primitive element α.

To construct an LDPC code, we generate all m-flats and l-flats in PG(n, q) using the method described above. An incidence matrix of m-flats and l-flats in PG(n, q) is a binary matrix with $N_{\mathrm{PG}}(m,n,q)$ rows and $N_{\mathrm{PG}}(l,n,q)$ columns. The (i,j)th element of the incidence matrix $\mathbf{A}_{\mathrm{PG}}(m,l,n,q)$ is a one if and only if the jth l-flat is contained in the ith m-flat of the geometry. An LDPC code is constructed by considering the incidence matrix or its transpose as the parity-check matrix of the code. It is widely accepted that the performance of an LDPC code under an iterative decoder is adversely affected by the presence of four-cycles in the Tanner graph of the code. In order to avoid four-cycles, we impose an additional constraint that l should be one less than m. If l is one less than m, then no two distinct m-flats have more than one l-dimensional subspace in common. This guarantees that the girth (length of the shortest cycle) of the graph is 6.

Figure 10.11 shows the code length as a function of quantum code rate for EA LDPC codes derived from Designs 2 and 3, obtained for different values of parameter $s=1-9$. The results of simulations are shown in Figure 10.12 for 30 iterations using the sum-product algorithm. The EA quantum LDPC codes from Design 3 are compared against dual-containing quantum LDPC codes discussed in the previous chapter. We can see that EA LDPC codes significantly outperform dual-containing LDPC codes.

Before concluding this section, we provide a very general design method [64], which is based on the theory of mutually orthogonal Latin squares (MOLS) constructed using the MacNeish–Mann theorem [28]. This design allows us to determine a corresponding code for an arbitrary composite number q.

Design 5: Iterative decodable codes based on the MacNeish–Mann theorem [64]. Let the prime decomposition of an integer q be

$$q = p_1^{s_1} p_2^{s_2} \ldots p_m^{s_m}, \tag{10.29}$$

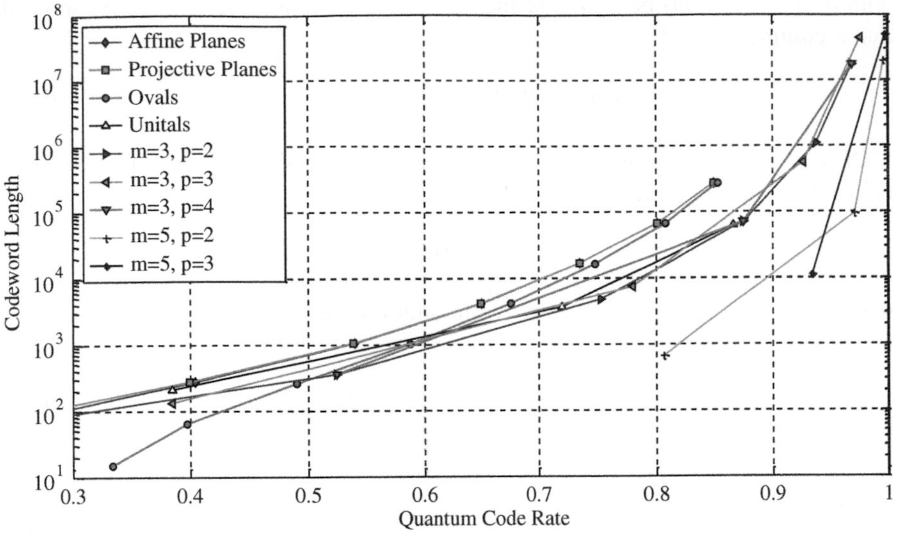

FIGURE 10.11

EA quantum LDPC codes from *m*-dimensional projective geometries.

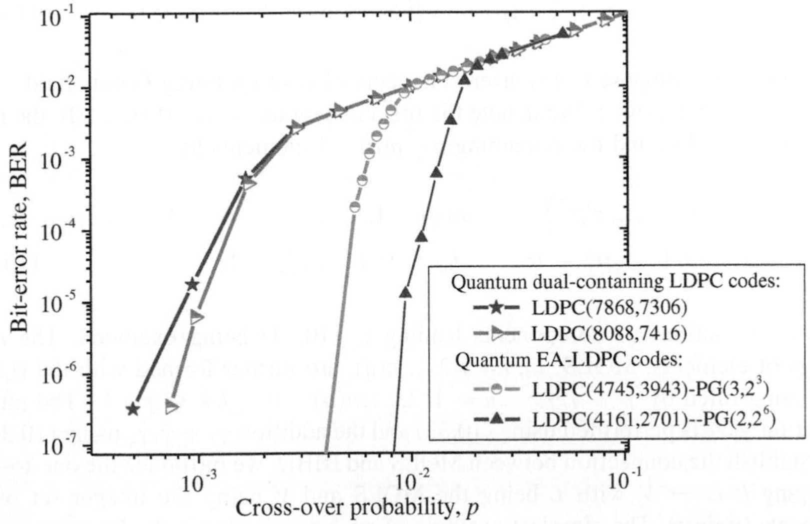

FIGURE 10.12

BERs against crossover probability on a binary symmetric channel.

(Modified from Ref. [49]; IEEE 2011; reprinted with permission.)

with p_i ($i = 1,2,\ldots,m$) being the prime numbers. MacNeish and Mann [28] showed that a complete set of

$$n(q) = \min(p_1^{s_1}, p_2^{s_2}, \ldots, p_m^{s_m}) - 1 \tag{10.30}$$

MOLS of order q can always be constructed. The construction algorithm is given below. Consider m-tuples

$$\gamma = (g_1, g_2, \ldots, g_m), \tag{10.31}$$

where $g_i \in GF(p_i^{s_i})$, $i = 1,2,\ldots,m$, with addition and multiplication operations defined by

$$\gamma_1 + \gamma_2 = (g_1, g_2, \ldots, g_m) + (h_1, h_2, \ldots, h_m) = (g_1 + h_1, g_2 + h_2, \ldots, g_m + h_m) \tag{10.32}$$

and

$$\gamma_1\gamma_2 = (g_1h_1, g_2h_2, \ldots, g_mh_m), \tag{10.33}$$

wherein the operations in brackets are performed in the corresponding Galois fields (e.g. g_1h_1 and $g_1 + h_1$ are performed in $\mathrm{GF}(p_1^{s_1})$).

We denote the elements from $\mathrm{GF}(p_i^{s_i})$, $i = 1, 2, \ldots, m$ as $0, g_i^{(1)} = 1, g_i^{(2)}, \ldots, g_i^{(p_i^{s_i}-1)}$. The following elements possess multiplicative inverses:

$$\gamma_i = \left(g_1^{(i)}, g_2^{(i)}, \ldots, g_m^{(i)}\right), \quad i = 1, 2, \ldots, n(q), \tag{10.34}$$

since they are composed of nonzero elements of corresponding Galois fields, and $\gamma_i - \gamma_j$ for $i \neq j$ as well. We denote the first element as $\gamma_0 = (0, 0, \ldots, 0)$, the next $n(q)$ as in (10.34), and the remaining $q - n(q) - 1$ elements as

$$\gamma_j = \left(g_1^{(i_1)}, g_2^{(i_2)}, \ldots, g_m^{(i_m)}\right), \; j = n(q) + 1, \ldots, q - 1; \;\; i_1 = 0, 1, \ldots, p_1^{s_1} - 1; \\ i_2 = 0, 1, \ldots, p_2^{s_2} - 1; \; \ldots; \;\; i_m = 0, 1, \ldots, p_m^{s_m} - 1, \tag{10.35}$$

with combinations of components leading to (10.34) being excluded. The $n(q)$ arrays of elements, MOLS, L_i, $i = 1,2,\ldots,n(q)$, are further formed with the (j,k)th cell being filled by $\gamma_j + \gamma_i\gamma_k$; $i = 1, 2, \ldots, n(q)$; $0 \leq j, k \leq q - 1$. (The multiplication $\gamma_i\gamma_k$ is performed using (10.33) and the addition $\gamma_j + \gamma_i\gamma_k$ using (10.32).) To establish the connection between MOLS and BIBD, we introduce the one-to-one mapping l: $L \rightarrow V$, with L being the MOLS and V being the integer set of q^2 elements (*points*). The simplest example of such a mapping is the linear mapping $l(x, y) = q(x - 1) + y$; $1 \leq x, y \leq q$ (therefore, $1 \leq l(x,y) \leq q^2$). (The numbers $l(x,y)$ are referred to as the cell *labels*.) Each L_i ($i = 1,\ldots,n(q)$) defines a set of q *blocks* (*lines*), $B_i = \{l(x, y) | L_i(x, y) = s, \;\; 1 \leq s \leq q\}$. Each line s ($1 \leq s \leq q$) in

B_i contains the labels of element s from L_i. These blocks are equivalent to the sets of lines of slopes $0 \leq s \leq q-1$ in the lattice design introduced in Ref. [19]. Labels of every line of a design specify the positions of ones in a row of the parity-check matrix. That is, the *incidence matrix* of a design is a $qn(q) \times q^2$ matrix $H = (h_{ij})$ defined by

$$h_{ij} = \begin{cases} 1 & \text{if the } i\text{th block contains the } j\text{th point} \\ 0 & \text{otherwise} \end{cases} \tag{10.36}$$

and represents the *parity-check matrix*. The code length is q^2, the code rate is determined by $R = (q^2 - \text{rank}(\boldsymbol{H}))/q^2$, and may be estimated as $R \cong (q^2 - n(q)q)/q^2 = 1 - n(q)/q$. The quantum code rate of the corresponding quantum code is given by

$$R_Q = \left(q^2 - 2\text{rank}(\boldsymbol{H}) + \text{rank}\left(\boldsymbol{H}\boldsymbol{H}^{\mathrm{T}}\right)\right)/q^2 \simeq \left(q^2 - 2n(q)q + \text{rank}\left(\boldsymbol{H}\boldsymbol{H}^{\mathrm{T}}\right)\right)/q^2. \tag{10.37}$$

10.4 ITERATIVE DECODING OF QUANTUM LDPC CODES

In Chapters 7 and 8, we defined the quantum error correction code (QECC), which encodes K qubits into N qubits, denoted as $[N,K]$, by an encoding mapping U from the K-qubit Hilbert space H_2^K onto a 2^K-dimensional subspace C_q of the N-qubit Hilbert space H_2^N. The $[N,K]$ quantum stabilizer code C_Q has been defined as the unique subspace of Hilbert space H_2^N that is fixed by the elements from stabilizer S of C_Q as follows:

$$C_Q = \bigcap_{s \in S} \left\{ |c\rangle \in H_2^N \,\middle|\, s|c\rangle = |c\rangle \right\}. \tag{10.38}$$

Any error in Pauli group G_N of N-qubit errors can be represented by

$$\begin{aligned} &E = j^{\lambda} X(\boldsymbol{a}) Z(\boldsymbol{b}); \quad \boldsymbol{a} = a_1 \ldots a_N; \quad \boldsymbol{b} = b_1 \ldots b_N; \quad a_i, b_i = 0,1; \; \lambda = 0,1,2,3 \\ &X(\boldsymbol{a}) \equiv X_1^{a_1} \otimes \ldots \otimes X_N^{a_N}; \quad Z(\boldsymbol{b}) \equiv Z_1^{b_1} \otimes \ldots \otimes Z_N^{b_N}. \end{aligned} \tag{10.39}$$

An important *property* of N-qubit errors is: $\forall\ E_1, E_2 \in G_N$ the following is valid, $[E_1, E_2] = 0$ or $\{E_1, E_2\} = 0$. In other words, the elements from the Pauli group either commute or anticommute:

$$\begin{aligned} E_1 E_2 &= j^{l_1 + l_2} X(\boldsymbol{a}_1) Z(\boldsymbol{b}_1) X(\boldsymbol{a}_2) Z(\boldsymbol{b}_2) = j^{l_1 + l_2} (-1)^{\boldsymbol{a}_1 \boldsymbol{b}_2 + \boldsymbol{b}_1 \boldsymbol{a}_2} X(\boldsymbol{a}_2) Z(\boldsymbol{b}_2) X(\boldsymbol{a}_1) Z(\boldsymbol{b}_1) \\ &= (-1)^{\boldsymbol{a}_1 \boldsymbol{b}_2 + \boldsymbol{b}_1 \boldsymbol{a}_2} E_2 E_1 = \begin{cases} E_2 E_1, & \boldsymbol{a}_1 \boldsymbol{b}_2 + \boldsymbol{b}_1 \boldsymbol{a}_2 = 0 \bmod 2 \\ -E_2 E_1, & \boldsymbol{a}_1 \boldsymbol{b}_2 + \boldsymbol{b}_1 \boldsymbol{a}_2 = 1 \bmod 2. \end{cases} \end{aligned} \tag{10.40}$$

In Chapters 7 and 8, we introduced the concept of syndrome of an error as follows. Let C_Q be a quantum stabilizer code with generators $g_1, g_2, \ldots, g_{N-K}$ and let $E \in G_N$ be an error. The *error syndrome* for error E is defined by the bit string $S(E) = [\lambda_1 \lambda_2 \ldots \lambda_{N-K}]^{\mathrm{T}}$ with component bits being determined by

$$\lambda_i = \begin{cases} 0, & [E, g_i] = 0 \\ 1, & \{E, g_i\} = 0 \end{cases} \qquad (i = 1, \ldots, N-K). \tag{10.41}$$

We have shown in Chapter 8 that the set of errors $\{E_i | E_i \in G_N\}$ has the same syndrome $\boldsymbol{\lambda} = [\lambda_1 \ldots \lambda_{N-K}]^{\mathrm{T}}$ if and only if they belong to the same coset $EC(S) = \{Ec | c \in C(S)\}$, where $C(S)$ is the *centralizer* of S (i.e. the set of errors $E \in G_N$ that commute with all elements from S). The Pauli encoded operators are introduced by the following mappings:

$$X_i \to \overline{X}_i = UX_{N-K+i}U^{\dagger} \quad \text{and} \quad Z_i \to \overline{Z}_i = UZ_{N-K+i}U^{\dagger}; \quad i = 1, \ldots, K, \tag{10.42}$$

where U is a Clifford operator acting on N qubits (an element of Clifford group $N(G_N)$, where N is the normalizer of G_N). By using this interpretation, the encoding operation is represented by

$$U(|\mathbf{0}\rangle_{N-K} \otimes |\psi\rangle), \qquad |\psi\rangle \in H_2^K. \tag{10.43}$$

An equivalent definition of syndrome elements is given by [65]

$$\lambda_i = \begin{cases} 1, & Eg_i = g_iE \\ -1, & Eg_i = -g_iE \end{cases} \qquad (i = 1, \ldots, N-K). \tag{10.44}$$

This definition can simplify verification of commutativity of two errors $E = E_1 \otimes \ldots \otimes E_N$ and $F = F_1 \otimes \ldots \otimes F_N$, where $E_i, F_i \in \{I, X, Y, Z\}$, as follows:

$$EF = \prod_{i=1}^{N} E_i F_i. \tag{10.45}$$

The Pauli channels can be described using the following error model, which has already been discussed in Section 7.5:

$$\xi(\rho) = \sum_{E \in G_N} p(E) E \rho E^{\dagger}, \quad p(E) \geq 0, \quad \sum_{E \in G_N} p(E) = 1. \tag{10.46}$$

A memoryless channel model can be defined in similar fashion to the discrete memoryless channel (DMC) models for classical codes (see Chapter 6) as follows [65]:

$$p(E = E_1 \ldots E_N) = p(E_1) \ldots p(E_N), \; E \in G_N. \tag{10.47}$$

The depolarizing channel is a very popular channel model for which $p(I) = 1 - p$, $p(X) = p(Y) = p(Z) = p/3$, for certain depolarizing strength $0 \leq p \leq 1$.

The probability of error E given syndrome λ can be calculated by

$$p(E|\lambda) \sim p(E) \prod_{i=1}^{N-K} \delta_{\lambda_i, Eg_i}. \quad (10.48)$$

The most likely error E_{opt} given the syndrome can be determined by

$$E_{\text{opt}}(\lambda) = \underset{E \in G_N}{\arg\max}\ p(E|\lambda). \quad (10.49)$$

Unfortunately, this approach is an NP-hard problem. We can use instead the qubit-wise most likely error evaluation as follows [65]:

$$E_{\text{q,opt}}(\lambda) = \underset{E_q \in G_1}{\arg\max}\ p_q(E_q|\lambda), \quad p_q(E_q|\lambda) = \sum_{E_1,\ldots,E_{q-1},E_{q+1},\ldots,E_N \in G_1} p(E_1 \ldots E_N|\lambda). \quad (10.50)$$

Notice that this marginal optimum does not necessarily coincide with the global optimum.

It is clear that from Eq. (10.50) that this marginal optimization involves summation of many terms exponential in N. A sum-product algorithm (SPA) similar to that discussed in Section 10.1 can be used here. For this purpose we can also use a *bipartite graph* representation of stabilizers. That is, we can identify stabilizers with function nodes and qubits with variable nodes. As an illustrative example let us consider (5,1) quantum cyclic code generated by the four stabilizers $g_1 = X_1Z_2Z_3X_4$, $g_2 = X_2Z_3Z_4X_5$, $g_3 = X_1X_3Z_4Z_5$, $g_4 = Z_1X_2X_4Z_5$. The corresponding quantum-check matrix is given by

$$\mathbf{A} = \begin{bmatrix} X \mid Z \\ 10010|01100 \\ 01001|00110 \\ 10100|00011 \\ 01010|10001 \end{bmatrix}.$$

The corresponding bipartite graph is shown in Figure 10.13, where dashed lines denote the X operator's action on particular qubits and solid lines denote the action of Z operators. This is an example of a *regular* quantum stabilizer code, with qubit node degree 3 and stabilizer node degree 4. For example, the qubit q_1 is involved in stabilizers g_1, g_3 and g_4, and therefore edges exist between this qubit and the corresponding stabilizers. The weight of edge is determined by action of the corresponding Pauli operator on the observed qubit:

$$g_1 = X_1Z_2Z_3X_4, \; g_2 = X_2Z_3Z_4X_5, \; g_3 = X_1X_3Z_4Z_5, \; g_4 = Z_1X_2X_4Z_5.$$

Let us denote the messages to be sent from qubit q to generator (check) node c as $m_{q \to c}$, which involve probability distributions over $E_q \in G_1 = \{I,X,Y,Z\}$ and

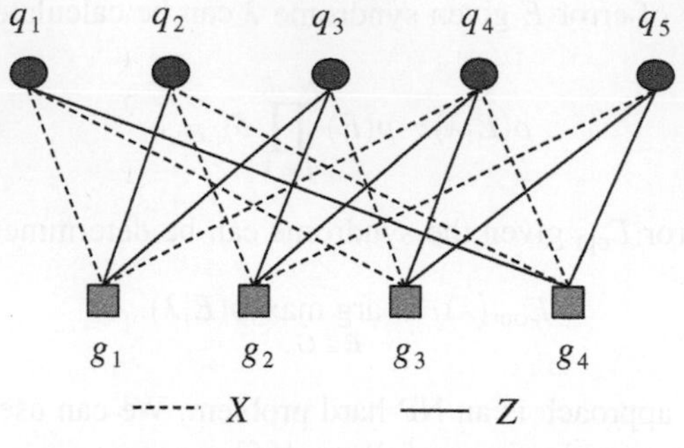

FIGURE 10.13

The bipartite graph of (5,1) quantum cyclic code with generators.

represent an array of four positive numbers, one for each value of E_q. Let us denote the neighborhood of q by $n(q)$ and the neighborhood of c by $n(c)$. The initialization is performed as $m_{q\to c}(E_q) = p_q(E_q)$, where $p_q(E_q)$ is given by (10.50). Each generator ("check") node c sends the message to its neighbor, qubit node q, by properly combining the extrinsic information from other neighbors as follows:

$$m_{c\to q}(E_q) \sim \sum_{E_{q'},\ q'\in n(c)\setminus q} \left(\delta_{\lambda_c, E_c g_c} \prod_{q'\in n(c)\setminus q} m_{q'\to c}(E_{q'})\right), \tag{10.51}$$

where $n(c)\backslash q$ denotes all neighbors of generator (check) node c except q. The proportionality factor can be determined from the normalization condition: $\sum_{E_i} m_{c\to q} = 1$. Clearly, the message $m_{c\to q}$ is related to the syndrome component λ_c associated with generator g_c and extrinsic information collected from all neighbors of c except q. This step can be called, in analogy with classical LDPC codes, the *generator (check) node update rule*.

The message to be sent from qubit q to generator (check) c is obtained by properly combining extrinsic information (message) from all generator neighbors of q, say c', except generator node c:

$$m_{q\to c}(E_q) \sim p_q(E_q) \prod_{c'\in n(q)\setminus c} m_{c'\to q}(E_q), \tag{10.52}$$

where $n(q)\backslash c$ denotes all neighbors of qubit node q except c. Clearly, $m_{q\to c}$ is a function of prior probability $p(E_q)$ and extrinsic messages collected from all neighbors of q except c. The proportionality constant in (10.52) can also be obtained from the normalization condition. Again, in analogy with classical LDPC codes, this step is called the *qubit node update rule*.

Steps (10.51) and (10.52) are iterated until a valid codeword is obtained or a predetermined number of iterations is reached. Now we have to calculate the beliefs $b_q(E_q)$, i.e. the marginal conditional probability estimation $p_q(E_q|\lambda)$, as follows:

$$b_q(E_q) \sim p_q(E_q) \prod_{c'\in n(q)} m_{c'\to q}(E_q). \tag{10.53}$$

Clearly, the beliefs are determined by taking the extrinsic messages of all neighbors of qubit q into account and the prior probability $p_q(E_q)$. The recovery operator can be obtained as the tensor product of qubit-wise maximum-belief Pauli operators:

$$E_{\mathrm{SPA}} = \otimes_{q=1}^{N} E_{\mathrm{SPA},q}, \quad E_{\mathrm{SPA},q} = \arg\max_{E_q \in G_1} \{ b_q(E_q) \}. \tag{10.54}$$

If the bipartite graph is a tree then the beliefs will converge to true conditional marginal probabilities $b_q(E_q) = p_q(E_q|\boldsymbol{\lambda})$ in a number of steps equal to the tree's depth. However, in the presence of cycles the beliefs will not converge to the correct conditional marginal probabilities in general. We will perform iterations until a trivial error syndrome is obtained, i.e. until $E_{\mathrm{MAP}} g_i = \lambda_i$ ($\forall\ i = 1,\ldots,N-K$) or until the predetermined number of iterations has been reached. This algorithm is applicable to both dual-containing and EA quantum codes.

The quantum SPA algorithm above is essentially very similar to the classical probability-domain SPA. In Section 10.1, we described the log-domain version of this algorithm, which is more suitable for hardware implementation as the multiplication of probabilities is replaced by addition of reliabilities.

Cycles of length 4, which exist in dual-containing codes, adversely affect the decoding BER performance. There are various methods to improve decoding performance, such as freezing, random perturbation, and collision methods [65]. The method described in Ref. [53], which was already described in Section 10.2, may also be used. In addition, by avoiding the so-called trapping sets, the BER performance of LDPC decoders can be improved [66,67].

10.5 SUMMARY

This chapter has considered quantum LDPC codes. In Section 10.1, classical LDPC codes, their design and decoding algorithms were described. Dual-containing quantum LDPC codes were explained in Section 10.2. Various design algorithms have been outlined. Section 10.3 covered entanglement-assisted quantum LDPC codes. Various classes of finite geometry codes are described. In Section 10.4, the probabilistic sum-product algorithm based on the quantum-check matrix was described. In the next section, a set of problems is given that will help the reader better understand the material in this chapter.

10.6 PROBLEMS

1. Consider an (8,4) product code composed of a (3,2) single-parity-check code:

c_0	c_1	c_2
c_3	c_4	c_5
c_6	c_7	

The parity-check equations of this code are given by

$$c_2 = c_0 + c_1, \quad c_5 = c_3 + c_4, \quad c_6 = c_0 + c_3, \quad c_7 = c_1 + c_4.$$

Write down the corresponding parity-check matrix and provide its Tanner graph. What is the minimum distance of this code? Can this classical code be used as a dual-containing quantum code? Explain your answer.

2. Consider a sequence of m independent binary digits $\boldsymbol{a} = (a_1, a_2, \ldots, a_m)$ in which $\Pr(a_k = 1) = p_k$.
   (a) Determine the probability that $\boldsymbol{a}$ contains an even number of ones.
   (b) Determine the probability that $\boldsymbol{a}$ contains an odd number of ones.

3. Here we will use the notation introduced in Section 10.1. Let $q_{ij}(b)$ denote the extrinsic information (message) to be passed from variable node v_i to function node f_j regarding the probability that $c_i = b$, $b \in \{0,1\}$, as illustrated in the left part of Figure 10.P3. This message concerns the probability that $c_i = b$ given extrinsic information from all check nodes, except node f_j, and given channel sample y_j. Let $r_{ji}(b)$ denote extrinsic information to be passed from node f_j to node v_i, as shown on the right of Figure 10.P3. This message concerns the probability that the jth parity-check equation is satisfied given $c_i = b$, and other bits have a separable distribution given by $\{q_{ij'}\}_{j' \neq j}$.
   (a) Show that $r_{ji}(0)$ and $r_{ji}(1)$ can be calculated from $q_{i'j}(0)$ and $q_{i'j}(1)$ by using Problem 2 as follows:

$$r_{ji}(0) = \frac{1}{2} + \frac{1}{2} \prod_{i' \in V_j \backslash i} (1 - 2q_{i'j}(1)),$$

$$r_{ji}(1) = \frac{1}{2} - \frac{1}{2} \prod_{i' \in V_j \backslash i} (1 - 2q_{i'j}(1)).$$

   (b) Show that $q_{ji}(0)$ and $q_{ji}(1)$ can be calculated as functions of $r_{ji}(0)$ and $r_{ji}(1)$ as follows:

$$q_{ij}(0) = (1 - P_i) \prod_{j' \in C_i \backslash j} r_{j'i}(0), \quad q_{ij}(1) = P_i \prod_{j' \in C_i \backslash j} r_{j'i}(1).$$

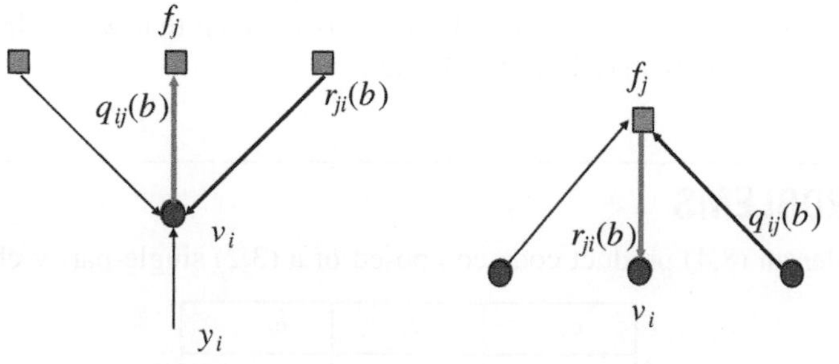

FIGURE 10.P3

Sum-product algorithm.

4. The probability domain sum-product algorithm "suffers" because (i) multiplications are involved (additions are less costly to implement) and (ii) in the calculation of the product many probabilities could become numerically unstable (especially for long codes with more than 50 iterations). The log-domain version of the sum-product algorithm is a preferable option. Instead of beliefs in the log domain we use log-likelihood ratios (LLRs), which for codeword bits c_i $(i = 2,\ldots,n)$ are defined as

$$L(c_i) = \log\left[\frac{\Pr(c_i = 0|y_i)}{\Pr(c_i = 1|y_i)}\right],$$

where y_i is a noisy sample corresponding to c_i. The LLRs corresponding to r_{ji} and q_{ij} are defined as follows:

$$L(r_{ji}) = \log\left[\frac{r_{ji}(0)}{r_{ji}(1)}\right], \quad L(q_{ij}) = \log\left[\frac{q_{ij}(0)}{q_{ij}(1)}\right].$$

Using $\tanh[\log(p_0/p_1)/2] = p_0 - p_1 = 1 - 2p_1$ show that:

(a) $L(r_{ji}) = 2\tanh^{-1}\{\prod_{i' \in V_j \backslash i} \tanh[\frac{1}{2}L(q_{i'j})]\}$.

(b) $L(q_{ij}) = L(c_i) + \sum_{j' \in C_i \backslash j} L(r_{j'i})$.

5. By using the following function:

$$\phi(x) = -\log\tanh\left(\frac{x}{2}\right) = \log\left[\frac{e^x + 1}{e^x - 1}\right],$$

show that $L(r_{ij})$ from Problem 4a can be written as

$$L(r_{ji}) = \left(\prod_{i'} \alpha_{i'j}\right)\phi\left[\sum_{i' \in C_j \backslash i} \phi(\beta_{i'j})\right],$$

where $\alpha_{ij} = \text{sign}[L(q_{ij})]$, $\beta_{ij} = |L(q_{ij})|$.

6. Show that $L(a_1 \oplus a_2) = L_1 \boxplus L_2$, $L_i = L(a_i)$ can be written as

$$L_1 \boxplus L_2 = \log\left(\frac{1 + e^{L_1 + L_2}}{e^{L_1} + e^{L_2}}\right).$$

7. Show that $L_1 \boxplus L_2$ from the previous problem can also be written as

$$L_1 \boxplus L_2 = \text{sign}(L_1)\text{sign}(L_2)\min(|L_1|,|L_2|) + s(L_1, L_2),$$

where $s(L_1, L_2) = \log(1 + e^{-|x+y|}) - \log(1 + e^{-|x-y|})$.

8. This problem concerns LDPC decoder implementation using the min-sum-with-correction-term algorithm. Implement the LDPC decoder using the min-sum-with-correction-term algorithm as described in Section 10.1.2.

(a) By assuming that the zero codeword was transmitted over an additive white Gaussian noise (AWGN) channel for PG(2,2^6)-based LDPC code, show the bit error rate (BER) dependence against signal-to-noise ratio (SNR) is dB scale for binary phase-shift keying (BPSK). Determine the coding gain at a BER of 10^{-6}.

(b) Now implement the corresponding encoder, generate a sufficiently long pseudo-random binary sequence, encode it, transmit it over the AWGN channel, and repeat the simulation from (a). Discuss the differences.

(c) Repeat the procedure for a binary symmetric channel (BSC).

9. This problem concerns LDPC decoder implementation using the probabilistic-domain algorithm described in Section 10.4. Implement this algorithm as described in that Section. Repeat steps (a)–(c) from Problem 8. Compare the results.

10. Let us now design an EA code based on Steiner triple systems from Section 10.3 having length comparable to PG(2,2^6)-based EA code. Which one is performing better at a BER of 10^{-6}? Justify the answer. Discuss the decoder complexity for these two codes.

11. Let us again observe EA code based on Steiner triple systems and PG(2,2^6)-based EA code. Evaluate the performance of these two EA codes (of comparable lengths) for the following channel models:

(a) Depolarizing channel.

(b) Amplitude damping channel.

For channel descriptions please refer to Chapter 7 (and Section 7.5).

References

[1] W.E. Ryan, Concatenated convolutional codes and iterative decoding, in: J.G. Proakis (Ed.), Wiley Encyclopedia of Telecommunications, John Wiley, 2003.

[2] I.B. Djordjevic, M. Arabaci, L. Minkov, Next generation FEC for high-capacity communication in optical transport networks, IEEE/OSA J. Lightwave Technol. 27 (2009) 3518–3530 (invited paper).

[3] I. Djordjevic, W. Ryan, B. Vasic, Coding for Optical Channels, Springer, 2010.

[4] R.G. Gallager, Low Density Parity Check Codes, MIT Press, Cambridge, MA, 1963.

[5] I.B. Djordjevic, S. Sankaranarayanan, S.K. Chilappagari, B. Vasic, Low-density parity-check codes for 40 Gb/s optical transmission systems, IEEE/LEOS J. Sel. Top. Quantum Electron. 12 (4) (July/August 2006) 555–562.

[6] J. Chen, A. Dholakia, E. Eleftheriou, M. Fossorier, X.-Y. Hu, Reduced-complexity decoding of LDPC codes, IEEE Trans. Commun. 53 (2005) 1288–1299.

[7] I.B. Djordjevic, B. Vasic, Nonbinary LDPC codes for optical communication systems, IEEE Photon. Technol. Lett. 17 (10) (October 2005) 2224–2226.

[8] M. Arabaci, I.B. Djordjevic, R. Saunders, R.M. Marcoccia, Non-binary quasi-cyclic LDPC based coded modulation for beyond 100-Gb/s transmission, IEEE Photon. Technol. Lett. 22 (6) (March 2010) 434–436.
[9] M.C. Davey, Error-Correction Using Low-Density Parity-Check Codes. Ph.D. dissertation, University of Cambridge, Cambridge, UK, 1999.
[10] D. Declercq, M. Fossorier, Decoding algorithms for nonbinary LDPC codes over GF(q), IEEE Trans. Commun. 55 (4) (April 2007) 633–643.
[11] M. Arabaci, I.B. Djordjevic, R. Saunders, R.M. Marcoccia, Polarization-multiplexed rate-adaptive non-binary-LDPC-coded multilevel modulation with coherent detection for optical transport networks, Opt. Express 18 (3) (2010) 1820–1832.
[12] M. Arabaci, I.B. Djordjevic, An alternative FPGA implementation of decoders for quasi-cyclic LDPC codes, Proc. TELFOR 2008 (November 2008) 351–354.
[13] M. Arabaci, I.B. Djordjevic, R. Saunders, R. Marcoccia, Non-binary LDPC-coded modulation for high-speed optical metro networks with back propagation, in: Proc. SPIE Photonics West 2010, OPTO: Optical Communications: Systems and Subsystems, Optical Metro Networks and Short-Haul Systems II, January 2010, pp. 23–28. San Francisco, CA, paper no. 7621-17.
[14] M. Arabaci, Nonbinary-LDPC-Coded Modulation Schemes for High-Speed Optical Communication Networks. Ph.D. dissertation, University of Arizona, November 2010.
[15] I.B. Djordjevic, O. Milenkovic, B. Vasic, "Generalized Low-Density Parity-Check Codes for Long-Haul High-Speed Optical Communications," in Optical Fiber Communication Conference and Exposition and The National Fiber Optic Engineers Conference on CD-Rom (Optical Society of America, Washington, DC, 2005), paper no. OThW6.
[16] B. Vasic, I.B. Djordjevic, R. Kostuk, Low-density parity check codes and iterative decoding for long haul optical communication systems, IEEE/OSA J. Lightwave Technol. 21 (February 2003) 438–446.
[17] I.B. Djordjevic, et al., Projective plane iteratively decodable block codes for WDM high-speed long-haul transmission systems, IEEE/OSA J. Lightwave Technol. 22 (March 2004) 695–702.
[18] O. Milenkovic, I.B. Djordjevic, B. Vasic, Block-circulant low-density parity-check codes for optical communication systems, IEEE/LEOS J. Sel. Top. Quantum Electron. 10 (March/April 2004) 294–299.
[19] B. Vasic, I.B. Djordjevic, Low-density parity check codes for long haul optical communications systems, IEEE Photon. Technol. Lett. 14 (August 2002) 1208–1210.
[20] S. Chung, et al., On the design of low-density parity-check codes within 0.0045 dB of the Shannon limit, IEEE Commun. Lett. 5 (February 2001) 58–60.
[21] S. Lin, D.J. Costello, Error Control Coding: Fundamentals and Applications, second ed., Prentice-Hall, 2004.
[22] R.M. Tanner, A recursive approach to low complexity codes, IEEE Trans. Inform. Theory, IT-27 (September 1981) 533–547.
[23] I.B. Djordjevic, "Codes on graphs, coded Modulation and compensition of nonlinear impairments by turbo equalization," in: S. Kumar (Ed.), Impact of Nonlinearities on Fiber Optic Communication, Springer, pp. 451–505, March 2011.
[24] M.P.C. Fossorier, Quasi-cyclic low-density parity-check codes from circulant permutation matrices, IEEE Trans. Inform. Theory 50 (2004) 1788–1793.
[25] W.E. Ryan, An introduction to LDPC codes, in: B. Vasic (Ed.), CRC Handbook for Coding and Signal Processing for Recording Systems, CRC Press, 2004.

[26] H. Xiao-Yu, E. Eleftheriou, D.-M. Arnold, A. Dholakia, Efficient implementations of the sum-product algorithm for decoding of LDPC codes, Proc. IEEE Globecom vol. 2 (November 2001). 1036–1036E.

[27] D.J.C. MacKay, Good error correcting codes based on very sparse matrices, IEEE Trans. Inform. Theory 45 (1999) 399–431.

[28] D. Raghavarao, Constructions and Combinatorial Problems in Design of Experiments, Dover Publications, New York, 1988.

[29] Y. Kou, S. Lin, M.P.C. Fossorier, Low-density parity-check codes based on finite geometries: A rediscovery and new results, IEEE Trans. Inform. Theory 47 (7) (November 2001) 2711–2736.

[30] I.B. Djordjevic, S. Sankaranarayanan, B. Vasic, Projective plane iteratively decodable block codes for WDM high-speed long-haul transmission systems, J. Lightwave Technol. 22 (3) (March 2004) 695–702.

[31] I.B. Djordjevic, S. Sankaranarayanan, B. Vasic, Irregular low-density parity-check codes for long haul optical communications, IEEE Photon. Technol. Lett. 16 (1) (January 2004) 338–340.

[32] B. Vasic, I.B. Djordjevic, Low-density parity check codes for long haul optical communications systems, IEEE Photon. Technol. Lett. 14 (8) (2002) 1208–1210.

[33] I.B. Djordjevic, O. Milenkovic, B. Vasic, Generalized low-density parity-check codes for optical communication systems, IEEE/OSA J. Lightwave Technol. 23 (May 2005) 1939–1946.

[34] J. Boutros, O. Pothier, G. Zemor, Generalized low density (Tanner) codes, Proc. 1999 IEEE Int. Conf. Communication (ICC 1999) vol. 1 (1999) 441–445.

[35] M. Lentmaier, K.S. Zigangirov, On generalized low-density parity check codes based on Hamming component codes, IEEE Commun. Lett. 3 (8) (August 1999) 248–250.

[36] T. Zhang, K.K. Parhi, High-performance, low-complexity decoding of generalized low-density parity-check codes, Proc. IEEE Global Telecommunications Conf. 2001 (IEEE Globecom) vol. 1 (2001) 181–185.

[37] I.B. Djordjevic, L. Xu, T. Wang, M. Cvijetic, GLDPC codes with Reed–Muller component codes suitable for optical communications, IEEE Commun. Lett. 12 (September 2008) 684–686.

[38] T. Richardson, A. Shokrollahi, R. Urbanke, Design of capacity approaching irregular low-density parity-check codes, IEEE Trans. Inform. Theory 47 (2) (February 2001) 619–637.

[39] William Shieh and Ivan Djordjeric, OFDM for Optical Communications. Elsevier/ Academic Press, Oct. 2009.

[40] A.R. Calderbank, P.W. Shor, Good quantum error-correcting codes exist, Phys. Rev. A. 54 (1996) 1098–1105.

[41] A.M. Steane, Error correcting codes in quantum theory, Phys. Rev. Lett. 77 (1996) 793.

[42] D.J.C. MacKay, G. Mitchison, P.L. McFadden, Sparse-graph codes for quantum error correction, IEEE Trans. Inform. Theory 50 (2004) 2315–2330.

[43] T. Camara, H. Ollivier, J.-P. Tillich, Constructions and Performance of Classes of Quantum LDPC Codes (2005). quant-ph/0502086.

[44] M. Hagiwara, H. Imai, Quantum Quasi-Cyclic LDPC Codes (2007). quant-ph/ 0701020.

[45] I.B. Djordjevic, Quantum LDPC codes from balanced incomplete block designs, IEEE Commun. Lett. 12 (May 2008) 389–391.

[46] I.B. Djordjevic, Photonic quantum dual-containing LDPC encoders and decoders, IEEE Photon. Technol. Lett. 21 (13) (1 July 2009) 842–844.

[47] I.B. Djordjevic, On the photonic implementation of universal quantum gates, Bell states preparation circuit, quantum relay and quantum LDPC encoders and decoders, IEEE Photon. J. 2 (1) (February 2010) 81–91.

[48] I.B. Djordjevic, Photonic implementation of quantum relay and encoders/decoders for sparse-graph quantum codes based on optical hybrid, IEEE Photon. Technol. Lett. 22 (19) (1 October 2010) 1449–1451.

[49] I.B. Djordjevic, Cavity quantum electrodynamics based quantum low-density parity-check encoders and decoders, in: SPIE Photonics West 2011, Advances in Photonics of Quantum Computing, Memory, and Communication IV, The Moscone Center, San Francisco, CA, January 2011, pp. 22–27, paper no. 7948-38.

[50] I. Anderson, Combinatorial Designs and Tournaments, Oxford University Press, Oxford, 1997.

[51] B. Vasic, O. Milenkovic, Combinatorial constructions of low-density parity-check codes for iterative decoding, IEEE Trans. Inform. Theory 50 (June 2004) 1156–1176.

[52] R.C. Bose, On the construction of balanced incomplete block designs, Ann. Eugen. 9 (1939) 353–399.

[53] S. Sankaranarayanan, B. Vasic, Iterative decoding of linear block codes: A parity-check orthogonalization approach, IEEE Trans. Inform. Theory 51 (September 2005) 3347–3353.

[54] T. Brun, I. Devetak, M.-H. Hsieh, Correcting quantum errors with entanglement, Science 314 (2006) 436–439.

[55] M.-H. Hsieh, I. Devetak, T. Brun, General entanglement-assisted quantum error correcting codes, Phys. Rev. A. 76 (19 December 2007). 062313-1–062313-7.

[56] I.B. Djordjevic, Photonic entanglement-assisted quantum low-density parity-check encoders and decoders, Opt. Lett. 35 (9) (1 May 2010) 1464–1466.

[57] M.-H. Hsieh, Entanglement-Assisted Coding Theory. Ph.D. dissertation, University of Southern California, August 2008.

[58] I. Devetak, T.A. Brun, M.-H. Hsieh, Entanglement-assisted quantum error-correcting codes, in: V. Sidoravičius (Ed.), New Trends in Mathematical Physics, Selected Contributions of the XVth International Congress on Mathematical Physics, Springer, 2009, pp. 161–172.

[59] M.-H. Hsieh, W.-T. Yen, L.-Y. Hsu, High performance entanglement-assisted quantum LDPC codes need little entanglement, IEEE Inform. Theory 57 (3) (March 2011) 1761–1769.

[60] M.M. Wilde, Quantum Coding with Entanglement. Ph.D. dissertation, University of Southern California, August 2008.

[61] S.S. Shrikhande, Generalized Hadamard matrices and orthogonal arrays of strength two, Can. J. Math. 16 (1964) 736–740.

[62] S. Sankaranarayanan, I.B. Djordjevic, B. Vasic, Iteratively decodable codes on *m*-flats for WDM high-speed long-haul transmission, J. Lightwave Technol. 23 (November 2005) 3696–3701.

[63] I.F. Blake, R.C. Mullin, The Mathematical Theory of Coding, Academic Press, New York, 1975.

[64] I.B. Djordjevic, B. Vasic, MacNeish–Mann theorem based iteratively decodable codes for optical communication systems, IEEE Commun. Lett. 8 (August 2004) 538–540.

[65] D. Poulin, Y. Chung, On the iterative decoding of sparce quantum codes, Quantum Inform. Comput. 8 (10) (2008) 987–1000.

[66] Z. Zhang, L. Dolecek, B. Nikolic, V. Anantharam, M. Wainwright, Design of LDPC decoders for improved low error rate performance: Quantization and algorithm choices, IEEE Trans. Commun. 57 (11) (November 2009) 3258–3268.

[67] M. Ivkovic, S.K. Chilappagari, B. Vasic, Eliminating trapping sets in low-density parity-check codes by using Tanner graph covers, IEEE Trans. Inform. Theory 54 (8) (August 2008) 3763–3768.

CHAPTER

11 Fault-Tolerant Quantum Error Correction and Fault-Tolerant Quantum Computing

CHAPTER OUTLINE

This chapter considers one of the most important applications of quantum error correction, namely the protection of quantum information as it dynamically undergoes computation through so-called fault-tolerant quantum computing [1–6]. The chapter starts by introducing fault tolerance basics and traversal operations (Section 11.1). It continues with fault-tolerant quantum computation concepts and procedures (Section 11.2). In Section 11.2, the universal set of fault-tolerant quantum gates is introduced, followed by fault-tolerant measurement, fault-tolerant state preparation, and fault-tolerant encoded state preparation using Steane's code as an illustrative example. Section 11.3 provides a rigorous description of fault-tolerant

Quantum Information Processing and Quantum Error Correction. DOI: 10.1016/B978-0-12-385491-9.00011-3

quantum error correction. The section starts with a short review of some basic concepts from stabilizer codes. In subsection 11.3.1 fault-tolerant syndrome extraction is described, followed by a description of fault-tolerant encoded operations in subsection 11.3.2. Subsection 11.3.3 is concerned with the application of the quantum gates on a quantum register by means of a measurement protocol. This method, when applied transversally, is used in subsection 11.3.4 to enable fault-tolerant error correction based on an arbitrary quantum stabilizer code. The fault-tolerant stabilizer codes are described in subsection 11.3.4. In subsection 11.3.5, the [5,1,3] fault-tolerant stabilizer code is described as an illustrative example. Section 11.4 covers fault-tolerant computing, in particular the fault-tolerant implementation of the Tofolli gate. Finally, in Section 11.5, the accuracy threshold theorem is formulated and proved. After a summary in Section 11.6, in the final section (Section 11.7) a set of problems is provided to enable readers to gain a deeper understanding of fault-tolerance theory.

11.1 FAULT TOLERANCE BASICS

One of the most powerful applications of quantum error correction is the protection of quantum information as it dynamically undergoes quantum computation. Imperfect quantum gates affect quantum computation by introducing errors in computed data. Moreover, imperfect control gates introduce errors in a processed sequence since the wrong operations are applied. A quantum error correction coding (QECC) scheme now needs to deal not only with errors introduced by quantum channels, but also with errors introduced by imperfect quantum gates during the encoding/decoding process. Because of this, the reliability of data processed by quantum computers is not *a priori* guaranteed by QECC. The reason is threefold: (i) the gates used for encoders and decoders are composed of imperfect gates, including controlled imperfect gates; (ii) syndrome extraction applies unitary operators to entangle ancillary qubits with code block; and (iii) the error recovery action requires the use of controlled operation to correct for the errors. Nevertheless, it can be shown that arbitrary good quantum error protection can be achieved even with imperfect gates, providing that the error probability per gate is below a certain *threshold*; this claim is known as the accuracy threshold theorem and will be discussed later in the chapter.

From the discussion above it is clear that QECC does not *a priori* improve the reliability of a quantum computer. So the purpose of *fault-tolerant design* is to ensure the reliability of quantum computers given a threshold, by properly implementing the fault-tolerant quantum gates. Similarly as in previous chapters, we observe the following error model: (1) qubit errors occur independently; (2) X, Y, and Z qubit errors occur with the same probability; (3) the error probability per qubit is the same for all qubits; and (4) the errors introduced by imperfect quantum gates affect only the qubits acted on by that gate.

Condition (4) is added to ensure that errors introduced by imperfect gates do not propagate and cause catastrophic errors. Based on this error model, we can define fault-tolerant operation as follows.

Definition 11.1. We say that a quantum operation is *fault-tolerant* if the occurrence of a single gate/storage error during the operation does not produce more than one error on each encoded qubit block.

In other words, an operation is said to be fault-tolerant if, up to the first order in single-error probability p (i.e. $O(p)$), the operation generates no more than one error per block. The key idea of fault-tolerant computing is to construct fault-tolerant operations to perform the logic on encoded states. It may be argued that if we use a strong enough QECC we might allow for more than one error per code block. However, the QECC scheme is typically designed to deal with random errors introduced by quantum channels or storage devices. Allowing for too many errors due to imperfect gates may lead to the error correction capability of QECC being exceeded when both random and imperfect gate errors are present. By restricting the number of faulty gate errors to one, we can simplify fault-tolerant design.

Example. Let us observe the SWAP gate introduced in Chapter 3, which interchanges the states of two qubits: $U_{\text{SWAP}}^{(12)}|\psi_1\rangle|\psi_2\rangle = |\psi_2\rangle|\psi_1\rangle$. If the SWAP gate is imperfect it can introduce errors on both qubits, and based on Definition 10.1 above, this gate is faulty. However, if we introduce an additional auxiliary state, we can perform the SWAP operation fault-tolerantly as follows: $U_{\text{SWAP}}^{(23)}U_{\text{SWAP}}^{(12)}U_{\text{SWAP}}^{(13)}|\psi_1\rangle|\psi_2\rangle|\psi_3\rangle = |\psi_2\rangle|\psi_1\rangle|\psi_3\rangle$. For example, let us assume that a storage error occurred on qubit 1, leading to $|\psi_1\rangle \to |\psi_1'\rangle$, before the SWAP operation takes place. The resulting state will be $|\psi_2\rangle|\psi_1'\rangle|\psi_3\rangle$, so that a single error occurring during SWAP operation results in only one error per block, meaning that this modified gate operates fault-tolerantly. Another related operation commonly used in fault-tolerant quantum computing, namely *transversal operation*, can be shown to be fault-tolerant.

Definition 11.2. An operation that satisfies one of the following two conditions is a *transversal operation*: (1) it only employs one-qubit gates to the qubits in a code block; (2) the ith qubit in one code block interacts only with the ith qubit in a different code block or block of ancillary qubits.

Condition (1) is obviously consistent with the definition of fault tolerance. Let us now observe two-qubit operations, in particular the operation of gates involving the ith qubits of two different blocks. Based on the error model above, only these two qubits can be affected by this faulty gate. Since these two interacting qubits belong to two different blocks, the transversal operation is fault-tolerant. The definition of transversal operation is quite useful as it is much easier to verify than the definition of fault tolerance. In the next section, in order to familiarize readers with fault-tolerant quantum computing concepts, we will develop fault-tolerant gates assuming that the QECC scheme is based on Steane code [5].

11.2 FAULT-TOLERANT QUANTUM COMPUTATION CONCEPTS

In this section, we introduce basic fault-tolerant computation concepts. The purpose of this section is to gradually introduce the reader to the fault-tolerant concepts, before moving to the rigorous exposition in Section 11.3.

11.2.1 Fault-Tolerant Pauli Gates

The key idea of fault-tolerant computing is to construct fault-tolerant operations to perform the logic on encoded states. The encoded Pauli operators for Pauli X and Z operators for Steane [7,1] code are given as follows:

$$\overline{Z} = Z_1 Z_2 Z_3 Z_4 Z_5 Z_6 Z_7 \qquad \overline{X} = X_1 X_2 X_3 X_4 X_5 X_6 X_7. \tag{11.1}$$

The encoded Pauli gate Y can easily be implemented as

$$\overline{Y} = \overline{X}\,\overline{Z}. \tag{11.2}$$

Let us now check if the encoded Pauli gates are transversal gates. Based on the quantum channel model from the previous section, it is clear that single-qubit errors do not propagate. Because every encoded gate is implemented in bitwise fashion, the encoded Pauli gates are transversal and consequently fault-tolerant. The implementation of fault-tolerant Pauli gates is shown in Figure 11.1.

11.2.2 Fault-Tolerant Hadamard Gate

An encoded Hadamard gate H should interchange X and Z under conjugation, the same way as an unencoded Hadamard gate interchanges Z and X:

$$\overline{H} = H_1 H_2 H_3 H_4 H_5 H_6 H_7. \tag{11.3}$$

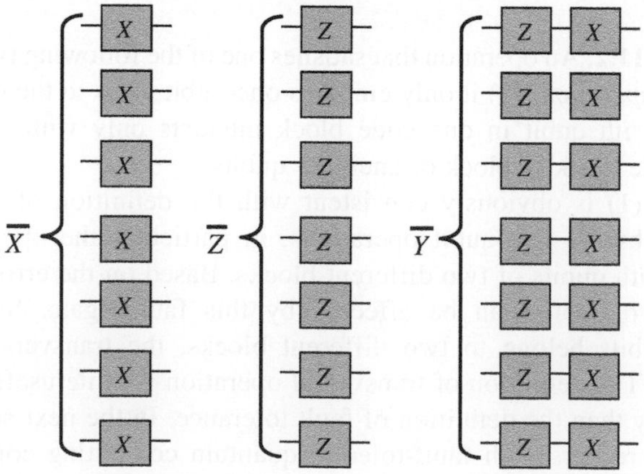

FIGURE 11.1

Transversal Pauli gates on a qubit encoded using a Steane [7,1] code.

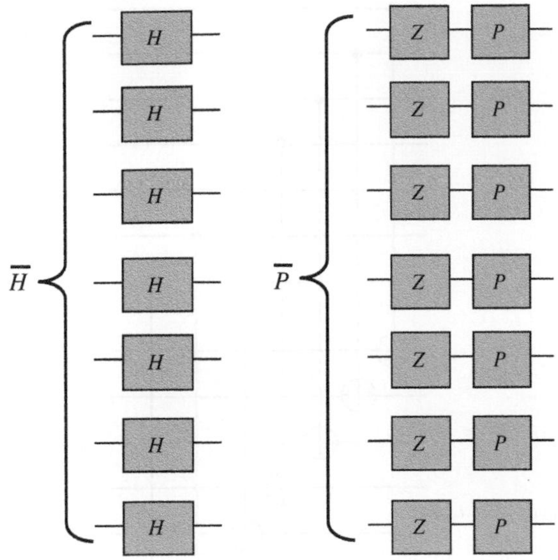

FIGURE 11.2

Transversal Hadamard and phase gates on a qubit encoded using a Steane [7,1] code.

Fault-tolerant Hadamard gate implementations is shown in Figure 11.2 (left). A single error, occurring on one qubit, does not propagate any further, indicating that an encoded Hadamard gate is transversal, and consequently fault-tolerant.

11.2.3 Fault-Tolerant Phase Gate (*P*)

Under conjugation the phase gate P maps Z to Z and X to Y as follows:

$$PZP^{\dagger} = Z \qquad PXP^{\dagger} = Y. \tag{11.4}$$

A fault-tolerant phase gate $\overline{P}$ should perform the same action on encoded Pauli gates. The bitwise operation of $\overline{P} = P_1P_2P_3P_4P_5P_6P_7$ on encoded Pauli Z and X gates gives

$$\overline{P}\,\overline{Z}\,\overline{P}^{\dagger} = \overline{Z} \qquad \overline{P}\,\overline{X}\,\overline{P}^{\dagger} = -\overline{Y}, \tag{11.5}$$

indicating that the sign in the conjugation operation of the encoded Pauli-X gate is incorrect, which can be fixed by inserting a Z operator in front of a single P operator, so that the fault-tolerant phase gates can be represented as shown in Figure 11.2 (right). A single error, occurring on one qubit, does not propagate any further, indicating that the operation is transversal and fault-tolerant.

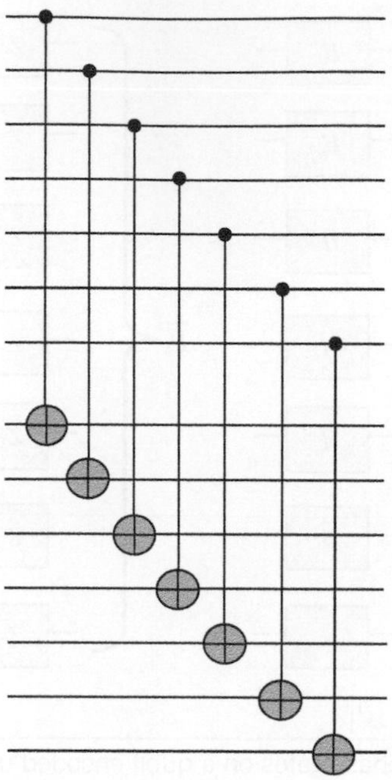

FIGURE 11.3

The fault-tolerant CNOT gate implementation.

11.2.4 Fault-Tolerant CNOT Gate

The fault-tolerant CNOT gate is shown in Figure 11.3. Let us assume that the ith qubit in the upper block interacts with the ith qubit of the lower block. From Figure 11.3 it is clear that an error in the ith qubit of the upper block will only affect the ith qubit of the lower block, so that condition (2) of Definition 10.2 is satisfied and the gate in Figure 11.3 is transversal, and consequently (based on Definition 10.1) fault-tolerant. The fault-tolerant quantum gates discussed so far are sufficient to implement an arbitrary encoder and decoder fault-tolerantly.

11.2.5 Fault-Tolerant $\pi/8$ (T) Gate

For a complete set of universal gates, the implementation of a nontransversal gate such as the $\pi/8$ (T) gate is needed. In order to implement the T gate fault-tolerantly, we need to apply the following three-step procedure [3,5]:

1. Prepare an ancillary, in base state, and apply a Hadamard followed by a nonfault-tolerant T gate:

$$|\phi\rangle = TH|0\rangle = T\frac{|0\rangle + |1\rangle}{\sqrt{2}} = \frac{|0\rangle + e^{j\pi/4}|1\rangle}{\sqrt{2}}. \qquad (11.6)$$

2. Apply a CNOT gate on the prepared state $|\phi\rangle$ as control qubit and on the state we want to transform $|\psi\rangle$ as the target qubit:

$$\begin{aligned}
U^{(12)}_{\mathrm{CNOT}}|\phi\rangle|\psi\rangle &= U^{(12)}_{\mathrm{CNOT}}\frac{|0\rangle + e^{j\pi/4}|1\rangle}{\sqrt{2}}\otimes(a|0\rangle + b|1\rangle)\\
&= \frac{1}{\sqrt{2}}U^{(12)}_{\mathrm{CNOT}}\left[|0\rangle(a|0\rangle + b|1\rangle) + e^{j\pi/4}|1\rangle(a|0\rangle + b|1\rangle)\right]\\
&= \frac{1}{\sqrt{2}}(a|0\rangle|0\rangle + b|0\rangle|1\rangle) + e^{j\pi/4}(a|1\rangle|1\rangle + b|1\rangle|0\rangle)\\
&= \frac{1}{\sqrt{2}}\left[(a|0\rangle + be^{j\pi/4}|1\rangle)|1\rangle + (b|0\rangle + ae^{j\pi/4}|1\rangle)|1\rangle\right], \qquad (11.7)
\end{aligned}$$

where we use superscript (12) to denote the action of a CNOT gate from qubit 1 (control qubit) to qubit 2 (target qubit).

3. In the third step, we need to measure the original qubit $|\psi\rangle$: If the result of a measurement is zero, the final state is $a|0\rangle + be^{j\pi/4}|1\rangle$; otherwise we have to apply a PX gate on the originally ancillary qubit as follows:

$$\begin{aligned}
PX\begin{bmatrix} b \\ ae^{j\pi/4}\end{bmatrix} &= \begin{bmatrix} 1 & 0 \\ 0 & j\end{bmatrix}\begin{bmatrix} 0 & 1 \\ 1 & 0\end{bmatrix}\begin{bmatrix} b \\ ae^{j\pi/4}\end{bmatrix} = \begin{bmatrix} 0 & 1 \\ j & 0\end{bmatrix}\begin{bmatrix} b \\ ae^{j\pi/4}\end{bmatrix}\\
&= \begin{bmatrix} ae^{j\pi/4} \\ be^{j\pi/2}\end{bmatrix} = e^{j\pi/4}\begin{bmatrix} a \\ be^{j\pi/4}\end{bmatrix} = e^{j\pi/4}T|\psi\rangle. \qquad (11.8)
\end{aligned}$$

Because $THZHT^{+} = TXT^{+} = e^{-j\pi/4}PX$, the state $|\phi\rangle$ can also be prepared from $|0\rangle$ fault-tolerantly by measuring PX: If the result is $+1$, preparation was successful, otherwise the procedure needs to be repeated. The fault-tolerant T gate implementation is shown in Figure 11.4. In this implementation, we assumed that the measurement procedure is perfect. The fault-tolerant measurement procedure is discussed in the next section.

11.2.6 Fault-Tolerant Measurement

The quantum circuit for performing measurement on a single-qubit operator U is shown in Figure 11.5a. Its operation has already been described in Chapter 3. This circuit is clearly faulty. The first idea of performing fault-tolerant measurement is to

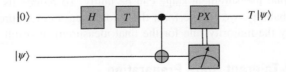

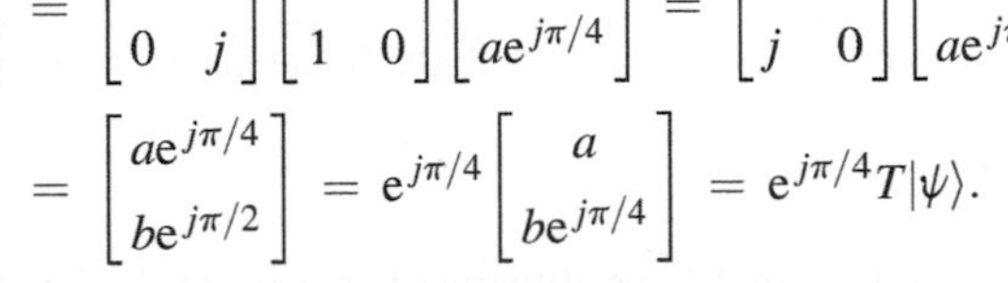

FIGURE 11.4

The fault-tolerant T gate implementation.

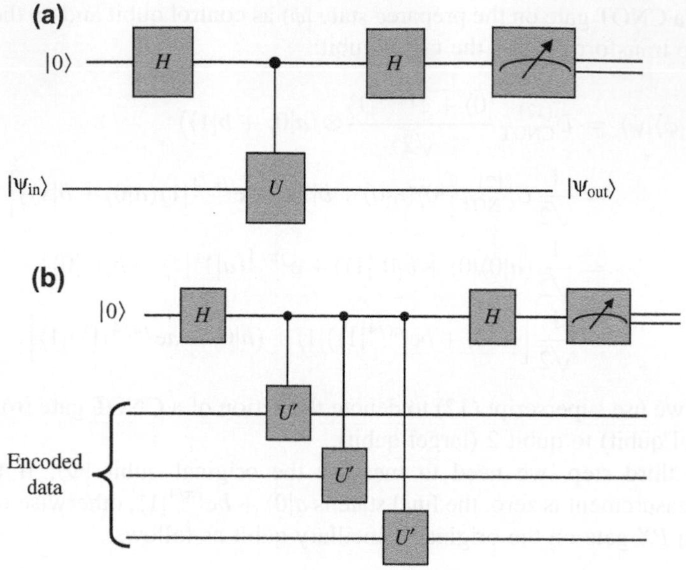

FIGURE 11.5

Faulty measurement circuits: (a) original circuit; (b) modified circuit obtained by putting U in transversal form.

put U in a transversal form U', as illustrated in Figure 11.5b, and then perform measurement in bitwise fashion with the same control ancillary. Unfortunately, such a measurement is not fault-tolerant, because an error in the ancillary qubit would affect several data qubits. We can convert the circuit shown in Figure 11.5b into fault-tolerant form by using three ancillary qubits as shown in Figure 11.6. This new circuit has several stages. The first stage is the cat-state $(|000\rangle+|111\rangle)/\sqrt{2}$ preparation circuit. The encoder in this stage is very similar to the three-qubit flip-code encoder described in Chapter 7. The second stage is the verification stage, which has certain similarities with the three-qubit flip-code error correction circuit. That is, this stage is used to verify if the state after the first stage is indeed the cat state. The third stage is the controlled-U stage, similarly as in Figure 11.5b. The fourth stage is the decoder stage, which is used to return ancillary states back to the original state. As expected, this circuit has similarities with a three-qubit code decoder. The final measurement stage can be faulty. To reduce its importance on qubit error probability, we can perform the whole measurement procedure several times and apply the majority rule for the final measurement result.

11.2.7 Fault-Tolerant State Preparation

The fault-tolerant state preparation procedure can be described as follows. We prepare the desired state and apply the verification procedure measurement as

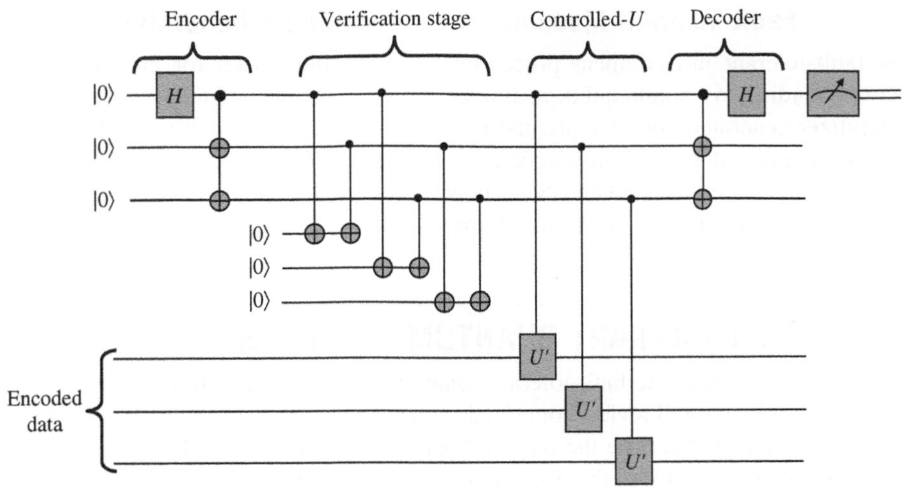

FIGURE 11.6

Fail-tolerant measurement circuit.

above. If the verification step is successful, that state is used for further quantum computation, otherwise we restart the verification procedure again, with fresh ancillary qubits.

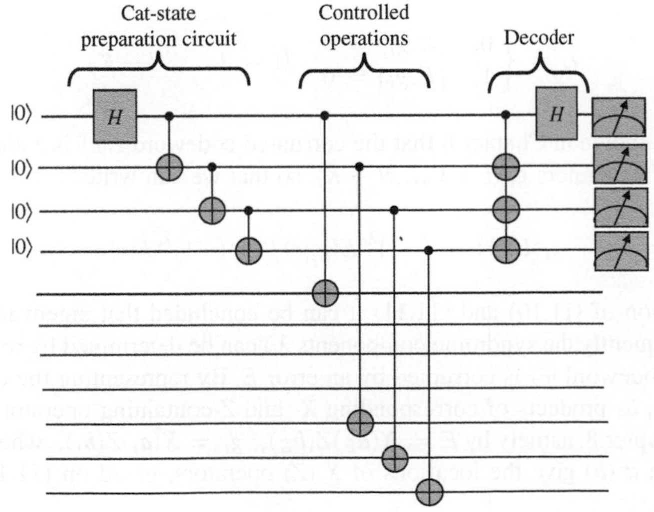

FIGURE 11.7

Fault-tolerant measurement for the generator $X_1X_5X_6X_7$.

11.2.8 Fault-Tolerant Measurement of Stabilizer Generators

A fault-tolerant measurement procedure has been introduced for an observable corresponding to a single qubit. A similar procedure can be applied to an arbitrary stabilizer generator. For example, the fault-tolerant measurement circuit of the first stabilizer generator of Steane's code $g_1 = X_1X_5X_6X_7$, from Section 8.4, is shown in Figure 11.7. After this elementary introduction of fault-tolerant concepts, gates and procedures, we turn our attention to a more rigorous description.

11.3 FAULT-TOLERANT QUANTUM ERROR CORRECTION

This section is devoted to fault-tolerant quantum error correction. Before we proceed with this topic, we will review some basic concepts of stabilizer codes introduced in Chapters 7 and 8. Let S be the largest Abelian subgroup of G_N (the N-qubit Pauli multiplicative error group) that fixes all elements $|c\rangle$ from quantum code C_Q, called the stabilizer group. The $[N,K]$ stabilizer code C_Q is defined as the K-dimensional subspace of the N-qubit Hilbert space H_2^N as follows:

$$C_Q = \bigcap_{s \in S} \left\{ |c\rangle \in H_q^N \middle| s|c\rangle = |c\rangle \right\}. \tag{11.9}$$

The very important concept of quantum stabilizer codes is the concept of syndrome of an error E from G_N. Let C_Q be a quantum stabilizer code with generators $g_1, g_2, \ldots, g_{N-K}$, then the error syndrome of $E \in G_N$ can be defined as the bit string $S(E) = [\lambda_1\lambda_2 \ldots \lambda_{N-K}]^T$ with component bits being determined by

$$\lambda_i = \begin{cases} 0, & [E, g_i] = 0 \\ 1, & \{E, g_i\} = 0 \end{cases} \quad (i = 1, \ldots, N-K). \tag{11.10}$$

It has been shown in Chapter 8 that the corrupted codeword $E|c\rangle$ is a simultaneous eigenket of generators g_i $(i = 1,\ldots,N - K)$, so that we can write:

$$g_i(E|c\rangle) = (-1)^{\lambda_i} E(g_i|c\rangle) = (-1)^{\lambda_i} E|c\rangle. \tag{11.11}$$

By inspection of (11.10) and (11.11) it can be concluded that eigenvalues $(-1)^{\lambda_i}$ (and consequently the syndrome components λ_i) can be determined by measuring g_i when the codeword $|c\rangle$ is corrupted by an error E. By representing the error E and generator g_i as products of corresponding X- and Z-containing operators as it was done in Chapter 8, namely by $E = X(\boldsymbol{a}_E)Z(\boldsymbol{b}_E)$, $g_i = X(\boldsymbol{a}_i)Z(\boldsymbol{b}_i)$, where nonzero positions in $\boldsymbol{a}$ ($\boldsymbol{b}$) give the locations of X (Z) operators, based on (11.11) we can write:

$$g_iE = (-1)^{\boldsymbol{a}_E \cdot \boldsymbol{b}_i + \boldsymbol{b}_E \cdot \boldsymbol{a}_i} E g_i, \tag{11.12}$$

where $\boldsymbol{a}_E \cdot \boldsymbol{b}_i + \boldsymbol{b}_E \cdot \boldsymbol{a}_i$ denotes the sympletic product introduced in Section 8.3 (the bitwise addition is mod 2). From Eqs (11.11) and (11.12) it is obvious that the ith component of the syndrome can be obtained by

$$\lambda_i = \boldsymbol{a}_E \cdot \boldsymbol{b}_i + \boldsymbol{b}_E \cdot \boldsymbol{a}_i \mod 2; \quad i = 1, \ldots, N - K. \tag{11.13}$$

By inserting identity operators, $I = U^\dagger U$, where U is a unitary operator, into (11.11) the following is obtained:

$$\begin{aligned} &g_i \overbrace{U^\dagger U}^{I} E \overbrace{U^\dagger U}^{I} |c\rangle = (-1)^{\lambda_i} E \overbrace{U^\dagger U}^{I} |c\rangle \\ &\overset{\cdot U(\text{from the left})}{\Leftrightarrow} (U g_i U^\dagger)(U E U^\dagger) U |c\rangle = (-1)^{\lambda_i} (U E U^\dagger) U |c\rangle. \end{aligned} \tag{11.14}$$

By introducing the notation $\overline{g}_i = U g_i U^\dagger$, $\overline{E} = U E U^\dagger$, and $|\overline{c}\rangle = U|c\rangle$, the second line of (11.14) can be rewritten as

$$\overline{g}_i \overline{E} |\overline{c}\rangle = (-1)^{\lambda_i} \overline{E} |\overline{c}\rangle. \tag{11.15}$$

Therefore, the corrupted image of the codeword $\overline{E}|\overline{c}\rangle$ is a simultaneous eigenket of the images of generators $\overline{g}_i$. Similarly to (11.14), we can show that

$$g_i|c\rangle = |c\rangle \Leftrightarrow g_i \overbrace{U^\dagger U}^{I} |c\rangle = |c\rangle \overset{\cdot U(\text{from the left})}{\Leftrightarrow} U g_i \overbrace{U^\dagger U}^{I} |c\rangle = U|c\rangle \Leftrightarrow \overline{g}_i |\overline{c}\rangle = |\overline{c}\rangle, \tag{11.16}$$

which will be used in the next section.

11.3.1 Fault-Tolerant Syndrome Extraction

To simplify further exposition of fault-tolerant quantum error correction, for the moment we will assume that generators $g_i = X(\boldsymbol{a}_i)Z(\boldsymbol{b}_i)$ do not contain the Y operators, in other words $a_{ij} b_{ij} = 0 \;\; \forall j$. In syndrome extraction, the ancillary qubits interact with a codeword in such a way to encode the syndrome $S(E)$ information into ancillary qubits and then perform appropriate measurements on ancillary qubits to reveal the syndrome vector $S(E)$. We will show that the syndrome information $S(E)$ can be incorporated by applying the transversal Hadamard $\overline{H}$ and transversal CNOT gate $\overline{U}_{\text{CNOT}}$ on the corrupted codeword $E|c\rangle$ as follows:

$$\overline{H}_i \overline{U}_{\text{CNOT}}(\boldsymbol{a}_i + \boldsymbol{b}_i) \overline{H}_i E|c\rangle \otimes |\boldsymbol{A}\rangle = E|c\rangle \otimes \sum_{\boldsymbol{A}_{\text{even}}} |\boldsymbol{A}_{\text{even}} + \boldsymbol{b}_E \cdot \boldsymbol{a}_i + \boldsymbol{a}_E \cdot \boldsymbol{b}_i\rangle, \tag{11.17}$$

where $|\boldsymbol{A}\rangle$ is the *Shor state* defined by

$$|\boldsymbol{A}\rangle = \frac{1}{2^{(\text{wt}(g_i)-1)/2}} \sum_{\boldsymbol{A}_{\text{even}}} |\boldsymbol{A}_{\text{even}}\rangle, \tag{11.18}$$

with summation being performed over all-even parity bit strings of length $\mathrm{wt}(g_i)$, denoted as $\boldsymbol{A}_{\mathrm{even}}$. The Shor state is obtained by first applying the cat-state preparation circuit on $|\mathbf{0}\rangle_{\mathrm{wt}(g_i)} = \underbrace{|0\rangle \otimes \ldots \otimes |0\rangle}_{\mathrm{wt}(g_i)\ \mathrm{times}}$ ancillary qubits, as explained in the corresponding text of Figure 11.7, followed by the set of $\mathrm{wt}(g_i)$ Hadamard gates, as shown in Figure 11.8. Therefore, we can represent the Shor state in terms of cat-state $|\Psi_{\mathrm{cat}}\rangle$ as follows:

$$|\boldsymbol{A}\rangle = \prod_{j=1}^{\mathrm{wt}(g_i)} H_j |\Psi_{\mathrm{cat}}\rangle, \quad |\Psi_{\mathrm{cat}}\rangle = \frac{1}{\sqrt{2}}[|0\ldots0\rangle + |1\ldots1\rangle]. \tag{11.19}$$

It is evident from Figure 11.8 that the stage of Hadamard gates is transversal as the ith error in the observed block introduces error only on the ith qubit of the corresponding interacting block. On the other hand, the error introduced in an arbitrary inner CNOT gate of the cat-state preparation circuit propagates and qubits below are affected, indicating that the cat-state preparation circuit is not fault-tolerant. We will return to this problem and Shor state verification later in this section, once we introduce the transversal Hadamard gate $\overline{H}$ and transversal CNOT gate $\overline{U}_{\mathrm{CNOT}}$. The transversal Hadamard gate $\overline{H}$ and transversal CNOT gate $\overline{U}_{\mathrm{CNOT}}$, related to generator

$$g_i = X(\boldsymbol{a}_i)Z(\boldsymbol{b}_i); \quad \boldsymbol{a}_i = [a_{i1}\ldots a_{iN}]^{\mathrm{T}}, \quad \boldsymbol{a}_i = [b_{i1}\ldots b_{iN}]^{\mathrm{T}} \quad (a_{ij}b_{ij} = 0 \ \forall j),$$

are defined respectively as

$$\overline{H}_i \doteq \prod_{j=1}^{N} (H_j)^{a_{ij}}, \quad \overline{U}_{\mathrm{CNOT}}(\boldsymbol{a}_i + \boldsymbol{b}_i) \doteq \prod_{j=1}^{N} (U_{\mathrm{CNOT}})^{a_{ij}+b_{ij}}. \tag{11.20}$$

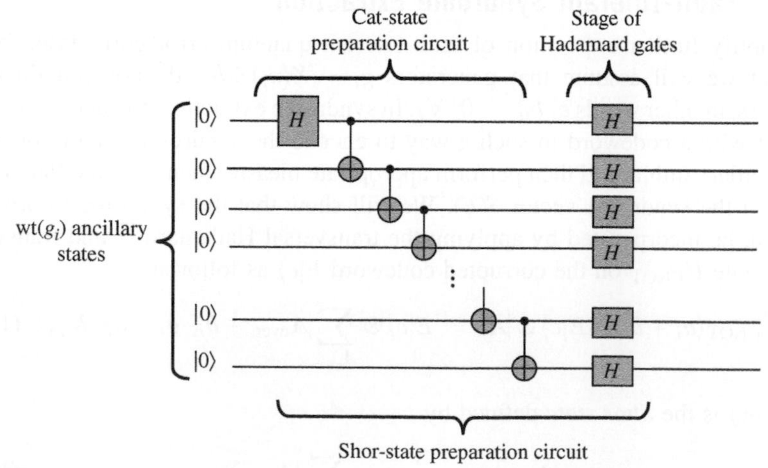

FIGURE 11.8

Shor-state preparation circuit.

In (11.20), the transversal CNOT gate $\overline{U}_{\text{CNOT}}$ is controlled by wt(g_i) qubits for which $a_{ij} + b_{ij} = 1$. From Eq. (11.17) it is evident that the encoded block and ancillary block are in a non-entangled state, and by performing measurement on the ancillary block we will not affect the encoded block (to collapse into one of the eigenkets). By conducting the measurement on the ancillary qubits, the modified Shor state will collapse into one of the eigenkets, say $|\boldsymbol{A}'_{\text{even}} + \boldsymbol{b}_E \cdot \boldsymbol{a}_i + \boldsymbol{a}_E \cdot \boldsymbol{b}_i\rangle$. Since $\text{wt}(\boldsymbol{A}'_{\text{even}}) \bmod 2 = 0$, the result of the measurement will be

$$(-1)^{\boldsymbol{b}_E \cdot \boldsymbol{a}_i + \boldsymbol{a}_E \cdot \boldsymbol{b}_i} = (-1)^{\lambda_i}, \tag{11.21}$$

and the ith syndrome component $\lambda_i = \boldsymbol{b}_E \cdot \boldsymbol{a}_i + \boldsymbol{a}_E \cdot \boldsymbol{b}_i$ mod2 will be revealed. By using a similar procedure for all components of the syndrome, we can identify the most probable error E based on syndrome $S(E) = [\lambda_1 \ldots \lambda_{N-K}]^{\text{T}}$ and perform the corresponding error recovery action $\boldsymbol{R} = R_1 \ldots R_N$.

As a reminder, the action of the Hadamard gate on the jth qubit is to perform the transformation $H_j X_j H_j = Z_j$. The application of transversal H gate on generator g_i yields:

$$\begin{aligned}\overline{g}_i &= \overline{H}_i g_i \overline{H}_i^{\dagger} \overset{\overline{H}_i^{\dagger} = \overline{H}_i}{=} \overline{H}_i X(\boldsymbol{a}_i) Z(\boldsymbol{b}_i) \overline{H}_i = \overline{H}_i X(\boldsymbol{a}_i) \overline{H}_i Z(\boldsymbol{b}_i) \overset{H_j X_j H_j = Z_j}{=} Z(\boldsymbol{a}_i) Z(\boldsymbol{b}_i) \\ &= Z(\boldsymbol{a}_i + \boldsymbol{b}_i).\end{aligned} \tag{11.22}$$

In similar fashion, we can show that the action of transversal H gate on error E gives

$$\begin{aligned}\overline{E} &= \overline{H}_i E \overline{H}_i^{\dagger} \overset{\overline{H}_i^{\dagger} = \overline{H}_i}{=} \overline{H}_i X(\boldsymbol{a}_E) Z(\boldsymbol{b}_E) \overline{H}_i \overset{\overline{\boldsymbol{a}}_i + \boldsymbol{a}_i = (1 \ldots 1)}{=} \overline{H}_i X(\boldsymbol{a}_E \cdot (\overline{\boldsymbol{a}}_i + \boldsymbol{a}_i)) \\ &\quad \times Z(\boldsymbol{b}_E \cdot (\overline{\boldsymbol{a}}_i + \boldsymbol{a}_i)) \overline{H}_i \\ &= \overline{H}_i \underbrace{X(\boldsymbol{a}_E \cdot \overline{\boldsymbol{a}}_i + \boldsymbol{a}_E \cdot \boldsymbol{a}_i)}_{X(\boldsymbol{a}_E \cdot \overline{\boldsymbol{a}}_i) X(\boldsymbol{a}_E \cdot \boldsymbol{a}_i)} \underbrace{Z(\boldsymbol{b}_E \cdot \overline{\boldsymbol{a}}_i + \boldsymbol{b}_E \cdot \boldsymbol{a}_i)}_{Z(\boldsymbol{b}_E \cdot \overline{\boldsymbol{a}}_i) Z(\boldsymbol{b}_E \cdot \boldsymbol{a}_i)} \overline{H}_i \\ &= \overline{H}_i X(\boldsymbol{a}_E \cdot \overline{\boldsymbol{a}}_i) \overline{H}_i \underbrace{\overline{H}_i X(\boldsymbol{a}_E \cdot \boldsymbol{a}_i) \overline{H}_i}_{Z(\boldsymbol{a}_E \cdot \boldsymbol{a}_i)} \overline{H}_i Z(\boldsymbol{b}_E \cdot \overline{\boldsymbol{a}}_i) \overline{H}_i \underbrace{\overline{H}_i Z(\boldsymbol{b}_E \cdot \boldsymbol{a}_i) \overline{H}_i}_{X(\boldsymbol{b}_E \cdot \boldsymbol{a}_i)} \\ &= X(\boldsymbol{a}_E \cdot \overline{\boldsymbol{a}}_i) Z(\boldsymbol{a}_E \cdot \boldsymbol{a}_i) Z(\boldsymbol{b}_E \cdot \overline{\boldsymbol{a}}_i) X(\boldsymbol{b}_E \cdot \boldsymbol{a}_i),\end{aligned} \tag{11.23}$$

where the "$\cdot$" operation is defined by $\boldsymbol{a} \cdot \boldsymbol{b} = (a_1 b_1, \ldots, a_N b_N)$ and $\overline{\boldsymbol{a}}_i$ is the bitwise complement $\overline{\boldsymbol{a}}_i = (a_{i1} + 1, \ldots, a_{iN} + 1)$. In the derivation above we used the property $\overline{\boldsymbol{a}}_i + \boldsymbol{a}_i = (1, \ldots, 1)$. The left-hand side of (11.17) can be rewritten, based on (11.23), as follows:

$$\begin{aligned}&\overline{H}_i \overline{U}_{\text{CNOT}}(\boldsymbol{a}_i + \boldsymbol{b}_i) \overline{H}_i E |c\rangle \otimes |\boldsymbol{A}\rangle \\ &= \overline{H}_i \overline{U}_{\text{CNOT}}(\boldsymbol{a}_i + \boldsymbol{b}_i) \overbrace{\overline{H}_i E \overline{H}_i^{\dagger}}^{\overline{E}} \overbrace{\overline{H}_i g_i \overline{H}_i g_i \overline{H}_i |c\rangle}^{g_i^2 |\overline{c}\rangle = |\overline{c}\rangle} \otimes |\boldsymbol{A}\rangle \\ &= \overline{H}_i \overline{U}_{\text{CNOT}}(\boldsymbol{a}_i + \boldsymbol{b}_i) \overline{E} |\overline{c}\rangle \otimes |\boldsymbol{A}\rangle.\end{aligned} \tag{11.24}$$

By substituting the last line of (11.23) into (11.24) we obtain:

$$\overline{H}_i \overline{U}_{\text{CNOT}}(\boldsymbol{a}_i+\boldsymbol{b}_i)\overline{H}_i E|c\rangle \otimes |\boldsymbol{A}\rangle \overset{|\overline{c}\rangle = \sum_m \overline{c}(m)|m\rangle}{=} \overline{H}_i \overline{U}_{\text{CNOT}}(\boldsymbol{a}_i+\boldsymbol{b}_i) X(\boldsymbol{a}_E\cdot\overline{\boldsymbol{a}}_i) \times Z(\boldsymbol{a}_E\cdot\boldsymbol{a}_i+\boldsymbol{b}_E\cdot\overline{\boldsymbol{a}}_i) X(\boldsymbol{b}_E\cdot\boldsymbol{a}_i) \sum_{\boldsymbol{m}} \overline{c}(\boldsymbol{m})|\boldsymbol{m}\rangle \otimes |\boldsymbol{A}\rangle, \tag{11.25}$$

where we used the expansion $|\overline{c}\rangle = \sum_{\boldsymbol{m}} \overline{c}(\boldsymbol{m})|\boldsymbol{m}\rangle$ in terms of computational basis (CB) kets. Upon applying the X-containing operators we obtain:

$$\begin{aligned}
&\overline{H}_i \overline{U}_{\text{CNOT}}(\boldsymbol{a}_i+\boldsymbol{b}_i)\overline{H}_i E|c\rangle \otimes |\boldsymbol{A}\rangle \\
&= \overline{H}_i \overline{U}_{\text{CNOT}}(\boldsymbol{a}_i+\boldsymbol{b}_i) X(\boldsymbol{a}_E\cdot\overline{\boldsymbol{a}}_i) Z(\boldsymbol{a}_E\cdot\boldsymbol{a}_i+\boldsymbol{b}_E\cdot\overline{\boldsymbol{a}}_i) \sum_{\boldsymbol{m}} \overline{c}(\boldsymbol{m})|\boldsymbol{m}+\boldsymbol{b}_E\cdot\boldsymbol{a}_i\rangle \otimes |\boldsymbol{A}\rangle \\
&= (-1)^{\lambda} \overline{H}_i \overline{U}_{\text{CNOT}}(\boldsymbol{a}_i+\boldsymbol{b}_i) Z(\boldsymbol{a}_E\cdot\boldsymbol{a}_i+\boldsymbol{b}_E\cdot\overline{\boldsymbol{a}}_i) X(\boldsymbol{a}_E\cdot\overline{\boldsymbol{a}}_i) \sum_{\boldsymbol{m}} \overline{c}(\boldsymbol{m})|\boldsymbol{m}+\boldsymbol{b}_E\cdot\boldsymbol{a}_i\rangle \otimes |\boldsymbol{A}\rangle \\
&= (-1)^{\lambda} \overline{H}_i \overline{U}_{\text{CNOT}}(\boldsymbol{a}_i+\boldsymbol{b}_i) Z(\boldsymbol{a}_E\cdot\boldsymbol{a}_i+\boldsymbol{b}_E\cdot\overline{\boldsymbol{a}}_i) \sum_{\boldsymbol{m}} \overline{c}(\boldsymbol{m})|\boldsymbol{m}+\boldsymbol{b}_E\cdot\boldsymbol{a}_i+\boldsymbol{a}_E\cdot\overline{\boldsymbol{a}}_i\rangle \otimes |\boldsymbol{A}\rangle,
\end{aligned} \tag{11.26}$$

where $\lambda = (\boldsymbol{b}_E\cdot\overline{\boldsymbol{a}}_i)\cdot(\boldsymbol{a}_E\cdot\overline{\boldsymbol{a}}_i)$. (Since $X(\boldsymbol{a}_E\cdot\overline{\boldsymbol{a}}_i)$ and $Z(\boldsymbol{a}_E\cdot\boldsymbol{a}_i)$ act on different qubits they commute.) By the application of a transversal CNOT gate on the Shor state from (11.26), we obtain:

$$\begin{aligned}
\overline{H}_i \overline{U}_{\text{CNOT}}(\boldsymbol{a}_i+\boldsymbol{b}_i)\overline{H}_i E|c\rangle \otimes |\boldsymbol{A}\rangle &= (-1)^{\lambda} \overline{H}_i Z(\boldsymbol{a}_E\cdot\boldsymbol{a}_i+\boldsymbol{b}_E\cdot\overline{\boldsymbol{a}}_i) \sum_{\boldsymbol{m}} \overline{c}(\boldsymbol{m})|\boldsymbol{m}+\boldsymbol{b}_E\cdot\boldsymbol{a}_i \\
&\quad +\boldsymbol{a}_E\cdot\overline{\boldsymbol{a}}_i\rangle \otimes \sum_{\boldsymbol{A}_{\text{even}}} \Big|\boldsymbol{A}_{\text{even}} + \underbrace{(\boldsymbol{a}_i+\boldsymbol{b}_i)(\boldsymbol{b}_E\cdot\boldsymbol{a}_i+\boldsymbol{a}_E\cdot\boldsymbol{b}_i)}_{\boldsymbol{a}_E\cdot\boldsymbol{b}_i+\boldsymbol{b}_E\cdot\boldsymbol{a}_i}\Big\rangle \\
&= E|c\rangle \otimes \sum_{\boldsymbol{A}_{\text{even}}} |\boldsymbol{A}_{\text{even}}+\boldsymbol{b}_E\cdot\boldsymbol{a}_i+\boldsymbol{a}_E\cdot\boldsymbol{b}_i\rangle,
\end{aligned} \tag{11.27}$$

therefore proving Eq. (11.17).

Example: The [5,1,3] code (revisited). The generators of this code are

$$g_1 = X_1Z_2Z_3X_4,\quad g_2 = X_2Z_3Z_4X_5,\quad g_3 = X_1X_3Z_4Z_5,\quad \text{and}\quad g_4 = Z_1X_2X_4Z_5.$$

Based on Eq. (11.20) and the generators above, we obtain the transversal Hadamard gates as follows:

$$\overline{H}_1 = H_1H_4,\quad \overline{H}_2 = H_2H_5,\quad \overline{H}_3 = H_1H_3,\quad \overline{H}_4 = H_2H_4.$$

Based on Eq. (11.22), the application of the transversal H gate on generators g_i ($i = 1,\ldots,4$) yields:

$$\begin{aligned}\overline{g}_1 &= Z(\boldsymbol{a}_1+\boldsymbol{b}_1) = Z_1Z_2Z_3Z_4, \quad \overline{g}_2 = Z(\boldsymbol{a}_2+\boldsymbol{b}_2) = Z_2Z_3Z_4Z_5,\\ \overline{g}_3 &= Z(\boldsymbol{a}_3+\boldsymbol{b}_3) = Z_1Z_3Z_4Z_5, \quad \overline{g}_4 = Z_1Z_2Z_4Z_5.\end{aligned}$$

The corresponding circuit for syndrome extraction is shown in Figure 11.9. The upper block corresponds to the action on codeword qubits and the lower block corresponds to the action on ancillary qubits (Shor state). It is clear that single-qubit errors in a codeword block can affect only one qubit in an ancillary block, suggesting that this implementation is transversal and consequently fault-tolerant. Let us observe the stage corresponding to generator g_1. Based on the transparent Hadamard gate $\overline{H}_1 = H_1H_4$ we have to place two Hadamard gates having action on codeword block qubits H_1 and H_4, respectively. The corresponding generator image $\overline{g}_1 = Z_1Z_2Z_3Z_4$ indicates that qubits 1, 2, 3, and 4 in the codeword block are control qubits, while the qubits in the Shor state block are target qubits. The purpose of this manipulation is to encode the syndrome information into ancillary qubits. We further perform measurements on the modified Shor state and the result of the measurement provides information about syndrome component λ_1. The other g_i ($i = 2,3,4$) stages operate in a similar fashion. Once the syndrome vector is determined, from the corresponding look-up table we determine the most probable error and apply the corresponding recovery action $\boldsymbol{R} = R_1R_2R_3R_4R_5$ to undo the action of imperfect gates.

As an alternative, Plenio et al. [7] proposed rearranging the sequence of CNOT gates in Eq. (11.17) as follows:

$$\overline{H}_i\overline{U}_{\mathrm{CNOT}}(\boldsymbol{a}_i)\overline{H}_i\overline{U}_{\mathrm{CNOT}}(\boldsymbol{b}_i)E|c\rangle \otimes |\boldsymbol{A}\rangle, \tag{11.28}$$

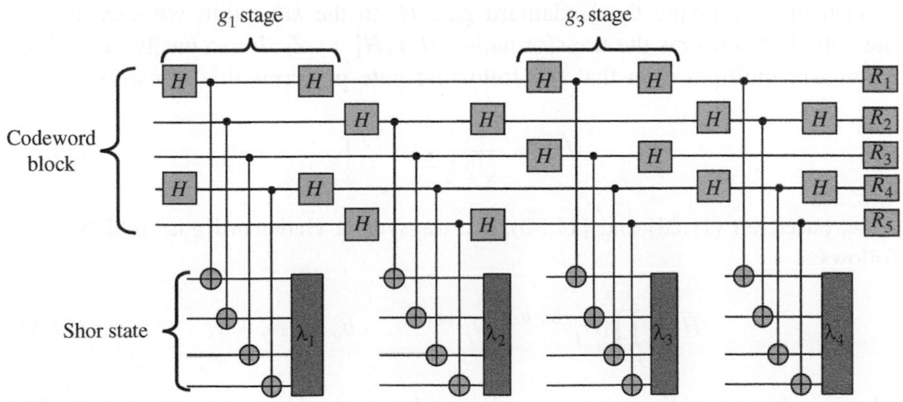

FIGURE 11.9

The syndrome extraction circuit for [5,1,3] stabilizer code.

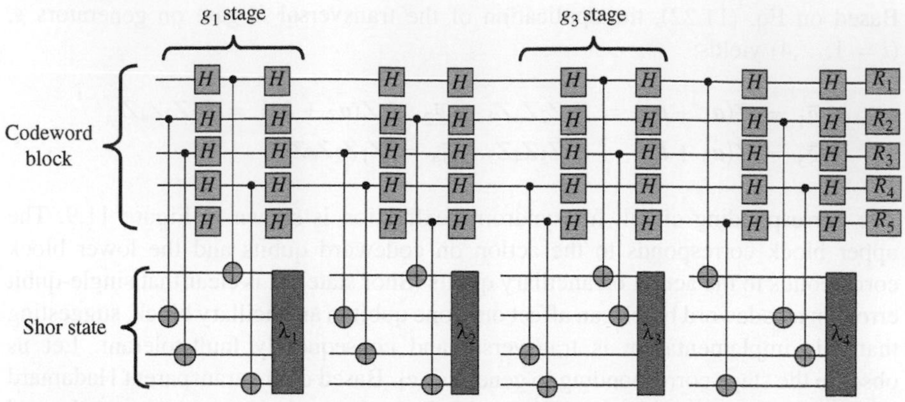

FIGURE 11.10

The syndrome extraction circuit for [5,1,3] stabilizer code based on the approach due to Plenio et al. [7].

where the transversal Hadamard gate is now applied on all qubits:

$$\overline{H}_i = \prod_{j=1}^{N} H_j. \tag{11.29}$$

Based on the generators above and Eqs (11.28) and (11.29), the corresponding syndrome extraction circuit is shown in Figure 11.10. The operating principle of this circuit is very similar to that shown in Figure 11.9.

In the discussion above we assumed that generators do not contain Y-operators. If generators do contain Y-operators, when constructing the images of generators $\overline{g}_i$ instead of applying the Hadamard gate H_k to the kth qubit, we need to find the gate that performs the transformation $\tilde{H}_k Y_k \tilde{H}_k^\dagger = Z_k$. It can easily be verified by matrix multiplication that the following gate performs this transformation:

$$\tilde{H}_k = \frac{1}{\sqrt{2}} \begin{bmatrix} 1 & -j \\ -j & 1 \end{bmatrix}. \tag{11.30}$$

Now, based on (11.20) and (11.30), the transversal Hadamard gate is defined as follows:

$$\overline{H}_i \doteq \prod_{j=1}^{N} (\tilde{H}_j)^{(\boldsymbol{a}_i \cdot \boldsymbol{b}_i)_j} (H_j)^{(\boldsymbol{a}_i \cdot \overline{\boldsymbol{b}}_i)_j}, \quad \overline{b}_{ij} = b_{ij} + 1, \tag{11.31}$$

where we have used the notation $(\boldsymbol{a}_i \cdot \boldsymbol{b}_i)_j$ to denote the jth component of $\boldsymbol{a}_i \cdot \boldsymbol{b}_i$. The images of generators now become

$$\begin{aligned}\overline{g}_i &= \overline{H}_i g_i \overline{H}_i^\dagger = \overline{H}_i X(\boldsymbol{a}_i) Z(\boldsymbol{b}_i) \overline{H}_i^\dagger = \overline{H}_i Y(\boldsymbol{a}_i \cdot \boldsymbol{b}_i) X(\boldsymbol{a}_i \cdot \overline{\boldsymbol{b}}_i) Z(\overline{\boldsymbol{a}}_i \cdot \boldsymbol{b}_i) \overline{H}_i^\dagger \\ &= \underbrace{\prod_{j=1}^{N} (\tilde{H}_j)^{(\boldsymbol{a}_i \cdot \boldsymbol{b}_i)_j} Y(\boldsymbol{a}_i \cdot \boldsymbol{b}_i) (\tilde{H}_j^\dagger)^{(\boldsymbol{a}_i \cdot \boldsymbol{b}_i)_j}}_{Z(\boldsymbol{a}_i \cdot \boldsymbol{b}_i)} \underbrace{\prod_{j=1}^{N} (H_j)^{(\boldsymbol{a}_i \cdot \overline{\boldsymbol{b}})_j} X(\boldsymbol{a}_i \cdot \overline{\boldsymbol{b}}_i) (H_j)^{(\boldsymbol{a}_i \cdot \overline{\boldsymbol{b}}_i)_j}}_{Z(\boldsymbol{a}_i \cdot \overline{\boldsymbol{b}})} Z(\overline{\boldsymbol{a}}_i \cdot \boldsymbol{b}_i) \\ &= Z(\boldsymbol{a}_i \cdot \boldsymbol{b}_i + \boldsymbol{a}_i \cdot \overline{\boldsymbol{b}}_i + \overline{\boldsymbol{a}}_i \cdot \boldsymbol{b}_i) = Z(\boldsymbol{a}_i \cdot \boldsymbol{b}_i + \boldsymbol{a}_i + \boldsymbol{a}_i \cdot \boldsymbol{b}_i + \boldsymbol{b}_i + \boldsymbol{a}_i \cdot \boldsymbol{b}_i) \\ &= Z(\boldsymbol{a}_i + \boldsymbol{b}_i + \boldsymbol{a}_i \cdot \boldsymbol{b}_i). \end{aligned} \tag{11.32}$$

In similar fashion, the image of the error E can be found by

$$\overline{E} = \overline{H}_i E \overline{H}_i^\dagger = (-1)^\lambda Z(\boldsymbol{a}_E \cdot (\boldsymbol{a}_i \cdot \overline{\boldsymbol{b}}_i) + \boldsymbol{b}_E \cdot \overline{\boldsymbol{a}_i \cdot \overline{\boldsymbol{b}}_i}) X(\boldsymbol{a}_E \cdot \boldsymbol{a}_i \cdot \overline{\boldsymbol{b}}_i + \boldsymbol{b}_E \cdot \boldsymbol{a}_i), \tag{11.33}$$

where $\lambda = (\boldsymbol{a}_E \cdot \overline{\boldsymbol{a}_i \cdot \overline{\boldsymbol{b}}_i}) \cdot (\boldsymbol{b}_E \cdot \overline{\boldsymbol{a}_i \cdot \overline{\boldsymbol{b}}_i})$. By application of the transversal CNOT gate on $E|c\rangle \otimes |\boldsymbol{A}\rangle$, we obtain:

$$\overline{U}_{\text{CNOT}}(\boldsymbol{a}_i + \boldsymbol{b}_i + \boldsymbol{a}_i \cdot \boldsymbol{b}_i) E|c\rangle \otimes |\boldsymbol{A}\rangle = \overline{E}|\overline{c}\rangle \otimes \sum_{\boldsymbol{A}_{\text{even}}} |\boldsymbol{A}_{\text{even}} + \boldsymbol{b}_E \cdot \boldsymbol{a}_i + \boldsymbol{a}_E \cdot \boldsymbol{b}_i\rangle. \tag{11.34}$$

Finally, by applying the transversal Hadamard gate given by Eq. (11.31) in (11.34), we derive:

$$\overline{H}_i \overline{U}_{\text{CNOT}}(\boldsymbol{a}_i + \boldsymbol{b}_i + \boldsymbol{a}_i \cdot \boldsymbol{b}_i) \overline{H}_i E|c\rangle \otimes |\boldsymbol{A}\rangle = E|c\rangle \otimes \sum_{\boldsymbol{A}_{\text{even}}} |\boldsymbol{A}_{\text{even}} + \boldsymbol{b}_E \cdot \boldsymbol{a}_i + \boldsymbol{a}_E \cdot \boldsymbol{b}_i\rangle, \tag{11.35}$$

which has a similar form to Eq. (11.17).

Example: The [4,2,2] code (revisited). The starting point, in this example, is generators of the code:

$$g_1 = X_1 Z_2 Z_3 X_4 \text{ and } g_2 = Y_1 X_2 X_3 Y_4.$$

Based on (11.31) we obtain the following transversal Hadamard gates corresponding to generators g_i $(i = 1,2)$:

$$\overline{H}_1 = H_1 H_4 \text{ and } \overline{H}_2 = \tilde{H}_1 H_2 H_3 \tilde{H}_4.$$

Based on the last line of (11.32) we obtain the images of generators as follows:

$$\overline{g}_1 = Z_1 Z_2 Z_3 Z_4 \text{ and } \overline{g}_2 = Z_1 Z_2 Z_3 Z_4.$$

Finally, based on (11.35) the syndrome extraction circuit for [4,2,2] code is shown in Figure 11.11.

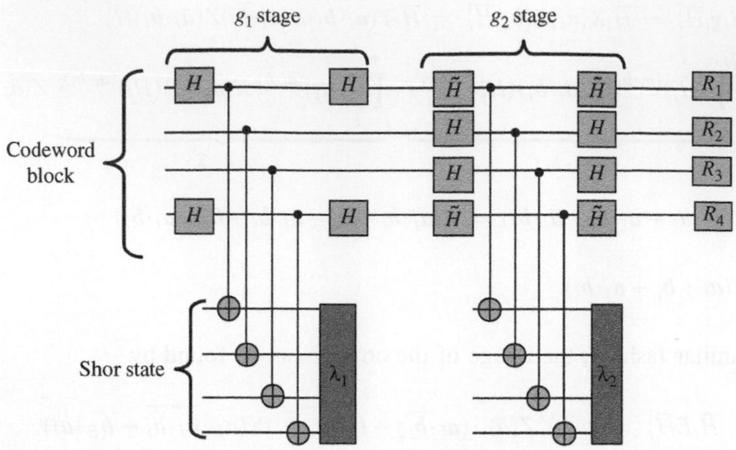

FIGURE 11.11

The syndrome extraction circuit for [4,2,2] stabilizer code.

The syndrome extraction circuits discussed above employ only transversal operations providing that the Shor state was prepared fault-tolerantly. Closer inspection of the Shor preparation circuit (see Figure 11.8) reveals that error introduced by the wrong qubit or faulty CNOT gate can cause the propagation of errors. From Figure 11.8, we can see that the first and last qubits have an even parity. Therefore, by adding an additional parity check on the first and last qubits, we can verify if the even parity condition is satisfied. If it is satisfied, we know that up to $O(p)$ no flip error has occurred. If the measurement of Z on an additional ancillary qubit is 1, we know that an odd number of errors occurred, and we have to repeat the verification procedure. The modified Shor preparation circuit enabling this verification is shown in Figure 11.12.

We learned earlier that if we perform measurement on the modified Shor state, this state will collapse into one of its eigenkets, say $|\boldsymbol{A}'_{\text{even}} + \boldsymbol{b}_E \cdot \boldsymbol{a}_i + \boldsymbol{a}_E \cdot \boldsymbol{b}_i\rangle$, and the result of the measurement will be $(-1)^{\boldsymbol{b}_E \cdot \boldsymbol{a}_i + \boldsymbol{a}_E \cdot \boldsymbol{b}_i + \text{wt}(\boldsymbol{A}'_{\text{e}})} = (-1)^{\lambda_i + \text{wt}(\boldsymbol{A}'_{\text{e}})}$. If an odd number of bit-flip errors has occurred during Shor state preparation, $\text{wt}(\boldsymbol{A}'_{\text{e}}) = 1$ and the syndrome value will be $\lambda_i + 1$. The other sources of errors are storage errors and imperfect CNOT gates used in syndrome calculation. The syndrome $S(E)$ can be used to identify the error introduced by various sources of imperfections discussed above. In an ideal world, the most probable error can be detected based on the syndrome and used to perform error correction. However, the syndrome extraction procedure, as mentioned above, can also be faulty. A wrongly determined syndrome during the syndrome extraction procedure can introduce additional errors during error correction. This problem can be solved by the following *syndrome verification protocol* due to Plenio et al. [7]:

1. If the result of measurements returns $S(E) = 0$, we accept the measurement result as correct and assume that no recovery operation is needed.

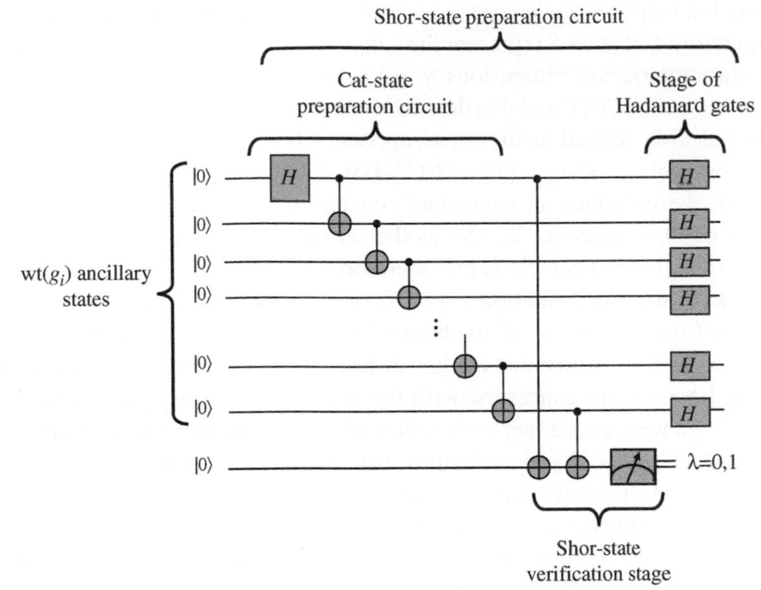

FIGURE 11.12

Modified Shor-state preparation circuit with verification stage.

2. On the other hand, if the result of measurements returns $S(E) \neq 0$, instead of using the syndrome to identify the most probable error followed by error correction, we repeat the whole procedure until the measurement returns $S(E) = 0$.

This protocol may be time consuming, but it ensures that a wrongly determined syndrome cannot propagate errors. Notice that even the $S(E) = 0$ result does not guarantee that storage error did not occur after the syndrome was determined, for example in the last transversal gates of Figures 11.9–11.11. However, this undetectable error will be present in the codeword block after error correction, indicating that the protocol is still fault-tolerant.

11.3.2 Fault-Tolerant Encoded Operations

We defined the encoded gates in Chapter 8 as unitary operators that perform mapping of one codeword to another codeword as given below:

$$(UsU^{\dagger})U|c\rangle = Us|c\rangle \overset{s|c\rangle=|c\rangle}{=} U|c\rangle; \quad \forall |c\rangle \in C_Q, \quad s \in S. \tag{11.36}$$

If $UsU^{\dagger} = s \quad \forall s \in S$, i.e. if stabilizer S is fixed by U under conjugation ($USU^{\dagger} = S$), then $U|c\rangle$ will be a codeword as, based on (11.36), it is fixed for any element from S. The operator U therefore represents an *encoded operation* (the operation that maps one codeword to another codeword). As shown in Chapter 8, the set of operators U that fix S represent the subgroup of the *unitary group* $U(N)$ (the set of all unitary matrices of dimensions $N \times N$). This subgroup is called the *normalizer* of S with respect to $U(N)$, and it is denoted here as $\mathcal{N}_U(S)$. The normalizer $\mathcal{N}_U(S)$ of the stabilizer S is defined as the set of operators $U \in U(N)$ that fix S under conjugation; therefore, $\mathcal{N}_U(S) = \{U \in U(N)|USU^\dagger = S\}$. In the Appendix (see also Chapter 8), we introduce an equivalent concept, namely the centralizer of S with respect to $U(N)$, denoted as $\mathcal{C}_U(S)$, as the set of all operators U (from $U(N)$) that commute for all stabilizer elements; therefore, $\mathcal{C}_U(S) = \{U \in U(N)|US = SU\}$. Clearly, the centralizer condition $US = SU$ can be converted to the normalizer by multiplying from the right by $U^\dagger$ to obtain $USU^\dagger = S$ (for a more formal proof, please refer to Section 8.1). Notice that the key difference between this section and Section 8.1 is that here we are concerned with the action of U from $U(N)$ on S, while in Section 8.1 we were concerned with action of an operator from Pauli multiplicative group G_N. As $G_N \subset U(N)$, it is obvious that $\mathcal{N}_G(S) \subset \mathcal{N}_U(S)$.

An important group for fault-tolerant quantum error correction is the *Clifford* group, which is defined as the normalizer of the Pauli multiplicative group G_N with respect to $U(N)$, and denoted as $\mathcal{N}_U(G_N)$. The *Clifford operator* U is therefore an operator that preserves the elements of the Pauli group under conjugation, namely $\forall\ O \in G_N:\ \ UOU^\dagger \in G_N$. The Clifford group elements can be generated based on simple sets of gates: Hadamard, phase, and CNOT gates, or equivalently Hadamard, phase, and controlled-Z gates. In this section, we are concerned with encoded gates $U \in \mathcal{N}(G_N) \cap \mathcal{N}(S)$ that can be constructed using these gates and encoded operations that are implemented in transversal fashion, and which are consequently fault-tolerant.

We learned in Chapter 8 that the quotient group $C(S)/S$ can be represented by

$$C(S)/S = \cup\{\overline{X}_m S, \overline{Z}_n S:\ m,n = 1,\ldots,K\} \wedge \{j^l:\ l = 0,1,2,3\}, \tag{11.37}$$

where $\{\overline{X}_m, \overline{Z}_n|m,n = 1,\ldots,K\}$ denotes the Pauli encoded operators. The elements of $C(S)/S$ (cosets), denoted as $\overline{O}S$, can be represented as follows:

$$\overline{O}S = j^l\overline{X}(\boldsymbol{a})\overline{Z}(\boldsymbol{b})S = j^l(\overline{X}_1 S)^{a_1}\ldots(\overline{X}_k S)^{a_K}(\overline{Z}_1 S)^{b_1}\ldots(\overline{Z}_1 S)^{b_K}, \tag{11.38}$$

as proved in Chapter 8. By denoting the conjugation mapping image of encoded Pauli operators as $\tilde{X}_m = U\overline{X}_m U^\dagger,\ \tilde{Z}_n = U\overline{Z}_n U^\dagger$, we can show that

$$\begin{aligned}
[\tilde{X}_m, g_n] &= \tilde{X}_m g_n - g_n\tilde{X}_m = U\overline{X}_m U^\dagger g_n - g_n U\overline{X}_m U^\dagger \overset{g_n = Us_nU^\dagger}{=} \\
&\quad U\overline{X}_m U^\dagger U s_n U^\dagger - U s_n U^\dagger U\overline{X}_m U^\dagger \\
&= U\overline{X}_m s_n U^\dagger - U s_n \overline{X}_m U^\dagger \overset{s_n\overline{X}_m = \overline{X}_m s_n}{=} U\overline{X}_m s_n U^\dagger - U\overline{X}_m s_n U^\dagger = 0,
\end{aligned} \tag{11.39}$$

where we represent the generators g_n in terms of the stabilizer element s_n by $g_n = Us_nU^\dagger$ (i.e. as U fixes the stabilizer under conjugation, any element $s \in S$, including generators, can be represented in the form $UsU^{\dagger}$). Clearly, the images of Pauli operators also commute with generators g_n and therefore belong to $\mathscr{C}(S)$ as well. Because the operator U fixes the stabilizer S under conjugation, based on (11.38) and the commutation relation (11.39), the operator U maps $\mathscr{C}(S)/S$ to $\mathscr{C}(S)/S$, as the coset $\overline{O}S$ gets mapped to

$$\tilde{O}S = j^l(\tilde{X}_1S)^{a_1}\ldots(\tilde{X}_kS)^{a_K}(\tilde{Z}_1S)^{b_1}\ldots(\tilde{Z}_1S)^{b_K}. \tag{11.40}$$

In the Appendix, we have defined *homomorphism* as the mapping f from group G onto group H if the image of a product ab equals the product of images of a and b. If f is a homomorphism and it is bijective (1:1 and onto), then f is said to be an *isomorphism* from G to H, and is commonly denoted by $G \cong H$. Finally, if $G = H$ we call the isomorphism *automorphism*. Let us now observe the mapping f from G_N to G_N that acts on elements from G_N by conjugation, i.e. $f(g) = UgU^{\dagger}\ \forall\ g \in G_N,\ U \in \mathscr{N}(G_N)$. It will be shown below that this mapping is an automorphism. We first need to show that this mapping is 1:1. Let us observe two *different* elements $a, b \in G_N$ and assume that $f(a) = f(b)$. The conjugation definition ($f(g) = UgU^{\dagger}$) function indicates that $f(a) = UaU^{\dagger} = UbU^{\dagger} = f(b)$. By multiplying by U from the right we obtain $Ua = Ub$, and by multiplying by $U^{\dagger}$ from the left we obtain $a = b$, which is a contradiction. Therefore, the mapping $f(g)$ is 1:1 mapping. In the second step, we need to prove that the mapping $f(g)$ is an onto mapping. We have shown in the previous paragraph that the range of mapping $f(g)$, denoted as $f(G_N)$, is contained in G_N. Since the mapping $f(g)$ is 1:1, $f(G_N)$ must equal G_N, indicating that the conjugation mapping is indeed onto mapping. In the final stage, we need to prove that the image of the product ab, where $a,b \in G_N$, must be equal to the product of images $f(a)f(b)$. This claim can easily be proved from the definition of conjugation mapping as follows:

$$f(ab) = UabU^{\dagger} = Ua\overbrace{U^{\dagger}U}^{I}bU^{\dagger} = \overbrace{(UaU^{\dagger})}^{f(a)}\overbrace{(UbU^{\dagger})}^{f(b)} = f(a)f(b). \tag{11.41}$$

Therefore, the mapping $f(g) = UgU^{\dagger}$ is an automorphism of G_N.

Because every Clifford operator U from $\mathscr{N}(G_N)$ is an automorphism of G_N, in order to determine the action of U on G_N, it is sufficient to determine the action of generators of $\mathscr{N}(G_N)$ on generators of G_N. We also know that every Clifford generator can be represented in terms of Hadamard, phase, and CNOT (or controlled-Z) gates, and it is essential to study the action of single-qubit Hadamard and phase gates on generators of G_1 and the action of two-qubit CNOT gates on generators of G_2.

The matrix representation of the Hadamard gate in the computational basis, based on Chapter 2, is given by

$$H = \frac{1}{\sqrt{2}}\begin{bmatrix} 1 & 1 \\ 1 & -1 \end{bmatrix}. \tag{11.42}$$

The Pauli gates X, Y, and Z under the conjugation operation $HOH^\dagger$ ($O \in \{X,Y,Z\}$) are mapped to

$$X \to HXH^\dagger = \begin{bmatrix} 1 & 1 \\ 1 & -1 \end{bmatrix} \begin{bmatrix} 0 & 1 \\ 1 & 0 \end{bmatrix} \begin{bmatrix} 1 & 1 \\ 1 & -1 \end{bmatrix} = Z,$$

$$Z \to HZH^\dagger = \begin{bmatrix} 1 & 1 \\ 1 & -1 \end{bmatrix} \begin{bmatrix} 1 & 0 \\ 0 & -1 \end{bmatrix} \begin{bmatrix} 1 & 1 \\ 1 & -1 \end{bmatrix} = X, \tag{11.43}$$

$$Y \to HYH^\dagger = \begin{bmatrix} 1 & 1 \\ 1 & -1 \end{bmatrix} \begin{bmatrix} 0 & -j \\ j & 0 \end{bmatrix} \begin{bmatrix} 1 & 1 \\ 1 & -1 \end{bmatrix} = -Y.$$

The matrix representation of the phase gate in the computational basis is given by

$$P = \begin{bmatrix} 1 & 0 \\ 0 & j \end{bmatrix}. \tag{11.44}$$

The Pauli gates X, Y, and Z under the conjugation operation $POP^\dagger$ ($O \in \{X,Y,Z\}$) are mapped to

$$X \to PXP^\dagger = \begin{bmatrix} 1 & 0 \\ 0 & j \end{bmatrix} \begin{bmatrix} 0 & 1 \\ 1 & 0 \end{bmatrix} \begin{bmatrix} 1 & 0 \\ 0 & -j \end{bmatrix} = Y,$$

$$Z \to PZP^\dagger = \begin{bmatrix} 1 & 0 \\ 0 & j \end{bmatrix} \begin{bmatrix} 1 & 0 \\ 0 & -1 \end{bmatrix} \begin{bmatrix} 1 & 0 \\ 0 & -j \end{bmatrix} = Z, \tag{11.45}$$

$$Y \to PYP^\dagger = \begin{bmatrix} 1 & 0 \\ 0 & j \end{bmatrix} \begin{bmatrix} 0 & -j \\ j & 0 \end{bmatrix} \begin{bmatrix} 1 & 0 \\ 0 & -j \end{bmatrix} = -X.$$

From Eqs (11.43) and (11.45) it is clear that both operators (H and G) permute the generators of G_1, and that the phase operator performs the rotation around the z-axis for 90°.

The last gate needed to implement any Clifford operator is the CNOT gate. Since the CNOT gate is a two-qubit gate we need to study its action on generators from G_2, namely $X_1 = XI$, X_2, Z_1, and Z_2. In computational basis, the CNOT gate U_{CNOT} and generators from G_2 can be represented as

$$U_{\text{CNOT}} = \begin{bmatrix} I & 0 \\ 0 & X \end{bmatrix}, \quad X_1 = X \otimes I = \begin{bmatrix} 0 & 1 \\ 1 & 0 \end{bmatrix} \otimes I = \begin{bmatrix} 0 & I \\ I & 0 \end{bmatrix},$$

$$X_2 = I \otimes X = \begin{bmatrix} 1 & 0 \\ 0 & 1 \end{bmatrix} \otimes X = \begin{bmatrix} X & 0 \\ 0 & X \end{bmatrix},$$

$$Z_1 = Z \otimes I = \begin{bmatrix} 1 & 0 \\ 0 & -1 \end{bmatrix} \otimes I = \begin{bmatrix} I & 0 \\ 0 & -I \end{bmatrix}, \tag{11.46}$$

$$Z_2 = I \otimes Z = \begin{bmatrix} 1 & 0 \\ 0 & 1 \end{bmatrix} \otimes Z = \begin{bmatrix} Z & 0 \\ 0 & Z \end{bmatrix}.$$

The generators of G_2 $\{X_1, X_2, Z_1$ and $Z_2\}$, under the conjugation operation $U_{CNOT}\, O U_{CNOT}^{\dagger}$ $(O \in \{X_1, X_2, Z_1, Z_2\})$ are mapped to

$$
\begin{aligned}
X_1 \rightarrow U_{CNOT}\, X_1 U_{CNOT}^{\dagger} &= \begin{bmatrix} I & 0 \\ 0 & X \end{bmatrix} \begin{bmatrix} 0 & I \\ I & 0 \end{bmatrix} \begin{bmatrix} I & 0 \\ 0 & X \end{bmatrix} = \begin{bmatrix} 0 & X \\ X & 0 \end{bmatrix} = X \otimes X = X_1 X_2 \\
X_2 \rightarrow U_{CNOT}\, X_2 U_{CNOT}^{\dagger} &= \begin{bmatrix} I & 0 \\ 0 & X \end{bmatrix} \begin{bmatrix} X & 0 \\ 0 & X \end{bmatrix} \begin{bmatrix} I & 0 \\ 0 & X \end{bmatrix} = I \otimes X = X_2 \\
Z_1 \rightarrow U_{CNOT} Z_1 U_{CNOT}^{\dagger} &= \begin{bmatrix} I & 0 \\ 0 & X \end{bmatrix} \begin{bmatrix} I & 0 \\ 0 & -I \end{bmatrix} \begin{bmatrix} I & 0 \\ 0 & X \end{bmatrix} = Z \otimes I = Z_1 \\
Z_2 \rightarrow U_{CNOT} Z_2 U_{CNOT}^{\dagger} &= \begin{bmatrix} I & 0 \\ 0 & X \end{bmatrix} \begin{bmatrix} Z & 0 \\ 0 & Z \end{bmatrix} \begin{bmatrix} I & 0 \\ 0 & X \end{bmatrix} = Z \otimes Z = Z_1 Z_2.
\end{aligned}
\tag{11.47}
$$

From Eq. (11.47) it is clear that an error introduced on the control qubit affects the target qubit, while the phase error introduced on the target qubit affects the control qubit.

For certain implementations, such as cavity quantum electrodynamics (CQED) implementation, the use of a controlled-Z U_{CZ} gate is more appropriate than the use of a CNOT gate [8]. We have shown in Chapter 3 that a CNOT gate can be implemented as shown in Figure 11.13. The gate shown in Figure 11.13 in computational basis can be expressed as

$$
\begin{bmatrix} I & 0 \\ 0 & HZH \end{bmatrix} = \begin{bmatrix} I & 0 \\ 0 & X \end{bmatrix} = U_{CNOT}, \tag{11.48}
$$

indicating that a controlled-Z gate can indeed be used instead of a CNOT gate. We now study the action of the controlled-Z gate U_{CZ} on generators of G_2. The

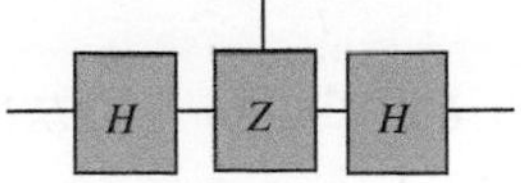

FIGURE 11.13

The CNOT gate expressed in terms of the controlled-Z gate.

generators of $G_2 - X_1$, X_2, Z_1, and Z_2 – under the conjugation operation $U_{CZ}OU_{CZ}^{\dagger}$ ($O \in \{X_1, X_2, Z_1, Z_2\}$) are mapped to

$$\begin{aligned}
X_1 \to U_{CZ}X_1U_{CZ}^{\dagger} &= \begin{bmatrix} I & 0 \\ 0 & Z \end{bmatrix}\begin{bmatrix} 0 & I \\ I & 0 \end{bmatrix}\begin{bmatrix} I & 0 \\ 0 & Z \end{bmatrix} = X \otimes Z = X_1 Z_2 \\
X_2 \to U_{CZ}X_2U_{CZ}^{\dagger} &= \begin{bmatrix} I & 0 \\ 0 & Z \end{bmatrix}\begin{bmatrix} X & 0 \\ 0 & X \end{bmatrix}\begin{bmatrix} I & 0 \\ 0 & Z \end{bmatrix} = Z \otimes X = Z_1 X_2 \\
Z_1 \to U_{CZ}Z_1U_{CZ}^{\dagger} &= \begin{bmatrix} I & 0 \\ 0 & Z \end{bmatrix}\begin{bmatrix} I & 0 \\ 0 & -I \end{bmatrix}\begin{bmatrix} I & 0 \\ 0 & Z \end{bmatrix} = Z \otimes I = Z_1 \\
Z_2 \to U_{CNOT}Z_2U_{CNOT}^{\dagger} &= \begin{bmatrix} I & 0 \\ 0 & X \end{bmatrix}\begin{bmatrix} Z & 0 \\ 0 & Z \end{bmatrix}\begin{bmatrix} I & 0 \\ 0 & X \end{bmatrix} = I \otimes Z = Z_2.
\end{aligned} \tag{11.49}$$

From Eq. (11.49) it is evident that the bit-flip error on one qubit introduces a phase error on the other qubit, while a phase error on either qubit does not propagate.

We turn our attention now to the determination of the unitary operator when the corresponding automorphism is known. We will use the example due to Gottesman [9]. The automorphism on G_1 in this example is given by cyclic permutation of Pauli gates [9]:

$$f_U : X \to Y \to Z \to X. \tag{11.50}$$

We know from Chapter 2 that the computational basis ket $|0\rangle$ is an eigenket of Pauli operator Z with eigenvalue of $+1$. Let us now apply the unknown gate U on basis ket $|0\rangle$:

$$U|0\rangle \overset{Z|0\rangle = |0\rangle}{=} UZ|0\rangle = UZ\overbrace{U^{\dagger}U}^{I}|0\rangle = (UZU^{\dagger})U|0\rangle = X(U|0\rangle), \tag{11.51}$$

where we used the mapping property (11.50). By comparing the first and last portions of Eq. (11.51) we conclude that $U|0\rangle$ is an eigenket of the Pauli operator X with eigenvalue $+1$, which is given by (see Chapter 2)

$$U|0\rangle = \frac{1}{\sqrt{2}}(|0\rangle + |1\rangle). \tag{11.52}$$

The elements of the first column in the matrix representation of U,

$$U = \begin{bmatrix} U_{00} & U_{01} \\ U_{10} & U_{11} \end{bmatrix}, \tag{11.53}$$

can be determined from (11.52) as follows: $U_{00} = \langle 0|U|0\rangle = 1/\sqrt{2}$ and $U_{10} = \langle 1|U|0\rangle = 1/\sqrt{2}$. The computational ket $|1\rangle$ can be obtained as $X|0\rangle = |1\rangle$, and by using a similar approach as in (11.51) we obtain:

$$\begin{aligned}
U|1\rangle &\overset{|1\rangle = X|0\rangle}{=} UX|0\rangle = UX\overbrace{U^{\dagger}U}^{I}|0\rangle = (UXU^{\dagger})U|0\rangle = Y(U|0\rangle) \\
&\overset{U|0\rangle = \frac{1}{\sqrt{2}}(|0\rangle + |1\rangle)}{=} \frac{1}{\sqrt{2}}Y(|0\rangle + |1\rangle) = \frac{-j}{\sqrt{2}}(|0\rangle - |1\rangle).
\end{aligned} \tag{11.54}$$

The elements of the second column of (11.53) can be found from (11.54) as follows. $U_{01} = \langle 0|U|1\rangle = -j/\sqrt{2}$, and $U_{11} = \langle 1|U|1\rangle = j/\sqrt{2}$, so that the matrix representation of U is given by

$$U_T = \frac{1}{\sqrt{2}}\begin{bmatrix} 1 & -j \\ 1 & j \end{bmatrix}. \tag{11.55}$$

The generators of G_1 – X, Y, and Z – under the conjugation operation $U_T O U_T^\dagger$ ($O \in \{X,Y,Z\}$) are mapped to

$$\begin{aligned} X \to U_T X U_T^\dagger &= \begin{bmatrix} 1 & -j \\ 1 & j \end{bmatrix}\begin{bmatrix} 1 & 0 \\ 0 & 1 \end{bmatrix}\begin{bmatrix} 1 & 1 \\ j & -j \end{bmatrix} = Y \\ Y \to U_T Y U_T^\dagger &= \begin{bmatrix} 1 & -j \\ 1 & j \end{bmatrix}\begin{bmatrix} 0 & -j \\ j & 0 \end{bmatrix}\begin{bmatrix} 1 & 1 \\ j & -j \end{bmatrix} = Z \\ Z \to U_T Z U_T^\dagger &= \begin{bmatrix} 1 & -j \\ 1 & j \end{bmatrix}\begin{bmatrix} 1 & 0 \\ 0 & -1 \end{bmatrix}\begin{bmatrix} 1 & 1 \\ j & -j \end{bmatrix} = X, \end{aligned} \tag{11.56}$$

which is consistent with Eq. (11.50).

11.3.3 Measurement Protocol

In this section we are concerned with applying the quantum gate on a quantum register by means of a measurement. This method, applied transversally, will be used in subsequent sections to enable fault-tolerant quantum computation by using arbitrary quantum stabilizer codes. DiVincenzo and Shor [10] have described a method that can be used to perform the measurements using any operator O from multiplicative Pauli group G_N (see also Refs [1,9]). There are three possible options involving the relationship of this operator O and stabilizer group S: (i) it can belong to S, when measuring O will not provide any useful information on the state of the system as the result will always be $+1$ for a valid codeword; (ii) it can commute with all elements from S, indicating that O belongs to $\mathcal{N}(S)/S$; and (iii) it can anticommute with certain elements from S, which represents a useful case in practice, as discussed in Chapters 7 and 8. Let us create a set of generators $S_0 = \{g_1, g_2, \ldots, g_M\}$ such that the operator O anticommutes with g_1 and commutes with the remaining generators. If the mth generator g_m anticommutes with O we can replace it with $g_1 g_m$, which will clearly commute with O. Because the operator O commutes with generators g_m ($m = 2,3,\ldots,M$), the measurement of O will not affect them (they can be simultaneously measured with complete precision). On the other hand, since g_1 anticommutes with O, the measurement of O will affect g_1. The measurement of O will perform a projection of an arbitrary quantum state $|\psi_0\rangle$ as $P_\pm|\psi_0\rangle$, with projection operators ($P_\pm^2 = P_\pm$) given by

$$P_\pm = \frac{1}{2}(I \pm O). \tag{11.57}$$

See Chapter 2 for a justification of this representation. Because the operator O has the second order ($O^2 = I$; see Chapter 8), its eigenvalues are ± 1. Therefore, the projection operators $P_{\pm}$ will project state $|\psi_0\rangle$ onto ± 1 eigenkets, respectively. The action of generators g_m on state $|\psi_0\rangle$ is given by $g_m|\psi_0\rangle = |\psi_0\rangle$, as expected. We come to the point where we can introduce the *measurement protocol*, which consists of three steps:

1. Perform the measurement of O fault-tolerantly, as described in Section 11.2.
2. If the measurement outcome is $+1$, do not take any further action. The final state will be $P_+|\psi_0\rangle = |\psi_+\rangle$.
3. If the measurement outcome is -1, apply the g_1 operator on the post-measurement state to obtain:

$$\begin{aligned} g_1 P_-|\psi_0\rangle &= g_1 \frac{1}{2}(I - O)|\psi_0\rangle = \frac{1}{2}(I + O) g_1|\psi_0\rangle = P_+ g_1|\psi_0\rangle \\ &= P_+|\psi_0\rangle = |\psi_+\rangle. \end{aligned} \tag{11.58}$$

In both cases, the measurement of O performs the mapping $|\psi_0\rangle \rightarrow |\psi_+\rangle$. The final state is fixed by stabilizer $S_1 = \{\, O, g'_1, \ldots, g'_M\}$, with generators given by

$$g'_m = \begin{cases} g_m, & [g_m, O] = 0 \\ g_1 g_m, & \{g_m, O\} = 0 \end{cases}; \quad m = 2, \ldots, M. \tag{11.59}$$

It can be shown, by observing the generators from S_0, that the measurement procedure above also performs the mapping $S_0 \rightarrow S_1$. In the Appendix, we define the centralizer of S_i ($i = 0,1$) as the set of operators from G_N that commute with all elements from S_i. We will show below that the measurement protocol performs the mapping: $\mathscr{C}(S_0)/S_0 \rightarrow \mathscr{C}(S_1)/S_1$. To prove this claim, let us observe an element n from $\mathscr{C}(S_0)$ and a state $|\psi_0\rangle$ that is fixed by S_0. The action of operator n on state $P_+|\psi_0\rangle$, during the measurement of operator O, must be the same as $P_+ n'|\psi_0\rangle$, where n' is the image of n (under measurement of O). If n commutes with O, then $n' = n$. On the other hand, if n anticommutes with O, $g_1 n$ will commute with O, and the image of n will be $n' = g_1 n$. Therefore, the measurement of O in the first case maps $n \rightarrow n$ and in the second case it maps $n \rightarrow g_1 n$. Since in the first case n commutes with O and $n' = n$, then n' must belong to the coset nS_1. In the second case, $g_1 n$ commutes with O and n' belongs to the coset $(g_1 n)S_1$. In conclusion, in the first case the measurement of O performs the mapping $nS_0 \rightarrow nS_1$ and in the second case $nS_0 \rightarrow g_1 n S_1$, proving the claim above. Let us denote the encoded Pauli operators for S_0 by $\hat{X}_i$, $\hat{Z}_i$. To summarize, in order to determine the action of measurement of O on the initial state $|\psi_0\rangle$, we need to perform the following procedure [2]:

1. Identify a generator g_1 from S_0 that anticommutes with O. Substitute any generator g_m ($m > 1$) that anticommutes with O with $g_1 g_m$.

2. Create the stabilizer S_1 from S_0 by replacing g_1 (from S_0) with operator O. The remaining generators for S_1 are created according to the following rule:

$$g'_m = \begin{cases} g_m, & [g_m, O] = 0 \\ g_1 g_m, & \{g_m, O\} = 0 \end{cases}; \quad m = 2, \ldots, M.$$

3. In the final stage, substitute any $\hat{X}_i$ $(\hat{Z}_i)$ that anticommutes with operator O with $g_1\hat{X}_i$ $(g_1\hat{Z}_i)$, to ensure that $\hat{X}_i$ $(\hat{Z}_i)$ belongs to $\mathscr{C}(S_1)$.

The following example due to Gottesman [9] (see also Ref. [2]) will be used to illustrate this measurement procedure. Consider a two-qubit system $|\psi\rangle \otimes |0\rangle$, where $|\psi\rangle$ is an arbitrary state of qubit 1. Let us now apply the CNOT gate using the rules we derived in Eq. (11.47). The stabilizer for $|\psi\rangle \otimes |0\rangle$ has a generator g given by $I \otimes Z$. The operators $\hat{X} = X \otimes I$, $\hat{Z} = Z \otimes I$ both commute with generator g and mutually anticommute, and satisfy properties of Pauli encoded operators. The CNOT gate with qubit 1 serving as control qubit and qubit 2 as target will perform the following mapping, based on (11.47):

$$\begin{aligned} \hat{X} &= X_1 \to \hat{X}' = X \otimes X = X_1 X_2 \\ \hat{Z} &= Z_1 \to \hat{Z}' = Z \otimes I = Z_1 \\ g &= Z_2 \to g' = Z \otimes Z = Z_1 Z_2. \end{aligned} \tag{11.60}$$

We now study the measurement of operator $O = I \otimes Y$. Clearly, the operators $\hat{X}$ and g anticommute with O (as they both differ in one position with O). The measurement of O will perform the following mapping, based on the three-step procedure given above:

$$\begin{aligned} \hat{X}' &= X \otimes X = X_1 X_2 \to \hat{X}'' = g'\hat{X}' = (Z \otimes Z)(X \otimes X) = -Y \otimes Y = -Y_1 Y_2 \\ \hat{Z}' &= Z \otimes I = Z_1 \to \hat{Z}'' = \hat{Z}' = Z_1 \\ g' &= Z \otimes Z = Z_1 Z_2 \to g'' = O = Y_2. \end{aligned} \tag{11.61}$$

The measurement protocol leaves the second qubit in the $+1$ eigenket of Y_2, and the second portions of g'', $\hat{X}''$ and $\hat{Z}''$ fix it. As the qubits 1 and 2 remain non-entangled after the measurement, qubit 2 can be discarded from further consideration. The left portions of operators $\hat{X}$, $\hat{X}'$, $\hat{Z}$, $\hat{Z}'$ will determine the operation carried out on qubit 1 during the measurement. From Eq. (11.61) it is evident that, during the measurement of O, X is mapped to $-Y$, and Z is mapped to Z, which is the same as the action of $P^\dagger$. Therefore, the measurement of O and the CNOT gate can be used to implement the phase gate.

The Hadamard gate can be represented in terms of the phase gate as follows:

$$PR_x(-\pi/2)^\dagger P = \begin{bmatrix} 1 & 0 \\ 0 & j \end{bmatrix} \frac{1}{\sqrt{2}} \begin{bmatrix} 1 & j \\ j & 1 \end{bmatrix}^\dagger \begin{bmatrix} 1 & 0 \\ 0 & j \end{bmatrix} = \frac{1}{\sqrt{2}} \begin{bmatrix} 1 & 1 \\ 1 & -1 \end{bmatrix} = H, \tag{11.62}$$

where $R_x(-\pi/2)$ denotes rotation about the x-axis by $-\pi/2$, which is an operator introduced in Chapter 3, with matrix representation given by

$$R_x(-\pi/2) = \frac{1}{\sqrt{2}}\begin{bmatrix} 1 & j \\ j & 1 \end{bmatrix}.$$

This maps the Pauli operators X and Z to X and Y respectively. What remains is to show that this rotation operator can be implemented by using the measurement protocol and CNOT gate. The initial two-qubit system will now be $|\psi\rangle \otimes (|0\rangle + |1\rangle)$, and the corresponding generator $g = X_2$. The encoded Pauli operator analogs are given by $\hat{X} = X_1$, $\hat{Z} = Z_1$, and the operator O as X_2. Let us now apply the CNOT gate where qubit 2 is the control qubit and qubit 1 is the target qubit. The CNOT gate will perform the following mapping, based on Eq. (11.47):

$$\begin{aligned} \hat{X} &= X_1 \rightarrow \hat{X}' = X \otimes I = X_1 \\ \hat{Z} &= Z_1 \rightarrow \hat{Z}' = Z \otimes Z = Z_1 Z_2 \\ g &= X_2 \rightarrow g' = X \otimes X = X_1 X_2. \end{aligned} \tag{11.63}$$

We then apply the phase gate on qubit 2, which performs the following mapping:

$$\begin{aligned} \hat{X}' &= X_1 \rightarrow \hat{X}'' = X_1 \\ \hat{Z}' &= Z_1 Z_2 \rightarrow \hat{Z}'' = Z_1 Z_2 \\ g' &= X_1 X_2 \rightarrow g'' = X_1 Y_2. \end{aligned} \tag{11.64}$$

In the final stage, we perform the measurement on operator O. It is clear from (11.64) that $\hat{Z}'' = Z_1 Z_2$ and g'' anticommute with operator O (because they both differ from O in one position). By applying the three-step measurement procedure above we obtain the following mapping:

$$\begin{aligned} \hat{X}'' &\rightarrow \hat{X}''' = X_1 \\ \hat{Z}'' &\rightarrow \hat{Z}''' = g''\hat{Z}'' = (X_1 Y_2)(Z_1 Z_2) = Y_1 X_1 \\ g'' &\rightarrow g''' = O = X_2. \end{aligned} \tag{11.65}$$

By disregarding qubit 2, we conclude that the measurement procedure maps X_1 to X_1 and Z_1 to Y_1, which is the same as the action of rotation operator $R_x(-\pi/2)$. In conclusion, we have just shown that any gate from the Clifford group can be implemented based on the CNOT gate, the measurement procedure described above, and a properly prepared initial state.

11.3.4 Fault-Tolerant Stabilizer Codes

We start our description of fault-tolerant stabilizer codes with a four-qubit unitary operation due to Gottesman [9], which when applied transversally can generate an encoded version of itself for all $[N,1]$ stabilizer codes. This four-qubit operation performs the following mapping [9]:

$$\begin{aligned} &X\otimes I\otimes I\otimes I\rightarrow X\otimes X\otimes X\otimes I\\ &I\otimes X\otimes I\otimes I\rightarrow I\otimes X\otimes X\otimes X\\ &I\otimes I\otimes X\otimes I\rightarrow X\otimes I\otimes X\otimes X\\ &I\otimes I\otimes I\otimes X\rightarrow X\otimes X\otimes I\otimes X\\ &Z\otimes I\otimes I\otimes I\rightarrow Z\otimes Z\otimes Z\otimes I\\ &I\otimes Z\otimes I\otimes I\rightarrow I\otimes Z\otimes Z\otimes Z\\ &I\otimes I\otimes Z\otimes I\rightarrow Z\otimes I\otimes Z\otimes Z\\ &I\otimes I\otimes I\otimes Z\rightarrow Z\otimes Z\otimes I\otimes Z. \end{aligned} \tag{11.66}$$

Clearly, the ith row ($i = 2,3,4$) is obtained by cyclic permutation of the $(i - 1)$th row, one position to the right. Similarly, the jth row ($j = 6,7,8$) is obtained by cyclic permutation of the $(j - 1)$th row, one position to the right. The corresponding quantum circuit to perform this manipulation is shown in Figure 11.14. For any element $s = e^{jl\pi/2}X(\boldsymbol{a}_s)X(\boldsymbol{b}_s)\in S$, where the phase factor $\exp(jl\pi/2)$ is employed to ensure the Hermitian properties of s, this quantum circuit maps $s\otimes I\otimes I\otimes I\rightarrow s\otimes s\otimes s\otimes I$ and its cyclic permutations as given below:

$$\begin{aligned} &s\otimes I\otimes I\otimes I\rightarrow s\otimes s\otimes s\otimes I\\ &I\otimes s\otimes I\otimes I\rightarrow I\otimes s\otimes s\otimes s\\ &I\otimes I\otimes s\otimes I\rightarrow s\otimes I\otimes s\otimes s\\ &I\otimes I\otimes I\otimes s\rightarrow s\otimes s\otimes I\otimes s. \end{aligned} \tag{11.67}$$

It can be easily verified that images of generators are themselves the generators of stabilizer $S\otimes S\otimes S\otimes S$. The encoded Pauli operators $\overline{X}_m = e^{jl_m\pi/2}X(\boldsymbol{a}_m)Z(\boldsymbol{b}_m)$, $\overline{Z}_m = e^{jl'_m\pi/2}X(\boldsymbol{c}_m)Z(\boldsymbol{d}_m)$ are mapped in a similar fashion:

$$\begin{aligned} &\overline{X}_m\otimes I\otimes I\otimes I\rightarrow \overline{X}_m\otimes \overline{X}_m\otimes \overline{X}_m\otimes I\\ &I\otimes \overline{X}_m\otimes I\otimes I\rightarrow I\otimes \overline{X}_m\otimes \overline{X}_m\otimes \overline{X}_m\\ &I\otimes I\otimes \overline{X}_m\otimes I\rightarrow \overline{X}_m\otimes I\otimes \overline{X}_m\otimes \overline{X}_m\\ &I\otimes I\otimes I\otimes \overline{X}_m\rightarrow \overline{X}_m\otimes \overline{X}_m\otimes I\otimes \overline{X}_m\\ &\overline{Z}_m\otimes I\otimes I\otimes I\rightarrow \overline{Z}_m\otimes \overline{Z}_m\otimes \overline{Z}_m\otimes I\\ &I\otimes \overline{Z}_m\otimes I\otimes I\rightarrow I\otimes \overline{Z}_m\otimes \overline{Z}_m\otimes \overline{Z}_m\\ &I\otimes I\otimes \overline{Z}_m\otimes I\rightarrow \overline{Z}_m\otimes I\otimes \overline{Z}_m\otimes \overline{Z}_m\\ &I\otimes I\otimes I\otimes \overline{Z}_m\rightarrow \overline{Z}_m\otimes \overline{Z}_m\otimes I\otimes \overline{Z}_m. \end{aligned} \tag{11.68}$$

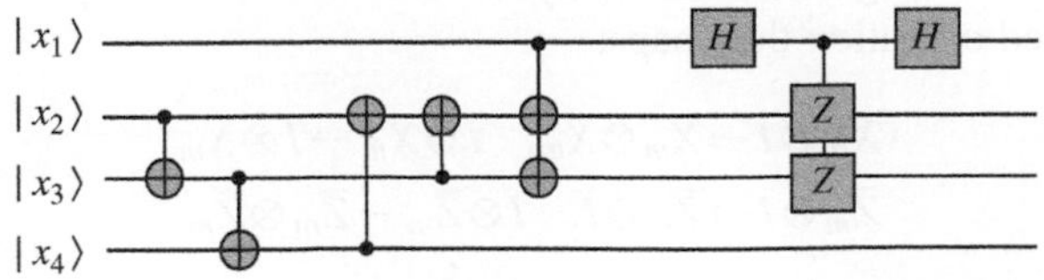

FIGURE 11.14

The quantum circuit to perform the mapping due to Gottesman [9], given by Eq. (11.66).

The transversal application of four-qubit operation provides the encoded version of operation itself, which is true for all [N,1] stabilizer codes. On the other hand, for [$N,K > 1$] stabilizer codes, the four-qubit operation maps the encoded Pauli operators $\overline{X}_m$, $\overline{Z}_m$ for all m positions simultaneously. In order to be suitable to map a subset of encoded Pauli operators, certain modifications are needed, which will be described later in this section.

Let us now provide a simple but illustrative example due to Gottesman [9] (see also Ref. [2]). We are concerned with the mapping of the four-qubit state $|\psi_1\psi_2\rangle \otimes |00\rangle$, where $|\psi_1\psi_2\rangle$ is an arbitrary two-qubit state. The stabilizer S_0 is described by the two generators $g_1 = Z_3$ and $g_2 = Z_4$. The encoded Pauli operators are given by

$$\tilde{X}_1 = X_1, \quad \tilde{X}_2 = X_2, \quad \tilde{Z}_1 = Z_1, \quad \tilde{Z}_2 = Z_2. \tag{11.69}$$

Applying the four-qubit operation given by (11.66), we obtain the following mapping:

$$g_1 \to g_1' = Z_1Z_3Z_4, \quad g_2 \to g_1' = Z_1Z_2Z_4, \quad \tilde{X}_1 \to \tilde{X}_1' = X_1X_2X_3,$$
$$\tilde{X}_2 \to \tilde{X}_2' = X_2X_3X_4, \quad \tilde{Z}_1 \to \tilde{Z}_1' = Z_1Z_2Z_3, \quad \tilde{Z}_2 \to \tilde{Z}_2' = Z_2Z_3Z_4.$$

We now perform measurements using operators $O_1 = X_3$ and $O_2 = X_4$. By applying the measurement protocol from the previous section, we obtain the following mapping as a result of measurements:

$$g_1'' \to O_1, \quad g_1' = Z_1Z_3Z_4, \quad g_2 \to O_2, \quad g_2' = Z_1Z_2Z_4, \quad \tilde{X}_1' \to \tilde{X}_1'' = X_1X_2X_3,$$
$$\tilde{X}_2' \to \tilde{X}_2'' = X_2X_3X_4, \quad \tilde{Z}_1' \to \tilde{Z}_1'' = g_1'g_2'\tilde{Z}_1' = Z_1, \quad \tilde{Z}_2' \to \tilde{Z}_2'' = g_1'g_2'\tilde{Z}_2' = Z_4. \tag{11.70}$$

By disregarding the measured ancillary qubits in (11.69) and (11.70) (qubits 3 and 4), we conclude that effectively we performed the following mapping on the first two qubits:

$$\begin{aligned} &X_1 \to X_1X_2, \quad X_2 \to X_2 \\ &Z_1 \to Z_1, \quad Z_2 \to Z_1Z_2, \end{aligned} \tag{11.71}$$

which is the same as the CNOT gate given by Eq. (11.47). Therefore, by using the four-qubit operation and measurements we can implement the CNOT gate. In the previous section, we learned how to implement an arbitrary Clifford operator by using the CNOT gate and measurements. If we apply a similar procedure to four code blocks and disregard the ancillary code blocks we can perform the fault-tolerant encoded operation that maps:

$$\begin{aligned} &\overline{X}_m \otimes I \to \overline{X}_m \otimes \overline{X}_m, \quad I \otimes \overline{X}_m \to I \otimes \overline{X}_m \\ &\overline{Z}_m \otimes I \to \overline{Z}_m \otimes I, \quad I \otimes \overline{Z}_m \to \overline{Z}_m \otimes \overline{Z}_m. \end{aligned} \tag{11.72}$$

Equation (11.72) is simply a block-encoded CNOT gate. For [$N,K > 1$] stabilizer codes, with this procedure the encoded CNOT gate is applied to all mth encoded qubits in two code blocks simultaneously. Another difficulty that arises is the fact that this procedure cannot be applied to two encoded qubits lying in the same code block or one lying in the first block and the other one in the second block.

We turn our attention to the description of a procedure that can be applied to $[N,K > 1]$ stabilizer codes as well. The first problem can be solved by moving the mth qubit to two ancillary blocks, by employing the procedure described below. During this transfer the mth encoded qubit has been transferred to two ancillary qubit blocks, while other qubits in ancillary blocks are initialized to $|\bar{0}\rangle$. We then apply the block-encoded CNOT gate given by Eq. (11.71) and the description above it. Because we transferred the mth encoded qubit to ancillary blocks, the other coded blocks and other ancillary block will not be affected. Once we apply the desired operation to the mth encoded qubit, we can transfer it back to the original code block. Let us consider a data qubit in an arbitrary state $|\psi\rangle$ and ancillary qubit prepared in the $+1$ eigenket of $X^{(a)}$ $((|0\rangle+|1\rangle)/\sqrt{2}$ state in computational basis), where we use superscript (a) to denote the ancillary qubit. The stabilizer for this two-qubit state is described by the generator $g = I^{(d)}X^{(a)}$, where we now use the superscript (d) to denote the data qubit. The encoded Pauli operator analogs are given by $\tilde{X} = X^{(d)}I^{(a)}$, $\tilde{Z} = Z^{(d)}I^{(a)}$. The CNOT gate is further applied assuming that the ancillary qubit is a control qubit and the data qubit is a target qubit. The CNOT gate performs the following mapping (see Eq. (11.47)):

$$g = I^{(d)}X^{(a)} \rightarrow g' = X^{(d)}X^{(a)}, \quad \tilde{X} = X^{(d)}I^{(a)} \rightarrow \tilde{X}' = X^{(d)}I^{(a)},$$
$$\tilde{Z} = Z^{(d)}I^{(a)} \rightarrow \tilde{Z}' = Z^{(d)}Z^{(a)}. \tag{11.73}$$

We then perform the measurement on operator $O = Z^{(d)}I^{(a)}$, and by applying the measurement procedure from the previous section we perform the following mapping:

$$g' = X^{(d)}X^{(a)} \rightarrow g'' = O = Z^{(d)}I^{(a)}, \quad \tilde{X}' \rightarrow \tilde{X}'' = g'\tilde{X}' = I^{(d)}X^{(a)},$$
$$\tilde{Z}' = Z^{(d)}I^{(a)} \rightarrow \tilde{Z}'' = Z^{(d)}Z^{(a)}. \tag{11.74}$$

By disregarding the measurement (data) qubit, we conclude that effectively we performed the mapping as follows:

$$X^{(d)} \rightarrow X^{(a)}, \quad Z^{(d)} \rightarrow Z^{(a)}, \tag{11.75}$$

indicating that the data qubit is transferred to the ancillary qubit. If, on the other hand, the initial ancillary qubit was in the $|0\rangle$ state, when the CNOT gate is applied with the ancillary qubit serving as control qubit, the target (data) qubit would be unaffected. The procedure discussed in this paragraph can therefore be used to transfer an encoded qubit to an ancillary block, with other qubits being initialized to $|\bar{0}\rangle$. To summarize, the procedure can be described as follows. The ancillary block is prepared in such a way that all encoded qubits n, different from the mth qubit, are initialized to the +1 eigenket $|\overline{0}\rangle$of $\overline{Z}_n^{(a)}$. The encoded qubit m is, on the other hand, initialized to the +1 eigenket of $\overline{X}_m^{(a)}$. The block-encoded CNOT gate, described by (11.72), is then applied by using the ancillary block as the control block. We further perform measurement on $\overline{Z}_m^{(d)}$. With this procedure, the mth encoded qubit is transferred to the ancillary block, while other encoded ancillary qubits are unaffected. Upon measurement, the mth qubit is left the +1 eigenket $|\overline{0}\rangle$of $\overline{Z}_m^{(d)}$, wherein the remaining encoded data qubits are unaffected. Clearly, with this procedure we are able to transfer an encoded qubit from one block to another. What remains now is to transfer the mth encoded qubit from the ancillary encoded block back to the original data block, which can be done using the procedure described in the next paragraph.

Let us assume now that our two-qubit data ancillary system is initialized as $|0\rangle^{(d)} \otimes |\psi\rangle^{(a)}$, with the generator of the stabilizer given by $g = Z^{(d)}I^{(a)}$. The corresponding encoded Pauli generator analogs are given by $\tilde{X} = I^{(d)}X^{(a)}$, $\tilde{Z} = I^{(d)}Z^{(a)}$. We then apply the CNOT gate using the ancillary qubit as a control qubit, and data qubit as a target qubit, which performs the following mapping (see again Eq. (11.47)):

$$\begin{aligned} g &= Z^{(d)}I^{(a)} \rightarrow g' = Z^{(d)}Z^{(a)}, \quad \tilde{X} = I^{(d)}X^{(a)} \rightarrow \tilde{X}' = X^{(d)}X^{(a)}, \\ \tilde{Z} &= I^{(d)}Z^{(a)} \rightarrow \tilde{Z}' = I^{(d)}Z^{(a)}. \end{aligned} \tag{11.76}$$

We further perform the measurement on operator $O = I^{(d)}X^{(a)}$, which performs the mapping given by

$$\begin{aligned} g' &= Z^{(d)}Z^{(a)} \rightarrow g'' = O = I^{(d)}X^{(a)}, \quad \tilde{X}' = X^{(d)}X^{(a)} \rightarrow \tilde{X}'' = X^{(d)}X^{(a)}, \\ \tilde{Z}' &= I^{(d)}Z^{(a)} \rightarrow \tilde{Z}'' = g'\tilde{Z}' = Z^{(d)}I^{(a)}. \end{aligned} \tag{11.77}$$

By disregarding the measurement (ancillary) qubit, we conclude that effectively we performed the mapping:

$$X^{(a)} \rightarrow X^{(d)}, \quad Z^{(a)} \rightarrow Z^{(d)}, \tag{11.78}$$

which is the opposite of the mapping given by (11.75). With this procedure, therefore, we can move the mth encoded qubit back to the original encoded data block.

We turn our attention now to the second problem, that of applying the CNOT gate from the mth to the nth qubit within the same block or between two different blocks. This problem can be solved by transferring the qubits under consideration to two ancillary blocks with all other positions being in the $|\overline{0}\rangle$ state. We further apply the block-encoded CNOT gate to the corresponding ancillary blocks, a procedure that will leave other ancillary qubits unaffected. When performing the desired CNOT operation we will need to transfer the correspond qubits back to the original data positions. The key device for this manipulation is the fault-tolerant SWAP gate. The SWAP gate has already been described in Chapter 3. The corresponding quantum circuit is provided in Figure 11.15a, to facilitate

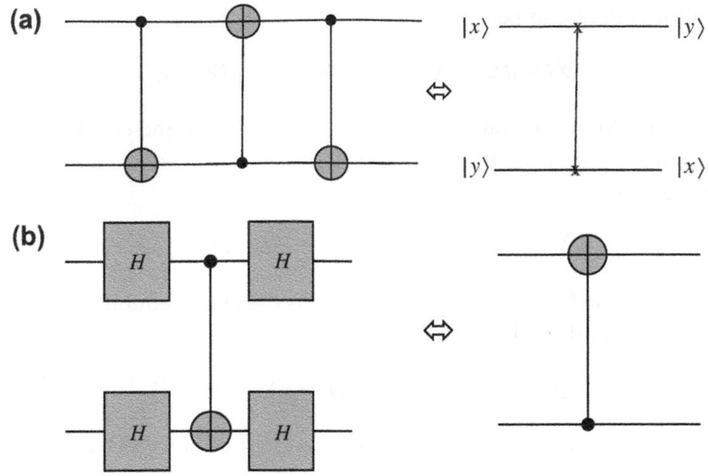

FIGURE 11.15

Quantum circuits to perform: (a) SWAP operation; (b) CNOT gate action from qubit 2 to qubit 1.

description of fault-tolerant SWAP gate implementation. In Figure 11.15b, we also describe a convenient method to implement the CNOT gate operation from qubit 2 to qubit 1, which has also been described in Chapter 3. Although we have already proved the correctness of equivalences given in Figure 11.15, it is interesting to prove the equivalencies by using the procedure we introduced in Section 11.3.2. In order to prove the equivalence given in Figure 11.15a, we apply the CNOT gate from qubit 1 to qubit 2, which performs the following mapping:

$$XI \rightarrow XX, \quad IX \rightarrow IX, \quad ZI \rightarrow ZI, \quad IZ \rightarrow ZZ. \tag{11.79}$$

In the second stage, we apply the CNOT gate using qubit 2 as the control qubit, and based on Eq. (11.47) we perform the following mapping:

$$\begin{aligned} XX &= (XI)(IX) \rightarrow (XI)(XX)^{X} \overset{X^2=I}{=} IX, \quad IX \rightarrow XX, \quad ZI \rightarrow ZZ, \\ ZZ &= (ZI)(IZ) \rightarrow (ZZ)(IZ)^{Z} \overset{Z^2=I}{=} ZI. \end{aligned} \tag{11.80}$$

In the third stage, we again apply the CNOT gate by using qubit 1 as the control qubit:

$$\begin{aligned} IX \rightarrow IX, \quad XX &= (XI)(IX) \rightarrow (XX)(IX) = XI, \\ ZZ &= (ZI)(IZ) \rightarrow (ZI)(ZZ) = IZ, \quad ZI \rightarrow ZI. \end{aligned} \tag{11.81}$$

The overall mapping can be written as

$$XI \rightarrow IX, \quad IX \rightarrow XI, \quad ZI \rightarrow IZ, \quad IZ \rightarrow ZI, \tag{11.82}$$

which is the SWAPing action of qubits 1 and 2. To prove the equivalence given in Figure 11.15b, we first apply Hadamard gates on both qubits, and based on Eq. (11.43) we can write:

$$XI \rightarrow ZI, \quad IX \rightarrow IZ, \quad ZI \rightarrow XI, \quad IZ \rightarrow IX. \tag{11.83}$$

We further apply the CNOT gate using qubit 1 as the control qubit, and based on Eq. (11.47) we perform the following mapping:

$$ZI \rightarrow ZI, \quad IZ \rightarrow ZZ, \quad XI \rightarrow XX, \quad IX \rightarrow IX. \tag{11.84}$$

Finally, we apply again the Hadamard gates on both qubits to obtain:

$$ZI \rightarrow XI, \quad ZZ \rightarrow XX, \quad XX \rightarrow ZZ, \quad IX \rightarrow IZ. \tag{11.85}$$

The overall mapping can be written as

$$XI \rightarrow XI, \quad IX \rightarrow XX, \quad ZI \rightarrow ZZ, \quad IZ \rightarrow IZ, \tag{11.86}$$

which is the same as the action of the CNOT gate from qubit 2 to qubit 1.

In certain implementations, it is difficult to implement the *automorphism group* of the stabilizer S, denoted as Aut(S), and defined as the group of functions performing the mapping $G_N \rightarrow G_N$. (Note that the automorphism group of the group G_N preserves the multiplication table.) In these situations it is instructive to design a SWAP gate based on one of its unitary elements, say U, which it is possible to implement. This operator is an automorphism of stabilizer S, it fixes it, it is an encoded operator that maps codewords to codewords, and also maps $\mathscr{C}(S)/S$ to $\mathscr{C}(S)/S$. Based on the fundamental isomorphism theorem (see Appendix and Section 8.2, Eq. (8.38)), we know that $\mathscr{C}(S)/S \cong G_K$, and $U \in \mathscr{N}(S)$, indicating that the action of U can be determined by studying its action on generators of G_K. The automorphism U must act on the first and nth encoded qubits, and by measuring the other encoded qubits upon action of U, we come up with an encoded operation in $\mathscr{N}(G_2)$ that acts only on the first and nth qubits. Gottesman [11] has shown that as far as single-qubit operations are concerned, the only two-qubit gates in $N(G_2)$ are: identity, CNOT-gate, SWAP-gate, and CNOT-gate followed by SWAP-gate. If the automorphism U yields action of the SWAP gate, our design is complete. On the other hand, if the automorphism U yields action of the CNOT gate, by using the equivalency in Figure 11.15a we can implement the SWAP gate. Finally, if the automorphism U yields the action of the CNOT gate followed by the SWAP gate, the desired SWAP gate that swaps the first qubit in the encoded state $|\overline{0}\rangle$ and the nth encoded qubit can be implemented based on the circuit shown in Figure 11.16a. Notice that the SWAP of two qubits in the block is not a transversal operation. An error introduced during the swapping operation can cause errors on both qubits being

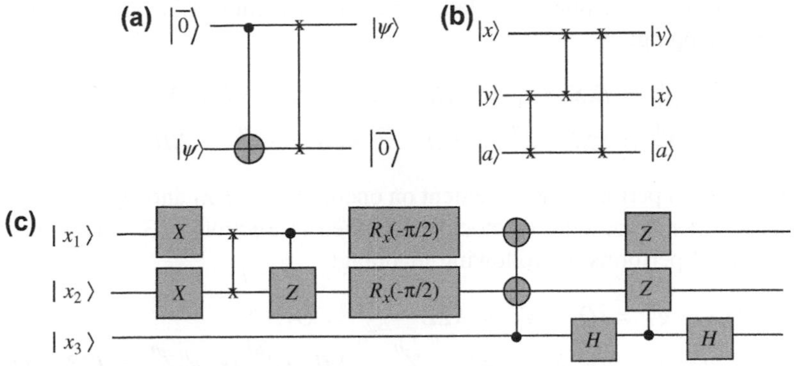

FIGURE 11.16

(a) The quantum circuit to swap encoded qubits 1 and n. (b) The quantum circuit to swap kets $|x\rangle$ and $|y\rangle$ by means of auxiliary ket $|a\rangle$. (c) The quantum circuit to perform the mapping due to Gottesman [9], given by Eq. (11.96).

swapped. This problem can be solved by using the circuit shown in Figure 11.16b, in which the two qubits to be swapped do not interact directly but by means of an ancillary qubit instead.

There are certain implementations, such as linear optics, in which the implementation of the CNOT gate is challenging, so we should keep the number of CNOT gates low. Typically, the CNOT gate in linear optics is implemented as a probabilistic gate [12–14], which performs a desired action with a certain probability. For such technologies, we can implement the SWAP gate by means of *quantum teleportation* [15], introduced in Chapter 3. Alternatively, highly nonlinear optical fibers (HNLFs) can be used in combination with the four-wave mixing effect for quantum teleportation. In quantum teleportation, entanglement in the Bell state, also known as the Einstein–Podolsky–Rosen (EPR) pair, is used to transport an arbitrary quantum state $|\psi\rangle$ between two distant observers A and B (often called Alice and Bob). The quantum teleportation system employs three qubits: qubit 1 is an arbitrary state to be transferred, while qubits 2 and 3 are in Bell state $(|00\rangle+|11\rangle)/\sqrt{2}$. Observer A has access to two qubits 1 and 2, while observer B has access to qubit 3. The observers also share an auxiliary classical communication channel. Therefore, the initial three-qubit state is $|\psi\rangle\otimes(|00\rangle+|11\rangle)/\sqrt{2}$. The corresponding generators of the stabilizer are $g_1 = IXX$ and $g_2 = IZZ$, while the encoded Pauli operator analogs are $\tilde{X} = XII$, $\tilde{Z} = ZII$. Observer A first applies the CNOT gate by using qubit 1 as control qubit and qubit 2 as target. Based on Eq. (11.47), we know that generators and encoded Pauli operator analogs will be mapped to

$$\begin{aligned} g_1 &= IXX \to g_1' = IXX, \quad g_2 = IZZ \to g_2' = ZZZ, \\ \tilde{X} &= XII \to \tilde{X}' = XXI, \quad \tilde{Z} = ZII \to \tilde{Z}' = ZII. \end{aligned} \tag{11.87}$$

Observer A further applies the Hadamard gate on qubit 1, which performs the following mapping:

$$g_1' = IXX \to g_1'' = IXX, \quad g_2' = ZZZ \to g_2'' = XZZ,$$
$$\tilde{X}' = XXI \to \tilde{X}'' = ZXI, \quad \tilde{Z}' = ZII \to \tilde{Z}'' = XII. \tag{11.88}$$

Observer A then performs measurement on operators $O_1 = Z_1$ and $O_2 = Z_2$, by using the measurement procedure (protocol) described in Section 11.3.3. This measurement protocol performs the following mapping:

$$g_1'' = IXX \to g_1''' = O_2, \quad g_2'' = XZZ \to g_2''' = O_1,$$
$$\tilde{X}'' = ZXI \to \tilde{X}''' = g_1''\tilde{X}'' = ZIX, \quad \tilde{Z}'' = XII \to \tilde{Z}''' = g_2''\tilde{Z}'' = IZZ. \tag{11.89}$$

By discarding the measured qubits, we conclude that we have effectively performed the mapping: $X_1 \to X_3$, $Z_1 \to Z_3$, meaning that qubit 1 has been transferred to qubit 3, which is equivalent to the SWAP operation.

We now explain how to use the quantum teleportation procedure to perform encoded qubit swapping. We first prepare an encoded block in which all qubits are in state $|\overline{0}\rangle$ except the nth encoded qubit, which is an arbitrary state $|\psi\rangle$. We further prepare another block in which all qubits are in state $|\overline{0}\rangle$ except the first and nth encoded qubits that are prepared into encoded Bell state $(|\overline{00}\rangle + |\overline{11}\rangle)/\sqrt{2}$. Quantum teleportation is then performed between the nth encoded qubits in two blocks, which is achieved by transferring the nth encoded qubit from the first block into the first encoded qubit in the second block, while leaving the nth encoded qubit in the second block in the state $|\overline{0}\rangle$. By discarding the first coded block, we have effectively performed an encoded SWAP operation. This procedure requires block-encoded CNOT gate, block-Hadamard gate, and $O_n = Z_n$ measurements, which is possible for the arbitrary stabilizer code. Therefore, we have just completed the description of how to fault-tolerantly apply encoded operations in the Clifford group $\mathcal{N}(G_N)$ for arbitrary stabilizer code. Before concluding this section, in the next subsection we provide an illustrative example of [5,1,3] stabilizer code, which is due to Gottesman [9] (see also Ref. [2]).

11.3.5 The [5,1,3] Fault-Tolerant Stabilizer Code

This is a stabilizer code that can correct single errors (because the minimum distance is 3), and it is also a perfect quantum code, as it satisfies the quantum Hamming bound with equality (see Section 7.4). This quantum code is generated by the following four generators of the stabilizer: $g_1 = XZZXI$, $g_2 = IXZZX$, $g_3 = XIXZZ$, and $g_4 = ZXIXZ$, and the corresponding quantum-check matrix is given by

$$A_Q = \begin{bmatrix} X & Z \\ 10010 & 01100 \\ 01001 & 00110 \\ 10100 & 00011 \\ 01010 & 10001 \end{bmatrix}. \tag{11.90}$$

Since every next row is the cyclic shift of the previous row, this code is a cyclic quantum code. The corresponding encoded Pauli operators, by using the standard form formalism, are obtained as follows:

$$\overline{X} = X_1X_2X_3X_4X_5, \quad \overline{Z} = Z_1Z_2Z_3Z_4Z_5. \tag{11.91}$$

In Section 11.3.2, we introduced the following Pauli gates cyclic permutation operator:

$$U: X \rightarrow Y \rightarrow Z \rightarrow X. \tag{11.92}$$

When this operator is applied transversally to the codeword of [5,1,3] code it leaves the stabilizer S invariant, as shown below:

$$\begin{aligned} g_1 &= XZZXI \rightarrow YXXYI = g_3g_4, \quad g_2 = IXZZX \rightarrow IYXXY = g_1g_2g_3, \\ g_3 &= XIXZZ \rightarrow YIYXX = g_2g_3g_4, \quad g_4 = ZXIXZ \rightarrow XYIYX = g_1g_2. \end{aligned} \tag{11.93}$$

Since the operator U fixes the stabilizer S, when applied transversally it can serve as an encoded fault-tolerant operation. To verify this claim, let us apply U to Pauli encoded operators:

$$\begin{aligned} &\overline{X} = X_1X_2X_3X_4X_5 \rightarrow Y_1Y_2Y_3Y_4Y_5 = \overline{Y}, \quad \overline{Z} = Z_1Z_2Z_3Z_4Z_5 \rightarrow X_1X_2X_3X_4X_5 = \overline{X}, \\ &\overline{Y} = Y_1Y_2Y_3Y_4Y_5 \rightarrow Z_1Z_2Z_3Z_4Z_5 = \overline{Z}. \end{aligned} \tag{11.94}$$

In Section 11.3.4, we introduced the four-qubit operation (see Eq. (11.66)) that can be used, when applied transversally, to generate an encoded version of itself for all stabilizer codes. Alternatively, the following three-qubit operation can serve the same role for [5,1,3] code [9]:

$$T_3 \doteq \begin{bmatrix} 1 & 0 & j & 0 & j & 0 & 1 & 0 \\ 0 & -1 & 0 & j & 0 & j & 0 & -1 \\ 0 & j & 0 & 1 & 0 & -1 & 0 & -j \\ j & 0 & -1 & 0 & 1 & 0 & -j & 0 \\ 0 & j & 0 & -1 & 0 & 1 & 0 & -j \\ j & 0 & 1 & 0 & -1 & 0 & -j & 0 \\ -1 & 0 & j & 0 & j & 0 & -1 & 0 \\ 0 & 1 & 0 & j & 0 & j & 0 & 1 \end{bmatrix}. \tag{11.95}$$

This three-qubit operation performs, with the corresponding quantum circuit shown in Figure 11.16c, the following mapping on unencoded qubits:

$$XII \rightarrow XYZ, \quad ZII \rightarrow ZXY, \quad IXI \rightarrow YXZ, \quad IZI \rightarrow XZY, \quad IIX \rightarrow XXX, \quad IIZ \rightarrow ZZZ. \tag{11.96}$$

The transversal application of T_3 performs the following mapping on the generators of [5,1,3] code:

$$\begin{aligned}
g_1 \otimes I \otimes I &= XZZXI \otimes IIIII \otimes IIIII \rightarrow XZZXI \otimes \underbrace{YXXYI}_{g_3 g_4} \otimes \underbrace{ZYYZI}_{g_1 g_3 g_4} \\
&= g_1 \otimes g_3 g_4 \otimes g_1 g_3 g_4 \\
g_2 \otimes I \otimes I &= IXZZX \otimes IIIII \otimes IIIII \rightarrow IXZZX \otimes \underbrace{IYXXY}_{g_1 g_2 g_3} \otimes \underbrace{IZYYZ}_{g_1 g_3} \\
&= g_2 \otimes g_1 g_2 g_3 \otimes g_1 g_3 \\
g_3 \otimes I \otimes I &\rightarrow g_3 \otimes g_2 g_3 g_4 \otimes g_2 g_4, \quad g_4 \otimes I \otimes I \rightarrow g_4 \otimes g_1 g_2 \otimes g_1 g_2 g_4 \\
I \otimes g_1 \otimes I &\rightarrow g_3 g_4 \otimes g_1 \otimes g_1 g_3 g_4, \quad I \otimes g_2 \otimes I \rightarrow g_1 g_2 g_3 \otimes g_2 \otimes g_1 g_3 \\
I \otimes g_3 \otimes I &\rightarrow g_2 g_3 g_4 \otimes g_3 \otimes g_2 g_4, \quad I \otimes g_4 \otimes I \rightarrow g_1 g_2 \otimes g_4 \otimes g_1 g_2 g_4 \\
I \otimes I \otimes g_n &\rightarrow g_n \otimes g_n \otimes g_n; \quad n = 1,2,3,4.
\end{aligned} \tag{11.97}$$

By closer inspection of Eq. (11.97) we conclude that the T_3 operator applies U and U^2 on the other two blocks, except for the last line. Since both of these operators fix S, it turns out that T_3 fixes $S \otimes S \otimes S$ as well, for all lines in (11.97) except the last one. On the other hand, by inspection of the last line in (11.97) we conclude that the T_3 operator also fixes $S \otimes S \otimes S$ in this case. Therefore, the T_3-operator applied transversally represents an encoded fault-tolerant operation for the [5,1,3] code. In similar fashion, by applying U and U^2 on the other two blocks in encoded Pauli operators we obtain the following mapping:

$$\begin{aligned}
&\bar{X} \otimes I \otimes I \rightarrow \bar{X} \otimes \bar{Y} \otimes \bar{Z}, \quad \bar{Z} \otimes I \otimes I \rightarrow \bar{Z} \otimes \bar{X} \otimes \bar{Y}, \quad I \otimes \bar{X} \otimes I \rightarrow \bar{Y} \otimes \bar{X} \otimes \bar{Z}, \\
&I \otimes \bar{Z} \otimes I \rightarrow \bar{X} \otimes \bar{Z} \otimes \bar{Y}, \quad I \otimes I \otimes \bar{X} \rightarrow \bar{X} \otimes \bar{X} \otimes \bar{X}, \quad I \otimes I \otimes \bar{Z} \rightarrow \bar{Z} \otimes \bar{Z} \otimes \bar{Z}.
\end{aligned} \tag{11.98}$$

Therefore, the T_3 operator, applied transversally, has performed the mapping to the encoded version of the operation itself. Now by combining the operations U and T_3 together with the measurement protocol, we are in position to implement fault-tolerantly all operators in $\mathcal{N}(G_N)$ for [5,1,3] stabilizer code. Because of the equivalence of Eqs (11.96) and (11.98), it is sufficient to describe the unencoded design procedures. We start our description with the phase gate implementation. Our starting point is the three-qubit system $|00\rangle \otimes |\psi\rangle$. The generators of the stabilizer are $g_1 = ZII$ and $g_2 = IZI$, while the corresponding encoded Pauli operator analogs are $\tilde{X} = IIX$, $\tilde{Z} = IIZ$. The application of the T_3 operation results in the following mapping:

$$\begin{aligned}
g_1 &= ZII \rightarrow g_1' = ZXY, \quad g_2 = IZI \rightarrow g_2' = XZY, \\
\tilde{X} &= IIX \rightarrow \tilde{X}' = XXX, \quad \tilde{Z} = IIZ \rightarrow \tilde{Z}' = ZZZ.
\end{aligned} \tag{11.99}$$

Next, we perform the measurement of operators $O_1 = IZI$ and $O_2 = IIZ$. We see that g_1' and $\tilde{X}'$ anticommute with O_1, while g_1', g_2', and $\tilde{X}'$ anticommute with O_2. Based on the measurement protocol, we perform the following mapping:

$$\begin{aligned} g_1' &= ZXY \rightarrow g_1'' = O_1, \quad g_2' \rightarrow g_2'' = O_2, \\ \tilde{X}' &= XXX \rightarrow \tilde{X}'' = g_1' \tilde{X}' = YIZ, \quad \tilde{Z}' \rightarrow \tilde{Z}'' = ZZZ. \end{aligned} \tag{11.100}$$

By disregarding the measurement qubits (qubits 2 and 3), we effectively performed the mappings $X_3 \rightarrow Y_1$ and $Z_3 \rightarrow Z_1$, which is in fact the phase gate combined with transfer from qubit 3 to qubit 1. Since

$$U^{\dagger} P = \frac{1}{\sqrt{2}} \begin{bmatrix} 1 & j \\ j & 1 \end{bmatrix} = R_x(-\pi/2),$$

and the Hadamard gate can be represented as $H = PR_x(-\pi/2)P^{\dagger}$, we can implement an arbitrary single-qubit gate in the Clifford group. For arbitrary Clifford operator implementation, the CNOT gate is needed. The starting point is the three-qubit system $|\psi_1\rangle \otimes |\psi_2\rangle \otimes |0\rangle$. The generator of the stabilizer is given by $g = IIZ$, while the encoded Pauli operator analogs are given by $\tilde{X}_1 = XII$, $\tilde{X}_2 = IXI$, $\tilde{Z}_1 = ZII$, $\tilde{Z}_2 = IZI$. We then apply the T_3 operator, which performs the following mapping:

$$\begin{aligned} g &= IIZ \rightarrow g' = ZZZ, \quad \tilde{X}_1 = XII \rightarrow \tilde{X}_1' = XYZ, \quad \tilde{X}_2 = IXI \rightarrow \tilde{X}_2' = YXZ, \\ \tilde{Z}_1 &= ZII \rightarrow \tilde{Z}_1' = ZXY, \quad \tilde{Z}_2 = IZI \rightarrow \tilde{Z}_2' = XZY. \end{aligned} \tag{11.101}$$

We further perform the measurement of $O = IXI$. Clearly g', $\tilde{X}_1'$, and $\tilde{Z}_2$ anticommute with O, and the measurement protocol performs the mapping:

$$\begin{aligned} g' \rightarrow g'' &= O, \quad \tilde{X}_1' = XYZ \rightarrow \tilde{X}_1'' = g'\tilde{X}_1' = YXI, \quad \tilde{X}_2' \rightarrow \tilde{X}_2'' = YXZ, \quad \tilde{Z}_1' \rightarrow \tilde{Z}_1'' \\ &= ZXY, \quad \tilde{Z}_2' = XZY \rightarrow \tilde{Z}_2'' = g'\tilde{Z}_2' = YIX. \end{aligned} \tag{11.102}$$

By disregarding the measurement qubit (namely qubit 2), we effectively performed the following mapping:

$$\begin{aligned} & X_1 I_2 \rightarrow Y_1 I_3, \quad \tilde{X}_2 = I_1 X_2 \rightarrow Y_1 Z_3, \quad \tilde{Z}_1 = Z_1 I_2 \rightarrow Z_1 Y_3, \\ & I_1 Z_2 \rightarrow Y_1 X_3. \end{aligned} \tag{11.103}$$

By relabeling qubit 3 in the images with qubit 2, mapping equation (11.103) can be described by the following equivalent gate U_e [9] (see also Ref. [2]):

$$U_e = (I \otimes U^2)(P \otimes I) U_{\text{CNOT}}^{(21)} (I \otimes R_x(-\pi/2)), \tag{11.104}$$

where $U_{\text{CNOT}}^{(21)}$ denotes the CNOT gate with qubit 2 as control qubit and qubit 1 as target qubit. By using the fact that the U gate is of order 3 ($U^3 = I \Leftrightarrow (U^2)^\dagger = U$), the $U_{\text{CNOT}}^{(21)}$ gate can be expressed in terms of the U_e gate as

$$U_{\text{CNOT}}^{(21)} = (P^\dagger \otimes I)(I \otimes U)U_e(I \otimes (R_x(-\pi/2))^\dagger). \tag{11.105}$$

The CNOT gate $U_{\text{CNOT}}^{(21)}$ can be implemented based on $U_{\text{CNOT}}^{(21)}$ and four Hadamard gates in a fashion similar to that shown in Figure 11.15b. This description of CNOT gate implementation completes the description of all gates needed for the implementation of an arbitrary Clifford operator. For the implementation of a universal set of gates required for quantum computation, we have to study the design of the Toffoli gate, which is the subject of the next section.

11.4 FAULT-TOLERANT QUANTUM COMPUTATION

As already discussed in Chapter 3, there exist various sets of universal quantum gates: (i) {Hadamard, phase, π/8 gate, CNOT} gates; (ii) {Hadamard, phase, CNOT, Toffoli} gates; (iii) the {Barenco} gate [16]; and (iv) the {Deutsch} gate [17]. To complete set (i), the fault-tolerant π/8 gate is needed, which has already been discussed in Section 11.2.5 (see also Ref. [5]). To complete set (ii) the Toffoli gate [2,3,11] is needed, which is the subject of interest in this section. The description of the fault-tolerant Toffoli gate below is based on Gottesman's proposal [11] (see also Ref. [2]).

11.4.1 Fault-Tolerant Toffoli Gate

The Toffoli gate has already been introduced in Chapter 2, and its operating principle is described in Figure 11.17. It can be considered as a generalization of the CNOT gate, which has two control qubits ($|x_1\rangle$ and $|x_2\rangle$) and one target qubit $|y\rangle$. The NOT gate operation is applied to the target qubit only when $x_1 = x_2 = 1$:

$$U_{\text{Toffoli}}|x_1x_2\rangle|y\rangle = |x_1x_2\rangle|y \oplus x_1x_2\rangle, \tag{11.106}$$

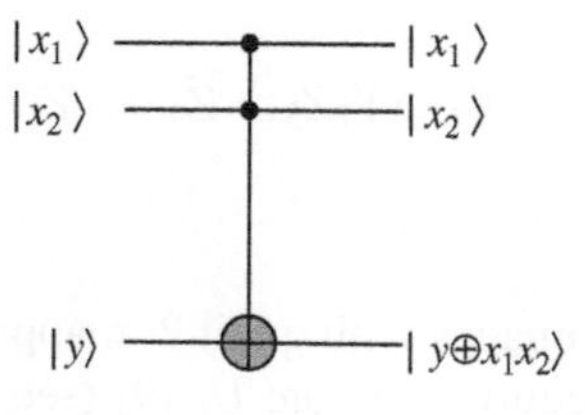

FIGURE 11.17

Toffoli gate description.

where $\oplus$ denotes mod 2 addition, as usual. For convenience, we can express the Toffoli gate's action in terms of projection operators. We have shown earlier (see Chapter 2) that, for arbitrary Hermitian operator O of order 2 ($O^2 = I$, with eigenvalues ± 1), the projection operators are defined as

$$P_{\pm}(O) = \frac{1}{2}(I \pm O). \tag{11.107}$$

By using the projection operators, we can express the Toffoli operator as follows [11] (see also Ref. [2]):

$$\begin{aligned} U_{\text{Toffoli}} &= \frac{1}{2}\{P_+(Z_1) + P_+(Z_2) + P_-(Z_1Z_2)\} \ + P_-(Z_1)P_-(Z_2)X_3 \\ &= \frac{1}{4}[3I + Z_1 + Z_2 - Z_1Z_2 + (I - Z_1)(I - Z_2)X_3]. \end{aligned} \tag{11.108}$$

Equation (11.108) can be interpreted as

$$\begin{aligned} U_{\text{Toffoli}}|x_1x_2y\rangle &= \frac{1}{2}(\delta_{x_1,0} + \delta_{x_2,0} + \delta_{x_1+x_2,1})|x_1x_2y\rangle + \delta_{x_1x_2,1}|x_1x_2y\rangle \\ &= \begin{cases} |00y\rangle; & x_1 = x_2 = 0 \\ |10y\rangle; & x_1 = 1,\ x_2 = 0 \\ |01y\rangle; & x_1 = 0,\ x_2 = 1 \\ |11(y\oplus 1)\rangle; & x_1 = 1,\ x_2 = 1, \end{cases} \end{aligned} \tag{11.109}$$

therefore proving its correctness. We are interested in the action of the Toffoli operator on elements from Pauli multiplicative group G_3, which is essentially the conjugation operation:

$$f(a) = U_{\text{Toffoli}}\, aU_{\text{Toffoli}}^{\dagger}, \quad \forall a \in G_3. \tag{11.110}$$

If this mapping is a homomorphism, then it will be sufficient to study the action of the Toffoli operator on generator elements from G_3. Let us first verify whether the multiplication table is preserved by conjugation:

$$\begin{aligned} f(ab) &= U_{\text{Toffoli}}\, abU_{\text{Toffoli}}^{\dagger} = U_{\text{Toffoli}}\, a\, \overbrace{U_{\text{Toffoli}}U_{\text{Toffoli}}^{\dagger}}^{I}\, bU_{\text{Toffoli}}^{\dagger} \\ &= \underbrace{U_{\text{Toffoli}}aU_{\text{Toffoli}}}_{f(a)}\ \underbrace{U_{\text{Toffoli}}^{\dagger}bU_{\text{Toffoli}}^{\dagger}}_{f(b)} = f(a)f(b), \quad \forall a,b \in G_3. \end{aligned} \tag{11.111}$$

Since the conjugation function of the product of two elements from G_3 equals the product of images, the mapping is clearly a homomorphism. It is evident that the image of arbitrary element form G_3 under conjugation by the Toffoli operator is contained in $\mathcal{N}(G_3)$, which indicates that the Toffoli operator does not fix G_3, and consequently it cannot belong to the normalizer of G_3, $\mathcal{N}(G_3)$. Since $G_3 \subset G_N$, the Toffoli operator cannot fix G_N and, consequently, the Toffoli operator does not belong to the Clifford group $N(G_N)$. Let us now study the action of the Toffoli operator on X_i, Z_i ($i = 1,2,3$). We start by mapping $f(X_1)$ [11]:

$$\begin{aligned} X_1 \rightarrow U_{\text{Toffoli}} X_1 U^{\dagger} &= \frac{1}{4}[3I + Z_1 + Z_2 - Z_1 Z_2 + (I - Z_1)(I - Z_2)X_3] X_1 \\ &\quad \frac{1}{4}[3I + Z_1 + Z_2 - Z_1 Z_2 + (I - Z_1)(I - Z_2)X_3] \\ &= \underbrace{\frac{1}{2}(I + Z_2 + (I - Z_2)X_3)}_{U_{\text{CNOT}}^{(23)}} X_1 \\ &= U_{\text{CNOT}}^{(23)} X_1. \end{aligned} \tag{11.112}$$

The mapping $f(X_2)$ can be obtained analogously to mapping (11.112) as follows:

$$\begin{aligned} X_2 \rightarrow U_{\text{Toffoli}} X_2 U_{\text{Toffoli}}^{\dagger} &= \frac{1}{4}[3I + Z_1 + Z_2 - Z_1 Z_2 + (I - Z_1)(I - Z_2)X_3] X_2 \\ &\quad \frac{1}{4}[3I + Z_1 + Z_2 - Z_1 Z_2 + (I - Z_1)(I - Z_2)X_3] \\ &= \underbrace{\frac{1}{2}(I + Z_1 + (I - Z_1)X_3)}_{U_{\text{CNOT}}^{(13)}} X_2 \\ &= U_{\text{CNOT}}^{(13)} X_2. \end{aligned} \tag{11.113}$$

The operators X_3, Z_1, and Z_2 do not change during the conjugation mapping, while the mapping $f(Z_3)$ is given by

$$\begin{aligned} Z_3 \rightarrow U_{\text{Toffoli}} Z_3 U_{\text{Toffoli}}^{\dagger} &= \frac{1}{4}[3I + Z_1 + Z_2 - Z_1 Z_2 + (I - Z_1)(I - Z_2)X_3] Z_3 \\ &\quad \frac{1}{4}[3I + Z_1 + Z_2 - Z_1 Z_2 + (I - Z_1)(I - Z_2)X_3] \\ &= \underbrace{\frac{1}{2}(I + Z_1 + (I - Z_1)Z_2)}_{U_{\text{CP}}^{(12)}} Z_3 \\ &= U_{\text{CP}}^{(12)} Z_3, \end{aligned} \tag{11.114}$$

where U_{CP} is the control-phase (or control-Z) gate. Based on (11.112)–(11.114), the stabilizer generators

$$g_1 = U_{CNOT}^{(23)} X_1, \quad g_2 = U_{CNOT}^{(13)} X_2, \quad g_3 = U_{CP}^{(12)} Z_3 \qquad (11.115)$$

fix the state:

$$A = \frac{1}{2}(|000\rangle + |010\rangle + |100\rangle + |111\rangle). \qquad (11.116)$$

We now create the composite state $|A\rangle \otimes \underbrace{|\psi_1\rangle \otimes |\psi_2\rangle \otimes |\psi_3\rangle}_{|\psi\rangle}$. The corresponding generators of the stabilizer for this composite state are

$$G_1 = U_{CNOT}^{(23)} X_1 III, \quad G_2 = U_{CNOT}^{(13)} X_2 III, \quad G_3 = U_{CP}^{(12)} Z_3 III, \qquad (11.117)$$

while the corresponding encoded Pauli operator analogs are

$$\tilde{X}_1 = X_4, \quad \tilde{X}_2 = X_5, \quad \tilde{X}_3 = X_6, \quad \tilde{Z}_1 = Z_4, \quad \tilde{Z}_2 = Z_5, \quad \tilde{Z}_3 = Z_6. \qquad (11.118)$$

By applying the CNOT gates $U_{CNOT}^{(14)}$, $U_{CNOT}^{(25)}$, $U_{CNOT}^{(63)}$, we perform the following mapping:

$$\begin{aligned} &G_1 \to G_1' = U_{CNOT}^{(23)} X_1 X_4, \quad G_2 \to G_2' = U_{CNOT}^{(13)} X_2 X_5, \quad G_3 \to G_3' = U_{CP}^{(12)} Z_3 Z_6, \\ &\tilde{X}_1 \to \tilde{X}_1' = X_4, \quad \tilde{X}_2 \to \tilde{X}_2' = X_5, \quad \tilde{X}_3 \to \tilde{X}_3' = X_3 X_6, \quad \tilde{Z}_1 \to \tilde{Z}_1' = Z_1 Z_4, \\ &\tilde{Z}_2 \to \tilde{Z}_2' = Z_2 Z_5, \quad \tilde{Z}_3 \to \tilde{Z}_3' = Z_6. \end{aligned} \qquad (11.119)$$

By measuring the operators $O_1 = Z_4$, $O_2 = Z_5$, and $O_3 = X_6$, since the operator O_1 (O_2) anticommutes with G_1', $\tilde{X}_1'$ (G_2', $\tilde{X}_2'$) and operator O_3 anticommutes with G_3', $\tilde{Z}_3'$, the measurement protocol performs the following mapping:

$$\begin{aligned} &G_1' \to G_1'' = O_1, \quad G_2' \to G_2'' = O_2, \quad G_3' \to G_3'' = O_3, \\ &\tilde{X}_1' \to \tilde{X}_1'' = G_1' \tilde{X}_1' = U_{CNOT}^{(23)} X_1 = G_1, \quad \tilde{X}_2' \to \tilde{X}_2'' = G_2' \tilde{X}_2 = U_{CNOT}^{(13)} X_2 = G_2, \\ &\tilde{X}_3' \to \tilde{X}_3'' = X_3 X_6, \\ &\tilde{Z}_1' \to \tilde{Z}_1'' = Z_1 Z_4, \quad \tilde{Z}_2' \to \tilde{Z}_2'' = Z_2 Z_5, \quad \tilde{Z}_3' \to \tilde{Z}_3'' = G_3' \tilde{Z}_3' = U_{CP}^{(12)} Z_3 = G_3. \end{aligned} \qquad (11.120)$$

By closer inspection of Eqs (11.118) and (11.120), and by disregarding the measured qubits (1, 2, and 3), we conclude that the procedure above has performed the following mapping:

$$X_4 \to g_1, \quad X_5 \to g_2, \quad X_6 \to X_3, \quad Z_4 \to Z_1, \quad Z_5 \to Z_2, \quad Z_6 \to g_3, \qquad (11.121)$$

which is the same as the homomorphism introduced by the Toffoli operator given by (11.112)–(11.114), in combination with transfer of data from qubits 1–3 to qubits 4–6.

The encoded Toffoli gate can now be implemented by using the encoded CNOT gate and measurement. The encoded g_3 operation, denoted as $\overline{g}_3$, can be represented as the product of the encoded controlled-phase gate on encoded qubits 1 and 3 and encoded Pauli operator $\overline{Z}_3$, i.e. $\overline{g}_3 = \overline{U}_{\mathrm{CP}}^{(12)}\overline{Z}_3$ (see Eq. (11.115)). From Figure 11.13 we know that the encoded controlled-phase (controlled-Z) gate can be expressed in terms of the encoded CNOT gate as follows:

$$\overline{U}_{\mathrm{CP}}^{(12)} = (I\otimes\overline{H})\overline{U}_{\mathrm{CNOT}}^{(12)}(I\otimes\overline{H}). \tag{11.122}$$

This equations allows us to implement $\overline{U}_{\mathrm{CP}}^{(12)}$ transversally since we have already learned how to implement $\overline{U}_{\mathrm{CNOT}}^{(12)}$ and $\overline{H}$ transversally. Because the encoded Z_3 operator can be represented by $\overline{Z}_3 = j^{l_3}X(\boldsymbol{a}_3)Z(\boldsymbol{b}_3)$, it contains only single-qubit operators and its action is by default transversal.

The previous description and discussion is correct providing that ancillary state $|A\rangle$ can be prepared fault-tolerantly. In order to be able to perform the Toffoli gate on encoded states, we must find a way to generate the encoded version of $|A\rangle$, denoted as $|\overline{A}\rangle$. To facilitate implementation, we introduce another auxiliary encoded state $|\overline{B}\rangle$ that is related to $|\overline{A}\rangle$ by $|\overline{B}\rangle = \overline{X}_3|\overline{A}\rangle$, where $\overline{X}_3$ is the encoded X_3 operator. We will assume that the corresponding coding blocks required are obtained from an [N,1,D] stabilizer code. For other [N,K,D] stabilizer codes, we will need to start with three ancillary blocks with encoded qubits being in the $|\overline{0}\rangle$ state. We then further need to apply the procedure leading to the encoded state $|\overline{A}\rangle$. The first portion in these blocks will be in the $|\overline{A}\rangle$ state, while the remaining qubits will stay in the $|\overline{0}\rangle$ state. The three encoded data qubits involved in the Toffoli gate will further be moved into three ancillary blocks initialized into $|\overline{0}\rangle$ states. Further, the Toffoli gate procedure will be applied to the six ancillary blocks by using block-encoded CNOT gates and the measurement protocol. Finally, the encoded data qubits will be transferred back to the original code blocks. Therefore, the study of encoding procedures using [N,1,D] stabilizer code will not affect the generality of procedures. By ignoring the normalization constants, the encoded $|\overline{A}\rangle$ and $|\overline{B}\rangle(=\overline{X}_3|\overline{A}\rangle)$ states can be represented as

$$\begin{aligned}|\overline{A}\rangle &= |\overline{000}\rangle + |\overline{010}\rangle + |\overline{100}\rangle + |\overline{111}\rangle,\\ |\overline{B}\rangle &= \overline{X}_3|\overline{A}\rangle = |\overline{001}\rangle + |\overline{011}\rangle + |\overline{101}\rangle + |\overline{110}\rangle.\end{aligned} \tag{11.123}$$

It is interesting to note that the action of the encoded g_3 operator on encoded states $|\overline{A}\rangle$ and $|\overline{B}\rangle$ is as follows:

$$\overline{g}_3|\overline{A}\rangle = |\overline{A}\rangle,\quad \overline{g}_3\,|\overline{B}\rangle = -|\overline{B}\rangle. \tag{11.124}$$

Therefore, the transversal application of the encoded g_3 operator maps $|\overline{A}\rangle \rightarrow |\overline{A}\rangle$ and $|\overline{B}\rangle \rightarrow -|\overline{B}\rangle$. From (11.123) it is clear that

$$|\overline{A}\rangle + |\overline{B}\rangle = |\overline{000}\rangle + |\overline{010}\rangle + |\overline{100}\rangle + |\overline{111}\rangle + |\overline{001}\rangle + |\overline{011}\rangle + |\overline{101}\rangle + |\overline{110}\rangle$$
$$= (|\overline{0}\rangle + |\overline{1}\rangle)^3 = \sum_{d_1,d_2,d_3=0}^{1} |\overline{d_1 d_2 d_3}\rangle, \tag{11.125}$$

which indicates that the state $|\overline{A}\rangle + |\overline{B}\rangle$ can be obtained by measuring $\overline{X}$ for each code block. Let an ancillary block be prepared in cat state $|0\ldots0\rangle + |1\ldots1\rangle$ and the code block in state $|\overline{A}\rangle + |\overline{B}\rangle$. By executing the encoded g_3 operator only when the ancillary is in state $|1\ldots1\rangle$, we will effectively perform the following mapping on the composite state:

$$(|0\ldots0\rangle + |1\ldots1\rangle)(|\overline{A}\rangle + |\overline{B}\rangle) \rightarrow (|0\ldots0\rangle + |1\ldots1\rangle)|\overline{A}\rangle + (|0\ldots0\rangle - |1\ldots1\rangle)|\overline{B}\rangle. \tag{11.126}$$

Now, we perform the measurement of $O = X_1\ldots X_N$ on the ancillary cat-state block. We know that

$$X_1\ldots X_N(|0\ldots0\rangle \pm |1\ldots1\rangle) = \pm(|0\ldots0\rangle \pm |1\ldots1\rangle), \tag{11.127}$$

so that if the result of the measurement is $+1$ the code block will be in state $|\overline{A}\rangle$, otherwise it will be in state $|\overline{B}\rangle$. When in state $|\overline{B}\rangle$, we need to apply the encoded operator $\overline{X}_3$ to obtain $\overline{X}_3\,|\overline{B}\rangle = |\overline{A}\rangle$. The only problem that remains is the cat-state preparation circuit that is not fault-tolerant, as we discussed in Section 11.3.1. The quantum circuit in Figure 11.12 (when Hadamard gates are omitted) generates the cat state where there is no error. It also verifies that up to the order of p (p is the single-qubit error probability) there is no bit-flip error present. However, the phase error in cat-state generation will introduce an error in the implementation of state $|\overline{A}\rangle$. Since

$$Z_i(|0\ldots0\rangle + |1\ldots1\rangle)(|\overline{A}\rangle + |\overline{B}\rangle) = (|0\ldots0\rangle - |1\ldots1\rangle)(|\overline{A}\rangle + |\overline{B}\rangle), \tag{11.128}$$

the application of mapping (11.126) in the presence of the phase-flip error yields:

$$Z_i(|0\ldots0\rangle + |1\ldots1\rangle)(|\overline{A}\rangle + |\overline{B}\rangle) \rightarrow (|0\ldots0\rangle - |1\ldots1\rangle)|\overline{A}\rangle + (|0\ldots0\rangle + |1\ldots1\rangle)|\overline{B}\rangle. \tag{11.129}$$

By measuring the operator $O = X_1\ldots X_N$ on the ancillary cat-state block the result $+1$ leads to $|\overline{B}\rangle$, while the result -1 leads to $|\overline{A}\rangle$. This problem can be solved by repeating the cat-state preparation procedure and measurement of O, every time with a fresh ancillary block, until we get the same result twice in a row. It can be shown that in this case, up to the order of p, the code block will be in state $|\overline{A}\rangle$.

11.5 ACCURACY THRESHOLD THEOREM

In our discussion about the Shor-state preparation circuit we learned that up to $O(P_e)$ no flip error has occurred, where P_e is the qubit error probability. Moreover, in our discussion of the fault-tolerant Toffoli gate we learned that up to the order of P_e we can prepare the state $|\overline{A}\rangle$ fault-tolerantly. Therefore, the encoding data using QECCs and applying fault-tolerant operations can also introduce new errors, if the qubit error probability P_e is not sufficiently low. It can be shown that the probability for a fault-tolerant circuit to introduce two or more errors into a single code block is proportional to P_e^2 for some constant of proportionality C. So the encoded procedure will be successful with probability $1 - CP_e^2$. If the probability of qubit error P_e is small enough, e.g. $P_e < 10^{-4}$, there is a real benefit of using fault-tolerant technology. For an arbitrary long quantum computation with the number of qubits involved of the order of millions, the failure probability per block CP_e^2 is still too high. That is, the success probability for an N-qubit computation is approximately $(1 - CP_e^{2N})^N$. Even with millions of qubits being involved with a failure probability $P_e < 10^{-4}$, we cannot make the computation error arbitrary small. Knill and Laflamme [18] (see also Refs [19–21]) solved this problem by introducing *concatenated codes*. We have already encountered the concept of concatenated codes in the discussion of Shor's nine-qubit code in Chapter 7. Moreover, the concept of quantum concatenated codes is very similar to the concept of classical concatenated codes introduced in Chapter 6. A concatenated code first encodes a set of K qubits by using an $[N,K,D]$ code. Every qubit from the codeword is encoded again using an $[N_1,1,D_1]$ code. The codeword qubits thus obtained are further encoded using an $[N_2,1,D_2]$ code, and so on. This procedure is illustrated in Figure 11.18. After k concatenation stages, the resulting quantum code has the parameters $[NN_1N_2\ldots N_k,K,DD_1D_2\ldots D_k]$. The obtained concatenated code has a failure probability of order $(CP_e)^{2^k}/C$. Therefore, the improvement with a quantum concatenated code is remarkable. With a quantum concatenated code, we are in position to design QECCs with arbitrary small probability of error. Unfortunately, the complexity increases as the number of concatenation stages increases. In practice, careful engineering is needed to satisfy these conflicting requirements. We are now ready to formulate the well-known accuracy threshold theorem.

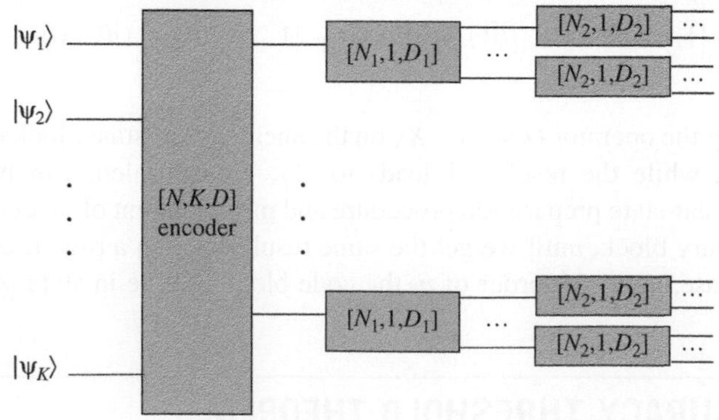

FIGURE 11.18

The principle of quantum concatenated code (two-stage example).

Accuracy threshold theorem [2]. A quantum computation operation of arbitrary duration can be performed with arbitrary small failure error probability in the presence of quantum noise and by using imperfect (faulty) quantum gates providing that following conditions are satisfied:

1. The computational operation is protected by using a concatenated QECC.
2. Fault-tolerant procedures are used for encoded quantum gates, error correction, and measurement protocols.
3. The storage quantum registers and a universal set of quantum gates are available with the failure error probabilities P_e that are smaller than a threshold value P_{tsh}, known as the threshold accuracy: $P_e < P_{tsh}$.

For a proof of the theorem, any interested reader can refer to Chapter 6 of Ref. [2] (see also Ref. [11]). Another interesting formulation is given by Neilsen and Chuang [3].

Threshold theorem for quantum computation [3]. A quantum circuit containing $p(N)$ gates, where $p(\cdot)$ is the polynomial function, can be simulated with a failure probability of at most P_f by using

$$O\left(\text{poly}\left(\frac{\log p(N)}{P_f}\right)p(N)\right), \tag{11.130}$$

quantum gates with quantum hardware whose components fail with probability at most P_e, provided P_e is below a constant threshold, $P_e < P_{tsh}$, and given reasonable assumptions about the quantum noise in the underlying quantum hardware. In Eq. (11.130), poly$(\cdot)$ denotes the polynomial complexity.

If the desired failure probability of a quantum computation circuit is P_f, the error probability per gate must be smaller than $P_f/p(N)$, where $p(N)$ is the number of gates. The number of stages needed in concatenated code for this accuracy can be determined from the following inequality:

$$\frac{(CP_e)^{2^k}}{C} \leq \frac{P_f}{p(N)}, \tag{11.131}$$

where $P_e < P_{tsh}$, $C = 1/P_{tsh}$. By solving (11.131) we obtain that the number of stages is lower bounded by

$$k \geq \log_2\left\{1 + \frac{\log[p(N)/P_f]}{\log(P_{tsh}/P_e)}\right\}. \tag{11.132}$$

The size of the corresponding quantum circuit is given by d^k, where d is the maximum number of gates used in the fault-tolerant procedure to implement an encoded gate. From (11.132) it is clear that the quantum circuit size per encoded gate is of order poly(log($p(N)/P_f$)), proving the claim of Eq. (11.130). This theorem is very important because it dispelled the doubts about the capabilities of quantum computers to perform arbitrary long calculation due to the error accumulation phenomenon. For a more rigorous proof, the interested reader is referred to Ref. [2] (see also Ref. [11]).

11.6 SUMMARY

This chapter has discussed fault-tolerant quantum computation concepts. After the introduction of fault tolerance basics and traversal operations (Section 11.1), the basics of fault-tolerant quantum computation concepts and procedures were given (Section 11.2) by using Steane's code as an illustrative example. Section 11.3 included a rigorous description of fault-tolerant QECCs. Section 11.4 covered fault-tolerant computing, in particular the fault-tolerant implementation of the Tofolli gate. In Section 11.5 the accuracy threshold theorem was formulated and proved. In the next section, a set of problems are provided to enable readers to gain a deeper understanding of fault-tolerance theory.

11.7 PROBLEMS

1. This problem concerns the fault-tolerant *T*-gate implementation. As the first step in the derivation, show that the three circuits in Figure 11.P1a are equivalent to each other. Then prove that the circuit in Figure 11.P1b is in fact the fault-tolerant *T* gate.
2. Determine the fault-tolerant Toffoli gate using the [7,1] Steane code.
3. In Section 11.3 we studied the fault-tolerant syndrome extraction circuit implementation of [5,1,3] code. The corresponding generators did not contain *Y* operators. For an efficient implementation the standard form described in Chapter 8 is desirable. However, the corresponding generators obtained from a standard form representation contain *Y* operators:

$$g_1 = Y_1Z_2Z_4Y_5, \; g_2 = X_2Z_3Z_4Z_5, \; g_3 = Z_1Z_2X_3X_5 \text{ and } g_4 = Z_1Z_3Y_4Y_5.$$

Determine the quantum circuit for syndrome extraction based on the generators above. Compare the complexity of the corresponding schemes. Also determine the syndrome extraction circuit using Plenio's approach. Discuss the complexity of this circuit with respect to previous ones.

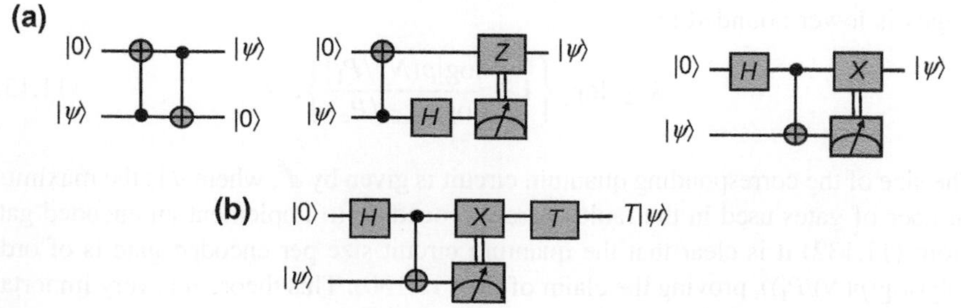

FIGURE 11.P1

Fault-tolerant *T*-gate implementation.

4. The generators for [8,3,3] quantum stabilizer code are given by

$$g_1 = X_1X_2X_3X_4X_5X_6X_7X_8, \quad g_2 = Z_1Z_2Z_3Z_4Z_5Z_6Z_7Z_8,$$
$$g_3 = X_2X_4Y_5Z_6Y_7Z_8, \quad g_4 = X_2Z_3Y_4X_6Z_7Y_8 \text{ and } g_5 = Y_2X_3Z_4X_5Z_6Y_8.$$

Determine the corresponding transversal syndrome extraction circuit. Provide the configuration of the Shor state preparation circuit as well. Discuss the Shor state verification procedure. Finally, describe the syndrome verification procedure.

5. Prove that the quantum circuit in Figure 11.P5 implements the following mapping:

$$\begin{aligned}
X\otimes I\otimes I\otimes I &\rightarrow X\otimes X\otimes X\otimes I\\
I\otimes X\otimes I\otimes I &\rightarrow I\otimes X\otimes X\otimes X\\
I\otimes I\otimes X\otimes I &\rightarrow X\otimes I\otimes X\otimes X\\
I\otimes I\otimes I\otimes X &\rightarrow X\otimes X\otimes I\otimes X\\
Z\otimes I\otimes I\otimes I &\rightarrow Z\otimes Z\otimes Z\otimes I\\
I\otimes Z\otimes I\otimes I &\rightarrow I\otimes Z\otimes Z\otimes Z\\
I\otimes I\otimes Z\otimes I &\rightarrow Z\otimes I\otimes Z\otimes Z\\
I\otimes I\otimes I\otimes Z &\rightarrow Z\otimes Z\otimes I\otimes Z.
\end{aligned}$$

This circuit, when applied transversally, can generate an encoded version of itself for any [N,1] stabilizer code as discussed earlier.

6. The one-qubit operator U performs the following Pauli operator cyclic shift:

$$U: X \rightarrow Y \rightarrow Z \rightarrow X.$$

Let us consider the two-qubit system $|\psi\rangle \otimes |0\rangle_Y = |\psi\rangle \otimes (|0\rangle + j|1\rangle)/\sqrt{2}$, where $|\psi\rangle$ is an arbitrary state. Let us now apply the CNOT gate by using qubit 2 as control qubit and qubit 1 as target qubit, followed by measurement of $O = Y_1$. Prove that this procedure implements the cyclic shift of Pauli operators introduced above.

The arbitrary state can be represented by $|\psi\rangle = a|0\rangle + b|1\rangle$. Determine the state of qubit 2 upon application of the measurement protocol described in the previous paragraph. Finally, determine the matrix representation of U.

7. Prove that the quantum circuit in Figure 11.P7 implements the following mapping:

$$XII \rightarrow XYZ, \quad ZII \rightarrow ZXY, \quad IXI \rightarrow YXZ, \quad IZI \rightarrow XZY, \quad IIX \rightarrow XXX, \quad IIZ \rightarrow ZZZ.$$

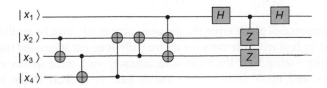

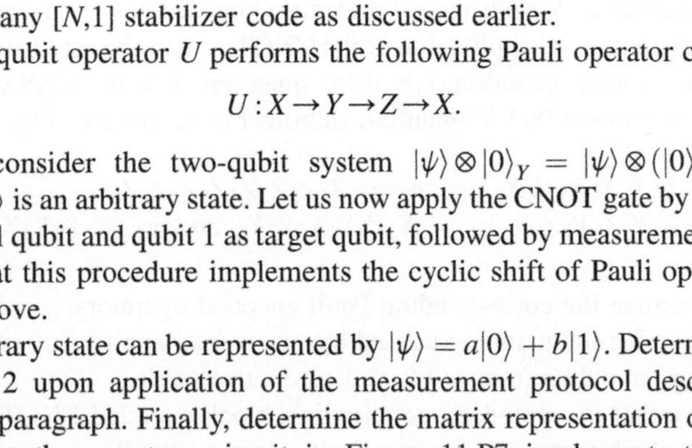

FIGURE 11.P5

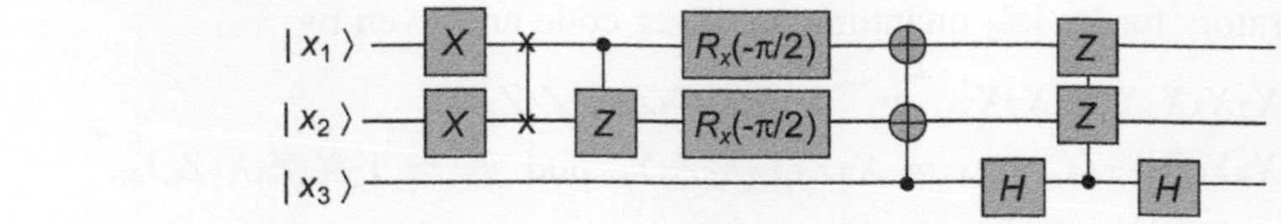

FIGURE 11.P7

This circuit, when applied transversally, can generate an encoded version of itself for the [5,1,3] stabilizer code as discussed earlier.

8. This problem concerns fault-tolerant encoded operations, [4,2,2] code, and fault-tolerant computing based on this code. The generators of this code are given by $g_1 = X_1X_2X_3X_4$ and $g_2 = Z_1Z_2Z_3Z_4$. The corresponding encoded Pauli operators are given by $\overline{X}_1 = X_1X_2$, $\overline{X}_2 = X_1Z_3$, $\overline{Z}_1 = Z_2Z_3$, $\overline{Z}_2 = Z_3Z_4$.

(a) Determine the mapping caused by transversal application of CNOT gates using the qubits in code block 1 as control qubits.

(b) Prove that the encoded operation is a block-encoded CNOT gate.

(c) Describe how to implement a fault-tolerant SWAP gate.

(d) Describe how to implement a fault-tolerant Toffoli gate.

9. Let the quantum [*N*,*K*,*D*] code be given. Consider now two encoded blocks. Prove that transversal application of the CNOT gate to these two encoded blocks yields an encoded operation only if the quantum code is a CSS code.

10. The generators for [8,3,3] quantum stabilizer code are given by

$$g_1 = X_1X_2X_3X_4X_5X_6X_7X_8, \quad g_2 = Z_1Z_2Z_3Z_4Z_5Z_6Z_7Z_8,$$
$$g_3 = X_2X_4Y_5Z_6Y_7Z_8, \quad g_4 = X_2Z_3Y_4X_6Z_7Y_8 \text{ and } g_5 = Y_2X_3Z_4X_5Z_6Y_8.$$

(a) Determine the corresponding Pauli encoded operators.

(b) Determine the mapping caused by transversal application of CNOT gates using the qubits in code-block 1 as control qubits.

(c) Prove that the encoded operation is a block-encoded CNOT gate.

(d) Describe how to implement a fault-tolerant SWAP gate.

(e) Describe how to implement a fault-tolerant Toffoli gate.

11. Show that in the absence of phase errors in the cat-state, the circuit in Figure 11.P11a can be used to generate the state:

$$|A\rangle = [|000\rangle + |010\rangle + |100\rangle + |111\rangle]/2.$$

In the presence of phase error, the measurement result must be verified. This can be done by replacing the cat-state above with three cat states and then applying a majority voting rule. Each cat-state interacts with three qubits as in Figure 11.P11a, and upon measurement the X gate is applied if the majority of measurement results is -1. Provide the corresponding circuit. Justify that the measurement result will be accurate up to the order of P_e (qubit error

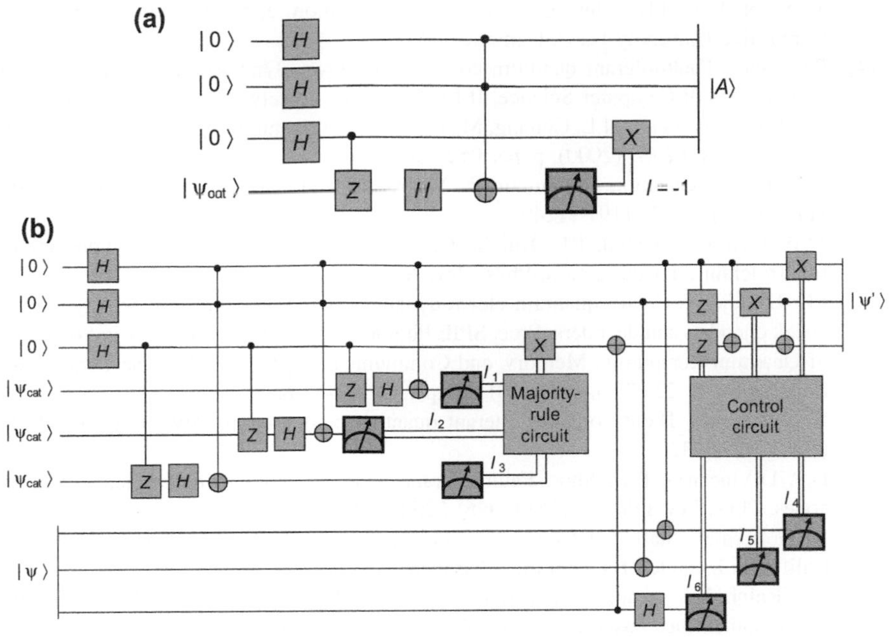

FIGURE 11.P11

probability). Prove that the circuit in Figure 11.P11b applies the Toffoli gate to the state $|A\rangle$. Describe the operating principle of this quantum circuit.

12. The quantum-check matrix $\boldsymbol{A}$ of a CSS code is given by

$$\boldsymbol{A} = \begin{bmatrix} \boldsymbol{H} & | & \boldsymbol{0} \\ \boldsymbol{0} & | & \boldsymbol{H} \end{bmatrix} \qquad \boldsymbol{H} = \begin{bmatrix} 1 & 0 & 0 & 1 & 1 & 1 \\ 1 & 1 & 1 & 0 & 0 & 1 \\ 0 & 1 & 1 & 1 & 1 & 0 \end{bmatrix}.$$

(a) Determine the mapping caused by transversal application of CNOT gates using the qubits in code block 1 as control qubits.
(b) Prove that the encoded operation is a block-encoded CNOT gate.
(c) Describe how to implement a fault-tolerant SWAP gate.
(d) Describe how to implement a fault-tolerant Toffoli gate.

References

[1] D. Gottesman, I.L. Chuang, Demonstrating the viability of universal quantum computation using teleportation and single-qubit operations, Nature 402 (25 November 1999) 390–393.

[2] F. Gaitan, Quantum Error Correction and Fault Tolerant Quantum Computing, CRC Press, 2008.

[3] M.A. Neilsen, I.L. Chuang, Quantum Computation and Quantum Information, Cambridge University Press, Cambridge, 2000.

[4] P.W. Shor, Fault-tolerant quantum computation. Proc. 37nd Annual Symposium on Foundations of Computer Science, IEEE Computer Society Press, 1996, pp. 56–65.

[5] X. Zhou, D.W. Leung, I.L. Chuang, Methodology for quantum logic gate construction, Phys. Rev. A. 62 (5) (2000), paper 052316.

[6] A.Y. Kitaev, Quantum computations: Algorithms and error correction, Russ. Math. Surv. 52 (6) (1997) 1191–1249.

[7] M.B. Plenio, V. Vedral, P.L. Knight, Conditional generation of error syndromes in fault-tolerant error correction, Phys. Rev. A. 55 (6) (1997) 4593–4596.

[8] I.B. Djordjevic, Cavity quantum electrodynamics based quantum low-density parity-check encoders and decoders. Proc. SPIE Photonics West 2011, Advances in Photonics of Quantum Computing, Memory, and Communication IV, The Moscone Center, San Francisco, CA, 22–27 January 2011, paper no. 7948-38.

[9] D. Gottesman, Theory of fault-tolerant quantum computation, Phys. Rev. A. 57 (1) (1998) 127–137.

[10] D.P. DiVincenzo, P.W. Shor, Fault-tolerant error correction with efficient quantum codes, Phys. Rev. Lett. 77 (15) (1996) 3260–3263.

[11] D. Gottesman, Stabilizer Codes and Quantum Error Correction. Ph.D. dissertation, California Institute of Technology, Pasadena, CA, 1997.

[12] T.C. Ralph, N.K. Langford, T.B. Bell, A.G. White, Linear optical controlled-NOT gate in the coincidence basis, Phys. Rev. A. 65 (2002) 062324-1–062324-5.

[13] E. Knill, R. Laflamme, G.J. Milburn, A scheme for efficient quantum computation with linear optics, Nature 409 (2001) 46–52.

[14] A. Politi, M. Cryan, J. Rarity, S. Yu, J.L. O'Brien, Silica-on-silicon waveguide quantum circuits, Science 320 (5876) (May 2008) 646–649.

[15] C.H. Bennett, G. Brassard, C. Crépeau, R. Jozsa, A. Peres, W.K. Wootters, Teleporting an unknown quantum state via dual classical and Einstein–Podolsky–Rosen channels, Phys. Rev. Lett. 70 (13) (1993) 1895–1899.

[16] A. Barenco, A universal two-bit quantum computation, Proc. R. Soc. Lond. A., Math. Phys. Sci. 449 (June 1995).

[17] D. Deutsch, Quantum computational networks, Proc. R. Soc. Lond. A., Math. Phys. Sci. 425 (1868) (September 1989) 73–90.

[18] E. Knill, R. Laflamme, Concatenated Quantum Codes. Available at: http://arxiv.org/abs/quant-ph/9608012, 1996.

[19] E. Knill, R. Laflamme, W.H. Zurek, Resilient quantum computation, Science 279 (5349) (16 January 1998) 342–345.

[20] C. Zalka, Threshold Estimate for Fault Tolerant Quantum Computation. Available at, http://arxiv.org/abs/quant-ph/9612028, 1996.

[21] M. Grassl, P. Shor, G. Smith, J. Smolin, B. Zeng, Generalized concatenated quantum codes, Phys. Rev. A. 79 (5) (2009), paper 050306(R).

CHAPTER

Quantum Information Theory

12

Shikhar Uttam

University of Pittsburgh (shf28@pitt.edu)

CHAPTER OUTLINE

Quantum information theory, similar to its classical counterpart, studies the meaning and limits of communicating classical and quantum information over quantum channels. This chapter introduces the basic concepts underlying this vast and fascinating area that is currently the subject of intense research.

In the classical context, information theory provides answers to two fundamental questions [1–3]: to what extent can information be compressed, and what is the maximum rate for reliable communication over a noisy channel. The former is the basis for source coding and data compression in communication theory, while the latter forms the basis for channel coding. Information theory also overlaps with many areas in other disciplines. Since its inception – with the 1948 seminal paper by Claude Shannon titled "A mathematical theory of communication" [1] – information theory has established links with many areas including, but not limited to, Kolmogorov complexity in computer science [4], large deviation theory within probability theory [5], hypothesis testing (e.g. Fisher information [6]) in statistics, and Kelly's rule in investment and portfolio theory [7].

In the last two decades classical information theory has provided the foundation for the development of its quantum analog. This is not to say that both are the same.

Quantum Information Processing and Quantum Error Correction. DOI: 10.1016/B978-0-12-385491-9.00012-5

In fact, as we will see in this chapter, quantum information theory has revealed consequences for the quantum world that cannot be predicted by classical information theory. Nevertheless, classical information theory provides the logical structure to coherently introduce the ideas of quantum information theory and this link will be exploited throughout this chapter. It is a challenge to do justice to quantum information theory in a single chapter. It is a comprehensive area with many consequences and whole books can be, and have been, devoted to it. Since space here is limited the focus is on telling a coherent story that, hopefully, will provide a helpful introduction to the subject. The references listed at the end of the chapter provide resources for the reader who decides to further pursue this field.

The chapter is divided into four sections. The first section introduces quantum entropy and discusses many of its properties and their interrelations. The second section expands on the first to introduce the important notion of Holevo information, its relation to accessible information, and the resulting Holevo bound. The third section deals with Schumacher's noiseless quantum channel coding theorem, which provides the condition under which it is possible to compress a quantum message. The fourth and final section of the chapter is devoted to understanding the classical capacity of a quantum channel and introduces the Holevo–Schumacher–Westmoreland (HSW) theorem. This theorem provides the condition under which communication over a noisy quantum channel is possible. For ease of exposition, each section broadly has two logical parts. In the first part classical information theory concepts are introduced that then form the basis of the development of the corresponding quantum theory ideas. Examples are also provided for clarity.

12.1 ENTROPY

The idea of entropy was formulated in statistical mechanics and in the works of Boltzmann and Gibbs in thermodynamics. In the context of information theory entropy quantifies the notion of information content of a source that represents some physical system. Entropy is an important underlying theme in all of information theory, and forms the basis of many of its results. We begin by introducing classical (or Shannon) entropy and then discuss quantum (or von Neumann) entropy.

12.1.1 Shannon Entropy

Consider a discrete memoryless stochastic source X that outputs a number x from a set of numbers $\mathcal{X}$, according to some probability distribution P. Each output is assumed to be independent of all other outputs. The uncertainty associated with the output x is given by the probability $P(x) = p_x$. Therefore, a precise description of uncertainty of the source X is given by listing the probabilities of all the elements of $\mathcal{X}$. This, however, can be an impractical task for many sources. More importantly, it fails to reveal the latent richness of the underlying stochastic process. We, therefore, try to encapsulate the uncertainty of the source, inherent in its probability distribution, with a single number. To define this number we first define a function f: $[0,1] \to \mathbb{R}_+$, such that for every $x_1, x_2 \in \mathcal{X}$, it satisfies

$$f(p_{x_1} p_{x_2}) = f(p_{x_1}) + f(p_{x_2}). \tag{12.1}$$

This condition says that the output of f will be the same whether we consider the joint probability $p_{x_1} p_{x_2}$ of the two independents events (output of numbers $x_1, x_2 \in \mathcal{X}$ by the source), or the marginal probabilities of the two events separately. We can interpret f as a measure of uncertainty about the two independent events. As a result, consistent with expectation, (12.1) indicates that the total amount of uncertainty remains the same whether we consider the two events jointly or separately. We can compute the average of this uncertainty over all outputs of X to get the single number

$$H(X) = \sum_{x \in \mathcal{X}} p_x f(p_x). \tag{12.2}$$

Noting that the function f that satisfies (12.1) is the logarithm function, we define the Shannon or classical entropy as

$$H(X) = -\sum_{x \in \mathcal{X}} P(x) \log P(x), \tag{12.3}$$

where the logarithm is taken with respect to base 2, and $\lim_{a \to 0} a \log a = 0$. The negative sign accounts for the fact that the base-2 logarithm of a probability is a negative number. Note that, in research literature, $H(X)$ is interchangeably denoted as $H(P)$. This is done to highlight the important fact that Shannon entropy is a function of the probability mass function. This can be seen from (12.3), where it is immaterial what the value of x is, as long as its probability is given by the number $P(x)$. We will employ both notations where their use will be clear from the context.

We have mentioned that Shannon entropy is a quantification of the average uncertainty of X with a probability distribution P. This interpretation applies when we have not yet seen what the output of the source is. Once we have the output we can interpret it as the average amount of information gained. These two interpretations – average uncertainty and average information gain – are complementary, and simply express the same idea from two different vantage points: the first before we have seen the source output and the second after we have seen what the source output is.

Example 12.1. Consider a source X that always outputs the number 10. We then have $\mathcal{X} = \{10\}$ with $P(10) = 1$. Using the definition of Shannon entropy in (12.3), we get $H(P) = -1 \log 1 = 0$. We can make two statements: First, the average uncertainty of the source X is zero, and second, the average information gain, once we have seen the output, is zero. Both interpretations are equivalent. Finally, because the base-2 logarithm is used the average information content or average uncertainty is expressed in units of binary digits or bits. Thus for this trivial source the Shannon entropy is 0 bit.

Example 12.2. Consider a source X with alphabet $\mathcal{X} = \{x_1, x_2\}$ and

$$X = \begin{cases} x_1 & \text{with } P(x_1) = p \\ x_2 & \text{with } P(x_2) = 1 - p. \end{cases} \tag{12.4}$$

Again using the Shannon entropy definition, we have for this source

$$H(P) = -p \log p - (1-p) \log(1-p). \tag{12.5}$$

We leave it as an exercise (see Problem 1) to show that $H(P)$ is a concave function of p with a maximum value of 1 bit for $p = \frac{1}{2}$. In other words, the Shannon entropy as a function of p is upper bounded by 1 bit,

$$H(P) \leq 1 = \log 2 = \log|\mathcal{X}|, \tag{12.6}$$

where $|\mathcal{X}|$ denotes the size of $\mathcal{X}$. This is a general result that shows that the entropy of a source with probability P and outputs drawn from $\mathcal{X}$ is upper bounded by $\log|\mathcal{X}|$, with equality if and only if the probability distribution is a uniform distribution. The proof of this result requires us to define relative entropy. Relative entropy, also known as Kullback–Leibler (KL) distance, is defined for two distributions P and Q on $\mathcal{X}$ as

$$D(P\,\|\,Q) = \sum_{x\in\mathcal{X}} P(x) \log \frac{P(x)}{Q(x)}, \tag{12.7}$$

where we define $\log 0/Q(x) = 0$ and $\log P(x)/0 = \infty$. Note that relative entropy is finite only when the support of P is a subset of the support of Q. KL distance is a measure of the increase in uncertainty when we assume Q to be the distribution when x is generated based on the distribution P. A result from classical information theory states that $D(P\,\|\,Q) \geq 0$ with equality if and only if $P = Q$. This result proceeds from a direct application of strict concavity of the logarithm to (12.7). Based on this result and the assumption $Q(x) = 1/|\mathcal{X}|$, we have

$$\begin{aligned} 0 \leq D(P\,\|\,Q) &= \sum_{x\in\mathcal{X}} P(x) \log \frac{P(x)}{\frac{1}{|\mathcal{X}|}} = \sum_{x\in\mathcal{X}} P(x) \log P(x) + \log|\mathcal{X}| \sum_{x\in\mathcal{X}} P(x) \\ &= -H(P) + \log|\mathcal{X}|, \end{aligned} \tag{12.8}$$

and, consequently, the desired result $H(P) \leq \log|\mathcal{X}|$.

Inherent in the discussion of Shannon entropy is the fundamental assumption that we can distinguish the source outputs from each other. This is not much of an assumption in the classical setting, where the source outputs are numbers, letters, strings, etc. In a quantum system, however, this is not necessarily true. For example, two non-orthogonal quantum states cannot be reliably distinguished. This distinction between the classical and quantum worlds requires a quantum definition of entropy.

12.1.2 Von Neumann Entropy

In the previous section we defined X to be a classical source that outputs classical symbols drawn from $\mathcal{X}$ according to some probability distribution P. We now define X to be a quantum source whose output symbol is a quantum state, described by the density operator $\boldsymbol{\rho}_x$. The probability that the output is $\boldsymbol{\rho}_x$ is given by $P(x) = p_x$. Our source is therefore defined by the ensemble $\{\boldsymbol{\rho}_x, p_x\}$. Given this ensemble, we can completely characterize the state of the quantum system Q at the source output – given that no measurement has been made – by the mixed density operator

$$\boldsymbol{\rho} = \sum_{x \in \mathcal{X}} p_x \boldsymbol{\rho}_x. \tag{12.9}$$

For this quantum system Q in state $\boldsymbol{\rho}$, the von Neumann or quantum entropy is defined as

$$S(\boldsymbol{\rho}) = -\mathrm{Trace}(\boldsymbol{\rho} \log \boldsymbol{\rho}). \tag{12.10}$$

If all the quantum states $\boldsymbol{\rho}_x$ in the ensemble are pure and mutually orthogonal, then the mixed density operator $\boldsymbol{\rho}$ is diagonalized by these pure states, and quantum entropy reduces to

$$S(\boldsymbol{\rho}) = -\mathrm{Trace}(\boldsymbol{\rho} \log \boldsymbol{\rho}) = -\sum_{x \in \mathcal{X}} p_x \log p_x = H(P), \tag{12.11}$$

the classical Shannon entropy. This result also agrees with the property that pure orthogonal states can be distinguished from each other, just like classical symbols. We should mention that, although for pure orthogonal states $S(\boldsymbol{\rho}) = H(P)$, they do not have the same meaning. $S(\boldsymbol{\rho})$ is a measure of the quantum uncertainty before any measurement is made, while $H(P)$ is a measure of the classical uncertainty of the outcomes after quantum measurements. To understand this point let us suppose that we measure the observable $\boldsymbol{A}$ denoted by

$$\boldsymbol{A} = \sum_i a_i |a_i\rangle\langle a_i|. \tag{12.12}$$

This measurement for the quantum system Q in a state described by $\boldsymbol{\rho}$ yields a_i with probability given by $p_i = \mathrm{Trace}(|a_i\rangle\langle a_i|\boldsymbol{\rho}) = \langle a_i|\boldsymbol{\rho}|a_i\rangle$. If the observable $\boldsymbol{A}$ commutes with $\boldsymbol{\rho}$, i.e. $[\boldsymbol{A}, \boldsymbol{\rho}] = 0$, then the same basis diagonalizes both $\boldsymbol{A}$ and $\boldsymbol{\rho}$, and the classical and quantum entropies are equal. This is the case for pure orthogonal states. In general, however, the two do not commute and $S(\boldsymbol{\rho}) < H(P_{\boldsymbol{A},\boldsymbol{\rho}})$. Here, $P_{\boldsymbol{A},\boldsymbol{\rho}}$ is the probability of the measurement outcomes when the observable is $\boldsymbol{A}$ and the density operator is $\boldsymbol{\rho}$.

An interesting property of quantum entropy is that it is invariant under a unitary change of basis,

$$S(\boldsymbol{U}\boldsymbol{\rho}\boldsymbol{U}^\dagger) = S(\boldsymbol{\rho}), \tag{12.13}$$

where U is a unitary transformation. An immediate consequence of this property allows us to easily compute the von Neumann entropy: diagonalize $\boldsymbol{\rho}$ using a unitary transform U_1 such that

$$\boldsymbol{\rho} = U_1 \boldsymbol{\rho}_{\mathrm{D}} U_1^{\dagger}, \tag{12.14}$$

where $\boldsymbol{\rho}_{\mathrm{D}}$ is a diagonal matrix whose diagonal entries are the eigenvalues of $\boldsymbol{\rho}$. Denoting the ith eigenvalue by λ_i, we can compute the quantum entropy as

$$S(\boldsymbol{\rho}) = -\sum_i \lambda_i \log \lambda_i. \tag{12.15}$$

Furthermore, since any density matrix is non-negative Hermetian with trace 1, we have

$$\sum_i \lambda_i = 1, \tag{12.16}$$

with $0 \leq \lambda_i \leq 1$. As a consequence, if our ensemble consists of a single pure state, then $S(\boldsymbol{\rho}) = 1 \log 1 = 0$. Another consequence is that $S(\boldsymbol{\rho}) \geq 0$, with the equality being satisfied if and only if the quantum system Q is in a pure state $\boldsymbol{\rho} = |\psi\rangle\langle\psi|$.

Having seen that quantum entropy is bounded by zero from below, can we also bound it from above? The answer to this question is yes, and the argument is similar to the way we upper bounded Shannon entropy. Let us start by defining the quantum relative entropy. Given two quantum systems in states specified by $\boldsymbol{\rho}$ and $\boldsymbol{\sigma}$, the relative entropy of $\boldsymbol{\rho}$ to $\boldsymbol{\sigma}$ is given by

$$S(\boldsymbol{\rho} \,\|\, \boldsymbol{\sigma}) = \mathrm{Trace}(\boldsymbol{\rho} \log \boldsymbol{\rho}) - \mathrm{Trace}(\boldsymbol{\rho} \log \boldsymbol{\sigma}). \tag{12.17}$$

This definition is analogous to the definition of classical relative entropy between two probability distributions P and Q we discussed earlier. We noted that its information theoretic interpretation said that wrongly assuming the distribution to be Q when it is actually P increases the uncertainty by $D(P \,\|\, Q)$. From the definition of the classical relative entropy it is clear that in certain cases it can be infinite — for example, when the support of P exceeds that of Q. A similar scenario also exists for quantum relative entropy. Define the kernel of any density matrix to be the eigenspace spanned by the eigenvectors of the density matrix corresponding to the eigenvalue 0. Additionally, define the support of any density matrix to be the eigenspace spanned by the eigenvectors of the density matrix corresponding to nonzero eigenvalues. Then, for cases when the nontrivial intersection between the support of $\boldsymbol{\rho}$ and the kernel of $\boldsymbol{\sigma}$ is empty, and the support of $\boldsymbol{\rho}$ is a subspace of the support of $\boldsymbol{\sigma}$, $S(\boldsymbol{\rho} \,\|\, \boldsymbol{\sigma})$ is finite. Otherwise it is infinite. An important property of $S(\boldsymbol{\rho} \,\|\, \boldsymbol{\sigma})$, known as the Klein inequality, states that

$$S(\boldsymbol{\rho} \,\|\, \boldsymbol{\sigma}) \geq 0, \tag{12.18}$$

with equality if and only if $\boldsymbol{\rho} = \boldsymbol{\sigma}$. Defining $\boldsymbol{\sigma} = \boldsymbol{I}/d$, for a quantum system in a d-dimensional Hilbert space, and using the Klein inequality and the definition of relative entropy, we have

$$S(\boldsymbol{\rho}) \leq \log d. \tag{12.19}$$

The equality is satisfied if and only if state $\boldsymbol{\rho}$ is chosen randomly, or in other words the quantum entropy is maximum when the quantum system is in a completely mixed state given by $\boldsymbol{I}/d$.

We turn back for a moment to see how the Klein inequality arises. Let us suppose that the eigen decompositions of $\boldsymbol{\rho}$ and $\boldsymbol{\sigma}$ are denoted by

$$\boldsymbol{\rho} = \sum_i \rho_i |\rho_i\rangle\langle\rho_i| \quad \text{and} \tag{12.20}$$

$$\boldsymbol{\sigma} = \sum_j \sigma_j |\sigma_j\rangle\langle\sigma_j|, \tag{12.21}$$

where $0 \leq \rho_i \leq 1$ and $0 \leq \sigma_i \leq 1$. We note that the decomposition can be any orthonormal decomposition because any unitary transform can be applied to the eigen decomposition without changing the basic proof. Substituting the decomposition of $\boldsymbol{\rho}$ in the relative entropy definition results in

$$S(\boldsymbol{\rho} \,\|\, \boldsymbol{\sigma}) = \text{Trace}(\boldsymbol{\rho} \log \boldsymbol{\rho}\,) - \text{Trace}(\boldsymbol{\rho} \log \boldsymbol{\sigma}), \tag{12.22}$$

$$= \sum_i \rho_i \log \rho_i - \text{Trace}\left(\sum_i \rho_i |\rho_i\rangle\langle\rho_i| \log \boldsymbol{\sigma} \right). \tag{12.23}$$

Rearranging (12.23) results in

$$S(\boldsymbol{\rho} \,\|\, \boldsymbol{\sigma}) = \sum_i \rho_i \log \rho_i - \sum_i \rho_i \langle\rho_i| \log \boldsymbol{\sigma} |\rho_i\rangle. \tag{12.24}$$

Further substituting (12.21) in (12.24) we get

$$S(\boldsymbol{\rho} \,\|\, \boldsymbol{\sigma}) = \sum_i \rho_i \log \rho_i - \sum_i \rho_i \langle\rho_i| \left(\sum_j |\sigma_j\rangle \log \sigma_j \langle\sigma_j| \right) |\rho_i\rangle \tag{12.25}$$

$$S(\boldsymbol{\rho} \,\|\, \boldsymbol{\sigma}) = \sum_i \rho_i \log \rho_i - \sum_i \sum_j \rho_i (\log \sigma_j) \langle\rho_i|\sigma_j\rangle\langle\sigma_j|\rho_i\rangle \tag{12.26}$$

$$S(\boldsymbol{\rho} \,\|\, \boldsymbol{\sigma}) = \sum_i \rho_i \left(\log \rho_i - \sum_j (\log \sigma_j) \langle\rho_i|\sigma_j\rangle\langle\sigma_j|\rho_i\rangle \right). \tag{12.27}$$

Note that since $\{|\rho_i\rangle\}$ and $\{|\sigma_j\rangle\}$ are two orthonormal bases, we have $\sum_i \langle\rho_i|\sigma_j\rangle \langle\sigma_j|\rho_i\rangle = 1$ and $\sum_j \langle\rho_i|\sigma_j\rangle\langle\sigma_j|\rho_i\rangle = 1$. Also, due to the nature of the inner product, we have $\langle\rho_i|\sigma_j\rangle\langle\sigma_j|\rho_i\rangle \geq 0$. Therefore, by exploiting the definition of strict concavity applied to the logarithm function, we easily see that

$$\sum_j (\log\sigma_j)\langle\rho_i|\sigma_j\rangle\langle\sigma_j|\rho_i\rangle \leq \log\left(\sum_j \sigma_j\langle\rho_i|\sigma_j\rangle\langle\sigma_j|\rho_i\rangle\right). \tag{12.28}$$

Substituting (12.28) in (12.27) we get

$$S(\boldsymbol{\rho}\,\|\,\boldsymbol{\sigma}) \geq \sum_i \rho_i(\log\rho_i - \log\gamma_i) \tag{12.29}$$

$$= \sum_i \rho_i \log\frac{\rho_i}{\gamma_i}, \tag{12.30}$$

where, $\gamma_i = \sum_j \sigma_j\langle\rho_i|\sigma_j\rangle\langle\sigma_j|\rho_i\rangle$. The right-hand side of (12.30) is identical to the definition of classical relative entropy for two probability distributions $P(i) = \rho_i$ and $Q(i) = \gamma_i$. We noted earlier that the classical relative entropy is bounded from below by zero. Using that result, we get

$$S(\boldsymbol{\rho}\,\|\,\boldsymbol{\sigma}) \geq 0, \tag{12.31}$$

with equality if and only if $\rho_i = \gamma_i$ for all i.

12.1.2.1 Composite Systems

We now turn to look at quantum systems with two or more subsystems and interrelations between the quantum entropies of the subsystems and the overall composite system. Let us begin by considering a bipartite system Q composed of uncorrelated subsystems A and B. This pure joint system is described by the product state density operator $\boldsymbol{\rho}_{AB} = \boldsymbol{\rho}_A \otimes \boldsymbol{\rho}_B$ and has the interesting property that uncorrelated subsystems have the same quantum entropy

$$S(\boldsymbol{\rho}_A) = S(\boldsymbol{\rho}_B). \tag{12.32}$$

To see this, we employ the Schmidt decomposition. A pure state $|\psi\rangle$ of a bipartite system AB can be represented as

$$|\psi\rangle = \sum_i c_i|i_A\rangle \otimes |i_B\rangle, \tag{12.33}$$

where $|i_A\rangle$ and $|i_B\rangle$ are the orthonormal bases of the subsystems A and B respectively, and $c_i \in \mathbb{R}_+$ are Schmidt coefficients that satisfy $\sum_i c_i^2 = 1$. We can use the Schmidt decomposition to express the density operator $\boldsymbol{\rho}_{AB}$ as

$$\boldsymbol{\rho}_{AB} = |\psi\rangle\langle\psi| = \sum_i\sum_j c_i\, c_j|i_A\rangle\langle j_A| \otimes |i_B\rangle\langle j_B|. \tag{12.34}$$

The partial trace of $\boldsymbol{\rho}_{AB}$ over subsystem A results in

$$\text{Trace}_A(\boldsymbol{\rho}_{AB}) = \sum_i \sum_j c_i\, c_j \langle i_A | j_A \rangle | i_B \rangle \langle j_B |. \tag{12.35}$$

Since $\langle i_A|$ and $\langle j_A|$ are orthonormal, we have

$$\langle i_A | j_A \rangle = \delta_{ij}, \tag{12.36}$$

whose substitution in (12.35) gives

$$\text{Trace}_A(\boldsymbol{\rho}_{AB}) = \sum_i c_i^2 | i_B \rangle \langle i_B | = \boldsymbol{\rho}_B. \tag{12.37}$$

Following the same procedure, and tracing over subsystem B, we get

$$\text{Trace}_A(\boldsymbol{\rho}_{AB}) = \sum_i c_i^2 | i_A \rangle \langle i_A | = \boldsymbol{\rho}_A. \tag{12.38}$$

From (12.37) and (12.38) we get the surprising result that for a pure joint state given by $\boldsymbol{\rho}_{AB}$, we have $\boldsymbol{\rho}_A = \boldsymbol{\rho}_B$, and consequently the desired result $S(\boldsymbol{\rho}_A) = S(\boldsymbol{\rho}_B)$.

For a general bipartite system AB the relation between the quantum entropies of the subsystem and composite system is given by the *subadditivity inequality*,

$$S(\boldsymbol{\rho}_{AB}) \leq S(\boldsymbol{\rho}_A) + S(\boldsymbol{\rho}_B), \tag{12.39}$$

with the equality satisfied when $\boldsymbol{\rho}_{AB} = \boldsymbol{\rho}_A \otimes \boldsymbol{\rho}_B$. Thus, the quantum entropy of the composite system is bounded from above by the sum of the entropies of the constituent subsystems, except when the subsystems are uncorrelated, in which case it is purely additive. The subadditivity inequality can be thought of as the quantum analog of the Shannon inequality for classical sources X and Y,

$$H(X,Y) \leq H(X) + H(Y). \tag{12.40}$$

This inequality states that the joint entropy of XY with a joint distribution $p(x,y)$ is less than the sum of the entropy of sources X and Y with marginal distributions $p(x)$ and $p(y)$ respectively, except for the case when the outputs of the two sources are independent of each other. In this case equality prevails. This result, however, should not make the reader draw a false equivalence between the classical and quantum worlds. In fact, the difference between them is starkly demonstrated by considering the lower bound on $S(\boldsymbol{\rho}_{AB})$ given by the Araki–Lieb triangle inequality

$$S(\boldsymbol{\rho}_{AB}) \geq |S(\boldsymbol{\rho}_A) - S(\boldsymbol{\rho}_B)|, \tag{12.41}$$

with the equality satisfied when $\boldsymbol{\rho}_{AB}$ is a pure state. Let us look at a consequence of this inequality through an example.

Example 12.3. We consider here subsystems of a bipartite system that are entangled. Entanglement implies that the subsystems have nonlocal quantum correlations. Let us consider one such system whose state is specified by the EPR pair

$$|\psi\rangle = \frac{1}{\sqrt{2}}(|0_A\rangle \otimes |0_B\rangle + |1_A\rangle \otimes |1_B\rangle). \tag{12.42}$$

The corresponding density operator is given by

$$\begin{aligned}\rho_{AB} = |\psi\rangle\langle\psi| = \frac{1}{2}((|0_A\rangle \otimes |0_B\rangle)(\langle 0_A| \otimes \langle 0_B| + (|0_A\rangle \otimes |0_B\rangle)(\langle 1_A| \otimes \langle 1_B| \\ + (|1_A\rangle \otimes |1_B\rangle)(\langle 0_A| \otimes \langle 0_B| + (|1_A\rangle \otimes |1_B\rangle)(\langle 1_A| \otimes \langle 1_B|)\end{aligned} \tag{12.43}$$

since the bipartite system is in a pure state $S(\rho_{AB}) = 0$. However, tracing over the subsystem B we get

$$\rho_A = \frac{1}{2}(|0_A\rangle\langle 0_A| + |1_A\rangle\langle 1_A|), \tag{12.44}$$

whose quantum entropy is given by $S(\rho_A) = -\left(\frac{1}{2}\log\frac{1}{2} + \frac{1}{2}\log\frac{1}{2}\right) = 1$. Similarly tracing over the subsystem A we get

$$\rho_B = \frac{1}{2}(|0_B\rangle\langle 0_B| + |1_B\rangle\langle 1_B|), \tag{12.45}$$

whose quantum entropy is given by $S(\rho_B) = -\left(\frac{1}{2}\log\frac{1}{2} + \frac{1}{2}\log\frac{1}{2}\right) = 1$. Using the Araki–Lieb triangle inequality we get $|S(\rho_A) - S(\rho_B)| = S(\rho_{AB}) = 0$ as expected. If we look closely something entirely nonclassical has happened here. The result indicates that the quantum entropy of a composite system with entangled subsystems is *less* than the quantum entropy of either subsystem. Thus, even though the composite system is in a pure state, measurements of the observables of the subsystems result in outcomes that have uncertainty associated with them. This result has considerable application in quantum error correction codes. It is important to note that there is no classical analog for this observation. In fact, classical considerations provide exactly the opposite result,

$$H(X,Y) \geq H(X), \text{ and } H(X,Y) \geq H(Y), \tag{12.46}$$

implying that joint Shannon entropy of the classical composite system is lower bounded by the Shannon entropy of either subsystem!

To prove the Araki–Lieb triangle inequality, we consider a purifying quantum system C, such that the resulting tripartite sytem ABC is in a pure state. Employing the subadditivity inequality for a composite system we have

$$S(\rho_{AC}) \leq S(\rho_A) + S(\rho_C) \tag{12.47}$$

$$S(\rho_{BC}) \leq S(\rho_B) + S(\rho_C). \tag{12.48}$$

Furthermore, because ABC is in a pure state, using (12.32) we note that

$$S(\boldsymbol{\rho}_{AB}) = S(\boldsymbol{\rho}_C) \tag{12.49}$$

$$S(\boldsymbol{\rho}_{AC}) = S(\boldsymbol{\rho}_B) \tag{12.50}$$

$$S(\boldsymbol{\rho}_{BC}) = S(\boldsymbol{\rho}_A). \tag{12.51}$$

Substituting (12.49) and (12.50) into (12.47) we get

$$S(\boldsymbol{\rho}_B) \leq S(\boldsymbol{\rho}_A) + S(\boldsymbol{\rho}_{AB}) \tag{12.52}$$

$$\Rightarrow S(\boldsymbol{\rho}_B) - S(\boldsymbol{\rho}_A) \leq S(\boldsymbol{\rho}_{AB}), \tag{12.53}$$

and substituting (12.49) and (12.51) into (12.48) we get

$$S(\boldsymbol{\rho}_A) - S(\boldsymbol{\rho}_B) \leq S(\boldsymbol{\rho}_{AB}). \tag{12.54}$$

Combining (12.53) and (12.54) gives us the desired triangle inequality. The equality is satisfied only when the bipartite system AB is in a pure state.

Before ending this subsection, we state without proof a very important inequality in quantum information theory known as strong subadditivity. For a tripartite system ABC in a state specified by $\boldsymbol{\rho}_{ABC}$ the following inequalities hold:

$$S(\boldsymbol{\rho}_A) + S(\boldsymbol{\rho}_B) \leq S(\boldsymbol{\rho}_{AC}) + S(\boldsymbol{\rho}_{BC}) \tag{12.55}$$

$$S(\boldsymbol{\rho}_{ABC}) + S(\boldsymbol{\rho}_B) \leq S(\boldsymbol{\rho}_{AB}) + S(\boldsymbol{\rho}_{BC}). \tag{12.56}$$

Strong subadditivity has important consequences in quantum information theory related to conditioning of quantum entropy, bounds on quantum mutual information, and the effect of quantum operations on quantum mutual information.

12.2 HOLEVO INFORMATION, ACCESSIBLE INFORMATION, AND HOLEVO BOUND

Let $\boldsymbol{\rho}_x$ be a set of mutually orthogonal mixed states, and let Alice prepare a quantum state given by the density operator

$$\boldsymbol{\rho} = \sum_x p_x \boldsymbol{\rho}_x. \tag{12.57}$$

If all that is available to Bob is $\boldsymbol{\rho}$, then the associated uncertainty is given by $S(\boldsymbol{\rho})$. If, however, Bob also knows how Alice prepared the state, i.e. he knows the probability p_x associated with each mixed state $\boldsymbol{\rho}_x$, then the uncertainty is given by $\sum_x p_x S(\boldsymbol{\rho}_x)$. We would intuitively expect that

$$S\left(\sum_x p_x \boldsymbol{\rho}_x\right) \geq \sum_x p_x S(\boldsymbol{\rho}_x), \tag{12.58}$$

because ignorance about how $\boldsymbol{\rho}$ was prepared is reduced in the latter case. Our intuition, in fact, is correct. To see this, let us choose a basis in which $\boldsymbol{\rho}$ is block diagonalized. This is easy to do because due to the mutual orthogonality of the mixed states, for $x_1 \neq x_2$ we have $\text{Trace}(\boldsymbol{\rho}_{x_1}\boldsymbol{\rho}_{x_2}) = 0$. We therefore have

$$S(\boldsymbol{\rho}) = -\sum_x \text{Trace}(p_x\boldsymbol{\rho}_x)\log(p_x\boldsymbol{\rho}_x) \tag{12.59}$$

$$= -\sum_x p_x \text{Trace}(\boldsymbol{\rho}_x)(\log p_x + \log \boldsymbol{\rho}_x) \tag{12.60}$$

$$= -\sum_x p_x \text{Trace}(\boldsymbol{\rho}_x)(\log p_x) - \sum_x p_x \text{Trace}(\boldsymbol{\rho}_x)(\log \boldsymbol{\rho}_x). \tag{12.61}$$

Invoking the basic property of a density operator, $\text{Trace}(\boldsymbol{\rho}_x) = 1$, (12.61) reduces to

$$S(\boldsymbol{\rho}) = -\sum_x p_x \log p_x - \sum_x p_x \log \boldsymbol{\rho}_x. \tag{12.62}$$

The first term on the right-hand side is simply the classical Shannon entropy, while the second term is $\sum_x p_x S(\boldsymbol{\rho}_x)$. We therefore have

$$S(\boldsymbol{\rho}) = H(X) + \sum_x p_x S(\boldsymbol{\rho}_x). \tag{12.63}$$

Further noting that Shannon entropy is bounded from below by zero, we get the desired result $S(\boldsymbol{\rho}) \geq \sum_x p_x S(\boldsymbol{\rho}_x)$. This result illustrates the concavity of quantum entropy.

On rearranging (12.63) we get

$$H(X) = S(\boldsymbol{\rho}) - \sum_x p_x S(\boldsymbol{\rho}_x) = \chi. \tag{12.64}$$

χ is known as the Holevo information for the ensemble $\{\boldsymbol{\rho}_x, p_x\}$. Equation (12.64) bears a close resemblance to the definition of mutual information in classical information theory. Given two random variables X and Y, the mutual information

$$I(X;Y) = H(X) - H(X|Y) \tag{12.65}$$

is the amount of information Y has about X on average. Stated differently, it represents the amount by which the average uncertainty about X, in terms of its Shannon entropy, is reduced given that we know Y. A similar interpretation can be given to the Holevo information

$$\chi = S(\boldsymbol{\rho}) - \sum_x p_x S(\boldsymbol{\rho}_x). \quad (12.66)$$

This tells us the average reduction in quantum entropy given we know how $\boldsymbol{\rho}$ was prepared. As shown above, as long as $\boldsymbol{\rho}_x$ have mutually orthogonal support, the Holevo information is equal to the Shannon entropy of a classical source X with probability distribution given by $P(x) = p_x$. In general, however, when the ensemble of mixed states does not satisfy the mutually orthogonal condition, Shannon entropy upper bounds the Holevo information

$$\chi \leq H(X). \quad (12.67)$$

Again, equality is for the above-stated case.

Suppose now that Alice wants to send classical information to Bob over a quantum channel. Alice has a classical source X that outputs the letter x drawn from an alphabet $\mathcal{X}$ with probability p_x. For each x Alice prepares the state $\boldsymbol{\rho}_x$ and sends it over the channel. For the sake of simplicity we assume that the quantum operation of the channel is characterized by an identity operator. When Bob receives $\boldsymbol{\rho}_x$, he performs generalized positive operator valued measure (POVM) $\boldsymbol{M}_y$, resulting in outcome y with probability

$$p(y|x) = \text{Trace}(\boldsymbol{M}_y \boldsymbol{\rho}_x). \quad (12.68)$$

Bob's goal in performing this measurement is to maximize the mutual information $I(X;Y)$. This maximized mutual information is the best that Bob can hope to do, and is defined as the information accessible to Bob. Since Bob can perform all the necessary measurements to maximize the mutual information, accessible information is defined as

$$H(X{:}\ Y) = \max_{\boldsymbol{M}_y} I(X;Y). \quad (12.69)$$

If the quantum states $\{\boldsymbol{\rho}_x\}$ are pure and mutually orthogonal, then instead of POVMs we can consider projective measurements $\boldsymbol{P}_y$ such that $p(y|x) = \text{Trace}(\boldsymbol{P}_y \boldsymbol{\rho}_x) = 1$, if and only if $x = y$, and zero otherwise. In this case $H(X : Y) = H(X)$, and Bob is accurately able to reconstruct the information Alice sent him. However, when the states are pure but non-orthogonal, the accessible information is bounded as

$$H(X{:}\ Y) \leq S(\boldsymbol{\rho}) \leq H(X), \quad (12.70)$$

where $\boldsymbol{\rho} = \sum_x p_x \boldsymbol{\rho}_x$. This result is a consequence of the fact that non-orthogonal states are not completely distinguishable.

An important interpretation of (12.70) is that it is a consequence of the no-cloning theorem. To see this, suppose two non-othogonal states $|\psi_1\rangle$ and $|\psi_2\rangle$ can be cloned. We can then perform repeated cloning of the states to obtain the states

$|\psi_1\rangle \otimes |\psi_1\rangle \otimes \ldots \otimes |\psi_1\rangle$ and $|\psi_2\rangle \otimes |\psi_2\rangle \otimes \ldots \otimes |\psi_2\rangle$. As the number of clones tends to infinity these new states start becoming orthogonal, so that Bob can unambiguously determine them, resulting in $H(X\colon Y) = H(X)$. Since this is not possible we have the desired result (12.70).

Finally, when $\boldsymbol{\rho}_x$ are not pure but mixed states the accessible information is bounded by the Holevo information

$$H(X\colon Y) \leq \chi. \tag{12.71}$$

This generalized result is known as the Holevo bound. The equality holds when the mixed states have mutually orthogonal support. Combining (12.67) and (12.71) we have

$$H(X\colon Y) \leq \chi \leq H(X), \tag{12.72}$$

implying that apart from the case when $\boldsymbol{\rho}_x$ have mutually orthogonal support there is no possibility for Bob to completely recover the classical information, characterized by $H(X)$, that Alice sent him over the quantum channel!

Example 12.4. Let us consider two orthogonal pure states $|0\rangle$ and $|1\rangle$ with the corresponding density operators

$$\boldsymbol{\rho}_0 = \begin{pmatrix} 1 & 0 \\ 0 & 0 \end{pmatrix} \tag{12.73}$$

$$\boldsymbol{\rho}_1 = \begin{pmatrix} 0 & 0 \\ 0 & 1 \end{pmatrix}. \tag{12.74}$$

Furthermore, let us assume that Alice selects each state with probability 1/2, resulting in

$$\boldsymbol{\rho} = \frac{1}{2}\begin{pmatrix} 1 & 0 \\ 0 & 1 \end{pmatrix}. \tag{12.75}$$

As a result $S(\boldsymbol{\rho}) = -\left(\frac{1}{2}\log\frac{1}{2} + \frac{1}{2}\log\frac{1}{2}\right) = 1$, while $\sum_x p_x S(\boldsymbol{\rho}_x) = \frac{1}{2}S(\boldsymbol{\rho}_0) + \frac{1}{2}S(\boldsymbol{\rho}_1) = 0$ because $S(\boldsymbol{\rho}_0) = S(\boldsymbol{\rho}_1) = 0$. Consequently, $\chi = 1 - 0 = 1$, and as expected is equal to the Shannon entropy $H(X) = -\left(\frac{1}{2}\log\frac{1}{2} + \frac{1}{2}\log\frac{1}{2}\right) = 1$. Now let us suppose that we still have the same pure orthogonal states, but Alice is uncertain whether a given preparation is $|0\rangle$ or $|1\rangle$, resulting in mixed density operators

$$\boldsymbol{\rho}_0 = \frac{1}{2}\begin{pmatrix} 1 & 0 \\ 0 & 0 \end{pmatrix} + \frac{1}{2}\begin{pmatrix} 0 & 0 \\ 0 & 1 \end{pmatrix} = \frac{1}{2}\begin{pmatrix} 1 & 0 \\ 0 & 1 \end{pmatrix} \tag{12.76}$$

$$\boldsymbol{\rho}_1 = \frac{1}{2}\begin{pmatrix} 1 & 0 \\ 0 & 0 \end{pmatrix} + \frac{1}{2}\begin{pmatrix} 0 & 0 \\ 0 & 1 \end{pmatrix} = \frac{1}{2}\begin{pmatrix} 1 & 0 \\ 0 & 1 \end{pmatrix}. \tag{12.77}$$

As in the previous case Alice selects each with probability 1/2, resulting in

$$\rho = \frac{1}{2}\rho_0 + \frac{1}{2}\rho_1 = \frac{1}{2}\begin{pmatrix} 1 & 0 \\ 0 & 1 \end{pmatrix} \tag{12.78}$$

and

$$\chi = S(\rho) - \sum_x p_x S(\rho_x) = 1 - 1 = 0. \tag{12.79}$$

Thus, the Holevo information is zero and there is no hope of Bob recovering any information Alice sent him. Note that although ρ_0 and ρ_1 are mixed states they are not mutually orthogonal. In fact, they are identical and therefore Bob's measurement outcomes are completely uncertain.

12.3 DATA COMPRESSION

Having studied the basic building blocks of quantum information theory we now turn to the quantum version of the first central question in information theory: to what extent can a quantum message be compressed without significantly affecting the quantum message fidelity. The answer to this question, given by Schumacher's quantum noiseless channel coding theorem, is the topic of this section. Schumacher's theorem is the quantum analog of Shannon's noiseless channel coding theorem that establishes the condition under which data compression of classical information can be done reliably. We, therefore, begin by discussing it and outlining its main ideas. This discussion will form the basis of our development of Schumacher's theorem and quantum data compression.

12.3.1 Shannon's Noiseless Channel Coding Theorem

Let us consider a classical source X that outputs a 1 or a 0 with probability p and $q = 1 - p$ respectively. Let us also assume that the output is independent of any other output. Then the probability of the output sequence $x_1, x_2, \ldots, x_n$ is given by

$$P(x_1, x_2, \ldots, x_n) = p^{\sum_{x_i=1}} q^{\sum_{x_i=0}(1-x_i)}. \tag{12.80}$$

Furthermore, as $n \to \infty$, we can use the frequentist approach to state that the probability of the output sequence is approximately given by

$$P(x_1, x_2, \ldots, x_n) \approx p^{np}(1-p)^{n-pn}. \tag{12.81}$$

Thus, as n increases, with high probability the output sequence will have np ones and $n(1-p)$ zeros. Note that this will not be true for *all* output sequences, but we can say with high probability it will be true for most sequences. These sequences are referred to as typical sequences, while the sequences for which the statement is not true and which rarely occur are called atypical sequences. Taking the base-2 logarithm of (12.81) we get

$$\log P(x_1, x_2, \ldots, x_n) \approx np \log p + n(1-p)\log(1-p) = -nH(X) \qquad (12.82)$$

or

$$P(x_1, x_2, \ldots, x_n) \approx 2^{-nH(X)}. \qquad (12.83)$$

$H(X)$ here denotes the average entropy per symbol, or the entropy rate. Equation (12.83) indicates that the probability of occurrence of the typical output sequence is approximately $2^{-nH(X)}$ with high probability. Furthermore, it also states that such typical sequences are at most $nH(X)$ in number. This tells us that for a binary source, with high probability, not much information will be lost if we only consider $2^{nH(X)}$ typical sequences instead of all the possible 2^n sequences. We can encode the typical sequences by first ordering them and assigning them an index value. Whenever we get a typical sequence as the source output we store it as an index from our listing. The index values can be represented by $nH(X)$ bits, thereby leading to data compression. If, however, the source output is an atypical sequence, we store it as some fixed index. Thus, for atypical sequences our scheme results in loss of information. But for increasing n the probability that a given sequence will be an atypical sequence approaches 0. As a result the loss of information can be made arbitrarily small.

Let us extend this idea to a nonbinary source whose outputs are independently and identically distributed (i.i.d.) random variables $X_1, X_2, \ldots, X_n$. For such random variables the law of large numbers states that as n $\to \infty$, the estimate of the expected value of the source X approaches the true expected value $E(X)$ in probability. More precisely, for sufficiently large n, and for $\epsilon, \delta > 0$:

$$P\left(\left|\frac{1}{n}\sum_i X_i - E(X)\right| \leq \epsilon\right) > 1 - \delta. \qquad (12.84)$$

If we replace the random variable X_i with the function of the random variable, $(-\log P(X_i))$, we get

$$2^{-n(H(X)+\epsilon)} \leq \log P(x_1, x_2, \ldots, x_n) \leq 2^{-n(H(X)-\epsilon)}. \qquad (12.85)$$

The sequences x_1, x_2, ..., x_n that satisfy (12.85) are typical sequences. The set of these typical sequences is the typical set $T(n, \epsilon)$. Thus, as a consequence of (12.84), for sufficiently large n, we have

$$P(T(n, \epsilon)) > 1 - \delta. \qquad (12.86)$$

Furthermore, based on (12.85) and (12.86) we can bound the size of the typical set to

$$(1-\epsilon)2^{n(H(X)-\epsilon)} \leq |T(n, \epsilon)| \leq 2^{n(H(X)+\epsilon)}. \qquad (12.87)$$

To see the right-hand side inequality, we note that the sum of probabilities of all typical sequences cannot be more than 1. Therefore, we have

$$1 \geq \sum_{(x_1,x_2,\ldots,x_n)\epsilon T(n,\epsilon)} P(x_1, x_2, \ldots, x_n) \geq \sum_{(x_1,x_2,\ldots,x_n)\epsilon T(n,\epsilon)} 2^{-n(H(X)+\epsilon)}$$

$$= |T(n,\epsilon)|2^{-n(H(X)+\epsilon)}, \tag{12.88}$$

resulting in

$$|T(n,\epsilon)| \leq 2^{n(H(X)+\epsilon)}. \tag{12.89}$$

To prove the left inequality, we invoke (12.86) to get

$$1 - \delta < P(T(n,\epsilon)) \leq \sum_{(x_1,x_2,\ldots x_n) \in T(n,\epsilon)} 2^{-n(H(X)-\epsilon)} = |T(n,\epsilon)|2^{-n(H(X)-\epsilon)}, \tag{12.90}$$

resulting in

$$|\mathrm{T}(n,\epsilon)| > (1-\delta)2^{n(H(X)-\epsilon)}. \tag{12.91}$$

We, therefore, have a couple of powerful results: With $n \to \infty$, the probability that an output sequence will be a typical sequence approaches 1 and the probability of occurrence of any one particular typical sequence is approximately given by $2^{-nH(X)}$. Furthermore, the total number of such typical sequences, i.e. the size of the typical set, is approximately $2^{nH(X)}$.

Suppose now we have a compression rate $R > H(X)$. We can choose $\epsilon > 0$ that satisfies $R > H(X) + \epsilon$. Then on the basis of the above results for sufficiently large n the total number of typical sequences is bounded from above as

$$|T(n,\epsilon)| \leq 2^{n(H(X)+\epsilon)} \leq 2^{nR}, \tag{12.92}$$

and the probability of occurrence of these typical sequences is bounded from below by $1-\delta$, where δ can be chosen to be arbitrarily small. Thus, the coding strategy is straightforward: divide the set of all possible sequences into sets of typical and atypical sequences. From (12.92), the sequences in the typical set can be represented by at most nR bits. If an atypical sequence occurs on rare occasions we assign it a fixed index. Note that with increasing n the probability of such atypical sequence occurrences becomes arbitrarily small, resulting in virtually no loss of information. We can, therefore, code and decode typical sequences with high reliability. Reliability implies that the probability of decoding error is very small.

Now let us consider what happens if $R < H(X)$. In such a case we can encode at most 2^{nR} typical sequences. Let us label this set as $T_R(n,\epsilon)$. Since a typical sequence can occur with a probability upper bounded by $2^{-n(H(X)-\epsilon)}$, the typical sequences in $T_R(n,\epsilon)$ will occur with probability $2^{n(R-H(X)+\epsilon)}$. Therefore, if $R < H(X)$, this probability of occurrence of a sequence in $T_R(n,\epsilon)$ will tend to 0 for $n \to \infty$, and as

a result we are unable to design a reliable compression scheme. We, therefore, have Shannon's noiseless coding theorem:
For an i.i.d. source X with entropy rate $H(X)$, a reliable compression method exists if $R > H(X)$. If, however, $R < H(X)$, then no reliable compression method exists.

12.3.2 Schumacher's Noiseless Quantum Channel Coding Theorem

Having developed the classical information theoretic idea of data compression, we now ask the question whether the concept of compression can be extended to the quantum world. As we will see in this section the answer is a surprising yes. We will first discuss it generally, and then to make the idea concrete will give an elaborate example.

Let us begin by considering a quantum source that outputs a pure state $|\psi_x\rangle$ with probability p_x. The output state of this quantum source is therefore given by the mixed density operator

$$\boldsymbol{\rho} = \sum_x p_x |\psi_x\rangle\langle\psi_x|. \tag{12.93}$$

The quantum message comprises n quantum source outputs that are independent of each other. We can therefore write the quantum message as the tensor product

$$\boldsymbol{\rho}_{\otimes n} = \boldsymbol{\rho} \otimes \boldsymbol{\rho} \otimes \ldots \otimes \boldsymbol{\rho}. \tag{12.94}$$

Schumacher's theorem gives us the condition under which there exists a reliable compression scheme that can compress and decompress the quantum message with high *fidelity*. We explain fidelity later in this section.

The development of Schumacher's noiseless quantum channel coding theorem closely parallels Shannon's classical version. If we express $\boldsymbol{\rho}$ in its orthonormal eigenbasis, we have

$$\boldsymbol{\rho} = \sum_\alpha \lambda_\alpha |\lambda_\alpha\rangle\langle\lambda_\alpha|. \tag{12.95}$$

The quantum source can then effectively be thought of as a classical source, and we say $\boldsymbol{\lambda} = \{\lambda_{\alpha_1}, \lambda_{\alpha_2}, \ldots, \lambda_{\alpha_n}\}$ is a typical sequence if, for sufficiently large n, it satisfies

$$P\left(\left|\frac{1}{n}\log\left(\frac{1}{P(\lambda_{\alpha_1})P(\lambda_{\alpha_2})\ldots P(\lambda_{\alpha_n})}\right) - S(\boldsymbol{\rho})\right| \le \epsilon\right) > 1-\delta. \tag{12.96}$$

Note that the above equation is a direct consequence of the law of large numbers in a manner identical to (12.84) and (12.85). We can now define $|\lambda_{\alpha_1}\rangle \otimes |\lambda_{\alpha_2}\rangle \otimes \ldots \otimes |\lambda_{\alpha_n}\rangle$ as the typical state corresponding to the above typical sequence. Thus, similar to the concept of the set of typical sequences – the typical set – for a classical information source, we introduce the notion of a typical subspace spanned by the typical basis states$|\lambda_{\alpha_1}\rangle \otimes |\lambda_{\alpha_2}\rangle \otimes \ldots \otimes |\lambda_{\alpha_n}\rangle$. We can then define the projection operator onto the typical subspace as

$$\boldsymbol{P}_{\mathrm{T}} = \sum_{\lambda} |\lambda_{\alpha_1}\rangle\langle\lambda_{\alpha_1}| \otimes |\lambda_{\alpha_2}\rangle\langle\lambda_{\alpha_2}| \ldots \otimes |\lambda_{\alpha_n}\rangle\langle\lambda_{\alpha_n}|. \tag{12.97}$$

Thus, $\boldsymbol{P}_{\mathrm{T}}\boldsymbol{\rho}_{\otimes n}$ is the projection of our quantum message onto the typical subspace. We can compute the probability of $\boldsymbol{P}_{\mathrm{T}}\boldsymbol{\rho}_{\otimes n}$ by computing its trace. As we have expressed $\boldsymbol{\rho}$ in its eigenbasis and because of (12.97), this trace is simply given by $\sum_{\lambda} P(\lambda_{\alpha_1})P(\lambda_{\alpha_2})\ldots P(\lambda_{\alpha_n})$. Furthermore, as a direct consequence of (12.86), for large n, the probability that the state of the quantum message lies in the typical subspace is bounded as

$$P(\boldsymbol{P}_{\mathrm{T}}\boldsymbol{\rho}_{\otimes n}) > 1 - \delta. \tag{12.98}$$

Earlier we showed that the probability of the typical set approaches unity for sufficiently large n. Analogously, we now have the result that the probability that the quantum message lies in the typical subspace approaches unity for sufficiently large n.

In the last section we were able to bound the size of the typical set. Following a similar procedure we can bound the dimensionality of the typical subspace, characterized by Trace($\boldsymbol{P}_{\mathrm{T}}$). Denoting the typical subspace by $T_{\lambda}(n, \varepsilon)$, the bounds are given by

$$(1 - \epsilon)2^{n(S(\rho)-\epsilon)} \le |T_{\lambda}(n, \epsilon)| \le 2^{n(S(\rho)+\epsilon)}. \tag{12.99}$$

Analogous to the classical development of Shannon's noiseless coding theorem, we have a couple of powerful results: With $n \to \infty$, the probability that the state of the quantum message will lie in the typical subspace approaches unity and the dimensionality of the typical subspace is approximately $2^{nS(\rho)}$, with the probability of the quantum message being a typical state approaching $2^{-nS(\rho)}$.

We now have a recipe for performing quantum compression. We have the typical subspace projection operator $\boldsymbol{P}_{\mathrm{T}}$ and the operator, $\boldsymbol{P}_{\mathrm{T}}^{\perp}$, projecting onto its orthogonal complement. It is easy to see that $\boldsymbol{P}_{\mathrm{T}}^{\perp} = \boldsymbol{I} - \boldsymbol{P}_{\mathrm{T}}$, where $\boldsymbol{I}$ is the identity operator, is a projection onto the atypical subspace. For compression we perform a fuzzy measurement using the two orthogonal operators with the outcomes denoted by 1 and 0 respectively. If the output is 1, we know the message is in the typical state and we do nothing further. (In Example 12.5 we discuss unitary encoding.) If, however, the outcome is 0, we know the message is in the atypical subspace and we denote it by some fixed state in the typical subspace. Since the probability that the quantum message will lie in the atypical subspace can be made arbitrarily small for large n, we can compress the quantum message without loss of information.

Schumacher showed that when $R > S(\boldsymbol{\rho})$ the above-described compression scheme is reliable. Reliability implies that the compressed quantum message can be decompressed with high fidelity. We first explain the reconstructed message fidelity and then discuss the above condition.

Let us reconsider our i.i.d. quantum source defined by the ensemble $\{|\psi_x\rangle, p_x\}$, where $|\psi_x\rangle$ are pure states. Given this ensemble, the kth instance of a length n quantum message is

$$|\psi^k_{\otimes n}\rangle = |\psi_{x^k_1}\rangle \otimes |\psi_{x^k_2}\rangle \otimes \ldots \otimes |\psi_{x^k_n}\rangle. \qquad (12.100)$$

And the corresponding density operator is $\boldsymbol{\rho}^k_{\otimes n} = |\psi^k_{\otimes n}\rangle\langle\psi^k_{\otimes n}|$. Alice encodes the message by performing a fuzzy measurement whose outcome will be 0_{typical} with probability $\text{Trace}(\boldsymbol{\rho}^k_{\otimes n}\boldsymbol{P}_T) > 1-\delta$ and 1_{atypical} with probability $\text{tr}(\boldsymbol{\rho}^k_{\otimes n}\boldsymbol{P}^{\perp}_{\text{T}}) \leq \delta$. As mentioned above, the atypical state is coded as some fixed state $|\varphi\rangle$ in the typical subspace, with

$$\boldsymbol{E} = \sum_j \boldsymbol{E}_j = \sum_j |\varphi\rangle\langle j|, \qquad (12.101)$$

where $|j\rangle$ is the orthonormal basis of the atypical subspace. The coded state is therefore given by

$$\rho_{\text{C}} = \boldsymbol{P}_{\text{T}}\boldsymbol{\rho}^k_{\otimes n}\boldsymbol{P}_{\text{T}} + \sum_j \boldsymbol{E}_j\boldsymbol{\rho}^k_{\otimes n}\boldsymbol{E}^{\dagger}_j. \qquad (12.102)$$

On substituting (12.101) in (12.102) we get

$$\rho_{\text{C}} = \boldsymbol{P}_{\text{T}}\boldsymbol{\rho}^k_{\otimes n}\boldsymbol{P}_{\text{T}} + \sum_j |\varphi\rangle\langle j|\psi^k_{\otimes n}\rangle\langle\psi^k_{\otimes n}|j\rangle\langle\varphi|, \qquad (12.103)$$

which on rearranging results in

$$\rho_{\text{C}} = \boldsymbol{P}_{\text{T}}\boldsymbol{\rho}^k_{\otimes n}\boldsymbol{P}_{\text{T}} + |\varphi\rangle\langle\varphi| \sum_j \langle\psi^k_{\otimes n}|j\rangle\langle j|\psi^k_{\otimes n}\rangle \qquad (12.104)$$

$$= \boldsymbol{P}_{\text{T}}\boldsymbol{\rho}^k_{\otimes n}\boldsymbol{P}_{\text{T}} + |\varphi\rangle\langle\varphi|\langle\psi^k_{\otimes n}|\boldsymbol{P}^{\perp}_{\text{T}}|\psi^k_{\otimes n}\rangle. \qquad (12.105)$$

The decoding operator for the compressed Hilbert space of the typical subspace is the identity operator. Therefore, the decoding operation results in the state $\rho_{\text{D}\circ\text{C}}$ given by (12.105). The question we need to answer is what is the fidelity of the decoding when n is sufficiently large. For the kth quantum message we consider, the fidelity $F_k = \langle\psi^k_{\otimes n}|\rho_{\text{D}\circ\text{C}}|\psi^k_{\otimes n}\rangle$ to be unity if $\rho_{\text{D}\circ\text{C}} = \rho^k_{\otimes n}$. Thus fidelity is a measure of how close the decoded quantum message is to the original quantum message. Since the quantum state of the source output is derived from an ensemble of quantum messages, we compute the average fidelity as

$$F = \sum_k p_k\langle\psi^k_{\otimes n}|\rho_{\text{D}\circ\text{C}}|\psi^k_{\otimes n}\rangle \qquad (12.106)$$

where $p_k = p_{x_1^k} p_{x_2^k} \dots p_{x_n^k}$ is the probability of the quantum message $\boldsymbol{\rho}^k_{\otimes n}$. Substituting (12.105) in (12.106) we get

$$F = \sum_k p_k \langle \psi^k_{\otimes n} | \boldsymbol{P}_{\mathrm{T}} \boldsymbol{\rho}^k_{\otimes n} | \psi^k_{\otimes n} \rangle + \text{atypical contribution} \tag{12.107}$$

$$= \sum_k p_k \langle \psi^k_{\otimes n} | \boldsymbol{P}_{\mathrm{T}} | \psi^k_{\otimes n} \rangle \langle \psi^k_{\otimes n} | \boldsymbol{P}_{\mathrm{T}} | \psi^k_{\otimes n} \rangle + \text{atypical contribution} \tag{12.108}$$

$$= \sum_k p_k \mathrm{Trace}(\boldsymbol{P}_{\mathrm{T}} \boldsymbol{\rho}^k_{\otimes n}) \mathrm{Trace}(\boldsymbol{P}_{\mathrm{T}} \boldsymbol{\rho}^k_{\otimes n}) + \text{atypical contribution} \tag{12.109}$$

$$\geq \sum_x p_k \mathrm{Trace}(\boldsymbol{P}_{\mathrm{T}} \boldsymbol{\rho}^k_{\otimes n})^2 \tag{12.110}$$

$$= \mathrm{Trace}(\boldsymbol{P}_{\mathrm{T}} \boldsymbol{\rho}_{\otimes n})^2 \tag{12.111}$$

$$> (1 - \delta)^2 \tag{12.112}$$

$$= (1 - \delta)^2 \geq 1 - 2\delta. \tag{12.113}$$

The inequality in (12.110) follows from the atypical contribution being non-negative. The second last inequality in (12.112) is a consequence of (12.98), while the last inequality simply follows from basic algebra. Thus, we have shown that we can compress the quantum message to $n(S(\boldsymbol{\rho}) + \epsilon)$ qubits with arbitrarily small loss in fidelity for sufficiently large n.

What happens if the quantum message is encoded to $R < S(\boldsymbol{\rho})$ qubits? In this case the dimension of the Hilbert subspace Ω in which the compressed quantum message lies is 2^{nR}. Let the projection operator onto this subspace be $\boldsymbol{V}$ and let the decoded state be $\boldsymbol{\rho}_{\mathrm{D}\circ\mathrm{C}}$. This decoded state lies in Ω and consequently can be expanded in the basis of this subspace as

$$\boldsymbol{\rho}_{\mathrm{D}\circ\mathrm{C}} = \sum_i |v_i\rangle\langle v_i| \boldsymbol{\rho}_{\mathrm{D}\circ\mathrm{C}} |v_i\rangle\langle v_i|. \tag{12.114}$$

Using (12.106) and (12.114), we can now compute the average fidelity of reconstruction as

$$F = \sum_k p_k \langle \psi^k_{\otimes n} | \boldsymbol{\rho}_{\mathrm{D}\circ\mathrm{C}} | \psi^k_{\otimes n} \rangle \tag{12.115}$$

$$= \sum_k p_k \left\langle \psi^k_{\otimes n} \middle| \sum_i |v_i\rangle\langle v_i| \boldsymbol{\rho}_{\mathrm{D}\circ\mathrm{C}} |v_i\rangle \middle\langle v_i \middle| \psi^k_{\otimes n} \right\rangle \tag{12.116}$$

$$\leq \sum_k p_k \langle \psi^k_{\otimes n} | v_i \rangle \langle v_i | \psi^k_{\otimes n} \rangle \tag{12.117}$$

$$= \sum_k p_k \text{Trace}(\boldsymbol{V} \boldsymbol{\rho}^k_{\otimes n}) \tag{12.118}$$

$$= \text{Trace}(\boldsymbol{V} \boldsymbol{\rho}_{\otimes n}). \tag{12.119}$$

The inequality is a result of $\langle v_i | \boldsymbol{\rho}_{\text{D}\circ\text{C}} | v_i \rangle \geq 0$. Following our discussion of Shannon's noiseless coding theorem, the probability that a typical state lies in Ω is $2^{n(R-S(\boldsymbol{\rho})+\epsilon)}$. For $R < S(\boldsymbol{\rho})$ and large n, this probability approaches 0. Thus, the fidelity of a quantum message reconstructed after being compressed to $R < S(\boldsymbol{\rho})$ qubits is not good. We therefore have the following statement for Schumacher's quantum noiseless coding theorem:

For a $\{|\psi_x\rangle, p_x\}$ i.i.d. quantum source a reliable compression method exists if $R > S(\boldsymbol{\rho})$. If, however, $R < S(\boldsymbol{\rho})$, then no reliable compression method exists.

We now give an example to illustrate Schumacher's noiseless coding theorem.

Example 12.5. Let us consider a source that produces state $|\psi_1\rangle$ and $|\psi_2\rangle$ with probabilities p_1 and p_2 respectively. Let us define the states to be

$$|\psi_1\rangle = \frac{1}{\sqrt{2}}(|0\rangle + |1\rangle) \tag{12.120}$$

$$|\psi_2\rangle = \frac{\sqrt{3}}{2}|0\rangle + \frac{1}{2}|1\rangle, \tag{12.121}$$

with $p_1 = p_2 = 0.5$. Thus, the resulting density is given by

$$\boldsymbol{\rho} = p_1|\psi_1\rangle\langle\psi_1| + p_2|\psi_2\rangle\langle\psi_2| = \begin{pmatrix} 0.6250 & 0.4465 \\ 0.4665 & 0.3750 \end{pmatrix}. \tag{12.122}$$

The eigen decomposition of $\boldsymbol{\rho}$ is

$$\boldsymbol{\rho} = \begin{pmatrix} -0.7934 & -0.6088 \\ -0.6088 & 0.7934 \end{pmatrix} \begin{pmatrix} 0.9830 & 0 \\ 0 & 0.0170 \end{pmatrix} \begin{pmatrix} -0.7934 & -0.6088 \\ -0.6088 & 0.7934 \end{pmatrix}^\dagger \tag{12.123}$$

and therefore the eigenvectors are

$$|\lambda_1\rangle = \begin{pmatrix} -0.7934 \\ -0.6088 \end{pmatrix} \tag{12.124}$$

$$|\lambda_2\rangle = \begin{pmatrix} -0.6088 \\ 0.7934 \end{pmatrix}, \tag{12.125}$$

with the corresponding eigenvalues given by $\lambda_1 = 0.9830$ and $\lambda_2 = 0.0170$.

With this setup in mind, let us consider the compression of a three-qubit quantum message. With our source as defined above, the possible quantum messages are

$$|\psi_{111}\rangle = |\psi_1\rangle \otimes |\psi_1\rangle \otimes |\psi_1\rangle, \qquad |\psi_{112}\rangle = |\psi_1\rangle \otimes |\psi_1\rangle \otimes |\psi_2\rangle$$

$$|\psi_{121}\rangle = |\psi_1\rangle \otimes |\psi_2\rangle \otimes |\psi_1\rangle, \qquad |\psi_{122}\rangle = |\psi_1\rangle \otimes |\psi_2\rangle \otimes |\psi_2\rangle$$

$$|\psi_{211}\rangle = |\psi_2\rangle \otimes |\psi_1\rangle \otimes |\psi_1\rangle, \qquad |\psi_{212}\rangle = |\psi_2\rangle \otimes |\psi_1\rangle \otimes |\psi_2\rangle$$

$$|\psi_{221}\rangle = |\psi_2\rangle \otimes |\psi_2\rangle \otimes |\psi_1\rangle, \qquad |\psi_{222}\rangle = |\psi_2\rangle \otimes |\psi_2\rangle \otimes |\psi_2\rangle.$$

Let us also define the eigenvector-based basis states:

$$|\lambda_{111}\rangle = |\lambda_1\rangle \otimes |\lambda_1\rangle \otimes |\lambda_1\rangle, \qquad |\lambda_{112}\rangle = |\lambda_1\rangle \otimes |\lambda_1\rangle \otimes |\lambda_2\rangle$$

$$|\lambda_{121}\rangle = |\lambda_1\rangle \otimes |\lambda_2\rangle \otimes |\lambda_1\rangle, \qquad |\lambda_{122}\rangle = |\lambda_1\rangle \otimes |\lambda_2\rangle \otimes |\lambda_2\rangle$$

$$|\lambda_{211}\rangle = |\lambda_2\rangle \otimes |\lambda_1\rangle \otimes |\lambda_1\rangle, \qquad |\lambda_{212}\rangle = |\lambda_2\rangle \otimes |\lambda_1\rangle \otimes |\lambda_2\rangle$$

$$|\lambda_{221}\rangle = |\lambda_2\rangle \otimes |\lambda_2\rangle \otimes |\lambda_1\rangle, \qquad |\lambda_{222}\rangle = |\lambda_2\rangle \otimes |\lambda_2\rangle \otimes |\lambda_2\rangle.$$

These states can be used to define the projective operators onto the Hilbert space $\mathbb{H}^{\otimes 3}$ spanned by $|\lambda_{ijk}\rangle, i, j = 1, 2$. Using these measurement states, let us see if there exists a subspace of the Hilbert space where the quantum messages lie. In other words, can we find the typical subspace. To do this we first note that

$$\text{Trace}\big((|\psi_i\rangle\langle\psi_i|)(|\lambda_j\rangle\langle\lambda_j|)\big) = |\langle\psi_i|\lambda_j\rangle|^2 = \lambda_j; i, j = 1, 2. \tag{12.126}$$

Furthermore,

$$\text{Trace}\big((|\psi_{ijk}\rangle\langle\psi_{ijk}|)(|\lambda_{i'j'k'}\rangle\langle\lambda_{i'j'k'}|)\big) = |\langle\psi_i|\lambda_{i'}\rangle|^2|\langle\psi_j|\lambda_{j'}\rangle|^2|\langle\psi_k|\lambda_{k'}\rangle|^2 = \lambda_{i'}\lambda_{j'}\lambda_{k'}, \tag{12.127}$$

where,$i, j, k = 1, 2; i', j', k' = 1, 2$. We now compute $\lambda_{i'}\lambda_{j'}\lambda_{k'}$,

$\lambda_1\lambda_1\lambda_1 = 0.9499$, $\lambda_1\lambda_1\lambda_2 = 0.0164$, $\lambda_1\lambda_2\lambda_1 = 0.0164$, $\lambda_1\lambda_2\lambda_2 = 2.8409\text{E}(-4)$,

$\lambda_2\lambda_1\lambda_1 = 0.0164$, $\lambda_2\lambda_1\lambda_2 = 2.8409\text{E}(-4)$, $\lambda_2\lambda_2\lambda_1 = 2.8409\text{E}(-4)$,

$\lambda_2\lambda_2\lambda_2 = 4.9130\text{E}(-6)$.

From these values we can see that the probability that the quantum message lies in the subspace spanned by $|\lambda_{111}\rangle, |\lambda_{112}\rangle, |\lambda_{121}\rangle, |\lambda_{211}\rangle$ is $0.9499 + 0.0164 + 0.0164 + 0.0164 = 0.9991$, while the probability that the quantum message lies in the subspace spanned by $|\lambda_{122}\rangle, |\lambda_{212}\rangle, |\lambda_{221}\rangle, |\lambda_{212}\rangle$ is $2.8409\text{E}(-4) + 2.8409\text{E}(-4) + 2.8409\text{E}(-4) + 4.9130\text{E}(-6) = 8.5717\text{E}(-4)$. Thus, with probability close to 1, the quantum message will lie in the subspace spanned by the first set of quantum states, while the probability δ of the quantum message lying in subspace spanned by the second set of quantum states lies close to 0. This becomes even more apparent if n, which is equal to 3 right now, increases. Thus, $|\lambda_{111}\rangle, |\lambda_{112}\rangle, |\lambda_{121}\rangle, |\lambda_{211}\rangle$ span the typical subspace, while $|\lambda_{122}\rangle, |\lambda_{212}\rangle, |\lambda_{221}\rangle, |\lambda_{212}\rangle$ span the complement subspace.

We can now perform data compression using unitary encoding. Unitary encoding implies that a unitary operator $\boldsymbol{U}$ acts on the Hilbert space such that any quantum message state $|\psi_{ijk}\rangle$ that lies in the typical subspace goes to $|\xi_1\rangle \otimes |\xi_2\rangle \otimes |\text{fixed}_{\text{typical}}\rangle$, while the message that lies in the atypical subspace goes to $|\xi_1\rangle \otimes |\xi_2\rangle \otimes |\text{fixed}_{\text{atypical}}\rangle$. The unitary operator can be chosen such that, for example, $|\text{fixed}_{\text{typical}}\rangle = |0\rangle$ and $|\text{fixed}_{\text{atypical}}\rangle = |1\rangle$. Thus, for encoding the message, Alice measures the third qubit. If the output is 0, then the quantum message is in the typical subspace and Alice sends $|\xi_1\rangle \otimes |\xi_2\rangle$ to Bob. If the output is 1, then the quantum message is in the atypical subspace. In this case Alice assigns a fixed quantum state $|\varphi\rangle$ in the typical subspace whose unitary transformation is the state $|a\rangle \otimes |b\rangle \otimes |0\rangle$. Alice then sends $|a\rangle \otimes |b\rangle$ to Bob.

To decompress the message Bob first appends $|0\rangle$ to the quantum state received by him, and applies the transform $\boldsymbol{U}^{-1}$. The possible resulting states are $|\psi_{ijk}\rangle$ or $|\varphi\rangle$. Thus, Bob is able to decompress the message as desired. For the cases where we get $|\varphi\rangle$ we lose information. However, as shown above, the probability of this happening is very small and can be made smaller for larger n.

Let us also look at the fidelity of the retrieved message. For a given quantum message m, the fidelity is given by

$$F_m = \langle \psi^m_{\otimes n} | \boldsymbol{P}_{\text{T}} \boldsymbol{\rho}^m_{\otimes n} \boldsymbol{P}_{\text{T}} | \psi^m_{\otimes n} \rangle + \text{atypical contribution.} \tag{12.128}$$

Here $\boldsymbol{P}_{\text{T}}$ is the projection onto the typical subspace. For our example for any quantum message, fidelity reduces to

$$F = (1-\delta)^2 + \delta |\langle \varphi | \psi_{ijk} \rangle|^2. \tag{12.129}$$

We have already computed $\delta = 8.5717\text{E}(-4)$, and this gives us a fidelity greater than $(1-\delta)^2 = 0.9983$. We thus have achieved a rate of $R = 2/3 = 0.667$ with high reconstruction fidelity. Let us compare this rate with the quantum entropy of the source. The quantum entropy turns out to be

$$S(\boldsymbol{\rho}) = \lambda_1 \log \lambda_1 + \lambda_2 \log \lambda_2 = 0.1242. \tag{12.130}$$

We therefore have $R > S$, therefore satisfying Schumacher's condition. Equation (12.130) also hints at another interesting fact: we can do better than two qubits, and go down to a single qubit. For this case, $R = 1/3 = 0.333$ is still greater than $S(\boldsymbol{\rho})$. We note that here quantum entropy represents the incompressible information in the message generated by a quantum source in a manner similar to Shannon entropy representing the incompressible information of a classical source.

12.4 HOLEVO–SCHUMACHER–WESTMORELAND (HSW) THEOREM

We now turn to the second question we raised at the beginning of this chapter: What is the maximum rate at which we can transfer information over a noisy communication channel? Specifically, we will try to answer this question for classical information over quantum channels. As in the previous section, we begin by looking at the classical counterpart.

Let us consider a binary symmetric channel. Such a channel takes in as input a binary symbol that is flipped with probability $q > 0$, and remains unchanged with probability $1-q$. For example, if the input is a 0 (or a 1), it will be flipped to a 1 (or 0) with probability q or remain unchanged with probability $1-q$. Is it possible to reliably communicate a message over such a channel? The answer is a resounding yes, and the key lies in randomly coding the input message with long codewords. For example, if we want to transmit M messages over the binary symmetric channel, we first define the rate of such a transfer per channel use as

$$R = \frac{\log M}{n}, \tag{12.131}$$

where n is the length of the codewords. The total number of these codewords is given by 2^{nR}. We say that the code rate R is achievable if the probability of error in decoding the message after transmission over the channel approaches zero as n increases. The set of codewords is referred to as the codebook, and their construction is done through random coding. Random coding requires that we first define a binary stochastic source that outputs 1 with some probability p and 0 with probability $1-p$. Each output is independent of the previous one. Given such a source, the codeword for the mth message is given by the tensor product of n outputs of the source.

If any codeword from our codebook were sent over the binary symmetric channel, it would flip the kth bit with probability q. In other words, for a single codeword we can define a set of error words. As a consequence of modulo-2 addition, the number of ones in any given error word is the Hamming distance between the codeword that is sent over the channel and the one that is received. Thus, for a single codeword the set of error words can be thought of as defining a Hamming sphere around that codeword. From the discussion of typical sequences in the previous section, the error words in the Hamming sphere approach $2^{nH(q)}$ as n increases. Such a Hamming sphere exists for each codeword. If the spheres do not overlap, then the codewords can be reliably decoded. If, however, there is overlap between them then unreliable communication can result. Shannon provided the condition under which it is possible to have reliable communication. Specifically, he showed that if the rate R is below the capacity C of the channel, then there exists a codebook $(2^{nR}, n)$ for which reliable decoding is possible. The proof is based on the idea of jointly typical sequences. A rough outline of the proof will be given here.

Two n-length sequences $\boldsymbol{x}$ and $\boldsymbol{y}$ are jointly typical sequences if they satisfy

$$P\left(\left|\frac{1}{n}\log\frac{1}{P(\boldsymbol{x})} - H(X)\right| \leq \epsilon\right) > 1-\delta \tag{12.132}$$

$$P\left(\left|\frac{1}{n}\log\frac{1}{P(\boldsymbol{y})} - H(Y)\right| \leq \epsilon\right) > 1-\delta \tag{12.133}$$

and

$$P\left(\left|\frac{1}{n}\log\frac{1}{P(\boldsymbol{x},\boldsymbol{y})} - H(X,Y)\right| \leq \epsilon\right) > 1-\delta, \tag{12.134}$$

where $P(\boldsymbol{x},\boldsymbol{y})$ is the joint probability of the two sequences and $H(X,Y)$ is their joint entropy. For an n-length input codeword randomly generated according to the probability distribution of a source X, the number of input typical sequences is approximately $2^{nH(X)}$ and the output typical sequences is approximately $2^{nH(Y)}$. Furthermore, the total number of input and output sequences that are jointly typical is $2^{nH(X,Y)}$. Therefore, the total pairs of sequences that are $\boldsymbol{x}$-typical, $\boldsymbol{y}$-typical and also jointly typical are $2^{n(H(X)+H(Y)-H(X,Y))} = 2^{nI(X;Y)}$, where, $I(X;Y)$ is the mutual information between X and Y. These are the maximum number of codeword sequences that can be distinguished. One way of seeing this is to consider a single codeword. For this codeword, the action of the channel, characterized by the conditional probability $P(\boldsymbol{y}|\boldsymbol{x})$, defines the Hamming sphere in which this codeword can lie after the action of the channel. The size of this Hamming sphere is approximately $2^{nH(Y|X)}$. Given that the total number of output typical sequences is approximately $2^{nH(Y)}$, if we desire to have no overlap between two Hamming spheres the maximum number of codewords we can consider is $\frac{2^{nH(Y)}}{2^{nH(Y|X)}} = 2^{n(H(Y)-H(Y|X))} = 2^{nI(X;Y)}$.

To increase this number we need to maximize $I(X;Y)$ over the distribution of X. (Note that the conditional probability of the channel is not in our control.) This maximal mutual information is referred to as the capacity of the channel. If we have a rate $R < C$, then Shannon's noisy channel coding theorem tells us that we can construct 2^{nR} n-length codewords that can be sent over the channel with maximum probability of error approaching zero for large n.

Suppose now that instead of communicating over a classical channel, Alice decides to communicate over a quantum channel. In this case the same idea of random block coding is used, except that now Alice uses a quantum source, that for each message $m \in M = 2^{nR}$ generates a product state codeword

$$\rho^m = \rho_{m_1} \otimes \rho_{m_2} \otimes \ldots \otimes \rho_{m_n} = \rho_{\otimes n} \in \mathbb{H}_{\otimes n}, \tag{12.135}$$

where each ρ is drawn from the ensemble $\{\rho_x, p_x\}$. This quantum codeword is then sent over a quantum channel that is described by the trace-preserving quantum operation ε, resulting in the received codeword $\sigma_m = \varepsilon(\rho_{m_1}) \otimes \varepsilon(\rho_{m_2}) \otimes \ldots \otimes \varepsilon(\rho_{m_n}) = \sigma_{\otimes n}$. Bob performs measurements on σ_m to decode Alice's message. Our goal is to find the maximum rate at which Alice can communicate over the quantum channel such that the probability of decoding error is negligibly small. This maximum rate is the capacity $(C(\varepsilon))$ of the quantum channel for transmitting classical information. The key idea behind the HSW theorem is that for $R < C(\varepsilon)$ there exist quantum codes with n-length codewords such that the probability of incorrectly decoding the quantum codeword can be made arbitrarily small.

To show this, we perform the eigen decomposition of our received state σ_m. Denoting the eigen decomposition of $\sigma_{m_k} = \varepsilon(\rho_{m_k})$ by

$$\sigma_{m_k} = \sum_j \lambda^j_{m_k} |\lambda^j_{m_k}\rangle\langle\lambda^j_{m_k}|, \tag{12.136}$$

we have

$$\sigma_m = \sum_i \Lambda^i_m |\Lambda^i_m\rangle\langle\Lambda^i_m|, \tag{12.137}$$

where the eigenvalue is $\Lambda^i_m = \lambda^{i_1}_{m_1}\lambda^{i_2}_{m_2}\ldots\lambda^{i_n}_{m_n}$ and the eigenvector is $|\Lambda^i_m\rangle = |\lambda^{i_1}_{m_1}\rangle \otimes |\lambda^{i_2}_{m_2}\rangle \otimes \ldots \otimes |\lambda^{i_n}_{m_n}\rangle$. Furthermore, we can define the quantum entropy associated with σ as

$$\langle S(\sigma)\rangle = \sum_x p_x S(\sigma_x). \tag{12.138}$$

Therefore, the sequence $\lambda^{i_1}_{m_1}\lambda^{i_2}_{m_2}\ldots\lambda^{i_n}_{m_n}$ is a typical sequence if according to our earlier discussion on typical sequences,

$$P\left(\left|\frac{1}{n}\log\frac{1}{\lambda^{i_1}_{m_1}\lambda^{i_2}_{m_2}\ldots\lambda^{i_n}_{m_n}} - \langle S(\sigma)\rangle\right| \le \epsilon\right) > 1 - \delta, \tag{12.139}$$

and correspondingly $\{|\Lambda^i_m\rangle\}$ define the basis of the typical subspace. Let us now define a projection operator P_m onto the typical subspace. By invoking the conclusions of the typical subspace results discussed earlier, the dimensionality of the typical subspace given by the trace of the projection operator is bounded by

$$(1-\epsilon)2^{n(\langle S(\sigma)\rangle - \epsilon)} \le \text{Trace}(P_m) \le 2^{n(\langle S(\sigma)\rangle + \epsilon)}. \tag{12.140}$$

Furthermore, the average probability that σ_m lies in the typical subspace is

$$\langle \text{Trace}(P_m \sigma_m)\rangle > 1 - \delta. \tag{12.141}$$

The received state σ_m denotes one received codeword. We therefore have to average over all such possible codewords, hence the angled brackets on the left-hand side of (12.141). Based on (12.141) we can say that the received codeword with probability approaching 1 will lie in the typical space. Let $\{M_m\}$ represent the POVM measurement operators that Bob uses to decode the message. Note that these measurement operators are approximately given by the projection operators P_m. As a result, measurement outcomes indicate if the received codeword lies within the typical subspace, and the probability that Bob will successfully decode the received codeword σ_m using POVM measurement operator M_m is given by

$$p_m = \text{Trace}(\sigma_m M_m), \tag{12.142}$$

while the probability of error in decoding is given by

$$p_{m_e} = 1 - \text{Trace}(\sigma_m M_m). \tag{12.143}$$

If we assume that Alice has uniformly selected the codewords from the codebook, we can define the average probabilities of success and error as

$$p_m^{\text{avg}} = \frac{\sum_m p_m}{2^{nR}} \tag{12.144}$$

and

$$p_{m_e}^{\text{avg}} = \frac{\sum_m p_{m_e}}{2^{nR}} \tag{12.145}$$

respectively. Furthermore, by employing Fano's inequality and the Holevo bound – a technical proof that we will forego here – the expectation of the average probability of error over all possible input codewords is upper bounded by

$$\langle p_{m_e}^{\text{avg}} \rangle \leq 4\delta + (2^{nR} - 1)2^{-n(\chi - 2\epsilon)}, \tag{12.146}$$

where $\chi = S(\langle \sigma \rangle) - \langle S(\sigma) \rangle$ is the Holevo information. For large n, the right-hand side reduces to $4\delta + 2^{-n(\chi - R + 2\epsilon)}$ and if $R < \chi$, we have

$$\langle p_{m_e}^{\text{avg}} \rangle \leq 4\delta. \tag{12.147}$$

Thus, for large n the probability of incorrectly decoding the message is upper bounded by 4δ, where δ can be made arbitrarily small. So we have the main result of the HSW theorem that for rate $R < \chi$ there exist codes – here we have used random product state codewords – that can allow reliable error-free communication over a quantum channel. Note that the larger χ is, the higher R can be. Consequently, we define the product state capacity as

$$C(\epsilon) = \max_{\{p_x, \rho_x\}} \chi. \tag{12.148}$$

To make the idea of capacity concrete we compute the capacity of two example quantum channels: the depolarizing channel and the bit-flip channel. We assume quantum source outputs $\boldsymbol{\rho}_0$ with probability p_0, and $\boldsymbol{\rho}_1$ with probability p_1. We further assume that both states are orthogonal pure states given by

$$\boldsymbol{\rho}_0 = \begin{pmatrix} 1 & 0 \\ 0 & 0 \end{pmatrix}, \quad \boldsymbol{\rho}_1 = \begin{pmatrix} 0 & 0 \\ 0 & 1 \end{pmatrix}. \tag{12.149}$$

Example 12.6: Depolarizing channel. The depolarizing channel, defined by the relation

$$\varepsilon(\boldsymbol{\rho}) = q\frac{1}{2} + (1-q)\boldsymbol{\rho}, \tag{12.150}$$

either transforms the input state to a uniform mixed state – denoted by the identity operator I – with probability q, or leaves the state unchanged with probability $1-q$. The transformation to the mixed state is referred to as the depolarization action of the channel. Applying the quantum operation (12.150) to both $\boldsymbol{\rho}_0$ and $\boldsymbol{\rho}_1$ we get

$$\boldsymbol{\sigma}_0 = \varepsilon(\boldsymbol{\rho}_0) = \frac{q}{2}\begin{pmatrix} 1 & 0 \\ 0 & 1 \end{pmatrix} + (1-q)\begin{pmatrix} 1 & 0 \\ 0 & 0 \end{pmatrix} = \begin{pmatrix} 1-\frac{q}{2} & 0 \\ 0 & \frac{q}{2} \end{pmatrix} \tag{12.151}$$

$$\boldsymbol{\sigma}_1 = \varepsilon(\boldsymbol{\rho}_1) = \frac{q}{2}\begin{pmatrix} 1 & 0 \\ 0 & 1 \end{pmatrix} + (1-q)\begin{pmatrix} 1 & 0 \\ 0 & 0 \end{pmatrix} = \begin{pmatrix} \frac{q}{2} & 0 \\ 0 & 1-\frac{q}{2} \end{pmatrix}. \tag{12.152}$$

The corresponding quantum entropies are easily seen to be

$$S(\boldsymbol{\sigma}_0) = \left(1-\frac{q}{2}\right)\log\left(1-\frac{q}{2}\right) + \frac{q}{2}\log\frac{q}{2} \tag{12.153}$$

$$S(\boldsymbol{\sigma}_1) = \frac{q}{2}\log\frac{q}{2} + \left(1-\frac{q}{2}\right)\log\left(1-\frac{q}{2}\right) = S(\boldsymbol{\sigma}_0). \tag{12.154}$$

Consequently, we have

$$\begin{aligned}\langle S(\boldsymbol{\sigma})\rangle &= p_0\, S(\boldsymbol{\sigma}_0) + p_1 S(\boldsymbol{\sigma}_1) = (p_0 + p_1)\, S(\boldsymbol{\sigma}_1) \\ &= \frac{q}{2}\log\frac{q}{2} + \left(1-\frac{q}{2}\right)\log\left(1-\frac{q}{2}\right).\end{aligned} \tag{12.155}$$

This gives us the second term of Holevo information χ. To find the first term we first compute

$$\langle\sigma\rangle = \left\langle \varepsilon\left(\sum_i p_i\boldsymbol{\rho}_i\right)\right\rangle = \frac{q}{2}\begin{pmatrix} 1 & 0 \\ 0 & 1 \end{pmatrix} + (1-q)\begin{pmatrix} p_0 & 0 \\ 0 & p_1 \end{pmatrix}$$

$$= \begin{pmatrix} \frac{q}{2} + (1-q)p_0 & 0 \\ 0 & \frac{q}{2} + (1-q)p_1 \end{pmatrix}, \tag{12.156}$$

and therefore we have

$$S(\langle\boldsymbol{\sigma}\rangle) = \left(\frac{q}{2} + (1-q)p_0\right)\log\left(\frac{q}{2} + (1-q)p_0\right) + \left(\frac{q}{2} + (1-q)p_1\right)\log\left(\frac{q}{2} + (1-q)p_1\right). \tag{12.157}$$

Under the constraint that $p_0 + p_1 = 1$, $S(\langle\boldsymbol{\sigma}\rangle)$ is maximized for $p_0 = p_1 = 1/2$, and this maximum value is 1 bit. Therefore, the Holevo information reduces to

$$\chi = 1 - \left(1 - \frac{q}{2}\right)\log\left(1 - \frac{q}{2}\right) + \frac{q}{2}\log\frac{q}{2}. \tag{12.158}$$

We can see that χ is equal to 1 for $q = 0$, and is 0 when $q = 1$. Thus, the capacity $C(\varepsilon) = 1$ bit. It is interesting to note that for $q = 1$, χ vanishes and the channel is of no use. The reason can be deduced from the definition of the depolarization channel. When we set $q = 1$ in (12.150), the output of the channel is a mixed state $\boldsymbol{I}/2$, no matter what the input. Consequently, we deterministically know that the channel output is a mixed state (maximum uncertainty) and the channel is of no use!

Example 12.7: Phase-flip channel. As the next example let us consider the phase-flip channel. If the input to the channel is a state given by $\alpha|0\rangle + \beta|1\rangle$, the action of the phase-flip channel results in $\alpha|0\rangle - \beta|1\rangle$ with probability q, or leaves the state unchanged with probability $1-q$. The phase-flip action is performed using the Pauli matrix

$$\mathbf{Z} = \begin{pmatrix} 1 & 0 \\ 0 & -1 \end{pmatrix}. \tag{12.159}$$

We therefore define the phase-flip channel mathematically by

$$\varepsilon(\boldsymbol{\rho}) = q\mathbf{Z}\boldsymbol{\rho}\mathbf{Z} + (1-q)\boldsymbol{\rho}. \tag{12.160}$$

Following the steps in the previous example we have

$$\boldsymbol{\sigma}_0 = \varepsilon(\boldsymbol{\rho}_0) = q\begin{pmatrix} 1 & 0 \\ 0 & -1 \end{pmatrix}\begin{pmatrix} 1 & 0 \\ 0 & 0 \end{pmatrix}\begin{pmatrix} 1 & 0 \\ 0 & -1 \end{pmatrix} + (1-q)\begin{pmatrix} 1 & 0 \\ 0 & 0 \end{pmatrix} = \begin{pmatrix} 1 & 0 \\ 0 & 0 \end{pmatrix} \tag{12.161}$$

$$\boldsymbol{\sigma}_1 = \varepsilon(\boldsymbol{\rho}_1) = q\begin{pmatrix}1 & 0\\ 0 & -1\end{pmatrix}\begin{pmatrix}0 & 0\\ 0 & 1\end{pmatrix}\begin{pmatrix}1 & 0\\ 0 & -1\end{pmatrix} + (1-q)\begin{pmatrix}0 & 0\\ 0 & 1\end{pmatrix} = \begin{pmatrix}0 & 0\\ 0 & 1\end{pmatrix}. \tag{12.162}$$

The corresponding quantum entropies are

$$S(\boldsymbol{\sigma}_0) = (1)\log(1) + 0\log 0 = 0 \tag{12.163}$$

$$S(\boldsymbol{\sigma}_1) = 0\log 0 + (1)\log(1) = 0 = S(\boldsymbol{\sigma}_0), \tag{12.164}$$

and therefore the second term of the Holevo information is given by

$$\langle S(\boldsymbol{\sigma})\rangle = p_0\, S(\boldsymbol{\sigma}_0) + p_1 S(\boldsymbol{\sigma}_1) = 0. \tag{12.165}$$

To calculate the first term, as in the previous example we first find the expression for

$$\langle\boldsymbol{\sigma}\rangle = \left\langle \varepsilon\left(\sum_i p_i\boldsymbol{\rho}_i\right)\right\rangle = q\begin{pmatrix}1 & 0\\ 0 & -1\end{pmatrix}\begin{pmatrix}p_0 & 0\\ 0 & p_1\end{pmatrix}\begin{pmatrix}1 & 0\\ 0 & -1\end{pmatrix} + (1-q)\begin{pmatrix}p_0 & 0\\ 0 & p_1\end{pmatrix}$$
$$= \begin{pmatrix}p_0 & 0\\ 0 & p_1\end{pmatrix}. \tag{12.166}$$

This gives us

$$S(\langle\boldsymbol{\sigma}\rangle) = p_0\log p_0 + p_1\log p_1. \tag{12.167}$$

Therefore,

$$\chi = S(\langle\boldsymbol{\sigma}\rangle) - \langle S(\boldsymbol{\sigma})\rangle = p_0\log p_0 + p_1\log p_1. \tag{12.168}$$

Holevo information is maximized over $\{p_0, p_1\}$, and this maximization can be done by noting that the capacity of any channel is upper bounded by the entropy of the source

$$S(\boldsymbol{\rho}) = S\begin{pmatrix}p_0 & 0\\ 0 & p_1\end{pmatrix} = p_0\log p_0 + p_1\log p_1, \tag{12.169}$$

which is the same as χ! Under the constraint

$$p_0 + p_1 = 1, \tag{12.170}$$

the maximum χ is 1 bit for

$$p_0 = p_1 = \frac{1}{2}. \tag{12.171}$$

Therefore, the capacity of the phase-flip channel is 1 bit!

Example 12.8: Lossy bosonic channel. As a last example we consider a continuous variable quantum channel, the lossy bosonic channel. In quantum optical communication bosonic channels generally refer to photon-based communication channels like free-space optical fiber transmission links. They are an important area of research because they allow us to systematically explain and study quantum optical communication by exploiting system linearity coupled with certain nonlinear devices such as parametric down converters and squeezers [8].

For a lossy channel the effect of the channel on the input quantum states can be modeled by a quantum noise source. Taking this noise source into account, the evolution of a multimode lossy bosonic channel can be described as

$$\boldsymbol{A}_{o_k} = \sqrt{\eta_k}\boldsymbol{A}_{i_k} + \sqrt{1-\eta_k}\boldsymbol{B}_k, \tag{12.172}$$

where $\boldsymbol{A}_{i_k}$ and $\boldsymbol{A}_{o_k}$ are the annihilation ladder operators of the kth mode of the input and output electromagnetic field, $\boldsymbol{B}$ is the annihilation operator associated with the kth environmental noise mode, and η_k is the kth mode channel transmissivity or quantum efficiency. For such a channel the capacity is given by

$$C(\varepsilon) = \sum_k g(\eta_k N_k(\beta)), \tag{12.173}$$

with $N_k(\beta)$ being the optimal photon-count distribution

$$N_k(\beta) = \frac{1/\eta_k}{e^{\beta\hbar\omega_k/\eta_k} - 1}, \tag{12.174}$$

where $\hbar$ is the reduced Planck constant, ω_k is the frequency of the kth mode, and β is the Lagrange multiplier determined by the average transmitted energy. A detailed derivation of this result is given in Ref. [9] and the references therein. The function $g()$,

$$g(x) = (x+1)\log_2(x+1) - x\log_2 x, \tag{12.175}$$

is known as the Shannon entropy of the Bose–Einstein probability distribution [10].

An interesting consequence of (12.173) is that by constructing the encoded product state

$$\rho_{\otimes k} = \otimes_k \int \mathrm{d}u p_k(u) |u\rangle_k \langle u|, \tag{12.176}$$

where the kth modal state $\int \mathrm{d}u p_k(u)|u\rangle_k\langle u|$ is a mix of coherent states $|u\rangle_k$, and $p_k(u)$ is the Guassian probability distribution

$$p_k(u) = \frac{1}{\pi N_k} \mathrm{e}^{-|u|/N_k} \tag{12.177}$$

with N_k being the modal photon count and u being the continuous variable, the channel capacity can be realized with a single use of the channel without employing entanglement. More details can be found in Refs [8–13].

12.5 CONCLUSION

In this chapter fundamental ideas of quantum information theory have been introduced. The key concept of quantum entropy has been presented, and many of its properties have been discussed in the context of single and composite systems. Holevo information has been defined and the very important Holevo bound discussed. All these ideas arc brought together in the quantum versions of the classical noiseless and noisy channel coding theorem. Because quantum information theory is built on the foundation of classical information theory, at every stage it has been contrasted with, or built upon, classical concepts to highlight the similarities and differences between the classical and quantum worlds. Despite this, quantum information theory is a vast field, and this chapter does not do it complete justice. Readers who wish to delve further into the exciting and rich complexities of this field should consult the references listed at the end. References [1–7] provide background of classical information theory along with its relation to a variety other fields in computer science: statistics, probability, estimation and detection, and finance. Reference [14] is a classic book in the field of quantum information theory and computation. This is a must read for anybody interested in quantum computing. References [15–23] list important papers that led to breakthroughs in quantum information theory. Finally, Refs [8–13] detail interesting theoretical work done in quantifying the capacity of quantum optical channels.

12.6 PROBLEMS

1. For Example 2.2 show that $H(P)$ is a concave function of p.
2. Given two distributions P and Q on $\mathcal{X}$, prove that the relative entropy $D(P||Q) \geq 0$, with equality if and only if $P = Q$.
3. For two random variables X and Y, the classical mutual information is

$$I(X;Y) = H(X) - H(X|Y),$$

where $H(X|Y)$ is the conditional entropy defined as

$$H(X|Y) = \sum_{x\in\mathcal{X}} \sum_{y\in\mathcal{Y}} P(x,y)\log(y|x).$$

Show that

(a) $I(X;Y) = I(Y;X)$.

(b) $I(X;Y) \geq 0$ with equality if and only if X and Y are independent.

4. Consider the following state of a bipartite quantum system AB:

$$|\psi\rangle = \sqrt{\frac{1}{3}}(|1\rangle \otimes |1\rangle + |1\rangle \otimes |1\rangle + 0\rangle \otimes |1\rangle).$$

(a) Compute the quantum entropy of the joint system.
(b) Compute the quantum entropy of the subsystems.
(c) Verify the triangle inequality and the Araki–Lieb triangle inequality.
(d) Is the composite system in a pure state?

5. For the product state $\rho_x \otimes \rho_y$, show that $S(\rho_x \otimes \rho_y) = S(\rho_x) + S(\rho_y)$.

6. Given a quantum source characterized by the density operator

$$\rho = \frac{1}{4}\begin{pmatrix} 3 & -1 \\ -1 & 1 \end{pmatrix}.$$

(a) Find the maximum achievable rate for this quantum source.
(b) Given that we have three-qubit quantum messages, what is the corresponding typical subspace? Assume $\delta < 0.1$.
(c) What will be the typical subspace if the message length is four qubits?

7. Let $\{P_m\}$ be a set of projective measurements that satisfy the following properties:

$$P_m^{\dagger} = P_m,$$

$$\sum_m P_m = I, \quad \text{and}$$

$$P_{m_1} P_{m_2} = \delta_{m_1 m_2} P_{m_1}.$$

For $\sigma = \sum_m P_m \rho P_m$, show that $S(\sigma) \geq S(\rho)$, with the necessary and sufficient condition for the equality being $\sigma = \rho$.

8. Consider the depolarizing channel

$$\varepsilon(\rho_i) = (1-q)\rho + \frac{q}{3} X\rho X^{\dagger} + \frac{q}{3} Y\rho Y^{\dagger} + \frac{q}{3} Z\rho Z^{\dagger},$$

where X, Y, Z are the Pauli matrices, and the density operator for the two-symbol alphabet is as described in Examples 12.5 and 12.6. This channel leaves the state invariant with probability $1-q$ or results in a bit flip (X), phase flip (Z), or a bit and phase flip (Y) with equal probability $q/3$.

(a) Show ε is a trace-preserving quantum channel operation.
(b) Determine the capacity of this quantum channel.
(c) We write the quantum channel in the operator-sum representation

$$\varepsilon(\rho_i) = E_1 \rho_i E_1^{\dagger} + E_2 \rho_i E_2^{\dagger} + E_3 \rho_i E_3^{\dagger} + \frac{q}{3} E_4 \rho_i E_4^{\dagger}$$

where

$$E_1 = \sqrt{1-q}\, I, \qquad E_2 = \sqrt{\frac{q}{3}}\, X,$$

$$E_3 = -i\sqrt{\frac{q}{3}}Y, \quad E_4 = \sqrt{\frac{q}{3}}Z.$$

Determine the quality of the transmitted bit given by the channel fidelity

$$F = \sum_i (\text{Trace}(\rho E_i))(\text{Trace}(\rho E_i)).$$

9. Consider the bit-flip channel

$$\varepsilon(\rho) = qX\rho X + (1-q)\rho$$

that takes as input the qubit $\alpha|0\rangle + \beta|1\rangle$ and performs the bit-flip operation with probability q while leaving the qubit alone with probability $1-q$.

(a) Compute its Holevo information.

(b) What can you say about the capacity of this channel?

(c) Is it possible that under certain conditions the capacity is 0?

10. Repeat Problem 6, but with the input state given by

$$|\psi\rangle = \frac{1}{\sqrt{2}}\begin{pmatrix}1\\1\end{pmatrix} + \frac{1}{\sqrt{2}}\begin{pmatrix}1\\-1\end{pmatrix}.$$

References

[1] C.E. Shannon, A mathematical theory of communication, Bell Syst. Tech. J. 27 (1948) 379–423, 623–656.

[2] C.E. Shannon, W. Weaver, The Mathematical Theory of Communication, University of Illinois Press, Urbana, IL, 1949.

[3] T. Cover, J. Thomas, Elements of Information Theory, Wiley-Interscience, 1991.

[4] A.N. Kolmogorov, Three approaches to the quantitative definition of information, Prob. Inform. Transm. 1 (1) (1965) 1–7.

[5] H. Touchette, The large deviation approach to statistical mechanics, Phys. Rep. 478 (1–3) (2009) 1–69.

[6] B.R. Frieden, Science from Fisher Information: A Unification, Cambridge University Press, 2004.

[7] J.L. Kelly Jr., A new interpretation of information rate, Bell Syst. Tech. J. 35 (1956) 917–926.

[8] A.S. Holevo, R.F. Werner, Evaluating capacities of boson Gaussian channels, Phys. Rev. A 63 (2001), paper 032312.

[9] V. Giovannetti, S. Guha, S. Lloyd, L. Maccone, J.H. Shapiro, H.P. Yuen, Classical capacity of the lossy bosonic channel: The exact solution, Phys. Rev. Lett. 92 (2004), paper 027902.

[10] J.H. Shapiro, S. Guha, B.I. Erkmen, Ultimate channel capacity of free-space optical communications, J. Opt. Networking 4 (2005) 501–516.

[11] V. Giovannetti, S. Lloyd, L. Maccone, P.W. Shor, Broadband channel capacities, Phys. Rev. A 68 (2003), paper 062323.

[12] V. Giovannetti, S. Lloyd, L. Maccone, P.W. Shor, Entanglement assisted capacity of the broadband lossy channel, Phys. Rev. Lett. 91 (2003), paper 047901.

[13] S. Guha, J.H. Shapiro, B.I. Erkmen, Classical capacity of bosonic broadcast communication and a minimum output entropy conjecture, Phys. Rev. A 76 (2007), paper 032303.

[14] M.A. Nielsen, I.L. Chuang, Quantum Computation and Quantum Information, Cambridge University Press, 2000.

[15] J. Preskill, Reliable quantum computers, Proc. R. Soc. Lond. A 454 (1998) 385–410.

[16] A.S. Holevo, Problems in the mathematical theory of quantum communication channels, Rep. Math. Phys. 12 (1977) 273–278.

[17] A.S. Holevo, Capacity of a quantum communications channel, Prob. Inform. Transm. 15 (1979) 247–253.

[18] R. Jozsa, B. Schumacher, A new proof of the quantum noiseless coding theorem, J. Mod. Opt. 41 (1994) 2343–2349.

[19] B. Schumacher, M. Westmoreland, W.K. Wootters, Limitation on the amount of accessible information in a quantum channel, Phys. Rev. Lett. 76 (1997) 3452–3455.

[20] A.S. Holevo, The capacity of the quantum channel with general signal states, IEEE Trans. Inform. Theory 44 (1998) 269–273.

[21] M. Sassaki, K. Kato, M. Izutsu, O. Hirota, Quantum channels showing superadditivity in channel capacity, Phys. Rev. A 58 (1998) 146–158.

[22] C.A. Fuchs, Nonorthogonal quantum states maximize classical information capacity, Phys. Rev. Lett. 79 (1997) 1162–1165.

[23] C.H. Bennett, P.W. Shor, J.A. Smolin, A.V. Thapliyal, Entanglement enhanced classical capacity of noisy quantum channels, Phys. Rev. Lett. 83 (1999) 3081–3084.

CHAPTER

Physical Implementations of Quantum Information Processing

13

CHAPTER OUTLINE

This chapter, which is based on Ref. [1–62], describes several promising physical implementations of quantum information processing [1–62]. The chapter starts with a description of physical implementation basics, *di Vincenzo criteria* [50,60], and an overview of physical implementation concepts (Section 13.1). Section 13.2 is concerned with the nuclear magnetic resonance (NMR) implementation, whose basic concepts are used in various implementations. In Section 13.3, the use of ion traps in quantum computing is described. Next, the various photonic implementations (Section 13.4) are considered, followed by quantum relay implementation (Section 13.5) and the implementation of quantum encoders and decoders (Section 13.6). The implementation based on optical cavity electrodynamics is further described in Section 13.7. Finally, the use of quantum dots in quantum computing is discussed in Section 13.8. After a short summary (Section 13.9), a set of problems (Section 13.10) is provided for self-study.

13.1 PHYSICAL IMPLEMENTATION BASICS

We are still in the early stages of implementation of quantum information processing devices and quantum computers. In subsequent sections, several successful implementations are described. However, at this time, it is difficult to predict which of

Quantum Information Processing and Quantum Error Correction. DOI: 10.1016/B978-0-12-385491-9.00013-7

them will actually be used in the foreseeable future. The main requirements, also known as *di Vincenzo criteria* [50,51,60], to be fulfilled before a particular physical system can be considered a viable candidate for quantum information processing can be summarized as follows:

1. **Initialization capability.** This concerns the ability to prepare the physical system into the desired initial state, say $|00\ldots0\rangle$.
2. **Quantum register.** This represents a physical-state system with well-defined states and qubits making up the storage device.
3. **Universal set of gates.** This requirement relates to the ability to perform a universal family of unitary transformations. The most popular sets of universal quantum gates, as discussed in Chapter 3, are: {Hadamard (H), phase (S), CNOT, Toffoli (U_T)} gates, {H, S, $\pi/8$ (T), CNOT} gates, the Barenco gate [1], and the Deutsch gate [2].
4. **Low error and decoherence rate criterion.** This requirement is related to high-fidelity quantum gates, which operate with reasonably low probability of error P_e, say $P_e < 10^{-3}$. In addition, the quantum decoherence time must be much longer than gate operation duration so that meaningful quantum computation is possible. Decoherence is enemy number one of quantum information processing. Because of decoherence, the computations must be performed in a time interval shorter than the decoherence time $\tau_{\text{Decoherence}}$. Let $\tau_{\text{operation}}$ denote the duration of an elementary operation performed by a given quantum gate. The maximum number of operations $N_{\text{operations}}$ that a quantum computer can perform is given by $N_{\text{operations}} = \tau_{\text{Decoherence}} / \tau_{\text{operation}}$.
5. **Read-out capability.** This requirement refers to the ability to perform measurements on qubits in computational basis in order to determine the state of qubits at the end of the computation procedure.

   In addition to the requirements above, in quantum teleportation, quantum superdense coding, quantum key distribution (QKD), and distributing quantum computation applications, one additional requirement is needed. This requirement is:
6. **Communication capability.** This is the ability to teleport qubits to distant locations.

In subsequent sections of this chapter, we describe several possible technologies, including: (1) nuclear magnetic resonance, which is discussed from a conceptual point of view [4–6,15,51], (2) trapped ion-based implementations [7–9,15,49,51], (3) photonic quantum implementations [15–17,29–33], (4) cavity quantum electrodynamics-based quantum computation [15,41–49] and (5) quantum dot-based implementations [47,51]. We also discuss the physical implementation of quantum relays and quantum encoders and decoders.

In many of these implementations, the theory of harmonic oscillators, described in Chapter 2 (see Section 2.9), is used. With this approach, the energy levels $|0\rangle, |1\rangle, \ldots, |2^N\rangle$ of a single quantum harmonic oscillator can be used to represent N qubits. An arbitrary unitary operation U can be implemented by matching the eigenvalue spectrum of U to that of the harmonic oscillator Hamiltonian $H = \hbar\omega a^\dagger a$, where $a^\dagger$ and a are creation and annihilation operators respectively (see Section 2.9 for additional details). The *annihilation a* and *creation* $a^\dagger$ operators are related to the momentum p and position z operators as follows:

$$a = \sqrt{\frac{M\omega}{2\hbar}}\left(z + \frac{jp}{M\omega}\right) \qquad a^\dagger = \sqrt{\frac{M\omega}{2\hbar}}\left(z - \frac{jp}{M\omega}\right). \tag{13.1}$$

The Hamiltonian of a particle in a one-dimensional parabolic potential well $V(z) = M\omega^2 z^2/2$ is given by

$$H = \frac{p^2}{2M} + \frac{M\omega^2 z^2}{2}. \tag{13.2}$$

The energy eigenvalues E_m can be determined from

$$H|m\rangle = \hbar\omega\left(a^\dagger a + \frac{1}{2}\right)|m\rangle = \hbar\omega\left(m + \frac{1}{2}\right)|m\rangle = E_m|m\rangle. \tag{13.3}$$

The time evolution of an arbitrary state is given by

$$|\psi(t)\rangle = \sum_m c_m(0)\mathrm{e}^{-jm\omega t}|m\rangle, \tag{13.4}$$

where $c_n(0)$ is the expansion coefficient for $|\psi(0)\rangle$. Let the logical states be defined as [15]

$$|00\rangle_{\bar{l}} = |0\rangle,\ \ |01\rangle_{\bar{l}} = |2\rangle,\ \ |10\rangle_{\bar{l}} = (|4\rangle + |1\rangle)/\sqrt{2},\ \ |11\rangle_{\bar{l}} = (|4\rangle - |1\rangle)/\sqrt{2}. \tag{13.5}$$

By setting $t = \pi/(\hbar\omega)$, the basis states evolve as follows:

$$|m\rangle \rightarrow \mathrm{e}^{-\frac{j}{\hbar}Ht}|m\rangle = \exp(-j\pi a^\dagger a)|m\rangle = (-1)^m|m\rangle. \tag{13.6}$$

The logic states, on the other hand, evolve as [15]

$$\begin{aligned} &|00\rangle_{\bar{l}} \rightarrow (-1)^0|00\rangle_{\bar{l}} = |00\rangle_{\bar{l}},\ \ |01\rangle_{\bar{l}} \rightarrow (-1)^2|01\rangle_{\bar{l}} = |01\rangle_{\bar{l}} \\ &|10\rangle_{\bar{l}} \rightarrow ((-1)^4|4\rangle + (-1)^1|1\rangle)/\sqrt{2} = (|4\rangle - |1\rangle)/\sqrt{2} = |11\rangle_{\bar{l}}, \\ &|11\rangle_{\bar{l}} \rightarrow ((-1)^4|4\rangle - (-1)^1|1\rangle)/\sqrt{2} = (|4\rangle + |1\rangle)/\sqrt{2} = |10\rangle_{\bar{l}}. \end{aligned} \tag{13.7}$$

By interpreting the first qubit as the control qubit and the second qubit as the target qubit, from (13.7) we conclude that when the control qubit is in the $|1\rangle$ state, the target qubit is flipped, which corresponds to CNOT gate operation.

13.2 NUCLEAR MAGNETIC RESONANCE (NMR) IN QUANTUM INFORMATION PROCESSING

In this section, we describe the use of nuclear magnetic resonance (NMR) in quantum information processing [4–6,15,51]. This approach is important from pedagogical point of view, as its basic concepts are used in ion trap and CQED applications. To describe the basic concepts of NMR and its use in quantum information processing, we follow the description similar to that of Le Bellac [51]. Let us assume that spin-1/2 systems are located in a strong time-invariant magnetic field B_0, of the order of a few teslas. The time-independent Hamiltonian is given by

$$\hat{H} = \begin{bmatrix} \hbar\omega' & 0 \\ 0 & \hbar\omega'' \end{bmatrix}, \tag{13.8}$$

where $\hbar\omega'$ and $\hbar\omega''$ represent the energy levels of the spin-1/2 systems. From the time-evolution equation:

$$j\hbar\frac{dU}{dt} = \hat{H}U, \tag{13.9}$$

since the Hamiltonian is time-invariant, by separation of variables we obtain the following expression for time-evolution operator $U(t,t_0)$:

$$U(t, t_0) = \exp[-j\hat{H}(t - t_0)/\hbar] = \begin{bmatrix} e^{-j\omega'(t-t_0)} & 0 \\ 0 & e^{-j\omega''(t-t_0)} \end{bmatrix}. \tag{13.10}$$

The time evolution of arbitrary state $|\psi(t=0)\rangle = a|0\rangle + b|1\rangle$ is then given by

$$\begin{aligned} |\psi(t)\rangle &= e^{-j\omega' t}a|0\rangle + e^{-j\omega'' t}b|0\rangle = a(t)|0\rangle + b(t)|0\rangle; \\ a(t) &= ae^{-j\omega' t}, \quad b(t) = be^{-j\omega'' t}. \end{aligned} \tag{13.11}$$

Given the arbitrariness of the absolute phase, what really matters is the phase difference, or equivalently the frequency difference $\omega' - \omega'' = \omega_0$, so that the Hamiltonian can be rewritten as

$$\hat{H} = \begin{bmatrix} \hbar\omega_0 & 0 \\ 0 & -\hbar\omega_0 \end{bmatrix}. \tag{13.12}$$

The energy $\hbar\omega_0$ is often referred to as the resonance energy and the frequency ω_0 as the *resonance frequency* (*Larmor frequency*). This Hamiltonian is also applicable to two-level atoms, where ω' represents the ground state and ω'' ($\omega'' > \omega'$) represents the excited state. If a two-level atom is raised to an excited state, it returns spontaneously to the ground state by emitting a photon of energy $\hbar(\omega'' - \omega') = \hbar\omega_0$. On the other hand, if the two-level atom is exposed to a laser beam of frequency $\omega \cong \omega_0$, a *resonance phenomenon* will be observed. That is, the closer ω is to ω_0, the stronger the absorption will be.

Let the proton now be placed in a magnetic field containing both constant and periodic components as follows:

$$\boldsymbol{B} = B_0\hat{z} + B_1(\hat{x}\cos\omega t - \hat{y}\sin\omega t), \tag{13.13}$$

which is the same as in NMR or (nuclear) *magnetic resonance imaging* (MRI). Its spin is associated with the operator $\hbar\boldsymbol{\sigma}/2$, where $\boldsymbol{\sigma} = [X\ Y\ Z]^T$ (X, Y, and Z are Pauli operators). The corresponding *magnetic moment* μ is given by $\mu = \gamma_p \sigma/2$, where γ_p is the *gyromagnetic ratio* to be determined experimentally ($\gamma_p \cong 5.59 q_p \hbar/(2m_p)$, where q_p is the proton charge and m_p is the proton mass). The proton Hamiltonian in this magnetic field can be written as

$$\hat{H} = -\boldsymbol{\mu}\cdot\boldsymbol{B} = -\frac{1}{2}\gamma_p \boldsymbol{\sigma}\cdot\boldsymbol{B} = -\frac{1}{2}\gamma_p [B_0 Z + B_1(X\cos\omega t - Y\sin\omega t)]. \quad (13.14)$$

By substituting: $\gamma_p B_0 = \hbar\omega_0$ and $\gamma_p B_1 = \hbar\omega_1$, we obtain:

$$\hat{H} = -\frac{1}{2}\hbar\omega_0 Z - \frac{1}{2}\hbar\omega_1 (X\cos\omega t - Y\sin\omega t). \quad (13.15)$$

Clearly, the frequency ω_1, known as the *Rabi frequency*, is proportional to B_1. The corresponding matrix representation of the Hamiltonian, based on (13.15), is

$$\hat{H} = -\frac{\hbar}{2}\begin{bmatrix} \omega_0 & \omega_1 e^{j\omega t} \\ \omega_1 e^{-j\omega t} & -\omega_0 \end{bmatrix}. \quad (13.16)$$

For an arbitrary state $|\psi(t=0)\rangle = a|0\rangle + b|1\rangle$, we can use the Hamiltonian (13.16) to solve the evolution equation:

$$j\hbar\frac{d|\psi(t)\rangle}{dt} = \hat{H}|\psi(t)\rangle, \quad |\psi(t)\rangle = \begin{bmatrix} a(t) \\ b(t) \end{bmatrix}. \quad (13.17)$$

After the substitution $a(t) = \tilde{a}(t)e^{j\omega_0 t/2}$, $b(t) = \tilde{b}(t)e^{-j\omega_0 t/2}$, the solution of evolution equation (13.17) is obtained as

$$\tilde{a}(t) = A\cos\left(\frac{\omega_1 t}{2}\right) + B\sin\left(\frac{\omega_1 t}{2}\right), \quad \tilde{b}(t) = jA\sin\left(\frac{\omega_1 t}{2}\right) - jB\cos\left(\frac{\omega_1 t}{2}\right), \quad (13.18)$$

where the constants A and B are dependent on the initial state. For example, if the initial state at $t=0$ was $|0\rangle$, the constants are obtained as $A=1$, $B=0$. At time instance $t = \pi/(2\omega_1)$, the $|0\rangle$ state evolves to the superposition state:

$$|\psi(t)\rangle|_{t=\pi/(2\omega_1)} = \frac{1}{2}\left(e^{j\omega_0 t/2}\left|0\right\rangle + e^{-j\omega_0 t/2}\left|1\right\rangle\right). \quad (13.19)$$

By redefining the $|0\rangle$ and $|1\rangle$ states, we can obtain the conventional superposition state: $(|0\rangle + |1\rangle)/\sqrt{2}$. This transformation is often called a $\pi/2$ pulse transformation. On the other hand, if we allow t to be π/ω_1 the initial state $|0\rangle$ is transformed to the $|1\rangle$ state, and this transformation is known as π pulse (because $\omega_1 t = \pi$). In general, the probability of finding the initial state $|0\rangle$ in the $|1\rangle$ state, denoted as $p_{01}(t)$, is given by

$$p_{01}(t) = \left(\frac{\omega_1}{\Omega}\right)^2 \sin^2\left(\frac{\Omega t}{2}\right), \quad \Omega = \left[(\omega - \omega_0)^2 + \omega_1^2\right]^{1/2}, \quad \delta = \omega - \omega_0. \quad (13.20)$$

These oscillations between states $|0\rangle$ and $|1\rangle$ are known as *Rabi oscillations*. The oscillations have maximum amplitude when detuning δ is zero.

A similar manipulation, which is fundamental in quantum computing applications, can be performed on two-level atoms. In this case, $\hbar\omega_a$ represents the energy difference between excited and ground states, while the Rabi frequency ω_1 is proportional to the electric dipole moment of the atom $\boldsymbol{d}$ and the electric field $\boldsymbol{E}$ of the laser beam, i.e. $\omega_1 \approx \boldsymbol{d}\cdot\boldsymbol{E}$.

The Hamiltonian given by Eq. (13.15) can be rewritten as

$$\hat{H} = \underbrace{-\frac{1}{2}\hbar\omega_0 Z}_{\hat{H}_0} \underbrace{-\frac{1}{2}\hbar\omega_1(\sigma_+ e^{j\omega t} + \sigma_- e^{-j\omega t})}_{\hat{H}_1(t)} = \hat{H}_0 + \hat{H}_1(t), \tag{13.21}$$

$$\sigma_\pm = (X \pm jY)/2,$$

where σ_+ and σ_- denote the *lowering* and *raising* operators respectively (see Chapter 2), defined as $\sigma_\pm^{(i)} = (X_i \pm jY_i)/2;\ \ i = 1,2$. To facilitate the evolution study of state $|\psi(t)\rangle$, we redefine the state in the *rotating reference frame* as follows:

$$|\tilde{\psi}(t)\rangle = e^{-j\omega Zt/2}|\psi(t)\rangle. \tag{13.22}$$

Since

$$|\psi(t)\rangle = e^{-j\hat{H}_0 t/\hbar}|\psi(0)\rangle = e^{-j\omega_0 Zt/2}|\psi(0)\rangle, \tag{13.23}$$

when $\omega = \omega_0$ and $\omega_1 = 0$ we obtain:

$$|\tilde{\psi}(t)\rangle\big|_{\omega=\omega_0,\omega_1=0} = e^{-j\omega Zt/2}|\psi(t)\rangle = e^{-j\omega Zt/2}e^{j\omega_0 Zt/2}|\psi(0)\rangle = |\psi(0)\rangle, \tag{13.24}$$

indicating that, in this reference frame, the state is time-invariant as long as $\omega = \omega_0$ and $\omega_1 = 0$. Because in the reference frame the raising and lowering operators are given by

$$\tilde{\sigma}_\pm = e^{-j\omega Zt/2}\sigma e^{j\omega Zt/2} = e^{\mp j\omega t}\sigma_\pm, \tag{13.25}$$

the Hamiltonian in the reference frame can be related to the Hamiltonian given by (13.21) as

$$\tilde{H}(t) = \frac{1}{2}\hbar\omega_0 Z + e^{-j\omega Zt/2}\hat{H}(t)e^{j\omega Zt/2} = \frac{\hbar}{2}\delta Z - \frac{\hbar}{2}\omega_1 X, \quad \delta = \omega - \omega_0, \tag{13.26}$$

and it is time invariant. Notice that for $\omega = \omega_0$, the evolution operator becomes

$$e^{-j\tilde{H}t/\hbar} = e^{-j\omega_1 Xt/2}, \tag{13.27}$$

and represents the rotation by an angle $-\omega_1 t$ around the x-axis. Therefore, in order to perform rotation around the x-axis for a given angle, the duration of the radio-frequency (RF) pulse needs to be properly adjusted.

We are now concerned with the study of the interaction of two spins within the same molecule. As an example, the first spin can be carried by a proton and the second spin can be carried by a ^{13}C nucleus. As these two spins have different magnetic moments, their resonance $\omega_0^{(i)}$ $(i = 1, 2)$ and Rabi frequencies $\omega_1^{(i)}$ $(i = 1, 2)$ will be different. This type of interaction can be described by the operator $\hbar J_{12} Z_1 Z_2$, where $|J_{12}| \ll \omega_1^{(i)}$ $(i = 1, 2)$. The Hamiltonian can be obtained by generalizing Eq. (13.21) as follows:

$$\begin{aligned}\hat{H}_{12} = &-\frac{1}{2}\hbar\omega_0^{(1)} Z_1 - \frac{1}{2}\hbar\omega_0^{(2)} Z_2 \quad - \frac{1}{2}\hbar\omega_1^{(1)}\left(\sigma_+^{(1)} \mathrm{e}^{j\omega^{(1)}t} + \sigma_-^{(1)} \mathrm{e}^{-j\omega^{(1)}t}\right) \\ &- \frac{1}{2}\hbar\omega_1^{(2)}\left(\sigma_+^{(2)} \mathrm{e}^{j\omega^{(2)}t} + \sigma_-^{(2)} \mathrm{e}^{-j\omega^{(2)}t}\right) + \hbar J_{12} Z_1 Z_2. \end{aligned} \tag{13.28}$$

Given the fact that resonance and Rabi frequencies for two different spins are different, the fields applied to different spins should be properly adjusted as

$$\left|\omega^{(i)} - \omega_0^{(i)}\right| = \left|\delta^{(i)}\right| \ll \omega_1^{(i)}; \; i = 1, 2. \tag{13.29}$$

The state of this two-spin system can be described in the rotating reference frame by generalizing Eq. (13.22) in the following way:

$$\left|\tilde{\psi}_1(t)\right\rangle \otimes \left|\tilde{\psi}_2(t)\right\rangle = \mathrm{e}^{-j\omega^{(1)} Z_1 t/2} \mathrm{e}^{-j\omega^{(2)} Z_2 t/2} |\psi_1(t)\rangle \otimes |\psi_2(t)\rangle. \tag{13.30}$$

In this rotating reference frame, the Hamiltonian is again time invariant:

$$\tilde{H} = \frac{\hbar}{2}\delta^{(1)} Z_1 + \frac{\hbar}{2}\delta^{(2)} Z_2 - \frac{\hbar}{2}\omega_1^{(1)} X_1 - \frac{\hbar}{2}\omega_1^{(2)} X_2 + \hbar J_{12} Z_1 Z_2. \tag{13.31}$$

Our attention now shifts to *quantum logic gates*. The qubits are represented as 1/2-spin systems. To manipulate qubits, it is sufficient to apply an RF field for a suitable time interval, with the frequency of the field being in close proximity to the resonance frequency $\omega_0^{(i)}$, which corresponds to the *i*th qubit to be manipulated. The CNOT gate can be implemented by using $Z_1 Z_2$ interaction between two qubits (1/2-spin systems). This interaction is internal to the system, while in many other implementations this kind of interaction is introduced externally. Since this interaction is internal to the system it is always present, and in certain situations we need to suppress its presence. By using the fact that the $Z_1 Z_2$ opearator is second-order Hermitian, i.e. $(Z_1 Z_2)^2 = (Z_1 Z_2)(Z_1 Z_2) = Z_1^2 \otimes Z_2^2 = I$, the evolution operator can be represented as

$$\mathrm{e}^{-jJZ_1 Z_2 t} = \cos(J_{12} t) I - j \sin(J_{12} t) Z_1 Z_2. \tag{13.32}$$

Equation (13.32) suggests that the action of the evolution operator is in fact the rotation operation; the time needed to perform interaction is typically of the order of milliseconds, about two orders of magnitude greater than the time interval needed to perform rotation on a single qubit (~10 μs), which causes an interoperability problem.

In order to explain how the CNOT gate can be implemented based on NMR, let us introduce the following operator:

$$O_{12}(t) = \mathrm{e}^{jJZ_1 Z_2 t} R_z^{(1)}(\pi/2) R_z^{(2)}(\pi/2), \tag{13.33}$$

where $R_z(\theta)$ is the rotation operator for angle θ around the z-axis, introduced in Chapter 3 as follows:

$$R_z(\theta) = e^{-j\theta Z/2} = \cos\left(\frac{\theta}{2}\right)I - j\sin\left(\frac{\theta}{2}\right)Z, \tag{13.34}$$

which for $\theta = \pi/2$ becomes

$$R_z(\pi/2) = \cos\left(\frac{\pi}{4}\right)I - j\sin\left(\frac{\pi}{4}\right)Z = \frac{1}{\sqrt{2}}(I - jZ).$$

For $t = \pi/(4J)$ (or equivalently $Jt = \pi/4$), the operator given by Eq. (13.33) becomes

$$\begin{aligned}
O_{12}(\pi/(4J_{12})) &= e^{j\pi Z_1 Z_2/4} R_z^{(1)}(\pi/2) R_z^{(2)}(\pi/2) \\
&= \underbrace{[\cos(\pi/4)I + j\sin(\pi/4)Z_1Z_2]}_{\frac{1}{\sqrt{2}}(I + jZ_1Z_2)} \frac{1}{2}(I - jZ_1)(I - jZ_2) \\
&= \frac{1-j}{2^{3/2}}(I + Z_1I_2 + I_1Z_2 - Z_1Z_2) \\
&= \frac{1-j}{2^{3/2}}\left\{\begin{bmatrix}1 & 0\\ 0 & 1\end{bmatrix} \otimes \begin{bmatrix}1 & 0\\ 0 & 1\end{bmatrix} + \begin{bmatrix}1 & 0\\ 0 & -1\end{bmatrix} \otimes \begin{bmatrix}1 & 0\\ 0 & 1\end{bmatrix}\right. \\
&\quad \left. + \begin{bmatrix}1 & 0\\ 0 & 1\end{bmatrix} \otimes \begin{bmatrix}1 & 0\\ 0 & -1\end{bmatrix} - \begin{bmatrix}1 & 0\\ 0 & -1\end{bmatrix} \otimes \begin{bmatrix}1 & 0\\ 0 & -1\end{bmatrix}\right\} \\
&= \frac{1-j}{\sqrt{2}}\begin{bmatrix}1 & 0 & 0 & 0\\ 0 & 1 & 0 & 0\\ 0 & 0 & 1 & 0\\ 0 & 0 & 0 & -1\end{bmatrix} = \frac{1-j}{\sqrt{2}}\begin{bmatrix}I & 0\\ 0 & Z\end{bmatrix} \\
&= \frac{1-j}{\sqrt{2}}C(Z), \quad C(Z) = \begin{bmatrix}I & 0\\ 0 & Z\end{bmatrix},
\end{aligned} \tag{13.35}$$

which is equivalent to the controlled-Z gate ($C(Z)$). The Hadamard gate can be implemented by a rotation of π rad around the axis $(1/\sqrt{2}, 0, 1/\sqrt{2})$ as follows:

$$\begin{aligned}
R_{\hat{n}=(1/\sqrt{2},0,1/\sqrt{2})}(\pi) &= \exp(-j\pi\hat{n}\cdot\sigma/2) = \cos\left(\frac{\pi}{2}\right)I - j\sin\left(\frac{\pi}{2}\right)(n_xX + n_yY + n_zZ) \\
&= -\frac{j}{\sqrt{2}}(X + Z) = -\frac{j}{\sqrt{2}}\begin{bmatrix}1 & 1\\ 1 & -1\end{bmatrix} = -jH, \; H = \frac{1}{\sqrt{2}}\begin{bmatrix}1 & 1\\ 1 & -1\end{bmatrix}.
\end{aligned} \tag{13.36}$$

Further, the phase S gate can be obtained (up to the phase constant) by rotation of $\pi/2$ rad around the z-axis:

$$R_z(\pi/2) = \begin{bmatrix} e^{-j\pi/4} & 0 \\ 0 & e^{j\pi/4} \end{bmatrix} = e^{-j\pi/4}\begin{bmatrix} 1 & 0 \\ 0 & e^{j\pi/2} \end{bmatrix} = e^{-j\pi/4}\begin{bmatrix} 1 & 0 \\ 0 & j \end{bmatrix}$$
$$= e^{-j\pi/4}S, \quad S = \begin{bmatrix} 1 & 0 \\ 0 & j \end{bmatrix}. \tag{13.37}$$

Finally, by rotating by $\pi/4$ rad around the z-axis, the $\pi/8$ gate is obtained:

$$R_z(\pi/2) = \begin{bmatrix} e^{-j\pi/8} & 0 \\ 0 & e^{j\pi/8} \end{bmatrix} = T, \tag{13.38}$$

which completes the implementation of the universal set of gates based on NMR. If it is preferable to use the CNOT gate instead of the controlled-Z gate, two Hadamard gates must be applied on second qubit before and after the $C(Z)$ gate, as shown in Figure 3.6 of Chapter 3. In other words, we perform the following transformation of the $C(Z)$ gate:

$$(I\otimes H_2)C(Z)(I\otimes H_2) = \begin{bmatrix} H & 0 \\ 0 & H \end{bmatrix}\begin{bmatrix} I & 0 \\ 0 & Z \end{bmatrix}\begin{bmatrix} H & 0 \\ 0 & H \end{bmatrix} = \begin{bmatrix} H^2 & 0 \\ 0 & HZH \end{bmatrix} = \begin{bmatrix} I & 0 \\ 0 & X \end{bmatrix}$$
$$= \text{CNOT}. \tag{13.39}$$

Based on the discussion above, the overall implementation of the CNOT gate based on NMR is shown in Figure 13.1. Two molecules, chloroform for two-qubit operations and perfluorobutadienyl iron complex for seven-qubit operations [3], are shown in Figure 13.2. The atoms used in quantum computing are encircled. For additional examples, the interested reader is referred to Refs [4–6] (and references therein). The authors in Ref. [3] have used the perfluorobutadienyl iron complex (see Figure 13.2b) to implement the Shor factorization algorithm in factorizing integer 15. Since b in b^x mod N can take any value from {2,4,7,8,11,13,14}, the largest period is $r = 4$ for $b = 2, 7, 8$, and 13. In order to see two periods, x should take values 0, 1, ..., $2^3 - 1$, while the

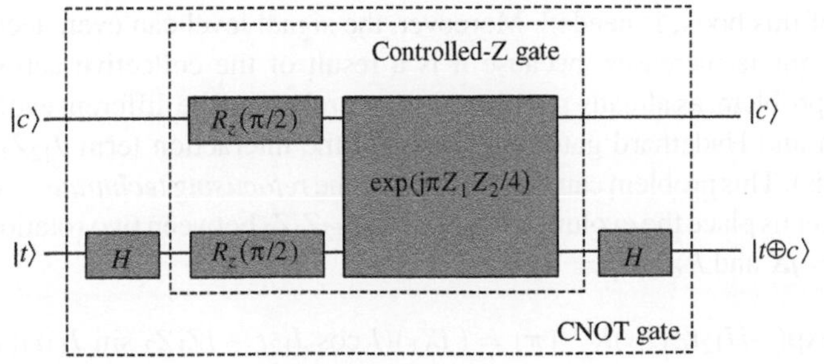

FIGURE 13.1

NMR-based controlled-Z gate and CNOT gate implementations.

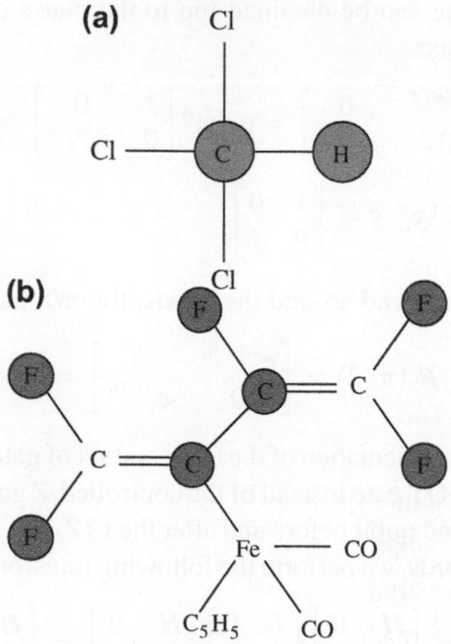

FIGURE 13.2

NMR-based quantum computing examples: (a) chloroform-based two-qubit example; (b) perfluorobutadienyl iron complex for seven-qubit operations.

corresponding $f(x)$ values are 0, 1,..., $2^4 - 1$. Therefore, a three-qubit input register and a four-qubit output quantum register are needed, indicating that seven-bit NMR-based molecules are sufficient. Although recently more than 10-qubit operations based on NMR have been demonstrated, there are many challenges to be solved. For example, the synthesis of a molecule with as many distinguishable sites as the number of qubits and the ability to select the appropriate frequencies to act on different qubits, for any meaningful computation (of the order of thousands of qubits) would be quite challenging. Further, NMR does not employ individual quantum systems (objects), but a collection of $>10^{18}$ active molecules diluted in a solvent. Therefore, the resulting signal is a result of collective action. Very complicated initialization, which is beyond the scope of this book, is needed. Moreover, the signal level can even decrease as the number of qubits increases, because it is a result of the collective action. Another interesting problem, as already mentioned above, relates to the different evolution times for rotation and Hadamard gates (~10 μs) and the interaction term $J_{12}Z_1Z_2$ (several milliseconds). This problem can be solved using the *refocusing technique*, used in NMR and MRI. Let us place the evolution term due to $J_{12}Z_1Z_2$ between two rotations of spin 1 $R_x^{(1)}(-\pi) = jX$ and $R_x^{(1)}(\pi) = -jX$ as follows:

$$\begin{aligned} R_x^{(1)}(-\pi)\exp(-jJ_{12}tZ_1Z_2)R_x^{(1)}(\pi) &= (jX_1)(I\cos J_{12}t - jZ_1Z_2\sin J_{12}t)(-jX_1) \\ &= I\cos J_{12}t + jZ_1Z_2\sin J_{12}t = \exp(jJ_{12}tZ_1Z_2). \end{aligned} \tag{13.40}$$

If two spins have evolved during a time interval t as $\exp(-jJ_{12}tZ_1Z_2)$, the overall result will be

$$\left[R_x^{(1)}(-\pi)\exp(-jJ_{12}tZ_1Z_2)R_x^{(1)}(\pi)\right]\exp(-jJ_{12}tZ_1Z_2)$$
$$= \exp(jJ_{12}tZ_1Z_2)\exp(-jJ_{12}tZ_1Z_2) = I, \qquad (13.41)$$

indicating that it is possible to cancel out the evolution of qubits different from those involved in the CNOT gate. Therefore, the purpose of refocusing methods is to effectively turn off undesired couplings, for a desired interval of time. For additional examples the interested reader is referred to Ref. [4].

Before concluding this section, it is interesting to describe how to determine the coupling coefficients in a Hamiltonian. The simplest way would be to observe the NMR spectrum and determine the coefficients by reading off the positions of peaks (on a frequency scale). For example, the three ^{13}C nuclei of alanine in 9.4 T are governed by the following Hamiltonian [4]:

$$\tilde{H}_{\text{alanine}} = \pi\left[10^8(Z_1 + Z_2 + Z_3) - 12580Z_2 + 3440Z_3\right] + \frac{\pi}{2}(53Z_1Z_2 + 38Z_1Z_3 + 1.2Z_2Z_3), \qquad (13.42)$$

where the frequencies are given in Hz.

13.3 TRAPPED IONS IN QUANTUM INFORMATION PROCESSING

In an ion-trap-like quantum computer [7–9,15,49,51], the qubits are represented by the internal electronic states of single alkali-like ions such as ions from group II of the periodical system of elements ($^{40}Ca^+$ or $^9Be^+$). Similarly to the previous section, to describe the concepts of trapped ions and corresponding quantum gates implementation, we follow interpretation due to Le Bellac [51]. The two states of a qubit are the ground state of an ion $|g\rangle \equiv |0\rangle$ and an excited state $|e\rangle \equiv |1\rangle$ with a very long lifetime (it could be ~1 s). The excited state is either a metastable state or a hyperfine state. For example, a $^{40}Ca^+$ ion with $S_{1/2}$ state as ground state and $D_{5/2}$ metastable state as the excited state (with a lifetime of the order of 1 s) can be used as a qubit in quantum computing applications. The individual qubits are subjected to a laser beam at a given frequency. A similar model to the previous section can be used, where now $\hbar\omega_0$ represents the energy difference between excited and ground states, while Rabi frequency ω_1 is proportional to the electric dipole moment of the ion $\boldsymbol{d}$ and the electric field $\boldsymbol{E}$ of the laser beam, in other words $\omega_1 \approx \boldsymbol{d} \cdot \boldsymbol{E}$. The transition between excited and ground state in a $^{40}Ca^+$ ion corresponds to a wavelength of 729 nm. The ions (representing the qubits) are placed in a linear Paul trap (named after their inventor), as shown in Figure 13.3. The ion trap is formed by using

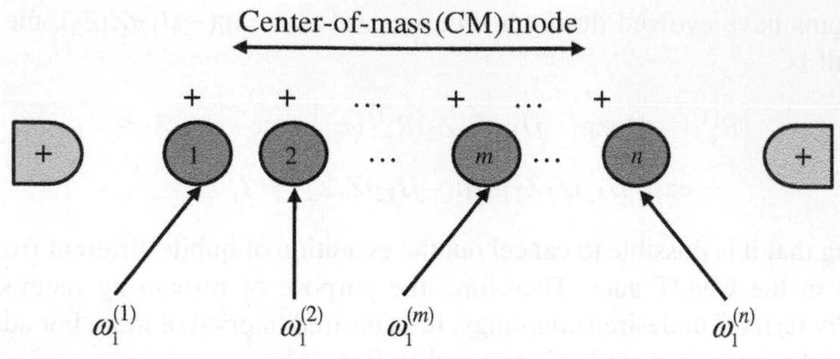

FIGURE 13.3

Operating principle of ion-trap-based quantum register with N qubits.

four parallel electrodes of radius comparable to the distance from the electrodes to the common axis. The sinusoidal RF voltage signal is applied to two opposite electrodes, while the other two electrodes are grounded. In order to confine ions in the axial direction, two positive static potentials are applied from opposite sides of the ion chain, as illustrated in Figure 13.3. The interaction among ions is performed by the *vibrating motion* of the ions, which represents the *external degree of freedom* of the ions. The chain of ions is kept at sufficiently low temperature so that ions can be trapped in the transversal direction. The overall result is the location of the ions in a harmonic potential:

$$V(x,y,z) = \frac{1}{2}M_{\mathrm{I}}\left(\omega_x^2 + \omega_y^2 + \omega_z^2\right), \tag{13.43}$$

where M_{I} is the ion mass and (x,y,z) denotes the position of the ion in the ion trap. In practice, the trap frequencies (of the order of MHz) satisfy the following conditions:

$$\omega_x^2 \approx \omega_y^2 \gg \omega_z^2, \tag{13.44}$$

so that the ion moves in the direction of the z-axis in a potential and we can write:

$$V(z) \cong \frac{1}{2}M_I\omega_z^2. \tag{13.45}$$

The corresponding Hamiltonian is given by

$$\hat{H} = \frac{p_z^2}{2M_I} + \frac{1}{2}M_I\omega_z^2, \tag{13.46}$$

where $p_z^2/(2M_I)$ is the kinetic energy term, with coordinate z and momentum p_z components being Hermitian and satisfying the commutation relation $[z,p_z] = j\hbar I$. By using Eq. (13.1), the Hamiltonian can be expressed in terms of creation $a^\dagger$ and annihilation a operators as follows:

$$\hat{H} = \hbar\omega_z(a^\dagger a + 1/2). \tag{13.47}$$

The eigenvalues of the Hamiltonian are $\hbar\omega(m+1/2)=E_m$ $(m=0,1,2,\ldots)$, as given by Eq. (13.3), and the corresponding eigenkets are denoted by $|m\rangle$. The action of a creation (annihilation) operator is to increase (decrease) the energy level (see Chapter 2 for more details):

$$a^\dagger|m\rangle=\sqrt{m+1}|m+1\rangle,\quad a|m\rangle=\sqrt{m}|m-1\rangle;\quad a|0\rangle=0. \tag{13.48}$$

The energy corresponding to the $|0\rangle$ state is nonzero, i.e. $E_0=\hbar\omega_z/2$, and the integer m is called the vibrational quantum number of the ion trap. If we use the Heisenberg inequality $\Delta z\Delta p_z\sim\hbar/2$ (see Chapter 2) and replace the coordinate z and momentum p_z operators by corresponding standard deviations Δz and Δp_z we can provide a heuristic interpretation as follows:

$$E\sim\frac{(\Delta p_z)^2}{2M_\mathrm{I}}+\frac{M_\mathrm{I}\omega_z^2}{2}(\Delta z)^2\overset{\Delta p_z\sim\hbar/(2\Delta z)}{\sim}\frac{\hbar^2}{8M_\mathrm{I}(\Delta z)^2}+\frac{M_\mathrm{I}\omega_z^2}{2}(\Delta z)^2. \tag{13.49}$$

By differentiating E with respect to Δz and setting the derivative to zero, we obtain the minimum energy as

$$E_\mathrm{opt}=\hbar\omega_z/2,\qquad(\Delta z)^2=\hbar/(2M_I\omega_z). \tag{13.50}$$

Therefore, the minimum energy is the best compromise between kinetic and potential energies, while the region where the ion is most probably going to be located is determined by $\Delta z=\sqrt{\hbar/(2M_I\omega_z)}$. This mode, known as normal mode, corresponds to the case when all ions oscillate together as a rigid body with frequency ω_z, i.e. it is a center-of-mass (CM) mode.

Let us now model the ion as a two-level system trapped in the potential $V(z)\cong M_I\omega_z^2/2$, and affected by an oscillating electric field of the form:

$$\mathbf{E}=2E_1\hat{x}\cos(\omega t-kz-\phi)=E_1\hat{x}\left[\mathrm{e}^{j(\omega t-\phi)}\mathrm{e}^{-jkz}+\mathrm{e}^{-j(\omega t-\phi)}\mathrm{e}^{jkz}\right]. \tag{13.51}$$

The corresponding Hamiltonian will have three terms: the first term $\hat{H}_0$ corresponds to the absence of an oscillating field ($E_1=0$), and the other two terms originate from the expansion $\exp(\pm jkz)\approx I\pm jkz$, which is valid when

$$\eta=k\Delta z_0=k\sqrt{\hbar/(2M_I\omega_z)}\ll 1,$$

where η is the so-called *Lamb–Dicke parameter.* It is common practice to redefine the zero of the vibrational energy such that the ground state has zero energy. The Hamiltonian operator in the absence of an electric field can be written as

$$\hat{H}_0=-\frac{\hbar\omega_0}{2}Z+\hbar\omega_z a^\dagger a. \tag{13.52}$$

The internal ion states are now the $|0\rangle$ state of energy $-\hbar\omega_0/2$ and the $|1\rangle$ state of energy $\hbar\omega_0/2$. Similarly to NMR-based quantum computation, we can use the rotating reference frame approach to facilitate our study. Given an operator A, the corresponding operator frame in the rotating reference frame will be

$$A(t) = e^{j\hat{H}_0 t/\hbar} A e^{-j\hat{H}_0 t/\hbar}, \tag{13.53}$$

which is essentially the Heisenberg picture representation (see Chapter 2 for more details). Following a similar procedure as in the previous section we can relate the annihilation, creation, lowering, and raising operators in the original and rotating frames as follows:

$$\tilde{a}(t) = a e^{-j\omega_z t}, \ \tilde{a}^\dagger(t) = a^\dagger e^{j\omega_z t}, \ \tilde{\sigma}_-(t) = \sigma_- e^{j\omega_0 t}, \ \tilde{\sigma}_+(t) = \sigma_+ e^{-j\omega_0 t}. \tag{13.54}$$

Based on Eq. (13.51) and the previous section, the interaction Hamiltonian can be represented as

$$\hat{H}_{\text{interaction}} = -\frac{\hbar\omega_1}{2} \underbrace{(\sigma_- + \sigma_+)}_{X} \underbrace{\left[e^{j(\omega t - \phi)} e^{-jkz} + e^{-j(\omega t - \phi)} e^{jkz}\right]}_{2\cos(\omega t - kz - \phi)}, \tag{13.55}$$

where ω_1 is the Rabi frequency, as indicated above. The approximation $\exp(\pm jkz) \approx I \pm jkz$ is now employed. The first term of the expansion (the identity operator term) yields the following contribution for the interaction Hamiltonian:

$$\hat{H}_1 = -\frac{\hbar\omega_1}{2} (\sigma_- + \sigma_+) \left[e^{j(\omega t - \phi)} + e^{-j(\omega t - \phi)}\right], \tag{13.56}$$

which in the rotating reference frame becomes

$$\tilde{H}_1 = -\frac{\hbar\omega_1}{2} (\sigma_- e^{j\omega_0 t} + \sigma_+ e^{j\omega_0 t}) \left[e^{j(\omega t - \phi)} + e^{-j(\omega t - \phi)}\right]. \tag{13.57}$$

By now employing the *rotating-wave approximation*, in which the terms of the form $\exp[\pm j(\omega + \omega_0)t]$ can be neglected as they average to zero because of their rapid oscillation in comparison with other terms, we can approximate (13.57) as

$$\begin{aligned} \tilde{H}_1 &\simeq -\frac{\hbar\omega_1}{2} [\sigma_- e^{-j(\overbrace{(\omega - \omega_0)}^{\delta} t - \phi)} + \sigma_+ e^{j(\overbrace{(\omega - \omega_0)}^{\delta} t - \phi)}] \\ &= -\frac{\hbar\omega_1}{2} [\sigma_- e^{-j(\delta t - \phi)} + \sigma_+ e^{j(\delta t - \phi)}], \ \delta = \omega - \omega_0. \end{aligned} \tag{13.58}$$

By comparison with (13.26) we conclude that the form is similar and therefore similar methods can be used to manipulate the qubits. At the resonance ($\delta = 0$), the time-evolution operator is the rotation operator:

$$e^{-j\theta(\sigma_- e^{j\phi} + \sigma_+ e^{-j\phi})} = e^{-j\theta(X\cos\phi + Y\sin\phi)/2}, \ \theta = -\omega_1 t, \tag{13.59}$$

which represents rotation around the axis: $\hat{\boldsymbol{n}} = (\cos\phi, \sin\phi, 0)$.

The second term in the expansion ($\pm jkz$) yields the following interaction Hamiltonian (based on Eq. (13.55)):

$$\hat{H}_2 \simeq -\frac{\hbar\omega_1}{2}(\sigma_- + \sigma_+)\left[-je^{j(\omega t-\phi)}kz + je^{-j(\omega t-\phi)}kz\right] \overset{z=\sqrt{\hbar/(2M_I\omega_z)}\,(a^\dagger+a)}{\simeq} \frac{j\eta\hbar\omega_1}{2}(\sigma_- + \sigma_+)(a^\dagger + a)\left[e^{j(\omega t-\phi)} - e^{-j(\omega t-\phi)}\right]. \quad (13.60)$$

In the rotating reference frame, the corresponding Hamiltonian term becomes

$$\begin{aligned}\tilde{H}_2 &\simeq \frac{j\eta\hbar\omega_1}{2}\left(\sigma_- e^{j\omega_0 t}a^\dagger e^{j\omega_z t} + \sigma_- e^{j\omega_0 t}ae^{-j\omega_z t} + \sigma_+ e^{-j\omega_0 t}a^\dagger e^{j\omega_z t}\right.\\ &\quad \left.+\sigma_+ e^{-j\omega_0 t}ae^{-j\omega_z t}\right)\left[e^{j(\omega t-\phi)} - e^{-j(\omega t-\phi)}\right]\\ &= \frac{j\eta\hbar\omega_1}{2}\left(\sigma_- a^\dagger e^{j(\omega_0+\omega_z)t} + \sigma_- ae^{j(\omega_0-\omega_z)t} + \sigma_+ a^\dagger e^{-j(\omega_0-\omega_z)t}\right.\\ &\quad \left.+\sigma_+ ae^{-j(\omega_0+\omega_z)t}\right)\left[e^{j(\omega t-\phi)} - e^{-j(\omega t-\phi)}\right]. \end{aligned} \quad (13.61)$$

By choosing the laser operating frequency to be $\omega = \omega_0 + \omega_z$, we select the upper ("blue") side-band frequency, while the lower side band can be neglected according to the rotating-wave argument above, so that the interaction Hamiltonian becomes

$$\tilde{H}_2 \simeq \frac{j\eta\hbar\omega_1}{2}(-e^{j\phi}\sigma_- a^\dagger + e^{-j\phi}\sigma_+ a). \quad (13.62)$$

The state of the single ion in the trap can be represented as $|n,m\rangle$, where $n = 0,1$ is related to the spin (internal) state and $m = 0,1$ is related to the vibrational state of the harmonic oscillator. The action of Hamiltonian terms in (13.62) is to introduce transitions from $|00\rangle$ to $|11\rangle$, and vice versa (see Figure 13.4), as follows:

$$\sigma_- a^\dagger|00\rangle = |11\rangle, \quad \sigma_+ a|11\rangle = |00\rangle. \quad (13.63)$$

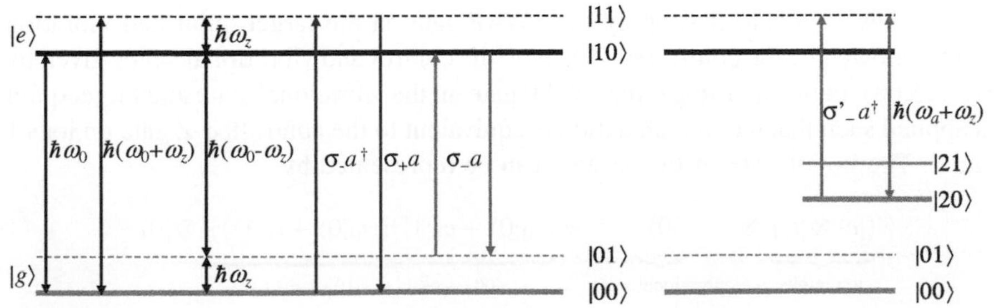

FIGURE 13.4

Energy levels and transitions describing coupling of spin and vibrational degrees of freedom.

On the other hand, by choosing the laser operating frequency to be $\omega = \omega_0 - \omega_z$, we select the lower ("red") side-band frequency, so that the interaction Hamiltonian (13.61) becomes

$$\tilde{H}_2 \simeq \frac{j\eta\hbar\omega_1}{2}\left(-\mathrm{e}^{j\phi}\sigma_- a + \mathrm{e}^{-j\phi}\sigma_+ a^\dagger\right). \tag{13.64}$$

The action of operator terms in (13.64) is to introduce the transformation $|01\rangle \leftrightarrow |10\rangle$ (see Figure 13.4):

$$\sigma_- a|01\rangle = |10\rangle, \quad \sigma_+ a^\dagger|10\rangle = |01\rangle. \tag{13.65}$$

The four basis states for quantum computation are $|00\rangle$, $|01\rangle$, $|10\rangle$, and $|11\rangle$. In order to facilitate the construction of two-qubit gates, for spin we will introduce an auxiliary state, say $|2\rangle$, which is illustrated in Figure 13.4. A laser tuned to the frequency $\omega = \omega_a + \omega_z$ stimulates the transitions $|20\rangle \leftrightarrow |11\rangle$:

$$\sigma'_- a^\dagger|20\rangle = |11\rangle, \quad \sigma'_+ a|11\rangle = |20\rangle, \tag{13.66}$$

and the corresponding Hamiltonian is given by

$$\tilde{H}_{\text{auxilary}} \simeq \frac{j\eta\hbar\omega'_1}{2}\left(-\mathrm{e}^{-j\phi}\sigma'_- a^\dagger + \mathrm{e}^{j\phi}\sigma'_+ a\right). \tag{13.67}$$

The laser beam is applied for a duration to introduce the $R_x(2\pi)$ rotation, which corresponds to the mapping $|11\rangle \rightarrow -|11\rangle$ as $R_x(2\pi) = \cos\pi I - j\sin\pi X = -1$. Since the other states are not affected, the matrix representation of this operation is

$$\begin{bmatrix} I & 0 \\ 0 & Z \end{bmatrix} = C(Z), \tag{13.68}$$

which implements the controlled-Z gate. However, given the fact that information is imposed on spin states and not on vibrational states, this type of gate does not appear to be useful. On the other hand, if this gate is combined with a SWAP gate, the vibration states can be used as intermediate states, and the controlled-Z gate on two ion qubits can be implemented. Let ion 1 serve as control qubit and ion 2 as target qubit, while the vibrational state used is an external state.

The key idea is to first apply the SWAP gate on the target qubit and vibration state, followed by the controlled-Z gate on the control and vibrational states given by Eq. (13.68). In the final stage, the SWAP gate on the vibrational state and target qubit is applied such that the overall action is equivalent to the controlled-Z gate on ions 1 and 2. The initial state of this system can be represented by

$$\begin{aligned} \underbrace{(|c\rangle \otimes |t\rangle)}_{\text{ion qubits}} \otimes \underbrace{|0\rangle}_{\text{vibrational state}} &= \underbrace{(c_0|0\rangle + c_1|1\rangle)((t_0|0\rangle + t_1|1\rangle))}_{a|00\rangle + b|01\rangle + c|10\rangle + d|11\rangle} \otimes |0\rangle \\ &= a|000\rangle + b|010\rangle + c|100\rangle + d|110\rangle; \\ a = c_0 t_0,\ b &= c_0 t_1,\ c = c_1 t_0,\ d = c_1 t_1. \end{aligned} \tag{13.69}$$

By applying a SWAP gate on the target and vibrational state, we obtain:

$$\begin{aligned}&\mathrm{SWAP}^{(23)}(a|000\rangle + b|010\rangle + c|100\rangle + d|110\rangle)\\&= a|000\rangle + b|001\rangle + c|100\rangle + d|101\rangle,\end{aligned} \tag{13.70}$$

where the superscript is used to denote the swapping of qubits 2 and 3. In the second stage, we apply the $C(Z)$ gate, with qubit 1 serving as control and vibrational qubit 3 serving as target, and obtain:

$$\begin{aligned}&\mathrm{C}^{(13)}(\mathrm{Z})(a|000\rangle + b|001\rangle + c|100\rangle + d|101\rangle)\\&= a|000\rangle + b|001\rangle + c|100\rangle - d|101\rangle.\end{aligned} \tag{13.71}$$

In the third stage, the SWAP gate on the second and third qubits is applied, leading to

$$\begin{aligned}&\mathrm{SWAP}^{(23)}(a|000\rangle + b|001\rangle + c|100\rangle - d|101\rangle)\\&= a|000\rangle + b|010\rangle + c|100\rangle - d|110\rangle = (a|00\rangle + b|01\rangle + c|10\rangle - d|11\rangle) \otimes |0\rangle,\end{aligned} \tag{13.72}$$

and the overall action of ion qubits 1 and 2 is that of a controlled-Z gate. The CNOT gate can be implemented based on Eq. (13.39), while the Hadamard gate can be implemented by rotation of π rad around the axis $(1/\sqrt{2}, 0, 1/\sqrt{2})$, as given by Eq. (13.36). What remains is to describe how to implement the SWAP gate on two ions in the trap. By tuning the laser to the lower (red) side-band frequency ($\omega = \omega_0 - \omega_z$), allowing the duration of the laser beam to introduce a rotation of π rad, and by setting ϕ in (13.51) to $-\pi/2$, the overall action can be represented in matrix form as follows:

$$\begin{bmatrix} 1 & 0 & 0 & 0 \\ 0 & 0 & -1 & 0 \\ 0 & 1 & 0 & 0 \\ 0 & 0 & 0 & 1 \end{bmatrix} = U'_{\mathrm{SWAP}}, \tag{13.73}$$

which is the same as that of SWAP gate, except for the sign. By redefining the state of the target qubit as $|1\rangle \to -|1\rangle$, the conventional SWAP gate is obtained.

The ion-trap-based N-qubit register has already been shown in Figure 13.3. The potential energy of this register is given by

$$V(z_1, \ldots, z_n) = \frac{M_I \omega_z^2}{2} \sum_{n=1}^{N} z_n^2 + \frac{q^2}{4\pi\varepsilon_0} \sum_{m \neq n} \frac{1}{|z_n - z_m|}, \tag{13.74}$$

under the assumption that the ion trap is linear. To ensure that the ion trap is confined in the transversal direction, the temperature must be kept sufficiently low so that ions are in the vibrational state with $m = 0$. If $k_B T \geq \hbar\omega_z$ (k_B is the Boltzmann constant),

$m \neq 0$ and the ions must be cooled down, which can be achieved by *Doppler cooling*. That is, the ion is placed between two laser beams directed in opposite directions. When the ion moves in a direction opposite to the direction of one of the beams, it experiences more energetic photons due to the Doppler effect, the transition comes closer to the resonance, and absorption becomes more pronounced. On the other hand, when it moves in the opposite direction it experiences the same effect, which leads to the slowing down of ions, until temperature $T \simeq \hbar 2\pi\Delta\nu/k_B$, where $\Delta\nu$ is the laser linewidth. The lowest mode, CM mode, corresponding to ω_z, refers to the motion of ensembles of ions. The first excited mode, known as the *breathing* mode, corresponds to ions oscillating at frequency $\sqrt{3}\omega_z$ with amplitude proportional to the distance from the axial axis. That is, by setting $z_i = z_0 + u_i$ and by using a Taylor expansion up to two terms, Eq. (13.74) for $N=2$ can be written as

$$\begin{aligned} V(z_1, z_2) &\simeq \frac{M_I\omega_z^2}{2}\left(2z_0^2 + 2z_0(u_1 - u_2) + u_1^2 + u_2^2\right) \\ &+ \frac{q^2}{4\pi\varepsilon_0 z_0}\left[1 - \frac{u_1 - u_2}{2z_0} + \frac{(u_1 - u_2)^2}{4z_0^2}\right]. \end{aligned} \tag{13.75}$$

The equilibrium is obtained by setting the linear terms to zero:

$$M_I\omega_z^2 z_0 - \frac{q^2}{4\pi\varepsilon_0 z_0}\frac{1}{2z_0} = 0, \tag{13.76}$$

which yields:

$$z_0 = 2^{-1/3} l, \quad l = \left(\frac{q^2}{4\pi\varepsilon_0 M_I \omega_z^2}\right)^{1/3}. \tag{13.77}$$

The normal modes are obtained by studying the quadratic terms, which leads to the potential:

$$V_2 \simeq \frac{M_I\omega_z^2}{2}\left(u_1^2 + u_2^2\right) + \frac{q^2}{4\pi\varepsilon_0 z_0}\frac{(u_1 - u_2)^2}{4z_0^2}. \tag{13.78}$$

The corresponding equations of motion are given by

$$M_I\ddot{u}_1 = M_I\omega_z^2 u_1 + \frac{q^2}{4\pi\varepsilon_0 z_0}\frac{u_1 - u_2}{2z_0^2}, \quad M_I\ddot{u}_2 = M_I\omega_z^2 u_2 - \frac{q^2}{4\pi\varepsilon_0 z_0}\frac{u_1 - u_2}{2z_0^2}. \tag{13.79}$$

The center of mass mode, $(u_1 + u_2)/2$, oscillates at frequency ω_z as

$$\ddot{u}_1 + \ddot{u}_2 = -\omega_z^2(u_1 + u_2). \tag{13.80}$$

On the other hand, the breathing mode, $u_1 - u_2$, oscillates at frequency $\sqrt{3}\omega_z$ as

$$\ddot{u}_1 - \ddot{u}_2 = -3\omega_z^2(u_1 - u_2). \tag{13.81}$$

One very important problem is how to accurately illuminate a single ion using a laser beam. Currently, acousto-optic modulators (AOMs) [10,11] are used, driven by appropriate RF signals, to perform controlled-Z, SWAP, and Hadamard gate operations. The AOMs are unfortunately slow devices and are highly temperature sensitive. Moreover, the extinction ratio of AOMs is low and power consumption is high. A possible solution is to use an electro-optical switch such as an optical cross-point switch (OXS) implemented by means of active vertical coupler (AVC) structures [12–14]. The use of this kind of device will provide ample bandwidth (not the main problem here) with high reliability. Moreover, this approach is able to significantly reduce the overall power consumption. This is possible as power is consumed only when the photonic switching cell is on. The AVC-based switching technique is a low loss technique and, more importantly, insensitive to temperature when compared to similar devices such as AOM-based devices. Finally, the extinction ratio is significantly better than that of AOMs.

13.4 PHOTONIC QUANTUM IMPLEMENTATIONS

Photonic technologies can be used in many different ways to perform quantum computation, including [15–17]: (i) the cavity superposition of zero or one photon $|\psi\rangle = c_0|0\rangle + c_1|1\rangle$; (ii) the photon can be located in the first cavity $|01\rangle$ or in the second cavity $|10\rangle$ (dual-rail representation), while the qubit representation is given by $|\psi\rangle = c_0|01\rangle + c_1|10\rangle$; and (iii) spin-angular momentum can be used to carry quantum information, while the qubit is represented as $|\psi\rangle = c_0|0\rangle + c_1|1\rangle$, where the logical "0" is represented by a horizontal (H) photon $|H\rangle \equiv |0\rangle = (1\ 0)^{\mathrm{T}}$ and the logical "1" is represented by a vertical (V) photon $|V\rangle \equiv |1\rangle = (0\ 1)^{\mathrm{T}}$. The bulky optics implementation [15] is first described not because of its practicality, but the simplicity of its description.

13.4.1 Bulky Optics Implementation

The laser output state is known as the *coherent state* and represents the right eigenket of the *annihilation operator a* (the operator that decreases the number of photons by one):

$$a|\alpha\rangle = \alpha|\alpha\rangle, \tag{13.82}$$

where α is the complex eigenvalue. The coherent state ket $|\alpha\rangle$ can be represented in terms of orthonormal eigenkets $|n\rangle$ (the number or Fock state) of the number operator $a^{\dagger}a$ as follows (see Chapter 2):

$$|\alpha\rangle = \exp\left[-|\alpha|^2/2\right]\sum_{n=0}^{+\infty}(n!)^{-1/2}\alpha|n\rangle, \tag{13.83}$$

and the mean energy $\langle\alpha|n|\alpha\rangle = |\alpha|^2$ has a Poisson distribution. When the laser beam is sufficiently attenuated, the coherent state becomes weak. By properly choosing the eigenvalue α, the weak coherent state behaves as a single-photon state with a high probability. For example, by setting $\alpha = 0.1$, the weak coherent state $|0.1\rangle = \sqrt{0.90}|0\rangle + \sqrt{0.09}|1\rangle + \sqrt{0.002}|2\rangle + \ldots$ behaves as a single-photon state (with high probability). Moreover, the implementation of single-photon sources is an active research topic (see Refs [18–21] and references therein).

The spontaneous parametric down-conversion (SPDC) and Kerr effect are now briefly described as they are key effects in the implementation of optical quantum computers. By sending photons at frequency ω_p (pump signal) into a nonlinear optical medium, say a KH_2PO_4 or BBO (β-barium borate) crystal, we generate photon pairs at frequencies ω_s (signal) and ω_i (idler), satisfying the energy conservation principle $\hbar(\omega_s + \omega_i) = \hbar\omega_p$, and momentum conservation principle $\boldsymbol{k}_s + \boldsymbol{k}_i = \boldsymbol{k}_p$, which is illustrated in Figure 13.5. Namely, a nonlinear crystal is used to split photons into pairs of photons that satisfy the law of conservation of energy, have combined energies and momenta equal to the energy and momentum of the original photon, phase matched in the frequency domain, and have correlated polarizations. If the photons have the same polarization this is called Type I correlation, otherwise they have perpendicular polarizations and belong to Type II. As the SPDC is stimulated by random vacuum fluctuations, the photon pairs are created at random time points. The output of a Type I down-converter is known as a squeezed vacuum and it contains only even numbers of photons. The output of a Type II down-converter is a two-mode squeezed vacuum. Modern methods are based on periodically poled $LiNbO_3$ (PPLN) and highly nonlinear fiber (HNLF) [22,23]. PPLN and HNLF are also used in optical phase conjugation and parametric amplifiers. PPLN employs parametric difference frequency generation (DFG), while HLNF employs the four-wave mixing (FWM) effect [24]. FWM occurs in HNLFs during the propagation of composite optical signals, such as the WDM signal. The three optical signals with different carrier frequencies f_i, f_j, and f_k ($i,j,k = 1,\ldots,M$) interact and generate a new optical signal at frequency $f_{ijk} = f_i + f_j - f_k$, providing that the phase-matching condition is satisfied, $\beta_{ijk} = \beta_i + \beta_j - \beta_k$, where β_m is the propagation constant. The phase-matching condition is in fact a requirement of momentum conservation. The FWM process can be considered as the annihilation of two photons with energies $\hbar\omega_i$ and $\hbar\omega_j$, and the generation of two new photons with energies $\hbar\omega_k$ and $\hbar\omega_{ijk}$. In the FWM process the indices i and j are not necessarily

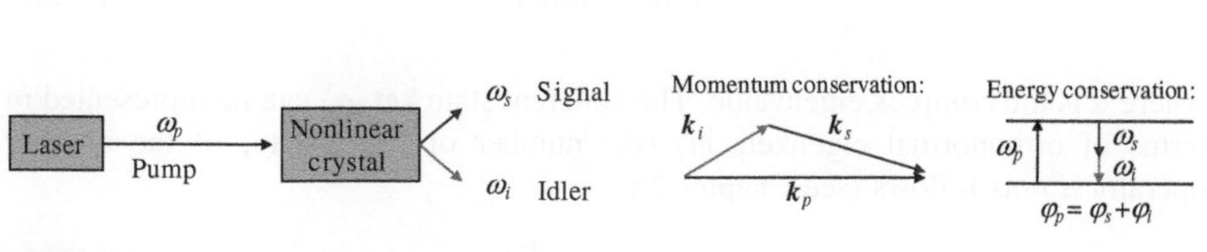

FIGURE 13.5

Spontaneous parametric down-conversion process.

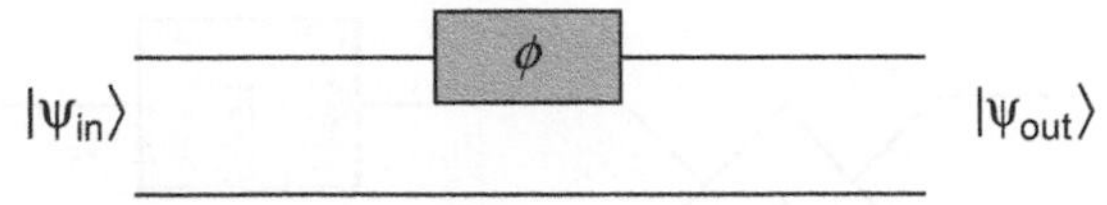

FIGURE 13.6

The equivalent scheme of the phase shifter.

distinct, which indicates that only two modes can interact to create the new one; this case is known as a *degenerate*.

For sufficiently high intensity of light, the index of reflection is not only a function of frequency (responsible for dispersion), but also a function of the intensity of light traveling through nonlinear media: $n(\omega,I) = n(\omega) + n_2 I$, where n_2 is the Kerr coefficient. The phase shift by a nonlinear waveguide of length L is determined by the complex factor $\exp(jn_2 IL\omega/c_0)$ (c_0 is the speed of the light in a vacuum).

Regarding quantum level photodetectors, avalanche photodiodes (APDs) and photomultiplier tubes can be used. Moreover, single photodetector development is an active research topic (see Refs [25–28] and references therein).

Regarding the implementation of photonic quantum gates, the most commonly used devices to manipulate the photon states are: (i) mirrors, which are used to change the direction of propagation; (ii) phase shifters, which are used to introduce a given phase shift; and (iii) beam splitters, which are used to implement various quantum gates. The *phase shifter* is a slab of transparent medium, say borosilicate glass, with index of refraction n being higher than that of air, and it is used to perform the following operation:

$$|\psi_{out}\rangle = \begin{bmatrix} e^{j\phi} & 0 \\ 0 & 1 \end{bmatrix} |\psi_{in}\rangle; \quad \phi = kL, \; k = n\omega/c. \tag{13.84}$$

The corresponding equivalent scheme is shown in Figure 13.6. The *beam splitter* is a partially silvered piece of glass with reflection coefficient parameterized as $R = \cos\theta$, so that its action is to perform the following operation:

$$|\psi_{out}\rangle = \begin{bmatrix} \cos\theta & \sin\theta \\ -\sin\theta & \cos\theta \end{bmatrix} |\psi_{in}\rangle. \tag{13.85}$$

A 50:50 beam splitter is obtained for $\theta = \pi/4$, and its implementation is shown in Figure 13.7.

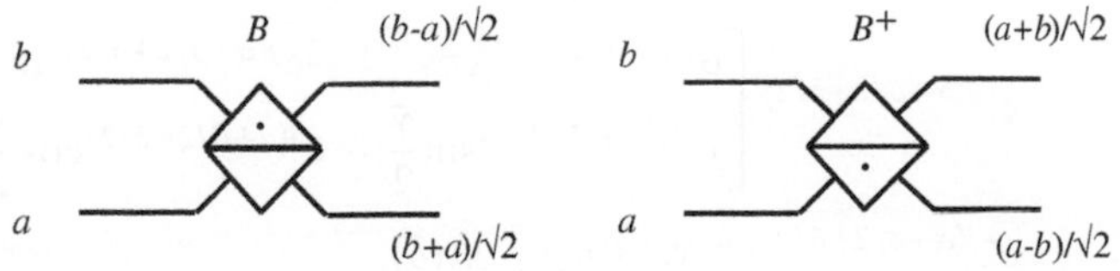

FIGURE 13.7

Operating principle of 50/50 beam splitter (B) and its Hermitian conjugate ($B^{\dagger}$).

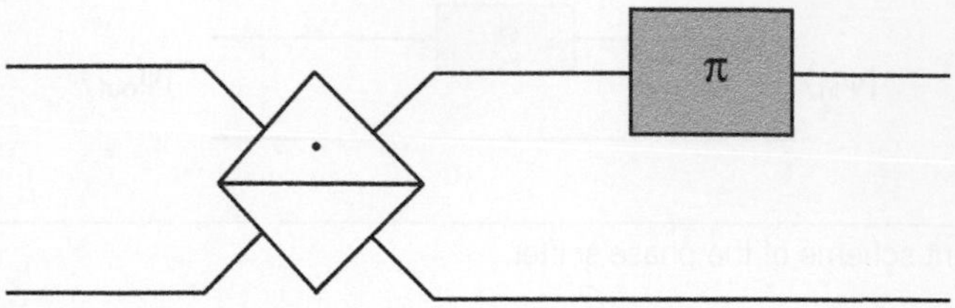

FIGURE 13.8

Hadamard gate implementation in bulky optics based on 50:50 beam splitter and π rad phase shifter.

By using these basic building blocks (the phase shifter and beam splitter), we can implement various single-qubit gates. As an illustration, in Figure 13.8 Hadamard gate implementation is shown, by concatenating a 50:50 beam splitter and π rad phase shifter. The operating principle of the gate in Figure 13.8 is as follows:

$$
\begin{aligned}
|\psi_{\text{out}}\rangle &= \begin{bmatrix} e^{j\pi} & 0 \\ 0 & 1 \end{bmatrix} \begin{bmatrix} \cos(\pi/4) & \sin(\pi/4) \\ -\sin(\pi/4) & \cos(\pi/4) \end{bmatrix} |\psi_{\text{in}}\rangle = -\frac{1}{\sqrt{2}} \begin{bmatrix} 1 & 1 \\ 1 & -1 \end{bmatrix} |\psi_{\text{in}}\rangle \\
&= -H|\psi_{\text{in}}\rangle, \quad H = \frac{1}{\sqrt{2}} \begin{bmatrix} 1 & 1 \\ 1 & -1 \end{bmatrix}.
\end{aligned} \tag{13.86}
$$

Let us observe the concatenation of two phase shifters and one beam splitter as shown in Figure 13.9. The overall action of this gate is given by

$$
\begin{aligned}
U_3U_2U_1 &= \begin{bmatrix} e^{-j\beta} & 0 \\ 0 & 1 \end{bmatrix} \begin{bmatrix} \cos(\gamma/2) & -\sin(\gamma/2) \\ \sin(\gamma/2) & \cos(\gamma/2) \end{bmatrix} \begin{bmatrix} e^{j[\alpha-\delta/2-(\alpha+\delta/2)]} & 0 \\ 0 & 1 \end{bmatrix} \\
&= e^{-j(\alpha+\beta/2+\delta/2)} \begin{bmatrix} e^{-j\beta}/2 & 0 \\ 0 & e^{j\beta/2} \end{bmatrix} \begin{bmatrix} \cos(\gamma/2) & -\sin(\gamma/2) \\ \sin(\gamma/2) & \cos(\gamma/2) \end{bmatrix} \\
&\quad \times \begin{bmatrix} e^{j(\alpha-\delta/2)} & 0 \\ 0 & e^{j(\alpha+\delta/2)} \end{bmatrix} \\
&= e^{-j(\alpha+\beta/2+\delta/2)} \begin{bmatrix} e^{j(\alpha-\beta/2-\delta/2)}\cos\frac{\gamma}{2} & -e^{j(\alpha-\beta/2+\delta/2)}\sin\frac{\gamma}{2} \\ e^{j(\alpha+\beta/2-\delta/2)}\sin\frac{\gamma}{2} & e^{j(\alpha+\beta/2+\delta/2)}\cos\frac{\gamma}{2} \end{bmatrix} \\
&= e^{-j(\alpha+\beta/2+\delta/2)} e^{j\alpha} R_z(\beta) R_y(\gamma) R_z(\delta),
\end{aligned} \tag{13.87}
$$

which up to the global phase constant is the same as that of the Y–Z decomposition theorem in Chapter 3. Therefore, by using the circuit in Figure 13.9, an arbitrary

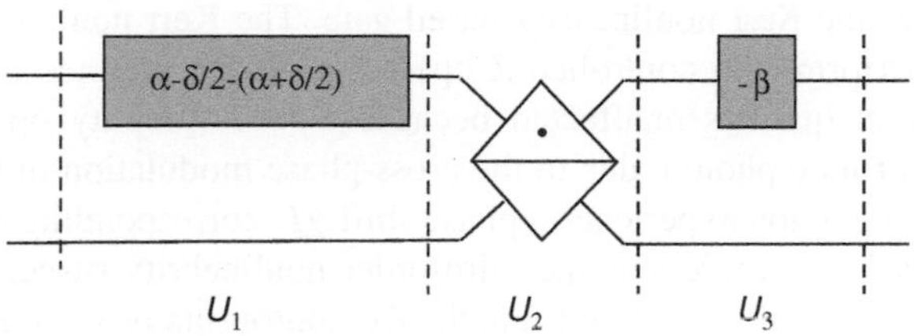

FIGURE 13.9

The *Y–Z* decomposition theorem by bulky optics devices.

single-qubit gate can be implemented. In order to complete the set of universal quantum gates, the implementation of the controlled-*Z* gate or CNOT gate is needed.

Let the logical "0" be represented by a horizontal (H) photon $|H\rangle \equiv |0\rangle = (1\ 0)^{\mathrm{T}}$ and the logical "1" by a vertical (V) photon $|V\rangle \equiv |1\rangle = (0\ 1)^{\mathrm{T}}$. The implementation of the CNOT gate by bulky optics is shown in Figure 13.10. One polarization beam splitter (PBS) is used per input port, while one polarization beam combiner (PBC) is used per output port. The output control $|C_{\mathrm{o}}\rangle = [c_{\mathrm{H,o}}\ c_{\mathrm{V,o}}]^{\mathrm{T}}$ and target qubits $|T_{\mathrm{o}}\rangle = [t_{\mathrm{H,o}}\ t_{\mathrm{V,o}}]^{\mathrm{T}}$ are related to the corresponding input qubits by

$$\begin{bmatrix} c_{\mathrm{H},o} \\ c_{\mathrm{V},o} \\ t_{\mathrm{H},o} \\ t_{\mathrm{V},o} \end{bmatrix} = \underbrace{\frac{1}{\sqrt{2}}\begin{bmatrix} 1 & 1 & 0 & 0 \\ 1 & -1 & 0 & 0 \\ 0 & 0 & 1 & 1 \\ 0 & 0 & 1 & -1 \end{bmatrix}}_{I\otimes H} \underbrace{\begin{bmatrix} 1 & 0 & 0 & 0 \\ 0 & 1 & 0 & 0 \\ 0 & 0 & 1 & 0 \\ 0 & 0 & 0 & -1 \end{bmatrix}}_{K} \underbrace{\frac{1}{\sqrt{2}}\begin{bmatrix} 1 & 1 & 0 & 0 \\ 1 & -1 & 0 & 0 \\ 0 & 0 & 1 & 1 \\ 0 & 0 & 1 & -1 \end{bmatrix}}_{I\otimes H} \begin{bmatrix} c_{\mathrm{H}} \\ c_{\mathrm{V}} \\ t_{\mathrm{H}} \\ t_{\mathrm{V}} \end{bmatrix}$$

$$= U_{\mathrm{CNOT}} \begin{bmatrix} c_{\mathrm{H}} \\ c_{\mathrm{V}} \\ t_{\mathrm{H}} \\ t_{\mathrm{V}} \end{bmatrix}; \quad U_{\mathrm{CNOT}} = \begin{bmatrix} 1 & 0 & 0 & 0 \\ 0 & 1 & 0 & 0 \\ 0 & 0 & 0 & 1 \\ 0 & 0 & 1 & 0 \end{bmatrix}, \tag{13.88}$$

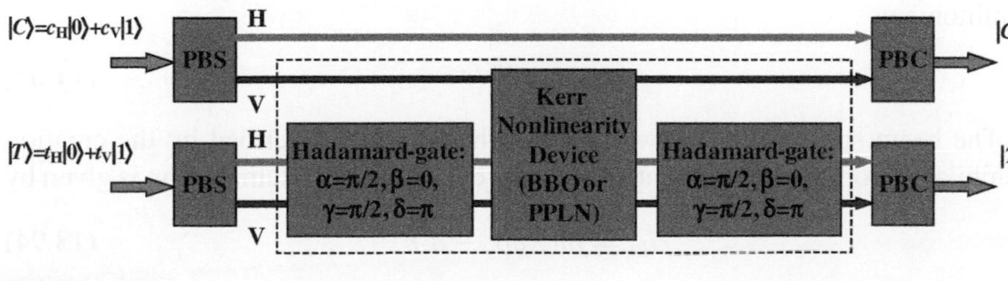

FIGURE 13.10

The implementation of the CNOT gate using bulky optics when photon polarization is used to carry the quantum information.

where K denotes the Kerr nonlinearity-based gate. The Kerr nonlinearity device in Figure 13.10 performs the controlled-Z operation. In the absence of a control c_V photon, the target qubit is unaffected because $H^2=I$ (identity operator). In the presence of control c_V photon, due to the cross-phase modulation in HNLF/crystal, the target vertical photon experience a phase shift χL, corresponding to the complex phase term $\exp(j\chi L)$, where χ is the third-order nonlinearity susceptibility coefficient and L is the nonlinear crystal length. By appropriately selecting the crystal length, we obtain $\chi L=\pi$ and the overall action on the target qubit is $HZH=X$, which corresponds to the action of the CNOT gate. Therefore, when only both control and qubit photons are vertical, a phase shift of π rad is introduced. By omitting the Hadamard gates from Figure 13.10, the corresponding circuit operates as a controlled-Z gate.

In the previous discussion, it was assumed that the matrix representation given by Eqs (13.84), (13.85) and the representation of the K operator in Eq. (13.88) are correct, without providing the quantum mechanical justification. The rest of this section is concerned with the quantum mechanical interpretation of various bulky optics devices used above. The phase shifter P of length L introduces delay in the light propagation by an amount $\Delta=(n-n_0)L/c$. The action of the phase-shift circuit P on a vacuum state and a single-photon state is as follows:

$$P|0\rangle=|0\rangle,\quad P|1\rangle=e^{j\Delta}|1\rangle. \tag{13.89}$$

On the other hand, the action of P on a dual-rail state is

$$P|\psi\rangle=P(c_{01}|01\rangle+c_{10}|10\rangle)=c_{01}e^{-j\Delta/2}|01\rangle+c_{10}e^{j\Delta/2}|10\rangle. \tag{13.90}$$

The action of P can be described in terms of a rotation operator around the z-axis by

$$R_z(\Delta)=e^{-jZ\Delta/2}. \tag{13.91}$$

The corresponding Hamiltonian is given by

$$\hat{H}=\hbar(n_0-n)Z, \tag{13.92}$$

so that the action of P can be interpreted as the evolution described by the Hamiltonian:

$$P=e^{-j\hat{H}L/\hbar c_0} \tag{13.93}$$

The beam splitter acts on two modes, which can be described by the creation (annihilation) operators a ($a^\dagger$) and b ($b^\dagger$). The corresponding Hamiltonian is given by

$$\hat{H}_{\mathrm{bs}}=j\hbar\omega(ab^\dagger-a^\dagger b). \tag{13.94}$$

The action of a beam splitter can be represented by the evolution operation as

$$B=e^{-j\hat{H}t/\hbar}=e^{\theta(ab^\dagger-a^\dagger b)}=e^{\theta G},\quad G=ab^\dagger-a^\dagger b,\theta=\omega t. \tag{13.95}$$

The Baker–Campbell–Hausdorf formula is then applied:

$$e^{\lambda G} A e^{-\lambda G} = \sum_{n=0}^{\infty} \frac{\lambda^n}{n!} C_n; \quad C_0 = A, \ C_n = [G, C_{n-1}], \ n = 1, 2, \ldots, \tag{13.96}$$

By noticing that

$$[a, a^\dagger] = [b, b^\dagger] = 1, \quad [G, a] = -b, \quad [G, b] = a, \tag{13.97}$$

it can be concluded that

$$\begin{aligned} &C_0 = a, \ C_1 = [G, a] = -b, \ C_2 = [G, C_1] = -a, \\ &G_3 = [G, C_2] = -[G, C_0] = b, \end{aligned} \tag{13.98}$$

and consequently:

$$C_n = \begin{cases} j^n a, & n - \text{even} \\ j^{n+1} b, & n - \text{odd}. \end{cases} \tag{13.99}$$

By using the Baker–Campbell–Hausdorf formula, the following is obtained:

$$\begin{aligned} BaB^\dagger &= e^{\theta G} a e^{-\theta G} = \sum_{n=0}^{\infty} \frac{\theta^n}{n!} C_n \\ &= \underbrace{\sum_{n-\text{even}} \frac{(j\theta)^n}{n!}}_{\cos\theta} a + j \underbrace{\sum_{n-\text{odd}} \frac{(j\theta)^n}{n!}}_{\sin\theta} b = a\cos\theta - b\sin\theta. \end{aligned} \tag{13.100}$$

In similar fashion it can be shown that

$$Bb^\dagger B = a\cos\theta + b\sin\theta, \tag{13.101}$$

so that the matrix representation of B is

$$B = \begin{bmatrix} \cos\theta & -\sin\theta \\ \sin\theta & \cos\theta \end{bmatrix} = e^{j\theta Y}, \tag{13.102}$$

which is the same as the rotation operator around the y-axis. The action of operator B on dual-rail states is given by

$$\begin{aligned} B|01\rangle &= Ba^\dagger|00\rangle = Ba^\dagger \underbrace{B^\dagger B}_{I} |00\rangle \overset{B|00\rangle=|00\rangle}{=} Ba^\dagger B^\dagger |00\rangle \\ &= (a^\dagger \cos\theta + b^\dagger \sin\theta)|00\rangle = \cos\theta|01\rangle - \sin\theta|10\rangle \\ B|10\rangle &= \cos\theta|10\rangle + \sin\theta|01\rangle, \end{aligned} \tag{13.103}$$

which leads to the same matrix representation as given by Eq. (13.102).

Cross-phase modulation (XPM) is another effect caused by the intensity dependence of the refractive index, and occurs during propagation of a composite signal. The nonlinear phase shift of a specific optical mode is affected not only by the intensity of the observed mode, but also by the intensity of the other optical modes. Quantum mechanically, the XPM between two modes a and b is described by the following Hamiltonian:

$$\hat{H}_{\mathrm{XPM}} = -\hbar\chi a^\dagger a b^\dagger b. \tag{13.104}$$

The propagation of these two modes over nonlinear media of length L is governed by the following evolution operator:

$$K = \mathrm{e}^{-j\hat{H}L/\hbar} = \mathrm{e}^{j\chi L a^\dagger a b^\dagger b}. \tag{13.105}$$

Therefore, the operator K is a unitary transform of nonlinear media of length L. The action of K on single-photon states is as follows:

$$\begin{aligned} &K|00\rangle = |00\rangle, \quad K|01\rangle = |01\rangle, \quad K|10\rangle = |10\rangle, \quad K|11\rangle = \mathrm{e}^{j\chi L}|11\rangle, \\ &\chi L = \pi: \quad K|11\rangle = -|11\rangle, \end{aligned} \tag{13.106}$$

and the corresponding matrix representation is given by

$$K = \begin{bmatrix} 1 & 0 & 0 & 0 \\ 0 & 1 & 0 & 0 \\ 0 & 0 & 1 & 0 \\ 0 & 0 & 0 & -1 \end{bmatrix} = \begin{bmatrix} I & 0 \\ 0 & Z \end{bmatrix} = C(Z), \tag{13.107}$$

which is the same as that of the controlled-Z gate ($C(Z)$). Since the Pauli X operator can be implemented as $X = HZH$ (see Figure 13.10), this implementation of controlled-Z gates completes the implementation of the universal set of gates $\{H, S, \pi/8\ (T), C(Z)\}$ or equivalently $\{H, S, T, \mathrm{CNOT}\}$. On the other hand, for a dual-rail representation, if the base kets are selected as follows [15]:

$$|e_{00}\rangle = |1001\rangle, \; |e_{01}\rangle = |1010\rangle, \; |e_{10}\rangle = |0101\rangle, \; |e_{11}\rangle = |0110\rangle, \tag{13.108}$$

then the action of the K operator on base kets is given by

$$K|e_i\rangle = |e_i\rangle, \; i \neq 11; \; K|e_{11}\rangle = -|e_{11}\rangle, \tag{13.109}$$

and the corresponding matrix representation is given by Eq. (13.107). Before concluding this section, we describe the implementation of the controlled-SWAP gate, also known as the Fredkin gate. Fredkin gate implementation for single-photon states using bulky optics is shown in Figure 13.11. The operating principle of this circuit is given by $U = B^\dagger K B$, where B is the 50:50 beam splitter, $B = B(\pi/2)$. Based on Eqs (13.95) and (13.105), the operator U can be represented as

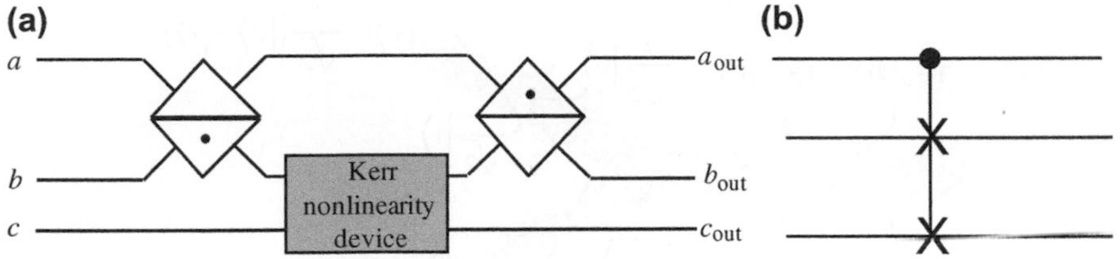

FIGURE 13.11

The implementation of the Fredkin gate using bulky optics: (a) implementation architecture; (b) quantum circuit representation.

$$\begin{aligned} U &= \exp\left(j\chi L c^\dagger c \frac{b^\dagger - a^\dagger}{2}\frac{b-a}{2}\right) \\ &= e^{j\frac{\pi}{2}b^\dagger b}\underbrace{e^{\frac{\chi L}{2}c^\dagger c\left(ab^\dagger - b^\dagger a\right)}}_{U_{\text{Fredkin}(\chi L)}} e^{-j\frac{\pi}{2}b^\dagger b} e^{j\frac{\chi L}{2}a^\dagger a c^\dagger c} e^{j\frac{\chi L}{2}b^\dagger b c^\dagger c}. \end{aligned} \tag{13.110}$$

The second term in (13.110),

$$U_{\text{Fredkin}}(\chi L) = e^{\frac{\chi L}{2}c^\dagger c(ab^\dagger - b^\dagger a)}, \tag{13.111}$$

operates as a Fredkin gate for $\chi L = \pi$ on single-photon states. In the absence of a photon in input port c, the output ports are related to the input ports as $a_{\text{out}} = a$, $b_{\text{out}} = b$, since $B^\dagger B = I$. On the other hand, in the presence of a photon in input port c, $a_{\text{out}} = b$, $b_{\text{out}} = a$, which is the same as the action of the SWAP gate. Therefore, this circuit behaves as a controlled-SWAP (Fredkin) gate with input port c serving as control qubit. The corresponding matrix representation is

$$U_{\text{Fredkin}} = \begin{bmatrix} 1 & 0 & & \cdots & & & & 0 \\ 0 & 1 & 0 & & & & & \\ & 0 & 1 & 0 & & & & \vdots \\ & & 0 & 1 & 0 & & & \\ \vdots & & & 0 & 1 & 0 & 0 & 0 \\ & & & & 0 & 0 & 1 & 0 \\ & & & & 0 & 1 & 0 & 0 \\ 0 & & \cdots & & 0 & 0 & 0 & 1 \end{bmatrix}. \tag{13.112}$$

13.4.2 Integrated Optics Implementation

This section covers integrated optics and all-fiber implementations of universal quantum gates [29–32]. If the waveguides supporting two modes are used, the Hadamard gate can be implemented based on a Y-junction, as shown in Figure 13.12.

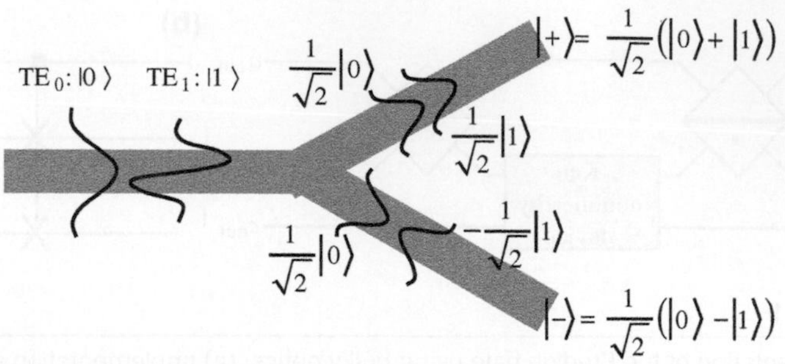

FIGURE 13.12

Hadamard gate implementation in integrated optics based on Y-junction.

However, since dual-mode waveguides are required for this implementation, we restrict our attention to the single-mode solution, as integrated optics for single-mode devises are more mature. Further study of dual-mode universal quantum gates is postponed until the Problems section. In what follows, the logical "0" is represented by a horizontal (H) photon $|H\rangle \equiv |0\rangle = (1\ 0)^{\mathrm{T}}$ and the logical "1" is represented by a vertical (V) photon $|V\rangle \equiv |1\rangle = (0\ 1)^{\mathrm{T}}$. An arbitrary single-qubit gate can be implemented based on a *directional coupler*, as shown in Figure 13.13. We use

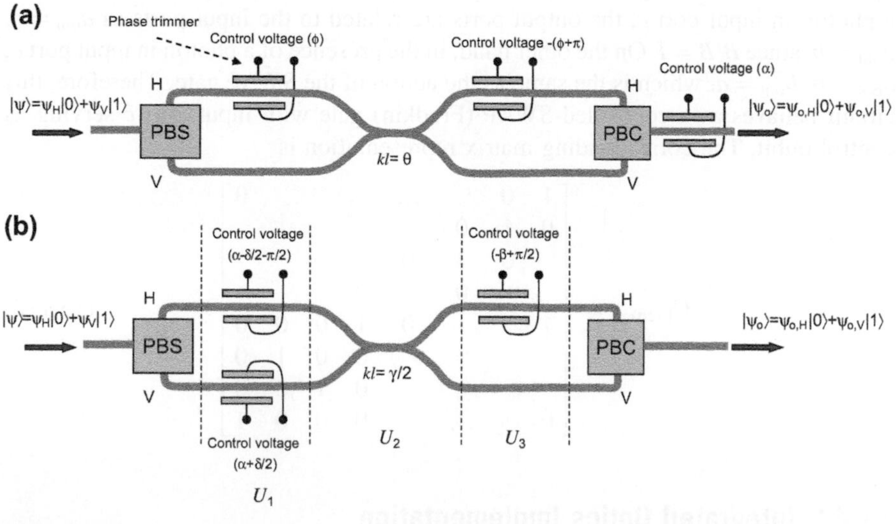

FIGURE 13.13

Integrated optics implementation of an arbitrary single-qubit gate based on a single directional coupler: (a) Barenco-type gate; (b) *Z*–*Y* decomposition theorem-based type. PBS/C, polarization beam splitter/combiner.

a polarization beam splitter (PBS) at the input of a quantum gate and a polarization beam combiner (PBC) at the output of the gate. The horizontal output (input) of PBS (PBC) is denoted by H, while the vertical output (input) of PBS (PBC) is denoted by V. The input qubit is denoted by $|\psi\rangle = \psi_H|0\rangle + \psi_V|1\rangle = [\psi_H\ \psi_V]^T$, while the output qubit is denoted by $|\psi_o\rangle = \psi_{o,H}|0\rangle + \psi_{o,V}|1\rangle = [\psi_{o,H}\ \psi_{o,V}]^T$. In Figure 13.13a, we show an implementation based on the Barenco-type proposal [1]. It can be shown that the output qubit is related to the input qubit by

$$\begin{bmatrix} \psi_{H,o} \\ \psi_{V,o} \end{bmatrix} = U_B \begin{bmatrix} \psi_H \\ \psi_V \end{bmatrix}, \quad U_B = \begin{bmatrix} e^{j\alpha}\cos(\theta) & -je^{j(\alpha+\phi)}\sin(\theta) \\ -je^{j(\alpha-\phi)}\sin(\theta) & e^{j\alpha}\cos(\theta) \end{bmatrix}. \tag{13.113}$$

In Eq. (13.113) $\theta = kl$, where k is the coupling coefficient and l is the coupling region length. By setting appropriately phase trimmer voltages, we can perform the arbitrary single-qubit operation. For example, by setting $\alpha = \phi = 0$ rad and $\theta = 2\pi$ we obtain an identity gate, while by setting $\alpha = 0$ rad, $\phi = \pi$, and $\theta = \pi/2$ we obtain the X gate. In Figure 13.13b, we show an implementation based on $Z-Y$ decomposition theorem. The output qubit is related to the input qubit by

$$\begin{bmatrix} \psi_{H,o} \\ \psi_{V,o} \end{bmatrix} = U \begin{bmatrix} \psi_H \\ \psi_V \end{bmatrix}, \quad U = \begin{bmatrix} \cos\left(\frac{\gamma}{2}\right)e^{j(\alpha-\beta/2-\delta/2)} & -\sin\left(\frac{\gamma}{2}\right)e^{j(\alpha-\beta/2+\delta/2)} \\ \sin\left(\frac{\gamma}{2}\right)e^{j(\alpha+\beta/2-\delta/2)} & \cos\left(\frac{\gamma}{2}\right)e^{j(\alpha+\beta/2+\delta/2)} \end{bmatrix}. \tag{13.114}$$

In Eq. (13.114) $\gamma = 2kl$. By setting $\alpha = \pi/4$, $\beta = \pi/2$, $\gamma = 2\pi$, and $\delta = 0$ rad, the U gate described by (13.114) operates as the phase gate:

$$S = \begin{bmatrix} 1 & 0 \\ 0 & j \end{bmatrix}. \tag{13.115}$$

By setting $\alpha = \pi/8$, $\beta = \pi/4$ rad, $\gamma = 2\pi$, and $\delta = 0$ rad, the U gate operates as $\pi/8$ gate:

$$T = \begin{bmatrix} 1 & 0 \\ 0 & e^{j\pi/4} \end{bmatrix}. \tag{13.116}$$

Finally, by setting $\alpha = \pi/2$, $\beta = 0$ rad, $\gamma = \pi/2$, and $\delta = \pi$, the U gate given by Eq. (13.114) operates as a Hadamard gate:

$$H = \frac{1}{\sqrt{2}} \begin{bmatrix} 1 & 1 \\ 1 & -1 \end{bmatrix}. \tag{13.117}$$

To complete the implementation of a set of universal quantum gates, the implementation of the CNOT gate is needed. The authors in Refs [32,62] proposed the use of directional couplers to implement the CNOT gate. For completeness of presentation, Figure 13.14a shows a simplified version of the CNOT gate proposed

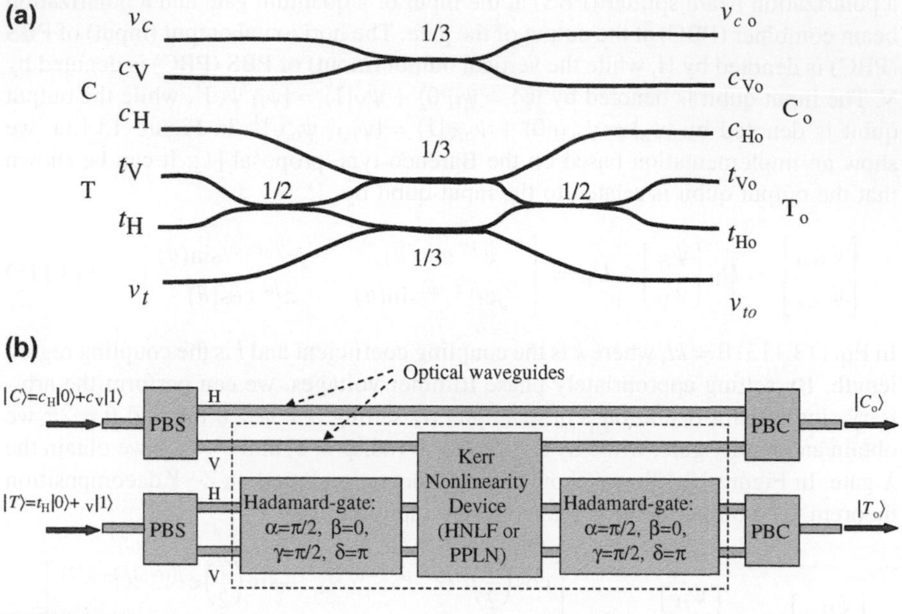

FIGURE 13.14

Integrated optics (or all-fiber) implementation of the CNOT gate: (a) probabilistic gate proposed in Ref. [32]; (b) deterministic gate. PBS/C: polarization beam splitter/combiner.

in Refs [32,62]. We see that the control output qubit $[c_{H,o},c_{V,o}]^T$ is related to the input control qubit $[c_H,c_V]^T$ and the input target qubit $(t_H,t_V)^T$ by [62]

$$\left[c_{H,o} = \left(1/\sqrt{3}\right)\left(\sqrt{2}v_c + c_H\right),\ c_{V,o} = \left(1/\sqrt{3}\right)(-c_V + t_H + t_V)\right]^T. \quad (13.118)$$

Because the output control qubit is affected by the input target qubit, the definition of CNOT gate operation (the control qubit must be unaffected by the target qubit) is violated. This gate operates correctly only with probability of 1/9, and is essentially a probabilistic gate. Figure 13.14b shows the deterministic implementation of the CNOT gate based on the single-qubit gate shown in Figure 13.13 and HNLF. Notice that different implementations of the CNOT gate in Refs [32,33] employ multi-rail/dual-rail representations, while the implementation in the figure is based on single-photon polarization (spin angular momentum). Because the different quantum gate implementations in Refs [32,33] require either two modes for dual-rail representation [32,62] or $2n$ modes for multi-rail representation [33] they cannot be implemented at all using conventional single-mode fibers (SMFs) or single-mode devices. This is a key disadvantage of the multi-rail representation compared to Figure 13.14b. It can be shown that output control $|C_o\rangle = [c_{H,o}\ c_{V,o}]^T$ and target qubits $|T_o\rangle = [t_{H,o}\ t_{V,o}]^T$ are related to the corresponding input qubits by

$$\begin{bmatrix} c_{H,o} \\ c_{V,o} \\ t_{H,o} \\ t_{V,o} \end{bmatrix} = \underbrace{\frac{1}{\sqrt{2}}\begin{bmatrix} 1 & 1 & 0 & 0 \\ 1 & -1 & 0 & 0 \\ 0 & 0 & 1 & 1 \\ 0 & 0 & 1 & -1 \end{bmatrix}}_{I\otimes H} \underbrace{\begin{bmatrix} 1 & 0 & 0 & 0 \\ 0 & 1 & 0 & 0 \\ 0 & 0 & 1 & 0 \\ 0 & 0 & 0 & -1 \end{bmatrix}}_{K} \underbrace{\frac{1}{\sqrt{2}}\begin{bmatrix} 1 & 1 & 0 & 0 \\ 1 & -1 & 0 & 0 \\ 0 & 0 & 1 & 1 \\ 0 & 0 & 1 & -1 \end{bmatrix}}_{I\otimes H} \begin{bmatrix} c_H \\ c_V \\ t_H \\ t_V \end{bmatrix}$$

$$= U_{CNOT} \begin{bmatrix} c_H \\ c_V \\ t_H \\ t_V \end{bmatrix}, \quad U_{CNOT} = \begin{bmatrix} 1 & 0 & 0 & 0 \\ 0 & 1 & 0 & 0 \\ 0 & 0 & 0 & 1 \\ 0 & 0 & 1 & 0 \end{bmatrix}. \tag{13.119}$$

The Kerr nonlinearity device in Figure 13.14b performs the controlled-Z operation. In the absence of a control c_V photon, the target qubit is unaffected because $H^2 = I$ (identity operator). In the presence of a control c_V photon, due to the cross-phase modulation in HNLF, the target vertical photon experiences a phase shift χL, where χ is the third-order nonlinearity susceptibility coefficient and L is the HNLF length. By appropriately selecting the fiber length, we obtain $\chi L = \pi$ and the overall action on the target qubit is $HZH = X$, which corresponds to the action of the CNOT gate. The Toffoli gate can be obtained in straightforward manner as a generalization of the CNOT gate above by adding an additional control qubit, while the Deutsch gate can be obtained by employing three control qubits, instead of the one used above.

The implementation of Pauli gates X, Y, and Z in integrated optics based on the single-qubit gate shown in Figure 13.13b is now described. By appropriately setting the phase shifts of the U gate α, β, γ, and δ, we can obtain the corresponding Pauli gates. The Y gate is obtained by setting $\alpha = \pi/2$, $\beta = \delta = 0$ rad, and $\gamma = \pi$:

$$Y = \begin{bmatrix} 0 & -j \\ j & 0 \end{bmatrix}. \tag{13.120}$$

The Z gate is obtained by setting $\alpha = \pi/2$, $\beta = \pi$, $\gamma = 2\pi$, and $\delta = 0$ rad:

$$Z = \begin{bmatrix} 1 & 0 \\ 0 & -1 \end{bmatrix}. \tag{13.121}$$

Finally, the X gate is obtained by setting $\alpha = \pi/2$, $\beta = -\pi$, $\gamma = \pi$, and $\delta = 0$ rad:

$$X = \begin{bmatrix} 0 & 1 \\ 1 & 0 \end{bmatrix}. \tag{13.122}$$

Another device that can be used for single-qubit manipulation, when information is imposed on the photon by using spin angular momentum, is an *optical hybrid*. The optical hybrid is shown in Figure 13.15a. Let the electrical fields at the input ports be denoted as $E_{i,1}$ and $E_{i,2}$ respectively. The output electrical fields, $E_{o,1}$ and $E_{o,2}$, are related to the input electrical fields by

$$\begin{aligned} E_{o,1} &= (E_{i,1} + E_{i,2})\sqrt{1-k} \\ E_{o,2} &= (E_{i,1} + E_{i,2}\exp(-j\phi))\sqrt{k}, \end{aligned} \tag{13.123}$$

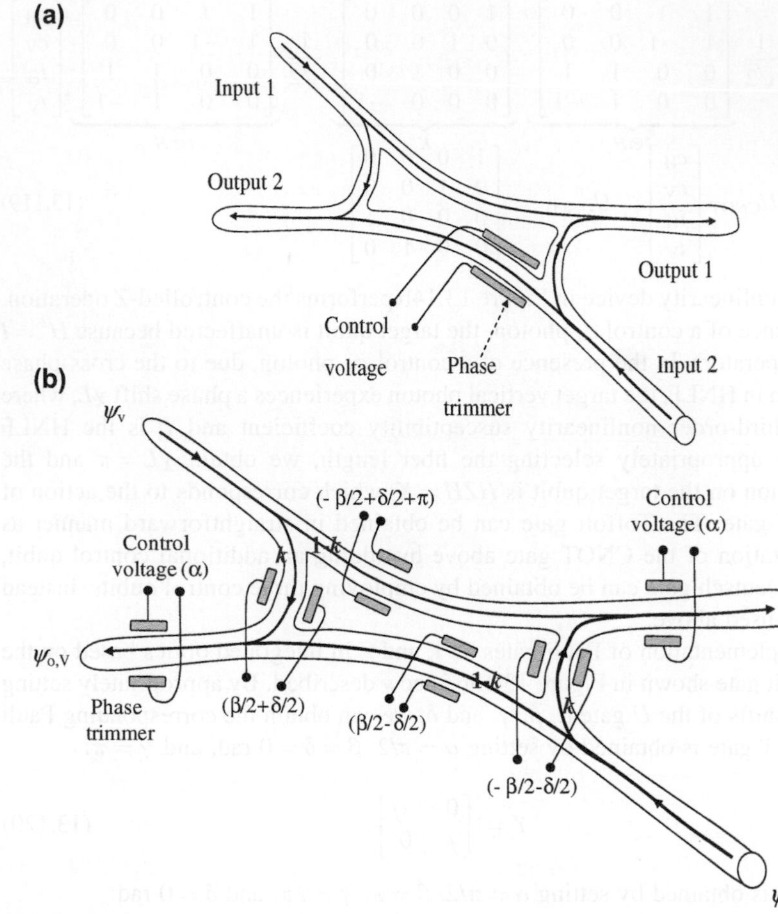

FIGURE 13.15

The optical hybrid as a basic building block to implement various single-qubit gates: (a) implementation of the Hadamard gate ($k = 1/2$, $\phi = \pi$); (b) implementation of an arbitrary single-qubit quantum gate in integrated optics based on a single optical hybrid. (α, β, and δ are the phase shifts introduced by corresponding phase trimmers).

where k is the power-splitting ratio ($0 \le k \le 1$) between the Y-junction output ports of the optical hybrid, and ϕ is the phase shift introduced by the phase trimmer (see Figure 13.15). Equation (13.123) can be rewritten in matrix form as follows:

$$\begin{bmatrix} E_{o,1} \\ E_{o,2} \end{bmatrix} = \begin{bmatrix} \sqrt{1-k} & \sqrt{1-k} \\ \sqrt{k} & e^{-j\phi}\sqrt{k} \end{bmatrix} \begin{bmatrix} E_{i,1} \\ E_{i,2} \end{bmatrix} = U \begin{bmatrix} E_{i,1} \\ E_{i,2} \end{bmatrix}, \quad U = \begin{bmatrix} \sqrt{1-k} & \sqrt{1-k} \\ \sqrt{k} & e^{-j\phi}\sqrt{k} \end{bmatrix}. \tag{13.124}$$

By setting the power-splitting ratio to $k = 1/2$ and phase shift to $\phi = \pi$ rad, the matrix U becomes

$$U(k = 1/2, \phi = \pi) = \frac{1}{\sqrt{2}} \begin{bmatrix} 1 & 1 \\ 1 & -1 \end{bmatrix} = H, \tag{13.125}$$

which is the same as the matrix representation of the Hadamard gate, providing that the polarization beam splitter and combiner are used at the input and output ports respectively. The optical hybrid can be generalized to the circuit shown in Figure 13.15b, which can be used to implement an arbitrary single-qubit gate. In this scheme, the output qubit $[\psi_{o,H}\ \psi_{o,V}]^T$ is related to the input qubit $[\psi_H\ \psi_V]^T$ by

$$\begin{bmatrix} \psi_{o,H} \\ \psi_{o,V} \end{bmatrix} = U \begin{bmatrix} \psi_H \\ \psi_V \end{bmatrix}, \quad U = \begin{bmatrix} \cos\left(\frac{\gamma}{2}\right) e^{j(\alpha-\beta/2-\delta/2)} & -\sin\left(\frac{\gamma}{2}\right) e^{j(\alpha-\beta/2+\delta/2)} \\ \sin\left(\frac{\gamma}{2}\right) e^{j(\alpha+\beta/2-\delta/2)} & \cos\left(\frac{\gamma}{2}\right) e^{j(\alpha+\beta/2+\delta/2)} \end{bmatrix}. \tag{13.126}$$

The matrix U in (13.126) represents a matrix representation of an arbitrary single-qubit quantum gate according to the Z–Y decomposition theorem. For optical hybrid, the corresponding phase shifts α, β, δ can be introduced by a phase trimmer either thermally or electro-optically, while the proper power-splitting ratio $k = \cos^2(\gamma/2)$ should be set in the fabrication phase. By setting $\gamma = \delta = 0$ rad, $\alpha = \pi/4$, and $\beta = \pi/2$ rad, the U gate described by (13.126) operates as the phase gate; by setting $\gamma = \delta = 0$ rad, $\alpha = \pi/8$, and $\beta = \pi/4$ rad, the U gate operates as a $\pi/8$ gate, while by setting $\gamma = \pi/2$, $\alpha = \pi/2$, $\beta = 0$ rad, and $\delta = \pi$, the U gate given by Eq. (13.126) operates as a Hadamard gate. To complete the set of universal quantum gates, the controlled-Z or CNOT gates are needed, which can be implemented as shown in Figure 13.14b, but now with Hadamard gates implemented by using the optical hybrid shown in Figure 13.15b.

Single-qubit gates can also be implemented based on a Mach–Zehnder interferometer (MZI), as shown in Figure 13.16. By using a similar approach as above, the output and input ports can be related by Eq. (13.126), proving that an arbitrary single-qubit gate can indeed be based on a single MZI.

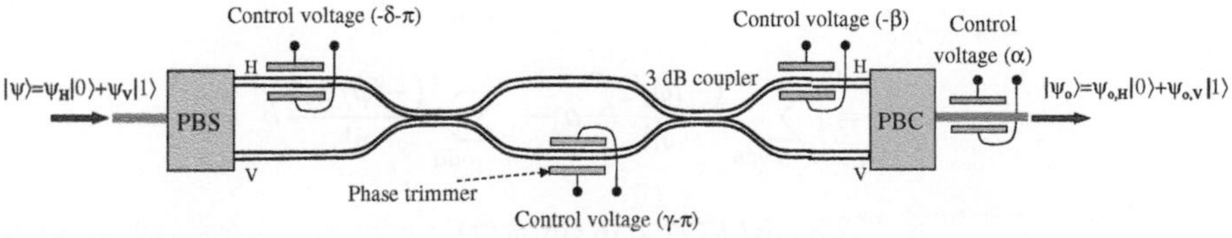

FIGURE 13.16

Photonic implementation of an arbitrary single-qubit gate based on a Mach–Zehnder interferometer. PBS/C, polarization beam splitter/combiner.

In the previous discussion, we assumed that directional coupler theory and the S matrix approach can also be applied on a quantum level, without a formal proof. As an illustration, in the rest of this section the derivation of (13.126) is provided by using quantum mechanical concepts for the directional coupler-based gate in Figure 13.13b. The unitary operator U given by Eq. (13.126) can be written as the product of three unitary operators: (i) U_1 corresponding to the first phase section; (ii) U_2 corresponding to the directional coupler section; and (iii) U_3 corresponding to the second phase section. When the photon is present in the upper branch of the first phase section, it will experience phase shift $\exp[i(-\alpha - \delta/2 - \pi/2)]$, and phase shift $\exp[i(\alpha + \delta/2)]$ when in the lower branch, so that the matrix representation of this section is

$$U_1 = \begin{bmatrix} \mathrm{e}^{i(\alpha-\delta/2-\pi/2)} & 0 \\ 0 & \mathrm{e}^{i(\alpha+\delta/2)} \end{bmatrix}. \tag{13.127}$$

In similar fashion, when the photon is present in the upper branch of the second phase section, it will experience phase shift $\exp[i(-\beta + \pi/2)]$, and no phase shift when in the lower branch, indicating that the matrix representation of this section is

$$U_3 = \begin{bmatrix} \mathrm{e}^{i(-\beta+\pi/2)} & 0 \\ 0 & 1 \end{bmatrix}. \tag{13.128}$$

The directional coupler action on H and V photons can be described by the creation (annihilation) operators a ($a^\dagger$) and b ($b^\dagger$). The action of a directional coupler is given by

$$U_2 = \mathrm{e}^{-i(\phi/2)(a^\dagger b + ab^\dagger)} = \mathrm{e}^{-i(\phi/2)G}, \quad G = a^\dagger b + ab^\dagger. \tag{13.129}$$

By using the Baker–Campbell–Hausdorf formula:

$$\mathrm{e}^{\lambda G} A \mathrm{e}^{-\lambda G} = \sum_{n=0}^{\infty} \frac{\lambda^n}{n!} C_n; \quad C_0 = A, \quad C_n = [G, C_{n-1}], \quad n = 1, 2, \ldots;$$
$$C_n = \begin{cases} a, & n - \text{even} \\ -b, & n - \text{odd}, \end{cases} \tag{13.130}$$

we can show that

$$\begin{aligned} U_2 a U_2^\dagger &= \mathrm{e}^{-i(\phi/2)G} a \mathrm{e}^{+i(\phi/2)G} = \sum_{n=0}^{\infty} \frac{(-i\phi/2)^n}{n!} C_n \\ &= \sum_{n-\text{even}} \frac{(-i\phi/2)^n}{n!} a - \sum_{n-\text{odd}} \frac{(-i\phi/2)^n}{n!} b \\ &= a\cos(\phi/2) + ib\sin(\phi/2) \\ U_2 b U_2^\dagger &= \mathrm{e}^{-i(\phi/2)G} b \mathrm{e}^{+i(\phi/2)G} \\ &= i\, a\sin(\phi/2) + b\cos(\phi/2). \end{aligned} \tag{13.131}$$

The corresponding matrix representation is

$$U_2 = \begin{bmatrix} \cos(\phi/2) & \text{i}\sin(\phi/2) \\ \text{i}\sin(\phi/2) & \cos(\phi/2) \end{bmatrix}, \tag{13.132}$$

which is the same as the corresponding expression derived from directional coupler theory. The overall operation of the gate from Figure 13.13b is then

$$\begin{bmatrix} \psi_{\text{o,H}} \\ \psi_{\text{o,V}} \end{bmatrix} = U_3 U_2 U_1 \begin{bmatrix} \psi_{\text{H}} \\ \psi_{\text{V}} \end{bmatrix} = \begin{bmatrix} \text{e}^{i(-\beta+\pi/2)} & 0 \\ 0 & 1 \end{bmatrix}$$

$$\begin{bmatrix} \cos(\phi/2) & i\sin(\phi/2) \\ i\sin(\phi/2) & \cos(\phi/2) \end{bmatrix} \begin{bmatrix} \text{e}^{i(\alpha-\delta/2-\pi/2)} & 0 \\ 0 & \text{e}^{i(\alpha+\delta/2)} \end{bmatrix} \begin{bmatrix} \psi_{\text{H}} \\ \psi_{\text{V}} \end{bmatrix}, \tag{13.133}$$

therefore proving Eq. (13.126).

13.5 PHOTONIC IMPLEMENTATION OF QUANTUM RELAY

In this section, we first describe the implementation of the Bell states preparation circuit in integrated optics, required in quantum teleportation systems, which is shown in Figure 13.17. Among many possible versions of Hadamard and CNOT gates, we have chosen two with similar propagation times. The upper circuit operates as a Hadamard gate, while the rest of the circuit operates as a CNOT gate, as

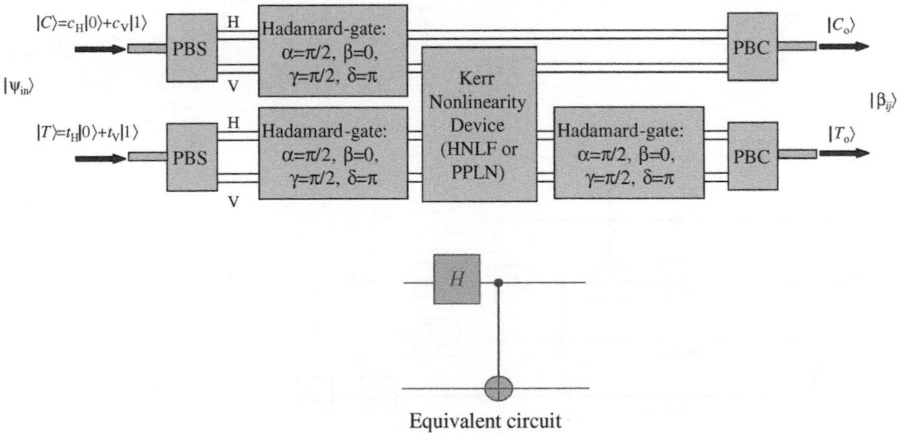

FIGURE 13.17

Photonic implementation of Bell states preparation circuit. PPLN, periodically poled $LiNbO_3$.

already explained in the previous section. It can be shown that the output quantum state $|\beta_{ij}\rangle$ is related to the input state $|\psi_{in}\rangle$ by

$$|\beta_{ij}\rangle = \frac{1}{\sqrt{2}}\begin{bmatrix} 1 & 0 & 0 & 0 \\ 0 & 1 & 0 & 0 \\ 0 & 0 & 0 & 1 \\ 0 & 0 & 1 & 0 \end{bmatrix}\begin{bmatrix} 1 & 0 & 1 & 0 \\ 0 & 1 & 0 & 1 \\ 1 & 0 & -1 & 0 \\ 0 & 1 & 0 & -1 \end{bmatrix}|\psi_{in}\rangle = \frac{1}{\sqrt{2}}\begin{bmatrix} c_H t_H + c_V t_V \\ c_H t_V + c_V t_V \\ c_V t_H - c_V t_V \\ c_H t_H - c_V t_H \end{bmatrix}. \tag{13.134}$$

For example, by setting $c_H = t_H = 1$ and $c_V = t_V = 0$, we obtain the Bell state $|\beta_{00}\rangle = [1 \;\; 0 \;\; 0 \;\; 1]^T/\sqrt{2} = (|00\rangle + |11\rangle)/\sqrt{2}$. In Figure 13.18, we describe how to implement the quantum relay based on the Bell states preparation circuit (shown in Figure 13.17), Hadamard, controlled-X, and controlled-Z gates described above. We employ the principle of differed measurement and perform corresponding measurements only on the last intermediate node. The measurement circuits in Figure 13.18 represent the avalanche photodiodes (APDs), which are used to detect the presence of c_V photons in corresponding control qubits. The detection of c_V photons triggers the application of required control voltages on phase trimmers to perform controlled-X and controlled-Z operations at the destination node.

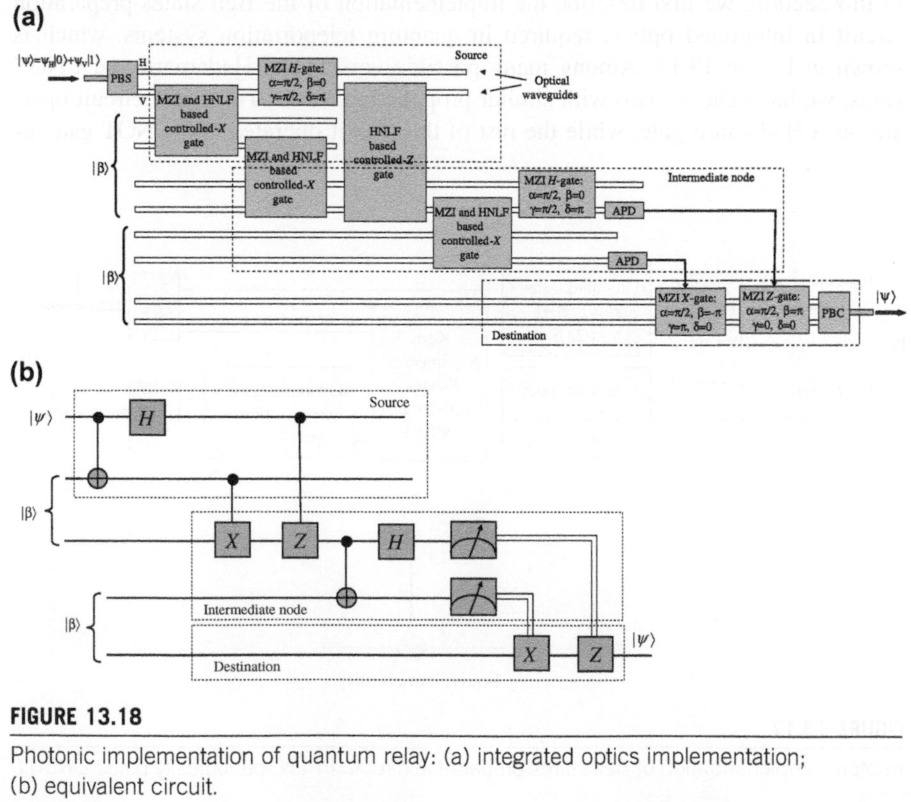

FIGURE 13.18

Photonic implementation of quantum relay: (a) integrated optics implementation; (b) equivalent circuit.

13.6 IMPLEMENTATION OF QUANTUM ENCODERS AND DECODERS

In this section, we describe two classes of sparse-graph quantum codes – (i) quantum dual-containing LDPC codes and (ii) entanglement-assisted LDPC codes – and show how the corresponding encoders and decoders can be implemented. For more details of various classes of quantum LDPC codes, the interested reader is referred to Chapter 10.

The block scheme of entanglement-assisted quantum code, which requires a certain number of entangled qubits to be shared between the source and destination, is shown in Figure 13.19. The number of necessary pre-existing entanglement qubits (also known as ebits [34]) can be determined by $e = \text{rank}(\boldsymbol{H}\boldsymbol{H}^{\text{T}})$, where $\boldsymbol{H}$ is the parity-check matrix of a classical code (and rank( ) is the rank of a given matrix). The source encodes quantum information in state $|\psi\rangle$ with the help of local ancillary qubits $|0\rangle$ as well as the source's half of shared ebits, and then sends the encoded qubits over a noisy quantum channel (e.g. a free-space or fiber-optic channel). The receiver performs decoding on all qubits to diagnose the channel error and performs a recovery unitary operation to reverse the action of the channel. Notice that the channel does not affect at all the receiver's half of the shared ebits. By omitting the ebits, the conventional quantum coding scheme is obtained.

Most practical quantum codes belong to the class of CSS codes [15,34–37], and can be designed via a pair of conventional linear codes satisfying the twisted property (one code includes the dual of another code). Their quantum-check matrix has the form

$$\boldsymbol{A} = \begin{bmatrix} \boldsymbol{H} & | & \boldsymbol{0} \\ \boldsymbol{0} & | & \boldsymbol{G} \end{bmatrix}, \qquad \boldsymbol{H}\boldsymbol{G}^{\text{T}} = \boldsymbol{0}, \tag{13.135}$$

where $\boldsymbol{H}$ and $\boldsymbol{G}$ are $M \times N$ matrices. The condition $\boldsymbol{H}\boldsymbol{G}^{\text{T}} = \boldsymbol{0}$ ensures that twisted product condition is satisfied. Each row in (13.135) represents a stabilizer, with those in the left half of $\boldsymbol{A}$ corresponding to the positions of X operators, and those in the right half ($\boldsymbol{G}$) corresponding to the positions of Z operators. As there are $2M$ stabilizer conditions applying to N qubit states, $N - 2M$ qubits are encoded in N

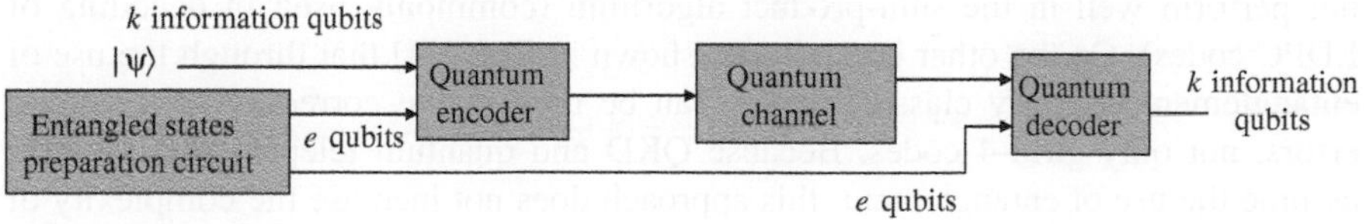

FIGURE 13.19

Entanglement-assisted quantum code.

qubits. The commutativity of stabilizers now appears as the *orthogonality of rows* with respect to a *twisted* (*sympletic*) *product*, formulated as follows: If the kth row in $\boldsymbol{A}$ is $r_k = (x_k; z_k)$, where x_k is the X binary string and z_k is the Z binary string, then the twisted product of rows k and l is defined by [35]

$$r_k \odot r_l = x_k \cdot z_l + x_l \cdot z_k \bmod 2, \tag{13.136}$$

where $x_k \cdot z_l$ is a dot (scalar) product defined by $x_k \cdot z_l = \Sigma_j x_{kj} z_{lj}$. The twisted product is zero if and only if there is an even number of places where the operators corresponding to rows k and l differ (and neither are the identity), i.e. if the operators commute. The CSS codes based on dual-containing codes are simplest to implement. Their (quantum) check matrix can be represented by [15,34–37]

$$\boldsymbol{A} = \left[\begin{array}{c|c} \boldsymbol{H} & \boldsymbol{0} \\ \boldsymbol{0} & \boldsymbol{H} \end{array}\right], \tag{13.137}$$

where $\boldsymbol{H}\boldsymbol{H}^{\mathrm{T}} = 0$, which is equivalent to $C^{\perp}(\boldsymbol{H}) \subset C(\boldsymbol{H})$, where $C(\boldsymbol{H})$ is the code having $\boldsymbol{H}$ as the parity-check matrix and $C^{\perp}(\boldsymbol{H})$ is its corresponding dual code. The quantum LDPC codes have many advantages over other classes of quantum codes, due to the sparseness of their parity-check matrices [17,35,36]. From Eq. (13.137) it follows that, providing that the $\boldsymbol{H}$ matrix of a dual-containing code is sparse, the corresponding $\boldsymbol{A}$ matrix will be sparse as well, while corresponding stabilizers will be of low weight. For example, the $\boldsymbol{H}$ matrix given below satisfies the condition $\boldsymbol{H}\boldsymbol{H}^{\mathrm{T}} = 0$, and can be used in quantum-check matrix (13.135) as dual-containing code:

$$\boldsymbol{H} = \begin{bmatrix} 0 & 0 & 1 & 0 & 1 & 0 & 1 & 1 \\ 1 & 0 & 0 & 1 & 1 & 0 & 0 & 1 \\ 0 & 1 & 0 & 1 & 0 & 0 & 1 & 1 \\ 1 & 0 & 0 & 0 & 0 & 1 & 1 & 1 \\ 0 & 0 & 1 & 1 & 0 & 1 & 0 & 1 \\ 1 & 1 & 1 & 0 & 0 & 0 & 0 & 1 \\ 0 & 1 & 0 & 0 & 1 & 1 & 0 & 1 \end{bmatrix}.$$

The main drawback of dual-containing LDPC codes is the fact that they are essentially girth-4 codes (girth represents the shortest cycle in the corresponding bipartite graph representation of a parity-check matrix of classical code), which do not perform well in the sum-product algorithm (commonly used in decoding of LDPC codes). On the other hand, it was shown in Ref. [34] that through the use of entanglement arbitrary classical codes can be used in the correction of quantum errors, not only girth-4 codes. Because QKD and quantum teleportation systems assume the use of entanglement, this approach does not increase the complexity of the system at all.

The number of entanglement qubits (ebits) needed in EA LDPC codes is $e = \text{rank}(\boldsymbol{H}\boldsymbol{H}^{\text{T}})$, as indicated above, so that the minimum number of required EPR pairs (Bell states) is 1, exactly the same as already in use in certain QKD schemes. For example, the LDPC code given below has $\text{rank}(\boldsymbol{H}_1\boldsymbol{H}_1{}^{\text{T}}) = 1$ and girth 6:

$$\boldsymbol{H}_1 = \begin{bmatrix} 0 & 0 & 1 & 0 & 1 & 0 & 1 \\ 1 & 0 & 0 & 1 & 1 & 0 & 0 \\ 0 & 1 & 0 & 1 & 0 & 0 & 1 \\ 1 & 0 & 0 & 0 & 0 & 1 & 1 \\ 0 & 0 & 1 & 1 & 0 & 1 & 0 \\ 1 & 1 & 1 & 0 & 0 & 0 & 0 \\ 0 & 1 & 0 & 0 & 1 & 1 & 0 \end{bmatrix}$$

and requires only one ebit to be shared between source and destination. Since arbitrary classical codes can be used with this approach, including LDPC codes of girth $g \geq 6$, the performance of quantum LDPC codes can be significantly improved. Notice that the $\boldsymbol{H}$ matrix above is obtained from the matrix $\boldsymbol{H}_1$ by adding an all-ones column.

Because two Pauli operators on N qubits commute if and only if there is an even number of places in which they differ (neither of which is the identity I operator), we can extend the generators in $\boldsymbol{A}$ (for $\boldsymbol{H}_1$) by adding an $e = 1$ column so that they can be embedded into a larger Abelian group; this procedure is known as Abelianization in abstract algebra. The stabilizer version of the Gram–Schmidt orthogonalization algorithm may be used to simplify this procedure, as indicated in Ref. [34].

For example, by performing Gauss–Jordan elimination, the quantum-check matrix (13.137) can be put in standard form [37] (see also Chapter 8):

$$\boldsymbol{A} = \begin{bmatrix} 1 & 0 & 0 & 0 & 0 & 1 & 1 & 1 & 0 & 0 & 0 & 0 & 0 & 0 & 0 & 0 \\ 0 & 1 & 0 & 0 & 1 & 1 & 0 & 1 & 0 & 0 & 0 & 0 & 0 & 0 & 0 & 0 \\ 0 & 0 & 1 & 0 & 1 & 0 & 1 & 1 & 0 & 0 & 0 & 0 & 0 & 0 & 0 & 0 \\ 0 & 0 & 0 & 1 & 1 & 1 & 1 & 0 & 0 & 0 & 0 & 0 & 0 & 0 & 0 & 0 \\ 0 & 0 & 0 & 0 & 0 & 0 & 0 & 0 & 0 & 1 & 0 & 1 & 0 & 1 & 1 \\ 0 & 0 & 0 & 0 & 0 & 0 & 0 & 0 & 1 & 0 & 0 & 0 & 0 & 1 & 1 & 1 \end{bmatrix}.$$

The corresponding generators in standard form are

$$g_1 = X_1X_6X_7X_8 \qquad g_2 = X_2X_5X_6X_8 \qquad g_3 = X_3X_5X_7X_8$$
$$g_4 = X_4X_5X_6X_7 \qquad g_5 = Z_3Z_5Z_7Z_8 \qquad g_6 = Z_1Z_6Z_7Z_8,$$

where the subscripts are used to denote the positions of corresponding X and Z operators. The encoding circuit is shown in Figure 13.20. We use the efficient implementation of encoders introduced by Gottesman [37]. It is clear from Figure 13.20 that for encoder implementation of quantum LDPC codes Hadamard (H), CNOT ($\oplus$), and controlled-Z gates are sufficient, whose implementation in integrated optics, NMR, and ion traps has already been discussed in previous

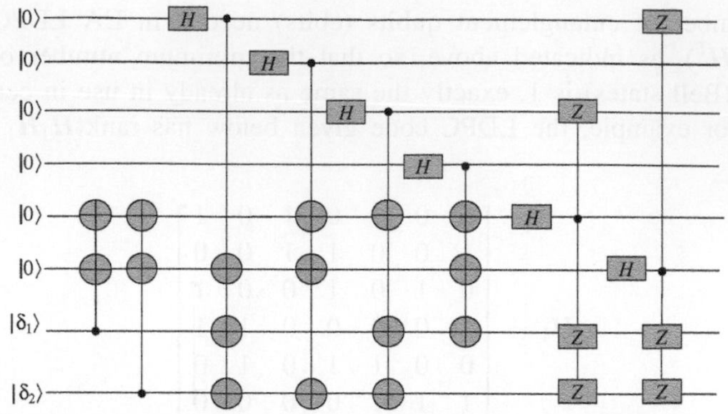

FIGURE 13.20

Encoding circuit for quantum (8,2) LDPC code. $|\delta_1\delta_2\rangle$ are information qubits.

sections. In the next section, another implementation based on cavity quantum electrodynamics will be described as well. For example, in Figure 13.14b it is clear that the controlled-X gate can be implemented based on one controlled-Z and two H gates. Therefore, by using Eq. (13.39), the quantum LDPC encoder can be implemented based only on Hadamard and controlled-Z gates, and their implementation is therefore compatible with the various technologies described in previous sections. Because $H^2 = I$, the corresponding circuit based on H gates and controlled-Z gates can be simplified. It can be shown in similar fashion that the corresponding decoder can be implemented based only on H gates and controlled-Z gates or CNOT gates.

As another illustrative example, the $\boldsymbol{H}$ matrix given below satisfies the condition $\boldsymbol{HH}^{\mathrm{T}} = 0$, and can be used in quantum-check matrix (13.137) as dual-containing code:

$$\boldsymbol{H} = \begin{bmatrix} 1 & 0 & 0 & 1 & 1 & 1 \\ 1 & 1 & 1 & 0 & 0 & 1 \\ 0 & 1 & 1 & 1 & 1 & 0 \end{bmatrix}.$$

By performing Gauss–Jordan elimination the quantum-check matrix (13.137) can be put in standard form as follows:

$$\boldsymbol{A} = \begin{bmatrix} 1 & 0 & 0 & 1 & 1 & 1 & 0 & 0 & 0 & 0 & 0 & 0 \\ 0 & 1 & 1 & 1 & 1 & 0 & 0 & 0 & 0 & 0 & 0 & 0 \\ 0 & 0 & 0 & 0 & 0 & 0 & 1 & 1 & 1 & 0 & 0 & 1 \\ 0 & 0 & 0 & 0 & 0 & 0 & 1 & 0 & 0 & 1 & 1 & 1 \end{bmatrix}.$$

The corresponding generators in standard form are

$$g_1 = X_1X_4X_5X_6 \qquad g_2 = X_2X_3X_4X_5 \qquad g_3 = Z_1Z_2Z_3Z_6 \qquad g_4 = Z_1Z_4Z_5Z_6.$$

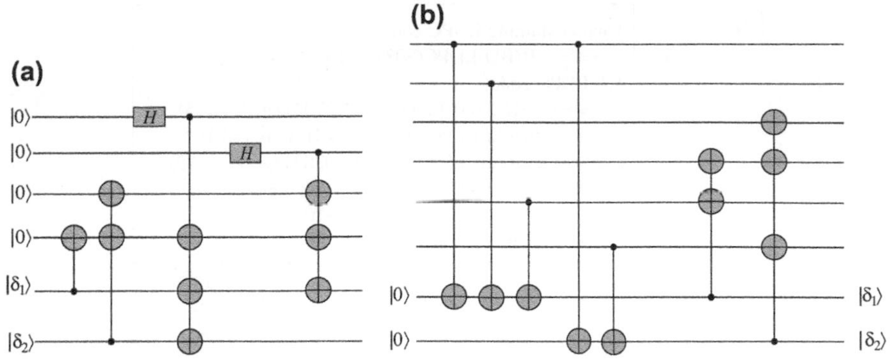

FIGURE 13.21

Encoding and decoding circuits for quantum (6,2) LDPC code: (a) encoder configuration; (b) decoder configuration.

The encoding circuit is shown in Figure 13.21a, while the decoding circuit is shown in Figure 13.21b. We use the efficient implementation of encoders and decoders introduced by Gottesman [37] (see also Chapter 8). Again, it is clear from Figure 13.21 that for encoder and decoder implementation of quantum LDPC codes, Hadamard (H) and CNOT ($\oplus$) gates are sufficient.

By closer inspection of Eq. (13.137) we conclude that generators for CSS design employ exclusively either X or Z gates, but not both. This is also evident from the example above (generators g_1, g_2, g_3, and g_4 contain only X operators, while generators g_5 and g_6 contain only Z operators). Therefore, quantum LDPC decoders of CSS type can be implemented based on controlled-Z gates only, as shown in Figure 13.22. For entanglement-assisted LDPC decoder implementation, we need to follow the procedure described above, in the text relating to Figure 13.20.

We now turn our attention to the design of quantum LDPC codes based on Steiner triple systems (STSs) [38], which is provided here for completeness of presentation. The STS represents a particular instance of a balanced incomplete

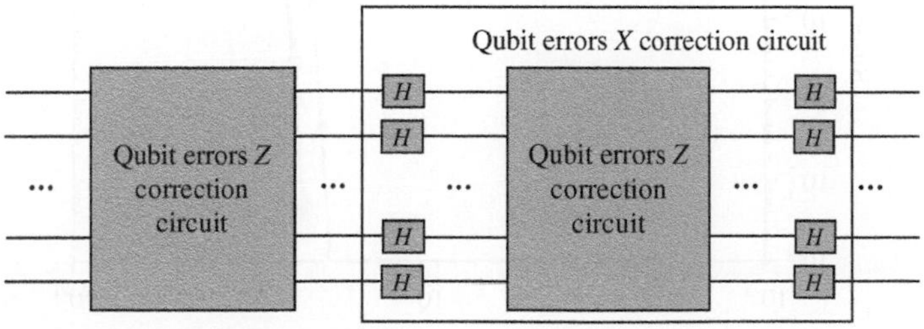

FIGURE 13.22

Quantum LDPC decoder of CSS-type implementation based only on controlled-Z gates.

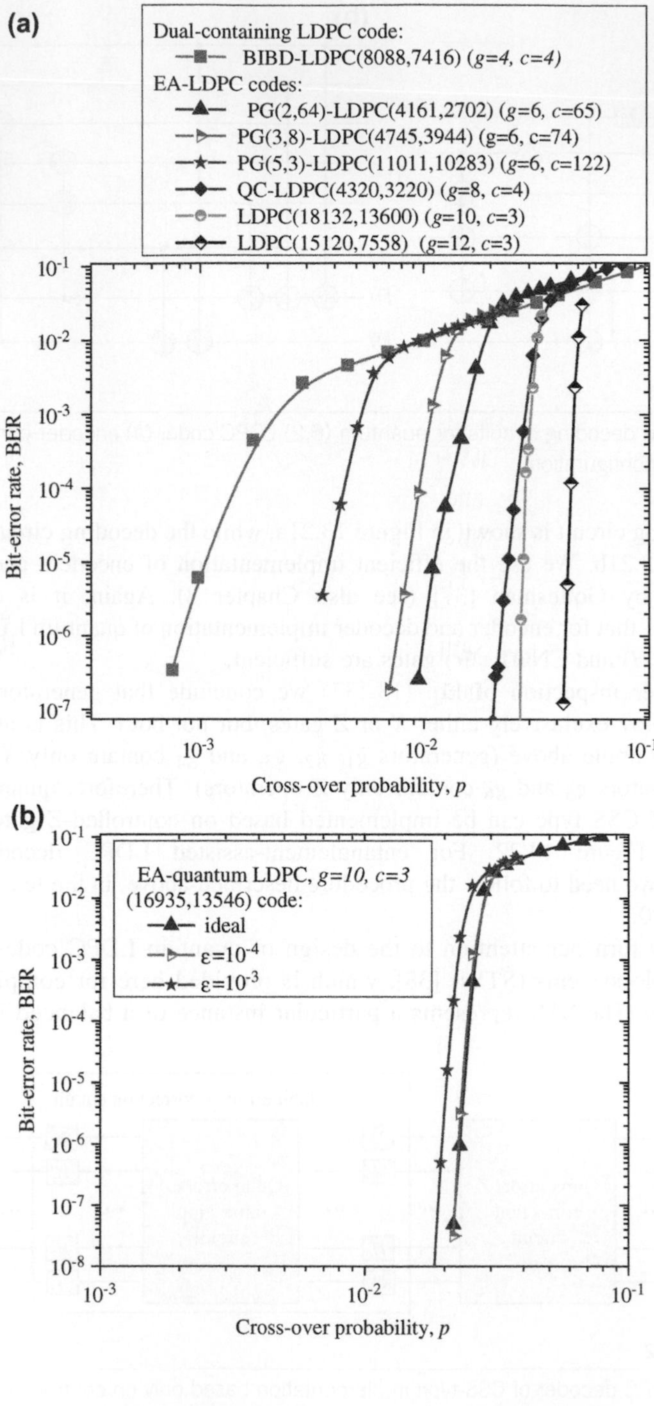
(a)
Dual-containing LDPC code:
BIBD-LDPC(8088,7416) (g=4, c=4)
EA-LDPC codes:
PG(2,64)-LDPC(4161,2702) (g=6, c=65)
PG(3,8)-LDPC(4745,3944) (g=6, c=74)
PG(5,3)-LDPC(11011,10283) (g=6, c=122)
QC-LDPC(4320,3220) (g=8, c=4)
LDPC(18132,13600) (g=10, c=3)
LDPC(15120,7558) (g=12, c=3)
Bit-error rate, BER
Cross-over probability, p
(b)
EA-quantum LDPC, g=10, c=3
(16935,13546) code:
ideal
ε=10^-4
ε=10^-3
Bit-error rate, BER
Cross-over probability, p

block design (BIBD) [36,38]. BIBD(v,k,λ) is defined as a collection of blocks of length k for the set V of integers of size v, such that each pair of elements of V occur together in exactly λ of the blocks. STS(v) is defined as BIBD(v,3,1). It has been shown in Ref. [36] that BIBD of even λ can be used to design quantum LDPC codes belonging to CSS codes, by using dual-containing classical codes. It has also been shown in Ref. [30] that BIBDs of unitary index ($\lambda = 1$) can be used to design entanglement-assisted LDPC codes that require only one ebit to be shared between source and destination, since rank($\boldsymbol{HH}^{\mathrm{T}}$) = 1. For example, STS(7) is given by the following collection of blocks of length $k = 3$: {2,4,6}, {6,3,7}, {5,6,1}, {3,2,5}, {1,7,2}, {7,5,4}, {4,1,3}. By identifying these blocks with nonzero positions of the corresponding parity-check matrix we obtain an LDPC code of girth-6 satisfying the property rank($\boldsymbol{HH}^{\mathrm{T}}$) = 1. For example, matrix $\boldsymbol{H}_1$ above is obtained from STS(7). By adding an all-ones column to the obtained matrix we obtain a dual-containing code, since rank($\boldsymbol{HH}^{\mathrm{T}}$) = 0. The matrix $\boldsymbol{H}$ above is obtained using this simple approach. Therefore, both entanglement-assisted codes and dual-containing quantum codes can be obtained by using STSs. By selectively removing the blocks from STSs we can increase the girth of the corresponding LDPC code, and therefore improve BER performance, at the expense of increasing the complexity of an equivalent entanglement-assisted LDPC code, since now rank($\boldsymbol{HH}^{\mathrm{T}}$) > 1. For more details of various STSs, the interested reader is referred to Ref. [38]. Notice that the codes from STSs are easy to implement because the column weight of the corresponding classical parity-check matrices is only 3, while the parity-check matrices' column weight of projective geometry (PG)-based codes [38–40], which also satisfy the rank($\boldsymbol{HH}^{\mathrm{T}}$) = 1 property, is huge (see Figure 13.23).

In Figure 13.23, we provide a comparison of EA LDPC codes of girth $g = 6$, 8, 10, and 12 against dual-containing LDPC code ($g = 4$). Parameter c denotes the column weight of the corresponding parity-check matrix. The dual-containing code of girth 4 is designed based on BIBDs [36]. The entanglement-assisted codes from PGs are designed as described in Refs [30,39]. These codes are of girth 6, but require only one ebit to be shared between source and destination. The entanglement-assisted codes of girth 10 and 12 are obtained in a similar fashion to STS described above, by selectively removing the blocks from the design.

From Figure 13.23a it is clear that EA LDPC codes outperform by more than an order in magnitude, in terms of crossover probability, the corresponding dual-

FIGURE 13.23

BER performance of various quantum codes: (a) assuming that gates are perfect; (b) assuming that gates are imperfect. BIBD, balanced incomplete block design; EA, entanglement-assisted codes; PG, projective geometry; QC, quasi-cyclic code. The parameters g and c denote girth and column weight of the corresponding parity-check matrix; respectively. (After Ref. [41]; IEEE 2011; reprinted with permission)

containing LDPC code. It is also evident that as we increase the girth we get better performance, at the expense of increased entanglement complexity, since rank($\boldsymbol{HH}^{\mathrm{T}}$) > 1 for $g > 6$ codes, and increases as girth increases. Notice, however, that finite geometry codes [39,40] typically have large column weight (see Figure 13.23), so that although they require a small number of ebits their decoding complexity is high because of the huge column weight. In practice, we will need to make a compromise between decoding complexity, number of required ebits, and BER performance. The results shown in Figure 13.23a are obtained by assuming that quantum gates are perfect. In Figure 13.23b, we study the influence of imperfect quantum gates on BER performance for girth-8 EA LDPC (16935,13546) code. Some practical problems such as decoherence will affect the operation of gates, causing the controlled-Z gate (or equivalently the CNOT gate) to fail with certain probability. When gates fail with probability $\varepsilon = 10^{-4}$, the BER performance loss is negligible. On the other hand, when the gates fail with probability $\varepsilon = 10^{-3}$, the BER performance loss is small but noticeable.

13.7 CAVITY QUANTUM ELECTRODYNAMICS (CQED)-BASED QUANTUM INFORMATION PROCESSING

CQED techniques can be used in many different ways to perform quantum information processing, including [15,41–46]: (i) quantum information can be represented by photon states wherein the cavities with atoms are used to provide nonlinear interaction between photons [42,43]; (ii) quantum information can be represented using atoms wherein the photons can be used to communicate between atoms [44]; and (iii) quantum information can be represented using a quantum interface between a single photon and the spin state of an electron trapped in a quantum dot [45].

Before describing one particular example of CQED-based quantum information processing, a simple model for atom–field interaction based on the Jaynes–Cummings model (proposed by Edwin Jaynes and Fred Cummings in 1963) is introduced (see Ref. [46] for more details). This model describes the interaction between a two-level atom (only two relevant levels in the atom are observed) and a quantized field. Let the ground state and excited state of the atom be denoted by $|g\rangle$ and $|e\rangle$ respectively, and let the photon number states $|0\rangle$ and $|1\rangle$ represent logic 0 and 1 respectively. The atom's energy levels are separated by $\hbar\omega_a$, the atomic transition energy. The photon's energy levels are separated by $\hbar\omega_0$. The two atom states are of opposite parity and mutually orthogonal, and can be used to form the basis $\{|g\rangle\langle g|, |g\rangle\langle e|, |e\rangle\langle g|, |e\rangle\langle e|\}$ An arbitrary operator O can be expanded in this basis as follows:

$$O = O_{gg}\sigma_{gg} + O_{ge}\sigma_{ge} + O_{eg}\sigma_{eg} + O_{ee}\sigma_{ee}; \quad O_{mn} = \langle m|O|n\rangle,$$
$$\sigma_{mn} = |m\rangle\langle n|, \quad m,n \in \{e,g\}. \tag{13.138}$$

Because the two atom states have opposite parity, the dipole operator can be represented by

$$\hat{d} = \hat{d}_{ge}\sigma_{ge} + \hat{d}_{eg}\sigma_{eg}; \quad \langle g|\hat{d}|g\rangle = \langle e|\hat{d}|e\rangle = 0. \tag{13.139}$$

Initially, the atom and photon do not interact, and the atom–field state can be represented as the tensor product of atom and photon states. The corresponding Hamiltonian is given by

$$\hat{H} = \underbrace{\hbar\omega_a|e,0\rangle\langle e,0|}_{\hat{H}_{\text{atom}}} + \underbrace{\hbar\omega_0|g,1\rangle\langle g,1|}_{\hat{H}_{\text{photon}}} = \hbar\omega_a\sigma_{ee} + \hbar\omega_0 a^\dagger a = \hat{H}_{\text{atom}} + \hat{H}_{\text{photon}}, \tag{13.140}$$

or in matrix form as

$$\hat{H} \doteq \begin{bmatrix} \hbar\omega_0 & 0 \\ 0 & \hbar\omega_a \end{bmatrix}, \tag{13.141}$$

which is the same as that given by Eq. (13.8). In (13.140), we use a ($a^\dagger$) to denote the photon annihilation (creation) operator, as before. The energy levels of the atom–photon system before interaction are shown in Figure 13.24a. The interaction Hamiltonian is a function of dipole moment $\boldsymbol{d}$ and electrical field $\boldsymbol{E}$ as follows:

$$\begin{aligned} \hat{H}_{\text{int}} &= -\hat{\boldsymbol{d}}\cdot\boldsymbol{E} = -(\hat{\boldsymbol{d}}_{ge}\sigma_{ge} + \hat{\boldsymbol{d}}_{eg}\sigma_{eg})\sqrt{\frac{\hbar\omega_0}{2\varepsilon_0 V}}[u(\boldsymbol{r})a + \text{h.c.}] \\ &= -\hbar\Omega_0(\sigma_{ge} + \sigma_{eg})(a + a^\dagger), \end{aligned} \tag{13.142}$$

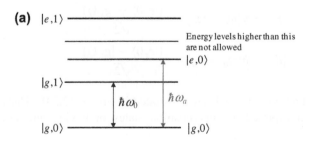

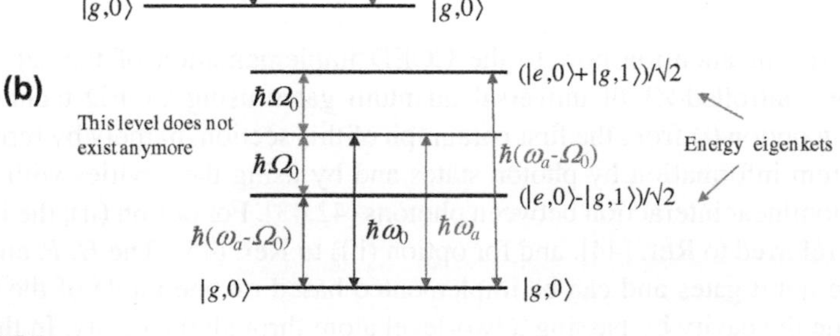

FIGURE 13.24

Atom–photon energy levels: (a) before interaction; (b) after interaction.

where V is the "photon mode volume" (the volume of space occupied by the photon) and $u(\boldsymbol{r})$ is the mode field at the center of mass $\boldsymbol{r}$. The interaction Hamiltonian (13.142) can be simplified as follows:

$$\begin{aligned}\hat{H}_{\text{int}} &= -\hbar\Omega_0(\sigma_{ge}+\sigma_{eg})(a+a^\dagger) \\ &= -\hbar\Omega_0(\sigma_{ge}a+\sigma_{ge}a^\dagger+\sigma_{eg}a+\sigma_{eg}a^\dagger) \simeq -\hbar\Omega_0(\sigma_{ge}a+\sigma_{eg}a^\dagger), \end{aligned} \tag{13.143}$$

as the terms $\sigma_{ge}a^\dagger$, $\sigma_{eg}a$ can be neglected by using the rotating wave approximation. The operator $\sigma_{ge}a$ raises the atom to its excited state, while annihilating a photon. On the other hand, the operator $\sigma_{eg}a^\dagger$ brings the atom to the ground state, while creating a photon. The overall Hamiltonian can now be represented by

$$\hat{H} = \hbar\omega_a\sigma_{ee} + \hbar\omega_0 a^\dagger a - \hbar\Omega_0\left(\sigma_{ge}a+\sigma_{eg}a^\dagger\right), \tag{13.144}$$

with the corresponding matrix representation given by

$$\hat{H} \doteq \begin{bmatrix} \hbar\omega_0 & \hbar\Omega_0 \\ \hbar\Omega_0 & \hbar\omega_a \end{bmatrix}. \tag{13.145}$$

To make this interaction stronger, we have to tune the light frequency to atomic transition frequency. In addition, since $\Omega_0 \sim d\sqrt{\omega_0/V}$, we can make Ω_0 larger by increasing the dipole moment (the proper choice of the atom and corresponding states) and by decreasing the volume of the cavity. The energy eigenkets of Hamiltonian (13.145) are given by

$$\begin{aligned}|E_+ = \hbar(\omega_a+\Omega_0)\rangle &= \frac{(|e,0\rangle+|g,1\rangle)}{\sqrt{2}} \\ |E_- = \hbar(\omega_a-\Omega_0)\rangle &= \frac{(|e,0\rangle-|g,1\rangle)}{\sqrt{2}}, \end{aligned} \tag{13.146}$$

and the corresponding energy levels are illustrated in Figure 13.24b. These energy eigenkets can be interpreted as superposition states of having and not having a photon.

We turn our attention now to the CQED implementation of the set $\{H, P, T, U_{\text{CNOT}}$ or controlled-$Z\}$ of universal quantum gates using CQED technology by employing option (i) from the first paragraph of this section, namely by representing the quantum information by photon states and by using the cavities with atoms to provide nonlinear interaction between photons [42,43]. For option (ii), the interested reader is referred to Ref. [44], and for option (iii) to Ref. [45]. The H, P, and T gates are single-qubit gates and can be implemented based on one mode of the radiation field inside the cavity by passing a two-level atom through the cavity. In the middle of the passage of an atom through the cavity a short classical pulse of amplitude A_{p} is to be applied. Let the ground state and excited state of the atom be denoted by $|g\rangle$

and $|e\rangle$ respectively, and let the photon states $|0\rangle$ and $|1\rangle$ represent logic 0 and 1 respectively. The interaction Hamiltonian can be represented by [46]

$$H_{\text{int}} = \hbar\Omega(a|e\rangle\langle g| + a^{\dagger}|g\rangle\langle e|), \tag{13.147}$$

where a and $a^{\dagger}$ denote the photon annihilation and creation operators and Ω is the corresponding vacuum Rabi frequency associated with the interaction of the cavity mode with atom states. Based on (13.147), the time-evolution operator can be derived [46]:

$$U(t) = \cos(\Omega t\sqrt{a^{\dagger}a+1})\,|g\rangle\langle g| + \cos(\Omega t\sqrt{a^{\dagger}a+1})\,|e\rangle\langle e| - i\frac{\sin(\Omega t\sqrt{a^{\dagger}a+1})}{\sqrt{a^{\dagger}a+1}}a|e\rangle\langle g|$$
$$-ia^{\dagger}\frac{\sin(\Omega t\sqrt{a^{\dagger}a+1})}{\sqrt{a^{\dagger}a+1}}|g\rangle\langle e|, \tag{13.148}$$

and the time evolution of the initial state $|\psi(0)\rangle$ can be described by $|\psi(t)\rangle = U(t)|\psi(0)\rangle$. The atom–field state $|\psi(0)\rangle = |e,0\rangle$ is unaffected after $\Delta t = \pi/2\Omega$, while $|\psi(0)\rangle = |e,1\rangle$ moves to $-i|g,0\rangle$. After the initial time $\Delta t = \pi/2\Omega$, the pulse of amplitude A_p is applied, which prepares the atom in the superposition state [43]:

$$\begin{aligned}&|g\rangle \to \cos\theta|g\rangle + ie^{-i\phi}\sin\theta|e\rangle, \quad |e\rangle \to ie^{i\phi}\sin\theta|g\rangle + \cos\theta|e\rangle;\\ &\theta = \omega t/2, \quad \omega = |d|A_p/\hbar,\end{aligned} \tag{13.149}$$

where $d = |d|e^{i\phi}$ is the dipole moment. The atom again interacts with the cavity field for the same duration $\Delta t = \pi/2\Omega$ so that the initial cavity modes are transformed to [43]

$$|0\rangle \to \cos\theta|0\rangle + e^{i\phi}\sin\theta|1\rangle, \quad |1\rangle \to e^{-i\phi}\sin\theta|0\rangle - \cos\theta|1\rangle, \tag{13.150}$$

which is equivalent to the one-qubit unitary operator $U(\theta,\phi)$:

$$U(\theta,\phi) = \begin{bmatrix}\cos\theta & e^{i\phi}\sin\theta\\ e^{-i\phi}\sin\theta & -\cos\theta\end{bmatrix}. \tag{13.151}$$

For example, by setting $\theta = \pi/4$ and $\phi = 0$, the unitary gate $U(\pi/4,0)$ becomes the Hadamard gate H:

$$U(\pi/4,0) = \frac{1}{\sqrt{2}}\begin{bmatrix}1 & 1\\ 1 & -1\end{bmatrix} = H. \tag{13.152}$$

The Z gate is obtained by setting $\theta = \phi = 0$:

$$U(0,0) = \begin{bmatrix}1 & 0\\ 0 & -1\end{bmatrix} = Z. \tag{13.153}$$

The P and T gates and other Pauli gates can be obtained by properly selecting θ and ϕ and/or by concatenation of two U gates with properly chosen parameters. The quantum phase-shift gate based on CQED, in which two qubits are represented as two radiation modes inside the cavity in combination with a three-level atom that provides the desired control interaction, is described in Ref. [43]. That is, the quantum phase-shift gate operation can be described by

$$C(U_\alpha) = |0_1 0_2\rangle\langle 0_1 0_2| + |0_1 1_2\rangle\langle 0_1 1_2| + |1_1 0_2\rangle\langle 1_1 0_2| + e^{i\alpha}|1_1 1_2\rangle\langle 1_1 1_2|. \quad (13.154)$$

By setting $\alpha = \pi$, the controlled-Z, $C(Z)$, gate can be obtained.

To complete the implementation of the set $\{H, P, T, U_{\mathrm{CNOT}}$ or controlled-$Z\}$ of universal quantum gates, the implementation of the CNOT gate/controlled-Z gate is needed. One possible implementation based on CQED, compatible with photon polarization states, is shown in Figure 13.25. To enable the interaction of vertical photons we use an optical cavity with a single trapped three-level atom, as illustrated in Figure 13.25 [41]. The atom has three relevant levels: the ground $|g\rangle$, the intermediate $|i\rangle$, and the excited $|e\rangle$ states. The ground and intermediate states are close to each other and can be hyperfine states. The atom has initially been prepared in superposition state $|\psi_A\rangle = (|g\rangle + |i\rangle)/\sqrt{2}$. The transition $|i\rangle \rightarrow |e\rangle$ is coupled to a cavity mode in vertical polarization and it is resonantly driven by the vertical photon from the input. When the incoming photon is in vertical polarization and the atom is in the ground state, the incoming photon is resonant with the cavity mode; it interacts with the atom and after interaction the atom goes back to the initial state while the V photon acquires a phase shift of π rad. If, on the other hand, the atom was in the intermediate state, the frequency corresponding to the entangled mode is significantly detuned from the frequency of the input photon, and the photon leaves the cavity without any phase change. The operation of the gate shown in Figure 13.25, by ignoring the Hadamard gates, can be described by

$$U_{\mathrm{AC}} R_{\mathrm{A}}(-\theta) U_{\mathrm{AT}} R_{\mathrm{A}}(\theta) U_{\mathrm{AC}} |CT\rangle |\psi_{\mathrm{A}}\rangle, \quad (13.155)$$

where $|\psi_{\mathrm{A}}\rangle$ denotes the initial atom state $(|g\rangle + |i\rangle)/\sqrt{2}$, $|CT\rangle$ is the input two-qubit state, U_{AC} (U_{AT}) denotes the operator describing atom–control photon

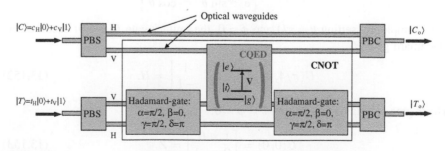

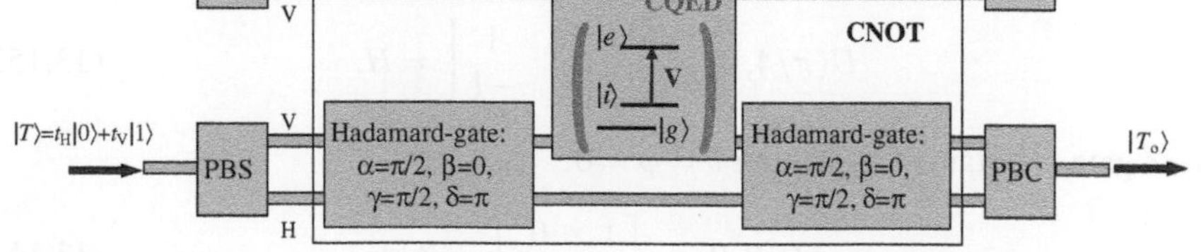

FIGURE 13.25

Deterministic CNOT gate implementation based on CQED. PBS/C, polarization beam splitter/combiner.

(atom−target photon) interaction, and $U_A(\theta)$ is the atom rotation operator performed by applying the θ pulse on the atom. In the absence of a control vertical photon, the action of the operators above is simply the identity operator, since $R_A(-\theta)R_A(\theta) = I$ and $U_{AC}^2 = I$. In the presence of a control vertical photon, the sequence of operators is as follows: (i) the vertical control photon interacts with the atom; (ii) the rotation operator is applied on the atom; (iii) the target vertical photon interacts with the atom; (iv) the de-rotation operator is applied on the atom; and (v) the vertical target photon interacts with the atom. After this sequence of operations, the control photon and atom go back to their initial states, while the target photon achieves a π phase shift. Therefore, the overall action is controlled-Z operation.

The two additional Hadamard gates are used to perform the transformation $HZH = X$, resulting in CNOT gate operation. A more rigorous derivation can be obtained by applying a similar procedure to that provided in Ref. [43]. From Figure 13.25 it is evident that the use of the controlled-Z gate instead of the CNOT gate for quantum computing applications and quantum teleportation is more appropriate in CQED technology, since the controlled-Z gate implementation is simpler (two Hadamard gates from Figure 13.25 are not needed for controlled-Z gate implementation). The quantum gate shown in Figure 13.25 is suitable for implementation in photon crystal technology [47,48]. For proper operation, on-chip integration is required. The first step towards this implementation would be achieving the coherent control of quantum $|CT\rangle$ states from Eq. (13.155). The quantum error correction-based fault-tolerant concepts should be used to facilitate this implementation.

An important practical problem to be addressed in future relates to the trapping of an atom within a cavity that requires ultra-cold conditions [49]. There are several research groups performing research in this area: R. Schoelkopf and S. Girvin at Yale, J. Kimble at Caltech, S. Harocheat at ENS, D. Stamper-Kurn at Berkley, P. S. Jessen at the University of Arizona, to name just a few. As an alternative solution, the quantum dot approach can be used, as indicated in Ref. [47].

13.8 QUANTUM DOTS IN QUANTUM INFORMATION PROCESSING

A quantum dot is the structure on a semiconductor that is able to confine electrons in three dimensions such that discrete energy levels are obtained. The quantum dot behaves as an artificial atom, whose properties can be controlled. Quantum dots can be formed spontaneously by depositing a semiconductor material on a substrate with different lattice spacing (this configuration is known as heterostructure). The quantum dots have a bowl-like shape with a diameter ~100 nm and height ~30 nm, as illustrated in Figure 13.26. A layer of AlGaAs of ~5 nm thickness is placed between two thicker layers of GaAs. The lower GaAs layer is n-doped to provide free electrons at the upper GaAs−AlGaAs interface, forming so-called two-dimensional

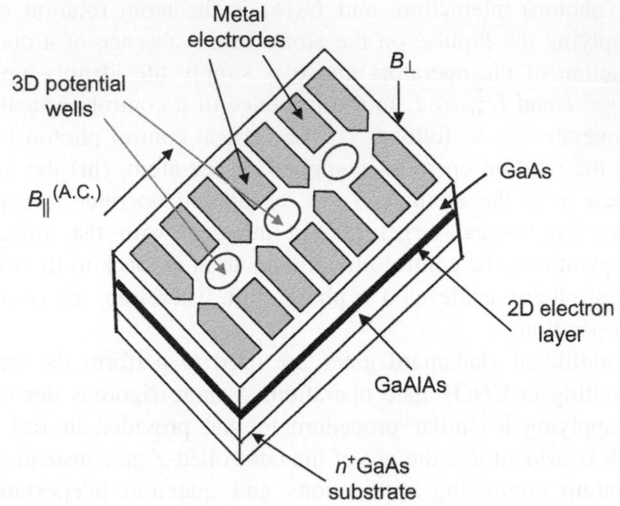

FIGURE 13.26

Semiconductor heterostructure for implementation of quantum information processing by using spins of electrons, confined in individual quantum dots, as qubits.

(2D) "electron gas". This process of creation of 2D electron gas is equivalent to total internal reflection, as two interface layers serve as boundaries of layers with different refractive indices. Further, an array of metal contacts (gates) is placed on the top of the upper GaAs layer (10–50 nm above the 2D electron gas). External voltages are applied on these gates to restrict the movement of the electrons in the horizontal plane. The three-dimensional potential wells are induced in the region among the metal electrodes.

There are many different schemes that can be used by employing this concept. One approach is based on *electron–hole pairs*, also known as *excitons*, whose energy is given by $E_{excitons} = E_{bg} - E_{be}$, where E_{bg} is the band-gap energy and E_{be} is the binding energy of electron–hole pairs. Now by using two quantum dots, the entangled qubits can be formed. The electron and hole can be in the first dot, representing state $|0\rangle$, or in the second dot, representing the state $|1\rangle$. Let the first qubit correspond to the electron and the second qubit to the hole. The state $|00\rangle$ denotes that the electron–hole pair is located in the first dot, while the $|11\rangle$ state denotes that the electron–hole pair is located in the second dot. The state $|01\rangle$ denotes that the electron is in the first dot and the hole is in the second dot. Finally, the state $|10\rangle$ denotes that the hole is in the first dot and the electron is in the second dot. By ensuring that the distance between interacting dots is ~5 nm, the electron and holes can tunnel from one dot to the other dot. The electron–hole recombines at a time interval of ~1 ns, and the wavelength is dependent on the state occupied by the particles before recombination.

Another approach is to use the *spin of an electron* as the qubit. There are two key advantages of this approach: (i) the Hilbert space is two-dimensional and (ii) the decoherence time is of the order of miliseconds. The Hamiltonian for a set of electron spins localized in a coupler array of quantum dots can be represented by [51]

$$\hat{H}=\sum_{\langle m,n\rangle} J_{mn}(t)\boldsymbol{\sigma}_m\cdot\boldsymbol{\sigma}_n+\frac{1}{2}\mu_B\sum_m g_m(t)\boldsymbol{B}_m(t)\cdot\boldsymbol{\sigma}_m,\quad \mu_B=e\hbar/2m_e, \tag{13.156}$$

where $\boldsymbol{\sigma}_m$ denotes the Pauli spin matrix associated with the mth electron, $\boldsymbol{\sigma}_m=(X_m,Y_m,Z_m)$, and J_{mn} denotes the coupling interaction between the mth and nth spins, which can be used to build two-qubit gates. $\boldsymbol{\sigma}_m\cdot\boldsymbol{\sigma}_n$ is the simultaneous scalar and tensor product and $\Sigma_{<m,n>}$ denotes the summation over closest neighbors. The second term corresponds to the coupling related to the external magnetic field $\boldsymbol{B}_m$. The coupling interaction $J_{mn}(v)$ (v is the parameter used to control the coupling interaction) can be switched on and off adiabatically, provided that $|v^{-1}\mathrm{d}v/\mathrm{d}t| << \Delta E/\hbar$, where ΔE is energy separation. The single-qubit gates can be implemented by properly varying the magnetic field $\boldsymbol{B}_m$ or Landé factor g_m so that the resonance frequency is dependent on qubit position. The two-qubit Hamiltonian, based on Eq. (13.156), can be written as [51]

$$\hat{H}_{12}(t)=J_{12}(\tau)\boldsymbol{\sigma}_1\cdot\boldsymbol{\sigma}_2. \tag{13.157}$$

In Problem 9 in Chapter 3, it was shown that the action of the operator $\boldsymbol{\sigma}_1\cdot\boldsymbol{\sigma}_2$ is to swap the qubits:

$$\begin{aligned}&\boldsymbol{\sigma}_1\cdot\boldsymbol{\sigma}_2|00\rangle=|00\rangle,\quad \boldsymbol{\sigma}_1\cdot\boldsymbol{\sigma}_2|11\rangle=|11\rangle,\quad \boldsymbol{\sigma}_1\cdot\boldsymbol{\sigma}_2|01\rangle=|10\rangle,\\&\boldsymbol{\sigma}_1\cdot\boldsymbol{\sigma}_2|10\rangle=|01\rangle.\end{aligned} \tag{13.158}$$

By allowing the evolution to last such that

$$\int_0^{T_s} J(t)\mathrm{d}t=J_0T_s=\pi \bmod 2\pi, \tag{13.159}$$

the SWAP gate operation is performed. The controlled-Z gate, $C(Z)$, can be implemented by proper combination of rotation and SWAP gates as follows [51]:

$$\begin{aligned}&C(Z)=\mathrm{e}^{j\pi Z_1/4}\mathrm{e}^{-j\pi Z_2/4}U_{\mathrm{SWAP}}^{1/2}\mathrm{e}^{j\pi Z_1/2}U_{\mathrm{SWAP}}^{1/2},\\&U_{\mathrm{SWAP}}=\frac{1}{1+j}\begin{bmatrix}1+j&0&0&0\\0&1&j&0\\0&j&1&0\\0&0&0&1+j\end{bmatrix}.\end{aligned} \tag{13.160}$$

If it is required that the CNOT gate be used instead of the controlled-Z gate, the transformation given by Eq. (13.39) should be employed.

13.9 SUMMARY

This chapter has covered several promising physical implementations of quantum information processing. The introductory section provided *di Vincenzo criteria* and an overview of physical implementation concepts. Section 13.2 was concerned with the nuclear magnetic resonance (NMR) implementation. In Section 13.3, the use of ion traps in quantum computing was described. Section 13.4 described various photonic implementations, including the bulky optics implementation (subsection 13.4.1) and integrated optics implementation (subsection 13.4.2). Section 13.5 was concerned with quantum relay implementation. The implementation of quantum encoders and decoders was discussed in Section 13.6. Further, the implementation of quantum computing based on optical cavity electrodynamics was considered in Section 13.7. The use of quantum dots in quantum computing was discussed in Section 13.8. In the next section, several interesting problems are provided for self-study.

13.10 PROBLEMS

1. Prove that the circuit shown in Figure 13.P1 for dual-rail representation operates as a CNOT gate.
2. Using quantum mechanical arguments, prove that the circuit shown in Figure 13.P2, for photon polarization states, behaves as the CNOT gate.
3. The Hadamard gate can be implemented in integrated optics with two-mode waveguides as shown in Figure 13.P3. By using a quantum mechanical interpretation, prove this claim. By using this two-mode technology, describe how

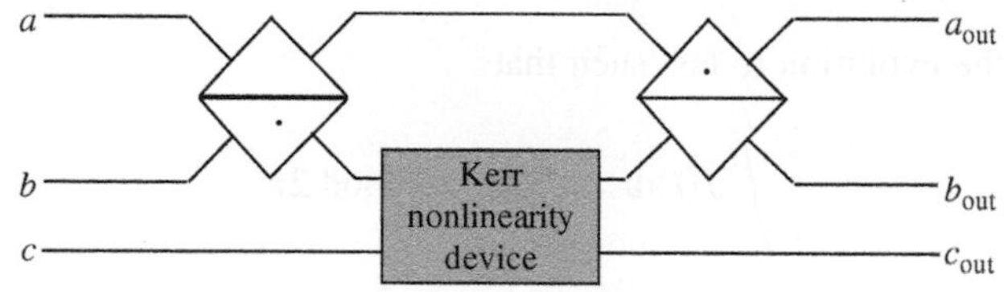

FIGURE 13.P1

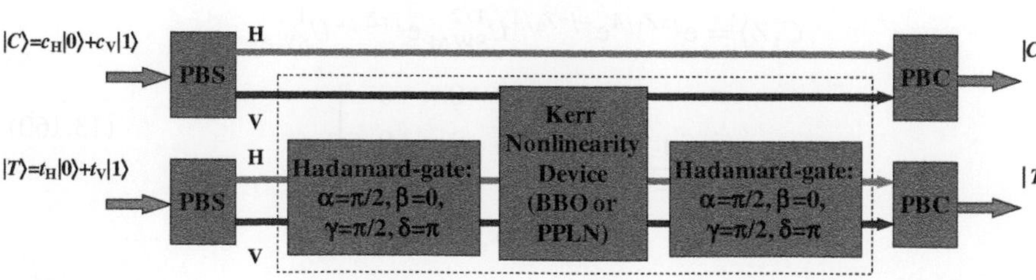

FIGURE 13.P2

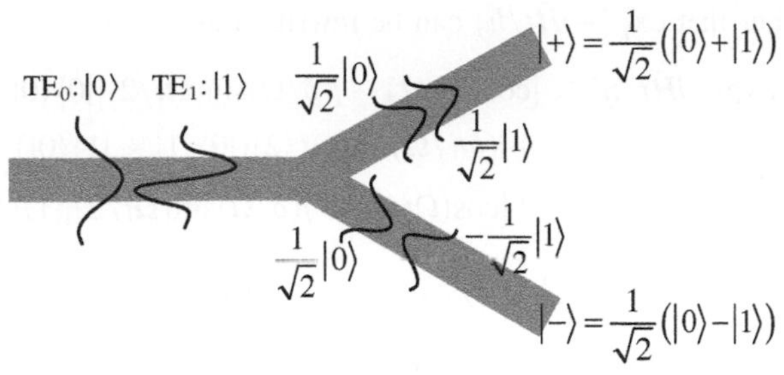

FIGURE 13.P3

the following universal set of quantum gates can be implemented: $\{H, S, \pi/8, \text{controlled-}Z\}$.

4. Determine the energy eigenkets of the Hamiltonian given by

$$\hat{H} \doteq \begin{bmatrix} \hbar\omega_0 & \hbar\Omega_0 \\ \hbar\Omega_0 & \hbar\omega_a \end{bmatrix}.$$

5. The Hamiltonian used in CQED can be represented by

$$\hat{H} = \underbrace{\hbar\omega_a\sigma_{ee} + \hbar\omega_0 a^\dagger a}_{\hat{H}_0} \underbrace{-\hbar\Omega_0(\sigma_{ge}a + \sigma_{ge}a^\dagger + \sigma_{eg}a + \sigma_{eg}a^\dagger)}_{\hat{H}_{\text{int}}},$$

where corresponding operators are defined in Section 13.7. In the interaction picture, the Hamiltonian can be represented as

$$\hat{H}_{\text{int}}^{(I)} = e^{j\hat{H}_0 t/\hbar}\hat{H}_{\text{int}}e^{-j\hat{H}_0 t/\hbar}.$$

By using the rotating wave approximation, prove that the interaction Hamiltonian can be approximated as

$$\hat{H}_{\text{int}} \simeq -\hbar\Omega_0\left(\sigma_{ge}a + \sigma_{eg}a^\dagger\right).$$

6. Let the Hamiltonian be given by

$$\tilde{H} = \frac{\hbar}{2}\delta Z - \frac{\hbar}{2}\omega_1 X, \quad \delta = \omega - \omega_0.$$

Show that $\exp(-j\tilde{H}t/\hbar)$ can be written as

$$\exp(-j\tilde{H}t/\hbar) = \exp[-j\Omega t(\delta Z/\Omega - \omega_1 X/\Omega)/2], \quad \Omega = \sqrt{\delta^2 + \omega_1^2}.$$

Show further that $\exp(-j\tilde{H}t/\hbar)$ can be rewritten as

$$\begin{aligned}\exp(-j\tilde{H}t/\hbar) = &[\cos(\Omega t/2) - j(\delta/\Omega)\sin(\Omega t/2)]|0\rangle\langle 0| \\ &+j(\omega_1/\Omega)\sin(\Omega t/2)(|0\rangle\langle 1| + |1\rangle\langle 0|) \\ &+[\cos(\Omega t/2) + j(\delta/\Omega)\sin(\Omega t/2)]|1\rangle\langle 1|.\end{aligned}$$

7. The Hamiltonian for a CQED system can be written as

$$\hat{H} = \hbar\omega_a\sigma_{ee} + \hbar\omega_0 a^\dagger a - \hbar\Omega_0\left(\sigma_{eg}a + \sigma_{ge}a^\dagger\right),$$

which possesses the constant of motion:

$$N = \langle\sigma_{ee}\rangle + \left\langle a^\dagger a\right\rangle.$$

The state with $N = n$ contains only two states $|n-1,e\rangle$ and $|n,g\rangle$, which are the eigenkets of the uncoupled system, described by the Hamiltonian $\hat{H}_0 = \hbar\omega_a\sigma_{ee} + \hbar\omega_0 a^\dagger a$. For strong atom–field coupling, these states no longer represent a proper description of an interacting atom–field system. We can decompose the Hamiltonian matrix into 2×2 submatrices. The nth submatrix is spanned by $|n-1,e\rangle$ and $|n,g\rangle$. Provide its matrix representation. Determine the energy eigenvalues and corresponding eigenkets. Also determine how the initial state $|e,0\rangle$ evolves over time. What would be the probability that a photon will be emitted in an empty cavity?

8. Provide the sequence of operations to be performed using ion-trap technology to implement the Toffoli gate and controlled-Z gate, which is controlled with two qubits.

9. Provide the sequence of operation to be performed using NMR to implement four-qubit GHZ states.

10. Let us observe quantum computation based on single photons and assume the following representation of states: $|0\rangle = |0\ldots0\rangle$, $|1\rangle = |0\ldots1\rangle$, $|2\rangle = |0\ldots10\rangle$, ..., $|2^n-1\rangle = |11\ldots1\rangle$. Show that arbitrary uniform transformation can be implemented by using only beam splitters and phase shifters, without any Kerr nonlinearity medium.

References

[1] A. Barenco, A universal two-bit quantum computation, Proc. R. Soc. Lond. A 449 (1937) (1995) 679–683.

[2] D. Deutsch, Quantum computational networks, Proc. R. Soc. Lond. A 425 (1868) (1989) 73–90.

[3] L.M.K. Vandersypen, M. Steffen, G. Breyta, C.S. Yannoni, M.H. Sherwood, I.L. Chuang, Experimental realization of Shor's quantum factoring algorithm using nuclear magnetic resonance, Nature 414 (20 December 2001) 883–887.

[4] D.G. Cory, N. Boulant, G. Boutis, E. Fortunato, M. Pravia, Y. Sharf, R. Laflamme, E. Knill, R. Martinez, C. Negrevergne, W.H. Zurek, T.F. Havel, L. Viola, S. Lloyd,

Y.S. Weinstein, G. Teklemariam, NMR based quantum information processing: Achievements and prospects, in: S.L. Braunstein, H.-K. Lo, P. Kok (Eds.), Scalable Quantum Computers: Paving the Way to Realization, Wiley-VCH, Weinheim, 2005 (doi: 10.1002/3527603182.ch8).

[5] S.J. Glaser, NMR quantum computing, Angew. Chem. Int. Ed. 40 (1) (2001) 147–149.

[6] L.M.K. Vandersypen, I. Chuang, NMR techniques for quantum control and computation, Rev. Mod. Phys. 76 (4) (2004) 1037–1069.

[7] M.G. Raizen, J.M. Gilligan, J.C. Bergquist, W.M. Itano, D.J. Wineland, Ionic crystals in a linear Paul trap, Phys. Rev. A 45 (1992) 6493.

[8] J.I. Cirac, P. Zoller, Quantum computations with cold trapped ions, Phys. Rev. Lett. 74 (1995) 4091.

[9] A. Steane, The ion trap quantum information processor, Appl. Phys. B 64 (1997) 623.

[10] G.P. Agrawal, Lightwave Technology: Components and Devices, John Wiley, 2004.

[11] Acousto-optic components, Brimrose Co. Available at: http://www.brimrose.com/.

[12] I.B. Djordjevic, R. Varrazza, M. Hill, S. Yu, Packet switching performance at 10 Gb/s across a 4×4 optical crosspoint switch matrix, IEEE Photon. Technol. Lett. 16 (January 2004) 102–104.

[13] R. Varrazza, I.B. Djordjevic, S. Yu, Active vertical-coupler based optical crosspoint switch matrix for optical packet-switching applications, IEEE/OSA J. Lightwave Technol. 22 (September 2004) 2034–2042.

[14] R. Varrazza, I.B. Djordjevic, M. Hill, S. Yu, 4×4 optical crosspoint packet switch matrix with minimized path-dependent optical gain, Opt. Lett. 28 (22) (November 2003) 2252–2254.

[15] M.A. Neilsen, I.L. Chuang, Quantum Computation and Quantum Information, Cambridge University Press, Cambridge, 2000.

[16] J.L. O'Brien, Optical quantum computing, Science 318 (7 December 2007) 1567–1570.

[17] I.B. Djordjevic, Photonic quantum dual-containing LDPC encoders and decoders, IEEE Photon. Technol. Lett. 21 (13) (1 July 2009) 842–844.

[18] T.M. Babinec, B.J.M. Hausmann, M. Khan, Y. Zhang, J.R. Maze, P.R. Hemmer, M. Loncar, A diamond nanowire single-photon source, Nature Nanotechnol. 5 (2010) 195–199.

[19] I. Friedler, C. Sauvan, J.P. Hugonin, P. Lalanne, J. Claudon, J.M. Gérard, Solid-state single photon sources: The nanowire antenna, Opt. Express 17 (2009) 2095–2110.

[20] B. Lounis, W.E. Moerner, Single photons on demand from a single molecule at room temperature, Nature 407 (28 September 2000) 491–493.

[21] A. Beveratos, R. Brouri, T. Gacoin, A. Villing, J.-P. Poizat, P. Grangier, Single photon quantum cryptography, Phys. Rev. Lett. 89 (2002), paper 187901.

[22] S. Radic, C.J. McKinistrie, Optical amplification and signal processing in highly nonlinear optical fiber, IEICE Trans. Electron. E88-C (May 2005) 859–869.

[23] S.L. Jansen, D. van den Borne, P.M. Krummrich, S. Spälter, G.-D. Khoe, H. de Waardt, Long-haul DWDM transmission systems employing optical phase conjugation, IEEE J. Sel. Top. Quantum Electron. 12 (July/August 2006) 505–520.

[24] M. Cvijetic, Optical Transmission Systems Engineering, Artech House, 2004.

[25] R.H. Hadfield, Single-photon detectors for optical quantum information applications, Nature Photon. 3 (2009) 696–705.

[26] G.N. Gol'tsman, O. Okunev, G. Chulkova, A. Lipatov, A. Semenov, K. Smirnov, B. Voronov, A. Dzardanov, C. Williams, R. Sobolewski, Picosecond superconducting single-photon optical detector, Appl. Phys. Lett. 79 (6) (2001) 705.

[27] E.J. Gansen, M.A. Rowe, D. Rosenberg, M. Greene, T.E. Harvey, M.Y. Su, R.H. Hadfield, S.W. Nam, R.P. Mirin, Single-photon detection using a semiconductor quantum dot, optically gated, field-effect transistor. Conference on Lasers and Electro-Optics/Quantum Electronics and Laser Science Conference and Photonic Applications Systems Technologies, Technical Digest (CD), paper JTuF4, Optical Society of America, 2006.
[28] S. Komiyama, Single-photon detectors in the terahertz range, IEEE Sel. Top. Quantum Electron. 17 (1) (January–February 2011) 54–66.
[29] I.B. Djordjevic, On the photonic implementation of universal quantum gates, Bell states preparation circuit, quantum relay and quantum LDPC encoders and decoders, IEEE Photon. J. 2 (1) (February 2010) 81–91.
[30] I.B. Djordjevic, Photonic entanglement-assisted quantum low-density parity-check encoders and decoders, Opt. Lett. 35 (9) (1 May 2010) 1464–1466.
[31] I.B. Djordjevic, Photonic implementation of quantum relay and encoders/decoders for sparse-graph quantum codes based on optical hybrid, IEEE Photon. Technol. Lett. 22 (19) (1 October 2010) 1449–1451.
[32] A. Politi, M. Cryan, J. Rarity, S. Yu, J.L. O'Brien, Silica-on-silicon waveguide quantum circuits, Science 320 (2008) 646.
[33] G.J. Milburn, Quantum optical Fredking gate, Phys. Rev. Lett. 62 (1988) 2124.
[34] T. Brun, I. Devetak, M.-H. Hsieh, Correcting quantum errors with entanglement, Science 314 (2006) 436–439.
[35] D.J.C. MacKay, G. Mitchison, P.L. McFadden, Sparse-graph codes for quantum error correction, IEEE Trans. Inform. Theory 50 (2004) 2315–2330.
[36] I.B. Djordjevic, Quantum LDPC codes from balanced incomplete block designs, IEEE Commun. Lett. 12 (2008) 389–391.
[37] D. Gottesman, Stabilizer Codes and Quantum Error Correction. Ph.D. dissertation, California Institute of Technology, Pasadena, CA, 1997.
[38] I. Anderson, Combinatorial Designs and Tournaments, Oxford University Press, 1997.
[39] I.B. Djordjevic, S. Sankaranarayanan, B. Vasic, Projective plane iteratively decodable block codes for WDM high-speed long-haul transmission systems, IEEE/OSA J. Lightwave Technol. 22 (March 2004) 695–702.
[40] S. Sankaranarayanan, I.B. Djordjevic, B. Vasic, Iteratively decodable codes on *m*-flats for WDM high-speed long-haul transmission, IEEE/OSA J. Lightwave Technol. 23 (November 2005) 3696–3701.
[41] I.B. Djordjevic, Cavity quantum electrodynamics (CQED) based quantum LDPC encoders and decoders, IEEE Photon. J. 3 (August 2011) 727–738.
[42] Q.A. Turchette, C.J. Hood, W. Lange, H. Mabuchi, H.J. Kimble, Measurement of conditional phase shifts for quantum logic, Phys. Rev. Lett. 75 (1995) 4710–4713.
[43] M.S. Zubairy, M. Kim, M.O. Scully, Cavity-QED-based quantum phase gate, Phys. Rev. A 68 (2003), paper 033820.
[44] C.-H. Su, A.D. Greentree, W.J. Munro, K. Nemoto, L.C.L. Hollenberg, High-speed quantum gates with cavity quantum electrodynamics, Phys. Rev. A 78 (2008), paper 062336.
[45] C. Bonato, F. Haupt, S.S.R. Oemrawsingh, J. Gudat, D. Ding, M.P. van Exter, D. Bouwmeester, CNOT and Bell-state analysis in the weak-coupling cavity QED regime, Phys. Rev. Lett. 104 (2010), paper 160503.
[46] M.O. Scully, M.S. Zubairy, Quantum Optics, Cambridge University Press, 1997.

[47] I. Fushman, D. Englund, A. Faraon, N. Stoltz, P. Petroff, J. Vuckovic, Controlled phase shifts with a single quantum dot, Science 320 (5877) (2008) 769–772.
[48] A. Faraon, A. Majumdar, D. Englund, E. Kim, M. Bajcsy, J. Vuckovic, Integrated quantum optical networks based on quantum dots and photonic crystals, New J. Phys. 13 (2011), paper 055025.
[49] K.L. Moore, Ultracold Atoms, Circular Waveguides, and Cavity QED with Millimeter-Scale Electromagnetic Traps. Ph.D. dissertation, University of California, Berkley, 2007.
[50] D.P. DiVincenzo, The physical implementation of quantum computation, Fortschr. Phys. 48 (2000) 771.
[51] M. Le Bellac, An Introduction to Quantum Information and Quantum Computation, Cambridge University Press, 2006.
[52] G. Jaeger, Quantum Information: An Overview, Springer, 2007.
[53] D. Petz, Quantum Information Theory and Quantum Statistics, Theoretical and Mathematical Physics, Springer, Berlin, 2008.
[54] P. Lambropoulos, D. Petrosyan, Fundamentals of Quantum Optics and Quantum Information, Springer-Verlag, Berlin, 2007.
[55] G. Johnson, A Shortcut Through Time: The Path to the Quantum Computer, Knopf, New York, 2003.
[56] J. Preskill, Quantum Computing (1999). Available at: http://www.theory.caltech.edu/~preskill/.
[57] J. Stolze, D. Suter, Quantum Computing, Wiley, New York, 2004.
[58] R. Landauer, Information is physical, Phys. Today 44 (5) (May 1991) 23–29.
[59] R. Landauer, The physical nature of information, Phys. Lett. A 217 (1991) 188–193.
[60] D.P. DiVincenzo, Two-bit gates are universal for quantum computation, Phys. Rev. A 51 (2) (1995) 1015–1022.
[61] A. Barenco, C.H. Bennett, R. Cleve, D.P. DiVincenzo, N. Margolus, P. Shor, T. Sleator, J.A. Smolin, H. Weinfurter, Elementary gates for quantum computation, Phys. Rev. A 52 (5) (1995) 3457–3467.
[62] T.C. Ralph, N.K. Langford, T.B. Bell, A.G. White, Linear optical controlled-NOT gate in the coincidence basis, Phys. Rev. A 65 (2002), paper 062324.

This page intentionally left blank

Abstract Algebra Fundamentals

This appendix presents abstract algebra fundamentals [1−12]. Only the most important topics needed for a better understanding of quantum information processing and quantum error correction are covered. This appendix is organized as follows. In Section A.1, the concept of groups is introduced and their basic properties provided. The group action on the set is discussed in Section A.2. Group mapping, in particular homomorphism, is described in Section A.3. In Section A.4, the concept of fields is introduced, while in Section A.5 the concept of vector spaces is considered. Further, Section A.6 provides basic facts on character theory, which is important in nonbinary quantum error correction. Section A.7 covers the algebra of finite fields, which is very important in both classical and quantum error correction. In Section A.8 the concept of metric spaces is introduced, while in Section A.9 the concept of Hilbert space is described.

A.1 GROUPS

Definition D1. A group is the set G that together with an operation, denoted by "+", satisfies the following axioms:

1. *Closure*: $\forall\ a, b \in G \Rightarrow a + b \in G$.
2. *Associative law*: $\forall\ a, b, c \in G \Rightarrow a + (b + c) = (a + b) + c$.
3. *Identity element*: $\exists\ e \in G$ such that $a + e = e + a = a\ \forall\ a \in G$.
4. *Inverse element*: $\exists\ a^{-1} \in G\ \forall\ a \in G$, such that $a + a^{-1} = a^{-1} + a = e$.

We call a group an *Abelian* or a *commutative* group if the operation "+" is also commutative: $\forall\ a, b \in G \Rightarrow a + b = b + a$.

Theorem T1. The identity element in a group is the unique one, and each element in the group has a unique inverse element.

Proof. If the identity element e is not unique, then there exists another one, e': $e' = e' + e = e$, implying that e' and e are identical. In order to show that a group element a has a unique inverse a^{-1}, let us assume that it has another inverse element $a_1^{-1} : a_1^{-1} = a_1^{-1} + e = a_1^{-1} + (a + a^{-1}) = (a_1^{-1} + a) + a^{-1} = e + a^{-1} = a^{-1}$, thus implying that the inverse element in the group is unique.

Examples:

- $F_2 = \{0,1\}$ and "+" operation is in fact the modulo-2 addition, defined by: $0 + 0 = 0$, $0 + 1 = 1$, $1 + 0 = 1$, $1 + 1 = 0$. The closure property is satisfied, 0 is the identity element, 0 and 1 are their own inverses, and operation "+" is associative. The set F_2 with modulo-2 addition therefore forms a group.

- The set of integers forms a group under the usual addition operation on which 0 is the identity element, and for any integer n, $-n$ is its inverse.
- Consider the set of codewords of *binary linear (N,K) block code.* Any two codewords added per modulo-2 form another codeword (according to the definition of the linear block code). An all-zero codeword is the identity element, and each codeword is its own inverse. The associative property also holds. Therefore, the all group properties are satisfied, and the set of codewords forms a group; this is the reason why this is called *group code.*

The number of elements in a group is typically called the *order* of the group. If the group has a finite number of elements, it is called a *finite* group.

Definition D2. Let H be a subset of elements of group G. We call H a *subgroup* of G if the elements of H themselves form the group under the same operation as that defined on elements from G.

In order to verify that the subset S is the subgroup it is sufficient to check the closure property and that an inverse element exists for every element of the subgroup.

Theorem T2 (Lagrange theorem). The order of a finite group is an integer multiple of any of its subgroup order.

Example. Consider the set Y of N-tuples (received words on the receiver side of a communication system), each element of the N-tuple being 0 or 1. It can easily be verified that elements of Y form a group under modulo-2 addition. Consider a subset C of Y with elements being the codewords of a binary (N,K) block code. Then C forms a subgroup of Y, and the order of group Y is divisible by the order of subgroup C.

There exist groups, all of whose elements can be obtained as a power of some element, say a. Such a group G is called a *cyclic group*, and the corresponding element a is called the *generator* of the group. The cyclic group can be denoted by $G = <a>$.

Example. Consider an element α of a finite group G. Let S be the set of elements $S = \{\alpha, \alpha^2, \alpha^3, \ldots, \alpha^i, \ldots\}$. Because G is finite, S must be finite as well, and therefore not all powers of α are distinct. There must be some l, m $(m > l)$ such that $\alpha^m = \alpha^l$ so that $\alpha^m \alpha^{-l} = \alpha^l \alpha^{-l} = 1$. Let k be the smallest such power of α of which $\alpha^k = 1$, meaning that $\alpha, \alpha^2, \ldots, \alpha^k$ are all distinct. We can now verify that set $S = \{\alpha, \alpha^2, \ldots, \alpha^k = 1\}$ is a subgroup of G. S contains the identity element, and for any element α^i, the element α^{k-i} is its inverse. Given that any two elements $\alpha^i, \alpha^j \in S$, the corresponding element obtained as their product $\alpha^{i+j} \in S$ if $i + j \leq k$. If $i + j > k$ then $\alpha^{i+j} \cdot 1 = \alpha^{i+j} \cdot \alpha^{-k}$, and because $i + j - k \leq k$, the closure property is clearly satisfied. S is therefore a subgroup of group G. Given the definition of a cyclic group, the subgroup S is also cyclic. The set of codewords of a cyclic (N,K) block code can be obtained as a cyclic subgroup of the group of all N-tuples.

Theorem T3. Let G be a group and $\{H_i | i \in I\}$ be a non-empty collection of subgroups with index set I. The intersection $\bigcap_{i \in I} H_i$ is a subgroup.

Definition D3. Let G be a group and X be a subset of G. Let $\{H_i | i \in I\}$ be the collection of subgroups of G that contain X. Then the intersection $\bigcap_{i \in I} H_i$ is called the *subgroup of G generated by X* and is denoted by $<X>$.

Theorem T4. Let G be a group and X a non-empty subset of G with elements $\{x_i | i \in I\}$ (I is the index set). The subgroup of G generated by X, $<X>$, consists of all finite product of x_i. The x_i elements are known as generators.

Example. A *quantum stabilizer code* with stabilizer group S and generators g_1, g_2, ..., g_{N-K} is a good illustration of this theorem. Since S is an Abelian group, and the generators have order 2 (meaning that ${g^2}_i = I$, where I is the identity operator), any element $s \in S$ can be represented by $s = g_1^{c_1} \ldots g_{N-K}^{c_{N-K}}$, $c_i \in \{0, 1\}$. For $c_i = 1$ ($c_i = 0$) the corresponding generator is included (excluded).

Definition D4. Let G be a group and H be a subgroup of G. For any element $a \in G$, the set $aH = \{ah | h \in H\}$ is called the *left coset* of H in G. Similarly, the set $Ha = \{ha | h \in H\}$ is called the *right coset* of H in G.

Theorem T5. Let G be a group and H be a subgroup of G. The collection of right cosets of H, $Ha = \{ha | h \in H\}$, forms a partition of G.

Instead of formal proof we provide the following justification. Let us create the following table. In the first row we list all elements of subgroup H, beginning with the identity element e. The second column is obtained by selecting an arbitrary element from G, not used in the first row, as the leading element of the second row. We then complete the second row by "multiplying" this element by all elements of the first row from the right. Of the not previously used elements we arbitrarily select an element as the leading element of the third row. We then complete the third row by multiplying this element by all elements of the first row from the right. We continue this procedure until we exploit all elements from G. The resulting table is as follows:

$h_1 = e$	h_2	...	h_{m-1}	h_m
g_2	$h_2 g_2$	...	$h_{m-1} g_2$	$h_m g_2$
g_3	$h_2 g_3$	...	$h_{m-1} g_3$	$h_m g_3$
...	...	...	...	...
g_n	$h_2 g_n$	...	$h_{m-1} g_n$	$h_m g_n$

Each row in this table represents a coset, and the first element in each row is a coset leader. The number of cosets of H in G is in fact the number of rows, and it is called the *index of H in G*, typically denoted by $[G{:}H]$. It follows from the table above that $|H| = m$, $[G{:}H] = n$, and $|G| = nm = [G{:}H]|H|$, which can be used as the proof of Lagrange's theorem. In other words, $[G{:}H] = |G|/|H|$.

Definition D5. Let G be a group and H be a subgroup of G. H is a *normal subgroup* of G if it is invariant under conjugation, i.e. $\forall\ h \in H$ and $g \in G$, the element $ghg^{-1} \in H$.

In other words, H is *fixed* under conjugation by the elements from G, i.e. $gHg^{-1} = H$ for any $g \in G$. Therefore, the left and right cosets of H in G coincide: $\forall\ g \in G$, $gH = Hg$.

Theorem T6. Let G be a group and H a normal subgroup of G. If G/H denotes the set of cosets of H in G, then the set G/H with coset multiplication forms a group, known as the *quotient group* of G by H.

The coset multiplication of aH and bH is defined as $aH \cdot bH = abH$. It follows from Lagrange's theorem that $|G/H| = [G:H]$. It is straightforward to prove this theorem by using the table above.

A.2 GROUP ACTING ON THE SET

We first study the action of a group G on an arbitrary set S, then consider the action of G on itself. A group G is said to act on a set S if:

1. Each element $g \in G$ implements a map $s \rightarrow g(s)$, where $s, g(s) \in S$.
2. The identity element e in G produces the *identity map*: $e(s) = s$.
3. The map produced by g_1g_2 is the *composition* of the maps produced by g_1 and g_2, i.e. $g_1g_2(s) = g_1(g_2(s))$.

The set of elements from S that are images of s under the action of G is called the *orbit* of $s \in S$, and is often denoted as orb(s). The orbit is formally defined as orb(s) $= \{g(s)|g \in G\}$. Another important concept is that of a stabilizer. The *stabilizer* of $s \in S$, denoted as S_s, is the set of elements from G that *fix* s: $S_s = \{g \in G|g(s) = s\}$. It can be shown that (i) each orbit defines an equivalence class on S so that the *collection of orbits partitions* S; and (ii) the *stabilizer* S_s *is a subgroup of* G.

Consider a Pauli group G_N on N qubits (consisting of all Pauli matrices together with factors $\pm 1, \pm j$) and a set S with subspace spanned by the basis codewords. Each basis codeword $|i\rangle$ defines its own stabilizer S_i, and the code stabilizer is obtained as intersection of all S_i.

If $S = G$, in quantum mechanical applications the action is usually a *conjugation* action, i.e. $g \rightarrow xgx^{-1}$. The corresponding orbit of g is called the *conjugacy class* of g, and the corresponding stabilizer is called the *centralizer* of g in G, denoted as $C_G(g)$. The *centralizer of* g in G is formally defined as $C_G(g) = \{x \in G|xg = gx\} = \{x \in G|xgx^{-1} = g\}$. Therefore, the centralizer of g in G, $C_G(g)$, contains all elements in G that commute with g.

If $S \subseteq G$, the *centralizer of* S in G, denoted as $C_G(S)$, is defined by $C_G(S) = \{x \in G|xS = Sx\}$.

The *center of* G, denoted as $Z(G)$, is the set of all elements from G that commutes with all elements from G, i.e. $Z(G) = \{x \in G|xg = gx, \ \forall \ g \in G\}$.

The *normalizer of* S in G, denoted as $N_G(S)$, is defined by $N_G(S) = \{x \in G|xSx^{-1} = S\}$. Let H be a subgroup of G. The *self-normalizing* subgroup of G is one that satisfies $N_G(H) = H$.

A.3 GROUP MAPPING

Suppose A and B are two sets, and f is a *function* or mapping (map) from $A \rightarrow B$ that assigns to each element of A a corresponding element of B. The set A is called a *domain* and the set of images B is called the *range*. If each element of B is the *image of no more than one element from A*, then *f* is said to be *injective* or *one-to-one*. In other words, the mapping is injective if distinct elements from *A* have distinct images, i.e. the function f: $A \rightarrow B$ is injective if and only if $f(x_1) = f(x_2)$ implies $x_1 = x_2$. If each element of *B* is the *image of at least one element from A*, then *f* is said to be *surjective* or *onto*. In other words, the range is the whole set *B*. If *f* is both injective and surjective, then *f* is said to be *bijective* (*one-to-one* correspondence). The function f: $A \rightarrow B$ has an inverse only if it is bijective. The inverse function f^{-1} is a bijective function from *B* to *A*. The bijective mapping from *A* to *A* is typically called a *permutation*. After this short reminder, in the rest of this section we study the mapping from group *G* to group *H*.

Homomorphism

Definition D6. Let *G* and *H* be groups and *f* be a function that maps $G \rightarrow H$. The function *f* is a *homomorphism* from G to *H* if, $\forall\ a,b \in G$: $f(ab) = f(a)f(b)$.

If *f* maps *G* onto *H*, *H* is said to be a *homomorphic image* of G. In other words, *f* is said to be homomorphic if the *image of a product ab is the product of images of a and b*. If *f* is a homomorphism and it is bijective, then *f* is said to be an *isomorphism* from *G* to *H*, and is commonly denoted by $G \cong H$. Finally, if $G = H$ we call the isomorphism an *automorphism*.

Let *f* be a homomorphism from $G \rightarrow H$. The *kernel of f*, denoted as Ker(*f*), is the set of elements of *G* that are mapped to the identity element e_H of *H*:

$$\mathrm{Ker}(f) = \{x \in G | f(x) = e_H\}.$$

The *image of f*, denoted as Im(*f*), is the set of elements in *H* that are the image of some $a \in G$: $\mathrm{Im}(f) = \{b \in H | b = f(a) \text{ for some } a \in G\}$.

Theorem T7. Let f: $G \rightarrow H$ be a homomorphism with kernel *K*, then $f(a) = f(b)$ if and only if $Ka = Kb$.

Instead of a formal proof of this theorem we provide the following justification. Clearly, $f(a) = f(b)$ iff $f(a)[f(b)]^{-1} = e$, which is equivalent to $f(ab^{-1}) = e$. Based on the definition of a kernel we conclude that $ab^{-1} \in K$, which is equivalent to $Ka = Kb$. The theorem above can be reformulated as follows: Any two elements *a* and *b* from *G* have the same image if and only if they belong to the same kernel. This theorem will be used to prove the following important theorem.

Theorem T8 (fundamental homomorphism theorem). Let *G* and *H* be groups. If f: $G \rightarrow H$ is a homomorphism, then *f* induces an isomorphism between *H* and the quotient group $G/\mathrm{Ker}(f)$:

$$H \cong G/\mathrm{Ker}(f).$$

Proof. Let the function that matches $\mathrm{Ker}(f)x$ with $f(x)$ be called ϕ, i.e. we can write $\phi(\mathrm{Ker}(f)x) = f(x)$. This function must ensure that if $\mathrm{Ker}(f)a$ is the same coset as $\mathrm{Ker}(f)b$ then $\phi(\mathrm{Ker}(f)a) = \phi(\mathrm{Ker}(f)b)$, which is correct because of Theorem T7. If $\phi(\mathrm{Ker}(f)a) = \phi(\mathrm{Ker}(f)b)$ then $f(a) = f(b)$ and based on Theorem T7 we conclude that $\mathrm{Ker}(f)a = \mathrm{Ker}(f)b$, which indicates that function ϕ from $G/\mathrm{Ker}(f)$ to *H* is injective. Because every element of H has the form $f(x) = \phi(\mathrm{Ker}(f)x)$, the function ϕ is clearly surjective. Finally, from the coset multiplication definition we obtain: $\phi(\mathrm{Ker}(f)a \cdot \mathrm{Ker}(f)b) = \phi(\mathrm{Ker}(f)ab) = f(ab) = f(a)f(b) = \phi(\mathrm{Ker}(f)a)\phi(\mathrm{Ker}(f)b)$, which proves that the function ϕ is an isomorphism from $G/\mathrm{Ker}(f)$ to H.

A.4 FIELDS

Definition D7. A *field* is a set of elements F with two operations, addition "+" and multiplication "·", such that:

1. F is an Abelian group under addition operation, with 0 being the identity element.
2. The nonzero elements of F form an Abelian group under the multiplication operation, with 1 being the identity element.
3. The *multiplication operation is distributive over the addition operation*:

$$\forall a, b, c \in F \Rightarrow a \cdot (b + c) = a \cdot b + a \cdot c.$$

Examples:

- The set of real numbers, with operation + as ordinary addition and operation · as ordinary multiplication, satisfy the above three properties and is therefore a field.
- The set consisting of two elements {0,1}, with modulo-2 multiplication and addition, given in the table below, constitutes a field known as the *Galois field*, and is denoted by GF(2).

+	0	1
0	0	1
1	1	0

•	0	1
0	0	0
1	0	1

- The set of integers modulo-p, with modulo-p addition and multiplication, forms a field with p elements, denoted by GF(p), providing that p is a prime.
- For any q that is an integer power of prime number p ($q = p^m$, m is an integer), there exists a field with q elements, denoted as GF(q). (The arithmetic is not modulo-q arithmetic, except when $m = 1$.) GF(p^m) contains GF(p) as a subfield.

Addition and multiplication in GF(3) are defined as follows:

+	0	1	2
0	0	1	2
1	1	2	0
2	2	0	1

•	0	1	2
0	0	0	0
1	0	1	2
2	0	2	1

Addition and multiplication in GF(2^2) are defined as:

+	0	1	2	3
0	0	1	2	3
1	1	0	3	2
2	2	3	0	1
3	3	2	1	0

•	0	1	2	3
0	0	0	0	0
1	0	1	2	3
2	0	2	3	1
3	0	3	1	2

Theorem T9. Let Z_p denote the set of integers $\{0,1,\ldots,p-1\}$, with addition and multiplication defined as ordinary addition and multiplication modulo-p. Then Z_p is a field if and only if p is a prime.

A.5 VECTOR SPACES

Definition D8. Let V be a set of elements with a binary operation "+" and let F be a field. Further, let an operation "·" be defined between the elements of V and the elements of F. Then V is said to be a *vector space* over F if the following conditions are satisfied $\forall\ a, b \in F$ and $\forall\ \boldsymbol{x}, \boldsymbol{y} \in V$:

1. V is an Abelian group under the addition operation.
2. $a \in F$ and $\forall\ \boldsymbol{x} \in V$, then $a \cdot \boldsymbol{x} \in V$.
3. *Distributive law*:

$$a\cdot(x+y) = a\cdot x + a\cdot y$$

$$(a+b)\cdot x = a\cdot x + b\cdot x.$$

4. *Associative law*:

$$(a\cdot b)\cdot x = a\cdot(b\cdot x).$$

If 1 denotes the identity element of F, then $1 \cdot \boldsymbol{x} = \boldsymbol{x}$. Let 0 denote the zero element (identity element under +) in F, $\mathbf{0}$ the additive element in V, and -1 the additive inverse of 1, the multiplicative identity in F. It can easily be shown that the following two properties hold:

$$0\cdot x = 0$$

$$x + (-1)\cdot x = 0.$$

Examples:

- Consider the set V, whose elements are n-tuples of the form $\boldsymbol{v} = (v_0, v_1, \ldots, v_{n-1})$, $v_i \in F$. Let us define the addition of any two n-tuples as another n-tuple obtained by component-wise addition and multiplication of an n-tuple by an element of F. Then V forms vector space over F, and it is commonly denoted by F^n. If $F = R$ (R is the field of a real number), then R^n is called the Euclidean n-dimensional space.

- The set of n-tuples whose elements are from GF(2), again with component-wise addition and multiplication by an element from GF(2), forms a vector space over GF(2).
- Consider the set V of polynomials whose coefficients are from GF(q). Addition of two polynomials is the usual polynomial addition, addition being performed in GF(q). Let the filed F be GF(q). Scalar multiplication of a polynomial by a field element of GF(q) corresponds to the multiplication of each polynomial coefficient by the field element, carried out in GF(q). V is then a vector space over GF(q).

Consider a set V that forms a vector space over a field F. Let $\boldsymbol{v}_1,\boldsymbol{v}_2,\ldots,\boldsymbol{v}_k$ be vectors from V, and $a_1,a_2, \ldots,a_k$ be field elements from F. The *linear combination* of the vectors $\boldsymbol{v}_1,\boldsymbol{v}_2,\ldots,\boldsymbol{v}_k$ is defined by

$$a_1\boldsymbol{v}_1 + a_2\boldsymbol{v}_2 + \ldots + a_k\boldsymbol{v}_k$$

The set of vectors $\{\boldsymbol{v}_1,\boldsymbol{v}_2,\ldots,\boldsymbol{v}_k\}$ is said to be *linearly independent* if there does not exist a set of field elements $a_1,a_2,\ldots,a_k$, not all $a_i = 0$, such that

$$a_1\boldsymbol{v}_1 + a_2\boldsymbol{v}_2 + \ldots + a_k\boldsymbol{v}_k = \mathbf{0}.$$

Example. The vectors (0 0 1), (0 1 0), and (1 0 0) (from F^3) are linearly independent. However, the vectors (0 0 2), (1 1 0), and (2 2 1) are linearly dependent over GF(3) because they sum to zero vector.

Let V be a vector space and S be subset of the vectors in V. If S is itself a vector space over F under the same vector addition and scalar multiplication operations applicable to V and F, then S is said to be *a subspace* of V.

Theorem T10. Let $\{\boldsymbol{v}_1,\boldsymbol{v}_2,\ldots,\boldsymbol{v}_k\}$ be a set of vectors from a vector space V over a field F. Then the set consisting of all linear combinations of $\{\boldsymbol{v}_1,\boldsymbol{v}_2,\ldots,\boldsymbol{v}_k\}$ forms a vector space over F, and is therefore a subspace of V.

Example. Consider the vector space V over GF(2) given by the set {(0 0 0), (0 0 1),(0 1 0),(0 1 1),(1 0 0),(1 0 1),(1 1 0),(1 1 1)}. The subset $S =$ {(0 0 0),(1 0 0), (0 1 0),(1 1 0)} is a subspace of V over GF(2).

Example. Consider the vector space V over GF(2) given by the set {(0 0 0), (0 0 1),(0 1 0),(0 1 1),(1 0 0),(1 0 1),(1 1 0),(1 1 1)}. For the subset $B =$ {(0 1 0), (1 0 0)}, the set of all linear combinations is given by $S=${(0 0 0),(1 0 0),(0 1 0), (1 1 0)}, and it forms a subspace of V over F. The set of vectors B is said to *span S*.

Definition D9. A *basis* of a vector space V is a set of linearly independent vectors that spans the space. The number of vectors in a basis is called the *dimension* of the vector space.

In the example above, the set {(0 0 1),(0 1 0),(1 0 0)} is the basis of vector space V, and has dimension 3.

A.6 CHARACTER THEORY

Character theory is often used to facilitate the explanation of nonbinary quantum codes, Clifford codes, and subsystems codes.

The *general linear group* of vector space V (over a field F), denoted as GL(V), is the group of all automorphisms of V. In other words, GL(V) is the set of all bijective linear transformations $V \rightarrow V$ together with functional composition as the group operation.

A *representation of a group* G on a vector space V (over a field F) is a group homomorphism from G to GL(V). In other words, the representation is a mapping:

$$\rho : G \rightarrow \mathrm{GL}(V) \text{ such that } \rho(g_1 g_2) = \rho(g_1)\rho(g_2) \forall g_1, g_2 \in G.$$

The vector space V is called the *representation space* and the dimension of V is called the *dimension of the representation*.

Definition D10. Let V be a finite-dimensional vector space over a field F and let ρ: $G \rightarrow$ GL(V) be a representation of a group G on V. The *character of* ρ is the function χ_ρ: $G \rightarrow F$ given by $\chi_\rho(g) = \mathrm{Tr}(\rho(g))$.

A character χ_ρ is called *irreducible* if ρ is an irreducible representation. The character χ_ρ is called *linear* if dim(V) = 1.

The kernel of χ_ρ is defined by ker $\chi_\rho = \{g \in G | \chi_\rho(g) = \chi_\rho(1)\}$, where $\chi_\rho(1)$ is the value of $\chi_\rho(1)$ on the group identity. If ρ is a representation of G of dimension k and 1 is the identity of G then

$$\chi_\rho(1) = \mathrm{Tr}(\rho(1)) = \mathrm{Tr}\, I = k = \dim \rho.$$

Definition D11. Let N be a group, $Z(N)$ be the center of N, and Irr(N) be the set of irreducible characters. The scalar product of two characters $\chi, \psi \in N$ is defined by

$$(\chi, \psi)_N = \frac{1}{|N|} \sum_{n \in N} \chi(n)\psi(n^{-1}).$$

It can be shown that Irr(N) forms an orthonormal basis in this space.

The group representation can also be related to the *matrix group*, i.e. the set of complex square $n \times n$ matrices, denoted by M_n, that satisfies the properties of a group under the matrix multiplication. In this formalism, the *representation* ρ of a group G is the mapping ρ: $G \rightarrow M_n$, which preserves the matrix multiplication. The *character* is introduced as the function defined by $\chi_\rho(g) = \mathrm{Tr}(\rho(g))$.

A.7 ALGEBRA OF FINITE FIELDS

To facilitate the description of finite fields, we introduce several definitions related to rings and congruences. A *ring* is a set of elements R with two operations, addition "+" and multiplication "·", such that (i) R is an Abelian group under addition operation, (ii) multiplication operation is associative, and (iii) multiplication is associative over addition.

The quantity a is said to be *congruent to b to modulus n*, denoted as $a \equiv b \pmod{n}$, if $a - b$ is divisible by n. If $x \equiv a \pmod{n}$, then a is called a *residue* to x to modulus n. *A class of residues to modulus n* is the class of all integers congruent to a given residue (mod n), and every member of the class is called a representative of the class. There are n classes, represented by (0), (1), (2), ..., (n − 1), and the representatives of these classes are called *a complete system of incongruent residues* to modulus n. If i and j are two members of a complete system of incongruent residues to modulus n, then addition and multiplication between i and j are defined by

$$i + j = (i + j)(\text{mod } n) \text{ and } i \cdot j = (i \cdot j)(\text{mod } n).$$

A complete system of residues (mod n) forms a commutative ring with unity element. Let s be a nonzero element of these residues. Then *s possesses an inverse if and only if n is prime, p*. Therefore, if p is a prime, a complete system of residues (mod p) forms a *Galois (or finite) field*, and is denoted by GF(p).

Let $P(x)$ be any given polynomial in x of degree m with coefficients belonging to GF(p), and let $F(x)$ be any polynomial in x with integral coefficients. Then $F(x)$ may be expressed as

$$F(x) = f(x) + pq(x) + P(x)Q(x), \text{ where } f(x)$$
$$= a_0 + a_1x + a_2x^2 + \ldots + a_{m-1}x^{m-1}, a_i \in \text{GF}(p)$$

This relationship may be written as $F(x) \equiv f(x) \bmod\{p, P(x)\}$, and we say that *$f(x)$ is the residue of $F(x)$ modulus p and $P(x)$*. If p and $P(x)$ are kept fixed but $f(x)$ varied, p^m classes may be formed, because each coefficient of $f(x)$ may take p values of GF(p). The classes defined by $f(x)$ form a commutative ring, which will be a field if and only if $P(x)$ is *irreducible* over GF(p) (not divisible with any other polynomial of degree $m - 1$ or less).

The finite field formed by p^m classes of residues is called *a Galois field of order p^m* and is denoted by GF(p^m). The function $P(x)$ is said to be *minimum polynomial* for generating the elements of GF(p^m) (the smallest degree polynomial over GF(p) having a field element $\beta \in$ GF(p^m) as a root). The nonzero elements of GF(p^m) can be represented as polynomials of degree at most $m - 1$ or as powers of a *primitive root* α such that

$$\alpha^{p^m-1} = 1, \quad \alpha^d \neq 1 \quad (\text{for } d \text{ dividing } p^m - 1).$$

A *primitive element* is a field element that generates all nonzero field elements as its successive powers. A *primitive polynomial* is an irreducible polynomial that has a primitive element as its root.

Theorem T11. Two important properties of GF(q), $q = p^m$ are:

1. The roots of polynomial $x^{q-1} - 1$ are all nonzero elements of GF(q).
2. Let $P(x)$ be an irreducible polynomial of degree m with coefficients from GF(p) and β be a root from the extended field GF($q = p^m$). Then all the m roots of $P(x)$ are

$$\beta, \beta^p, \beta^{p^2}, \ldots, \beta^{p^{m-1}}.$$

To obtain a minimum polynomial we divide $x^q - 1$ ($q = p^m$) by the least common multiple of all factors like $x^d - 1$, where d is a divisor of $p^m - 1$. Then we get the *cyclotomic equation* – the equation having for its roots all primitive roots of the equation $x^{q-1} - 1 = 0$. The order of this equation is $\phi(p^m - 1)$, where $\phi(k)$ is the number of positive integers less than k and relatively prime to it. By replacing each coefficient in this equation by the least nonzero residue to modulus p, we get a *cyclotomic polynomial* of order $\phi(p^m - 1)$. Let $P(x)$ be an irreducible factor of this polynomial, then $P(x)$ is a minimum polynomial, which is in general not unique.

Example. Let us determine the minimum polynomial for generating the elements of GF(2^3). The cyclotomic polynomial is

$$(x^7 - 1)/(x - 1) = x^6 + x^5 + x^4 + x^3 + x^2 + x + 1 = (x^3 + x^2 + 1)(x^3 + x + 1).$$

Hence, $P(x)$ can be either $x^3 + x^2 + 1$ or $x^3 + x + 1$. Let us choose $P(x) = x^3 + x^2 + 1$:

$$\phi(7) = 6, \ \deg[P(x)] = 3.$$

Three different representations of GF(2^3) are given below:

Power of α	Polynomial	Three-tuple
0	0	000
α^0	1	001
α^1	α	010
α^2	α^2	100
α^3	$\alpha^2 + 1$	101
α^4	$\alpha^2 + \alpha + 1$	111
α^5	$\alpha + 1$	011
α^6	$\alpha^2 + \alpha$	110
α^7	1	001

A.8 METRIC SPACES

Definition D12. A linear space X, over a field K, is a *metric* space if there exists a non-negative function $u \rightarrow \|u\|$, called the *norm*, such that ($\forall$ $u,v \in X$, $c \in K$):

1. $\|u\| = 0 \Leftrightarrow u = 0$ (0 is the fixed point)
2. $\|cu\| = |c| \cdot \|u\|$ (homogeneity property)
3. $\|u + v\| \leq \|u\| + \|v\|$ (triangle inequality).

Definition D13. The metric is defined as $\rho(u,v) = \|u - v\|$.
The metric has the following important properties:

1. $\rho(u,v) = 0 \Leftrightarrow u = v$ (reflectivity)
2. $\rho(u,v) = \rho(v,u)$ (symmetry)
3. $\rho(u,v) + \rho(v,w) \geq \rho(u,w)$ (triangle inequality); for every u,v,w from X.

Examples. $K = R$ (the field of real numbers), $X = \mathrm{R}^n$ (Euclidean space). The various norms can be defined as follows:

$$\|\boldsymbol{x}\|_p = \left(\sum_{k=1}^{n} |x_k|^p\right)^{1/p} \qquad 1 \le p \le \infty \quad (p - \text{norm})$$

$$\|\boldsymbol{x}\|_\infty = \max_k |x_k| \quad \text{(absolute value norm)}$$

$$\|\boldsymbol{x}\|_2 = \left(\sum_{k=1}^{n} |x_k|^2\right)^{1/2} \qquad \text{(Euclidean norm).}$$

Example. For $X = C[a,b]$ (complex signals on interval $[a,b]$), the norm can be defined by either form below:

$$\|u\| = \max_{a \le t \le b} |u(t)|$$

$$\|u\| = \int_a^b |u(t)| \mathrm{d}t.$$

Example. For $X = L^r[a,b]$, the norm can be defined by

$$\|u\| = \left(\int_a^b |u(t)|^r \mathrm{d}t\right)^{1/r}.$$

Definition D14. Let $\{u_n\}_{n \in N}$ be a sequence of points in the metric space X and let $u \in X$ be such that

$$\lim_{n \to \infty} \|u_n - u\| = 0.$$

We say that u_n *converges to u*.

Definition D15. A series $\{u_n\}_{n \in N}$ satisfying

$$\lim_{n,m \to \infty} \|u_n - u_m\| = 0$$

is called the *Cauchy series*.

Definition D16. Metric space is *complete* if every Cauchy series converges.

Definition D17. Complete metric space is called *Banach* space.

A.9 HILBERT SPACE

Definition D18. A vector space X over a complex field C is called *unitary* (or space with a dot (scalar) product) if there exists a function $(u, v) \rightarrow C$ satisfying the following properties:

1. $(u,u) \geq 0$
2. $(u,u) = 0,\ u = 0$
3. $(u + v,w) = (u,w) + (v,w)$
4. $(cu,v) = c(u,v)$
5. $(u,v) = (v,u)^*$

The properties above are valid $\forall\ u,v,w \in X$ and $c \in C$. The function (u,v) is often called a dot (scalar) product.

The dot product has the following properties:

$$(u, cv) = c^*(u, v)$$

$$(u, v_1 + v_2) = (u, v_1) + (u, v_2)$$

$$|(u, v)|^2 \leq (u, u)(v, v).$$

Definition D19. Let us define a norm as follows:

$$\|u\| = \sqrt{(u, u)}.$$

Unitary space with this norm is called a pre-Hilbert space. If it is complete then it is called a *Hilbert space*.

Example. R^n is a Hilbert space with dot product introduced by

$$(\boldsymbol{x}, \boldsymbol{y}) = \sum_{k=1}^{n} x_k y_k = (\boldsymbol{y}^{\mathrm{T}}, \boldsymbol{x}^{\mathrm{T}}).$$

Example. C^n is a Hilbert space with dot product introduced by

$$(\boldsymbol{x}, \boldsymbol{y}) = \sum_{k=1}^{n} x_k y_k^* = (\boldsymbol{y}^{\dagger}, \boldsymbol{x}^{\dagger})*$$

In this space, the following important inequality, known as the Cauchy–Schwarz(–Buniakowsky) inequality, is valid:

$$\left| \sum_{k=1}^{n} x_k y_k^* \right|^2 \leq \left(\sum_{k=1}^{n} |x_k|^2 \right) \left(\sum_{k=1}^{n} |y_k|^2 \right).$$

Example. $L^2(a,b)$ is a Hilbert space with dot product introduced by

$$(u, v) = \int_a^b u(t)v^*(t)\mathrm{d}t.$$

Orthogonal systems in Hilbert space

Definition D20. A set of vectors $\{u_k\}_{k\in I}$ in the Hilbert space forms *an orthogonal system* if:

$$(u_n, u_k) = \delta_{n,k}\|u_k\|^2 \qquad \forall n, k\in I$$

$$\delta_{n,k} = \begin{cases} 1, & n = k \\ 0, & n\neq k \end{cases} \qquad \|u_k\| = \sqrt{(u_k, u_k)}.$$

The set of indices I can be finite, countable, or uncountable. If $\|u_k\| = 1$ the system is *orthonormal*.

Gram–Schmidt orthogonalization procedure

The Gram–Schmidt procedure assigns an orthogonal system of vectors $\{u_0, u_1, \ldots\}$ starting from a countable many linearly independent vectors $\{v_0, v_1, \ldots\}$. For the first basis vector we select $u_0 = v_0$. We then express the next basis vector as v_1 plus projection along basis u_0:

$$u_1 = v_1 + \lambda_{10}u_0.$$

The projection λ_{10} can be obtained by multiplying the previous equation by u_0 and by determination of the dot product as follows:

$$(u_1, u_0) = (v_1, u_0) + \lambda_{10}(u_0, u_0) = 0.$$

By solving this equation per λ_{10} we obtain:

$$\lambda_{10} = -(v_1, u_0)/(u_0, u_0).$$

Suppose that we have already determined the basis vectors $u_0, u_1, \ldots, u_{k-1}$. We can express the next basis vector u_k in terms of vector v_k and projections along already determined basis vectors as follows:

$$u_k = v_k + \lambda_{k0}u_0 + \lambda_{k1}u_1 + \ldots + \lambda_{k,k-1}u_{k-1}.$$

By multiplying the previous equation by u_i from the right side and by evaluating the dot product we obtain:

$$(u_k, u_i) = (v_k, u_i) + \sum_{j=0}^{k-1} \lambda_{k,j}(u_j, u_i) = 0\ ; \quad i = 0, 1, \ldots, k-1.$$

By invoking the principle of orthogonality, the summation in the equation above is nonzero only for $j = i$, so that the projection along the ith basis function can be obtained by

$$\lambda_{k,i} = -\frac{(v_k, u_i)}{(u_i, u_i)}.$$

Finally, the kth basis vector can be obtained by

$$u_k = v_k - \sum_{j=0}^{k-1} \frac{(v_k, u_j)}{(u_j, u_j)} u_j; \quad k = 1, 2, \ldots.$$

The *orthonormal basis* is obtained simply by normalizing u_k with $\|u_k\| = \sqrt{(u_k, u_k)}$ as follows:

$$\phi_k = \frac{u_k}{\|u_k\|}.$$

Example. Let us observe the following vectors:

$$v_1 = (1,1,1,1) \qquad v_2 = (1,2,4,5) \qquad v_3 = (1,-3,-4,-2).$$

By applying the Gram–Schmidt procedure we obtain the following basis:

$$u_1 = v_1 = (1,1,1,1)$$

$$u_2 = v_2 - \frac{(v_2, u_1)}{(u_1, u_1)} u_1 = (-2,-1,1,2)$$

$$u_3 = v_3 - \frac{(v_3, u_1)}{(u_1, u_1)} u_1 - \frac{(v_3, u_2)}{(u_2, u_2)} u_2 = \left(\frac{8}{5}, -\frac{17}{10}, -\frac{13}{10}, \frac{7}{5}\right).$$

The orthonormal basis can be obtained by normalization as follows:

$$\phi_1 = \frac{u_1}{\|u_1\|} = \frac{1}{2}(1,1,1,1)$$

$$\phi_2 = \frac{u_2}{\|u_2\|} = \frac{1}{\sqrt{10}}(-2,-1,1,2)$$

$$\phi_3 = \frac{u_3}{\|u_3\|} = \frac{1}{\sqrt{910}}(16,-17,-13,14).$$

The following table is a nice illustration of the similarities of conventional vector space with signal space, often used in communication theory, and Hilbert space, often used in quantum mechanics:

	Vector Space	Signal Space	Hilbert Space
	$\vec{u} = \sum_{k=1}^{n} u_k \vec{b}_k$	$u(t) = \sum_{k=1}^{n} u_k \phi_k(t)$	$u = \sum_{k=1}^{n} u_k b_k$
Projection along basis	$u_k = \vec{u} \cdot \vec{b}_k$	$u_k = (u(t), \phi_k(t))$	$u_k = (u, b_k)$
Dot product	$\vec{u} \cdot \vec{v} = \sum_{k=1}^{n} u_k v_k$	$(u(t), v(t)) = \int_a^b u(t) v^*(t) dt$	(u, v) – defined
Norm	$\Vert\vec{u}\Vert = \sqrt{\sum_{k=1}^{n} u_k^2}$	$\Vert\vec{u}\Vert = \sqrt{\int_a^b \vert u(t)\vert^2 \, dt}$	$\Vert u\Vert = \sqrt{(u, u)}$
Metric (distance)	$d_E(\vec{u}, \vec{v}) = \Vert\vec{u} - \vec{v}\Vert$	$d_E(u(t), v(t)) = \Vert u(t) - v(t)\Vert$	$\Vert u - v\Vert = \rho(u, v)$
	$= \sqrt{\sum_{k=1}^{n} (u_k - v_k)^2}$	$= \sqrt{\int_a^b \vert u(t) - v(t)\vert^2 \, dt}$	
Triangle inequality	$\Vert\vec{u} + \vec{v}\Vert \le \Vert\vec{u}\Vert + \Vert\vec{v}\Vert$	$\Vert u(t) + v(t)\Vert \le \Vert u(t)\Vert + \Vert v(t)\Vert$	$\Vert u + v\Vert \le \Vert u\Vert + \Vert v\Vert$
Schwartz inequality	$\Vert\vec{u} \cdot \vec{v}\Vert \le \Vert\vec{u}\Vert\Vert\vec{v}\Vert$	$\vert(u(t), v(t))\vert \le \Vert u(t)\Vert\Vert v(t)\Vert$	$\vert(u, v)\vert \le \Vert u\Vert\Vert v\Vert$

References

[1] C.C. Pinter, A Book of Abstract Algebra, Dover Publications, 2010 (reprint).

[2] J.B. Anderson, S. Mohan, Source and Channel Coding: An Algorithmic Approach, Kluwer Academic, 1991.

[3] S. Lin, D.J. Costello, Error Control Coding: Fundamentals and Applications, Prentice-Hall, 2004.

[4] F. Gaitan, Quantum Error Correction and Fault Tolerant Quantum Computing, CRC Press, 2008.

[5] P.A. Grillet, Abstract Algebra, Springer, 2007.

[6] A. Chambert-Loir, A Field Guide to Algebra, Springer, 2005.

[7] B. Vasic, Digital Communications I, Lecture Notes, University of Arizona, 2005.

[8] M.A. Neilsen, I.L. Chuang, Quantum Computation and Quantum Information, Cambridge University Press, Cambridge, 2000.

[9] D. Raghavarao, Constructions and Combinatorial Problems in Design of Experiments, Dover Publications, New York, 1988 (reprint).

[10] I.B. Djordjevic, W. Ryan, B. Vasic, Coding for Optical Channels, Springer, 2010.

[11] D.B. Drajic, P.N. Ivanis, Introduction to Information Theory and Coding, Akademska Misao, 2009 (in Serbian).

[12] S. Lang, Algebra, Addison-Wesley, Reading, MA, 1993.

This page intentionally left blank

Index

Note: Page numbers with "f" denote figures; "t" tables.

A

B

C

R

S

Printed and bound by CPI Group (UK) Ltd, Croydon, CR0 4YY
03/10/2024
01040313-0006